PRACTICAL
MACHINE
SHOP

PRACTICAL MACHINE SHOP

John E. Neely

Machine Technology
Lane Community College

1807 · 1982
175 YEARS OF PUBLISHING

John Wiley & Sons

New York □ Chichester □ Brisbane □ Toronto □ Singapore

Designer: Laura C. Ierardi
Editor: Judith A. Green
Manuscript editor: Dex Ott
Production supervisor: Sherry Berg

Library of Congress Cataloging in Publication Data:

Neely, John, 1920–
    Practical machine shop.

    Includes index.
    1. Machine-shop practice.   I. Title.
TJ1160.N3386   670.42′3   81–4537
ISBN 0–471–08000–4   AACR2

Printed in the United States of America

10 9 8 7 6 5 4 3 2 1

PREFACE

Practical Machine Shop was written for students of mechanical fields such as automotive, diesel, agricultural implement mechanics, mechanical engineering, and drafting. Although the book is useful for a short basic course in machine shop, its major purpose is to acquaint students of related mechanical fields with manufacturing processes and the use of the machine shop—its advantages and limitations.

The book is not intended to teach a course in mechanics. It does teach basic machine shop and some basic principles that will help students cope with mechanical problems and difficulties not usually covered in their own course of study. A damaged thread or broken stud, for instance, may be treated as a nuisance and a spurious problem in their mechanics class; in this book these problems are presented as specific training projects.

Students will learn how machining processes are used to rebuild, repair, and manufacture mechanical parts that they install. They will learn how to identify and solve a problem, making correct decisions whether to purchase or repair a worn part or to have a new one made locally. A strong emphasis is placed on problem solving.

The format of this book is patterned after *Machine Tool Practices*. Section and unit divisions are made with a statement of purpose, objectives, an information section, and a self-test in each unit. This similarity to *Machine Tool Practices* makes it possible to use both textbooks in the same area and on the same machinery for both a major machine shop and a related machine shop course. Many of the projects and exercises are interchangeable for both classes. *Practical Machine Shop* has a highly visual approach in which drawings and photographs illustrate and clarify the mechanical principles taught in the book.

A separate Workbook for students is available. The Workbook contains worksheets with projects and exercises; some alternative projects; and a number of tables for the students' convenience. An Instructor's Manual is also provided, which includes post-tests and Keys as well as information on using the text in various teaching systems.

Practical Machine Shop is equally adaptable to lecture-lab or totally individualized instruction, or even a hybrid form of part lecture and part self-paced instruction. This is because of its unique format, which provides a flexible, yet complete, learning program.

It is my hope that *Practical Machine Shop* will help students and others in the metals trades to understand the materials of their trades more fully, and to be able to solve more problems encountered in their particular fields.

John E. Neely

ACKNOWLEDGMENTS

I wish to express sincere appreciation and gratitude to my wife, Doris, for the long hours of typing the manuscript, organizing details, and handling the correspondence. Without her help, this book could not have been written.

My special thanks go to those who have served as models for photographs, and to John Allan, Jr., for supplying photographs. I would also like to thank the reviewers of the manuscript who were most helpful.

I wish to thank the following for their contributions to this textbook:

Aloris Tool Company, Inc.
Aluminum Company of America
American Iron & Steel Institute
American Society for Metals
The American Society of Mechanical Engineers
Armco Inc.
Bay State Abrasives Division, Dresser Industries, Inc.
The Bendix Corporation
Bethlehem Steel Corporation
Boyar–Schultz Corporation
Brodhead–Garrett Company
Brown & Sharpe Mfg. Company
Buck Tool Company
Carothers Company, Eugene, OR.
Cincinnati Milacron
Dake Corporation
The Desmond–Stephan Manufacturing Company
DoAll Company
Dover Publications
The duMont Corporation
El Jay Inc., Eugene, OR.
Elox, Division Colt Industries Corporation
Enco Manufacturing Company
Floturn, Inc., Division of the Lodge & Shipley Co.
FMC Corporation
Foothill–DeAnza Community College District
Giddings & Lewis, Inc.
Great Lakes Screw
Hardinge Brothers, Inc.
Harig Manufacturing Corporation
Heath Tool & Die, Eugene, OR.

Hitachi Magna–Lock Corporation
Hitchner Manufacturing Company, Inc.
Huffman's Auto Grinding, Eugene, OR.
Illinois/Eclipse, Division Illinois Tool Works, Inc.
The International Nickel Company, Inc.
John Wiley & Sons, Inc.
Kasto–Racine, Inc.
Kennametal, Inc.
Lane Community College
Locktite Corporation
The Lodge & Shipley Company
Louis Levin & Son, Inc.
The L. S. Starrett Company
Magnaflux Corporation
Monarch–Sidney, Division of Monarch Machine Tool Company
MTI Corporation
National Twist Drill & Tool Div. Lear Siegler, Inc.
Orchard Auto Parts Company, Eugene, OR.
Pacific Machinery & Tool Steel Company
PMC Industries
Ralmike's Tool-A-Rama
Rank Scheer–Tumico, Inc.
Reliance Electric Company
Republic Steel Corporation
Sandvik Madison, Inc.
Sellstrom Manufacturing Company
SKF Industries, Inc.
Snap-On Tools Corporation
Southwestern Industries, Inc.
Surface Finishes, Inc.

Sweetland Archery Products, Eugene, OR.
Syclone Products Inc.
Taper Micrometer Corporation
TRW Inc.
Victor Equipment Company
Waldes Kohinoor, Inc.
The Warner & Swasey Company
Warren Tool Corporation
The Weldon Tool Company
Weyerhaeuser Company
Whitnon Spindle, Division of Mite Corporation

J. E. N.

CONTENTS

CONTENTS

SECTION A

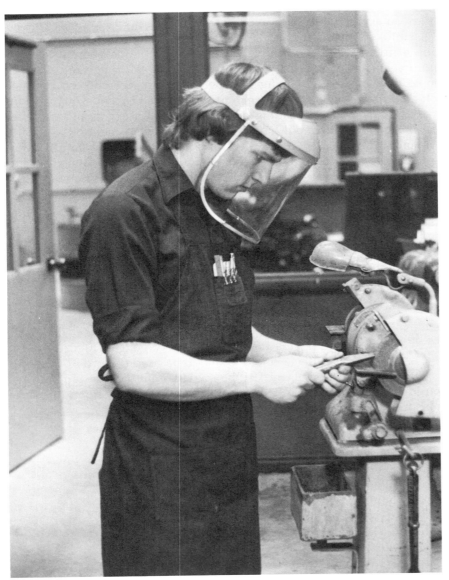

SHOP TECHNIQUES AND INFORMATION

UNIT 1   SHOP SAFETY
UNIT 2   INTRODUCTION TO MECHANICAL HARDWARE
UNIT 3   SELECTION AND IDENTIFICATION OF METALS
UNIT 4   USE OF PEDESTAL AND PORTABLE GRINDERS

UNIT 1 SHOP SAFETY

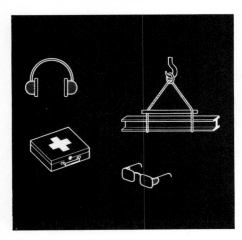

Safety is not often considered as you proceed through your daily tasks. Often you expose yourself to needless risk because you have experienced no harmful effects in the past. Unsafe habits become almost automatic. None of us really likes to think about the possible consequences of an unsafe act. However, safety can and does have an important effect on people who make their living in a potentially dangerous environment such as a machine shop. An accident can reduce or end your working career. Years spent in training and gaining experience can be wasted in an instant if you should have an accident, not to mention a possible permanent physical handicap for you and hardship on your family. Safety is an attitude that should extend far beyond the machine shop and into every facet of your life. You must constantly think about safety in everything you do.

OBJECTIVES

After completing this unit, you should be able to:
1. Identify hazards in the machine shop.
2. Identify and use safety equipment designed for a machinist.
3. Describe safe procedures around the shop and around a machine tool.

PERSONAL SAFETY

Eye Protection

Eye protection is a primary safety consideration around the machine shop. Machine tools produce metal chips, and there is always a possibility that these may be ejected from a machine at high velocity. Sometimes they can fly many feet. Furthermore, most cutting tools are made from hard materials. They can occasionally break or shatter from the stress applied to them during a cut. The result can be more flying metal particles.

Eye protection must be worn **at all times** in the machine shop. There are several types of eye protection available. Plain safety glasses are all that is required in most shops. These have shatterproof lenses that may be changed if they become scratched. The lenses have a high resistance to impact. Figure 1 shows the high impact resistance of these shatterproof lenses. Probably the most comfortable types are the fixed bow safety glasses (Figure 2).

Side shield safety glasses must be worn around any grinding operation. The side shield protects the side of the eye from flying particles. Side shield safety glasses may be of the solid or perforated type (Figure 3). The perforated side shield fits closer to the eye.

If you wear prescription glasses, you may want to cover them with a safety goggle (Figure 4). The full face shield may also be used (Figure 5). Prescription glasses can be made as safety glasses. In industry, prescription safety glasses are sometimes provided free to employees.

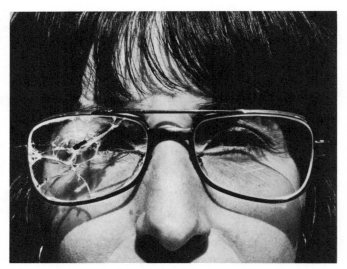

Figure 1. These shatterproof lenses saved the eye of the wearer when they were struck by a flying piece of metal. (Photo courtesy of John Allan, Jr.)

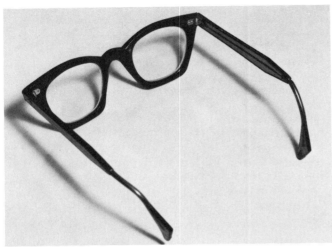

Figure 2. Fixed bow safety glasses. Their only disadvantage is that particles are able to enter from the sides. (Lane Community College)

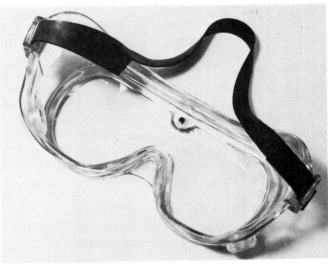

Figure 4. Safety goggles are adequate except in the vicinity of heavy flying metal objects. The relatively soft plastic on these goggles can be pierced. (Lane Community College)

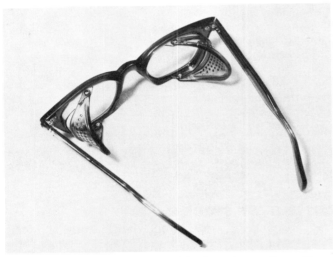

Figure 3. Perforated side shield safety glasses. (Lane Community College)

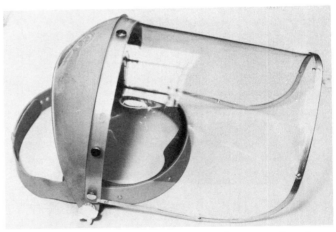

Figure 5. Safety face shield protects the entire face from flying particles or liquids. (Lane Community College)

Foot Protection

Generally, the machine shop does not present too great a hazard to the feet. However, there is always a possibility that you could drop something on your foot. A safety shoe is available. This will have a steel toe shield designed to resist impacts (Figure 6). Some safety shoes also have an instep guard. Shoes must be worn at all times in the machine shop. A solid leather shoe is recommended. Tennis shoes and san-

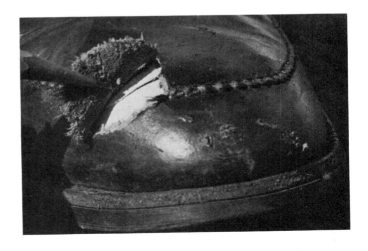

Figure 6. This safety-toe shoe is an example of the protection it can give your toes from falling objects. The steel liner is exposed. (Photo courtesy of John Allan, Jr.)

dals should not be worn. You must never even enter a machine shop with bare feet. Remember that the floor is often covered with razor sharp metal chips.

Ear Protection

The instructional machine shop usually does not present a noise problem. However, an industrial machine shop may be adjacent to a fabrication or punch press facility. New safety regulations are quite strict regarding exposure to noise. Several types of sound suppressors and noise reducing ear plugs may be worn.

Excess noise can cause a permanent hearing loss. Usually this occurs over a period of time, depending on the intensity of the exposure. Noise is considered an industrial hazard if it is continuously above 85 *decibels,* the units used in measuring sound waves. If it is over 115 decibels for short periods of time, ear protection must be worn (Figure 7). Ear muffs or plugs should be used wherever high intensity noise is likely. A considerate workman will not create excessive noise when it is not necessary. Table 1 shows the decibel level of various sounds; sudden sharp or high intensity noises are the most disturbing to your eardrums.

Figure 7. Ear muffs are designed to protect the ears from damage caused by loud noises. (White, Neely, Kibbe, and Meyer, *Machine Tools and Machining Practices,* John Wiley and Sons, Inc., Copyright © 1977, New York)

Table 1
The Decibel Level of Various Sounds

130—Painful sounds; jet engine on ground
120—Airplane on ground: reciprocating engine
110—Boiler factory
    —Pneumatic riveter
100—
    —Maximum street noise
    —Roaring lion
 90—
    —Loud shout
 80—Diesel truck
    —Piano practice
    —Average city street
 70—
    —Dog barking
    —Average conversation
 60—
    —Average city office
 50—
    —Average city residence
 40—One typewriter
    —Average country residence
 30—Turning page of newspaper
    —Purring cat
 20—
    —Rustle of leaves in breeze
    —Human heartbeat
 10—
  0—Faintest audible sound

Source: Warren T. White, John E. Neely, Richard R. Kibbe, and Roland O. Meyer, *Machine Tools and Machining Practices,* John Wiley and Sons, Inc., Copyright © 1977, New York.

CUTTING OFF STOCK

Care in handling is required when cutting off stock in a power saw. The stock should be brought to the saw on a rollcase (Figure 8) or a simple rollstand. The pieces being cut off can sometimes be several feet long and should be similarly supported. Sharp burrs left from the cutting should be removed immediately with a file. You can acquire a nasty cut by sliding your hand over one of these burrs.

Carrying Objects

Do not carry long stock in the vertical position because of the chance of hitting light fixtures and ceilings. A better way is to have someone carry each end of a long piece of material. Do not carry sharp tools in your pockets. They can injure you or someone else.

HOT METAL SAFETY

Oxy-acetylene torches are often used for cutting shapes, circles, and plates in machine shops. Safety

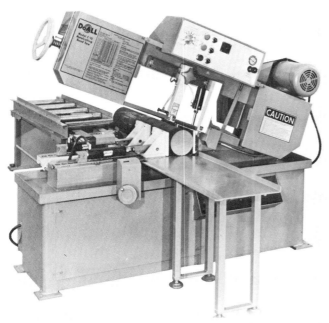

Figure 9. Face shield and gloves are worn for protection while heat treating and grinding. (Lane Community College)

Figure 8. The material is brought into the saw on the rollcase (opposite side) and, when pieces are cut off, they are supported by the stand (this side of the saw). The stand prevents the part from falling to the floor. (The DoAll Company)

when burning them requires proper clothing, gloves, and eye protection. It is also very important that any metal that has been heated by burning or welding be plainly marked, especially if it is left unattended. The common practice is to write the word *HOT* with soapstone on such items. Wherever arc welding is performed in a shop, the arc flash should be shielded from the other workers. *Never* look toward the arc because if the light enters your eye even from the side, the eye can be burned.

When handling and pouring molten metals such as babbitt, aluminum, or bronze, wear a face shield and gloves. Do not pour molten metals where there is a concrete floor unless it is covered with sand. When molten metal spills onto a concrete floor, it will cause a piece of the concrete to explode upward along with hot metal. If the molten metal falls onto a water puddle or even on wet sand, it will explode upward possibly causing injury to anyone in the vicinity.

When heat treating, always wear a face shield and heavy gloves (Figure 9). There is a definite hazard to the face and eyes when cooling tool steel by oil quenching, that is, submerging it in oil. The oil, hot from the steel, tends to fly upward, so you should stand to one side of the oil tank.

Certain metals, when divided finely as a powder or even as coarse as machining chips, can ignite with a spark or just by the heat of machining. Magnesium and zirconium are two such metals. The fire, once started, is difficult to extinguish, and if water or a water-based fire extinguisher is used, the fire will only increase in intensity. Chloride-based power fire extinguishers are commercially available. These are effective for such fires as they prevent water absorption and form an air-excluding crust over the burning metal. Sand is also used to smother fires in magnesium.

Hazardous Fumes

Some metals such as zinc give off toxic fumes when heated above their boiling point. Some of these fumes when inhaled cause temporary sickness, but other fumes can be severe or even fatal. The fumes of mercury and lead are especially dangerous, as their effect is cumulative in your body and can cause irreversible damage. Cadmium and beryllium compounds are also very poisonous. Therefore, when welding, burning, or heat treating metals, adequate ventilation is an absolute necessity. This is also true when parts are being carburized with compounds containing potassium cyanide. These *cyanogen compounds* are deadly poisonous and every precaution should be taken when using them. Kasenite, a trade name for a carburizing compound that is not toxic, is often found in school shops and in machine shops. Uranium salts are toxic and all radioactive materials are extremely dangerous.

Clothing, Hair, and Jewelry

Wear a short sleeve shirt or roll up long sleeves above the elbow. Keep your shirt tucked in and remove your necktie. It is recommended that you wear a shop apron. If you do, keep it tied behind you. If apron strings become entangled in the machine, you may be reeled in as well. A shop coat may be worn as long as you roll up long sleeves. Do not wear fuzzy sweaters around machine tools.

If you have long hair, keep it secured properly. In industry, you may be required to wear a hair net so that your hair cannot become entangled in a moving machine. The result of this can be disastrous.

Remove your wristwatch and rings before operating any machine tool. These can cause serious injury if they should be caught in a moving machine part.

Hand Protection

There is really no device that will totally protect your hands from injury. Next to your eyes, your hands are the most important tools that you have. It is up to you to keep them out of danger. Use a brush to remove chips from a machine (Figure 10). Do not use your hands. Chips are not only razor sharp; they are often extremely hot. Resist the temptation to grab chips as they come from a cut. Long chips are extremely dangerous. These can often be eliminated by properly sharpening your cutting tools. Chips should *not* be removed with a rag. The metal particles become imbedded in the cloth and they may cut you. Furthermore, the rag may be caught in a moving machine. Gloves must not be worn around most machine tools, although they are acceptable when working with a band saw blade or when cleaning chip pans while the machine is stopped. If a glove should be caught in a moving part, it will be pulled in, along with the hand inside it.

Various cutting oils, coolants, and solvents may affect your skin. The result may be a rash or possible infection. Avoid direct contact with these products as much as possible and wash your hands as soon as possible after contact.

Figure 10. Use a brush to clear chips.

Figure 11. Vacuum dust collector on grinders.

Grinding Dust

Grinding dust is produced by abrasive wheels and consists of extremely fine metal particles and abrasive wheel particles. These should not be inhaled. In the machine shop, most grinding machines have a vacuum dust collector (Figure 11). Grinding may be done with coolants that aid in dust control. A machinist may be involved in portable grinding operations. This is common in such industries as shipbuilding. You should wear an approved respirator if you are exposed to grinding dust. Change the respirator filter at regular intervals. Grinding dust can present a great danger to health. Examples include the dust of such metals as beryllium, or the presence of radioactivity in nuclear systems. In these situations, the spread of grinding dust must be carefully controlled.

Lifting

Improper lifting can result in a permanent back injury that can limit or even end your career. Back injury can be avoided if you lift properly at all times. If you must lift a large or heavy object, get some help or make use of a hoist or forklift. Don't try to be a "superman" and lift something that you know is too heavy. It is not worth the risk.

Objects within your lifting capability can be lifted safely by the following procedure (Figures 12*a* and 12*b*):

1. Keep your back straight.
2. Squat down, bending your knees.
3. Lift smoothly; let the muscles in your legs do the work. Keep your back straight. Bending over the load puts an excessive stress on your spine.
4. Position the load so that it is comfortable to carry.

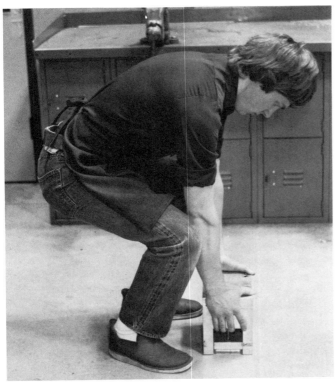

Figure 12a. Proper way of lifting. (Lane Community College)

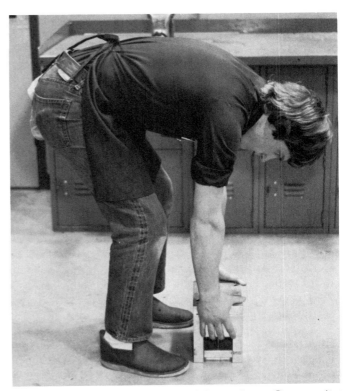

Figure 12b. Improper way of lifting. (Lane Community College)

Watch where you are walking when carrying a load.

5. If you are placing the load back to floor level, lower it in the same manner you picked it up.

Scuffling and Horseplay

The machine shop is no place for scuffling and horseplay. This activity can result in a serious injury to you, a fellow student, or worker. Practical joking is also very hazardous. What might appear to be a comical situation to you could result in a disastrous accident to someone else. In industry, horseplay and practical joking are often grounds for dismissal of an employee.

Injuries

If you should be injured, report it immediately to your instructor.

IDENTIFYING SHOP HAZARDS

A machine shop is a potentially dangerous place. One of the best ways to be safe is to be able to identify shop hazards before they involve you in an accident. By being aware of potential danger, you can better make safety part of your work in the machine shop.

Although industrial hazards exist, they are controlled by safety programs and employee–employer cooperation. The machine shop is a relatively safe occupation, but safety rules must be observed.

Compressed Air

Most machine shops have compressed air. This is needed to operate certain machine tools. Often flexible air hoses are hanging about the shop. Few people realize the large amount of energy that can be stored in a compressed gas such as air. When this energy is released, extreme danger may be present. You may be tempted to blow chips from a machine tool using compressed air. This is not good practice. The air will propel metal particles at high velocity. They can injure you or someone on the other side of the shop. Use a brush to clean chips from the machine. Do not blow compressed air on your clothing or skin. The air can be dirty and the force can implant dirt and germs into your skin. Air can be a hazard to ears as well. An eardrum can be ruptured.

Should an air hose break, or the nozzle on the end come unscrewed, the hose will whip wildly. This can result in an injury if you happen to be standing nearby. When an air hose is not in use, it is good practice to shut off the supply valve. The air trapped in the hose should be vented. When removing an air hose, even one that has a quick disconnect, from its supply valve, be sure that the supply is turned off and the hose has been vented. Removing a charged air hose will result in a sudden venting of air. This can surprise you and an accident might result if you are not careful.

Housekeeping

Keep floor and aisles clear of stock and tools. This will insure that all exits are clear if the building should have to be evacuated. Material on the floor, especially round bars, can cause falls. Clean oils or coolants that may have spilled on the floor. Several preparations designed to absorb oil are available. These may be used from time to time in the shop. Keep oily rags in an approved safety can (Figure 13). This will prevent possible fire from spontaneous combustion. Rag containers should be emptied every night.

Electrical

Electricity is another potential danger in a machine shop. Your exposure to electrical hazard will be minimal unless you become involved with machine maintenance. A machinist is mainly concerned with the on and off switch on a machine tool. However, if you are adjusting the machine or accomplishing maintenance, you should unplug it from the electrical service. If it is permanently wired, the circuit breaker may be switched off and tagged with an appropriate warning.

MACHINE HAZARDS

There are many machine hazards. Each section of this book will discuss the specific hazards applicable to that type of machine tool. Remember that a machine has no intelligence of its own. It cannot distinguish between cutting metal and cutting fingers. Do not think that you are strong enough to stop a machine should you become tangled in moving parts. You are not. When operating a machine, think about what you are going to do before you do it. Go over a checklist.

Safety Checklist

1. Do I know how to operate this machine?
2. What are the potential hazards involved?
3. Are all guards in place?
4. Are my procedures safe?
5. Am I doing something that I probably should not do?

Figure 13. Store oil soaked rags in an approved safety can.

6. Have I made all the proper adjustments and tightened all locking bolts and clamps?
7. Is the workpiece secured properly?
8. Do I have proper safety equipment?
9. Do I know where the stop switch is?
10. Do I think about safety in everything that I do?

Safety in Material Handling

Machinists were once expected to lift pieces of steel weighing a hundred pounds or more into awkward positions. This was a dangerous practice that resulted in too many injuries. Hoists and cranes are used to lift all but the smaller parts. Steel weighs about 487 pounds per cubic foot; water weighs 62.5 pounds per cubic foot, thus steel is a very heavy material for its size. You can easily be misled into thinking that a small piece of steel does not weigh much. Follow these two rules in all lifting you do: don't lift more than you can *easily* handle, and bend your knees and keep your back straight. If a material is too heavy or awkward for you to position it on a machine such as a lathe, use a hoist. Once the workpiece has been hoisted to the required level, it can hang in that posi-

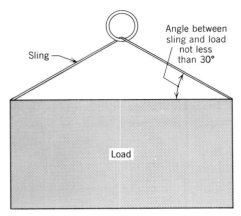

Figure 14. Load sling. *(Machine Tools and Machining Practices)*

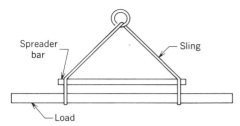

Figure 15. Sling for lifting long bars. *(Machine Tools and Machining Practices)*

tion until the clamps or chuck jaws on the machine have been secured.

HOISTING

When lifting heavy metal parts with a mechanical or electric hoist, always stand in a safe position, no matter how secure the slings and hooks seem to be. They don't often break, but it can and does happen, and if a careless foot is under the edge, a painful or crippling experience is sure to follow. Slings should not have less than a 30-degree angle with the load (Figure 14). It should be noted that when a sling, as shown in Figure 14, is at 30 degrees, it will support only one-half the load that the cable or chain will in a vertical lifting position. At 45 degrees, a sling will support about two-thirds, and at 60 degrees about three-fourths of a vertical rigging arrangement. When hoisting long bars or shafts, a spreader bar (Figure 15) should be used so the slings cannot slide together and unbalance the load. When operating a crane, be careful that no one is standing in the way of the load or hook.

Fire Extinguishers

It is an important safety consideration to know the correct fire extinguisher to use for a particular fire. For example, if you should use a water-based extinguisher on an electrical fire, you could receive a severe or fatal electrical shock. Fires are classed according to types as given in Table 2.

Table 2

Types of Extinguishers Used on the Classes of Fire (Brodhead–Garrett Company)

	Pressurized Water	Loaded Stream	CO$_2$	Regular Dry Chemical	All Use Dry Chemical
CLASS A FIRES Paper, wood, cloth, etc. where quenching by water or insulating by general purpose dry chemical is effective.	Yes Excellent	Yes Excellent	Small surface fires only	Small surface fires only	Yes Excellent Forms smothering film, prevents reflash
CLASS B FIRES Burning liquids (Gasoline, oils, cooking fats, etc.) where smothering action is required.	No Water will spread fire	Yes Has limited capability	Yes Carbon Dioxide has no residual effects on food or equipment	Yes Excellent Chemical Smothers fire	Yes Excellent Smothers fire, prevents reflash
CLASS C FIRES Fire in live electrical equipment (motors, switches, appliances, etc.) where a nonconductive extinguishing agent is required.	No Water is a conductor of electricity	No Water is a conductor of electricity	Yes Excellent CO$_2$ is a nonconductor, leaves no residue	Yes Excellent Nonconducting smothering film. Screens operator from heat	Yes Excellent Nonconducting smothering film. Screens operator from heat

There are four basic types of fire extinguishers used other than the use of tap water:

1. The dry chemical type is effective on Classes B and C fires.
2. The pressurized water and loaded stream types are safe only on Class A fires. This type may actually spread an oil or gasoline fire.
3. The dry chemical multi-purpose extinguisher may be safely used on Classes A, B, and C fires.

4. Pressurized carbon dioxide (CO_2) can be used on Classes B and C fires.

You should always make yourself aware of the locations of fire extinguishers in your working area. Take time to look at them closely and note their types and capabilities. This way, if there should ever be an oil-based fire or electrical fire in your area, you will know how to safely put it out.

SELF-TEST

1. What is the primary piece of safety equipment in the machine shop?
2. What can you do if you wear prescription glasses?
3. Describe proper dress for the machine shop.
4. What can be done to control grinding dust?
5. What hazards exist from coolants, oils, and solvents?
6. Describe proper lifting procedure.
7. Describe at least two compressed air hazards.
8. Describe good housekeeping procedures.
9. How should long pieces of material be carried?
10. List at least five points from the safety checklist for a machine tool.
11. Should you lift a 100 pound workpiece by hand to put it into a lathe chuck? Explain.
12. (a) Locate and identify all of the fire extinguishers in your area.
    (b) Name the types of extinguishers that can be safely used on electrical fires.

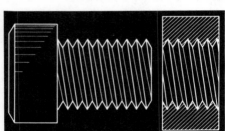

UNIT 2 INTRODUCTION TO MECHANICAL HARDWARE

Mechanical parts are useless until they are assembled into a structure or machine. Assembly requires a great variety of fasteners, each with a particular use. For example, capscrews and studs are used in an automobile engine, while various types of thread cutting screws are used on the body. In your career in any area of metalworking, you will use most or all of the fasteners and mechanical hardware items explained in this unit. It is important for you to know how to name a particular fastener. You would have a difficult time getting a simple bolt from a tool room or store if you didn't know what to call it. For example, a 10-32 machine screw is smaller in diameter than a $\frac{1}{4}$-28 and has a finer pitch or threads per inch. You will need an understanding of fasteners and screw threads as given in this unit when you cut threads on the lathe in Section H.

OBJECTIVE

After completing this unit, you should be able to:
Identify most common mechanical hardware used to assemble tools, machines, and other mechanisms.

THREADS

The **thread** is an extremely important mechanical device. It derives its usefulness from the inclined plane, one of the six simple machines. Almost every mechanical device is assembled with threaded fasteners. A straight thread is a helical groove that is formed on the outside or inside diameter of a cylinder (Figure 1). These spiral grooves take several forms. Furthermore, they have specific and even spacing. One of the fundamental tasks of a machinist is to produce both external and internal threads using several machine tools and hand tools. The majority of threads appear on **threaded fasteners.** These include many types of **bolts, screws,** and **nuts.** However, threads are used for a number of other applications aside from fasteners. These include threads for adjustment purposes, measuring tool applications, and the transmission of power. A close relative to the thread, the helical auger, is used to transport material.

THREAD FORMS

There are a number of thread forms. In later units, you will examine these in detail, and you will have the opportunity to make several of them on a machine tool. As far as the study of machined hardware is concerned, you will be most concerned with the **unified thread form** (Figure 2). The unified thread form was an outgrowth of the American National Standard form. In order to help standardize manufacturing in the United States, Canada, and Great Britain, the unified form was developed. Unified threads are a combination of the American National and the British Standard Whitworth forms. Unified threads are divided into the following series:

UNC—Unified Coarse Series
UNF—Unified Fine Series
UNS—Unified Special Series

IDENTIFYING THREADED FASTENERS

Unified coarse and unified fine refers to the number of threads per inch of length on standard threaded fasteners. A specific diameter of bolt or nut will have a specific number of threads per inch of length. For example, a $\frac{1}{2}$ in. diameter Unified Coarse Series bolt will have 13 threads per inch of length. This bolt will be identified by the following marking:

$$\frac{1}{2} \text{ in.}-13 \text{ UNC}$$

One-half is the **major diameter** and **13** is the **number of threads per inch of length.** A $\frac{1}{2}$ in. diameter Unified

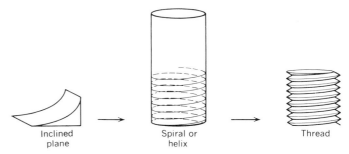

Figure 1. Thread helix. *(Machine Tools and Machining Practices)*

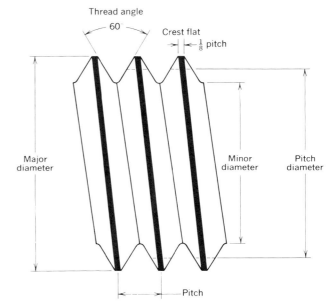

Figure 2. Unified thread form. *(Machine Tools and Machining Practices)*

Fine Series bolt will be identified by the following marking:

$$\frac{1}{2} \text{ in.}-20 \text{ UNF}$$

One-half is the major diameter and 20 is the number of threads per inch.

The Unified Special Threads are identified in the same manner. A $\frac{1}{2}$ in. diameter UNS bolt may have 12, 14, or 18 threads per inch. These are less common than the standard UNC and UNF. However, you may see them in machining technology. There are many other series of threads used for different applications. Information and data on these can be found in machinist's handbooks. You might wonder why there needs to be a UNC and UNF series. This has to do with thread applications. For example, an adjusting screw might require a fine thread, while a common bolt may require only a coarse thread.

CLASSES OF THREAD FITS

The preceding information was necessary for an understanding of **thread fit classes.** Some thread applications can tolerate loose threads, while other applications require tight threads.

Unified **thread fits** are classified as **1A, 2A, 3A,** or, **1B, 2B, 3B.** The **A symbol** indicates an **external** thread. the **B symbol** indicates an **internal** thread. This notation is added to the thread size and number of threads per inch. Let us consider the $\frac{1}{2}$ in. diameter bolt discussed previously. The complete notation reads:

$$\frac{1}{2}\text{--}13 \text{ UNC } 2A$$

On this particular bolt, the class of fit is 2. The symbol A indicates an external thread. If the notation had read:

$$\frac{1}{2}\text{--}13 \text{ UNC } 3B$$

this would indicate an internal thread with a class 3 fit. This could be a nut or a hole threaded with a tap. Taps are a very common tool for producing an internal thread.

Class 1A and 1B have the greatest manufacturing tolerance. They are used where ease of assembly is desired and a loose thread is not objectionable. Class 2 fits are used on the largest percentage of threaded fasteners. Class 3 fits will be tight when assembled. Each class of fit has a specific tolerance on **major diameter** and **pitch diameter.** These data may be found in machinist's handbooks and are required for the manufacture of threaded fasteners.

Threads can be either left hand or right hand. The hand of a thread can be determined by looking at the thread from the side with the bolt in a horizontal position. A right-hand thread slants to the right from top to bottom (example is shown in Figure 2) and a left-hand thread slants from top to bottom toward the left. Most fasteners are right hand.

STANDARD SERIES OF THREADED FASTENERS

Threaded fasteners, including all common bolts and nuts, range from quite small machine screws to quite large bolts. Below a diameter of $\frac{1}{4}$ in., threaded fasteners are given a number. Common UNC and UNF se-

Table 1
UNC and UNF Threaded Fasteners

UNC		UNF	
Size	Threads/in.	Size	Threads/in.
		0	80
1	64	1	72
2	56	2	64
3	48	3	56
4	40	4	48
5	40	5	44
6	32	6	40
8	32	8	36
10	24	10	32
12	24	12	28

Table 2
UNC and UNF Threaded Fasteners

UNC		UNF	
Size	Threads/in.	Size	Threads/in.
$\frac{1}{4}$ in.	20	$\frac{1}{4}$ in.	28
$\frac{5}{16}$ in.	18	$\frac{5}{16}$ in.	24
$\frac{3}{8}$ in.	16	$\frac{3}{8}$ in.	24
$\frac{7}{16}$ in.	14	$\frac{7}{16}$ in.	20
$\frac{1}{2}$ in.	13	$\frac{1}{2}$ in.	20
$\frac{9}{16}$ in.	12	$\frac{9}{16}$ in.	18
$\frac{5}{8}$ in.	11	$\frac{5}{8}$ in.	18
$\frac{3}{4}$ in.	10	$\frac{3}{4}$ in.	16
$\frac{7}{8}$ in.	9	$\frac{7}{8}$ in.	14
1 in.	8	1 in.	12

ries threaded fasteners are listed in Table 1. Table 2 lists common UNC and UNF threaded fasteners where the major diameter is expressed in fractional form. The fractional series continues up to about 4 in.

All of the sizes listed in the tables are very common fasteners found in all types of machines, automobiles, and other mechanisms. Your contact with these common sizes will be so frequent that you will soon begin to recall them from memory. *Note:* Mechanics often refer to coarse threads as "standard" and fine threads as "SAE." This refers to an obsolete thread series that is more or less equivalent to UNC (coarse) and UNF (fine).

COMMON EXTERNALLY THREADED FASTENERS

Common mechanical hardware includes threaded fasteners such as bolts, screws, nuts, and thread inserts. All of these are used in a variety of ways to hold parts and assemblies together. Complex assemblies such as an airplane, ship, or automobile may have many thousands of fasteners taking many forms.

External threads are made either by cutting them with a die or a single point tool in a lathe or by a grinding process. They also are formed by thread rolling, which displaces metal to form thread rings on a blank. Internal threads are cut in a hole with a tap or, in the case of larger diameter threads, in a lathe with a single point tool. Internal threads may also be formed by using a form tap that does not cut but, instead, displaces metal to make the thread.

Formed threads such as external rolled threads are 10 to 20 percent stronger than cut threads and have a much higher fatigue resistance. Another advantage of the rolling process is that the thread is much harder than a cut thread of the same material; thus wear resistance is increased and no stock is wasted in forming the thread.

Bolts and Screws

A general definition of a **bolt** is an externally threaded fastener that is inserted through holes in an assembly. A bolt is tightened with a **nut** (Figure 3, right). A **screw** is an externally threaded fastener that is inserted into a threaded hole and tightened or released by turning the head (Figure 3, left). From these definitions, it is apparent that a bolt can become a screw or the reverse can be true. This depends on the application of the hardware. Bolts and screws are the most common of the threaded fasteners. These fasteners are used to assemble parts quickly; and they make disassembly possible.

The strength of an assembly of parts depends to a large extent on the diameter of the screws or bolts used. In the case of screws, strength depends on the amount of **thread engagement**. Thread engagement is the distance that a screw extends into a threaded hole. The minimum thread engagement should be a distance equal to the diameter of the screw used; preferably it would be one and one-half times the screw diameter. Should an assembly fail, it is better that the screw break than to have the internal thread stripped from the hole. It is generally easier to remove a broken screw than to drill and tap for a larger screw size. With a screw engagement of $1\frac{1}{2}$ times its diameter, the screw will usually break rather than strip the thread in the hole.

Machine bolts (Figure 4) are made with **hexagonal** or **square** heads. These bolts are often used in the assembly of parts that do not require a precision bolt. The body diameter of machine bolts is usually slightly larger than the nominal or standard size of the bolt. Body diameter is the diameter of the un-

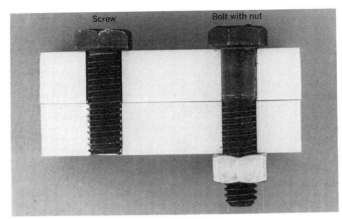

Figure 3. Screw and bolt with nut. (*Machine Tools and Machining Practices*)

Figure 4. Square head bolt and hex head bolt. Square head nut and hex nut. (*Machine Tools and Machining Practices*)

Figure 5. Stud bolt. (*Machine Tools and Machining Practices*)

threaded portion of a bolt below the head. A hole that is to accept a common bolt must be slightly larger than the body diameter. When machine bolts are purchased, nuts are frequently included. Common bolts are made with a class 2A unified thread and come in both UNC and UNF series. Sizes in hexagonal head machine bolts range from $\frac{1}{4}$ in. diameter to 4 in. diameter. Square head machine bolts are standard to $1\frac{1}{2}$ in. diameter.

Stud bolts (Figure 5) have threads on both ends. Stud bolts are used where one end is semipermanently screwed into a threaded hole. The end of the stud bolt screwed into the tapped hole has a class 3A thread, while the nut end is a class 2A thread.

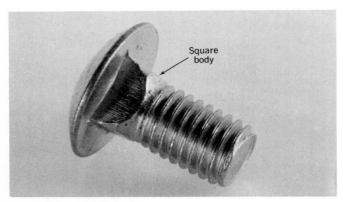

Figure 6. Carriage bolt. *(Machine Tools and Machining Practices)*

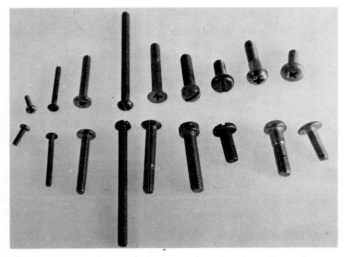

Figure 7. Machine screws. *(Machine Tools and Machining Practices)*

Carriage bolts (Figure 6) are used to fasten wood and metal parts together. Carriage bolts have round heads with a square body under the head. The square part of the carriage bolt, when pulled into the wood or a square hole punched in a metal part, keeps the bolt from turning while the nut is being tightened. Carriage bolts are manufactured with class 2A coarse threads.

Machine screws are made with either coarse or fine thread and are used for general assembly work. The heads of most machine screws are slotted to be driven by screwdrivers. Machine screws are available in many sizes and lengths (Figure 7). Several head styles are also available (Figure 8). Standard hexagonal head capscrews, button capscrews, and carriage bolts are all measured for length from the screw end to the shoulder of the head, not including the head. Countersink, flat head, and oval head screws are measured overall (head included) for length. See a machinist's handbook for bolt and screw lengths. Machine screw sizes fall into two categories. Fraction sizes range from diameters of $\frac{1}{4}$ to $\frac{3}{4}$ in. Below $\frac{1}{4}$ in. diameter, screws are identified by numbers from 0 to 12. A number 0 machine screw has a diameter of .060 in. (sixty thousandths of an inch). For each number above zero add .013 in. to the diameter.

EXAMPLE
Find the diameter of a number 6 machine screw:

$$\#0 \text{ diameter} = .060 \text{ in.}$$
$$\#6 \text{ diameter} = .060 \text{ in.} + (6 \times .013 \text{ in.})$$
$$= .060 \text{ in.} + .078 \text{ in.}$$
$$= .138 \text{ in.}$$

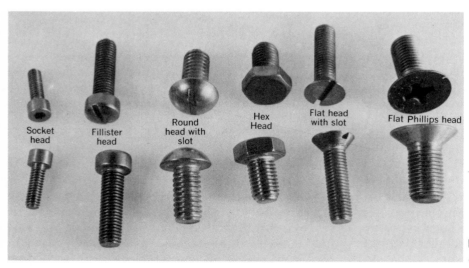

Figure 8. *(opposite)* Machine screw head styles. (Courtesy of Great Lakes Screw)

Figure 9. Capscrews. *(Machine Tools and Machining Practices)*

PAN Low large diameter with high outer edges for maximum driving power. With slotted or Phillips recess for machine screws. Available plain for driving screws.		**FLAT UNDERCUT** Standard 82° flat head with lower 1/3 of countersink removed for production of short screws. Permits flush assemblies in thin stock.	
TRUSS Similar to round head, except with shallower head. Has a larger diameter. Good for covering large diameter clearance holes in sheet metal. For machine screws and tapping screws.		**FLAT, 100°** Has larger head than 82° design. Use with thin metals, soft plastics, etc. Slotted or Phillips driving recess.	
BINDER Undercut binds and eliminates fraying of wire in electrical work. For machine screws, slotted or Phillips driving recess.		**FLAT TRIM** Same as 82° flat head except depth of countersink has been reduced. Phillips driving recess only.	
ROUND Used for general-purpose service. Used for bolts, machine screws, tapping screws and drive screws. With slotted or Phillips driving recess.		**OVAL** Like standard flat head. Has outer surface rounded for added attractiveness. Slotted, Phillips or clutch driving recess.	
ROUND WASHER Has integral washer for bearing surface. Covers larger bearing area than round or truss head. For tapping screws only; with slotted or Phillips driving recess.		**OVAL UNDERCUT** Similar to flat undercut. Has outer surface rounded for appearance. With slotted or Phillips driving recess.	
FLAT FILLISTER Same as standard fillister but without oval top. Used in counter bored holes that require a flush screw. With slot only for machine screws.		**OVAL TRIM** Same as oval head except depth of countersink is less. Phillips driving recess only.	
FILLISTER Smaller diameter than round head, higher, deeper slot. Used in counterbored holes. Slotted or Phillips driving recess. Machine screws and tapping screws.		**ROUND COUNTERSUNK** For bolts only. Similar to 82° flat head but with no driving recess.	
HEXAGON Head with square, sharp corners, and ample bearing surface for wrench tightening. Used for machine screws and bolts.		**SQUARE (SET-SCREW)** Square, sharp corners can be tightened to higher torque with wrench than any other set-screw head.	
HEXAGON WASHER Same as Hexagon except with added washer section at base to protect work surface against wrench disfigurement. For machine screws and tapping screws.		**SQUARE (BOLT)** Square, sharp corners, generous bearing surface for wrench tightening.	
FLAT, 82° Use where flush surface is desired. Slotted, clutch, Phillips, or hexagon-socket driving recess.		**SQUARE COUNTERSUNK** For use on plow bolts, which are used on farm machinery and heavy construction equipment.	

Machine screws 2 in. long or shorter have threads extending all the way to the head. Longer machine screws have a $1\frac{3}{4}$ in. thread length.

Capscrews (Figure 9) are made with a variety of different head shapes and are used where precision bolts or screws are needed. Capscrews are manufactured with close tolerances and have a finished appearance. Capscrews are made with coarse, fine, or special threads, and are normally made with a class 2A thread. The strength of screws depends mainly on the kind of material used to make the screw. Different screw materials are aluminum, brass, bronze, low carbon steel, medium carbon steel, alloy steel, stainless steel, and titanium. Steel hex head capscrews

Bolt head marking	SAE — Society of Automotive Engineers ASTM — American Society for Testing and Materials SAE — ASTM Definitions	Material	Minimum tensile strength in pounds per square inch (PSI)
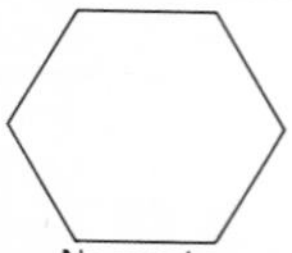 No marks	SAE grade 1 SAE grade 2 Indeterminate quality	Low carbon steel Low carbon steel	65,000 PSI
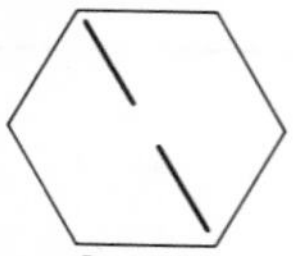 2 marks	SAE grade 3	Medium carbon steel, cold worked	110,000 PSI
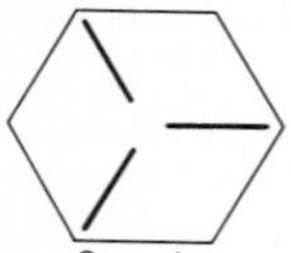 3 marks	SAE grade 5 ASTM — A 325 Common commercial quality	Medium carbon steel, quenched and tempered	120,000 PSI
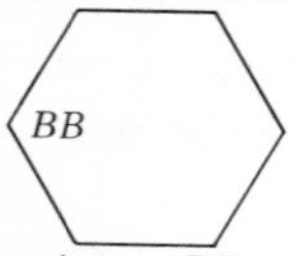 Letters *BB*	ASTM — A 354	Low alloy steel or medium carbon steel, quenched and tempered	105,000 PSI
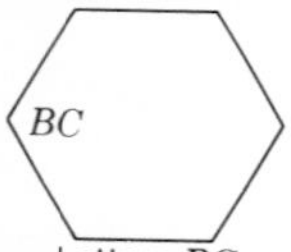 Letters *BC*	ASTM — A 354	Low alloy steel or medium carbon steel, quenched and tempered	125,000 PSI
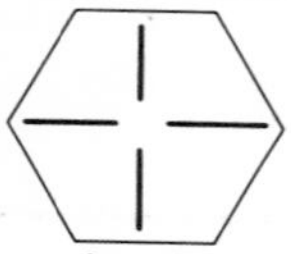 4 marks	SAE grade 6 Better commercial quality	Medium carbon steel, quenched and tempered	140,000 PSI
 5 marks	SAE grade 7	Medium carbon alloy steel, quenched and tempered, roll threaded after heat treatment	133,000 PSI
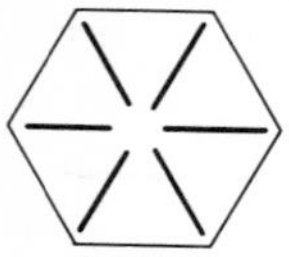 6 marks	SAE grade 8 ASTM — A 345 Best commercial quality	Medium carbon alloy steel, quenched and tempered	150,000 PSI

Figure 10. Grade markings for bolts. (*Machine Tools and Machining Practices*)

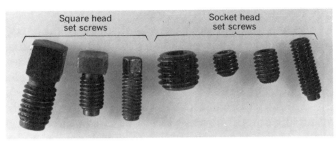

Figure 11. Socket and square head set screws. *(Machine Tools and Machining Practices)*

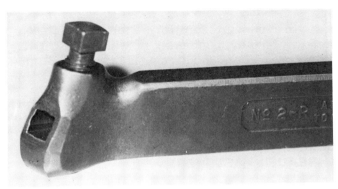

Figure 12. Square head set screws are found in tool holders. (Lane Community College)

Figure 13. Square head jack screw. (Lane Community College)

come in diameters from $\frac{1}{4}$ to 3 in., and their strength is indicated by symbols on the hex head (Figure 10). Slotted head capscrews can have flat heads, round heads, or fillister heads.

Socket head capscrews are also made with socket flat heads and socket button heads. The socket in the head used for turning the screw may be hexagonal or a spline type, which has four or six splines. Of course, the correct hexagonal or spline type wrench must be used in the socket. These screws are made of high grade steel, hardened and tempered to around 40Rc. The heads on these capscrews may be plain or knurled. Socket head capscrews are replacing hex head screws in many machinery applications and assemblies because of their ease of application and the fact that the head can be hidden in a counterbore. This style of capscrew is used almost exclusively in the tool and die and machine tool industries.

Set screws (Figure 11) are used to lock pulleys or collars on shafts. Set screws can have square heads with the head extending above the surface or, more often, the set screws are slotted or have socket heads. **Slotted or socket head set screws** usually disappear below the surface of the part to be fastened. A pulley or collar where the set screws are below the surface is much safer for persons working around them. Socket head set screws may have hex socket heads or spline socket heads. Set screws are manufactured in number sizes from 0 to 10 and in fractional sizes from $\frac{1}{4}$ to 2 in. Set screws are usually made from carbon or alloy steel and hardened.

Square head set screws are often used on tool holders (Figure 12) or as jackscrews in leveling machine tools (Figure 13). Set screws have several different points (Figure 14). The flat point set screw will make the least amount of indentation on a shaft and is used where frequent adjustments are made. A flat point set screw is also used to provide a jam screw action when a second set screw is tightened on another set screw to prevent its release through vibra-

Figure 14. Set screw points. *(Machine Tools and Machining Practices)*

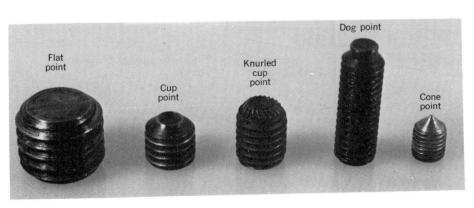

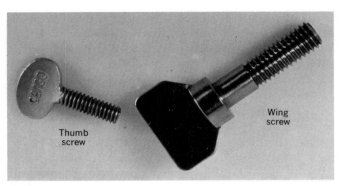

Figure 15. Thumb screw and wingscrew. *(Machine Tools and Machining Practices)*

tion. The oval point set screw will make a slight indentation as compared with the cone point. With a half dog or full dog point set screw holding a collar to a shaft, alignment between shaft and collar will be maintained even when the parts are disassembled and reassembled. This is because the shaft is drilled with a hole of the same diameter as the dog point. Cup pointed set screws will make a ring-shaped depression in the shaft and will give a very slip-resistant connection. Square head set screws have a class 2A thread and are usually supplied with a coarse thread. Slotted and socket head set screws have a class 3A UNC or UNF thread.

Thumbscrews and wingscrews (Figure 15) are used where parts are to be fastened or adjusted rapidly without the use of tools.

Thread forming screws (Figure 16) form their own threads and eliminate the need for tapping. These screws are used in the assembly of sheet metal parts, plastics, and nonferrous material. Thread forming screws form threads by displacing material with no cutting action. These screws require an existing hole of the correct size.

Thread cutting screws (Figure 17) make threads by cutting and producing chips. Because of the cutting action these screws need less driving torque than thread forming screws, Applications are similar as those for thread forming screws. These include fastening sheetmetal, aluminum brass, diecastings, and plastics.

Drive screws (Figure 18) are forced into the correct size hole by hammering or with a press. Drive screws make permanent connections and are often used to fasten name plates or identification plates on machine tools.

COMMON INTERNALLY THREADED FASTENERS

Nuts

Common nuts (Figure 19) are manufactured in as many sizes as there are bolts. Most nuts are either hex or square in shape. Nuts are identified by the size of the bolt they fit and not by their outside size. Common hex nuts are made in different thicknesses. A **thin hex nut** is called a **jam nut.** They are used where space is limited or where the strength of a

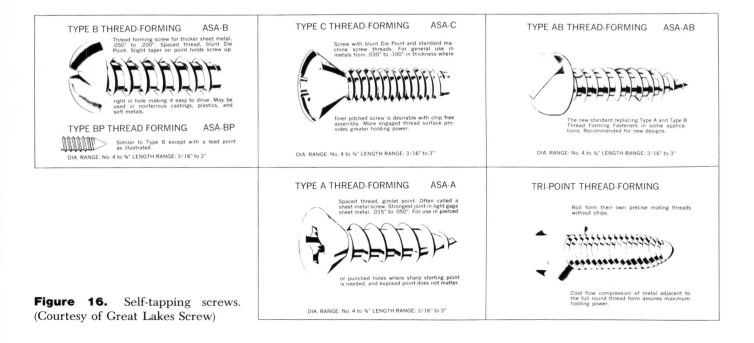

Figure 16. Self-tapping screws. (Courtesy of Great Lakes Screw)

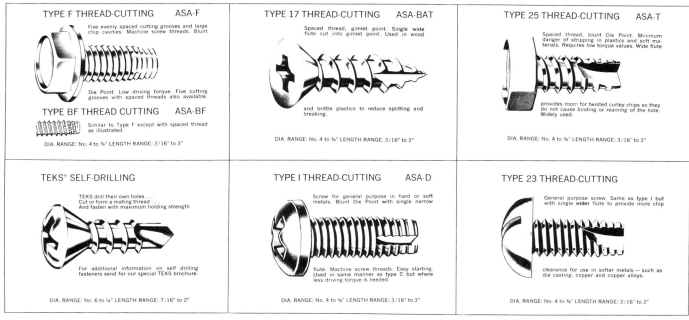

Figure 17. Thread cutting screws. (Courtesy of Great Lakes Screw)

Figure 18. Drive screw. *(Machine Tools and Machining Practices)*

regular nut is not required. Jam nuts are often used to lock other nuts (Figure 20). **Regular hex nuts** are slightly thinner than their size designation. A $\frac{1}{2}$ in. regular hex nut is $\frac{7}{16}$ in. thick. A $\frac{1}{2}$ in. heavy hex nut is $\frac{31}{64}$ in. thick. A $\frac{1}{2}$ in. light duty hex nut measures $\frac{11}{16}$ in. thick. Other common nuts include various stop or lock nuts. Two common types are the **elastic stop nut** and the **compression stop nut.** They are used in applications where the nut might vibrate off the bolt. **Wing nut** and **thumb nuts** are used where quick assembly or disassembly by hand is desired. Two other hex nuts are **slotted** and **castle nuts.** These nuts have slots cut into them. When the slots are aligned with holes in a bolt, a cotter pin may be used to prevent

Figure 19. Common nuts. *(Machine Tools and Machining Practices)*

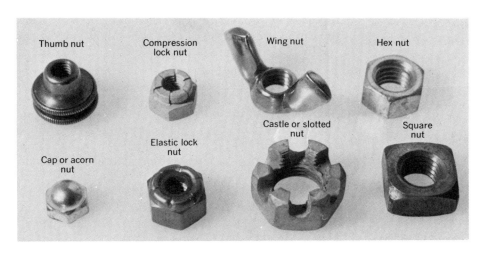

Thumb nut   Compression lock nut   Wing nut   Hex nut

Cap or acorn nut   Elastic lock nut   Castle or slotted nut   Square nut

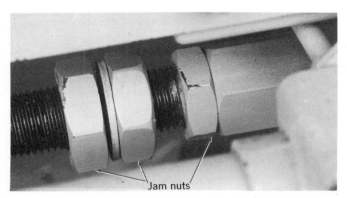

Figure 20. Jam nuts.

the nut from turning. Axle and spindles on vehicles have slotted nuts to prevent wheel bearing adjustments from slipping.

Cap or **acorn nuts** are often used where decorative nuts are needed. These nuts also protect projecting threads from accidental damage. Nuts are made from many different materials, depending on their application and strength requirements.

WASHERS, PINS, RETAINING RINGS, AND KEYS

Washers
Flat washers (Figure 21) are used under nuts and bolt heads to distribute the pressure over a larger area. Washers also prevent the marring of a finished surface when nuts or screws are tightened. Washers can be manufactured of many different materials. The nominal size of a washer is intended to be used with the same nominal size bolt or screw. Standard series of washers are narrow, regular, and wide. For example, the outside diameter of a $\frac{1}{4}$ in. narrow washer is $\frac{1}{2}$ in., the outside diameter of a $\frac{1}{4}$ in. regular washer is almost $\frac{3}{4}$ in., and the diameter of a wide $\frac{1}{4}$ in. washer measures 1 in.

Lock washers (Figure 22) are manufactured in

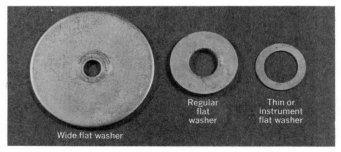

Figure 21. Wide, regular, and thin (or instrument) flat washers. *(Machine Tools and Machining Practices)*

many styles. The helical spring lock washer provides hardened bearing surfaces between a nut or bolt head and the components of an assembly. The spring-type construction of this lock washer will hold the tension between a nut and bolt assembly even if a small amount of looseness should develop. Helical spring lock washers are manufactured in five series: light, regular, heavy, extra duty, and hi-collar. The hi-collar lock washer has an outside diameter equal to the same nominal size socket head capscrew. This makes the use of these lock washers in a counterbored bolt hole possible. Counterbored holes have the end enlarged to accept the bolt head (Figure 23). A variety of standard tooth lock washers are produced, the external type providing the greatest amount of friction or locking effect between fastener and assembly. For use with small head screws and where a smooth appearance is desired, an internal tooth lock washer is used. When large bearing area is desired or where the assembly holes are oversized, an internal-external tooth lock washer is available. A countersunk tooth lock washer is used for a locking action with flat head screws.

Pins
Pins (Figure 24) find many applications in the assembly of parts. **Dowel pins** are heat treated and precision ground. Their diameter varies from the nominal dimension by only plus or minus .0001 in. ($\frac{1}{10.000}$ of an inch). Dowel pins are used where very accurate alignments must be maintained between two or more parts. Holes for dowel pins are reamed to provide a slight press fit. Reaming is a machining process during which a drilled hole is slightly enlarged to provide a smooth finish and accurate diameter. Dowel pins only locate. Clamping pressure is supplied by the screws. Dowel pins may be driven into a blind hole. A blind hole is closed at one end. When this kind of hole is used, provision must be made to let the air that is displaced by the pin escape. This can be done by drilling a small through hole or by grinding a narrow flat the full length of the pin. Always use the correct lubricant when screw and pin assemblies are made.

One disadvantage of dowel pins is that they tend to enlarge the hole in an unhardened workpiece if they are driven in and out several times. When parts are intended to be disassembled frequently, **taper pins** will give accurate alignment. Taper pins have a taper of $\frac{1}{4}$ in. per foot of length and are fitted into reamed taper holes. If a taper pin hole wears larger because of frequent disassembly, the hole can be reamed larger to receive the next larger size of taper pin. Diameters of taper pins range in size from $\frac{1}{16}$ in. to $\frac{11}{16}$ in. measured at the large end. Taper pins are identified by a number from 7/0 (small diameter) to number

EXTERNAL TYPE
External type lock washers provide greater torsional resistance due to teeth being on largest radius. Screw heads should be large enough to cover washer teeth. Available with left hand or alternate twisted teeth.

INTERNAL TYPE
For use with small screw heads or in applications where it is necessary to hide washer teeth for appearance or snag prevention.

EXTERNAL-INTERNAL TYPE
For use where a larger bearing surface is needed such as extra large screw heads or between two large surfaces. More biting teeth for greater locking power. Excellent for oversize or elongated screw holes.

HEAVY DUTY INTERNAL TYPE
Recommend for use with larger screws and bolts on heavy machinery and equipment.

DOME TYPE PLAIN PERIPHERY
For use with soft or thin materials to distribute holding force over larger area. Used also for oversize or elongated holes. Plain periphery is recommended to prevent surface marring.

DOME TYPE TOOTHED PERIPHERY
For use with soft or thin materials to distribute holding force over larger area. Used also for oversize or elongated holes. Toothed periphery should be used where additional protection against shifting is required.

COUNTERSUNK TYPE
Countersunk washers are used with either flat or oval head screws in recessed countersunk applications. Available for 82° and 100° heads and also internal or external teeth.

DISHED TYPE PLAIN PERIPHERY
Recommended for the same general applications as the dome type washers but should be used where more flexibility rather than rigidity is desired. Plain periphery for reduced marring action on surfaces.

DISHED TYPE TOOTHED PERIPHERY
Recommended for the same general applications as the dome type washers but should be used where more flexibility rather than rigidity is desired. Toothed periphery offers additional protection against shifting.

PYRAMIDAL TYPE
Specially designed for situations requiring very high tightening torque. The pyramidal washer offers bolt locking teeth and rigidity yet is flexible under heavy loads. Available in both square and hexagonal design.

FINISH TYPE
Recommended where marring or tearing of surface material by turning screw head must be prevented and for decorative use.

HELICAL SPRING LOCK TYPE
Spring lock washers may be used to eliminate annoying rattles and provide tension at fastening points.

CONE SPRING TYPE

CONE SPRING TYPE SERRATED PERIPHERY
Same general usage as the cone type with plain periphery but with the added locking action of a serrated periphery. Takes high tightening torque.

FLAT TYPE
For use with oversize and elongated screw holes. Spreads holding force over a larger area. Used also as a spacer. Available in all metals.

SPECIAL TYPES
Special washers with irregular holes, cup types, plate types with multiple holes or tab types may be supplied upon request. Consult our engineering department for any of your special needs.

FIBER AND ASBESTOS
In cases where insulation or corrosion resistance is more important than strength, fiber or asbestos washers are available.

DOUBLE SEMS
Two washers securely held from slipping off, yet free to spin and lock. Prevents gouging of soft metals.

Figure 22. Lock washers. (Courtesy of Great Lakes Screw)

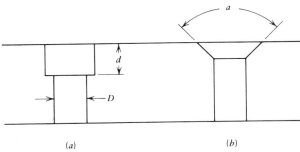

Figure 23. (a) Counterbore. Note that the depth of the counterbore (d) is always the same as the bolt diameter (D) of socket head capscrews. (b) Countersink. Included angles (a) of common screws are 82 degrees. They are also made 80, 90, 92, and 100 degrees included angle.

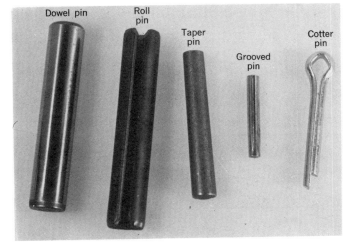

Figure 24. Pins.

10 (large diameter) as well as by their length. The large end diameter is constant for a given size pin, but the small diameter changes with the length of the pin. Taper pins are sometimes made with a threaded end to hold them in place even when subjected to severe vibration (Figure 25).

A **grooved pin** is either a cylindrical or tapered pin with longitudinal grooves pressed into the pin body. This causes the pin to deform. A groove pin will hold securely in a drilled hole even after repeated removal.

Roll pins can also be used in drilled holes with

Figure 25. Threaded taper pin. (Lane Community College)

no reaming required. These pins are manufactured from flat steel bands and rolled into cylindrical shape. Roll pins, because of their spring action, will stay tight in a hole even after repeated disassemblies.

Cotter pins are used to retain parts on a shaft or to lock a nut or a bolt as a safety precaution. Cotter pins make quick assembly and disassembly possible.

Retaining Rings

Retaining rings (Figure 26) are fasteners used in many assemblies. Retaining rings can easily be installed in machined grooves, internally in housings, or externally on shafts or pins (Figure 27). Some types of retaining rings do not require grooves but have a self-locking spring-type action. The most common application of a retaining ring is to provide a shoulder to hold and retain a bearing or other part on an otherwise smooth shaft. They may also be used in a bearing housing (Figure 28). Special pliers are used to install and remove retaining rings.

Keys

Keys (Figure 29) are used to prevent the rotation of gears or pulleys on a shaft. Keys are fitted into key seats in both the shaft and the external part. Keys should fit the key seats rather snugly. **Square keys,** where the width and the height are equal, are preferred on shaft sizes up to a $6\frac{1}{2}$ in. diameter. Above a $6\frac{1}{2}$ in. diameter rectangular keys are recommended. **Woodruff keys,** which are almost in the shape of a half circle, are used where relatively light loads are transmitted. One advantage of woodruff keys is that they cannot change their location on a shaft because they are retained in a pocket. A key fitted into an

Figure 26. Retaining rings. (Waldes Kohinoor, Inc.)

INTERNAL	**BASIC N5000** For housings and bores — Size Range 250—10.0 in. / 6.4—254.0 mm.	EXTERNAL	**BOWED 5101** For shafts and pins — Size Range .188—1.750 in. / 4.8—44.4 mm.	EXTERNAL	**REINFORCED 5115** For shafts and pins — Size Range .094—1.0 in. / ●	EXTERNAL	**TRIANGULAR NUT 5300** For threaded parts — Size Range 6-32 and 8-32 10-24 and 10-32 1/4-20 and 1/4-28
INTERNAL	**BOWED N5001** For housings and bores — Size Range .250—1.750 in. / 6.4—44.4 mm.	EXTERNAL	**BEVELED 5102** For shafts and pins — Size Range 1.0—10.0 in. / 25.4—254.0 mm.	EXTERNAL	**BOWED E-RING 5131** For shafts and pins — Size Range .110—1.375 in. / 2.8—34.9 mm.	EXTERNAL	**KLIPRING 5304** T-5304 For shafts and pins — Size Range .156—1.000 in. / 4.0—25.4 mm.
INTERNAL	**BEVELED N5002** For housings and bores — Size Range 1.0—10.0 in. / 25.4—254.0 mm.	EXTERNAL	**CRESCENT® 5103** For shafts and pins — Size Range .125—2.0 in. / 3.2—50.8 mm.	EXTERNAL	**E-RING 5133** For shafts and pins — Size Range 040—1.375 in. / 1.0—34.9 mm.	EXTERNAL	**TRIANGULAR 5305** For shafts and pins — Size Range .062—.438 in. / ●
INTERNAL	**CIRCULAR 5005** For housings and bores — Size Range .312—2.0 in. / ●	EXTERNAL	**CIRCULAR 5105** For shafts and pins — Size Range .094—1.0 in. / ●	EXTERNAL	**PRONG-LOCK® 5139** For shafts and pins — Size Range .092—.438 in. / ●	EXTERNAL	**GRIPRING® 5555** For shafts and pins — Size Range .079—.750 in. / 2.0—19.0 mm.
INTERNAL	**INVERTED 5008** For housings and bores — Size Range .750—4.0 in. / 19.0—101.6 mm.	EXTERNAL	**INTERLOCKING 5107** For shafts and pins — Size Range .469—3.375 in. / 11.9—85.7 mm.	EXTERNAL	**REINFORCED E-RING 5144** For shafts and pins — Size Range .094—.562 in. / 2.4—14.3 mm.	EXTERNAL	**HIGH-STRENGTH 5560** For shafts and pins — Size Range .101—.328 in. / ●
EXTERNAL	**BASIC 5100** For shafts and pins — Size Range .125—10.0 in. / 3.2—254.0 mm.	EXTERNAL	**INVERTED 5108** For shafts and pins — Size Range .500—4.0 in. / 12.7—101.6 mm.	EXTERNAL	**HEAVY-DUTY 5160** For shafts and pins — Size Range .394—2.0 in. / 10.0—50.8 mm.	EXTERNAL	**PERMANENT SHOULDER 5590** For shafts and pins — Size Range .250-.750 / 6.4-19.0 mm.

Figure 27. External retaining ring used on a shaft. (Courtesy of Waldes Kohinoor, Inc.)

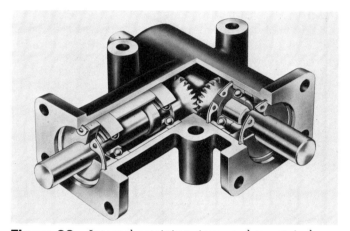

Figure 28. Internal retaining rings used to retain bearings. (Courtesy of Waldes Kohinoor, Inc.)

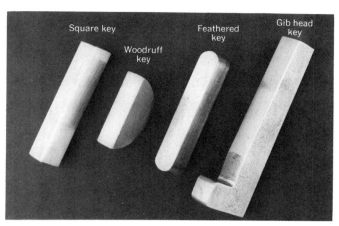

Figure 29. Keys. (Lane Community College)

endmilled groove will also retain its axial position on the shaft. Most of these keys are held under tension with one or more set screws threaded through the hub of the pulley or gear. Where extremely heavy shock loads or high torques are encountered, a **taper key** is used. Taper keys have a taper of $\frac{1}{8}$ in. per foot. Where a tapered key is used, the key seat in the shaft is parallel to the shaft axis and a taper to match the key is in the hub. Where only one side of an assembly is accessible, a **gib head taper key** is used instead of a plain taper key. When a gib head taper key is driven into the key seat as far as possible, a gap remains between the gib and the hub of the pulley or gear. The key is removed for disassembly by driving a wedge into the gap to push the key out. A **feathered key** is a key that is secured in a key seat with screws. A feathered key is often a part of a sliding gear or sliding pulley.

SELF-TEST

1. What is the difference between a bolt and a screw?
2. How much thread engagement is recommended when a screw is used in an assembly?
3. When are class 2 threads used?
4. What is the difference between a machine bolt and a capscrew?
5. What is the outside diameter of a No. 8 machine screw?
6. Where are set screws used?
7. When are stud bolts used?
8. Explain the difference between thread forming and thread cutting screws.
9. Where are castle nuts used?
10. Where are cap nuts used?
11. Explain two reasons why flat washers are used.
12. What is the purpose of a helical spring lock washer?
13. When is an internal-external tooth lock washer used?
14. When are dowel pins used?
15. When are taper pins used?
16. When are roll pins used?
17. What are retaining rings?
18. What is the purpose of a key?
19. When is a woodruff key used?
20. When is a gib head key used?

UNIT 3 SELECTION AND IDENTIFICATION OF METALS

When the village smithy plied his trade, there were only wrought iron and carbon steel for making tools, implements, and horseshoes, so the task of separating metals was relatively simple. As industry grew, more alloy steels and special metals were needed, and thus gradually developed. Today many hundreds of these metals are used. Without some means of reference to or identification of metals, work in the machine shop would be chaotic. Therefore, this unit introduces you to several systems used for marking steels, known as ferrous metals, and some ways to choose between them. You will also learn how to identify many nonferrous metals, metals other than iron and steel.

OBJECTIVE

After completing this unit, you should be able to:
Identify different types of metals by various means of shop testing.

STEEL IDENTIFICATION SYSTEMS

Color coding is used as one means of identifying a particular type of steel. Its main disadvantage is that there is no universal color coding system. Manufacturers and local shops all have different codes.

The most common systems in the United States used to classify steels by chemical composition were developed by the Society of Automotive Engineers (SAE) and American Iron and Steel Institute (AISI). The SAE and AISI systems use a four or five digit number (Table 1). The first number indicates the type of steel. Carbon, for instance, is denoted by the number 1, 2 is a nickel steel, 3 is a nickel-chromium steel, and so on. The second digit indicates the approximate percentage of the predominant alloying element. The third and fourth digits, represented by x, always denote the percentage of carbon in hundredths of one percent. For example, SAE 1040 denotes plain carbon steel with 0.40 percent carbon; SAE 4140 denotes a chromium-molybdenum steel containing 0.40 percent carbon and about 1.0 percent of the major alloy (molybdenum). Plain carbon steels, alloys, and tool steels can contain anywhere from 0.06 to 1.70 percent carbon. Steels having over one percent carbon require a five digit number. Certain corrosion and heat resisting alloys are classified with a five digit number in order to identify the approximate alloy composition of the steel.

The AISI numerical system is basically the same as the SAE system with certain capital letter prefixes. These prefixes designate the process used to make the steel. The lowercase letters from a to i as a suffix denote special conditions in the steel.

AISI prefixes:

B—Acid Bessemer, carbon steel

C—Basic open hearth carbon steel

CB—Either acid Bessemer or basic open hearth carbon steel at the option of the manufacturer

D—Acid open hearth carbon steel

E—Electric furnace alloy steel

STAINLESS STEEL

It is the element chromium (Cr) that makes stainless steels stainless. Steel must contain a minimum of about 11 percent chromium in order to gain resistance to atmospheric corrosion. Higher percentages of chromium make steel even more resistant to corrosion and high temperatures. Nickel is added to improve ductility, corrosion resistance, and other properties.

Table 1
SAE-AISI Numerical Designation of Alloy Steels
(x Represents Percent of Carbon in Hundredths)

Carbon Steels	
Plain carbon	10xx
Free-cutting, resulfurized	11xx
Manganese Steels	13xx
Nickel Steels	
0.50% nickel	20xx
1.50% nickel	21xx
3.50% nickel	23xx
5.00% nickel	25xx
Nickel-Chromium Steels	
1.25% nickel, 0.65% chromium	31xx
1.75% nickel, 1.00% chromium	32xx
3.50% nickel, 1.57% chromium	33xx
3.00% nickel, 0.80% chromium	34xx
Corrosion and Heat-Resisting Steels	303xx
Molybdenum Steels	
Chromium	41xx
Chromium-nickel	43xx
Nickel	46xx and 48xx
Chromium Steels	
Low-chromium	50xx
Medium-chromium	51xx
High-chromium	52xx
Chromium-Vanadium Steels	6xxx
Tungsten Steels	7xxx
Triple Alloy Steels	8xxx
Silicon-Manganese Steels	9xxx
Leaded Steels	11Lxx (example)

Source: John E. Neely, *Practical Metallurgy and Materials of Industry*, John Wiley and Sons, Inc., Copyright © 1979, New York.

Excluding the precipitation hardening types that harden over a period of time after solution heat treatment, there are three basic types of stainless steels: the martensitic and ferritic types of the 400 series, and the autenitic types of the 300 series.

The martensitic, hardenable type has carbon content up to 1 percent or more, so it can be hardened by heating to a high temperature and then quenching (cooling) in oil or air. The cutlery grades of stainless are to be found in this group. The ferritic type contains little or no carbon. It is essentially soft iron that has 11 percent or more chromium content. It is the least expensive of the stainless steels and is used for such things as building trim, pots, and pans. Both ferritic and martensitic types are magnetic.

Austenitic stainless steel contains chromium and nickel, little or no carbon, and cannot be hardened by quenching, but it readily work hardens while retaining much of its ductility. For this reason it can be work hardened until it is almost as hard as a hardened martensitic steel. Austenitic stainless steel is

somewhat magnetic in its work hardened condition, but nonmagnetic when annealed or soft. See the Glossary for definitions of these metallurgical terms.

Table 2 illustrates the method of classifying the stainless steels. Only a very few of the basic types are given here. You should consult a manufacturer's catalog for further information.

TOOL STEELS
Special carbon and alloy steels called tool steels have their own classification. There are six major tool steels for which one or more letter symbols have been assigned:

1. Water hardening tool steels
   W—High carbon steels
2. Shock resisting tool steels
   S—Medium carbon, low alloy
3. Cold work tool steels
   O—Oil hardening types
   A—Medium alloy air hardening types
   D—High carbon, high chromium types
4. Hot work tool steels
   H—H1 to H19, chromium base types; H20 to H39, tungsten base types; H40 to H59, molybdenum base types
5. High speed tool steels
   T—Tungsten base types
   M—Molybdenum base types
6. Special purpose tool steels
   L—Low alloy types
   F—Carbon tungsten types
   P—Mold steels P1 to P19, low carbon types; P20 to P39, other types

Several metals can be classified under each group, so that an individual type of tool steel will also have a suffix number that follows the letter symbol of its alloy group. The carbon content is given only in those cases where it is considered an identifying element of that steel.

TYPE OF STEEL	EXAMPLES
Water hardening:	
straight carbon tool steel	W1, W2, W4
Manganese, chromium, tungsten:	
oil hardening tool steel	O1, O2, O6
Chromium (5.0%):	
air hardening die steel	A2, A5, A10
Silicon, manganese, molybdenum:	
punch steel	S1, S5
High speed tool steel	M2, M3, M30
	T1, T5, T15

Table 2
Classification of Stainless Steels

Class	Alloy Content	Metallurgical Structure	Ability to be Heat Treated
I	Chromium	Martensitic	Hardenable (types 410, 414, 416, 405, 403, 420, 440, 431)
II	Chromium	Ferritic	Nonhardenable (types 430, 442, 409, 400, 446, Armco 18 SR)
III	Chromium-nickel	Austenitic	Nonhardenable (types 302, Armco 18-9 LW, 301, 303, 304, 305, 308, 309, 310, 347, 316, 321, 348, 317)
	Chromium-nickel-manganese	Austenitic	Nonhardenable (Armco Nitronic 32, Armco Nitronic 33, Armco Nitronic 40, Armco Nitronic 50, Armco Nitronic 60)
IV	Chromium-nickel (copper added)	Martensitic	Precipitation-hardening (Armco 17-4 PH, Armco 15-5 PH, Armco 13-8 Mo)
	Chromium-nickel	Semi-austenitic	Precipitation-hardening (Armco 17-7 PH, Armco PH 15-7 Mo, Armco PH 14-8 Mo)
	Chromium-nickel	Austenitic	Precipitation-hardening (Armco 17-10 P)

Source: Armco Inc., Middletown, Ohio, *Development of the Stainless Steels*, Copyright © 1977, Armco.

Note: The precipitation hardening stainless steels (for example Armco 17-4 PH) are normally purchased in the solution treated condition and somewhat martensitic, but machinable (Rc34 to Rc38). Solution treatment consists of heating to 1875 to 1950°F (1024 to 1065°C) for one-half hour and oil quenching. Further hardening is done by reheating 900 to 1150°F (482 to 621°C) and air cooling. This produces a Rockwell hardness up to Rc44.

SHOP TESTS FOR IDENTIFYING STEELS

One of the disadvantages of steel identification systems is that the marking is often lost. The end of a shaft is usually marked. If the marking is obliterated or cut off and the piece is separated from its proper storage rack, it is very difficult to determine its carbon content and alloy group. This shows the necessity of returning stock material to its proper rack. It is also good practice to leave the identifying mark on one end of the stock material and always to cut off the other end.

Unfortunately, in the shop there are always some short ends and otherwise useful pieces that have become unidentified. In addition, when repairing or replacing parts for old or nonstandard machinery, there is usually no record available for material selection. There are many shop methods a worker may use to identify the basic type of steel in an unknown sample. By the process of elimination, the worker can then determine which of the several steels of that type in the shop is most comparable to his sample. The following are several methods of shop testing that you can use.

Visual

Some metals can be identified by visual observation of their surface finishes (Figure 1). Heat scale or black mill scale is found on all hot rolled (HR) steels. These can be either low carbon (0.05 to 0.30 percent), medium carbon (0.30 to 0.60 percent), high carbon (0.60 to 1.70 percent), or alloy steels. Other surface coatings that might be detected are the sherardized, plated, case hardened, or nitrided surfaces (Figure 2). Sherardizing is a process in which zinc vapor is inoculated into the surface of iron or steel.

Cold rolled (CR) steel usually has a metallic luster. Not all cold worked steel is rolled; some bars are cold drawn through a die. Cold finish (CF) is a term used to designate the finish on all types of cold worked steel. Ground and polished (G and P) steel (for example, drill rod) has a bright, shiny finish with closer dimensional tolerances than CF (Figure 2). Also, cold drawn ebonized, or black, finishes are sometimes found on alloy and resulfurized shafting.

Chromium nickel stainless steel, which is austenitic and nonmagnetic, usually has a white appearance. Straight 12 to 13 percent chromium is ferritic and

Figure 1. The contrast of surface textures of hot rolled (HR) and cold rolled (CR) steels can be seen in this photograph. HR is on the left and CR on the right. (Lane Community College)

Figure 3. The ferritic stainless steel is attracted to a magnet and the austenitic is not. (Lane Community College)

Figure 2. Round bars having various surface finishes. Left to right: aluminum, ground and polished, cold finished steel, hot rolled steel, sherardized surface, and zinc dip surface. (John E. Neely, *Practical Metallurgy and Materials of Industry*, John Wiley and Sons, Inc., Copyright © 1979, New York)

magnetic with a bluish-white color. Manganese steel is blue when polished, but copper colored when oxidized. White cast iron fractures will appear silvery or white. Gray cast iron fractures appear dark gray and will smear a finger with a gray graphite smudge when touched.

Magnet Test

All ferrous metals such as iron and steel are magnetic; that is, they are attracted to a magnet. With the exception of nickel, cobalt, and a few rare earth metals, nonferrous metals are generally not attracted to a

magnet. United States "nickel" coins contain about 25 percent nickel and 75 percent copper, so they do not respond to the magnet test, but Canadian "nickel" coins are attracted to a magnet because they contain a much larger proportion of nickel. Ferritic and martensitic (400 series) stainless steels are also attracted to a magnet and so cannot be separated from other steels by this method. Austenitic (300 series) stainless steel is not attracted to a magnet unless it is work hardened (Figure 3).

Hardness Test

Wrought iron is very soft since it contains almost no carbon or any other alloying element. Generally speaking, the more carbon (up to 2 percent) and other elements that steel contains, the harder, stronger and less ductile it becomes, even if in an annealed state. Thus, the hardness of a sample can help us to separate low carbon steel from an alloy steel or a high carbon steel.

The Rockwell Hardness Tester (Figure 4) and the Brinell Hardness Tester are the most commonly used types of hardness testers for industrial and metallurgical purposes. Heat treaters, inspectors, and many others in industry often use these machines. These instruments provide a quick, accurate method of determining hardness by the use of a given weight and a penetrator. The Brinell test gives a reading in BHN (Brinell hardness number). The Rockwell test has several scales of which the B and C scales are most used: the B scale for soft metals such as aluminum, and the C scale for hard metals such as steel. Mild steel

Figure 4. The Rockwell Hardness Tester. (Lane Community College)

Figure 5a. A piece of keystock (mild steel) is scratched across an unknown sample. Since the sample is not scratched, it is harder than the keystock and probably is an alloy or tool steel. (John E. Neely, *Practical Metallurgy and Materials of Industry,* John Wiley and Sons, Inc., Copyright © 1979, New York)

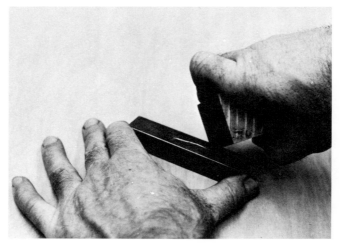

Figure 5b. The sample is now scratched against the keystock as a further test and it does scratch the keystock (*Practical Metallurgy and Materials of Industry*)

can be in the range of 10 to 30 Rc (Rockwell), 180 to 290 BHN. Spring steel is about 45 Rc, 430 BHN, and file hard is about 62 Rc, 685 BHN. See Table 1 in Appendix I.

Not all shops have hardness testers available, in which case the scratch test or the file test may be used.

Scratch Test

Geologists and "rock hounds" scratch rocks against items of known hardness for identification purposes. The same method can be used to check metals for relative hardness. Simply scratch one sample with another and the softer sample will be marked. Be sure all scale or other surface impurities have been removed before scratch testing (Figures 5a and 5b). A variation of this method is to strike two similar edges of two samples together. The one receiving the deepest indentation is the softer of the two.

File Test

Files can be used to establish the relative hardness between two samples, as in the scratch test, or they

can determine an approximate hardness of a piece on a scale of many steels. Table 3 gives the Rockwell and Brinell hardness numbers for this file test when using new files. This method, however, can only be as accurate as the skill that the user has acquired through practice.

Care must be taken not to damage the file, since filing on hard materials may ruin the file. Testing should be done on the tip or near the edge of the file.

Table 3
File Test and Hardness Table

Type Steel	Rockwell B	C	Brinell	File Reaction
Mild steel	65		100	File bites easily into metal. (Machines well but makes built up edge on tool.)
Medium carbon steel		16	212	File bites into metal with pressure. (Easily machined with high speed tools.)
High alloy steel				File does not bite into metal except with difficulty.
High carbon steel		31	294	(Readily machinable with carbide tools.)
Tool steel		42	390	Metal can only be filed with extreme pressure. (Difficult to machine even with carbide tools.)
Hardened tool steel		50	481	File will mark metal but metal is nearly as hard as the file, and machining is impractical; should be ground.
Case hardened parts and hardened tool steel		64	739	Metal is as hard as the file; should be ground.

Source: John E. Neely, *Practical Metallurgy and Materials of Industry*, John Wiley and Sons, Inc., Copyright © 1979, New York.
Note: Rockwell and Brinell hardness numbers are only approximations since file testing is not an accurate method of hardness testing.

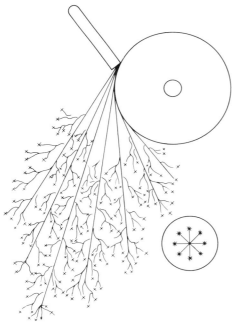

Figure 6. High carbon steel. Short, very white, or light yellow carrier lines with considerable forking, having many starlike bursts. Many of the sparks follow around the wheel.

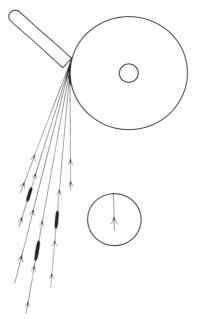

Figure 7. Low carbon steel. Straight carrier lines having a yellowish color with very small amount of branching and very few carbon bursts.

Spark Testing

Spark testing is a useful way to test for carbon content in many steels. The metal tested, when held against a grinding wheel, will display a particular spark pattern depending on its content. Spark testing provides a convenient means of distinguishing between tool steel (of medium or high carbon) and low carbon steel. High carbon steel (Figure 6) shows many more bursts than low carbon steel (Figure 7).

Almost all tool steel contains some alloying elements besides the carbon, which affects the carbon burst. Chromium, molybdenum, silicon, aluminum,

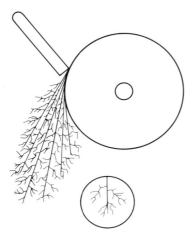

Figure 8. Cast iron. Short carrier lines with many bursts, which are red near the grinder and orange-yellow further out. Considerable pressure is required on cast iron to produce sparks.

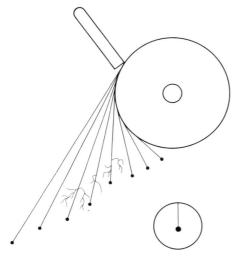

Figure 9. High speed steel. Carrier lines are orange, ending in pear-shaped globules with very little branching or carbon sparks. High speed steel requires moderate pressure to produce sparks.

and tungsten suppress the carbon burst. For this reason spark testing is not very useful in determining the content of an unknown sample of steel. It is useful, however, as a comparison test. Comparing the spark of a known sample to that of an unknown sample can be an effective method of identification for the trained observer. Cast iron may be distinguished from steel by the characteristic spark stream (Figure 8). High speed steel can also be readily identified by spark testing (Figure 9).

When spark testing, always wear safety glasses or a face shield. Adjust the wheel guard so the spark

Figure 10. Machinability test. When one of two samples having the same feed and speed shows a darker color (blue) on the chip, it can be assumed that it is a harder, stronger metal. (Lane Community College)

will fly outward and downward, and away from you. A coarse grit wheel that has been freshly dressed to remove contaminants should be used.

Machinability

Machinability will be studied further in another unit, but as a simple comparison test, it can be useful to determine a specific type of steel. For example, two unknown samples identical in appearance and size can be test cut in a machine tool, using the same speed and feed for both. The ease of cutting should be compared, and chips observed for heating color and curl (Figure 10).

Selection of Steel

Several properties should be considered when selecting a piece of steel for a job: strength, machinability, hardenability, weldability, fatigue resistance, and corrosion resistance. Manufacturer's catalogs and reference books are available for selection of standard structural shapes, bars, and other steel products. Others are available for the stainless steels, tool steels, and finished carbon steel and alloy shafting. Many of these steels are known by a trade name.

A worker is often called upon to select a shaft material from which to machine finish a part. Shafting is manufactured with two kinds of surface finish: cold finished (CF) found on low carbon steel, and ground and polished (G and P) found mostly on alloy steel shafts. Tolerances are kept much closer on ground and polished shafts. The following are some common alloy steels:

1. SAE 4140 is a chromium-molybdenum alloy with 0.40 percent carbon. It lends itself readily to heat treating, forging, and welding. It provides a high resistance to torsional and reversing stresses such as drive shafts.

2. SAE 1140 is a resulfurized, drawn, free machining bar stock. This material has good resistance to bending stresses because of its fibrous qualities and it has good tensile strength, but it has low resistance to high torque values. SAE 1140 is also useful where stiffness is required. It should not be heat treated or welded.

3. Leaded steels have all the free machining qualities and finishes of resulfurized steels. Leaded alloy steels such as SAE41L40 have the superior strength of 4140, but are much easier to machine.

4. SAE 1040 is a medium carbon steel that has a normalized tensile strength of about 85,000 PSI. It can be heat treated, but large sections will be hardened only on the surface and the core will still be in a normalized or relatively soft condition.

5. SAE 1020 is a low carbon steel that has good machining characteristics. It normally comes as CF shafting. It is very commonly used for shafting in industrial applications. It has a lower tensile strength than the alloy steels or higher carbon steels.

NONFERROUS METALS

Nonferrous metals such as gold, silver, copper, and tin were in use hundreds of years before the smelting of iron, and yet, some nonferrous metals have appeared relatively recently in common industrial use. For example, aluminum was first commercially extracted from ore in 1886 by the Hall–Heroult process; titanium, a space age metal, was produced in commercial quantities only after World War II.

In general, nonferrous metals are more costly than ferrous metals. It is not always easy to distinguish a nonferrous metal from a ferrous metal, nor to separate one from another, although they often vary widely in color and density or pounds per cubic inch (lb/in.^{3}).

Aluminum

Aluminum is white or white-gray in color and can have any surface finish from dull to shiny when polished. An **anodized** surface is frequently found on aluminum products. Aluminum weighs 168.5 pounds per cubic foot as compared to 487 pounds per cubic foot for steel, and has a melting point of 1220°F (660°C) when pure. It is readily machinable and can be manu-

Figure 11. Structural aluminum shapes used for building trim provides a pleasing appearance. *(Machine Tools and Machining Practices)*

factured into almost any shape or form (Figure 11).

Magnesium is also a much lighter metal than steel, as it weighs 108.6 pounds per cubic foot, and looks much like aluminum. In order to distinguish between the two metals, it is sometimes necessary to make a chemical test. A zinc chloride solution in water, or copper sulfate, will blacken magnesium immediately, but will not change aluminum.

There are several numerical systems used to identify aluminums, such as federal specifications, military specifications, the American Society for Testing Materials (ASTM), and SAE specifications. The system most used by manufacturers, however, is one adopted by the Aluminum Association in 1954.

From Table 4, you can see that the first digit of a number in the aluminum alloy series indicates the

Table 4
Aluminum and Aluminum Alloys

Code Number	Major Alloying Element
1xxx	None
2xxx	Copper
3xxx	Manganese
4xxx	Silicon
5xxx	Magnesium
6xxx	Magnesium and silicon
7xxx	Zinc
8xxx	Other elements
9xxx	Unused (not yet assigned)

Source: Warren T. White, John E. Neely, Richard R. Kibbe, and Roland O. Meyer, *Machine Tools and Machining Practices*, John Wiley and Sons, Inc., Copyright © 1977, New York.

alloy type. The second digit, represented by an x in the table, indicates any modifications that were made to the original alloy. The last two digits indicate the numbers of similar aluminum alloys of an older marking system, except in the 1100 series, where the last two digits indicate the amount of pure aluminum above 99 percent contained in the metal.

EXAMPLES

An aluminum alloy numbered 5056 is an aluminum-magnesium alloy, where the first 5 represents the alloy magnesium, the second digit represents modifications to the alloy, and 56 are numbers of a similar aluminum alloy of an older marking system. An aluminum numbered 1120 contains no major alloy, and has 0.20 percent pure aluminum above 99 percent.

Aluminum and its alloys are produced as castings or as wrought (cold worked) shapes such as sheets, bars, and tubing. They can be either cold worked by rolling, drawing, or extruding, or hot worked by forging. Cast aluminum is not generally as strong as wrought aluminum. Aluminum alloys are harder than pure aluminum and will scratch the softer (1100 series) aluminum. A 10 percent solution of sodium hydroxide (caustic soda) will not stain pure aluminum but will leave a dark stain on the aluminum alloy.

Pure aluminum and some of its alloys cannot be hardened by heat treating, but they can be annealed to soften them by heat treatment. One method of hardening aluminum is by cold working (strain hardening). These temper (hardness) designations are made by a letter that follows the four digit alloy series number:

- —F   As fabricated. No special control over strain hardening or temper designation is noted.
- —O   Annealed, recrystallized wrought products only. Softest temper.
- —H   Strain hardened, wrought products only. Strength is increased by work hardening.

This letter —H is always followed by two or more digits. The first digit, 1, 2, or 3, denotes the final degree of strain hardening.

- —H1   Strain hardened only
- —H2   Strain hardened and partially annealed
- —H3   Strain hardened and stabilized

The second digit denotes higher strength tempers obtained by heat treatment:

2   $\frac{1}{4}$ hard
4   $\frac{1}{2}$ hard
6   $\frac{3}{4}$ hard
8   Full hard
9   Extra hard

EXAMPLE

5056-H18 is an aluminum-magnesium alloy, strain hardened to a full hard temper.

Another method of hardening some aluminum alloys is a process called solution heat treatment and precipitation hardening or aging. This process involves heating the aluminum alloy to a temperature where the alloying element is dissolved into a solid solution. The aluminum alloy is then quenched in water and allowed to age or is artificially aged by heating slightly. The aging produces an internal strain that hardens and strengthens the aluminum. Some other nonferrous metals are also hardened by this process. For these aluminum alloys the letter —T follows the four digit series number. Numbers 2 to 10 follow this letter to indicate the sequence of treatment.

- —T2   Annealed (cast products only)
- —T3   Solution heat treated and cold worked
- —T4   Solution heat treated, but naturally aged
- —T6   Solution heat treated and artificially aged
- —T8   Solution heat treated, cold worked, and artificially aged
- —T9   Solution heat treated, artificially aged, and cold worked
- —T10 Artificially aged and then cold worked

EXAMPLE

2024-T6 Aluminum-copper alloy, solution heat treated and artificially aged.

Cast aluminum alloys generally have lower tensile strength than wrought alloys. Sand castings, permanent mold, and die casting alloys are of this group. They owe their mechanical properties to solution heat treatment and precipitation or to the addition of alloys. A classification system similar to that of wrought aluminum alloys is used (Table 5).

The cast aluminum 108F, for example, has an ultimate tensile strength of 24,000 PSI in the as-fabricated condition and contains no alloy. The 220.T4 copper aluminum alloy has a tensile strength of 48,000 PSI.

Cadmium

Cadmium has a blue-white color and is commonly used as a protective plating on parts such as screws, bolts, and washers. It is also used as an alloying ele-

Table 5

Cast Aluminum Alloy Designations

Code Number	Major Alloy Element
1xx.x	None, 99 percent aluminum
2xx.x	Copper
3xx.x	Silicon with Cu and/or Mg
4xx.x	Silicon
5xx.x	Magnesium
6xx.x	Zinc
7xx.x	Tin
8xx.x	Unused series
9xx.x	Other major alloys

Source: John E. Neely, *Practical Metallurgy and Materials of Industry,* John Wiley and Sons, Inc., Copyright © 1979, New York.

ment to make metal alloys that melt at low temperature, such as bearing metals, solder, type casting metals, and storage batteries. Cadmium compounds such as cadmium oxide are toxic and can cause illness when breathed. These toxic fumes can be produced by welding, cutting, or machining on cadmium plated parts. Breathing the fumes should be avoided by using adequate ventilation systems. The melting point of cadmium is 610°F (321°C). Its weight is 539.6 lb/ft³.

Copper and Copper Alloys

Copper is a soft, heavy metal that has a reddish color. It has high electrical and thermal conductivity when pure, but loses these properties to a certain extent when alloyed. It must be strain hardened when used for electric wire. Copper is very ductile and can be easily drawn into wire or tubular products. It is so soft that it is difficult to machine and it has a tendency to adhere to tools. Copper can be work hardened or hardened by solution heat treatment when alloyed with beryllium. The melting point of copper is 1981°F (1083°C). Its weight is 554.7 lb/ft³.

Beryllium Copper. Beryllium copper is an alloy of copper and beryllium that can be hardened by heat treating for making nonsparking tools and other products (Figure 12). Machining of this metal should be done after solution heat treatment and aging, and not when it is in the annealed state. Machining or welding beryllium copper can be very hazardous, if safety precautions are not followed. Machining dust or welding fumes should be removed by a heavy coolant flow or by a vacuum exhaust system, and a respirator type of face mask should be worn. The melting point of beryllium is 2345°F (1285°C), and its weight is 115 lb/ft³.

Figure 12. Beryllium copper chisel being used to cut a chip in mild steel. (Lane Community College)

Brass. Brass is an alloy of zinc and copper. Brass colors usually range from white to yellow, and in some alloys, red to yellow. Brasses range from gilding metal used for jewelry (95 percent copper, 5 percent zinc) to Muntz metal (60 percent copper, 40 percent zinc) used for bronzing rod and sheet stock. Brasses are easily machined. Brass is usually tougher than bronze and produces a stringy chip when machined. The melting point of brasses ranges from 1616 to 1820°F (880 to 993°C), and their weights range from 512 to 536 lb/ft³.

Bronze. Bronze is found in many combinations of copper and other metals, but copper and tin are the original elements combined to make bronze. Bronze colors usually range from red to yellow. Phosphor bronze contains 92 percent copper, 0.05 percent phosphorus, and 8 percent zinc. Aluminum bronze is often used in the shop for making bushings or bearings that support heavy loads (Figure 13). (Brass is not normally used for making antifriction bushings.)

Figure 13. Flanged bronze bushing. *(Practical Metallurgy and Materials of Industry)*

The melting point of bronze is about 1841°F (1005°C) and its weight is about 548 lb/ft³. Bronzes are usually harder than brasses, but are easily machined with sharp tools. The chip produced is often granular. Some bronze alloys are used as brazing rods.

Chromium

Chromium is a slightly gray metal that can take a high polish. It has a high resistance to corrosion by most reagents; exceptions are dilute hydrochloric and sulfuric acids. Chromium is widely used as a decorative plating on automobile parts and other products.

Chromium is not a very ductile or malleable metal and its brittleness limits its use as an unalloyed metal. It is commonly alloyed with steel to increase hardness and corrosion resistance. Chrome-nickel and chrome-molybdenum are two very common chromium alloys. Chromium is also used in electrical heating elements such as chromel or nichrome wire. The melting point of chromium is 2939°F (1615°C) and its weight is 432.4 lb/ft³.

Die Cast Metals

Finished castings are produced with various metal alloys by the process of die casting. Die casting is a method of casting molten metal by forcing it into a mold. After the metal has solidified, the mold opens and the casting is ejected. Carburetors, door handles, and many small precision parts are manufactured using this process (Figure 14). Die cast alloys, often called "pot metals," are classified in six groups:

1. Tin base alloys
2. Lead base alloys
3. Zinc base alloys
4. Aluminum base alloys
5. Copper, bronze, or brass alloys
6. Magnesium base alloys

The specific content of the alloying elements in each of the many die cast alloys may be found in handbooks or other references on die casting.

Lead and Lead Alloys

Lead is a heavy metal that is silvery when newly cut and gray when oxidized. It has a high density, low tensile strength, low ductility (cannot be easily drawn into wire), and high malleability (can be easily compressed into a thin sheet).

Lead has a high corrosion resistance and is alloyed with antimony and tin for various uses. It is used as shielding material for nuclear and X-ray radiation, for cable sheathing, and battery plates. Lead is added to steels, brasses, and bronzes to improve machinability. Lead compounds are very toxic and they are also cumulative in the body. Small amounts ingested over a period of time can be fatal. The melting point of lead is 621°F (327°C); its weight is 707.7 lb/ft³.

A **babbitt metal** is a soft, antifriction alloy metal often used for bearings and is usually tin or lead based (Figure 15). Tin babbitts usually contain from 65 to 90 percent tin with antimony, lead, and a small percentage of copper added. These are the higher grade and generally the more expensive of the two types. Lead babbitts contain up to 75 percent lead with antimony, tin, and some arsenic making up the difference.

Cadmium base babbitts resist higher temperatures than other tin and lead base types. These alloys contain from 1 to 15 percent nickel or a small percentage of copper and up to 2 percent silver. The melting point of babbitt is about 480°F (249°C).

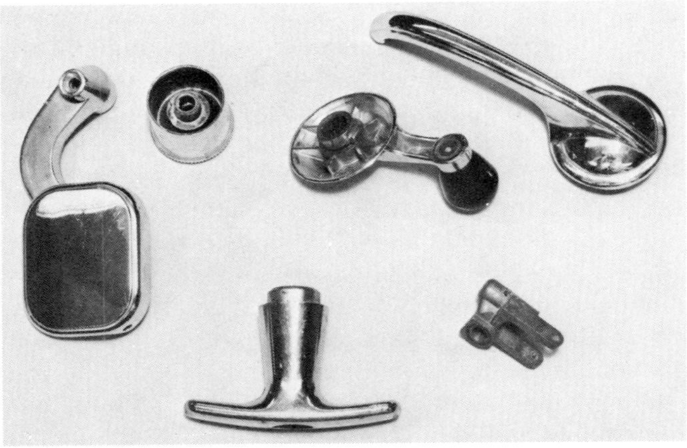

Figure 14. Die cast parts. *(Machine Tools and Machining Practices)*

Figure 15. Babbitted pillow block bearings. *(Machine Tools and Machining Practices)*

Magnesium

When pure, magnesium is a soft, silver-white metal that closely resembles aluminum, but is lighter in density. In contrast to aluminum, magnesium will readily burn with a brilliant white light. Cast and wrought magnesium alloys are designated by SAE and ASTM numbers, which may be found in metals reference handbooks such as *Machinery's Handbook.*

Magnesium, which is similar to aluminum in density and appearance, presents some quite different machining problems because magnesium chips can burn in air, and applying water will only cause the chips to burn more fiercely. Sand or special compounds should be used to extinguish these fires, and should always be on hand when machining magnesium. Thus, when working with magnesium, a water-based coolant should never be used. Magnesium can be machined dry when light cuts are taken and the heat is dissipated. Compressed air is sometimes used as a coolant. Anhydrous (containing no water) oils having a high flash point and low viscosity are used in most production work. Magnesium is machined with very high surface speeds and with tool angles similar to those used for aluminum. The melting point of magnesium is 1204°F (651°C), and its weight is 108.6 lb/ft³.

Molybdenum

As a pure metal, molybdenum is used for high temperature applications and, when machined, it chips like gray cast iron. It is used as an alloying element in steel to promote deep hardening and to increase its tensile strength and toughness. Pure molybdenum is used for filament supports in lamps and in electron tubes. The melting point of molybdenum is 4748°F (2620°C). Its weight is 636.5 lb/ft³.

Nickel

Nickel is noted for its resistance to corrosion and oxidation. It is a whitish metal used for electroplating and as an alloying element in steel and other metals to increase ductility and corrosion resistance. It resembles pure iron in some ways but has a greater corrosion resistance. Electroplating is the coating or covering of another material with a thin layer of metal, using electricity to deposit the layer.

When spark testing nickel, it throws short orange carrier lines with no sparks or sprigs (Figure 16). Nickel is attracted to a magnet, but becomes nonmagnetic near 680°F (360°C). The melting point of nickel is 2646°F (1452°C), and its weight is 549.1 lb/ft³.

Nickel Base Alloys

Monel is an alloy of 67 percent nickel and 28 percent

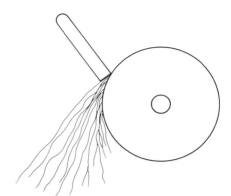

Figure 16. Nickel. Short carrier lines with no forks or sprigs. Average stream length is 10 inches, having an orange color. *(Machine Tools and Machining Practices)*

copper, plus impurities such as iron, cobalt, and manganese. It is a tough, but machinable, ductile, and corrosion resistant alloy. Its tensile strength (resistance of a metal to a force tending to tear it apart) is 70,000 to 85,000 lb/in.². Monel metal is used to make marine equipment such as pumps, steam valves, and turbine blades. On a spark test, monel shoots orange colored, straight sparks about 10 inches long, similar to those of nickel. K-monel contains 3 to 5 percent aluminum and can be hardened by heat treatment.

Chromel and nichrome are two nickel-chromium-iron alloys used as resistance wire for electric heaters and toasters. Nickel-silver contains nickel and copper in similar proportions to monel, but also contains 17 percent zinc. Other nickel alloys such as inconel are used for parts that are exposed to high temperatures for extended periods.

Inconel, a high temperature and corrosion resistant metal consisting of nickel, iron, and chromium, is often used for aircraft exhaust manifolds because of its resistance to high temperature oxidation (scaling) (Figure 17). The nickel alloys' melting point range is 2425 to 2950°F (1329 to 1621°C).

Precious Metals

Gold has a limited industrial value and is used in dentistry, electronic and chemical industries, and jewelry. In the past, gold has been used mostly for coinage. Gold coinage is usually hardened by alloying with about 10 percent copper. Silver is alloyed with 8 to 10 percent copper for coinage and jewelry. Sterling silver is 92.5 percent silver in English coinage and has been 90 percent silver for American coinage. Silver has many commercial uses such as an alloying element for mirrors, photographic compounds, and

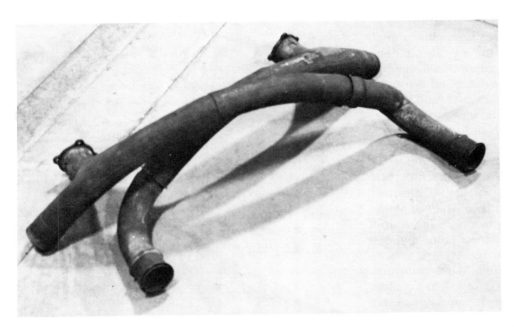

Figure 17. Inconel exhaust manifold for aircraft engines. *(Practical Metallurgy and Materials of Industry)*

Figure 18. The most familiar tin plate product is the steel based tin can. (American Iron & Steel Institute)

electrical equipment. It has a very high electrical conductivity. Silver is used in silver solders that are stronger and have a higher melting point than lead-tin solders.

Platinum, palladium, and iridium, as well as other rare metals are even more rare than gold. These metals are used commercially because of their special properties such as extremely high resistance to corrosion, high melting points, and high hardness. The melting points of some precious metals are: gold 1945°F (1063°C), iridium 4430°F (2443°C), platinum 3224°F (1773°C), and silver 1761°F (961°C). Gold has a weight of 1204.3 lb/ft³. The weight of silver is about 654 lb/ft³. Platinum is one of the heaviest of metals with a weight of 1333.5 lb/ft³. Iridium is also a heavy metal, weighing 1397 lb/ft³.

Tantalum

Tantalum is a bluish-gray metal that is difficult to machine because it is quite soft and ductile and the chip clings to the tool. It is immune to attack from all corrosive acids except hydrofluoric and fuming sulfuric acids. It is used for high temperature operations above 2000°F (1093°C). It is also used for surgical implants and in electronics. Tantalum carbides are combined with tungsten carbides for cutting tools that have extreme wear resistance. The melting point of tantalum is 5162°F (2850°C). Its weight is 1035.8 lb/ft³.

Tin

Tin has a white color with a slightly bluish tinge. It is whiter than silver or zinc. Since tin has a good corrosion resistance, it is used to plate steel, especially for the food processing industry (Figure 18). Tin is used as an alloying element for solder, babbitt, and pewter.

A popular solder is an alloy of 50 percent tin and 50 percent lead. Tin is alloyed with copper to make bronze. The melting point of tin is 449°F (232°C). Its weight is 454.9 lb/ft³.

Titanium

The strength and light weight of this silver-gray metal make it very useful in the aerospace industries for jet engine components, heat shrouds, and rocket parts. Pure titanium burns in air at 1156°F (610°C) and has a tensile strength of 60,000 to 110,000 PSI, similar to that of steel; by alloying titanium, its tensile strength can be increased considerably. Titanium weighs about half as much as steel and, like stainless steel, is a relatively difficult metal to machine. Machining can be accomplished with rigid setups, sharp tools, slower surface speed, and the use of proper coolants. When spark tested, titanium throws a brilliant white spark with a single burst on the end of each carrier. The melting point of titanium is 3272°F (1800°C), and its weight is 280.1 lb/ft³.

Tungsten

Typically, tungsten has been used for incandescent light filaments. It has the highest known melting point (6098°F or 3370°C) of any metal, but is not resistant to oxidation at high temperatures. Tungsten is used for rocket engine nozzles and welding electrodes and as an alloying element with other metals. Machining pure tungsten is very difficult with single point tools, and grinding is preferred for finishing operations. Tungsten carbide compounds are used to make extremely hard and heat resistant lathe tools and milling cutters by compressing the tungsten carbide powder into a briquette and sintering it in a furnace. Tungsten weighs about 1180 lb/ft³.

Zinc

The familiar galvanized steel is actually steel plated with zinc and is used mainly for its high corrosion resistance. The fumes of galvanized steel should be avoided when welding. Zinc alloys are widely used as die casting metals. Zinc and zinc based die cast metals conduct heat much more slowly than aluminum. The rate of heat transfer on similar shapes of aluminum and zinc is a means of distinguishing between them. The melting point of zinc is 787°F (419°C), and it weighs about 440 lb/ft³.

Zirconium

Zirconium is similar to titanium in both appearance and physical properties. It was once used as an explosive primer and as a flashlight powder for photography. Machining zirconium, like titanium, requires rigid setups and slow surface speeds. Pure zirconium will burn in air when heated to 958°F (500°C). Zirconium has an extremely high resistance to corrosion from acids and seawater. Zirconium alloys are used in nuclear reactors, flash bulbs, and surgical implants such as screws, pegs, and skull plates. When spark tested, it produces a spark that is similar to that of titanium. The melting point of zirconium is 3182°F (1750°C). Its weight is 399 lb/ft³.

SELF-TEST

1. By what universal coding system is carbon steel and alloy steel designated?
2. What are three basic types of stainless steels and what is the number series assigned to them? What are their basic differences?
3. If your shop stocked the following steel shafting, how would you determine the content of an unmarked piece of each, using shop tests as given in this unit?
   (a) AISI C1020 CF
   (b) AISI B1140 (G and P)
   (c) AISI C4140 (G and P)
   (d) AISI 8620 HR
   (e) AISI B1140 (Ebony)
   (f) AISI C1040
4. A small part has obviously been made by a casting process. How can you determine whether it is a ferrous or a nonferrous metal, or if it is steel, or white or gray cast iron?
5. What is the meaning of the symbols O1 and W1 when applied to tool steels?
6. When checking the hardness of a piece of steel with the file test, the file slides over the surface without cutting.
   (a) Is the steel piece readily machinable?
   (b) What type of steel is it most likely to be?
7. Steel that is nonmagnetic is called _____.
8. What nonferrous metal is magnetic?
9. List at least four properties of steel that should be kept in mind when you select the material for a job.
10. What advantage do aluminum and its alloys have over steel alloys? What disadvantages?
11. Describe the meaning of the letter "H" when it follows the four digit number that designates an aluminum alloy. The meaning of the letter "T."
12. Name two ways in which magnesium differs from aluminum.
13. What is the major use of copper? How can copper be hardened?
14. What is the basic difference between brass and bronze?
15. Name two uses for nickel.
16. Lead and tin have one useful property in common. What is it?
17. Molybdenum and tungsten are both used in _____ steels.
18. Babbitt metals, used for bearings, are made in what major basic types?
19. What type of metal can be injected under pressure into a permanent mold?
20. Which is stronger, cast or wrought (worked) aluminum?
21. What can be done to avoid building up an edge on the tool bit when machining aluminum?
22. Should a water based coolant be used when machining magnesium? Explain.
23. Should the rake angles on tools for brasses and bronzes be zero, positive, or negative? Explain.
24. How is tungsten used for cutting tools?

UNIT 4 USE OF PEDESTAL AND PORTABLE GRINDERS

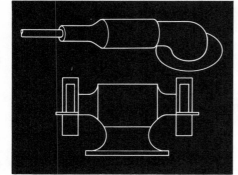

Workers in mechanical fields often find it necessary to use offhand and portable grinding equipment for sharpening tools, removing unwanted metal, and polishing and buffing. As a worker in any mechanical or machine shop area, you will need to know how to remove and mount wheels and true them by dressing. It is also necessary to know how to shape and sharpen certain frequently used tools such as chisels and screwdrivers. This unit will help you to use these offhand grinding tools safely and correctly for various operations.

OBJECTIVES

After completing this unit, you should be able to:
1. Safely mount and dress abrasive wheels on the pedestal grinder.
2. Resharpen punches and chisels and reshape screwdrivers.
3. Safely use hand grinding equipment.

SAFETY IN OFFHAND GRINDING

The pedestal grinder gets its name from the floor stand, or pedestal, that supports the motor and abrasive wheels. The pedestal grinder is a common machine tool that you will use almost daily in the machine shop and frequently in other shops. This grinding machine is used for general purpose offhand grinding where the workpiece is hand held and applied to the rapidly rotating abrasive wheel. One of the primary functions of the pedestal grinder is the shaping and sharpening of tools and drills. Pedestal grinders are also used for snagging, that is, rough grinding to smooth rough edges made in torch cutting or on castings, and to generally remove small amounts of unwanted metal. Grinders that are designated for tool grinding in a shop should **never** be used for rough grinding as this quickly ruins the wheel for accurate tool sharpening. There is usually one large pedestal grinder in a shop that is designated for snagging and rough grinding only. Rotary wire brushes or buffing wheels are often mounted on pedestal grinders to modify their use.

Under ordinary conditions and when reasonable care is taken, grinding is a relatively safe method of removing metal, but grinding wheels are breakable and they must be handled, mounted, and used carefully with adequate protection. The offhand grinding machine should always have a guard protecting the wheel which should never be removed when the grinder is in use. Excessive pressure should never be used when grinding. If this seems to be necessary, the improper grit or grade of wheel is being used. Grinding wheels, like other rotating machine tools, may catch loose clothing or long hair in the rotating abrasive wheel or wire wheel, as in the case of Figure 1. Long hair should be contained in an industrial-type hair net. **Always** use eye protection such as goggles or a face shield when doing any kind of grinding.

Nonferrous metals such as aluminum and brass should never be ground on aluminum oxide wheels, which are mostly found on pedestal grinders. These metals fill up the voids or spaces in the surface of the wheel so that more pressure is needed to remove the metal. This additional pressure sometimes causes the wheel to break and shatter. Silicon carbide wheels are used for grinding nonferrous metals because the

Figure 1. The long loose hair of the grinder operator was caught in the wire brush wheel and the force and speed of this action was such that the operator's head was jerked suddenly into the guard. Note that the cast aluminum guard was shattered as a result of the impact. (Lane Community College)

Figure 2. Using a piece of wood to hold the wheel while removing the spindle nut. (Lane Community College)

Figure 3. The ring test is made before mounting the wheel. (Lane Community College)

Figure 4. The wheel is mounted with the proper bushing in place. (Lane Community College)

abrasive breaks down more readily, keeping the wheel sharp. Silicon carbide wheels should always be used for this purpose.

When it becomes necessary to remove worn wheels, as shown in Figure 2, the side of the guard must be removed and the tool rest must be moved out of the way. A piece of wood may be used to keep the wheel from rotating so that the nut can be turned and removed. It may be well to remember that on the left side of the pedestal grinder the threads are left-handed and on the right side they are right-handed.

A new wheel should always be "ring tested" to determine if there are any cracks or imperfections (Figure 3). Support the wheel with a finger while using the handle of a screwdriver to strike the wheel near the rim. If there is a true ringing sound, like a bell, then the wheel should be solid and safe to use. But if it sounds like a dull thud, it may be cracked. The flanges and the spindle should be cleaned before mounting the wheel, and if a bushing is required, a proper size bushing should be used (Figure 4). Check to see if the wheel is rated at a safe operating speed for your grinder and check the condition of the blotters (papers on each side of the wheel) to see that

Figure 5. The tool rest is adjusted. (Lane Community College)

Figure 6. The spark guard is adjusted. (Lane Community College)

they are not damaged. The nut should be tightened only enough to hold the wheel firmly. The nut should not be tightened excessively as that could break the wheel.

After the guard and cover plate have been replaced, the tool rest should be brought up to the wheel between $\frac{1}{16}$ and $\frac{1}{8}$ in. as shown in Figure 5. If there is too much space between the tool rest and the wheel, a small piece such as a tool bit that is being ground may flip up and catch between the wheel and tool rest. Your finger may be caught between the workpiece and grinding wheel, resulting in a serious injury. It is an extremely important safety measure to keep this distance of a maximum of $\frac{1}{8}$ in. between the tool rest and wheel face. A spark guard on the upper side of the wheel guard, as shown in Figure 6, should also be brought up close to the wheel within $\frac{1}{16}$ in. This protects the operator if the wheel should shatter.

You are now ready to turn on the grinder. Stand

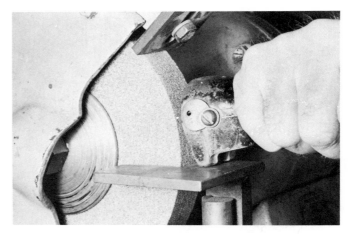

Figure 7. Although this is not the best method of using the Desmond dresser, it is the most common since it is more convenient. The correct or suggested method is to move the tool rest out far enough to hook the ears of the dresser behind so it can be used as a guide. (Lane Community College)

aside, out of the line of rotation of the wheel, and turn on the motor. Let the grinder run for one full minute without using it.

A new wheel does not always run exactly true and it must be dressed. If a wheel is considerably out-of-true, the entire grinder will vibrate. Figure 7 shows a Desmond dresser being used to true the face of the wheel. Wheels often get grooved, out of round, or misshapen and must frequently be dressed in order that proper results may be obtained. The following rules tell how to use wheels safely and with confidence.

Mounting the Wheel on the Pedestal Grinder

1. Select the correct wheel. "Ring" test the wheel and inspect it for cracks. **Never** use a cracked or chipped wheel. **(See Section K, Unit 1, for selecting wheels.)**
2. Do not exceed maximum safe speed for the wheel as noted on the blotter.
3. Use clean recessed matching flanges.
4. Make sure there is one clean, smooth blotter on each side of the wheel. If the existing blotters are not in good condition, that is, partially torn off or missing, replace them with new ones.
5. Never change the hole in the wheel or force the wheel on the spindle.
6. Do not tighten the nut excessively or it may break the wheel.

7. Adjust the spark guard and tool rest.

8. Put on safety glasses and stand to one side of the wheel before starting the grinder.

Safety Rules When Using Grinders

1. Start the grinder.

2. Stand aside and allow the wheel to run idle a full minute before starting to grind. **Always** stand to one side as much as possible when using the grinder.

3. Dress the wheel if it runs out of true.

4. Contact the grinding wheel smoothly without bumping.

5. Grind only on the face (outside circumference) of a straight wheel. An exception to this rule is when truing flat surfaces on small tools such as screwdrivers. Excessive side pressure on a straight wheel may break it.

6. Never exert force when grinding so that the motor slows noticeably.

7. Always wear appropriate eye protection when grinding or dressing wheels, or when watching others grind.

8. When grinding small objects such as small tool bits for lathes, use a locking plier to hold them instead of your fingers (Figure 8). Avoid having your fingers close to the wheel at any time. Grinders can cause severe abrasions to fingers, or worse, loss of hands or fingers.

SHARPENING TOOLS ON THE PEDESTAL GRINDER

Many kinds of small hand tools may be reshaped and resharpened by the offhand grinding process on a pedestal grinder. Twist drills are often sharpened by the offhand method. This process is explained in Section G, Unit 4, "Hand Grinding of Drills on the Pedestal Grinder." Likewise, high speed steel lathe tool bits are often sharpened offhand. For this process, see Section H, Unit 4, "Cutting Tools for the Lathe."

Chisels and punches need frequent sharpening when heavily used, mostly because of dullness or chipped or nicked cutting edges. The punch end may be flattened because it was used on steel with a higher hardness. Only just enough metal to reshape the point or edge should be removed by grinding since excessive grinding would shorten the life of the tool. Several shapes of chisels and punches may be seen in Figure 9.

Chisels should be resharpened at their correct angle on the periphery (outside circumference) of the

Figure 8. A locking plier being used to hold a small lathe tool bit while grinding it. Ordinary pliers do not hold work securely enough for this purpose. (Lane Community College)

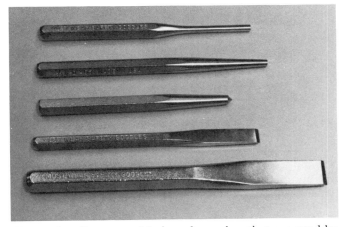

Figure 9. Common chisels and punches that are used by mechanics. (Lane Community College)

wheel (Figures 10*a* and 10*b*). Flat chisels should also be moved slightly side to side on the wheel to produce a slight curvature on the cutting edge (Figure 11). When a chisel or punch develops a mushroom head from hammering, it becomes a danger to anyone in the vicinity because small pieces of steel can break off and fly at a high velocity (Figure 12). When a mushroom head begins to form, it should be ground off

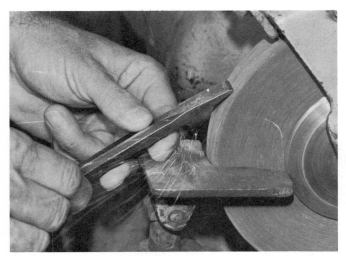

Figure 10a. Flat chisel being ground to produce a 60-degree angle. (Lane Community College)

Figure 10b. Diamond point chisel being sharpened on the end, which is the proper way. (Lane Community College)

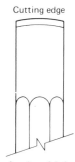

Figure 11. Flat chisels should be ground with a slight curvature on the cutting edge.

Figure 12. Chisels or wedges that develop mushroom heads from hammering, such as this one, are dangerous to use. Pieces of metal can fly off at high velocity. (Lane Community College)

Figure 13. The same tool as in Figure 12 has been corrected by grinding. (Lane Community College)

Figure 14. This punch is being correctly sharpened to produce a 90-degree angle for use as a center punch. It must be rotated while being ground. A sharper angle of 60 degrees may be ground to produce a prick punch. (Lane Community College)

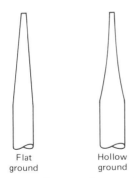

Flat
ground

Hollow
ground

Figure 15. The two shapes of screwdriver points—flat and hollow ground. Flat ground screwdriver points are stronger than hollow ground, but hollow ground points do not slip in the screw slot as easily as the flat ground.

Figure 16. Blade of screwdriver being shaped with the side of the wheel for a flat grind. (Lane Community College)

as shown in Figure 13. Punches should also be sharpened on the periphery at the correct angle (Figure 14). They should be carefully and evenly rotated while grinding to produce a true conical point.

It is also important not to overheat the tool since that would soften it and ruin its usefulness. Chisels and punches are made of carbon tool steel and are hardened and tempered. (See Section F, Unit 1, "Hardening, Case Hardening, and Tempering.") Tempering softens the hardened metal, so heating the tool above its correct tempering temperature will soften the cutting edge. Usually a blue color indicates the metal has been overheated beyond the range of normal tempering temperature and the edge has been made too soft for use. A corrective measure would be to have the tool heat treated, that is, hardened and tempered again. If the discolored area is very small, the tool can be ground back to the area where it was not overheated and then resharpen it. Frequent cooling in water will help to avoid damage by overheating.

Screwdrivers are probably the most misused of all tools, so they are often twisted or misshapen to the extent that they will no longer fit a screw slot. There are two ways of shaping a screwdriver point; the flat grind and the hollow grind (Figure 15). The flat grind can be made on the side of the wheel (Figure 16), and the hollow grind on the periphery of the wheel (Figure 17). The end of the screwdriver blade should be ground square to the axis of the tool as shown in Figure 18. The thickness of the point should be just under the standard screw slot that it will be used in (Figure 19).

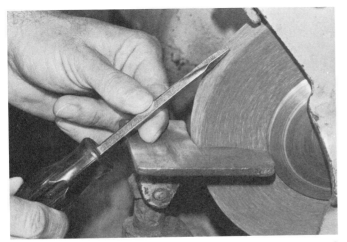

Figure 17. Hollow grind on a screwdriver being formed on the periphery of the wheel. (Lane Community College)

Figure 18. The end of the blade being squared on the wheel. (Lane Community College)

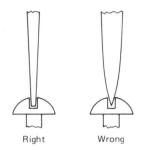

Right   Wrong

Figure 19. Proper and improper grinding of screwdriver blade.

Figure 20. Large and small straight grinders. (Lane Community College)

Figure 21. Offset hand grinder. (Lane Community College)

USE OF PORTABLE GRINDERS AND POLISHERS

Mechanics, welders, auto-body repair persons, and machinists will sometimes need to use hand held grinding and polishing tools. These can be powered either by air or electricity. If they are air operated, the regulator should be set at the correct pressure (PSI) for the tool and the lubricator should be functioning to prevent excessive wear in the air motor. Grinders with electric motors must have cords that are in good condition (not cut, frayed, or ground partly through). They should have the ground connector on the plug intact to prevent electric shock. Guards should be in place whenever possible.

Straight grinders are made in ranges from tiny, high speed tools for fine work to large heavy duty grinders (Figure 20). Straight abrasive wheels are mounted on the larger types of grinders while small, shaft mounted wheels are used on the high speed types. Cup or straight wire wheels can also be mounted on them.

Offset grinding tools are commonly used in welding and auto-body shops because of their versatility and rapid metal removal rate (Figure 21). Grinding discs, wire brushes, sanding discs, and polishing buffers can be mounted on these tools. Various sizes of these tools are extensively used for grinding and polishing in body shops (Figure 22).

Specific safety observations when using these off-hand grinders are as follows:

1. Wear eye and body protection. Gloves and, if possible, a leather apron should be used.
2. Avoid grinding in the presence of flammable or explosive liquids and gases, such as gasoline, batteries, paint thinner, and "empty" drums that once contained these substances. Hand grinders may throw sparks 20 to 30 feet.
3. After turning off a grinding tool, wait until it has stopped rotating before putting it down.
4. Never strike or bounce a wheel while it is revolving. This can cause it to shatter and explode into flying fragments.
5. Keep the rotating wheel as far away from your body as possible. If the wheel catches on a sharp projection, the grinder can bounce toward you and cause a very serious grinding gash.

Figure 22. Polishing a fender with an offset hand grinder. (Lane Community College)

SELF-TEST

1. What is the primary function of small pedestal grinders?
2. What kinds of personal protection should be observed when using a pedestal grinder?
3. Why should silicon carbide and not aluminum oxide wheels be used for nonferrous metals such as aluminum?
4. How can you tell if a wheel is cracked before you mount it on a grinder?
5. Why should the tool rest be kept within $\frac{1}{8}$ in. of the wheel?
6. How can you safely grind small pieces such as a boring bar tool bit less than an inch long?
7. Carbon steel tools should be frequently cooled in water while they are being ground. Why is this true?
8. How do you hollow grind a tool?
9. Hand held grinders with electric motors should be frequently checked for one important hazard. What is it?
10. Name at least three safety considerations to keep in mind when using a hand held grinder.

SECTION B

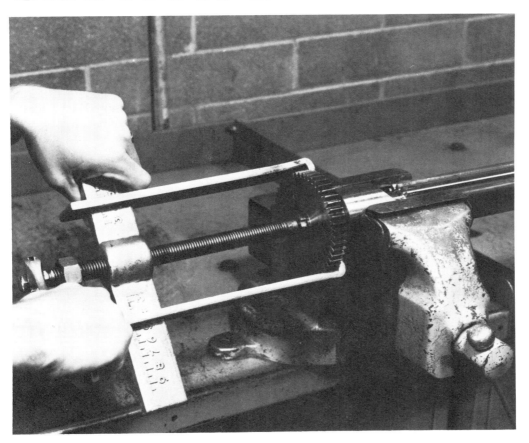

HAND TOOLS

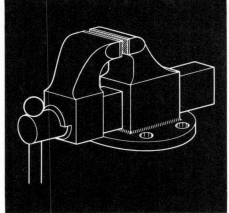

UNIT 1 WORKHOLDING FOR HAND OPERATIONS

The bench vise is a basic but very necessary tool in the shop. With proper care and use, this workholding tool will give many years of faithful service.

OBJECTIVES

After completing this unit, you should be able to:
1. Identify various types of vises and their uses.
2. Explain the procedures used for the care and maintenance of vises.

TYPES OF VISES

Vises of various types are used by machinists when doing hand or bench work. They should be mounted in such a way that a long workpiece can be held in a vertical position extending alongside the bench (Figure 1). Some bench vises have a solid base (Figure 2); others have a swivel base (Figure 3). The machinist's bench vise is measured by the width of the jaws (Figure 4).

Toolmakers often use small vises that pivot on a ball and socket for holding delicate work. Hand-held vises, called pin vises, are made for holding very small or delicate parts.

Most bench vises have hardened insert jaws that are serrated for greater gripping power (Figure 5). These criss-cross serrations are sharp and will dig into finished workpieces enough to mar them beyond repair. Soft jaws (Figure 6) made of copper, other soft metals, or wood are used to protect a finished surface on a workpiece. These soft jaws are made to slip over the vise jaws. Some vises used for sheet metal work have smooth, deep jaws (Figure 7).

USES OF VISES

Vises are used to hold work for filing, hacksawing, chiseling, and bending light metal. They are also used for holding work when assembling and disassembling parts.

Vises should be placed on the workbench at the correct working height for the individual. The top of the vise jaws should be at elbow height. Poor work is produced when the vise is mounted too high or too low. A variety of vise heights should be provided in the shop or skids made available to stand on (Figure 8).

Figure 1. When long work is clamped in the vise vertically, it should clear the workbench. (Lane Community College)

Figure 2. A solid base bench vise. (Lane Community College)

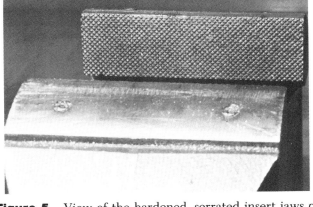

Figure 5. View of the hardened, serrated insert jaws on the vise. (Lane Community College)

Figure 3. A swivel base bench vise. (Lane Community College)

Figure 6. View of the soft jaws placed on the vise. (Lane Community College)

Figure 4. How to measure a vise. (Lane Community College)

Figure 7. Smooth-jawed vise for working with sheet metal. (Lane Community College)

Figure 8. The correct working height on vises. (Lane Community College)

Figure 9. Hammering on the slide bar should never be done to a vise. This may crack or distort it. (Lane Community College)

CARE OF VISES

Like any other tool, vises have limitations. "Cheater" bars or pipes should not be used on the handle to tighten the vise. Heat from a torch should not be applied to work held in the jaws as the hardened insert jaws will then become softened. There is usually one vise in a shop that is reserved for heating and bending.

Heavy hammering should not be done on a bench vise. The force of bending or pounding should be against the fixed jaw rather than the movable jaw of the vise. Bending light, flat stock or small round stock in the jaws is permissible if a light hammer is used. The movable jaw slide bar (Figure 9) *should never be hammered upon* as it is usually made of thin cast iron and can be cracked quite easily. An anvil is often provided behind the solid jaw for the purpose of light hammering.

Bench vises should occasionally be taken apart so that the screw, nut, and thrust collars may be cleaned and lubricated. The screw and nut should be cleaned in solvent. A heavy grease should be packed on the screw and thrust collars before reassembly (Figure 10).

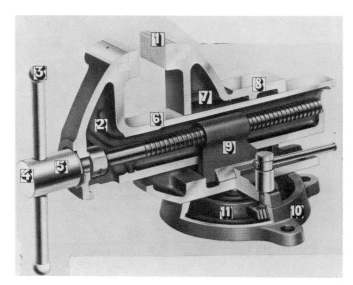

Figure 10. A cutaway view of a vise. (1) Replaceable hardened tool steel faces pinned to jaw. (2) Malleable iron front jaw. (3) Steel handle with ball ends. (4) Cold rolled steel screw. (5) Bronze thrust bearing. (6) Front jaw beam. (7) Malleable iron back jaw body. (8) Anvil. (9) Nut, mounted in back jaw keyseat for precise alignment. (10) Malleable iron swivel base. (11) Steel tapered gear and lock bolt. (The Warren Group, Division of Warren Tool Corporation)

SELF-TEST

1. What clamping position should be considered when mounting a vise on a workbench?
2. Name two types of bench vises.
3. How is the machinist's bench vise measured for size?
4. Small, delicate work may be held in a _____ or a _____ vise.
5. Explain two characteristics of the insert jaws on vises.
6. How can a finished surface be protected?
7. In what way are vises that are used for sheet metal work different from a machinist's vise?
8. What are vises usually used for?
9. Name three things that should never be done to a vise.
10. How should a vise be lubricated?

UNIT 2 ARBOR AND SHOP PRESSES

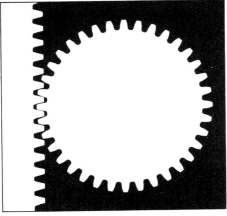

The arbor press and the small shop press are very common sights in most machine shops. It would be very difficult, indeed, to get along without these machines. The operator will find these machines extremely useful when he knows how to use them, but if he is not instructed in their use, they can be dangerous to him and destructive to the workpiece.

OBJECTIVES

After completing this unit, you should be able to:
1. Install and remove a bronze bushing using an arbor press.
2. Press on and remove a ball bearing from a shaft on an arbor press using the correct tools.
3. Press on and remove a ball bearing from a housing using an arbor press and correct tooling.
4. Install and remove a mandrel using an arbor press.
5. Install and remove a shaft with key in a hub using the arbor press.
6. Broach a $\frac{1}{4}$ inch keyway in a $1\frac{1}{4}$ inch bore.

TYPES

The arbor press is an essential piece of equipment in the small machine shop. Without it a machinist would be forced to resort to the use of a hammer or sledge to make any forced fit, a process that could easily damage the part.

Two basic types of arbor presses are manufactured and used: the hydraulic (Figure 1) and the mechanical (Figure 2). Both types are handpowered with a lever. The lever gives a "feel" or a sense of pressure applied, which is not possible with power-driven presses. This pressure sensitivity is needed when small delicate parts are being pressed so that a workman will know where to stop before collapsing the piece.

USES

The major uses of the arbor press are bushing installation and removal, ball and roller bearing installation and removal (Figure 3), pressing shafts into hubs (Figure 4), pressing mandrels into workpieces, broaching keyways (Figure 5), and straightening and bending (Figure 6).

Figure 1. Fifty ton capacity hydraulic shop press. (Lane Community College)

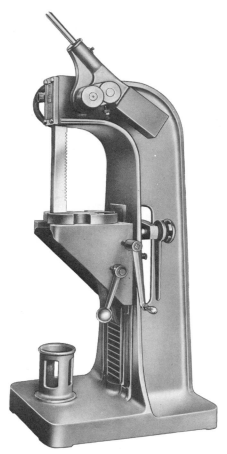

Figure 2. Simple ratchet floor type arbor press. (Dake Corporation)

Figure 3. Roller bearing being removed from axle. (Lane Community College)

Installing Bushings

A bushing is a short metal tube, machined inside and out to precision dimensions, and usually made to fit into a bore, or accurately machined hole. Many kinds of bushings are used for various purposes and are usually installed with an "interference fit" or press fit. This means that the bushing is slightly larger than the hole into which it is pressed. The amount of interference may be found in Table 9, Appendix I. There are many bushings made of many materials, including bronze and hardened steel, but they all have one thing in common: they must be lubricated with high pressure lubricant before they are pressed into the bore. Oil is not used as it will simply wipe off and cause the bushing to seize the bore. Seizing is the condition where two unlubricated metals tend to weld together under pressure. In this case, it may cause the bushing to be damaged beyond repair (Figure 7).

The bore should always have a strong chamfer, that is, an angled or beveled edge, since a sharp edge

Figure 4. Shaft being pressed into hub. (Lane Community College)

Figure 5. Keyseat being broached in gear hub with a mechanical arbor press. (The duMont Corporation, Greenfield, Mass.)

Figure 6. Bar being checked with a straightedge during straightening process. (Lane Community College)

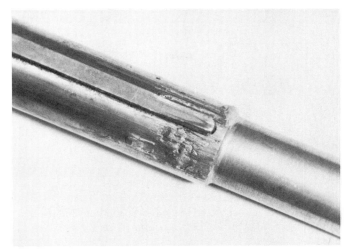

Figure 7. This shaft had just been made by a machinist and was forced into an interference fit bore for a press fit. No lubrication was used, and it immediately seized and welded to the bore, which was also ruined. (Lane Community College)

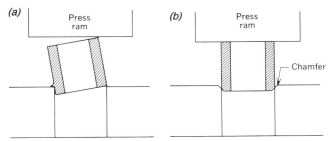

Figure 8. *(a)* Bushing being pressed where bore is not chamfered and bushing is misaligned. *(b)* Bushing being pressed into correctly chamfered hole in correct alignment. *(Machine Tools and Machining Practices)*

would cut into the bushing and damage it (Figures 8*a* and 8*b*). The bushing should also have a long tapered chamfer or "start" so it will not "dig in" and enter misaligned. Bushings are prone to go in crooked if there is a sharp edge, especially if it is a hardened steel bushing. Care should be taken to see that the bushing is straight as it enters the bore and that it continues into the bore in proper alignment. This should not be a problem if the tooling is right, that is, if the end of the press ram (the movable part that does the pressing) is square and if it is not loose and worn. The proper bolster plate (the stationary plate that supports the workpiece) should also be used under the part so that it cannot tilt out of alignment. Sometimes special tooling is used to guide the bushing (Figure 9*a*). Only the pressure needed to force the

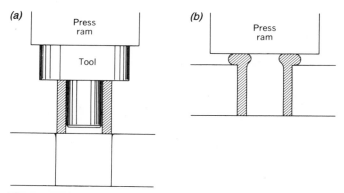

Figure 9. *(a)* Special tool to keep bushing square to press ram. *(b)* Effect of excessive pressure on bushing that exceeds bore length. *(Machine Tools and Machining Practices)*

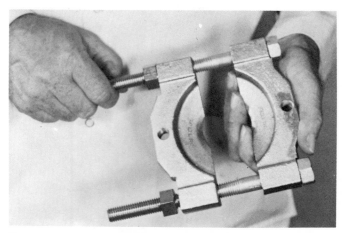

Figure 10. Special tool (bearing puller) for supporting inner bearing race. (See Figure 3.) (Lane Community College)

bushing into place should be applied, especially if the bushing is longer than the bore length. Excessive pressure might distort the bushing and cause it to be undersized (Figure 9*b*).

Ball and Roller Bearings

Ball and roller bearings pose special problems when they are installed and removed by pressing. This is because the pressure must be applied directly against the race and not through the balls or rollers, since this could destroy the bearing. Frequently, when removing ball bearings from a shaft, the inner race is hidden by a shoulder and cannot be supported in the normal way. In this condition a special tool is used (Figure 10). On the inner and outer races, bearings may be installed by pressing on the race with a steel tube of the proper diameter. As with bushings, high pressure lubricant should be used.

Sometimes there is no other way to remove an old ball bearing except by exerting pressure through the balls. When this is done, there is a real danger that the race may be violently shattered. In this case a scatter shield must be used. A scatter shield is a heavy steel tube about 8 to 12 in. long and is set up to cover the work. The shield is placed around the bearing during pressing to keep shattered parts from injuring the operator. It is a good safety practice to always use a scatter shield when ball bearings are removed from a shaft by pressing. Safety glasses should be worn during all pressing operations.

Bores and Shafts

Holes in the hubs of gears, sprockets, and other machine parts are also frequently designed for a force fit. In these instances, there is usually a key that needs to be aligned. The key is mounted in a groove, called a keyseat, in the shaft. If it is a round end keyseat,

Figure 11. Chamfer on key helps in alignment of parts being pressed together. (Lane Community College)

the key will not slip endwise when pressing. If the keyseat end curves out like a ski, the key may slip and bind in the slot. One solution to this problem is to occasionally tap the key in with a flat-end punch during the pressing operation. This key, in turn, also fits into a slot in the hub of a gear or pulley, and secures the part against the shaft, keeping it from rotating. When pressing shafts with keys into hubs with keyseats, it is sometimes helpful to chamfer the leading edge of the key so it will align itself properly (Figure 11). Seizing will occur in this operation, as with the installation of bushings, if high pressure lubricant is not used.

Figure 12. Mandrel being lubricated and pressed into part for further machining. (Lane Community College)

Figure 13. Hexagonal shape being push broached. Note this man is wearing no safety glasses. This is an unsafe practice. (The duMont Corporation, Greenfield, Mass.)

Mandrels

Mandrels, cylindrical pieces of steel with a slight taper, are pressed into bores (to secure the part for machining) in much the same way that shafts are pressed into hubs. There is one important difference, however. Since the mandrel is tapered about .006 of an inch per foot, it can be installed only with the small end in first. Determine which end is the small one by measuring with a micrometer or by trying the mandrel in the bore. The small end should start into the hole, but the large end should not. The large end often has a flat milled on it. Apply lubricant and press the mandrel in until definite resistance is felt (Figure 12). Do not use a hydraulic press as too much pressure may be applied; use an arbor press.

Keyseat Broaching

Although many types of keyseating machines are in use in many machine shops, keyseat broaching is often done on arbor presses. Broaching is the process of cutting out shapes on interior and exterior surfaces of a metal part by using a series of cutting teeth like a saw, only different in that each successive tooth is a few thousandths of an inch longer than the last one. Thus, each tooth cuts a certain amount as it passes across the work. The broaches can either be pulled through the work (pull broaches) or pushed through or past the work (push broaches).

Figure 14. A typical set of keyseat broaches. (The duMont Corporation, Greenfield, Mass.)

Keyseats are only one type of cutting that can be done by the push-type procedure. Such internal shapes as a square or hexagon can also be cut by this method (Figure 13). All that is needed for these procedures is the proper size of arbor press and a set of keyseat broaches (Figure 14), which are hardened cutters with stepped teeth so that each tooth cuts only a definite amount as pushed or pulled through a part. Bushings that fit standard bore diameters are provided to guide the broach. These are available in inch and metric dimensions.

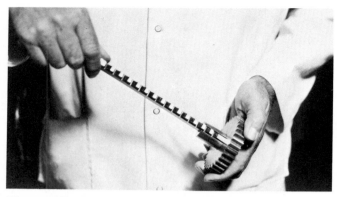

Figure 15. Broach with guide bushing inserted into gear. (Lane Community College)

Figure 17. Shims in place behind broach that is ready to lubricate and make final cut on part. (Lane Community College)

Figure 16. Broach, guide bushing placed in arbor press that is ready to lubricate and to perform first pass. (Lane Community College)

4. Lubricate.
5. Push the broach through.
6. Clean the broach.
7. Place second-pass shim in place.
8. Insert broach.
9. Lubricate.
10. Push the broach through.
11. If more than one shim is needed to obtain the correct depth, repeat the procedure (Figure 17).

The tools should be cleaned and returned to their box and the finished keyway should be deburred and cleaned.

Production or single-pass broaching requires no shims or second-pass cuts, and with some types no bushings need be used (Figure 18).

Two important things to remember when push broaching are alignment and lubrication. Misalignment, caused by a worn or loose ram, can cause the broach to hog (dig in) or break. Sometimes this can be avoided by facing the teeth of the broach toward the back of the press and permitting the bushing to

Steps in Broaching a Keyseat

1. Choose the bushing that fits the bore and the broach, and put it in place in the bore.
2. Insert the correct size broach into the bushing slot (Figure 15).
3. Place this assembly in the arbor press (Figure 16).

Figure 18. Production push broaching without bushings or shims. (The duMont Corporation, Greenfield, Mass.)

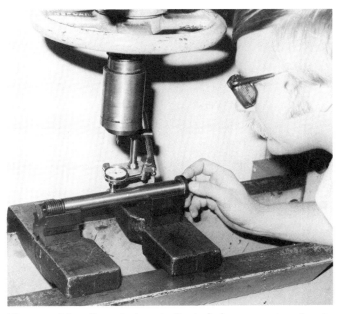

Figure 19. Part being indicated for runout prior to straightening. (Lane Community College)

protrude above the work to provide more support for the broach. After starting the cut, relieve the pressure to allow the broach to center itself. Repeat this procedure during each cut.

At least two or three teeth should be in contact with the work. If needed, stack two or more workpieces to lengthen the cut. The cut should never exceed the length of the standard bushing used with the broach. Never use a broach on material harder than Rockwell C35, one of many grades of a hardness test you will meet later in this book. If it is suspected that a part is harder than mild steel, its hardness should be determined before any broaching is attempted.

Use a good high pressure lubricant. Also apply a sulfur base cutting oil to the teeth of the broach. Always lubricate the back of the keyway broach to reduce friction, regardless of the material to be cut. Brass is usually broached dry, but bronzes cut better with oil or soluble oil. Cast iron is broached dry, and kerosene or cutting oil is recommended for aluminum.

Bending and Straightening

Bending and straightening are frequently done on hydraulic shop presses. Mechanical arbor presses are not usually used for this purpose. There is a definite safety hazard in this type of operation as a poor setup can allow pieces under pressure to suddenly fly out of the press. Brittle materials such as cast iron or hardened steel bearing races can suddenly break under pressure and explode into fragments.

A shaft to be straightened is placed between two nonprecision vee-blocks, steel blocks with a vee-shaped groove running the length of the blocks that support a round workpiece. In the vee blocks, the shaft is rotated to detect runout, or the amount of bend in the shaft. The rotation is measured on a dial indicator, which is a device capable of detecting very small mechanical movements, and read from a calibrated dial. The high point is found and marked on the shaft (Figure 19). After removing the indicator, a soft metal pad such as copper is placed between the shaft and the ram and pressure is applied (Figure 20). The shaft should be bent back to a straight position and then very slightly beyond that point. The pressure is then removed and the dial indicator is again put in position. The shaft is rotated as before, and the position of the mark noted, as well as the amount of runout. If improvement has been found, continue the process; but if the first mark is now opposite the high point, too much pressure has been applied. Repeat the same steps, applying less pressure.

Other straightening jobs on flat stock and other shapes are done in a similar fashion. Frequently, two or more bends will be found that may be opposite

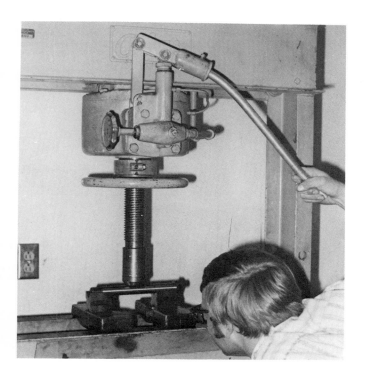

Figure 20. Pressure being applied to straighten shaft. (Lane Community College)

or are not in the same direction. This condition is best corrected by straightening one bend at a time and checking with a straightedge and feeler gage. Special shop press tooling is sometimes used for simple bending jobs in the shop.

Note: Be sure to stand to one side when applying pressure in all pressing operations. Be especially careful when using the press for straightening and always use safety glasses or an eye shield.

SELF-TEST

1. Why is it important to know how to use the arbor press properly and how to correctly set up pressing operations?
2. What kinds of shop presses are made? What makes them different from large commercial presses?
3. List several uses of the shop press.
4. A newly machined steel shaft with an interference fit is pressed into the bore of a steel gear. The result is a shaft is ruined beyond repair; the bore of the gear is also badly damaged. What has happened? What caused this failure?
5. The ram of an arbor press is loose in its guide and the pushing end is rounded off. What kind of problems could be caused by this?

6. A $\frac{1}{2}$ in. diameter bronze bushing is $\frac{1}{8}$ in. longer than the bore. Should you apply 30 tons of pressure to make sure it has seated on the press plate? If your answer is no, how much pressure should you apply?
7. If the inner race on a ball bearing is pressed onto a shaft, why should you not support the outer race while pushing the shaft off?
8. What difference is there in the way a press fit is obtained between mandrels and ordinary shafts?
9. Prior to installing a bushing with the arbor press, what two important steps must be taken?
10. Name five ways to avoid tool breakage and other problems when using push broaches for making keyways in the arbor press.

UNIT 3 NONCUTTING HAND TOOLS

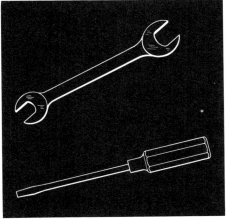

Hand tools are essential in all of the mechanical trades. This unit will help you learn the names and uses of most of the noncutting tools used by machinists.

OBJECTIVES

After completing this unit, you should be able to:
1. Identify the proper tool for a given job.
2. Determine the correct use of a selected tool.
3. Use torque wrenches and determine fastener strengths and torque limits.

CLAMPING DEVICES

A machinist has to do many jobs that require the use of clamping devices. These clamping devices are varied in types and tasks that they can perform, but they all serve one general purpose: to hold a workpiece while machining operations are being performed.

C-clamps are used to hold workpieces on machines such as drill presses, and are also used for clamping parts together. The size of the clamp is determined by the largest opening of the jaws. In Figure 1, a shielded screw clamp is shown at the top; a standard light duty C-clamp is shown at the bottom. The shield is designed to protect the screw from welding spatter and other damage. Machinist's heavy duty C-clamps (Figure 2) are used when heavy clamping is required, for example, to hold steel plates when drilling or welding on them.

Parallel clamps (Figure 3) are often used to hold precision parts to angle plates for layout or surface grinding operations. Since they have a large gripping area, they can hold workpieces very securely if they are properly tightened.

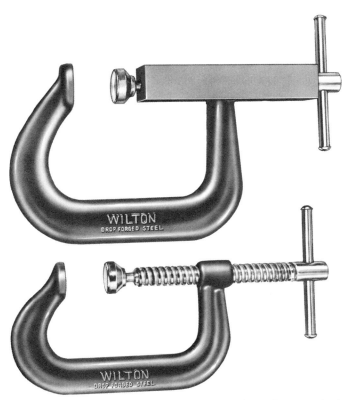

Figure 1. Two types of C-clamps. (Wilton Corporation)

Figure 2. Heavy duty C-clamp. (Lane Community College)

Figure 3. Single size parallel clamps.

Figure 4. Slip joint or combination pliers. (Snap-On Tools Corporation)

Figure 5. Interlocking joint or water pump pliers. (Snap-On Tools Corporation)

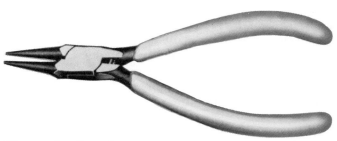

Figure 6. Round nose or wire looper pliers. (Snap-On Tools Corporation)

Figure 7. Needlenose pliers, straight. (Snap-On Tools Corporation)

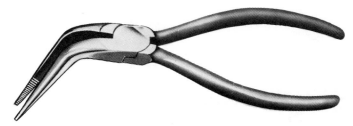

Figure 8. Needlenose pliers, bent. (Snap-On Tools Corporation)

PLIERS

Pliers come in several shapes and with several types of jaw action. Simple combination or slip joint pliers (Figure 4) will do most jobs for which you need pliers. The slip joint allows the jaws to expand to grasp a larger size work. They are measured by overall length and are made in 5-, 6-, 8-, and 10-in. sizes.

Interlocking joint pliers (Figure 5), or water pump pliers, were made to tighten packing gland nuts on water pumps on cars and trucks, but are useful for a variety of jobs. Pliers should never be used as a substitute for a wrench, as the nut or bolt head will be permanently deformed by the serrations in the plier jaws and the wrench will no longer fit properly. Round nose pliers (Figure 6) are used to make loops in wire and to shape light metal. Needle nose pliers are used for holding small delicate workpieces in tight spots. They are available in both straight (Figure 7) and bent nose (Figure 8) types. Linemen's pliers (Figure 9) can be used for wire cutting and bending. Some types have wire stripping grooves and insulated handles. Diagonal cutters (Figure 10) are only used for wire cutting.

The lever-jawed locking wrench, also called vise-grip, has an unusually high gripping power. The screw

Figure 9. Side cutting pliers. (Snap-On Tools Corporation)

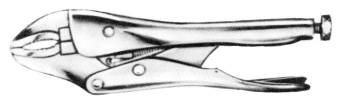

Figure 10. Diagonal cutters. (Snap-On Tools Corporation)

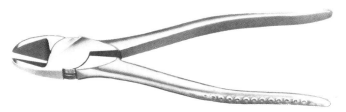

Figure 11. Vise grip wrench. (Snap-On Tools Corporation)

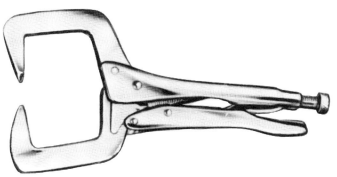

Figure 12. Vise grip C-clamp. (Snap-On Tools Corporation)

Figure 13. Maul. (*Machine Tools and Machining Practices*)

Figure 14. Ball peen hammer. (*Machine Tools and Machining Practices*)

Figure 15. Straight peen hammer.

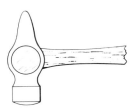

Figure 16. Cross peen hammer.

in the handle adjusts the lever action to the work size (Figure 11). They are made with special jaws for various uses such as the C-clamp type used in welding (Figure 12).

HAMMERS
Hammers are classified as either hard or soft. Hard hammers have steel heads such as blacksmith types or mauls made for heavy hammering (Figure 13). The ball peen hammer (Figure 14) is the one most fre-

quently used by machinists. It has a rounded surface on one end of the head, which is used for upsetting or riveting metal and a hardened striking surface on the other. Two hammers should never be struck together on the face, as pieces could break off. Hammers are specified according to the weight of the head. Ball peen hammers range from two ounces to three pounds. Those under 10 ounces are used for layout work. Two other shop hammers are the straight peen (Figure 15) and the cross peen (Figure 16).

Soft hammers are made of plastic (Figure 17), brass, copper, lead, rubber (Figure 18), or rawhide and are used to properly position workpieces with finishes that would be damaged by a hard hammer. The movable jaw on most machine tool vises tends

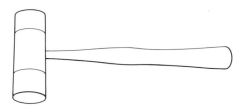

Figure 17.   Plastic hammer. *(Machine Tools and Machining Practices)*

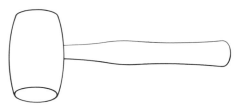

Figure 18.   Rubber mallet. *(Machine Tools and Machining Practices)*

to move slightly upward when tightened against the workpiece. Thus, the workpiece is moved upward and out of position. The machinist must then use a soft, heavy mallet to reposition it. Those having no rebound (such as lead hammers) work best for this purpose.

WRENCHES

A large variety of wrenches is made for different uses such as turning capscrews, bolts, and nuts. The adjustable wrench (Figure 19) is a general purpose tool but will not suit every job, especially those requiring work in close quarters. The wrench should be rotated toward the movable jaw and should fit the nut or bolt tightly. The size of the wrench is determined by its overall length in inches.

Open end wrenches (Figure 20) are best suited to square-headed bolts, and usually fit two sizes, one on each end. The ends on this type of wrench are also angled so they can be used in close quarters. Box wrenches (Figure 21) are also double ended and offset to clear the user's hand. The box completely surrounds the nut or bolt and usually has 12 points so that the wrench can be reset after rotating only a partial turn. Mostly used on hex-headed bolts, these wrenches have the advantage of precise fit. Combination and open end wrenches are made with a box at one end and an open end at the other (Figure 22).

Socket wrenches are similar to box wrenches in that they also surround the bolt or nut and usually are made with 12 points contacting the six-sided nut. Sockets are made to be detached from various types of drive handles (Figure 23).

Figure 19.   Adjustable wrench showing the correct direction of pull. Moveable jaw should always face the direction of rotation. (Snap-On Tools Corporation)

Figure 20.   Open end wrench. (Snap-On Tools Corporation)

Figure 21.   Box wrench. (Snap-On Tools Corporation)

Figure 22.   Combination wrench. (Snap-On Tools Corporation)

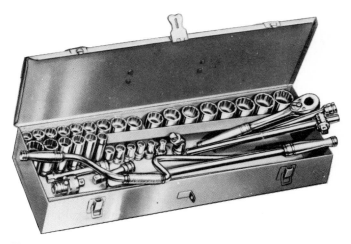

Figure 23.   Socket wrench set. (Snap-On Tools Corporation)

Pipe wrenches, as the name implies, are used for holding and turning pipe. These wrenches have sharp serrated teeth and will damage any finished part on which they are used (Figure 24). Strap wrenches (Fig-

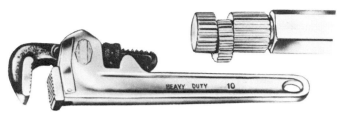

Figure 24. Pipe wrenches, external and internal. (Snap-On Tools Corporation)

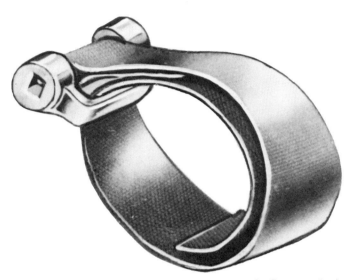

Figure 25. Strap wrench. (Snap-On Tools Corporation)

Figure 26. Fixed face spanner. *(Machine Tools and Machining Practices)*

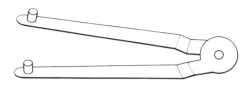

Figure 27. Adjustable face spanner. *(Machine Tools and Machining Practices)*

Figure 28. Hook spanner. *(Machine Tools and Machining Practices)*

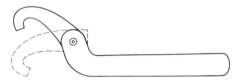

Figure 29. Adjustable hook spanner. *(Machine Tools and Machining Practices)*

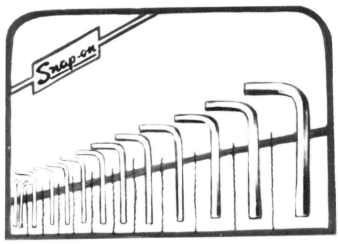

Figure 30. Socket head wrench set. (Snap-On Tools Corporation)

ure 25) are used for extremely large parts or to avoid marring the surface of tubular parts.

Spanner wrenches come in several basic types including face and hook. Face types are sometimes called pin spanners (Figure 26). Spanners are made in fixed sizes or adjustable types (Figures 27, 28, and 29).

Socket head wrenches (Figure 30) are six-sided bars having a 90-degree bend near one end. They are used with socket head capscrews and socket set screws.

The hand tap wrench (Figure 31) is used for medium and large size taps. The T-handle tap wrench (Figure 32) is used for small taps $\frac{1}{4}$ in. and under, as its more sensitive "feel" results in less tap breakage.

The following are safety hints for using wrenches:

1. Make sure the wrench you select fits properly. If

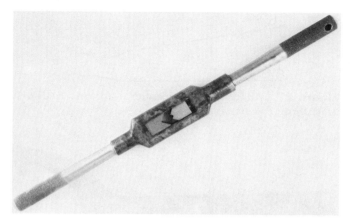

Figure 31. Hand tap wrench. *(Machine Tools and Machining Practices)*

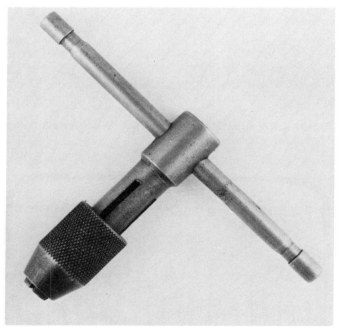

Figure 32. T-handle tap wrench. *(Machine Tools and Machining Practices)*

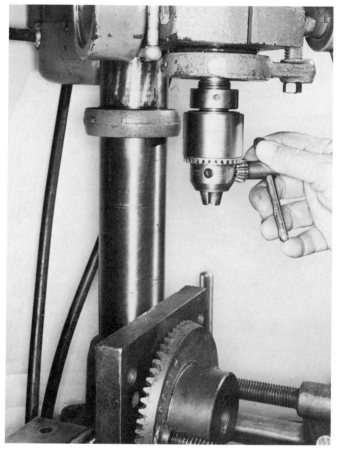

Figure 33. Chuck key and drill chuck in use. (Lane Community College)

it is a loose fit, it may round off the corners of the nut or bolt head.

2. Pull on a wrench instead of pushing to avoid injury.

3. Never use a wrench on moving machinery.

4. Do not hammer on a wrench or extend the handle for additional leverage. Use a larger wrench.

Chuck keys (Figure 33) are used to open and close chucks on drill presses and electric hand drills. Their size is determined by the size of the chuck. They should never be left in a chuck, as they can become a dangerous missile as soon as the machine is turned on. Safety chuck keys are made with a spring plunger that pushes them out of the chuck when no pressure is being applied on the key with the hand, and consequently cannot be left in the chuck.

SCREWDRIVERS

The two types of screwdrivers that are most used are the standard (Figure 34) and Phillips (Figure 35). Both types are made in various sizes and in several styles, straight, shank, and offset (Figures 36*a* and 36*b*). It is important to use the right width blade when installing or removing screws (Figure 37). The shape of the tip is important also. If the tip is badly worn or incorrectly ground, it will tend to jump out of the slot. The correct method of grinding a standard screwdriver is shown in Figure 38. Be careful not to overheat the tip when grinding, as it will become soft. Never use a screwdriver for a chisel or pry bar. Keep a screwdriver in proper shape by using it only on the screws for which it was meant.

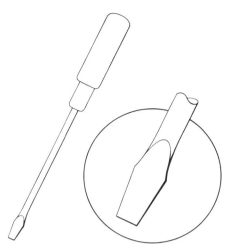

Figure 34. Screwdriver, standard. *(Machine Tools and Machining Practices)*

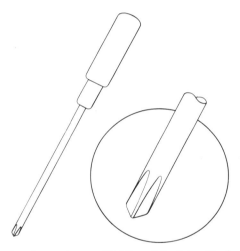

Figure 35. Screwdriver, Phillips. *(Machine Tools and Machining Practices)*

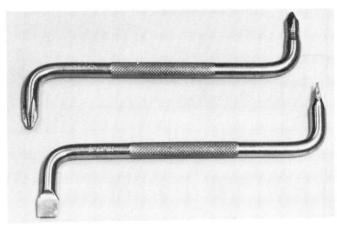

Figure 36a. Standard and Phillips offset screwdrivers. (Lane Community College)

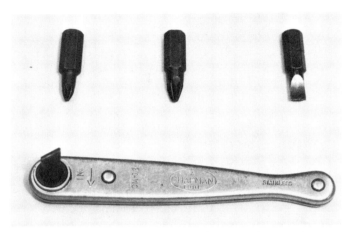

Figure 36b. Ratchet offset screwdriver with interchangeable points. (Lane Community College)

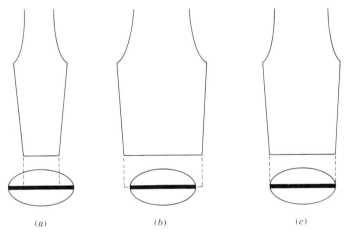

(a)   (b)   (c)

Figure 37. The proper width of a screwdriver blade: *(a)* Too narrow, *(b)* too wide, and *(c)* correct width. *(Machine Tools and Machining Practices)*

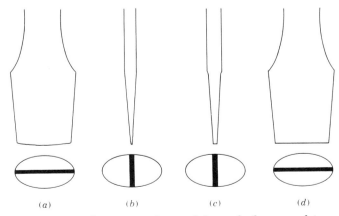

(a)   (b)   (c)   (d)

Figure 38. The proper shape of the end of a screwdriver blade. Blades *(a)* and *(b)* are badly worn; blades *(c)* and *(d)* are ground correctly. *(Machine Tools and Machining Practices)*

SNAP RING TOOLS

Snap ring pliers are made for both external and internal use (Figure 39). Since snap rings are made in many sizes, the points on snap ring tools are also made in various sizes and angles (Figure 40). Some types have interchangeable points.

TORQUE WRENCHES AND FASTENER STRENGTH

Materials may be subjected to three basic stresses: tensile, compressive, and shear. A fourth type is combined shear and tensile, called torsion shear (Figure 41). Fasteners, such as bolts and capscrews, generally have a direct tensile load applied when they are tightened with a wrench. Since a thread is really a circular inclined plane, the thread angle, and consequently the pitch, is involved in determining the tensile stress resulting from a given torque. It is sufficient to note that the finer the thread (smaller thread angle) the greater the tensile force exerted on the bolt for a given torque. (See Section A, Unit 2 for more information on fasteners.)

The tensile stress area of the threaded part of a bolt is the area within the minor diameter. This area and the ultimate strength of the material in PSI determine the strength of the bolt (Figure 42). The ultimate strength of a fastener is the amount of tensile load required to break it. This can be determined by the following formula where

$$P = SA$$

$P =$ the tensile load to break the screw

$S =$ the ultimate strength (tensile strength) of the material (This can be determined by grade markings from a table.)

$A =$ the tensile stress area in square inches (The minor diameter may be taken from screw thread tables and the area calculated with the formula πr^2.)

A more practical calculation is to substitute the yield point for the ultimate strength of the material. The yield point is the stress in PSI where the metal begins to permanently deform. When a fastener has deformed or stretched, it has been overloaded and damage has resulted. Safety factors are therefore used so that the fastener is from 2 to 15 times stronger (using its yield strength) than any expected load while it is in use. Whatever its safety factor may be, a mechanic can overtorque it past its yield point and cause a machine failure (Figure 43).

Undertorquing a bolt can also cause a fatigue failure from a vibrating (cyclic) stress. Undertightening

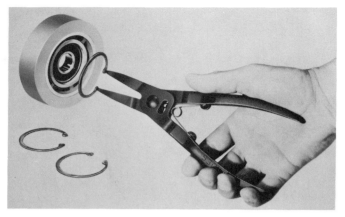

Figure 39. Inserting a snap ring with the proper tool. (Waldes Kohinoor Inc.)

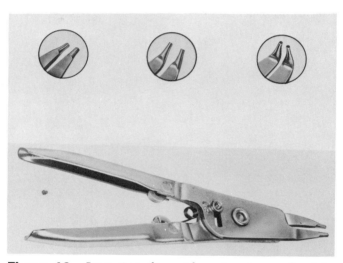

Figure 40. Snap ring plier with various types of points. (Waldes Kohinoor Inc.)

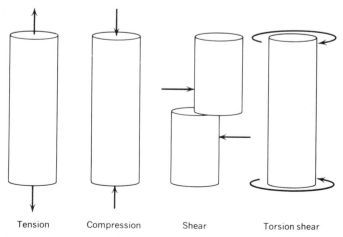

Tension   Compression   Shear   Torsion shear

Figure 41. The three types of stresses. (*Machine Tools and Machining Practices*)

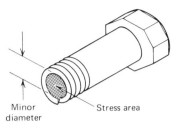

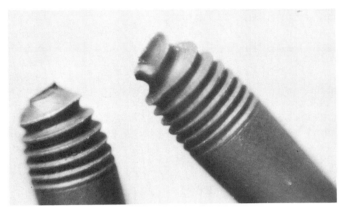

Figure 42. The effective stress area of a bolt.

Figure 43. These capscrews were overtorqued when the mechanic tightened them. (Lane Community College)

Figure 44a. Beam type torque wrench. (Lane Community College)

Figure 44b. Dial type torque wrench. (Lane Community College)

Figure 44c. Click type torque wrench. (Lane Community College)

parts on machinery, especially automobiles, can cause leaks and parts to come loose, while overtightening can cause distortion such as out of round brake drums, cylinders, and warped heads. A proper tightening sequence is needed to prevent distortion.

TORQUE WRENCHES

It is obvious that some kind of accurate method of measuring fastener stress is needed. This is done with a measuring tool called a torque wrench. Torque wrenches are made in three basic types, and read in English measure (inch pounds and foot pounds), or in metric measure (kilogram-centimeters and Newton-meters). Some models read in both inch and metric units. One type has a flexible beam and a stationary pointer on a scale (Figure 44a). A second type reads torque directly on a dial (Figure 44b), and the click type (Figure 44c) indicates the preset torque has been reached by an audible sound or "click" followed by a sudden release.

It has been found that a fastener should be tightened to a tension of about one-half of its elastic limit. At this point it will retain enough residual tension to resist loosening and yet the fastener will not tend to stretch or fatigue at this amount of load. Recommended foot and inch pounds of bolt torque is given

in Table 1. This is based on the ultimate tensile strength of the fastener in terms of grade markings, which are a SAE classification system for bolt strength. See also Figure 10 in Section A, Unit 2.

IMPACT WRENCHES

Power, or impact, wrenches that are either electric or air driven are found in most mechanics' shops. These tools are very useful for loosening stubborn bolts and nuts. Care must be taken when using these tools for assembling parts, such as automotive wheels on drums, so that even torquing is obtained. If one bolt is first overtorqued before tightening the others, the part, such as a brake drum, may be permanently distorted.

Table 1

Standard Bolt Torque Foot Pounds and Newton-Meters

Grade of Bolt	1 and 2		5		6		8	
Min. Tensile Strength-PSI	65,000		120,000		140,000		150,000	
Grade Markings on Head	⬡		⬡		⬡		⬡	
Bolt dia.	Ft. Lb	N-m	Ft. Lb	N-m	Ft. Lb	N-m	Ft. Lb	N-m
$\frac{1}{4}$–20	5	6.779	8	10.846	10	13.558	12	16.270
$\frac{1}{4}$–28	6	8.135	10	13.558			14	18.980
$\frac{5}{16}$–18	11	14.914	17	23.049	19	25.760	24	32.540
$\frac{5}{16}$–24	13	17.626	19	25.760			27	36.607
$\frac{3}{8}$–16	18	24.404	31	42.030	34	46.098	44	59.656
$\frac{3}{8}$–24	20	27.116	35	47.454			49	66.435
$\frac{7}{16}$–14	28	37.963	49	66.435	55	74.570	70	94.907
$\frac{7}{16}$–20	30	40.674	55	74.570			78	105.754
$\frac{1}{2}$–13	39	52.877	75	101.686	85	115.244	105	142.361
$\frac{1}{2}$–20	41	55.588	85	115.244			120	162.696
$\frac{9}{16}$–12	51	69.147	110	149.138	120	162.676	155	210.149
$\frac{9}{16}$–18	55	74.570	120	162.676			170	230.486
$\frac{5}{8}$–11	83	112.533	150	203.320	167	226.419	210	284.718
$\frac{5}{8}$–18	95	128.803	170	230.486			240	235.392
$\frac{3}{4}$–10	105	142.361	270	366.066	280	379.624	375	508.425
$\frac{3}{4}$–16	115	155.917	295	399.961			420	569.436
$\frac{7}{8}$–9	160	216.928	395	535.541	440	596.552	605	820.259
$\frac{7}{8}$–14	175	237.265	435	689.773			675	915.165
1–8	235	318.613	590	799.922	660	894.828	910	1,233.778
1–14	250	338.950	660	894.828			990	1,342.242

Note: Typical torque values for capscrews having clean, dry threads are shown. If fasteners are chrome or cadmium plated or have been oiled, reduce torque by 10 to 20 percent. Consult manufacturer's recommendations for torque values for any specific job. See also Figure 10, Section A, Unit 2. (N-m is the abbreviation for Newton-meters, a metric measure of torque.)

SELF-TEST

1. What are parallel clamps commonly used for?
2. How are C-clamps measured?
3. Compare the uses of light and heavy duty C-clamps.
4. In order to remove a nut or bolt, slip joint or water pump pliers make a good substitute for a wrench when a wrench is not handy.
   True _____   False _____
5. What advantage does the lever jawed wrench offer over other similar tools such as pliers?
6. Would you use a three-pound ball peen hammer for layout work? If not, what size do you think is right?
7. Some objects should never be struck with a hard hammer, such as a finished machine surface or the end of a shaft. What could you use to avoid damage?
8. A machine has a capscrew that needs to be tightened and released quite often. Which wrench would be best to use in this case: the adjustable or box-type wrench? Why?
9. Why should pipe wrenches never be used on bolts, nuts, or shafts?
10. What are two important things to remember about standard screwdrivers that will help you avoid problems in their use?
11. What can result from overtorquing or undertorquing fasteners?
12. How can you know how much torque to put on a bolt, even if you have a torque wrench?

UNIT 4 CUTTING HAND TOOLS: HACKSAWS

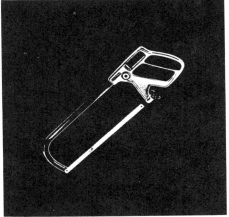

Hacksaws are one of the more frequently used hand tools. The hand hacksaw is a relatively simple tool to use, but the facts and rules contained in this unit will help you improve your use of the hacksaw.

OBJECTIVE

After completing this unit, you should be able to:
Identify, select, and use hand hacksaws.

The hacksaw consists of three parts: the frame, the handle, and the saw blade (Figure 1). Frames are either the solid or adjustable type. The solid frame can only be used with one length of saw blade. The adjustable frame can be used with hacksaw blades from 8 to 12 in. in length. The blade can be mounted to cut in line with the frame or at a right angle to the frame (Figures 2a and 2b). When the distance to be sawed is greater than the space between the blade and the frame, it may be necessary to set the blade at 90 degrees as in Figure 2b so the work can clear the frame.

Most hacksaw blades are made from high speed steel, and in standard lengths of 8, 10, and 12 in. Blade length is the distance between the centers of the holes at each end. Hand hacksaw blades are generally .5 in. wide and .025 in. thick. The kerf or cut produced by the hacksaw is wider than the .025 in. thickness of the blade because of the set of the teeth (Figure 3).

Figure 2a. Straight sawing with a hacksaw. (Lane Community College)

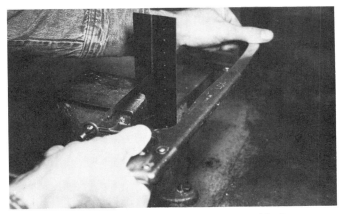

Figure 2b. Sawing with the blade set at 90 degrees to the frame. (Lane Community College)

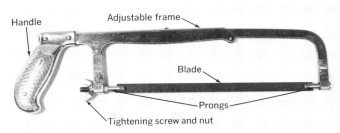

Figure 1. The parts of a hacksaw. (*Machine Tools and Machining Practices*)

Figure 3. The kerf is wider than the blade because of the set of the teeth. (Lane Community College)

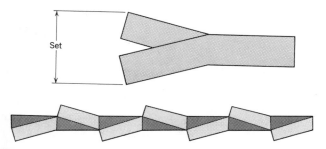

Figure 4. The straight (alternate) set. *(Machine Tools and Machining Practices)*

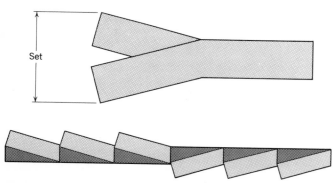

Figure 5. The wavy set. *(Machine Tools and Machining Practices)*

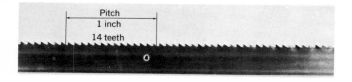

The set refers to the bending of teeth outward from the blade itself. Two kinds of set are found on hand hacksaw blades. First, with the straight or alternate set (Figure 4) one tooth is bent to the right and the next tooth to the left for the length of the blade. The second kind of set is the wavy set in which a number of teeth are gradually bent to the right and then to the left (Figure 5). A wavy set is found on most fine tooth hacksaw blades.

The spacing of the teeth on a hand hacksaw blade is called the pitch and is expressed in teeth per inch of length (Figure 6). Standard pitches are 14, 18, 24, and 32 teeth per inch, with the 18 pitch blade used as a general purpose blade.

The hardness and size or thickness of a workpiece determine to a great extent which pitch blade to use. As a rule, you should use a coarse tooth blade on soft materials, to have sufficient clearance for the chips, and a fine tooth blade on harder materials. But you also should have at least three teeth cutting at any time, which may require a fine tooth blade on soft materials with thin cross sections.

Hand hacksaw blades fall into two categories: soft-backed or flexible blades and all-hard blades. On the flexible blades only the teeth are hardened, the back being tough and flexible. The flexible blade is less likely to break when used in places that are difficult to get at such as in cutting off bolts on machinery, since it can bend or distort without breaking. The all-hard blade is, as the name implies, hard and very brittle and should be used only where the workpiece can be rigidly supported, as in a vise. On an all-hard blade even a slight twisting motion may break the blade. All-hard blades, in the hands of a skilled person, will cut true straight lines and give long service.

The blades are mounted in the frame with the teeth pointing away from the handle, so that the hacksaw cuts only on the forward stroke. No cutting pressure should be applied to the blade on the return stroke as this tends to dull the teeth. The sawing speed with a hacksaw should be from 40 to 60 strokes per minute. To get the maximum performance from a blade, make long, slow, and steady strokes using the full length of the blade. Sufficient pressure should be maintained on the forward stroke to keep the teeth cutting. Teeth on a saw blade will dull rapidly if too little or too much pressure is put on the saw. The teeth will dull also if too fast a cutting stroke is used; a speed in excess of 60 strokes a minute will dull the blade because friction will overheat the teeth.

Figure 6. The pitch of the blade is expressed as the number of teeth per inch. (Lane Community College)

Figure 7. A new blade must be started on the opposite side of the work, not in the same kerf as the old blade. (Lane Community College)

Figure 8. The workpiece is being sawed close to the vise to avoid vibration and chatter. (Lane Community College)

The saw blade may break if it is too loose in the frame or if the workpiece slips in the vise while sawing. Too much pressure may also cause the blade to break. A badly worn blade where the set has been worn down, will cut a too narrow kerf, which will cause binding and perhaps breakage of the blade. When this happens and a new blade is used to finish the cut, turn the workpiece over and start with the new blade from the opposite side and make a cut to meet the first one (Figure 7). The set on the new blade is wider than the old kerf. Forcing the new blade into an old cut will immediately ruin a new blade by wearing the set down.

A cut on a workpiece should be started with only light cutting pressure, with the thumb or fingers on one hand acting as a guide for the blade. Sometimes it helps to start a blade when a small vee-notch is filed in the workpiece. When a workpiece is supported in a vise, make sure that the cutting is done close to the vise jaws for a rigid setup free of chatter (Figure 8). Work should be positioned in a vise so that the saw cut is vertical. This makes it easier for the saw to follow a straight line. At the end of a saw cut, just before the pieces are completely parted, reduce the cutting pressure or you may be caught off balance when the pieces come apart and cut your hands on the sharp edges of the workpiece. While sawing, stand as close to the vise or workpiece as is comfortable, with your feet slightly apart to provide stability. If possible, the workpiece and your elbow should be at about the same level. To saw thin material, sandwich it between two pieces of wood for a straight cut. Avoid bending the saw blades because they are likely to break, and when they do, they usually shatter in all directions and could injure you or others nearby.

SELF-TEST

1. What is the kerf?
2. What is the set on a saw blade?
3. What is the pitch of a hacksaw blade?
4. What determines the selection of a saw blade for a job?
5. Hand hacksaw blades fall into two basic categories. What are they?
6. What speed should be used in hand hacksawing?
7. Give four causes that make saw blades dull.
8. Give two reasons why saw blades break.
9. A new hacksaw blade should not be used in a cut started with a blade that has been used. Why?
10. What dangers exist when a hacksaw blade breaks while it is being used?

UNIT 5 CUTTING HAND TOOLS: FILES

Files are often used to put the finishing touches on a machined workpiece, either to remove burrs or sharp edges or as a final fitting operation. Intricate parts or shapes are often entirely produced by skilled craftsmen using files. In this unit you are introduced to the types and uses of files in metalworking.

OBJECTIVE

After completing this unit, you should be able to:
Identify eight common files and some of their uses.

TYPES OF FILES

Files are tools that everyone in metalwork will use. Often, through lack of knowledge, these tools are misused. Files are made in many different lengths ranging from 4 to 18 in. (Figure 1). Files are manufactured in many different **shapes** and are used for many specific purposes. Figure 2 shows the parts of a file. When a file is measured, the **length** is taken from the heel to the point, and the tang is excluded. Most files are made from high carbon steel and are heat-treated to the correct hardness range. They are manufactured in four different **cuts:** single, double, curved tooth, and rasp. The single cut, double cut, and curved tooth are commonly encountered in machine shops (Figure 3). Rasps are usually used with wood. Curved tooth files will give excellent results with soft materials such as aluminum, brass, plastic, or lead.

Files also vary in their **coarseness:** rough, coarse, bastard, second cut, smooth, and dead smooth. The files most often used are the bastard, second cut, and smooth grades. Different sizes of files within the same coarseness designation will have varying sizes of teeth (Figure 4): the longer the file, the coarser the teeth. For maximum metal removal a double cut file is used. If the emphasis is on a smooth finish, a single cut file is recommended.

The face of most files is slightly convex (Figure 5) because they are made thicker in the middle than on the ends. Through this curvature only some teeth are cutting at any one time, which makes them penetrate better. If the face were flat, it would be difficult

to obtain an even surface because of the tendency to rock a file while filing. Some of this curvature is also offset by the pressure applied to make the file cut. New files do not cut as well as slightly used ones, since on new files some teeth are longer than most others and leave scratches on a workpiece.

Files are either blunt or tapered (Figure 6). A blunt file has the same cross-sectional area from heel

Figure 1. Files are made in several different lengths. (Lane Community College)

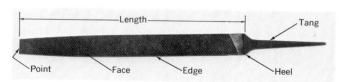

Figure 2. The parts of a file. (*Machine Tools and Machining Practices*)

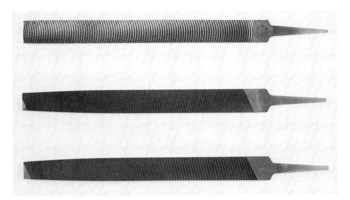

Figure 3. Three files that are frequently found in machine shops are the curved tooth, double cut, and single cut files.

Figure 4. These two files are both bastard cut, but since they are of different lengths, they have different coarsenesses. (Lane Community College)

Figure 5. The edge of this file shows the convex face. (Lane Community College)

to point, where a tapered file narrows toward the point.

Files fall into five basic categories: mill and saw files, machinists' files, Swiss pattern files, curved tooth files, and rasps. Machinists' and mill and saw files are classified as American pattern files. Mill files (Figure 7) were originally designed to sharpen large saws in lumber mills, but now they are used for draw filing, filing on a lathe (Figure 8), or filing a finish on a workpiece. Mill files are single cut and work well on brass and bronze. Mill files are slightly thinner than an

Figure 6. Blunt and tapered file shapes.

Figure 7. A mill file.

Figure 8. The lathe file has a longer angle on the teeth to clear the chips when filing on the lathe.

Figure 9. The flat file is usually a double cut file.

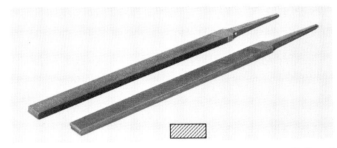

Figure 10. Two pillar files. (Lane Community College)

equal-sized flat file, which is a machinists' file (Figure 9) that is usually double cut. Double cut files are used when fast cutting is needed. The finish produced is relatively rough.

Pillar files (Figure 10) have a narrower but thicker cross section than flat files. Pillar files are parallel in width and taper slightly in thickness. They also have one or two safe edges that allow filing into a corner without damaging the shoulder. Square files (Figure 11) usually are double cut and are used to file in keyways, slots, or holes.

If a very thin file is needed with a rectangular cross section, a warding file (Figure 12) is used. This file is often used by locksmiths when filing notches into locks and keys. Another file that will fit into narrow slots is a knife file (Figure 13). The included angle between the two faces of this file is approximately 10 degrees.

Three-square files (Figure 14), also called three-cornered files, are triangular in shape with the faces at 60-degree angles to each other. These files are used for filing internal angles between 60 and 90 degrees as well as to make sharp corners in square holes. Half-round files (Figure 15) are available to file large internal curves. Half-round files, because of their tapered construction, can be used to file many different radii. Round files (Figure 16) are used to file small radii or to enlarge holes. These files are available in many diameter sizes.

Swiss pattern files (Figure 17) are manufactured to much closer tolerances than American pattern files, but are made in the same shapes. Swiss pattern files are more slender as they taper to finer points and their teeth extend to the extreme edges. Swiss pattern files range in length from 3 to 10 in. and their coarseness is indicated by numbers from 00 (coarse) to 6 (fine). Swiss pattern files are made with tangs to be used with file handles or as needle files with round or square handles that are part of the files. Another type of Swiss pattern files is die sinkers' rifflers (Figure 18). These files are double-ended with cutting surfaces on either end. Swiss pattern files are used primarily by tool and die makers, mold makers, and other craftsmen engaged in precision filing on delicate instruments.

Curved tooth files (Figure 19) cut very freely and remove material rapidly. The teeth on curved tooth files are all of equal height and the gullets or valleys between teeth are deep and provide sufficient room for the filings to curl and drop free. Curved tooth files are manufactured in three grades of cut: standard, fine, and smooth, and in length from 8 to 14 in. These files are made as rigid tang types for use with a file handle, or as rigid or flexible blade types used with special handles. Curved tooth file shapes are flat, half-round, pillar, and square.

The bastard cut file (Figure 20) has a safe edge that is smooth. Flat filing may be done up to the shoulders of the workpiece without fear of damage. Files of other cuts and coarseness are also available with safe edges on one or both sides.

Figure 11.   The square file. (Lane Community College)

Figure 12.   Warding file. (Lane Community College)

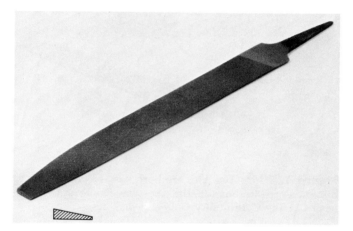

Figure 13.   Knife file. (Lane Community College)

Figure 14.   Three-square files are used for filing angles between 60 and 90 degrees. (Lane Community College)

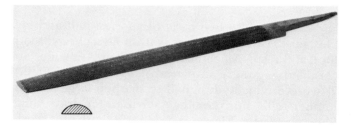

Figure 15.   Half-round files are used for internal curves. (Lane Community College)

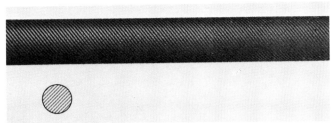

Figure 16. Round files are used to file a small radius or to enlarge a hole.

Figure 17. A set of Swiss pattern files. Since these small files are very delicate and can be broken quite easily, great care must be exercised in their use.

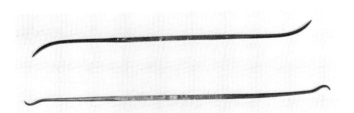

Figure 18. Die sinker's rifflers. *(Machine Tools and Machining Practices)*

Figure 19. Curved tooth files are used on soft metals.

Figure 20. A file with a safe edge will not cut into shoulders or corners when filing is being done.

Figure 21. Thread files. (Lane Community College)

Thread files (Figure 21) are used to clean up and reshape damaged threads. They are square in cross section and have eight different thread pitches on each file. The thread file of the correct pitch is most effectively used when held or stroked against the thread while it is rotating in a lathe. A thread can be repaired, however, even when it cannot be turned in a lathe.

CARE AND USE OF FILES

Files do an efficient job of cutting only while they are sharp. Files and their teeth are very hard and brittle. Do not use a file as a hammer or as a pry bar. When a file breaks, particles will fly quite a distance at high speed and may cause an injury. Files should be stored so that they are not in contact with any other file. The same applies to files on a workbench. Do not let files lie on top of each other because one file will break teeth on the other file (Figure 22). Teeth on files will also break if too much pressure is

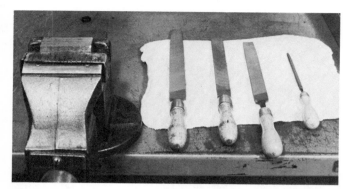

Figure 22. Files should be kept neatly arranged so that they will not strike each other and damage the cutting edges. (Lane Community College)

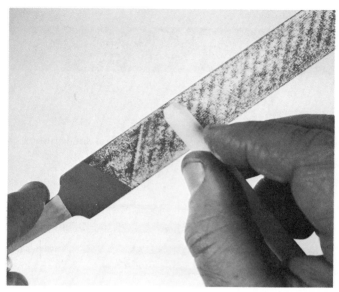

Figure 24. Using chalk on the file to help reduce pinning.

Figure 23. Using a file card to clean a file.

put on them while filing. On the other hand, if not enough pressure is applied while filing, the file only rubs the workpiece and dulls the teeth. A dull file can be identified by its shiny, smooth teeth and by the way it slides over the work without cutting. Dulling of teeth is also caused by the filing of hard materials or when filing too fast. A good filing speed is 40 to 50 strokes per minute, but remember that the harder the material is, the slower the strokes should be; the softer the materials, the coarser the file should be.

Too much pressure or a new file may cause "pinning," that is, filings wedged in the teeth; the result is deep scratches on the work surface. If the pins cannot be removed with a file card, try a piece of brass, copper, or mild steel and push it through the teeth (Figure 23). Do not use a scriber or other hard object for this operation. A file will not pin as much if some

blackboard chalk is applied to the face (Figure 24). *Never use a file without a file handle or the pointed tang may cause a serious hand or wrist injury* (Figures 25a and 25b).

Many filing operations are performed with the workpiece held in a vise. Clamp the workpiece securely, but remember to protect it from the serrated vise jaws with some soft piece of material such as copper, brass, wood, or paper (Figure 26). The workpiece should extend out of the vise only so the file clears the vise jaws by $\frac{1}{8}$ to $\frac{1}{4}$ inch. Since a file cuts only on the forward stroke, no pressure should be applied on the return stroke. Letting the file drag over the workpiece on the return stroke helps release the small chips so that they can fall from the file. Use a stroke as long as possible; this will make the file wear out evenly instead of just in the middle. To file a flat surface, change the direction of the strokes frequently to produce a cross-hatch pattern (Figure 27). By using a straightedge steel rule to test for flatness, we can easily determine where the high spots are that have to be filed away. It is best to make flatness checks often because, if any part is filed below a given layout line, the rest of the workpiece may have to be brought down just as far.

Figure 28 shows how a file should be held to file a flat surface. A smooth finish is usually obtained by draw filing (Figure 29), where a single cut file is held with both hands and drawn back and forth on a workpiece. The file should not be pushed over the ends of the workpiece as this would leave rounded edges.

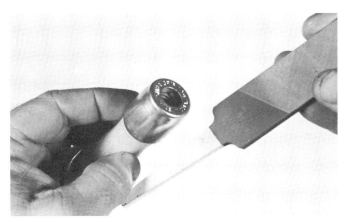

Figure 25a. A file should never be used without a file handle.

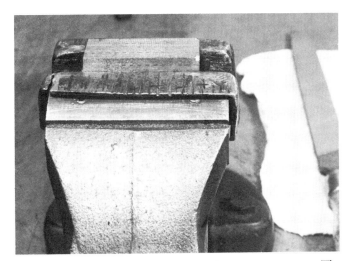

Figure 25b. This style of handle is designed to screw on rather than be driven on the tang.

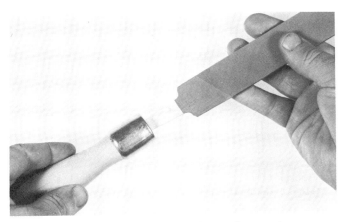

Figure 26. Workpiece in vise with protective jaws. The work extends only ¼ in. from the vise jaws for better rigidity. (Lane Community College)

Figure 27. The cross-hatch pattern shows that this piece has been filed from two directions thus producing a flatter surface. (Lane Community College)

Figure 28. Proper filing position. (Lane Community College) *Note:* safety glasses should be worn.

To get a smooth finish it sometimes helps to hold the file as shown in Figure 30, making only short strokes. The pressure is applied by a few fingers and does not extend over the ends of the workpiece. When a round file or half-round file is used, the forward stroke should also include a clockwise rotation for deeper cuts and a smoother finish. A tendency of people who are filing is to run their hands or fingers over a newly filed surface. This deposits a thin coat of skin oil on the surface. When filing is resumed, the file will not cut for several strokes, but will only slip over the surface causing the file to dull more quickly.

Figure 29. Draw filing. (Lane Community College)

Figure 30. Use this procedure to correct high spots on curvatures on the workpiece. Apply pressure with short strokes only where cutting is needed. (Lane Community College)

SELF-TEST

1. How is a file identified?
2. What are the four different cuts found on files?
3. Name four coarseness designations of files.
4. Which of the two kinds of files, single cut or double cut, is designed to remove more material?
5. Why are the faces of most files slightly convex?
6. What difference is there between a blunt and a tapered file?
7. What difference exists between a mill file and an equal-sized flat file?
8. What is a warding file?
9. An American pattern file differs in what way from a Swiss pattern file?
10. What are the coarseness designations for needle files?
11. Why should files be stored so they do not touch each other?
12. What happens if too much pressure is applied when filing?
13. What causes a file to get dull?
14. Why should a handle be used on a file?
15. Why should workpieces be measured often?
16. What happens when a surface being filed is touched with the hand or fingers?
17. How does the hardness of a workpiece affect the selection of a file?
18. How can rounded edges be avoided when a workpiece is drawfiled?
19. Should pressure be applied to a file on the return stroke?
20. Why is a round file rotated while it is being used?

UNIT 6 HAND REAMERS

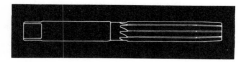

Holes produced by drilling are seldom accurate in size and often have rough surfaces. A reamer is used to finish a hole to an exact dimension with a smooth finish. This unit will describe some commonly used hand reamers and how they are used.

OBJECTIVES

After completing this unit, you should be able to:
1. Identify at least five types of hand reamers.
2. Hand ream a hole to a specified size.

Hand reamers are often used to finish a previously drilled hole to an exact dimension and a smooth surface. When parts of machine tools are aligned and fastened with capscrews or bolts, the final operation is often the hand reaming of a hole in which a dowel pin is placed to maintain the alignment. Hand reamers are designed to remove only a small amount of material from a hole—usually from .001 to .005 in. These tools are made from high carbon or high speed steel.

FEATURES OF HAND REAMERS

Figure 1 shows the major features of the most common design of hand reamer. Another design is available with a pilot ahead of the starting taper (see *Machinery's Handbook* for details). The square on the end of the shank permits the clamping of a tap wrench or T-handle wrench to provide the driving torque for reaming. The diameter of this square is between .004 and .008 in. smaller than the reamer size, and the shank of the reamer is between .001 and .006 in. smaller, to guide the reamer and to permit it to pass through a reamed hole without marring it. It is very important that these tools *not* be put into a drill chuck, because a burred shank can ruin a reamed hole as the shank is passed through it.

Hand reamers have a long starting taper that is usually as long as the diameter of the reamer, but may be as long as one-third of the fluted body. This starting taper is usually very slight and may not be apparent at a casual glance. Hand reamers do their cutting on this tapered portion. The gentle taper and length of the taper help to start the reamer straight and keep it aligned in the hole.

Details of the cutting end of the hand reamer are shown in Figure 2. The full diameter or actual size of the hand reamer is measured where the starting taper ends and the margin of the land appears. The diameter of the reamer should only be measured at this junction, as the hand reamer is generally back tapered or reduced in outside diameter by about .0005 to .001 inch per inch of length toward the shank. This back tapering is done to reduce tool contact with the workpiece. When hand reamers become dull, they are resharpened at the starting taper, using a tool and cutter grinder.

The function of the hand reamer is like that of a scraper, rather than an aggressive cutting tool like most drills and machine reamers. For this reason hand reamers typically have zero or negative radial rake on the cutting face, rather than the positive radial rake characteristic of most machine reamers. (See Section G, Unit 8, "Reaming in the Drill Press.") The right-hand cut with a left-hand helix is considered

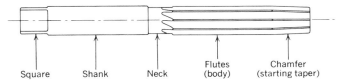

Figure 1. Major features of the hand reamer. *(Machine Tools and Machining Practices)*

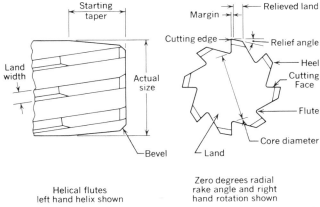

Figure 2. Functional details of the hand reamer. (Bendix Industrial Tools Division)

Figure 3. Straight flute hand reamer. (TRW Inc.)

Figure 4. Helical flute hand reamer. (TRW Inc.)

standard for hand reamers. The left-hand helix produces a negative axial rake for the tool, which contributes to a smooth cutting action. The left-hand helix will also prevent the reamers from being pulled into the hole.

Most reamers, hand or machine types, are made with staggered spacing, that is, the teeth are not uniform or symmetrical. Also, an odd number of teeth is sometimes used. This tends to reduce chatter by reducing harmonic effects between cutting edges. Harmonic chatter is especially a problem with adjustable hand reamers, which often leave a tooth pattern in the work.

Hand reamers are made with straight flutes (Figure 3) or helical flutes (Figure 4). Most hand reamers are manufactured with a right-hand cut, which means they will cut when rotated in a clockwise direction. Helical or spiral fluted reamers are available with a right-hand helix or a left-hand helix. Helical flute reamers are especially useful when reaming a hole having keyways or grooves cut into it, as the helical flutes tend to bridge the gaps and reduce binding or chattering.

Hand reamers for cylindrical holes are made as solid (Figures 3 and 4) or as expansion types (Figure 5). Expansion reamers are designed for use where it is necessary to enlarge a hole slightly for proper fit such as in maintenance applications. These reamers have an adjusting screw that allows limited expansion to an exact size. The maximum expansion of these reamers is approximately .006 in. for diameters up to $\frac{1}{2}$ in., .010 in. for diameters between $\frac{1}{2}$ and 1 in., and .012 in. for diameters for between 1 and $1\frac{1}{2}$ in. These tools are frequently broken by attempts to expand them beyond these limits.

Helical flute expansion reamers are especially adapted for the reaming of bushings or holes having a keyway or straight grooves because of their bridging and shearing cutting action. Expansion reamers have a slightly undersized pilot on the end that guides the reamer and helps to keep it in alignment.

The adjustable hand reamer (Figure 6) is different from the expansion reamer in that it has inserted blades. These cutting blades fit into tapered slots in the body of the reamer and are held in place by two locking nuts. The blades have a taper corresponding to the taper of the slots, which keeps them parallel at any setting. Adjustments in reamer size are made by loosening one nut while tightening the other. Adjustable hand reamers are available in diameters from $\frac{1}{4}$ to 3 in. The adjustment range varies from $\frac{1}{32}$ in. on the smaller diameter reamers to $\frac{5}{16}$ in. on the larger size reamers. Only a small amount of material should be removed at one time, as too large a cut will usually cause chatter.

Taper pin reamers (Figures 7 and 8) are used for reaming holes for standard taper pins used in the assembly of machine tools and other parts. Taper pin reamers have a taper of $\frac{1}{4}$ in. per foot of length, and are manufactured in 18 different sizes numbered from 8/0 to 0 and on up to size 10. The smallest size, number 8/0, has a large end diameter of .0514 in. and the largest reamer, a number 10, has a large end diameter of .7216 in. The sizes of these reamers are designed to allow the small end of each reamer to enter a hole reamed by the next smaller size reamer. As with other hand reamers, the helical flute reamer will cut with more shearing action and less chattering, especially on interrupted cuts.

Morse taper socket reamers are designed to produce holes for American Standard Morse taper shank

Figure 5. Straight flute expansion hand reamer. (TRW Inc.)

Figure 6. Adjustable hand reamers. The lower reamer is equipped with a pilot and tapered guide bushing for reaming in alignment with a second hole. (Lane Community College)

Figure 7. Straight flute taper pin hand reamer. (TRW Inc.)

Figure 8. Spiral flute taper pin hand reamer. (TRW Inc.)

Figure 9. Morse taper roughing reamer with taper shank. (TRW Inc.)

Figure 10. Morse taper finishing reamer with taper shank. (TRW Inc.)

tools. These reamers are available as roughing reamers (Figure 9) and as finishing reamers (Figure 10). The roughing reamer has notches ground at intervals along the cutting edges. These notches act as chipbreakers and make the tool more efficient at the expense of fine finish. The finishing reamer is used to impart the final size and finish to the socket. Morse taper socket reamers are made in sizes from number 0, with a large end diameter of .356 in. to number 5, with a large end diameter of 1.8005 in. There are two larger Morse tapers, but they are typically sized by boring rather than reaming.

SPECIAL PURPOSE REAMERS

Specialized reamers are used in various fields of metalworking. For example, gunsmiths use a special chambering reamer for rifle barrels to make the cartridge fit properly. Plumbers use a reamer to deburr the ends of pipe, and auto mechanics use a special tool called a ridge reamer (Figure 11) to remove a step in the cylinder caused by the wear of the piston rings on the cylinder wall.

USING HAND REAMERS

A hand reamer should be turned with a tap wrench or T-handle wrench rather than with an adjustable wrench. The use of a single end wrench makes it almost impossible to apply torque without disturbing the alignment of the reamer with the hole. A hand reamer should be rotated slowly and evenly, allowing the reamer to align itself with the hole to be reamed. Use a tap wrench large enough to give a steady torque and to prevent vibration and chatter. Use a steady and large feed; feeds up to one-quarter of the reamer diameter per revolution can be used. Small and lightweight workpieces can be reamed by fastening the reamer vertically in a bench vise and rotating the work over the reamer by hand (Figure 12).

In all hand reaming with solid, expansion, or adjustable reamers, never rotate the reamer backwards to remove it from the hole, as this will dull it rapidly. If possible, pass the reamer through the hole and remove it from the far side without stopping the forward rotation. If this is not possible, it should be withdrawn while maintaining the forward rotation.

The preferred stock allowance for hand reaming is between .001 and .005 in. Reaming more material than this would make it very difficult to force the reamer through the workpiece. Reaming too little, on the other hand, results in excessive tool wear because it forces the reamer to work in the zone of material work-hardened during the drilling operation. This stock allowance does not apply to taper reamers, for which a hole has to be drilled at least as large as the small diameter of the reamer. The hole size for a taper pin is determined by the taper pin number and its length. This data can be found in machinist's handbooks.

Since cylindrical hand reaming is restricted to small stock allowances, it is most important that you be able to drill a hole of predictable size and of a surface finish that will assure a finished cleanup cut by the reamer. It is a good idea to drill a test hole

Figure 11. Ridge reamer.

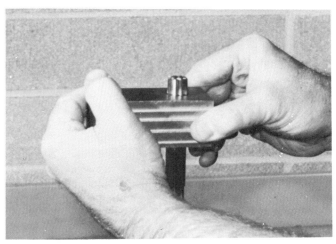

Figure 12. Hand reaming in a small workpiece with the reamer held in a vise. (Lane Community College)

in a piece of scrap of similar composition and carefully measure it both for size and for an enlarged or bellmouth entrance. You may find it necessary to drill a slightly smaller hole before drilling the correct reaming size to assure a more accurate hole size. Carefully spot drill the location before drilling the hole in your actual workpiece. The hole should then be lightly chamfered with a countersinking tool to remove burrs and to promote better reamer alignment.

The use of a cutting oil also improves the cutting action and the surface finish when reaming most metals. Exceptions are cast iron and brass, which should be reamed dry.

When a hand reamer is started it should be checked for squareness on two sides of the reamer, 90 degrees apart. Another way to assure alignment of the reamer with the drilled hole is to use the drill press as a reaming fixture. Put a piece of cylindrical stock with a 60 degree center in the drill chuck (Figure 13) and use it to guide and follow the squared end of the reamer as you turn the tool with the tap wrench. Be sure to plan ahead so that you can drill, countersink, and ream the hole without moving the table or head of the drill press between operations.

On deep holes, or especially on holes reamed with taper reamers, it becomes necessary to remove the chips frequently from the reamer flutes to prevent clogging. Remove these chips with a brush to avoid cutting your hands.

Reamers should be stored so they do not contact one another to avoid burrs on the tools that can damage a hole being reamed. They should be kept in their original shipping tubes or set up in a tool stand. Always check reamers for burrs or for pickup of previous material before you use them. Otherwise the reamed hole can be oversized or marred with a rough finish.

Note. Hand reamers should **never** be used in a drill press or lathe with the power on as this will either dull or break them.

Figure 13. Using the drill press as a reaming fixture. (Lane Community College)

SELF-TEST

1. How is a hand reamer identified?
2. What is the purpose of a starting taper on a reamer?
3. What is the advantage of a spiral flute reamer over a straight flute reamer?
4. How does the shank diameter of a hand reamer compare with the diameter measured over the margins?
5. When are expansion reamers used?
6. What is the difference between an expansion and an adjustable reamer?
7. What is the purpose of a taper pin reamer?
8. What is the purpose of coolant used while reaming?
9. Why should reamers not be rotated backwards?
10. How much reaming allowance is left for hand reaming?

UNIT 7 IDENTIFICATION AND USES OF TAPS

Most internal threads produced today are made with taps. These taps are available in a variety of styles, each designed to perform a specific type of tapping operation efficiently. Today's mass production of consumer goods depends to a large extent on the efficient and secure assembly of parts, for which we often rely on threaded fasteners. It takes skill to produce usable tapped holes, so a craftsman in the metal trades must have an understanding of the factors that affect the tapping of a hole, such as the work material and its cutting speed, the cutting fluid, and the size and condition of the hole. This unit will help you identify and select taps for threading operations, and you will learn about common tapping procedures.

OBJECTIVES

After completing this unit, you should be able to:
1. Identify common taps.
2. Select taps for specific applications.
3. Select the correct tap drill for a specific percentage of thread.
4. Determine the cutting speed for a given work material-tool combination.
5. Select the correct cutting fluid for tapping.
6. Tap holes by hand or with a drill press.
7. Identify and correct common tapping problems.

IDENTIFYING COMMON TAP FEATURES

Taps are used to cut internal threads into holes. This process is called tapping. Tap features are illustrated in Figures 1 and 2. The active cutting part of the tap is the chamfer, which is produced by grinding away the tooth form at an angle, with relief back of the cutting edge, so that the cutting action is distributed progressively over a number of teeth. The fluted portion of the tap provides a space for chips to accumulate and for the passage of cutting fluids. Two, three, and four flute taps are common.

The major diameter (Figure 2) is the outside diameter of the tool as measured over the thread crests at the first full thread behind the chamfer. This is the largest diameter of the cutting portion of the tap, as most taps are back tapered or reduced slightly in thread diameter toward the shank. This back taper reduces the amount of tool contact with the thread

during the tapping process, hence making the tap easier to turn.

The pitch diameter (Figure 2) is the diameter of an imaginary cylinder where the width of the spaces and the width of the threads are equal. The pitch of the thread is the distance between a point on one thread and the same point on the next thread.

Taps are made from either high carbon steel or high speed steel and have a hardness of about Rockwell C63. High speed steel taps are far more common in manufacturing plants than carbon steel taps. High speed steel taps typically are ground after heat treatment to ensure accurate thread geometry.

Another identifying characteristic of taps is the amount of chamfer at the cutting end of a tap (Figure 3). A set consists of three taps, taper, plug, and bottoming taps, which are identical except for the number of chamfered threads. The taper tap is useful in start-

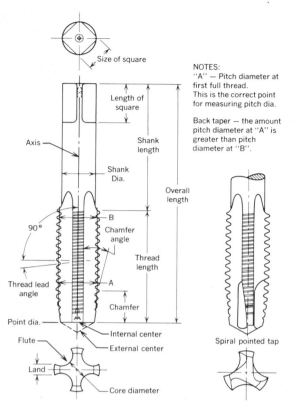

Figure 1. General tap terms. (Bendix Industrial Tools Division)

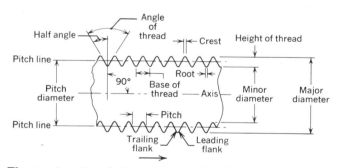

Figure 2. Detailed tap terms. (Bendix Industrial Tools Division)

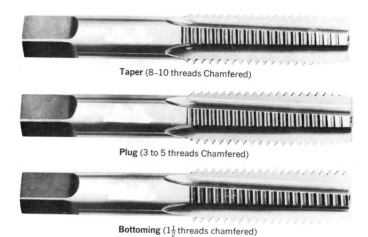

Taper (8–10 threads Chamfered)

Plug (3 to 5 threads Chamfered)

Bottoming (1½ threads chamfered)

Figure 3. Chamfer designations for cutting taps. (© TRW Inc., 1980)

Figure 4. Identifying marking on tap. *(Machine Tools and Machining Practices)*

ing a tapped thread square with the part. The most commonly used tap, both in hand and machine tapping, is a plug tap. Bottoming taps are used to produce threads that extend almost to the bottom of a blind hole. A blind hole is one that is not drilled clear through a part.

Figure 4 shows the identifying markings of a tap, where $\frac{5}{8}$ in. is the nominal size, 11 is the number of threads per inch, and NC refers to the standardized National Coarse thread series. G is the symbol used for ground taps. H3 identifies the tolerance range of the tap. HS means that the tap material is high speed steel. Left-handed taps will also be identified by an LH or left-hand marking on the shank.

OTHER KINDS AND USES OF TAPS

Although there are many kinds of taps used in manufacturing, only those most used in metal shops will be discussed here.

Spiral pointed taps (Figure 5), often called gun taps, are especially useful for machine tapping of through holes or blind holes with sufficient chip room below the threads. When turning the spiral point, the chips are forced ahead of the tap (Figure 6). Since the chips are pushed ahead of the tap, the problems caused by clogged flutes, especially breakage and dulling of taps, are eliminated. Also, since they are not needed for chip disposal, the flutes of gun taps can be made shallower, thus increasing the strength of the tap. Gun taps work best in a machine such as a drill press with a tapping attachment. They can be used for hand tapping, but do not work very well because they are extremely difficult to start and keep in alignment with the hole without some type of guide.

Spiral pointed taps can be operated at higher

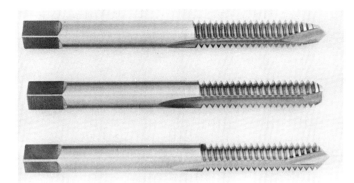

Figure 5. Set of spiral pointed (or "gun") taps.

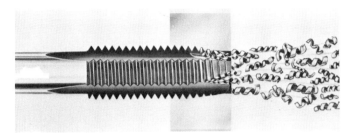

Figure 6. Cutting action of spiral pointed taps. (© TRW Inc., 1980)

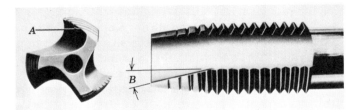

Figure 7. Detail of spiral pointed tap. (© TRW Inc., 1980).

Figure 8. Spiral fluted tap—fast spiral. The action of the tap lifts the chips out of hole to prevent binding. (© TRW Inc., 1980)

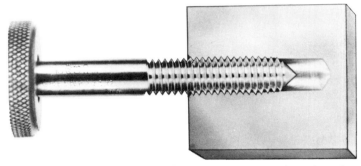

Figure 9. The thread forming action of a fluteless thread forming tap. (© TRW Inc., 1980)

speeds and require less torque to drive than ordinary hand taps. Figure 7 shows the design of the cutting edges. The cutting edges *(A)* at the point of the tap are ground at an angle *(B)* to the axis.

Spiral fluted taps are made with helical instead of straight flutes (Figure 8), which draw the chips out of the hole. This kind of tap is also used when tapping a hole that has a keyway or spline as the helical lands of the tap will bridge the interruptions. Spiral fluted taps are recommended for tapping deep blind holes in ductile materials such as aluminum, magnesium, brass, copper, and die-cast metals.

Thread forming taps are fluteless and do not cut threads in the same manner as conventional taps. They are forming tools and their action can be compared with external thread rolling. On ductile materials such as aluminum, brass, copper, die castings, lead, and leaded steels these taps give excellent results.

Thread forming taps are held and driven just as are conventional taps, but because they do not cut the threads no chips are produced. Problems of chip congestion and removal often associated with the tapping of blind holes are eliminated. Figure 9 shows how the forming tap displaces metal. The crests of the thread that are at the minor diameter may not be flat but will be slightly concave because of the flow

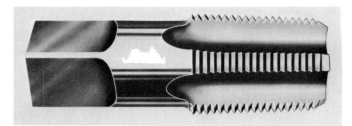

Figure 10. Taper pipe tap. (© TRW Inc., 1980)

Figure 11. Pulley tap. (© TRW Inc., 1980)

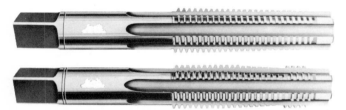

Figure 12. Set of Acme thread taps. The upper tap is used for roughing; the lower tap for finishing. (© TRW Inc., 1980)

Figure 13. A tandem Acme tap designed to rough and finish cut the thread in one pass. (Lane Community College)

Figure 14. Tap wrench. *(Machine Tools and Machining Practices)*

Figure 15. T-handle tap wrench. *(Machine Tools and Machining Practices)*

in which only the chamfered end and the first full thread are actually cutting.

A pulley tap (Figure 11) is used to tap setscrew and oilcup holes in the hubs of pulleys. The long shank also permits tapping in places that might be inaccessible for regular hand taps. When used for tapping pulleys, these taps are inserted through holes in the rims, which are slightly larger than the shanks of the taps. These holes serve to guide the taps and assure proper alignment with the holes to be tapped.

Figure 12 shows Acme taps for roughing and finishing. Acme threads are used to provide accurate movement such as lead screws on machine tools and for applying pressure in various mechanisms. These threads are generally coarser (fewer threads per inch) than 60-degree threads. Acme threads have a 29-degree included angle with a wide flat on the crest of the thread. On some Acme taps the roughing and finishing operation is performed with one tap (Figure 13). The length of this tap usually requires a through hole.

TAPPING BY HAND

A tap wrench (Figure 14) or a T-handle tap wrench (Figure 15) attached to the tap is used to provide driving torque while hand tapping. To obtain a greater accuracy in hand tapping, a hand tapper (Figure 16) may be used. This fixture acts as a guide for the tap to insure that it stays in alignment and cuts concentric threads.

The following procedure may be used for correct hand tapping:

of the displaced metal. Threads produced in this manner have improved surface finish and increased strength because of the cold working of the metal. The size of the hole to be tapped must be closely controlled, since too large a hole will result in a poor thread form and too small a hole will result in the breaking of the tap.

A tapered pipe tap (Figure 10) is used to tap holes with a $\frac{3}{4}$ in. per foot taper for pipes with a matching thread and to produce a leakproof fit. The nominal size of a pipe tap is that of the pipe fitting and not the actual size of the tap. When tapping taper pipe threads, every tooth of the tap engaged with the work is cutting until the rotation is stopped. This takes much more torque than does the tapping of a straight thread

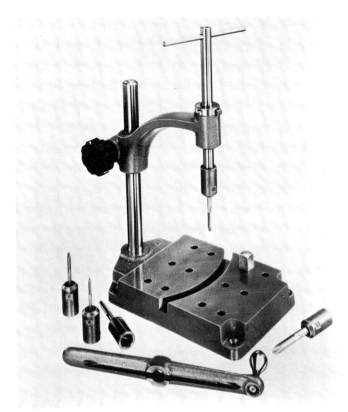

Figure 16. Hand tapper. (Ralmike's Tool-A-Rama)

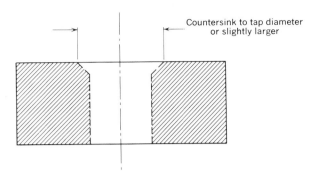

Countersink to tap diameter or slightly larger

Figure 17. Preparing the workpiece. *(Machine Tools and Machining Practices)*

Figure 18. Starting the tap. (Lane Community College)

1. Determine the size of the thread to be tapped and select the tap.
2. Select the proper tap drill with the aid of a tap drill chart. A taper tap should be selected for hand tapping; or if a drill press or tapping machine is to be used for alignment, use a plug tap.
3. Fasten the workpiece securely in a drill press vise. See Section G, Unit 5, Table 1 for cutting speeds and explanation of speed formula. Calculate the correct RPM for the drill used:

$$\text{RPM} = \frac{CS \times 4}{D}$$

Drill the hole using the recommended cutting fluid. Check the hole size.
4. Countersink the hole entrance to a diameter slightly larger than the major diameter of the threads (Figure 17). This allows the tap to be started more easily, and it protects the start of the threads from damage.
5. Mount the workpiece in a bench vise so that the hole is in a vertical position.
6. Tighten the tap in the tap wrench.
7. Cup your hand over the center of the wrench (Figure 18) and place the tap in the hole in a vertical position. Start the tap by turning two or three turns in a clockwise direction for a right-hand thread. At the same time keep a steady pressure downward on the tap. When the tap is started, it may be turned as shown in Figure 19.
8. After the tap is started for several turns, remove the tap wrench without disturbing the tap. Place the blade of a square against the solid shank of the tap to check for squareness (Figure 20). Check from two positions 90 degrees apart. If the tap is not square with the work, it will ruin the thread and possibly break in the hole if you continue tapping. Back the tap out of the hole and restart.
9. Use the correct cutting oil on the tap when cutting threads.
10. Turn the tap clockwise one-quarter to one-half turn and then turn it back a three-quarter turn

Figure 19. Tapping a thread by hand. (Lane Community College)

Figure 20. Checking the tap for squareness. (Lane Community College)

to break the chip. This is done with a steady motion to avoid breaking the tap.

11. When tapping a blind hole, use the taps in the order of starting, plug, and then bottoming. Remove the chips from the hole before using the bottoming tap and be careful not to hit the bottom of the hole with the tap.

12. Figure 21 shows a 60-degree point center chucked in a drill press to align a tap squarely with the previously drilled hole. Only very slight follow-up pressure should be applied to the tap. Too much downward pressure will cut a loose, oversize thread. The tap can also be aligned in the drill press by chucking the shank of the tap and turning the spindle by hand two or three revolutions to start the tap square to the work. The tap can then be released from the chuck and a tap wrench used to finish the thread. **Note:** Do not turn on the power when using this method.

MACHINE TAPPING

Holes can also be tapped in a drill press that has a spindle reverse switch, which is often foot operated for convenience. Drill presses without reversing switches can be used for tapping with a tapping attachment (Figure 22). Some of these tapping attachments have an internal friction clutch where downward pressure on the tap turns the tap forward and feeds it into the work. Releasing downward pres-

Figure 21. Using the drill press as a tapping fixture. (Lane Community College)

Figure 22. Drill press tapping attachment. (Lane Community College)

sure will automatically reverse the tap and back it out of the workpiece. Some tapping attachments have lead screws that provide tap feed rates equal to the lead of the tap. Most of these attachments also have an adjustment to limit the torque to match the size of tap, which eliminates most tap breakage.

THREAD PERCENTAGE AND HOLE STRENGTH

The strength of a tapped hole depends largely on the workpiece material, the percentage of full thread used, and the length of the thread. The workpiece material is usually selected by the designer, but the machinist can often control the percentage of thread produced and the depth of the thread. The percentage of thread produced is dependent on the diameter of the drilled hole. Tap drill charts generally give tap drill sizes to produce 75 percent thread.

An example will illustrate the relationships between the percentage of thread, torque required to drive the tap, and resulting thread strength. An increase in thread depth from 60 to 72 percent in AISI 1020 steel required twice the torque to drive the tap, but it increased the strength of the thread by only 5 percent. The practical limit seems to be 75 percent of full thread, since greater percentage of thread does not increase the strength of the threaded hole in most materials.

In some difficult-to-machine materials such as titanium alloys, high tensile steels, and some stainless steels, 50 to 60 percent thread depth will give sufficient strength to the tapped hole. Threaded assemblies are usually designed so that the bolt breaks before the threaded hole strips. Common practice is to have a bolt engage a tapped hole by 1 to $1\frac{1}{2}$ times its diameter.

DRILLING THE RIGHT HOLE SIZE

The condition of the drilled hole affects the quality of the thread produced, as an out-of-round hole leads to an out-of-round thread. Bell-mouthed holes will produce bell-mouthed threads. When an exact hole size is needed, the tap drill should be preceded by a drill just under the tap drill diameter. This ensures a fairly true hole diameter since the tap drill acts as a reamer. This is especially important for large diameter taps and when fine pitch threads are used. The size of the hole to be drilled is usually obtained from tap drill charts, which usually show a 75 percent thread depth. If a thread depth other than 75 percent is wanted, use the following formula to determine the proper hole size:

$$\text{Outside diameter of thread} - \frac{.01299 \times \text{percentage of thread}}{\text{Number of threads per inch}}$$

$$= \text{Hole size}$$

For example, calculate the hole size for a 1 in., 12-thread fastener with a 70 percent thread depth:

$$1 - \frac{.01299 \times 70}{12} = .924 \text{ in.}$$

Further information for calculating threads may be found in Section H, Unit 11, "Sixty Degree Thread Information and Calculations." Table 10 in the Appendix gives metric tap drill sizes.

A simple formula used by some machinists for finding tap drill sizes of both metric and unified 60-degree thread is as follows:

$$\text{Tap drill size} = \text{Thread diameter} - \text{Pitch}$$

This formula gives approximately 75 percent threads.

SPEEDS FOR TAPPING

The quality of the thread produced also depends on the speed at which a tap is operated. The selection of the best speed for tapping is limited, unlike the varying speeds and feeds possible with other cutting tools, because the feed per revolution is fixed by the lead of the thread. Excessive speed develops high temperatures that cause rapid wear of the tap's cutting edge. Dull taps produce rough or torn and off-size threads. High cutting speeds prevent adequate lubrication at the cutting edges and often create a problem of chip disposal.

When selecting the best speed for tapping, you should consider not only the material being tapped, but also the size of the hole, the kind of tap holder being used, and the lubricant being used. Table 1 gives some guidelines in selecting a cutting speed and a lubricant for some materials when using high speed steel taps.

These cutting speeds in feet per minute have to be translated into RPM to be useful. For example, calculate the RPM when tapping a $\frac{3}{8}$–24 UNF hole in free machining steel. The cutting speed chart gives a cutting speed between 60 and 80 feet per minute. Use the lower figure; you can increase the speed once you see how the material taps. The formula for calculating RPM is:

$$\frac{\text{Cutting Speed} \times 4}{\text{Diameter}} \quad \text{or} \quad \frac{60 \times 4}{\frac{3}{8}} = 640 \text{ RPM}$$

Lubrication is one of the most important factors in a tapping operation. Cutting fluids used when tap-

Table 1

Recommended Cutting Speeds and Lubricants for Machine Tapping

Material	Speeds in Feet per Minute	Lubricant
Aluminum	90–100	Kerosene and light base oil
Brass	90–100	Soluble oil or light base oil
Cast iron	70– 80	Dry or soluble oil
Magnesium	20– 50	Light base oil diluted with kerosene
Phosphor bronze	30– 60	Mineral oil or light base oil
Plastics	50– 70	Dry or air jet
Steels		
Low carbon	40– 60	Sulphur base oil
High carbon	25– 35	Sulphur base oil
Free machining	60– 80	Soluble oil
Molybdenum	10– 35	Sulphur base oil
Stainless	10– 35	Sulphur base oil

Source: Warren T. White, John E. Neely, Richard R. Kibbe, and Roland O. Meyer, *Machine Tools and Machining Practices,* John Wiley and Sons, Inc., Copyright © 1977, New York.

ping serve as coolants, but are more important as lubricants. It is important to select the correct lubricant because the use of a wrong lubricant may give results that are worse than if no lubricant was used. For lubricants to be effective, they should be applied in sufficient quantity to the actual cutting area in the hole.

SOLVING TAP PROBLEMS

In Table 2, common tapping problems are presented with some possible solutions.

Table 2

Common Tapping Problems and Possible Solutions

Causes of Tap Breakage	Solutions
Tap hitting bottom of hole or bottoming on packed chips	Drill hole deeper. Eject chips with air pressure. (*Caution:* Stand aside when you do this and always wear safety glasses.) Use spiral fluted taps to pull chips out of hole. Use a thread forming tap.
Chips are packing in flutes	Use tap style with more flute space. Tap to a lesser depth or use a smaller percentage of threads. Select a tap that will eject chips forward (spiral point) or backward (spiral fluted).
Hard materials or hard spots	Anneal the workpiece. Reduce cutting speed. Use longer chamfers on tap. Use taps with more flutes.
Inadequate lubricant	Use the correct lubricant and apply a sufficient amount of it under pressure at the cutting zone.
Tapping too fast	Reduce cutting speed.
Excessive wear:	
Abrasive materials	Improve lubrication. Use surface treated taps. Check the alignment of tap and hole to be tapped.
Chips clogging flutes:	
Insufficient lubrication	Use better lubricant and apply it with pressure at the cutting zone.
Excessive speed	Reduce cutting speed.
Wrong-style tap	Use a more free cutting tap such as spiral pointed tap, spiral fluted tap, interrupted thread tap, or surface treated taps.
Torn or rough threads:	
Dull tap	Resharpen.
Chip congestion	Use tap with more chip room. Use lesser percentage of thread. Drill deeper hole. Use a tap that will eject chips.
Inadequate lubrication and chips clogging flutes	Correct as previously suggested.
Hole improperly prepared	Torn areas on the surface of the drilled, bored, or cast hole will be shown in the minor diameter of the tapped thread.

Causes of Tap Breakage	Solutions
Undersize threads:	
Pitch diameter of tap too small	Use tap with a larger pitch diameter.
Excessive speed	Reduce tapping speed.
Thin wall material	Use a tap that cuts as freely as possible. Improve lubrication. Hold the workpiece so that it cannot expand while it is being tapped. Use an oversize tap.
Dull tap	Resharpen.
Oversize or bellmouth threads:	
Loose spindle or worn holder	Replace or repair spindle or holder.
Misalignment	Align spindle, fixture, and work.
Tap oversize	Use smaller pitch diameter tap.
Dull tap	Resharpen.
Chips packed in flutes	Use tap with deeper flutes, spiral flutes, or spiral points.
Buildup on cutting edges of tap	Use correct lubricant and tapping speed.

Source: Warren T. White, John E. Neely, Richard R. Kibbe, and Roland O. Meyer, *Machine Tools and Machining Practices*, John Wiley and Sons, Inc., Copyright © 1977, New York.

SELF-TEST

1. What difference exists between the taps in a set?
2. Where is a spiral pointed tap used?
3. When is a spiral fluted tap used?
4. How are thread forming taps different from conventional taps?
5. How are taper pipe taps identified?
6. Why are finishing and roughing Acme taps used?
7. What kind of tools are used to drive taps when hand tapping?
8. What is a hand tapper?
9. What is a tapping attachment?
10. Which three factors affect the strength of a tapped hole?
11. How deep should the usable threads be in a tapped hole?
12. What causes taps to break while tapping?
13. What causes rough and torn threads?
14. What causes oversize threads in a hole?
15. When should you be most careful to get an accurate diameter of a tap drilled hole?

UNIT 8 THREAD CUTTING DIES AND THEIR USES

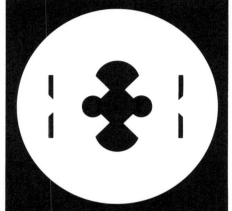

A die is used to cut external threads on the surface of a bolt or rod. Many machine parts and mechanical assemblies are held together with threaded fasteners, most of which are mass produced. Occasionally, however, a machinist has to make a bolt or extend the threads on a bolt for which he uses a die. In this unit you will be introduced to some dies and their hand threading uses.

OBJECTIVES

After completing this unit, you should be able to:
1. Identify dies used for hand threading.
2. Select and prepare a rod for threading.
3. Cut threads with a die.

Dies are used to cut external threads on round materials. Some dies are made from carbon steel, but most are made from high speed steel. Dies are identified by the markings on the face as to the size of thread, number of threads per inch, and form of thread, such as NC, UNF, or other standard designations (Figure 1). The characteristics of various thread forms are further discussed in Section H, "Turning Machines."

COMMON TYPES OF HAND THREADING DIES

The die shown in Figure 1 is an example of a round split adjustable die, also called a button die. These dies are made in all standardized thread sizes up to $1\frac{1}{2}$ in. thread diameters and $\frac{1}{2}$ in. pipe threads. The outside diameters of these dies vary from $\frac{5}{8}$ to 3 in.

Adjustments on these dies are made by turning a fine pitch screw that forces the sides of the die apart or allows them to spring together. The range of adjustment of round split adjustable dies is very small, allowing only for a loose or tight fit on a threaded part. Adjustments made to obtain threads several thousandths of an inch oversize will result in poor die performance because the heel of the cutting edge will drag on the threads. Excessive expansion may cause the die to break in two.

Some round split adjustable dies do not have the

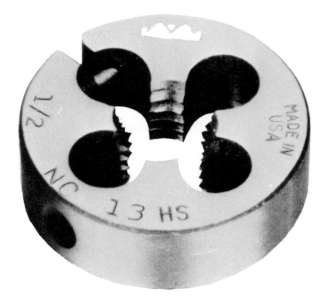

Figure 1. Markings on a die. (Example shown is a round split adjustable die.) *(Machine Tools and Machining Practices)*

built-in adjusting screw. Adjustments are then made with the three screws in the die stock (Figure 2). Two of these screws on opposite sides of the die stock hold the die in the die stock and also provide closing pres-

Figure 2. Diestock for round split adjustable dies. (© TRW Inc., 1980)

Figure 3. Die halves for two piece die. (© TRW Inc., 1980)

Cap Guide Collet

Figure 4. Components of a split adjustable die collet. (© TRW Inc., 1980)

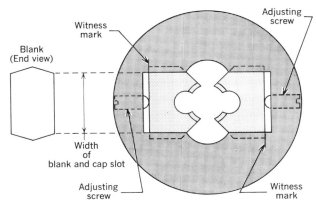

Figure 5. Setting the die position to the witness marks on the die and collet assembly. (© TRW Inc., 1980)

Figure 6. Diestock for adjustable die and collet assembly. (© TRW Inc., 1980)

sure. The third screw engages the split in the die and provides opening pressure. These dies are used in a die stock for hand threading or in a machine holder for machine threading.

Another type of threading die is the two piece die, whose halves (Figure 3) are called blanks. These blanks are assembled in a collet consisting of a cap and the guide (Figure 4). The normal position of the blanks in the collet is indicated by witness marks (Figure 5). The adjusting screws allow for precise control of the cut thread size. The blanks are inserted in the cap with the tapered threads toward the guide. Each of the two die halves is stamped with a serial number.

Make sure the halves you select have the same numbers. The guide used in the collet serves as an aid in starting and holding the dies square with the work being threaded. Each thread size uses a guide of the same nominal or indicated size. Collets are held securely in die stocks (Figure 6) by a knurled set screw that seats in a dimple in the cap.

Hexagon rethreading dies (Figure 7) are used to recut slightly damaged or rusty threads. These dies should not be used for cutting new threads since it is difficult to start the die square with the workpiece and to maintain the alignment. Rethreading dies are driven with a wrench large enough to fit the die. Solid square dies (Figure 8) are made to cut new threads, but they can be used as rethreading dies.

HAND THREADING PROCEDURES

Threading of a rod should always be started with the leading or throat side of the die. This side is identified by the chamfer on the first two or three threads and also by the size markings. The chamfer distributes

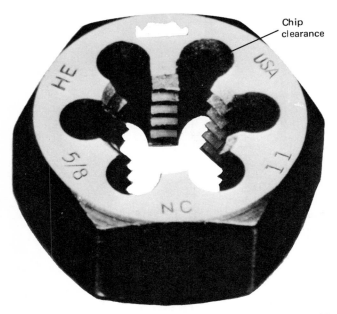

Figure 7. Hexagon rethreading die. (© TRW Inc., 1980)

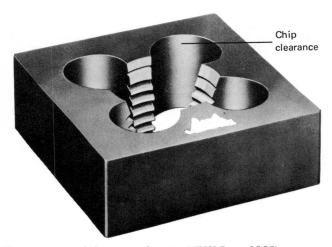

Figure 8. Solid square die. (© TRW Inc., 1980)

by the dies being relatively thin when compared to the diameter of thread that they cut. Only a few threads in the die can act as a guide on the already cut threads. This error usually does not cause problems when standard or thin nuts are used on the threaded part. However, when an item with a long internal thread is assembled with a threaded rod, it usually gets tight and then locks, not because the thread depth is insufficient, but because there is a lead error. This lead error can be as much as one-fourth of a thread in one inch of length.

The outside diameter of the material to be threaded should not be over the nominal size of the thread and preferably a few thousandths of an inch (.002 to .005 in.) undersize. After a few full threads are cut, the die should be removed so that the thread can be tested with a nut or thread ring gage. A thread ring gage set usually consists of two gages, a go and a no go gage. As the names imply, a go gage should screw on the thread, while the no go gage will not go more than $1\frac{1}{2}$ turns on a thread of the correct size. Do not assume that the die will cut the correct size thread; always check by gaging or assembling. Adjustable dies should be spread open for the first cut and set progressively smaller for each pass after checking the thread size.

It is very important that a die is started squarely on the rod to be threaded. A lathe can be used as a fixture for cutting threads with a die (Figure 9). The rod is fastened in a lathe chuck for rotation, while the die is held square because it is supported by the face of the tailstock spindle. The carriage or the compound rest prevents the die stock from turning while the chuck is rotated *by hand*. As the die advances, the tailstock spindle is also advanced to stay in contact with the die. Do not force the die with the tailstock spindle, or a loose thread may result. A die may be used to finish to size a long thread that has been rough threaded on the lathe.

Another method that is used on the lathe for more accurate threading with a button die is to use a guide or holder. A button die holder (Figure 10*a*) can be used to thread several inches without producing a "drunken thread" (Figure 10*b*).

Occasionally, a die is used to extend the thread on a bolt. Make certain that the bolt is not hardened or the die will be ruined. To cut full, usable threads close to a shoulder, first cut the thread normally until the die touches the shoulder, then reverse the die and use the unchamfered side to finish the last few threads.

It is always good practice to chamfer the end of a workpiece before starting a die (Figure 11). The chamfer on the end of a rod can be made by grinding

the cutting load over a number of threads, which produces better threads and less chance of chipping the cutting edges of the die. Cutting oil or other threading fluids are very important in obtaining quality threads and maintaining long die life. Once a cut is started with a die, it will tend to follow its own lead, but uneven pressure on the die stock will make the die cut variable helix angles or "drunken" threads.

Threads cut by hand often show a considerable accumulated lead error. The lead of a screw thread is the distance a nut will move on the screw if it is turned one full revolution. This problem is caused

Figure 9. Threading a rod with a hand die in a lathe. (Lane Community College)

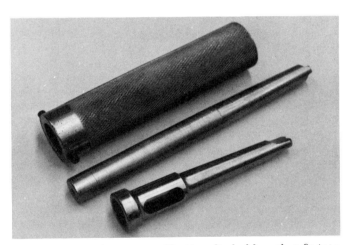

Figure 10a. Two types of button die holders that fit into the tailstock of a lathe. (Lane Community College)

Figure 10b. Button die holder being used to cut a three-inch long thread in a lathe. (Lane Community College)

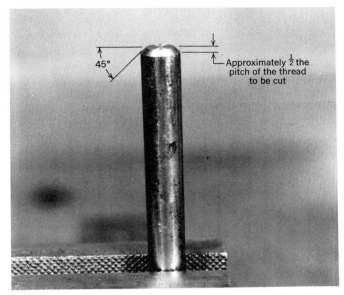

45°

Approximately ½ the pitch of the thread to be cut

Figure 11. Chamfer workpiece before using die. (Lane Community College)

on a pedestal grinder, by filing or with a lathe. This will help in starting the cut and it will also leave a finished thread end. While cutting threads with a hand die, the die rotation should be reversed after each full turn forward to break the chips into short pieces that will fall out of the die. Chips jammed in the clearance holes will tear the thread.

The following procedure is used for hand threading:

1. Select the workpiece to be threaded and measure its diameter. Then chamfer the end. This may be done on a grinder or with a file. The chamfer should be at least as deep as the thread to be cut.

2. Select the correct die and mount it in a die stock.

3. Mount the workpiece in a bench vise. Short workpieces are mounted vertically and the long pieces usually are held horizontally.

4. To start the thread, place the die over the workpiece. Holding the die stock with one hand (Figure 12), apply downward pressure and turn the die.

5. When the cut has started, apply cutting oil to the workpiece and die and start turning the die stock with both hands (Figure 13). After each complete revolution forward, reverse the die one-half turn to break the chips.

6. Check to see that the thread is started square, using a machinist's square. Corrections can be made by applying slight downward pressure on the high side while turning.

7. When several turns of the thread have been com-

Figure 12. Start the die with one hand. (Lane Community College)

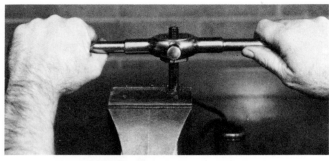

Figure 13. Use both hands to turn the threading die. (Lane Community College)

pleted, you should check the fit of the thread with a nut, thread ring gage, thread micrometer, or the mating part. If the thread fit is incorrect, adjust the die with the adjustment screws and take another cut with the adjusted die. Continue making adjustments until the proper fit is achieved.

8. Continue threading to the required thread length. To cut threads close to a shoulder, invert the die after the normal threading operation and cut the last two or three threads with the side of the die that has no chamfer.

SELF-TEST

1. What kind of threads does a die cut, internal or external?
2. What tool is used to drive a die?
3. How much adjustment is possible with a round split adjustable die?
4. What is the purpose of the guide in a two-piece adjustable die collet?
5. What are important points to watch when assembling two-piece dies in a collet?
6. Where are hexagon rethreading dies used?
7. Why do dies have a chamfer on the cutting end?
8. Why are cutting fluids used?
9. What diameter should a rod be before being threaded?
10. Why should a rod be chamfered before being threaded?

SECTION C

DIMENSIONAL
MEASUREMENT

UNIT 1 SYSTEMS OF MEASUREMENT

Throughout history there have been many systems of measurement. Prior to the era of national and international industrial operations, individual craftsmen were often responsible for the manufacture of a complete product. Since they made all the necessary parts and did the required assembly, they needed to conform only to their particular system of measurement. However, as machines replaced people and diversified mass production was established on a national and international basis, the need for standardization of measurement became readily apparent. Total standardization of measurement throughout the world still has not been fully realized. Most measurement in the modern world does, however, conform to either the English (inch-pound-second) or the metric (meter-kilogram-second) system. Metric measurement is now predominant in most of the industrialized nations of the world. The inch system is still used to a great extent in the United States. However, because of the interdependence of the world's industrial community, even the United States is turning more and more toward the use of metric measurement. This unit will introduce you to these systems of measurement and some methods of conversion from one system to the other.

OBJECTIVES

After completing this unit, you should be able to:
1. Identify common methods of measurement conversion.
2. Convert inch dimensions to metric equivalents and convert metric dimensions to inch equivalents.

THE ENGLISH SYSTEM OF MEASUREMENT

The English system of measurement uses the units of inches, pounds, and seconds to represent the measurement of length, mass, and time. Since we are primarily concerned with the measurement of length in the machine shop, we will simply refer to the English system as the **inch** system.

Subdivisions and Multiples of the Inch

The following table shows the common subdivisions and multiples of the inch that are used by the machinist.

Common Subdivisions

.000001	millionth
.00001	hundred thousandth
.0001	ten thousandth
.001	thousandth
.01	hundredth
.1	tenth
1.00	*Unit inch*

Common Multiples

12.00	= 1 foot
36.00	= 1 yard

Other common subdivisions of the inch are:

$\frac{1}{128}$	.007810 (decimal equivalent)
$\frac{1}{64}$	.015625
$\frac{1}{32}$	.031250
$\frac{1}{20}$	.050000
$\frac{1}{16}$	.062500
$\frac{1}{8}$	.125000
$\frac{1}{4}$	.250000
$\frac{1}{2}$	.500000

Multiples of Feet	*Multiples of Yards*
3 feet = 1 yard	1760 yards = 1 mile
5280 feet = 1 mile	

See Section G, Unit 2, Table 1 for fractional-decimal conversion. When measuring closer than $\frac{1}{64}$ in., thousandths of an inch is the standard subdivision in the metalworking trades.

THE METRIC SYSTEM AND THE INTERNATIONAL SYSTEM OF UNITS—SI

The basic unit of length in the metric system is the **meter.** Originally the length of the meter was defined by a natural standard, specifically, a portion of the earth's circumference. Later, more convenient metal standards were constructed. In 1886, the metric system was legalized in the United States, but its use was not mandatory. Since 1893 the yard has been defined in terms of the metric meter by the ratio

$$1 \text{ yard} = \frac{3600}{3937} \text{ meter}$$

Although the metric system has been in use for many years in many different countries, it still lacked complete standardization among its users. Therefore, an attempt was made to modernize and standardize the metric system. From this effort has come the **Systeme International d'Unites,** known as **SI** or the **International Metric System.**

The basic unit of length in SI is the meter or metre (in the common international spelling). The SI meter is defined by a physical standard that can be reproduced anywhere with unvarying accuracy.

1 meter = 1,650,763.73 wavelengths in a vacuum of the orange-red light spectrum of the Krypton-86 atom

Probably the primary advantage of the metric system is that of convenience in computation. All subdivisions and multiples use 10 as a divisor or multiplier. This can be seen in the following table.

.000001	(one-millionth meter or micrometer)
.001	(one-thousandth meter or millimeter)
.01	(one-hundredth meter or centimeter)
.1	(one-tenth meter or decimeter)
1.00	*Unit meter*
10	(ten meters or one dekameter)
100	(100 meters or one hectometer)
1000	(1000 meters or one kilometer)
1,000,000	(one million meters or one megameter)

METRIC SYSTEM EXAMPLES
1. One meter (m) = _____ millimeters (mm).
   Since a mm is $\frac{1}{1,000}$ part of an m, there are 1000 mm in a meter.

2. 50 mm = _____ centimeters (cm).
   Since 1 cm = 10 mm, $\frac{50}{10} = 5$ cm in 50 mm.
3. Four kilometers (km) = _____ m.
   Since 1 km = 1000 m then 4 km = 4000 m.
4. 582 mm = _____ cm.
   Since 10 mm = 1 cm, $\frac{582}{10} = 58.2$ cm.

Millimeters and hundredths of a millimeter are the standard metric measure in the metalworking trades.

CONVERSION BETWEEN SYSTEMS

Much of the difficulty with working in a two-system environment is experienced in converting from one system to the other. This can be of particular concern to the machinist as he must exercise due caution in making conversions. Arithmetic errors can be easily made. Therefore the use of a calculator and using metric measuring tools for metric work is recommended.

Conversion Factors and Mathematical Conversion

Since the historical evolution of the inch and metric systems is quite different, there are no obvious relationships between length units of the two systems. You simply have to memorize the basic conversion factors. We know from the preceding discussion that the yard has been defined in terms of the meter. Knowing this relationship, you can derive mathematically any length unit in either system. However, the conversion factor

$$1 \text{ yard} = \frac{3600}{3937} \text{ meter}$$

is a less common factor for the machinist. A more common factor can be determined by the following:

$$1 \text{ yard} = .91440 \text{ meter}$$

$$\left(\frac{3600}{3937} \text{ expressed in decimal form}\right)$$

Then

$$1 \text{ inch} = \frac{1}{36} \text{ of } .91440 \text{ meter}$$

So

$$\frac{.91440}{36} = .025400 \text{ meter}$$

We know that

$$1 \text{ m} = 1000 \text{ mm}$$

Therefore

$$1 \text{ inch} = .025400 \times 1000$$

Or

$$1 \text{ in.} = 25.4000 \text{ mm}$$

The conversion factor 1 in. = 25.4 mm is very common and should be memorized. From the example shown it should be clear that in order to find inches knowing millimeters, you must divide millimeters by 25.4.

$$1000 \text{ mm} = \underline{\hspace{1cm}} \text{ in.}$$

$$\frac{1000}{25.4} = 39.37 \text{ in.}$$

In order to simplify the arithmetic, any conversion can always take the form of a multiplication problem.

EXAMPLE

Instead of $\frac{1000}{25.4}$, multiply by the reciprocal of 25.4, which is $\frac{1}{25.4}$ or .03937.

Therefore, $1000 \times .03937 = 39.37$ in.

EXAMPLES OF CONVERSIONS [INCH TO METRIC]
1. 17 in.= _____ cm.
   Knowing inches, to find centimeters multiply inches by 2.54: 2.54 × 17 in. = 43.18 cm.
2. .807 in. = _____ mm.
   Knowing inches, to find millimeters multiply inches by 25.4: 25.4 × .807 in. = 20.49 mm.

EXAMPLES OF CONVERSIONS [METRIC TO INCH]
1. .05 mm = _____ in.
   Knowing millimeters, to find inches multiply millimeters by .03937: .05 × .03937 = .00196 in.
2. 1.63 m = _____ in.
   Knowing meters, to find inches, multiply meters by 39.37: 1.63 × 39.37 m = 64.173 in.

Conversion Factors to Memorize

$$1 \text{ in.} = 25.4 \text{ mm or } 2.54 \text{ cm}$$
$$1 \text{ mm} = .03937 \text{ in.}$$

Other Methods of Conversion

The **conversion chart** (Figure 1) is a popular device for making conversions between systems. Conversion charts are readily available from many manufacturers. However, most conversion charts give equivalents for whole millimeters or standard fractional inches. If you must find an equivalent for a factor that does not appear on the chart, you must interpolate. In this instance, knowing the common conversion factors and

MILLIMETERS TO INCHES
(Basis: 1 inch = 25.4 millimeters)

Millimeters	Inches	Millimeters	Inches	Millimeters	Inches	Millimeters	Inches
1	0.039370	26	1.023622	51	2.007874	76	2.992126
2	.078740	27	1.062992	52	2.047244	77	3.031496
3	.118110	28	1.102362	53	2.086614	78	3.070866
4	.157480	29	1.141732	54	2.125984	79	3.110236
5	.196850	30	1.181102	55	2.165354	80	3.149606
6	.236220	31	1.220472	56	2.204724	81	3.188976
7	.275591	32	1.259843	57	2.244094	82	3.228346
8	.314961	33	1.299213	58	2.283465	83	3.267717
9	.354331	34	1.338583	59	2.322835	84	3.307087
10	.393701	35	1.377953	60	2.362205	85	3.346457
11	.433071	36	1.417323	61	2.401575	86	3.385827
12	.472441	37	1.456693	62	2.440945	87	3.425197
13	.511811	38	1.496063	63	2.480315	88	3.464567
14	.551181	39	1.535433	64	2.519685	89	3.503937
15	.590551	40	1.574803	65	2.559055	90	3.543307
16	.629921	41	1.614173	66	2.598425	91	3.582677
17	.669291	42	1.653543	67	2.637795	92	3.622047
18	.708661	43	1.692913	68	2.677165	93	3.661417
19	.748031	44	1.732283	69	2.716535	94	3.700787
20	.787402	45	1.771654	70	2.755906	95	3.740157
21	.826772	46	1.811024	71	2.795276	96	3.779528
22	.866142	47	1.850394	72	2.834646	97	3.818898
23	.905512	48	1.889764	73	2.874016	98	3.858268
24	.944882	49	1.929134	74	2.913386	99	3.897638
25	.984252	50	1.968504	75	2.952756	100	3.937008

Note: The above table is approximate: 1/25.4 0.039370078740

Figure 1.   Metric conversion table. (MTI Corporation)

determining the equivalent mathematically is more efficient.

Several electronic calculators designed to convert directly from system to system are available. Of course, any calculator can and should be used to do a conversion problem. The direct converting calculator does not require that any conversion constant be remembered. These constants are permanently programmed into the calculator memory.

Converting Machine Tools

With the increase in metric measurement in industry, which predominantly uses the inch system, several devices have been developed that permit a machine tool to function in either system. These conversion devices eliminate the need to convert all dimensions prior to beginning a job.

Conversion equipment includes **conversion dials** (Figure 2) that can be attached to lathe cross slide screws as well as milling machine saddle and table screws. The dials are equipped with gear ratios that permit a direct metric reading to appear on the dial.

Metric mechanical and electronic travel indicators can also be used. The mechanical dial travel indicator (Figure 3) uses a roller that contacts a moving part of a machine tool. Travel of the machine component is indicated on the dial. This type of travel indicator discriminates to .01 millimeter. Whole millimeters are counted on the one millimeter counting wheel. Mechanical dial travel indicators are used in many

Figure 2. Inch/metric conversion dials for machine tools. (The Monarch Machine Tool Company, Sidney, Ohio)

Figure 3. Metric mechanical dial travel dial indicator. (Southwestern Industries, Inc. Trav-A-Dial® is a registered trademark of Southwestern Industries, Inc., Los Angeles, California)

applications such as reading the travel of a milling machine saddle and table (Figure 4).

The electronic travel indicator uses a sensor that is attached to the machine tool. Machine tool component travel is indicated on an electronic digital display. The equipment can be switched to read travel in inch or metric dimensions.

Metric conversion devices can be fitted to existing machine tools for a moderate expense. Many new machine tools, especially those built abroad, have dual system capability built into them.

HOW THINGS ARE MEASURED

The measurement of **length** is the distance along a line between two points (Figure 5). It is also length that defines the longer or longest dimension of an object. Depth is often called length when the object is long. **Width** is the dimension taken at right angles to the length. **Height** is the distance from the bottom to the top of an object standing upright. **Depth** is the direct linear measurement from the point of viewing, usually from the front to back of an object or the perpendicular measurement downward from a surface.

Length is measured in such basic linear units as **inches, millimeters,** and in advanced metrology,

Figure 4. Metric mechanical dial travel indicators reading milling machine saddle and table movement. (Southwestern Industries, Inc. Trav-A-Dial® is a registered trademark of Southwestern Industries, Inc., Los Angeles, California)

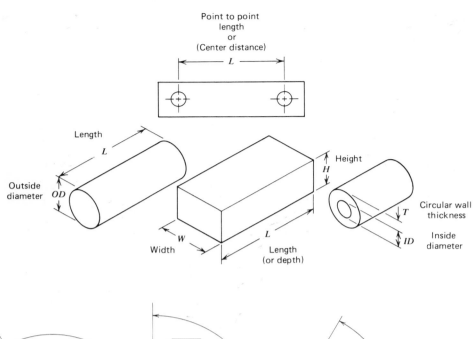

Point to point
length
or
(Center distance)

Length

Outside
diameter

Height

Circular wall
thickness

Inside
diameter

Width

Length
(or depth)

Figure 5. The measurement of length may appear under several different names.

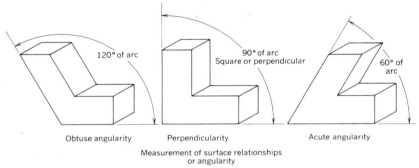

120° of arc

Obtuse angularity

90° of arc
Square or perpendicular

Perpendicularity

60° of
arc

Acute angularity

Measurement of surface relationships
or angularity

Figure 6. Measurement of surface relationships or angularity. *(Machine Tools and Machining Practices)*

Figure 7. Visual surface finish comparator gage. (The DoAll Company)

wavelengths of light. In addition, measurements are sometimes made to measure the relationship of one surface to another, which is commonly called **angularity** (Figure 6). **Squareness,** which is closely related to angularity, is the measure of deviation from true perpendicularity. A craftsman will measure angularity in the basic units of angular measure, **degrees, minutes,** and **seconds of arc.**

In addition to the measure of length and angular-

ity, you also need to measure such things as surface finish (Figure 7), concentricity, straightness, and flatness. Occasionally, you will come in contact with measurements that involve circularity, sphericity, and alignment (Figure 8). However, many of these more specialized measurement techniques are in the realm of the inspector or laboratory metrologist and appear infrequently in general shop work. To be reliable, measurements must be taken in line with the axis

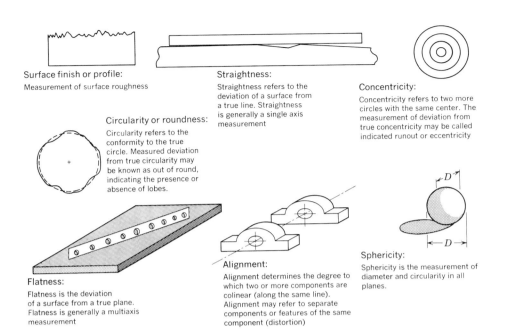

Surface finish or profile:
Measurement of surface roughness

Straightness:
Straightness refers to the deviation of a surface from a true line. Straightness is generally a single axis measurement

Concentricity:
Concentricity refers to two more circles with the same center. The measurement of deviation from true concentricity may be called indicated runout or eccentricity

Circularity or roundness:
Circularity refers to the conformity to the true circle. Measured deviation from true circularity may be known as out of round, indicating the presence or absence of lobes.

Flatness:
Flatness is the deviation of a surface from a true plane. Flatness is generally a multiaxis measurement

Alignment:
Alignment determines the degree to which two or more components are colinear (along the same line). Alignment may refer to separate components or features of the same component (distortion)

Sphericity:
Sphericity is the measurement of diameter and circularity in all planes.

Figure 8. Other measurements encountered by the machinist. *(Machine Tools and Machining Practices)*

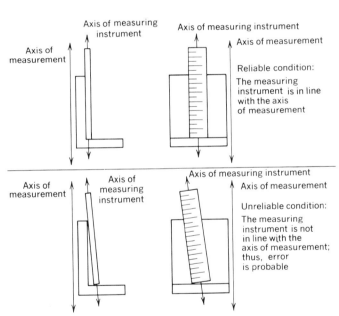

Figure 9. The axis of a linear measuring instrument must be in line with the axis of measurement. *(Machine Tools and Machining Practices)*

of measurement (Figure 9). This is true for all types of measuring tools, rules, calipers, micrometers, and transfer measuring tools.

CONSIDERATIONS IN MEASUREMENT

The units in this section discuss most of the common measuring tools available to a worker. The capabilities, discrimination, and reliability as well as proce-

dures for use are examined. It is, of course, the responsibility of a worker to select the proper measuring instrument for the job at hand. When faced with a need to measure, the following questions should be considered.

1. What degree of accuracy and precision must this measurement meet?
2. What degree of measuring tool discrimination does this required accuracy and precision demand?
3. What is the most reliable tool for this application?

Calibration

Accurate and reliable measurement places a considerable amount of responsibility on the worker, who must keep the measuring tools being used to an acceptable standard. This is the process known as calibration. Most of the common measuring instruments provide a factory standard. Even though calibration cannot be carried out under laboratory conditions, the instruments should at least be checked periodically against available standards.

Variables in Measuring

You should be further aware that any measurement is relative to the conditions under which it is taken. A common expression that is often used around the machine shop is: "The measurement is right on." There is, of course, little probability of obtaining a measurement that is truly exact. Each measurement has a certain degree of deviation from the theoretical exact size. This degree of error is dependent on many variables, including the measuring tool selected, the procedure used, the temperature of the part, the temperature of the room, the cleanliness of the air in the room, and the cleanliness of the part at the time of measurement. Of course, the **discrimination** of the measuring tool must be taken into account. Discrimination refers to the extent to which a unit of length has been divided. If a micrometer is divided to read thousandths of an inch, then it has a .001 in. discrimination.

The deviation of a measurement from exact size is taken into consideration by the designer. Every measurement has a tolerance, meaning that the measurement is acceptable within a specific range. Tolerance can be quite small depending on design requirements. When this condition exists, reliable measurement becomes more difficult because it is more heavily influenced by the many variables present. Therefore, before you make any measurement, you should stop for a moment and consider the possible variables involved. You should then consider what might be done to control as many of these variables as possible.

In practical terms, in most machine shops measurements are made to three or four decimal places in inch measure since micrometers and vernier calipers will measure to .001 in. and, if the micrometer has a vernier calibration, to .0001 in. It is common shop practice to avoid confusion over decimal numbers, to refer to thousandths only, and add the ten thousandths. Thus, the decimal equivalent of $\frac{7}{16}$ in. = .4375 would be read four hundred thirty-seven and five-tenths, instead of four thousand three hundred seventy-five ten-thousandths. This would be easily confused with $\frac{3}{8}$ = .375 in.

Metric micrometers measure to .01 mm; therefore, only two decimal places are necessary when rounding off a calculation. Most metric measurements in machine shops are made in millimeters. Ten-thousandths inch measurements are somewhat finer than one-hundredth millimeter measurements; .01 mm = .0003937 in., almost four ten-thousandths of an inch, while .0001 in. = .00254 mm, a dimension smaller than a standard metric micrometer can measure.

SELF-TEST

1. Convert 35 mm into in.
2. Convert .125 in. into mm.
3. Convert 6.273 in. into cm.
4. The greatest dimension of a rectangular object is called its _____.
5. Angularity is expressed in _____, _____, and _____.
6. What is meant by calibration of a measuring instrument?
7. Express the tolerance ±.050 in. in metric terms to the nearest $\frac{1}{100}$ mm.
8. To find cm knowing mm, do you multiply or divide by 10?
9. Express the tolerance ±.02 mm in terms of inches to the nearest $\frac{1}{10,000}$ in.
10. Can an inch machine tool be converted to work in metric units?

UNIT 2 USING STEEL RULES

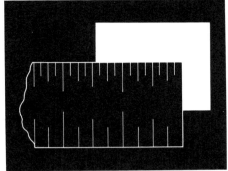

One of the most practical and common measuring tools available in the machining and inspection of parts is the steel rule. It is a tool that metal workers use daily in different ways. It is important that anyone engaged in machining be able to select and use steel rules.

OBJECTIVES

After completing this unit, you should be able to:
1. Identify various kinds of rules and their applications.
2. Measure and record the dimensions of 10 objects that you have obtained from your instructor, using a fractional rule, decimal rule, and a metric rule.

SCALES AND RULES

The terms **scale** and **rule** are often used interchangeably and often incorrectly. A rule is a linear measuring instrument whose graduations represent **real units** of lengths and their subdivisions. In contrast a **scale** is graduated into **imaginary** units that are either smaller or larger than the real units they represent. This is done for convenience where proportional measurements are needed. For example, an architect uses a scale that has graduations representing feet and inches. However, the actual length of the graduations on the architect's scale are quite different from full-sized dimensions.

DISCRIMINATION OF STEEL RULES

The general concept of **discrimination** was discussed in Unit 1 of this section. Discrimination refers to the extent to which a unit of length has been divided. If the smallest graduation on a specific rule is $\frac{1}{32}$ in., then the rule has a discrimination of, or discriminates to $\frac{1}{32}$ in. Likewise, if the smallest graduation of the rule is $\frac{1}{64}$ in., then this rule discriminates to $\frac{1}{64}$ in.

The maximum discrimination of a steel rule is generally $\frac{1}{64}$ in. or in the case of the decimal inch rule, $\frac{1}{100}$ in. The metric rule has a discrimination of $\frac{1}{2}$ mm (millimeter). Remembering that a measuring tool should never be used beyond its discrimination, the steel rule will not be reliable in trying to ascertain a measurement increment smaller than $\frac{1}{64}$ or $\frac{1}{100}$ in. If a specific measurement falls between the markings on the rule, only this can be said of this reading: it

is more or less than the amount of the nearest mark. No further data as to how much more or less can be reliably determined. It is not recommended practice to attempt to read between the graduations on a steel rule with the intent of obtaining reliable readings.

RELIABILITY AND EXPECTATION OF ACCURACY IN STEEL RULES

For reliability, great care must be taken if the steel rule is to be used at its maximum discrimination. Remember that the markings on the rule occupy a certain width. A good quality steel rule has engraved graduations. This means that the markings are actually fine cut in the metal from which the rule is made. Of all types of graduations, engraved ones occupy the least width along the rule. Other rules, graduated by other processes, may have markings that occupy greater width. These rules are not necessarily any less accurate, but they may require more care in reading. Generally, the reliability of the rule will diminish as its maximum discrimination is approached. The smaller graduations are more difficult to see without the aid of a magnifier. Of particular importance is the point from which the measurement is taken. This is the **reference point** and must be carefully aligned at the point where the length being measured begins.

From a practical standpoint, the steel rule finds widest application for measurements no smaller than $\frac{1}{32}$ in. on a fractional rule or $\frac{1}{50}$ in. on a decimal rule. This does not mean that the rule cannot measure to

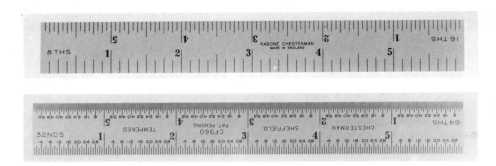

Figure 1. Six inch rigid steel rule (front side). *(Machine Tools and Machining Practices)*

Figure 2. Six inch rigid steel rule (back side). *(Machine Tools and Machining Practices)*

its maximum discrimination, because under the proper conditions it certainly can. However, at or very near maximum discrimination, the time consumed to insure reliable measurement is really not justified. You will be more productive if you make use of a type of measuring instrument with considerably finer discrimination for measurements below the nearest $\frac{1}{32}$ or $\frac{1}{50}$ in. It is good practice to take more than one reading when using a steel rule. After determining the desired measurement, apply the rule once again to see if the same result is obtained. By this procedure, the reliability factor is increased.

TYPES OF RULES

Rules may be selected in many different shapes and sizes, depending on the need. The common **rigid steel rule** is six inches long, $\frac{3}{4}$ in. wide and $\frac{3}{64}$ in. thick. It is engraved with number 4 standard rule graduations. A number 4 graduation consists of $\frac{1}{8}$ and $\frac{1}{16}$ in. on one side (Figure 1) and $\frac{1}{32}$ and $\frac{1}{64}$ in. divisions on the reverse side (Figure 2). Other common graduations are summarized in the following table.

Graduation Number	Front Side	Back Side
Number 3	32nds / 64ths	10ths / 50ths
Number 16	50ths / 100ths	32nds / 64ths

The number 16 graduated rule is often found in the aircraft industry where dimensions are specified in decimal fraction notations, based on 10 or a multiple of 10 divisions of an inch rather than 32 or 64 divisions as found on common rules. Many rigid rules are one inch wide.

Another common rule is the **flexible type** (Figure 3). This rule is 6 in. long, $\frac{1}{2}$ in. wide, and $\frac{1}{64}$ in. thick. Flexible rules are made from hardened and tempered spring steel. One advantage of a flexible rule is that it will bend, permitting measurements to be made in a space shorter than the length of the rule. Most flexible rules are 6 or 12 in. long.

The **narrow rule** (Figure 4) is very convenient when measuring in small openings, slots, or holes. Most narrow rules have only one set of graduations on each side. These can be number 10, which is 32nds and 64ths, or number 11, which is 64ths and 100ths.

The **standard hook rule** (Figure 5) makes it possible to reach through an opening; the rule is hooked on the far side in order to measure a thickness or the depth of a slot (Figure 6). When a workpiece has a chamfered edge, a hook rule will be advantageous over a common rule. Providing that the hook is not loose or excessively worn, it will provide an easy to locate reference point.

The **short rule set** (Figure 7) consists of a set of rules with a holder. Short rule sets have a range of $\frac{1}{4}$ to 1 in. They can be used to measure shoulders in holes or steps in slots, where space is extremely limited. The holder will attach to the rules at any angle, making these very versatile tools.

The **slide caliper rule** (Figure 8) is a versatile tool used to measure round bars, tubing, and other objects where it is difficult to measure at the ends and difficult to estimate the diameter with a rigid steel rule. The small slide caliper rule can also be used to measure internal dimensions from $\frac{1}{4}$ in. up to the capacity of the tool.

The **rule depth gage** (Figure 9) consists of a slotted steel head in which a narrow rule slides. For depth measurements the head is held securely against the surface with the rule extended into the cavity or hole to be measured (Figure 10). The locking nut is tightened and the rule depth gage can then be removed and the dimension determined.

CARE OF RULES

Rules are precision tools, and only those that are properly cared for will provide the kind of service they are designed to give. A rule should not be used as a screwdriver. Rules should be kept separate from hammers, wrenches, files, and other hand tools to protect them from possible damage. An occasional wiping of a rule with a lightly oiled shop towel will keep it clean and free from rust.

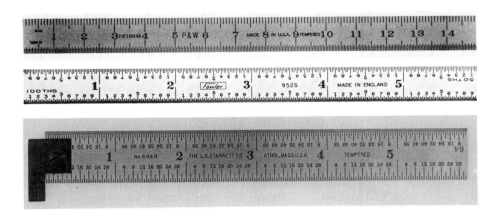

Figure 3. Flexible steel rule (metric). *(Machine Tools and Machining Practices)*

Figure 4. Narrow rule (decimal inch). *(Machine Tools and Machining Practices)*

Figure 5. Standard hook rule. *(Machine Tools and Machining Practices)*

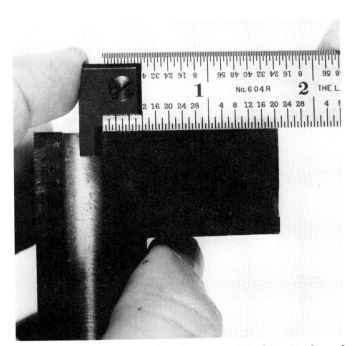

Figure 6. Standard hook rule in use. *(Machine Tools and Machining Practices)*

Figure 7. Short rule set with holder. (The L.S. Starrett Co.)

APPLYING STEEL RULES

When using a steel rule in close proximity to a machine tool, **always keep safety in mind.** Stop the machine before attempting to make any measurements of the workpiece. Attempting to measure with the machine running may result in the rule being caught by a moving part. This may damage the rule, but worse, may result in serious injury to the operator.

One of the problems associated with the use of rules is that of **parallax error.** Parallax error is the error that results when the observer making the measurement is not in line with the workpiece and the rule. You may see the graduation either too far left or too far right of its real position (Figure 11). Parallax

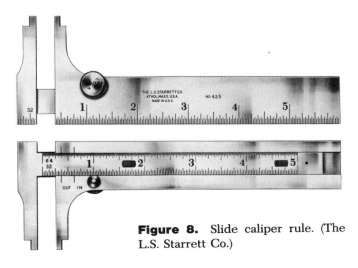

Figure 8. Slide caliper rule. (The L.S. Starrett Co.)

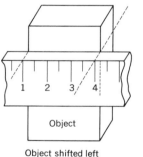
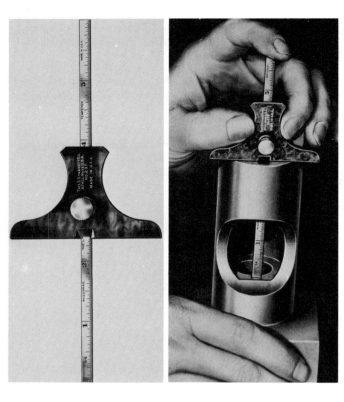

Figure 9. *(left)* Rule depth gage. (The L.S. Starrett Co.)

Figure 10. *(right)* Rule depth gage in use. (The L.S. Starrett Co.)

error occurs when the rule is read from a point other than one directly above the point of measurement. The **point of measurement** is the point at which the measurement is read. It may or may not be the true reading of the size depending on what location was used as the reference point on the rule. Parallax can be controlled by always observing the point of measurement from directly above. Furthermore, the graduations on a rule should be placed as close as possible to the surface being measured. In this regard, a thin rule is preferred over a thick rule.

As a rule is used it becomes worn, usually on the ends. The outside inch markings on a worn rule are less than one inch from the end. This has to be considered when measurements are made. A reliable way to measure (Figure 12) is to use the one inch mark on the rule as the reference point. In the figure, the measured point is at $2\frac{1}{32}$. Subtracting one inch results in a size of $1\frac{1}{32}$ for the part.

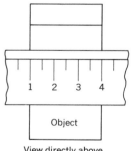
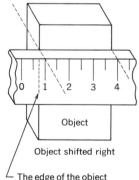

Figure 11. Parallax error. *(Machine Tools and Machining Practices)*

Object shifted left

View directly above proper view point for minimizing parallax

Object shifted right

The edge of the object appears to be at this point on the rule

When viewed from directly above, the rule graduations are exactly in line with the edge of the object being measured. However, when the object is shifted right or left of a point directly above the point of measurement, the alignment of the object edge and the rule graduations appears to no longer coincide

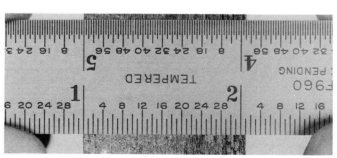

Figure 12. Using the one inch mark as the reference point. *(Machine Tools and Machining Practices)*

Figure 13. Measuring round objects. *(Machine Tools and Machining Practices)*

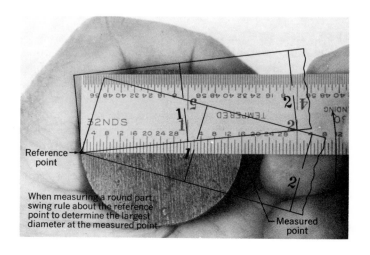

When measuring a round part, swing rule about the reference point to determine the largest diameter at the measured point

Reference point

Measured point

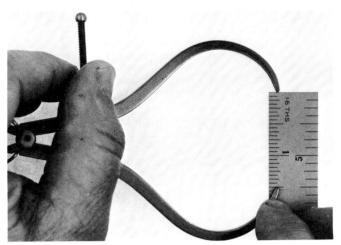

Figure 14. Using a rule to set an outside caliper. *(Machine Tools and Machining Practices)*

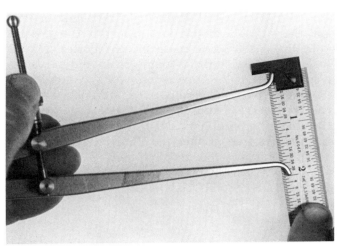

Figure 15. Using a rule to set an inside caliper. *(Machine Tools and Machining Practices)*

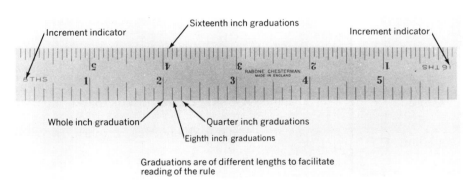

Figure 16. Front side graduations of the typical machinist's rule. *(Machine Tools and Machining Practices)*

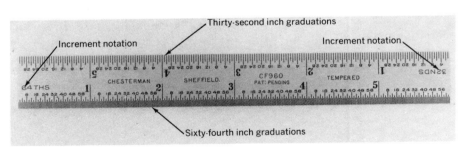

Figure 17. Back side graduations of the typical machinist's rule. *(Machine Tools and Machining Practices)*

Round bars and tubing should be measured with the rule applied on the end of the tube or bar (Figure 13). Select a reference point and set it carefully at a point on the circumference of the round part to be measured. Using the reference point as a pivot, move the rule back and forth slightly to find the largest distance across the diameter. When the largest distance is determined, read the measurement at that point.

Rules are also used for transfer measurements with calipers. The caliper is set to the part, and the reading is obtained by use of the rule. Both inside and outside calipers can be used in this manner (Figures 14 and 15).

READING FRACTIONAL INCH RULES

Most dimensions are expressed in inches and fractions of inches. These dimensions are measured with fractional inch rules. The typical machinist's rule is broken down into 1, $\frac{1}{2}$, $\frac{1}{4}$, $\frac{1}{8}$, $\frac{1}{16}$, $\frac{1}{32}$, and $\frac{1}{64}$ in. graduations. In order to facilitate reading, the 1, $\frac{1}{2}$, $\frac{1}{4}$, $\frac{1}{8}$, and $\frac{1}{16}$ in. graduations appear on one side of the rule (Figure 16). The reverse side of the rule has one edge graduated in $\frac{1}{32}$ in. increments and the other edge graduated in $\frac{1}{64}$ in. increments. On the $\frac{1}{32}$ in. side, every fourth mark is numbered and on the $\frac{1}{64}$ in. side, every eighth mark is numbered (Figure 17). This eliminates the need to count graduations from the nearest whole inch mark. On these rules, the length of the graduation line varies with the one inch line being the longest, the $\frac{1}{2}$ in. line being next in length, the $\frac{1}{4}$, $\frac{1}{8}$, and $\frac{1}{16}$ in. lines each being consecutively shorter. The difference in line lengths is an important aid in reading a rule. The smallest graduation on any edge of a rule is marked by small numbers on the end. Note that the words 8THS and 16THS appear at the ends of the rule. The numbers 32NDS and 64THS appear on the reverse side of the rule, thus indicating thirty-seconds and sixty-fourths of an inch.

EXAMPLES OF FRACTIONAL INCH READINGS:
FIGURE 18

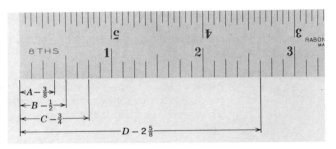

Figure 18. Examples of readings on the ⅛ in. discrimination edge. (*Machine Tools and Machining Practices*)

Distance *A* falls on the third $\frac{1}{8}$ in. graduation. This reading would be $\frac{3}{8}$ in.

Distance *B* falls on the longest graduation between the end of the rule and the first full inch mark. The reading is $\frac{1}{2}$ in.

Distance *C* falls on the sixth $\frac{1}{8}$ in. graduation making it $\frac{6}{8}$ or $\frac{3}{4}$ in.

Distance *D* falls at the fifth $\frac{1}{8}$ in. mark beyond the 2 in. graduation. The reading is $2\frac{5}{8}$ in.

FIGURE 19

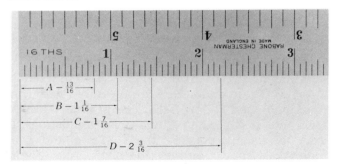

Figure 19. Examples of readings on the 1/16 in. discrimination edge. (*Machine Tools and Machining Practices*)

Distance *A* falls at the thirteenth $\frac{1}{16}$ in. mark making the reading $\frac{13}{16}$ in.

Distance *B* falls at the first $\frac{1}{16}$ in. mark past the 1 in. graduation. The reading is $1\frac{1}{16}$ in.

Distance *C* falls at the seventh $\frac{1}{16}$ in. mark past the 1 in. graduation. The reading is $1\frac{7}{16}$ in.

Distance *D* falls at the third $\frac{1}{16}$ in. mark past the 2 in. graduation. The reading is $2\frac{3}{16}$ in.

FIGURE 20

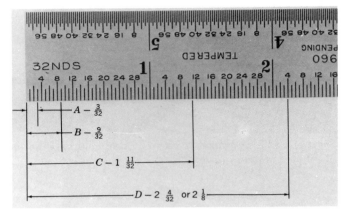

Figure 20. Examples of readings on the 1/32 in. discrimination edge. (*Machine Tools and Machining Practices*)

Distance *A* falls at the third $\frac{1}{32}$ in. mark. The reading is $\frac{3}{32}$ in.

Distance *B* falls at the ninth $\frac{1}{32}$ in. mark. The reading is $\frac{9}{32}$ in.

Distance *C* falls at the eleventh $\frac{1}{32}$ in. mark past the 1 in. graduation. The reading is $1\frac{11}{32}$ in.

Distance *D* falls at the fourth $\frac{1}{32}$ in. mark past the 2 in. graduation. The reading is $2\frac{4}{32}$ in., which reduced to lowest terms becomes $2\frac{1}{8}$ in.

FIGURE 21

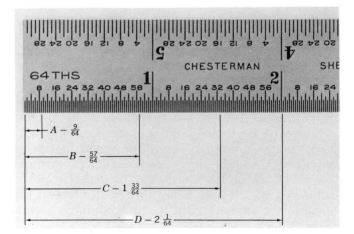

Figure 21. Examples of readings on the 1/64 in. discrimination edge. (*Machine Tools and Machining Practices*)

Distance *A* falls at the ninth $\frac{1}{64}$ in. mark making the reading $\frac{9}{64}$ in.

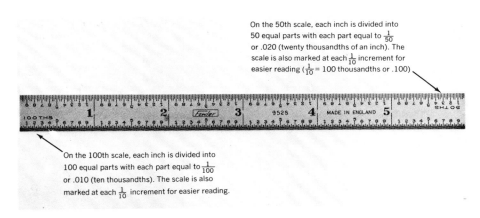

On the 50th scale, each inch is divided into 50 equal parts with each part equal to $\frac{1}{50}$ or .020 (twenty thousandths of an inch). The scale is also marked at each $\frac{1}{10}$ increment for easier reading ($\frac{1}{10}$ = 100 thousandths or .100)

On the 100th scale, each inch is divided into 100 equal parts with each part equal to $\frac{1}{100}$ or .010 (ten thousandths). The scale is also marked at each $\frac{1}{10}$ increment for easier reading.

Figure 22. Six inch decimal rule. *(Machine Tools and Machining Practices)*

Distance *B* falls at the fifty-seventh $\frac{1}{64}$ in. mark making the reading $\frac{57}{64}$ in.

Distance *C* falls at the thirty-third $\frac{1}{64}$ in. mark past the 1 in. graduation. The reading is $1\frac{33}{64}$ in.

Distance *D* falls at the first $\frac{1}{64}$ in. mark past the 2 in. graduation, making the reading $2\frac{1}{64}$ in.

READING DECIMAL INCH RULES

Many dimensions in the auto, aircraft, and missile industries are specified in **decimal notations,** which refers to the division of the inch into 10 parts or a multiple of 10 parts, such as 50 or 100 parts. In this case, a **decimal rule** would be used. Decimal inch dimensions are specified and read as thousandths of an inch. Decimal rules, however, do not discriminate to the individual thousandth because the width of an engraved or etched division on the rule is approximately .003 in. (three-thousandths of an inch). Decimal rules are commonly graduated in increments of $\frac{1}{10}$ in., $\frac{1}{50}$ in., or $\frac{1}{100}$ in.

A typical decimal rule may have $\frac{1}{50}$ in. divisions on the top edge and $\frac{1}{100}$ in. divisions on the bottom edge (Figure 22). The inch is divided into 10 equal parts, making each numbered division $\frac{1}{10}$ in. or .100 in. (100-thousandths of an inch). On the top scale each $\frac{1}{10}$ increment is further subdivided into five equal parts, which makes the value of each of these divisions .020 in. (20-thousandths of an inch).

EXAMPLES OF DECIMAL INCH READINGS: FIGURE 23

Distance *A* falls on the first marked graduation. The reading is $\frac{1}{10}$ or .100 thousandths in. This can also be read on the 50ths in. scale, as seen in the figure.

Distance *B* can only be read on the 100th in. scale, as it falls at the seventh graduation beyond the .10 in. mark. The reading is .100 in. plus .070 in., or .170 in. This distance cannot be read on the 50th in. scale because discrimination of the 50th in. scale is not sufficient.

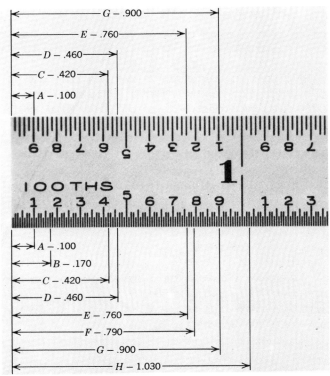

Figure 23. Examples of decimal rule readings. *(Machine Tools and Machining Practices)*

Distance *C* falls at the second mark beyond the .400 in. line. This reading is .400 in. plus .020 in., or .420 in. Since .020 in. is equal to $\frac{1}{50}$ in., this can also be read on the 50th in. scale as shown in the figure.

Distance *D* falls at the sixth increment beyond the .400 in. line. The reading is .400 in. plus .060 in., or .460 in. This can also be read on the 50th in. scale, as seen in the figure.

Distance *E* falls at the sixth division beyond the .700 in. mark. The reading is .700 in. plus .060 in., or .760 in. This can also be read on the 50th in. scale.

Distance *F* falls at the ninth mark beyond the .700 in. line. The reading is .700 in. plus .090 in., or .790 in. This cannot be read on the 50th in. scale.

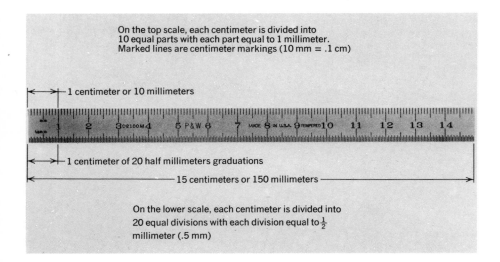

On the top scale, each centimeter is divided into
10 equal parts with each part equal to 1 millimeter.
Marked lines are centimeter markings (10 mm = .1 cm)

1 centimeter or 10 millimeters

1 centimeter of 20 half millimeters graduations

15 centimeters or 150 millimeters

On the lower scale, each centimeter is divided into
20 equal divisions with each division equal to ½
millimeter (.5 mm)

Figure 24. 150 millimeter metric rule. *(Machine Tools and Machining Practices)*

Distance *H* falls three marks past the first full inch mark. The reading is 1.00 in. plus .030 in., or 1.030 in. This cannot be read on the 50th in. scale.

READING METRIC RULES

Many products are made in metric dimensions requiring a machinist to use a **metric rule**. The typical metric rule has millimeter (mm) and half millimeter graduations (Figure 24).

EXAMPLES OF READING METRIC RULES:
FIGURE 25

Distance *A* falls at the 53rd graduation on the mm scale. The reading is 53 mm.

Distance *B* falls at the 22nd graduation on the mm scale. The reading is 22 mm.

Distance *C* falls at the sixth graduation on the mm scale. The reading is 6 mm.

Distance *D* falls at the 17th half mm mark. The reading is 8 mm plus an additional half mm, giving a total of 8.5 mm.

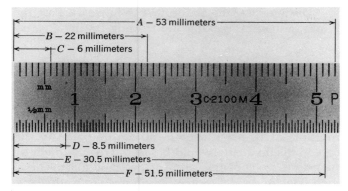

A — 53 millimeters

B — 22 millimeters

C — 6 millimeters

D — 8.5 millimeters

E — 30.5 millimeters

F — 51.5 millimeters

Figure 25. Examples of metric rule readings. *(Machine Tools and Machining Practices)*

Distance *E* falls one-half mm beyond the 3 centimeter (cm) graduation. Since 3 cm is equal to 30 mm, the reading is 30.5 mm.

Distance *F* falls one-half mm beyond the 51 mm graduation. The reading is 51.5 mm. In machine design, all dimensions are specified in mm. Hence, 1.5 meters (m) would be 1500 mm.

SELF-TEST

Read and record the dimensions indicated by the letters
A to H in Figure 26.

A _____   D _____   G _____

B _____   E _____   H _____

C _____   F _____

Read and record the dimensions indicated by the letters
A to E in Figure 27.

A _____

B _____

C _____

D _____

E _____

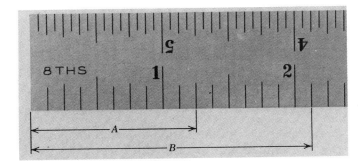

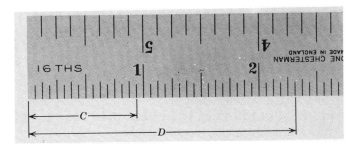

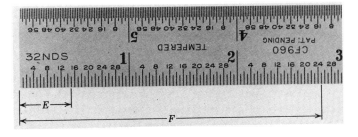

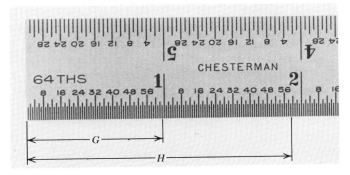

Figure 26. A to H.

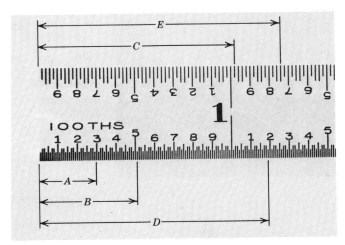

Figure 27. A to E.

Read and record the dimensions indicated by the letters
A to F in Figure 28.

A _____

B _____

C _____

D _____

E _____

F _____

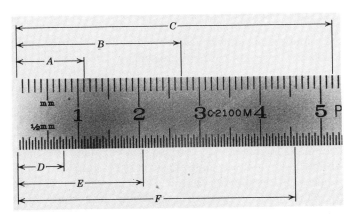

Figure 28. A to F.

UNIT 3 USING VERNIER CALIPERS AND VERNIER DEPTH GAGES

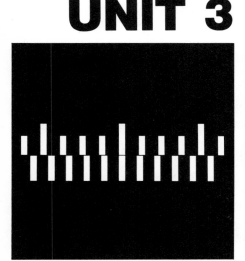

The inspection and measurement of machined parts requires various kinds of measuring tools. Often the discrimination of a rule is sufficient, but, in many cases the discrimination of a rule with a vernier scale is required. This unit explains the types, use, and applications of common vernier instruments.

OBJECTIVES

After completing this unit, you should be able to:

1. Measure and record the dimensions of 10 objects to an accuracy of plus or minus .001 in. with a vernier caliper.
2. Measure and record the dimensions of 10 objects to an accuracy of plus or minus .02 mm using a metric vernier caliper.
3. Measure and record the dimensions of 10 objects using a vernier depth gage.

PRINCIPLE OF THE VERNIER

The principle of the **vernier** may be used to increase the discrimination of all graduated scale measuring tools used by a machinist. A vernier system consists of a **main scale** and a **vernier scale**. The vernier scale is placed adjacent to the main scale so that graduations on both scales can be observed together. The spacing of the vernier scale graduations is shorter than the spacing of the main scale graduations. For example, consider a main scale divided as shown (Figure 1*a*). It is desired to further subdivide each main scale division into 10 parts with the use of a vernier. The spacing of each vernier scale division is made $\frac{1}{10}$ of a main scale division shorter than the spacing of a main scale division. This may sound confusing but, think of it as 10 vernier scale divisions corresponding to nine main scale divisions (Figure 1*a*). The vernier now permits the main scale to discriminate to $\frac{1}{10}$ of its major divisions. Therefore, $\frac{1}{10}$ is known as the **least count** of the vernier.

The vernier functions in the following manner.

Assume that the zero line on the vernier scale is placed as shown (Figure 1*b*). The reading on the main scale is two, plus a fraction of a division. It is desired to know the amount of the fraction over two, to the nearest tenth or least count of the vernier. As you inspect the alignment of the vernier scale and the main scale lines, you will note that they move closer together until one line on the vernier scale coincides with a line on the main scale. This is the **coincident line** of the vernier which indicates the fraction in tenths that must be added to the main scale reading. The vernier is coincident at the sixth line. Since the least count of the vernier is $\frac{1}{10}$, the zero vernier line is six-tenths past two on the main scale. Therefore, the main scale reading is 2.6 (Figure 1*b*).

DISCRIMINATION AND APPLICATIONS OF VERNIER INSTRUMENTS

Vernier instruments used for linear measure in the inch system discriminate to .001 in. ($\frac{1}{1000}$). Metric verni-

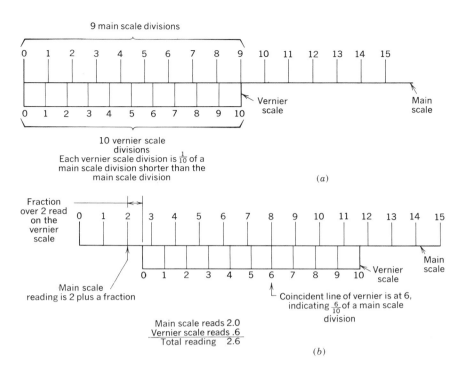

Figure 1. Principle of the vernier. *(Machine Tools and Machining Practices)*

ers generally discriminate to .02 $(\frac{1}{50})$ of a millimeter.

The most common vernier instruments include several styles of **calipers.** The common vernier caliper is used for outside and inside linear measurement. Another style of vernier caliper has the capability of depth measurement in addition to outside and inside capacity. The vernier also appears on a variety of depth gages.

Beyond its most common applications, the vernier also appears on a height gage, which is an extremely important layout tool for a machinist. The vernier is also used on the gear tooth caliper, a special vernier caliper used in gear measurement. As the principle of the vernier can be used to subdivide a unit of angular measure as well as linear measure, it appears on various types of protractors used for angular measurement.

RELIABILITY AND EXPECTATION OF ACCURACY IN VERNIER INSTRUMENTS

Reliability in vernier calipers and depth gages is highly dependent on proper use of the tool. The simple fact that the caliper or depth gage has increased discrimination over a rule does not necessarily provide increased reliability. The improved degree of discrimination in vernier instruments requires more than the mere visual alignment of a rule graduation against the edge of the object to be measured. The zero reference point of a vernier caliper is the positively placed contact of the solid jaw with the part to be measured. On the depth gage, the base is the zero reference point. Positive contact of the zero reference is an important consideration in vernier reliability. .

The vernier scale must be read carefully if a reliable measurement is to be determined. On many vernier instruments the vernier scale should be read with the aid of a magnifier. Without this aid, the coincident line of the vernier is difficult to determine. Therefore, the reliability of the vernier readings can be in question. The typical vernier caliper has very narrow jaws and thus must be carefully aligned with the axis of the measurement. On the plain slide (not having a fine adjustment screw) vernier caliper, no provision is made for the "feel" of the measuring pressure. Feel is obtained by moving the movable jaw against the work and determining the drag. Some calipers and the depth gage are equipped with a screw thread fine adjustment that gives them a slight advantage in determining the pressure applied during the measurement.

Generally, the overall reliability of vernier instruments for measurement at maximum discrimination of .001 in. is fairly low. The vernier should never be used in an attempt to discriminate below .001 in. The instrument does not have that capability. However, the vernier principle is also used on micrometers to .0001 in. and on protractors to discriminate to smaller units than degrees. Vernier instruments are a popular tool on the inspection bench, and they can serve very well for measurement in the range of plus or minus .005 of an inch. With proper use and an understanding of the limitations of a vernier instrument, this tool can be a valuable addition to the many measuring tools available to you.

VERNIER CALIPERS

With a rule, measurements can be made to the nearest $\frac{1}{64}$ or $\frac{1}{100}$ in., but often this is not sufficiently accurate. A measuring tool based on a rule but with much greater discrimination is the **vernier caliper** (Figure 2). Vernier calipers have a discrimination of .001 in. The **beam** or **bar** is engraved with the **main scale**. This is also called the **true scale,** as each inch marking is exactly one inch apart. Each inch is further broken down into minor divisions that can be either .025 in. or .050 in. apart. The beam and the solid jaw are square, or at 90 degrees to each other.

The movable jaw contains the **vernier scale**. This scale is located on the sliding jaw of a vernier caliper or it is part of the base on the vernier depth gage. The function of the vernier scale is to subdivide the minor division on the beam scale into the smallest increments that the vernier instrument is capable of measuring, which is usually .001 in. for inch reading vernier calipers.

Most of the longer vernier calipers have a fine adjustment clamp for precise adjustments of the movable jaw. Inside measurements are made over the **nibs** on the jaw and are read on the top scale of the vernier caliper (Figure 2). The top scale is a duplicate of the lower scale, with the exception that it is offset to compensate for the size of the nibs.

The "Mauser pattern" or "Cross Horn" vernier caliper is very common (Figure 3) though there are many other types that do not have this versatility. This is a versatile tool because of its capacity to make outside, inside, height, and depth measurements. Many different measuring applications are made with this particular design of vernier caliper (Figure 4).

VERNIER CALIPER PROCEDURES

To test a vernier caliper for accuracy, clean the contact surfaces of the two jaws. Bring the movable jaw with normal gaging pressure into contact with the solid jaw. Hold the caliper up to a light source and examine the alignment of the solid and movable jaws. If wear exists, a line of light will be visible between the jaw faces. A gap as small as .0001 ($\frac{1}{10,000}$) of an inch can be seen against a light. If the contact between the jaws is satisfactory, check the vernier scale alignment. The vernier scale zero mark should be in alignment with the zero on the main scale. Realignment of the vernier scale to adjust it to zero can be accomplished on some vernier calipers.

A vernier caliper is a delicate precision tool and should be treated as such. It is very important that the correct amount of pressure or feel is developed while taking a measurement. The measuring jaws should contact the workpiece firmly. However, excessive pressure will spring the jaws and give inaccurate readings. When measuring an object, use the solid

Figure 2. Typical inside-outside, 50 division vernier caliper.

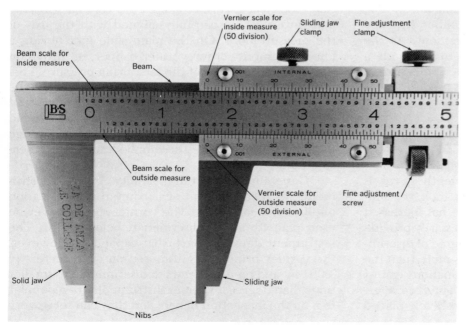

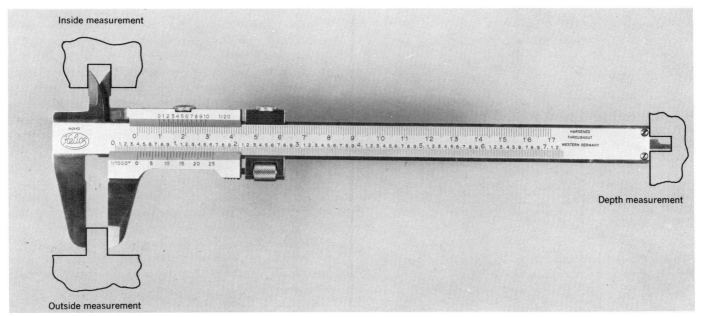

Figure 3. Typical "Cross Horn" or "Mauser Pattern" vernier caliper. *(Machine Tools and Machining Practices)*

Figure 4. This design of vernier caliper has many applications. (MTI Corporation)

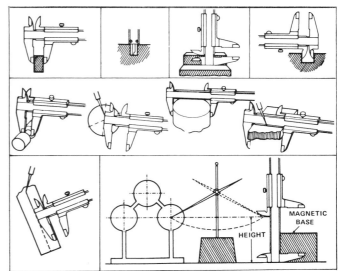

jaw as the reference point. Then move the sliding jaw until contact is made. When measuring with the vernier caliper make certain that the beam of the caliper is in line with the surfaces being measured. Whenever possible, read the vernier caliper while it is still in contact with the workpiece. Moving the instrument may change the reading. Any measurement should be repeated at least twice to assure reliability.

READING INCH VERNIER CALIPERS

Vernier scales are engraved with 25 or 50 divisions (Figures 5 and 6). On a 25 division vernier caliper, each inch on the main scale is divided into 10 major divisions numbered from 1 to 9. Each major division is .100 in. (one hundred-thousandths). Each major division has four minor divisions with a spacing of .025 in. (twenty-five thousandths). The vernier scale has 25 divisions with the zero line being the index.

To read the vernier caliper (Figure 7), read all of the graduations to the left of the index line. This would be one whole inch plus $\frac{2}{10}$ or .200 in. plus one

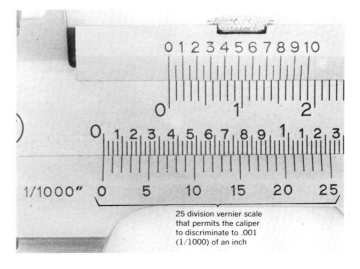

Figure 5. Lower scale is a 25 division vernier. *(Machine Tools and Machining Practices)*

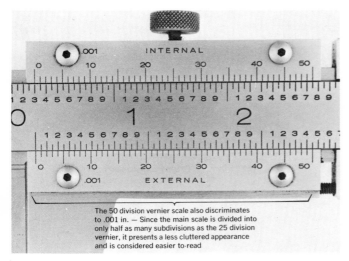

The 50 division vernier scale also discriminates to .001 in. — Since the main scale is divided into only half as many subdivisions as the 25 division vernier, it presents a less cluttered appearance and is considered easier to read

Figure 6. Fifty division vernier caliper. *(Machine Tools and Machining Practices)*

Figure 7. Reading a 25 division vernier caliper. *(Machine Tools and Machining Practices)*

minor division valued at .025 in. plus part of one minor division. The value of this partial minor division is determined by the coincidence of one line on the vernier scale with one line of the true scale. For this example, the coincidence is on line 13 of the vernier scale. This is the value in thousandths of an inch that has to be added to the value read on the beam. Therefore, 1 + .100 + .100 + .025 + .013 equals the total reading of 1.238 in. An aid in determining the coincidental line is to look at the lines adjacent to or on each side of the coincidental line. These lines should fall inside the lines on the true scale (Figure 8).

The 50 division vernier caliper is read as follows (Figure 9). The true scale has each inch divided into 10 major divisions of .100 in. each, with each major division subdivided into two minor divisions, thus being .050 in. each. The vernier scale has 50 divisions. The 50 division vernier caliper reading shown is read as follows:

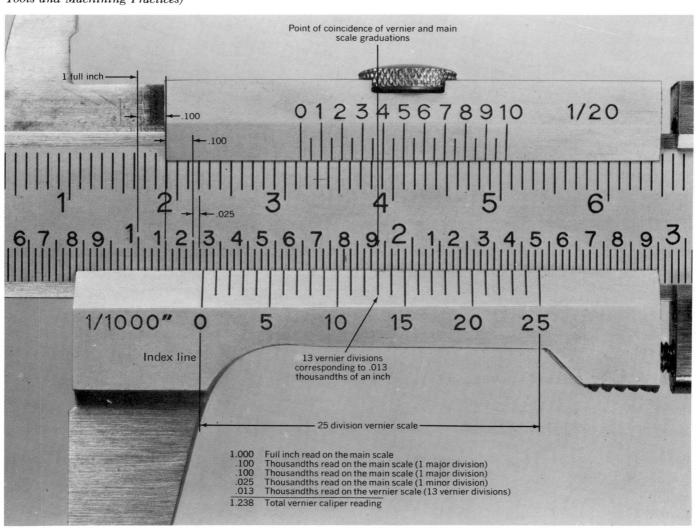

1.000	Full inch read on the main scale
.100	Thousandths read on the main scale (1 major division)
.100	Thousandths read on the main scale (1 major division)
.025	Thousandths read on the main scale (1 minor division)
.013	Thousandths read on the vernier scale (13 vernier divisions)
1.238	Total vernier caliper reading

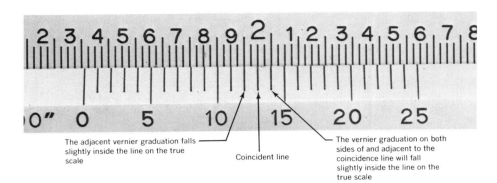

Figure 8. Determining the coincident line on a vernier. *(Machine Tools and Machining Practices)*

The adjacent vernier graduation falls slightly inside the line on the true scale

Coincident line

The vernier graduation on both sides of and adjacent to the coincidence line will fall slightly inside the line on the true scale

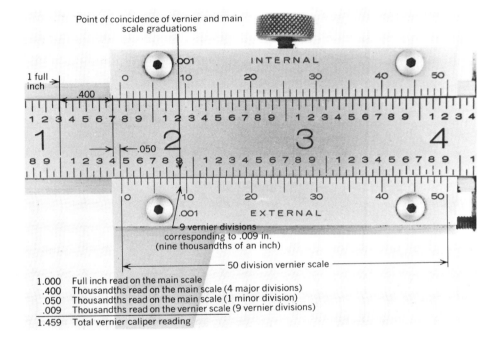

Figure 9. Reading a 50 division vernier. *(Machine Tools and Machining Practices)*

Point of coincidence of vernier and main scale graduations

1 full inch

.400

.050

.001

INTERNAL

.001   EXTERNAL

9 vernier divisions corresponding to .009 in. (nine thousandths of an inch)

50 division vernier scale

1.000   Full inch read on the main scale
 .400   Thousandths read on the main scale (4 major divisions)
 .050   Thousandths read on the main scale (1 minor division)
 .009   Thousandths read on the vernier scale (9 vernier divisions)
1.459   Total vernier caliper reading

Beam whole inch reading	1.000 in.
Additional major divisions	.400 in.
Additional minor divisions	.050 in.
Vernier scale reading	.009 in.
Total caliper reading	1.459 in.

READING METRIC VERNIER CALIPERS

The applications for a metric vernier caliper are exactly the same as those described for an inch system vernier caliper. The discrimination of metric vernier caliper models varies from .02 mm, .05 mm, or .1 mm. The most commonly used type discriminates to .02 mm. The main scale on a metric vernier caliper is divided into millimeters with every tenth millimeter mark numbered. The 10 millimeter line is numbered 1, the 20 millimeter line is numbered 2 and so on, up to the capacity of the tool (Figure 10). The vernier scale on the sliding jaw is divided into 50 equal spaces with every fifth space numbered. Each numbered division on the vernier represents one-tenth of a millimeter. The five smaller divisions between the numbered lines represent two-hundredths (.02 mm) of a millimeter.

To determine the caliper reading, read, on the main scale, whole millimeters to the left of the zero or the index line on the sliding jaw. The example (Figure 10) shows 27 mm plus part of an additional millimeter. The vernier scale coincides with the main scale at the 18th vernier division. Since each vernier scale spacing is equal to .02 mm, the reading on the vernier scale is equal to 18 times .02, or .36 mm. Therefore, .36 mm must be added to the amount showing on the main scale to obtain the final reading. The result is equal to 27 mm + .36 mm or 27.36 mm (Figure 10).

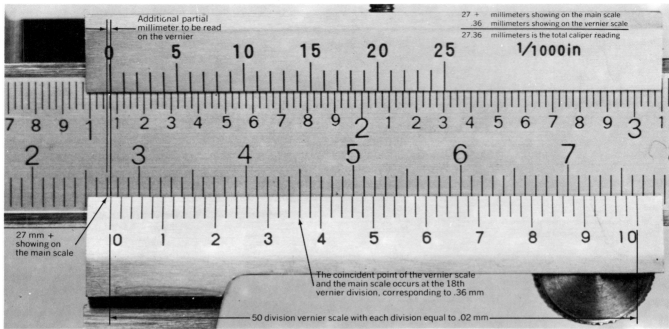

Figure 10. Reading a metric vernier caliper with .02 millimeter discrimination (lower scale). The upper scale is for inch reading. *(Machine Tools and Machining Practices)*

Figure 11. Dial caliper.

DIAL CALIPERS

Although not a vernier tool, the dial caliper (Figure 11) is a very popular measuring instrument. This instrument has a discrimination of .001 in. and can be used for outside, inside, and depth measurement. The caliper is easier to read because it does not require that the coincident line of a vernier be determined.

Figure 12. Vernier depth gage with .001 in. discrimination.

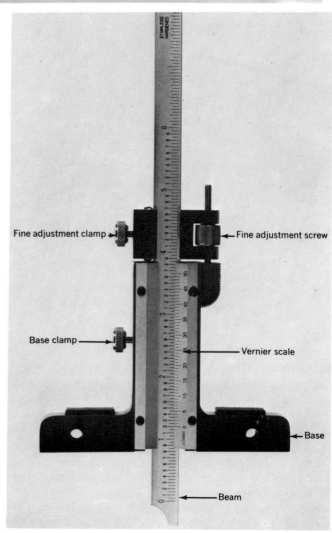

The caliper dial hand is operated by a pinion gear that engages a rack on the caliper beam. When using a dial caliper, remember what you have learned about the accuracy of caliper instruments. These instruments are more delicate than vernier calipers and can easily get out of adjustment when bumped or dropped.

READING VERNIER DEPTH GAGES

These measuring tools are designed to measure the depth of holes, recesses, steps, and slots. Basic parts of a vernier depth gage include the base or anvil with the vernier scale and the fine adjustment screw (Figure 12). Also shown is the graduated beam or bar that contains the true scale. To make accurate mea-

surements the reference surface needs to be flat and free from nicks and burrs. The base should be held firmly against the reference surface while the beam is brought in contact with the surface being measured. The measuring pressure should approximately equal the pressure exerted when making a light dot on a piece of paper with a pencil. On a vernier depth gage, dimensions are read in the same manner as on a vernier caliper.

VERNIER HEIGHT GAGES

Vernier height gages are used extensively for layout work and precision measurement. See Section E, Unit 2, "Layout Tools" and Unit 3, "Layout Practice" for an explanation and use of these tools.

SELF-TEST—READING INCH VERNIER CALIPERS

Determine the dimensions in the vernier caliper illustrations (Figures 13a to 13d).

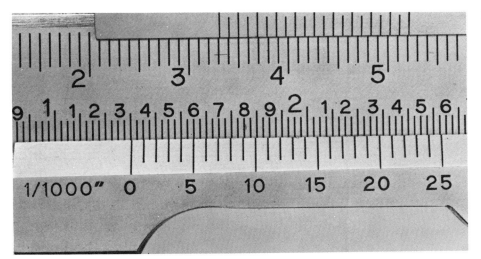

Figure 13a

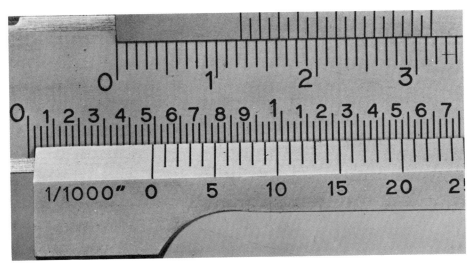

Figure 13b

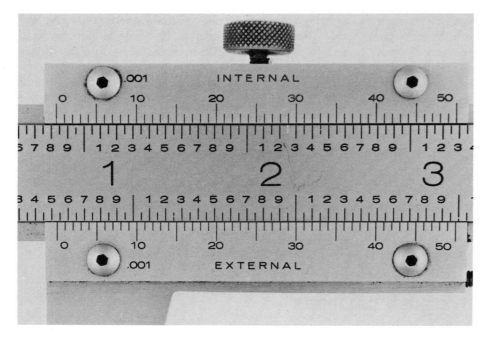

Figure 13c

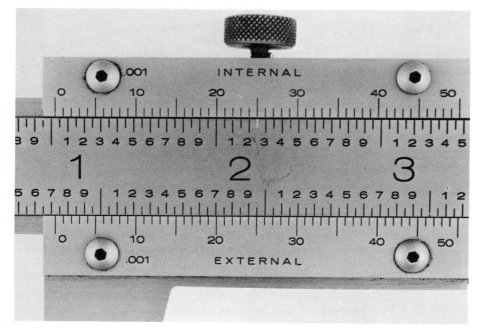

Figure 13d

SELF-TEST—READING METRIC VERNIER CALIPERS

Determine the metric vernier caliper dimensions illustrated
in Figures 14a to 14d.

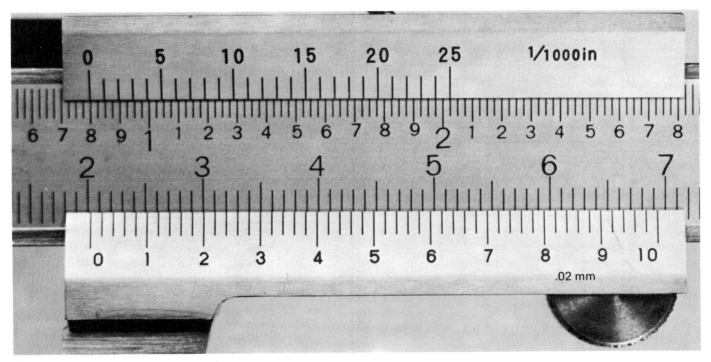

Figure 14a

Figure 14b

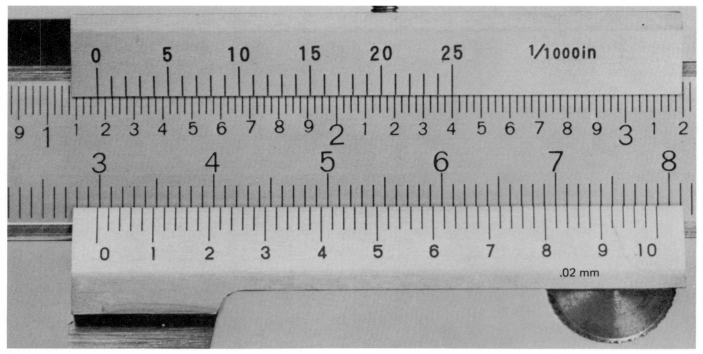

Figure 14c

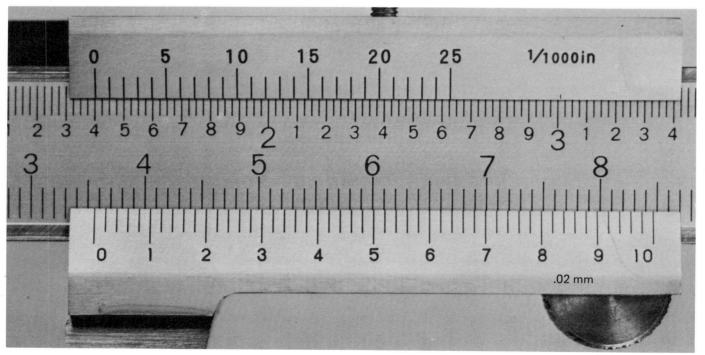

Figure 14d

SELF-TEST—READING VERNIER DEPTH GAGES

Determine the depth measurements illustrated in Figures
15a to 15d.

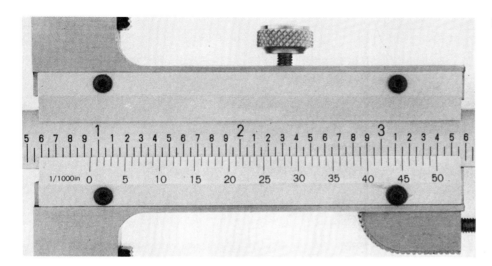

Figure 15a

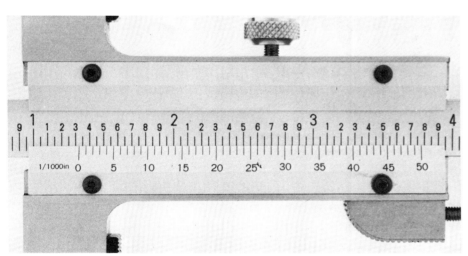

Figure 15b

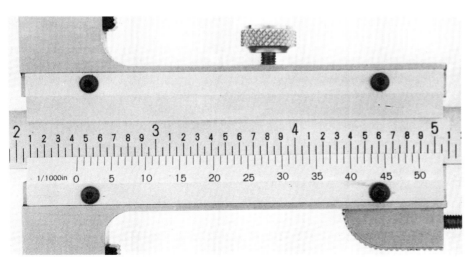

Figure 15c

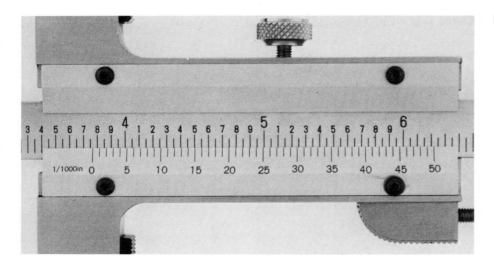

Figure 15d

UNIT 4 USING MICROMETER INSTRUMENTS

Micrometer measuring instruments are the most commonly used precision measuring tools found in industry. Correct use of them is essential to anyone engaged in making, inspecting, or assembling machined parts.

OBJECTIVES

After completing this unit and with the use of appropriate measuring kits, you should be able to:

1. Measure and record the dimensions of 10 objects, using outside micrometers, to an accuracy of plus or minus .001 of an inch.
2. Measure and record the diameters of five holes in test objects to an accuracy of plus or minus .001 in., using an inside micrometer.
3. Measure and record five depth measurements on a test object using a depth micrometer to an accuracy of plus or minus .001 in.
4. Measure and record the dimensions of 10 objects, using a metric micrometer, to an accuracy of plus or minus .01 mm.
5. Measure and record the dimensions of five objects, using a vernier micrometer, to an accuracy of plus or minus .0001 in. (assuming proper measuring conditions).

OUTSIDE MICROMETERS AND THEIR CARE

You should be familiar with the names of the major parts of the typical outside micrometer (Figure 1). The micrometer uses the movement of a precisely threaded rod turning in a nut for precision measurements. The accuracy of micrometer measurements is dependent on the quality of its construction, the care the tool receives, and the skill of the user. Con-

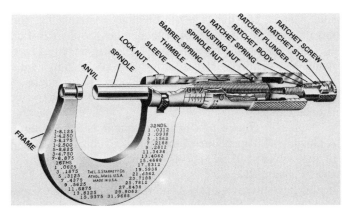

Figure 1. Parts of the outside micrometer. (The L.S. Starrett Co.)

Figure 2. Micrometers should always be kept on a tool board when used near a machine tool. (Lane Community College)

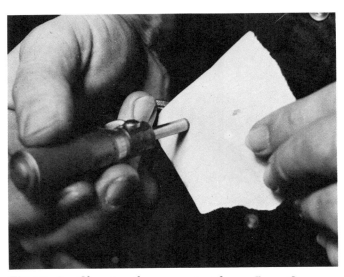

Figure 3. Cleaning the measuring faces. (Lane Community College)

sider some of the important factors in the care of the micrometer: A micrometer should be wiped clean of dust and oil before and after it is used. A micrometer should not be opened or closed by holding it by the thimble and spinning the frame around the axis of the spindle. Make sure that the micrometer is not dropped. Even a fall of a short distance can spring the frame. This will cause misalignment between the anvil and spindle faces and destroy the accuracy of this precision tool. A micrometer should be kept away from chips on a machine tool. The instrument should be placed on a clean tool board (Figure 2) or on a clean shop towel close to where it is needed.

Always remember that the machinist is responsible for any measurements that he may make. To excuse an inaccurate measurement on the grounds that a micrometer was not properly adjusted or cared for would be less than professional. When a micrometer is stored after use, make sure that the spindle face does not touch the anvil. Perspiration, moisture from the air, or even oils promote corrosion between the measuring faces with a corresponding reduction in accuracy.

Prior to using a micrometer, clean the measuring faces. The measuring faces of many newer micrometers are made from an extremely hard metal called tungsten carbide. These instruments are often known as carbide-tipped micrometers. If you examine the measuring faces of a carbide-tipped micrometer, you will see where the carbide has been attached to the face of the anvil and spindle. Carbide-tipped micrometers have very durable and long-wearing measuring faces. Screw the spindle down lightly against a piece of paper held between it and the anvil (Figure 3). Slide the paper out from between the measuring faces and blow away any fuzz that clings to the spindle or anvil. At this time, you should test the zero reading of the micrometer by bringing the spindle slowly into

contact with the anvil (Figure 4). Use the ratchet stop or friction thimble to perform this operation. The ratchet stop or friction thimble found on most micrometers is designed to equalize the gaging force. When the spindle and anvil contact the workpiece, the ratchet stop or friction thimble will slip as a predetermined amount of torque is applied to the micrometer thimble. If the micrometer does not have a ratchet device, use your thumb and index finger to provide a slip clutch effect on the thimble. Never use more pressure when checking the zero reading than when making actual measurements on the workpiece. If there is a small error, it may be corrected by adjusting the index line to the zero point (Figure

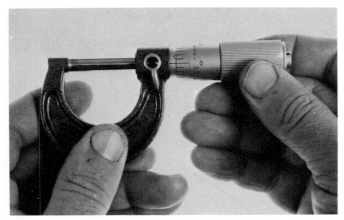

Figure 4. Checking the zero reading. *(Machine Tools and Machining Practices)*

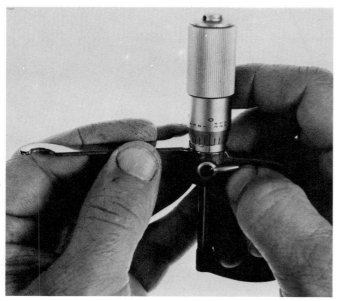

Figure 5. Adjusting the index line to zero. *(Machine Tools and Machining Practices)*

5). The manufacturer's instructions provided with the micrometer should be followed when making this adjustment. Also, follow the manufacturer's instructions for correcting a loose thimble to spindle connection or incorrect friction thimble or ratchet stop action. One drop of instrument oil applied to the micrometer thread at monthly intervals will help it to provide many years of reliable service.

READING INCH MICROMETERS

Dimensions requiring the use of micrometers will generally be expressed in decimal form to three deci-

mal places. In the case of the inch instrument, this would be the thousandths place. You should think in terms of thousandths whenever reading decimal fractions. For example, the decimal .156 of an inch would be read as one hundred and fifty-six thousandths of an inch. Likewise, .062 would be read as sixty-two thousandths.

On the **sleeve** of the micrometer is a graduated scale with 10 numbered divisions, each one being $\frac{1}{10}$ of one inch or .100 (100-thousandths) apart. Each of these major divisions is further subdivided into four equal parts, which makes the distance between these graduations $\frac{1}{4}$ of .100 or .025 (25-thousandths) (Figure 6). The **spindle screw** of a micrometer has 40 threads per inch. When the spindle is turned one complete revolution, it has moved $\frac{1}{40}$ of one inch, or expressed as a decimal, .025 (25-thousandths).

When we examine the **thimble**, we find 25 evenly spaced divisions around its circumference (Figure 6). Because each complete revolution of the thimble causes it to move a distance of .025 in., each thimble graduation must be equal to $\frac{1}{25}$ of .025, or .001 in. (one-thousandth). On most micrometers, each thimble graduation is numbered to facilitate reading the instrument. On older micrometers only every fifth line may be numbered.

When reading the micrometer (Figure 7), first determine the value indicated by the lines exposed on the sleeve. The edge of the thimble exposes three major divisions. This represents .300 in. (three hundred-thousandths). However, there are also two minor divisions showing on the sleeve. The value of these is .025, for a total of .050 in. (fifty-thousandths). The reading on the thimble is 9, which indicates .009 in. (nine thousandths). The final micrometer reading is determined by adding the total of the sleeve and thimble readings. In the example shown (Figure 7), the sleeve shows a total of .350 in. Adding this to the thimble, the final reading becomes .350 in. + .009 in., or .359 in.

USING THE MICROMETER

When used to measure small parts held in the hand, the micrometer should be gripped by the frame (Figure 8) leaving the thumb and forefinger free to operate the thimble. When possible, take micrometer readings while the instrument is in contact with the workpiece (Figure 9). Use only enough pressure on the **spindle** and **anvil** to yield a reliable result. This is what is referred to as *feel*. The proper feel of a micrometer will come only from experience. Obviously, excessive pressure will not only result in an inaccurate measurement, it will also distort the frame

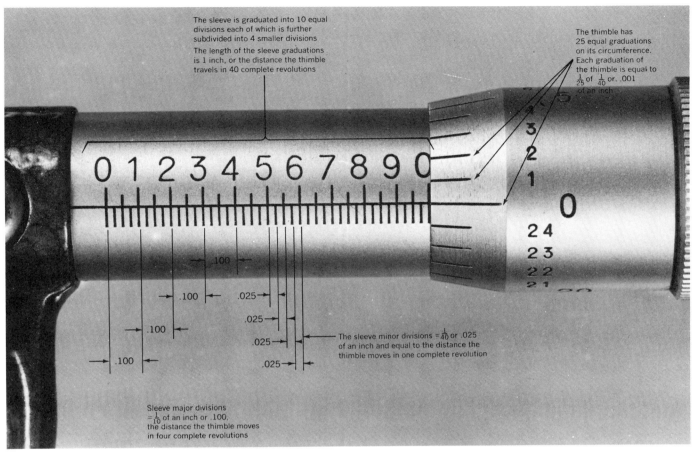

The sleeve is graduated into 10 equal divisions each of which is further subdivided into 4 smaller divisions

The length of the sleeve graduations is 1 inch, or the distance the thimble travels in 40 complete revolutions

The thimble has 25 equal graduations on its circumference. Each graduation of the thimble is equal to $\frac{1}{25}$ of $\frac{1}{40}$ or .001 of an inch

The sleeve minor divisions = $\frac{1}{40}$ or .025 of an inch and equal to the distance the thimble moves in one complete revolution

Sleeve major divisions = $\frac{1}{10}$ of an inch or .100, the distance the thimble moves in four complete revolutions

Figure 6. Graduations on the inch micrometer. *(Machine Tools and Machining Practices)*

Figure 7. Inch micrometer reading of .359, or three hundred fifty-nine thousandths. *(Machine Tools and Machining Practices)*

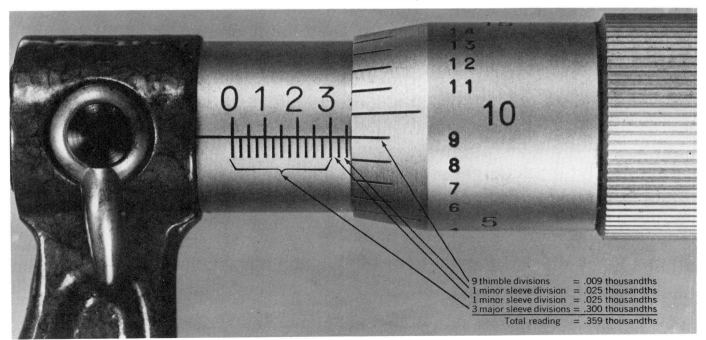

9 thimble divisions	= .009 thousandths
1 minor sleeve division	= .025 thousandths
1 minor sleeve division	= .025 thousandths
3 major sleeve divisions	= .300 thousandths
Total reading	= .359 thousandths

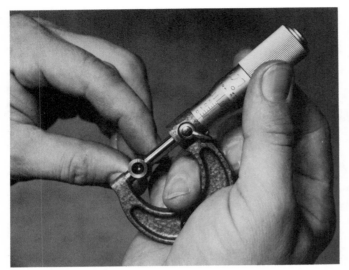

Figure 8. Proper way to hold a micrometer. (Lane Community College)

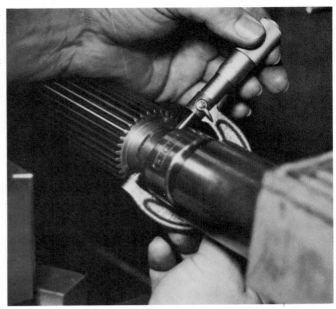

Figure 10. Hold a micrometer in both hands when measuring a round part. (Lane Community College)

Figure 9. Read a micrometer while still in contact with the workpiece. (Lane Community College)

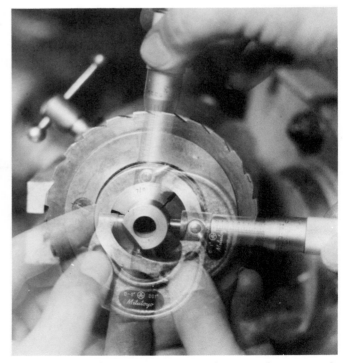

Figure 11. When measuring round parts, take two readings 90 degrees apart. (Lane Community College)

of the micrometer and possibly damage it permanently. You should also remember that too light a pressure on the part by the measuring faces can yield an unreliable result.

When used to measure parts in a machine, the micrometer should be held in both hands whenever possible. This is especially true when measuring cylindrical workpieces (Figure 10). Holding the instrument in one hand does not permit sufficient control for reliable readings. Furthermore, cylindrical workpieces should be checked at least twice with measurements made 90 degrees apart. This is to check for an out-

of-round condition (Figure 11). When critical dimensions are measured, that is, any dimension where a very small amount of tolerance is acceptable, make at least two consecutive measurements. Both readings should indicate identical results. If two identical readings cannot be determined, then the actual size of the part cannot be stated reliably. All critical measurements should be made at a temperature of 68°F (20°C). A workpiece warmer than this temperature will be larger because of heat expansion.

Outside English measure micrometers usually have a measuring range of one inch. They are identified by size as to the largest dimensions they measure. A two inch micrometer will measure from one to two inches. A three inch micrometer will measure from two to three inches. The capacity of the tool is increased by increasing the size of the frame. Typical outside micrometers range in capacity from 0 to 168 in. It requires a great deal more skill to get consistent measurements with large capacity micrometers.

READING VERNIER INCH MICROMETERS

When measurements must be made to a discrimination greater than .001 in., a standard micrometer is not sufficient. With a **vernier micrometer,** readings can be made to a **ten-thousandth part of an inch** (.0001

in.). This kind of micrometer is commonly known as a "tenth mike." A vernier scale is part of the sleeve graduations. The vernier scale consists of 10 lines parallel to the index line and located above it (Figure 12).

If the 10 spaces on the vernier scale were compared to the spacing of the thimble graduations, the 10 vernier spacings would correspond to 9 spacings on the thimble. Therefore, the vernier scale spacing must be smaller than the thimble spacing. That is, in fact, precisely the case. Since 10 vernier spacings compare to 9 thimble spacings, the vernier spacing is $\frac{1}{10}$ smaller than the thimble space. We know that the thimble graduations correspond to .001 in. (one-thousandth). Each vernier spacing must then be equal to $\frac{1}{10}$ of .001 in., or .0001 in. (one ten-thousandth). Thus, according to the principle of the vernier, each thousandth of the thimble is subdivided into 10 parts. This permits the vernier micrometer to discriminate to .0001 in.

To read a vernier micrometer, first read to the nearest thousandth as on a standard micrometer.

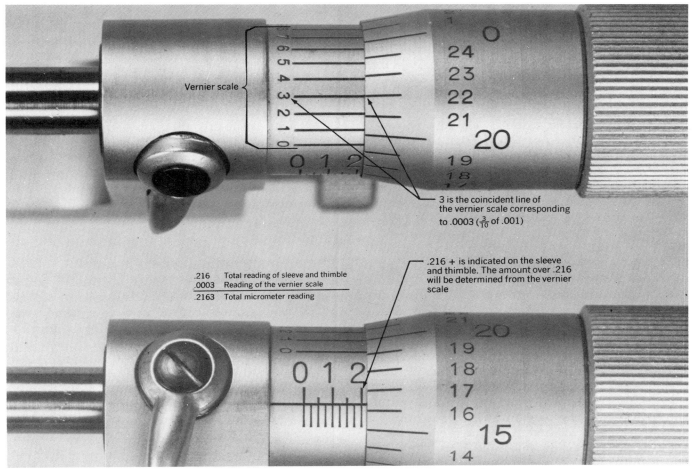

3 is the coincident line of the vernier scale corresponding to .0003 ($\frac{3}{10}$ of .001)

.216 + is indicated on the sleeve and thimble. The amount over .216 will be determined from the vernier scale

.216	Total reading of sleeve and thimble
.0003	Reading of the vernier scale
.2163	Total micrometer reading

Figure 12. Inch vernier micrometer reading of .2163 in. *(Machine Tools and Machining Practices)*

Then, find the line on the vernier scale that coincides with a graduation on the thimble. The value of this coincident vernier scale line is the value in ten-thousandths, which must be added to the thousandths reading thus making up the total reading. **Remember to add the value of the vernier scale line and not the number of the matching thimble line.**

In the lower view (Figure 12), a micrometer reading of slightly more than .216 in. is indicated. In the top view, on the vernier scale, the line numbered 3 is in alignment with the line on the thimble. This indicates that .0003 (three ten-thousandths) must be added to the .216 in. for a total reading of .2163. This number is read "two hundred sixteen-thousandths and three-tenths."

RELIABILITY AND EXPECTATION OF ACCURACY IN MICROMETER INSTRUMENTS

The standard inch micrometer will discriminate to .001 ($\frac{1}{1000}$) of an inch. In its vernier form, the discrimination is increased to .0001 ($\frac{1}{10,000}$) of an inch. The common metric micrometer discriminates to .01 ($\frac{1}{100}$) of a millimeter. The same rules apply to micrometers as apply to all measuring instruments. The tool should not be used beyond its discrimination. A standard micrometer with .001 discrimination should not be used in an attempt to ascertain measurements beyond that point. In order to measure to a discrimination of .0001 with the vernier micrometer, certain special conditions of accuracy must be met.

You must exercise cautious judgment when attempting to measure to a tenth of a thousandth using a vernier micrometer. The tenth measure should be carried out under controlled conditions if truly reliable results are to be obtained. The finish of the workpiece must be extremely smooth. Contact pressure of the measuring faces must be very consistent. The workpiece and instrument must be temperature stabilized. Furthermore, the micrometer must be carefully calibrated against a known standard. Only under these conditions can true reliability be realized.

The micrometer has increased reliability over the vernier. One reason for this is readability of the instruments. The .001 graduations that dictate the maximum discrimination of the micrometer are placed on the circumference of the thimble. The distance between the marks is therefore increased, making them easier to see.

The micrometer will yield very reliable results to .001 discrimination if the instrument is properly cared for, properly calibrated, and correct procedure for use is followed. **Calibration** is the process by which any measuring instrument is compared to a known

standard. If the tool deviates from the standard, it may then be adjusted to conformity. This is an additional advantage of the micrometer over the vernier. The micrometer must be periodically calibrated if reliable results are to be obtained.

Can a micrometer measure reliably to within .001? The answer is "no" for the standard micrometer since it cannot discriminate to smaller units than .001 in. You cannot be sure if it is "right on" when the thimble lines do not quite match up with the index line, for example. The answer is "yes" for the vernier micrometer, but only under controlled conditions. What then, is an acceptable expectation of accuracy that will yield maximum reliability? This is dependent to some degree on the tolerance specified and can be summarized in the following table.

Tolerance Specified	Acceptability of the Standard Micrometer	Acceptability of the Vernier Micrometer
±.0001	No	Yes (under controlled conditions)
±.001	Yes	Yes (vernier will not be required)

For a specified tolerance within .001 in., the vernier micrometer should be used. Plus or minus .001 in. is a total range of .002 in. or within the capability of the standard micrometers.

The micrometer is indeed a marvelous example of precision manufacturing. These rugged tools are produced in quantity with each one conforming to equally high standards. Micrometer instruments, in all their many forms, constitute one of the fundamental measuring instruments for the machinist.

READING METRIC MICROMETERS

The **metric micrometer** (Figure 13) has a spindle thread with a .5 mm lead. This means that the spindle will move .5 mm when the thimble is turned one complete revolution. Two revolutions of the thimble will advance the spindle one millimeter. In precision machining, metric dimensions are usually expressed in terms of .01 ($\frac{1}{100}$) of a millimeter. On the metric micrometer the thimble is graduated into 50 equal divisions with every fifth division numbered (Figure 14). If one revolution of the thimble is .5 mm, then each division on the thimble is equal to .5 mm divided by 50 or .01 mm. The sleeve of the metric micrometer is divided into 25 main divisions above the index line

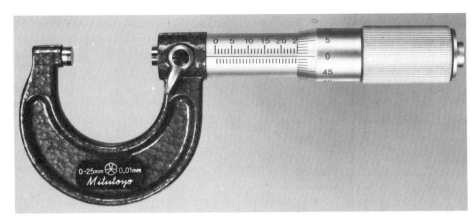

Figure 13. Metric micrometer. (Lane Community College)

Figure 14. Graduations on the metric micrometer. *(Machine Tools and Machining Practices)*

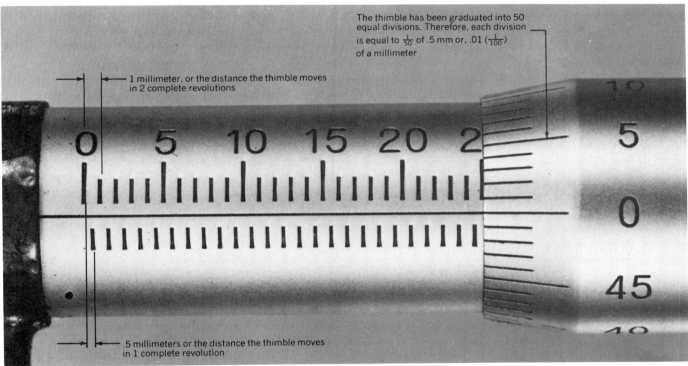

The thimble has been graduated into 50 equal divisions. Therefore, each division is equal to $\frac{1}{50}$ of .5 mm or, .01 ($\frac{1}{100}$) of a millimeter

1 millimeter, or the distance the thimble moves in 2 complete revolutions

.5 millimeters or the distance the thimble moves in 1 complete revolution

with every fifth division numbered. These are whole millimeter graduations. Below the index line are graduations which fall halfway between the divisions above the line. The lower graduations represent half or .5 mm values. The thimble edge (Figure 15) leaves the 12 mm line exposed with no .5 mm line showing. The thimble reading is 32, which is .32 mm. Adding the two figures results in a total of 12.32 mm.

The 15 mm mark (Figure 16) is exposed on the sleeve plus a .5 mm graduation below the index line. The thimble reads 20 or .20 mm. Adding these three values, 15.00 + .50 + .20, results in a total of 15.70 mm.

Any metric micrometer should receive the same care discussed in the section on outside micrometers.

Combination Metric/Inch or Inch/Metric Micrometer

The combination micrometer is designed for dual system use in metric and inch measurement. The tool has a digital reading scale for one system while the other system is read from the sleeve and thimble.

Direct Reading Micrometer

The direct reading micrometer, which may also be known as a high precision micrometer, reads directly to .0001 ($\frac{1}{10,000}$) of an inch.

USING INSIDE MICROMETERS

Inside micrometers are equipped with the same grad-

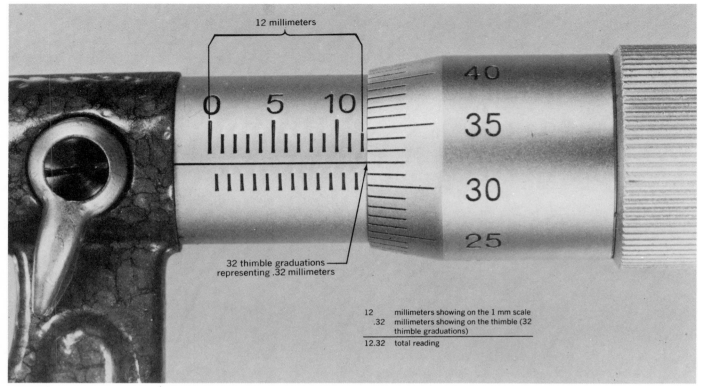

12   millimeters showing on the 1 mm scale
.32   millimeters showing on the thimble (32
      thimble graduations)
_____
12.32 total reading

Figure 15. Metric micrometer reading of 12.32 millimeters. *(Machine Tools and Machining Practices)*

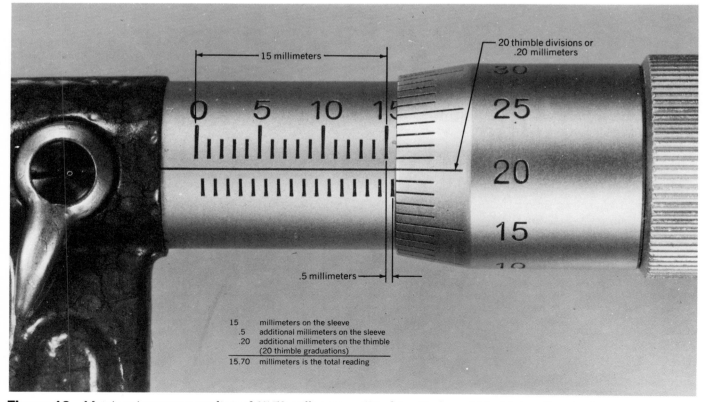

15   millimeters on the sleeve
.5   additional millimeters on the sleeve
.20   additional millimeters on the thimble
      (20 thimble graduations)
_____
15.70 millimeters is the total reading

Figure 16. Metric micrometer reading of 15.70 millimeters. *(Machine Tools and Machining Practices)*

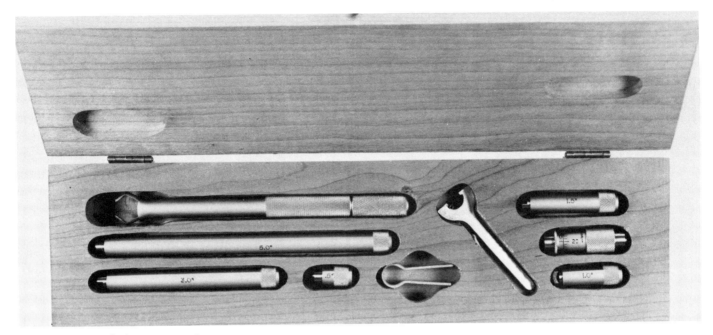

Figure 17. Tubular type inside micrometer set.

uations as outside micrometers. Inside inch micrometers discriminate to .001 in. and have a measuring capacity ranging from 1.5 to 20 in. or more. A typical **tubular type** inside micrometer set (Figure 17) consists of the **micrometer head** with detachable **hardened anvils** and several **tubular measuring rods** with **hardened contact tips.** The lengths of these rods differ in increments of .5 in. to match the measuring capacity of the micrometer head, which in this case is .5 in. A handle is provided to hold the instrument into places where holding the instrument directly would be difficult. Another common type of inside micrometer comes equipped with relatively small diameter solid rods that differ in inch increments, even though the head movement is .5 in. In this case, a .5 in. spacing collar is provided. This can be slipped over the base of the rod before it is inserted into the measuring head.

Inside micrometer heads have a range of .250, .500, 1.000, or 2.000 in. depending on the total capacity of the set. For example, an inside micrometer set with a head range of .500 in. will be able to measure from 1.500 to 12.500 in.

The measuring range of the inside micrometer is changed by attaching the extension rods. Extension rods may be solid or tubular. Tubular rods are lighter in weight and are often found in large range inside micrometer sets. Tubular rods are also more rigid. It is very important that all parts be extremely clean when changing extension rods (Figure 18). Even small

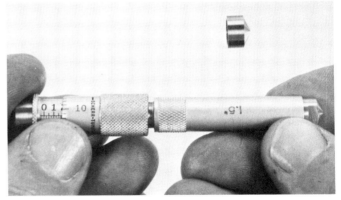

Figure 18. Attaching 1.5 in. extension rod to inside micrometer head.

dust particles can affect the accuracy of the instrument.

When making internal measurements, set one end of the inside micrometer against one side of the hole to be measured (Figure 19). An inside micrometer should not be held in the hands for extended periods, as the resultant heat may affect the accuracy of the instrument. A handle is usually provided, which eliminates the need to hold the instrument and also facilitates insertion of the micrometer into a bore or hole (Figure 20). One end of the micrometer will become the center of the arcing movement used when finding the centerline of the hole to be measured. The micrometer should then be adjusted to the size of the hole. When the correct hole size is reached,

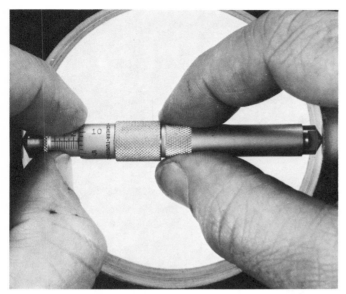

Figure 19. Placing the inside micrometer in the bore to be measured.

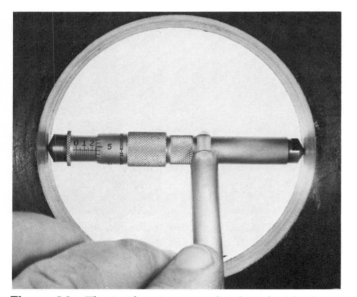

Figure 20. The inside micrometer head used with a handle.

there should be a very light drag between the measuring tip and the work when the tip is moved through the centerline of the hole. The size of the hole is determined by adding the reading of the micrometer head, the length of the extension rod, and the length of the spacing collar, if one was used. Read the micrometer **while it is still in place if possible.** If the instrument must be removed to be read, the correct range can be determined by checking with a rule (Figure 21). A skilled craftsman will usually use an accurate outside micrometer to verify a reading taken with an inside micrometer. In this case, the inside micrometer becomes an easily adjustable transfer measuring tool (Figure 22). Take at least two readings 90 degrees apart to obtain the size of a hole or bore. The readings should be identical unless the bore is out-of-round. Inside micrometers do not have a spindle lock. Therefore, to prevent the spindle from turning while establishing the correct feel, the adjusting nut should be maintained slightly tighter than normal.

USING DEPTH MICROMETERS

A depth micrometer is a tool that is used to precisely measure depths of holes, grooves, shoulders, and recesses. As other micrometer instruments, it will discriminate to .001 in. Depth micrometers usually come as a set with interchangeable rods to accommodate different depth measurements (Figure 23). The basic parts of the depth micrometer are the **base, sleeve, thimble, extension rod, thimble cap,** and frequently a **ratchet stop.** The bases of a depth micrometer can be of various widths. Generally the wider bases are more stable, but in many instances, space limitations dictate the use of narrower bases. Some depth micrometers are made with only a half base for measurements in confined spaces.

The extension rods are installed or removed by holding the thimble and unscrewing the thimble cap. Make sure that the seat between the thimble cap and rod adjusting nuts is clean before reassembling the

Figure 21. Confirming inside micrometer range using a rule.

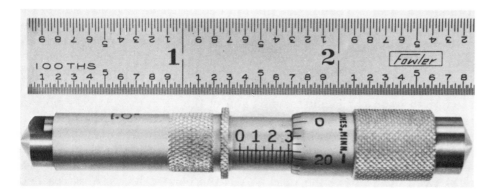

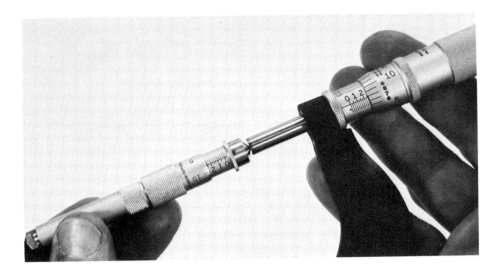

Figure 22. Checking the inside micrometer with an outside micrometer.

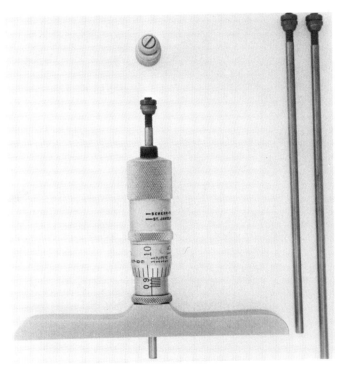

Figure 23. Depth micrometer set.

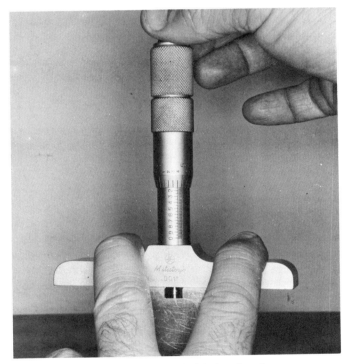

Figure 24. Proper way to hold the depth micrometer.

micrometer. Do not overtighten when replacing the thimble cap. Furthermore, do not attempt to adjust the rod length by turning the adjusting nuts. These rods are factory adjusted and matched as a set. The measuring rods from a specific depth micrometer set should always be kept with that set. Since these rods are factory adjusted and matched to a specific instrument, transposing measuring rods from set to set will usually result in incorrect measurements.

When making depth measurements, it is very important that the micrometer base has a smooth and flat surface on which to rest. Furthermore, sufficient pressure must be applied to keep the base in contact with the reference surface. When a depth micrometer is used without a ratchet, a slip clutch effect can be produced by letting the thimble slip while turning it between the thumb and index finger (Figure 24).

READING INCH DEPTH MICROMETERS

When a comparison is made between the sleeve of an outside micrometer and the sleeve of a depth mi-

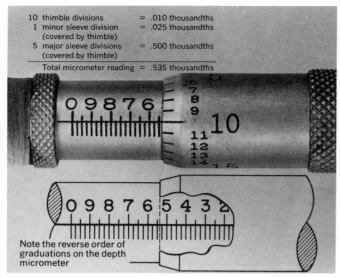

10 thimble divisions	= .010 thousandths
1 minor sleeve division	= .025 thousandths
(covered by thimble)	
5 major sleeve divisions	= .500 thousandths
(covered by thimble)	
Total micrometer reading	= .535 thousandths

Note the reverse order of
graduations on the depth
micrometer

Figure 25. Sleeve graduations on the depth micrometer are numbered in the opposite direction as compared to the outside micrometer. *(Machine Tools and Machining Practices)*

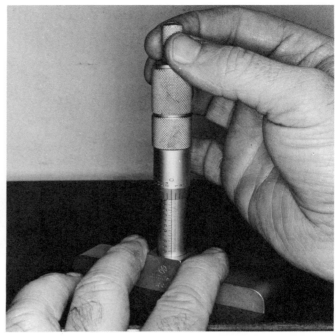

Figure 26. Checking a depth micrometer for zero adjustment using the surface plate as a reference surface.

crometer, note that the graduations are numbered in the opposite direction (Figure 25). When reading a depth micrometer, the distance to be measured is the value covered by the thimble. Consider the reading shown (Figure 25). The thimble edge is between the number 5 and 6. This indicates a value of at least .500 in. on the sleeve major divisions. The thimble also covers the first minor division on the sleeve. This has a value of .025 in. The value on the thimble circumference indicates .010 in. Adding these three values results in a total of .535 in., or the amount of extension of the rod from the base.

A depth micrometer should be tested for accuracy before it is used. When the 0 to 1 in. rod is used, retract the measuring rod into the base. Clean the base and contact surface of the rod. Hold the micrometer base firmly against a flat highly finished surface, such as a surface plate, and advance the rod until it contacts the reference surface (Figure 26). If the micrometer is properly adjusted, it should read zero. When testing for accuracy with the one inch extension rod, set the base of the micrometer on a one inch gage block and measure to the reference surface (Figure 27). Other extension rods can be tested in a like manner.

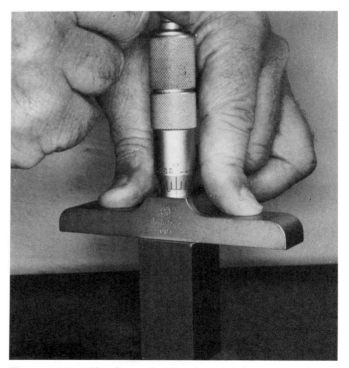

Figure 27. Checking the depth micrometer calibration at the 1.000 in. position using the 0–1 in. rod and a 1 in. square or "Hoke type" gage block.

SELF-TEST

1. Why should a micrometer be kept clean and protected?
2. Why should a micrometer be stored with the spindle out of contact with the anvil?
3. Why are the measuring faces of the micrometer cleaned before measuring?
4. How precise is the standard inch micrometer?
5. What affects the accuracy of a micrometer?
6. What is the difference between the sleeve and thimble?
7. Why should a micrometer be read while it is still in contact with the object to be measured?
8. How often should an object be measured to verify its actual size?

9. What effect has an increase in temperature on the size of a part?
10. What is the purpose of the friction thimble or ratchet stop on the micrometer?

Exercise

Read and record the five outside micrometer readings in Figures 28a to 28e.

Figure 28a. _____   Figure 28d. _____
Figure 28b. _____   Figure 28e. _____
Figure 28c. _____

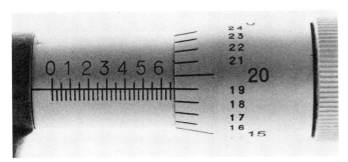

Figure 28a

Figure 28d

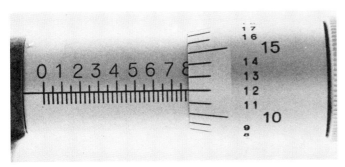

Figure 28b

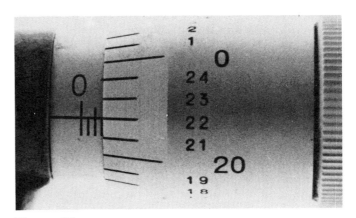

Figure 28e

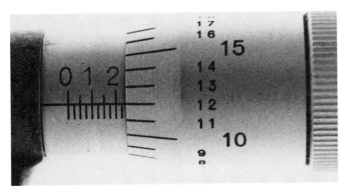

Figure 28c

SELF-TEST

Read and record the five vernier micrometer readings in Figures 29a to 29e.

Figure 29a. _____   Figure 29d. _____

Figure 29b. _____   Figure 29e. _____

Figure 29c. _____

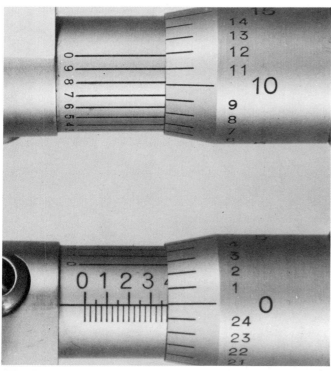

Figure 29a

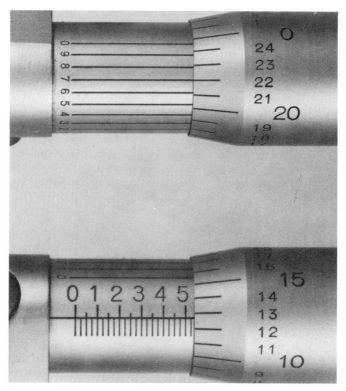

Figure 29b

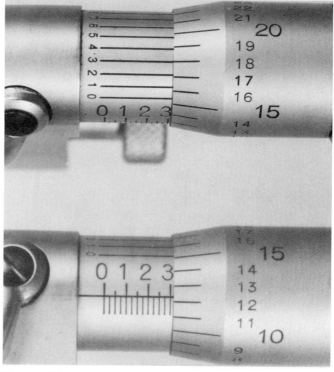

Figure 29c

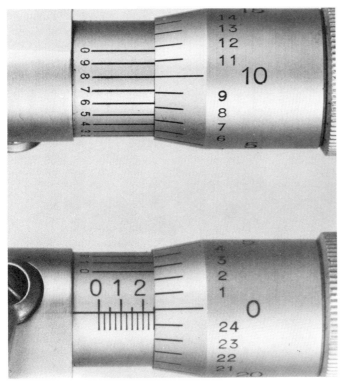

Figure 29d

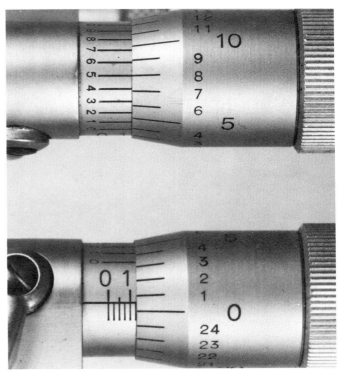

Figure 29e

SELF-TEST

Read and record the five metric micrometer readings in Figures 30a to 30e.

Figure 30a. _____   Figure 30d. _____
Figure 30b. _____   Figure 30e. _____
Figure 30c. _____

Exercise

Obtain a metric micrometer from your instructor and practice measuring objects around your laboratory.

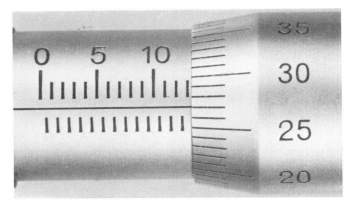

Figure 30b

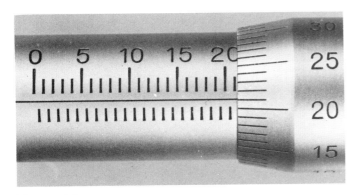

Figure 30a

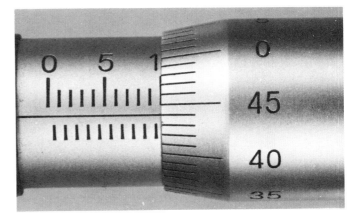

Figure 30c

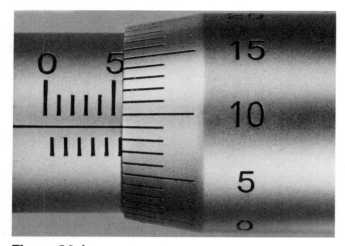

Figure 30d

Figure 30e

SELF-TEST

Read and record the five inside micrometer readings (Figures 31a to 31e). Micrometer head is 1.500 in. when zeroed.

Figure 31a. _____
Figure 31b. _____
Figure 31c. _____ (.5 in. extension)
Figure 31d. _____ (1.0 in. extension)
Figure 31e. _____ (1.0 in. extension)

Exercise

Obtain an inside micrometer set from your instructor and practice using the instrument on objects around your laboratory. Measure examples such as lathe spindle holes, bushings, bores of roller bearings, hydraulic cylinders, and tubing.

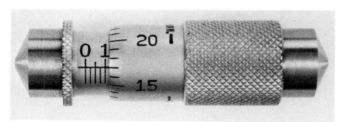

Figure 31a

Figure 31b

Figure 31c .5 in. extension.

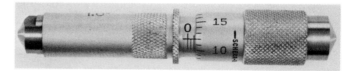

Figure 31d 1.0 in. extension.

Figure 31e 1.0 in. extension.

SELF-TEST

Read and record the five depth micrometer readings in Figures 32a to 32e.

Figure 32a. _____

Figure 32b. _____

Figure 32c. _____

Figure 32d. _____

Figure 32e. _____

Exercise

Obtain a depth micrometer from your instructor and practice measuring objects in your laboratory.

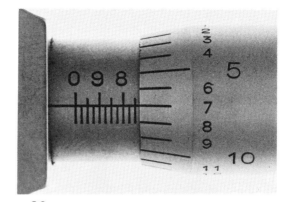

Figure 32c

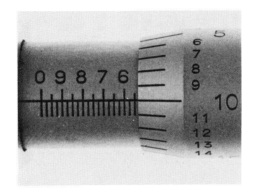

Figure 32a

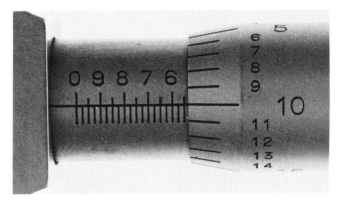

Figure 32d

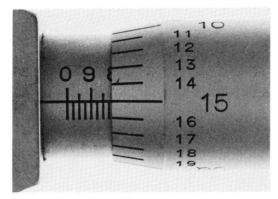

Figure 32b

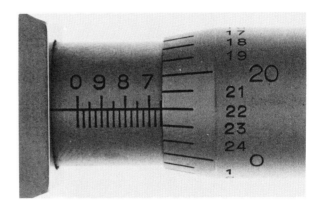

Figure 32e

UNIT 5 USING COMPARISON AND ANGULAR MEASURING INSTRUMENTS

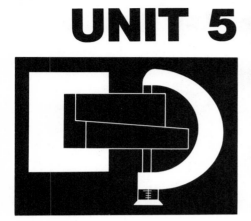

As a machinist, you will use a large number of measuring instruments that have no capacity within themselves to show a measurement. These tools will be used in comparison measurement applications where they are compared to a known standard or used in conjunction with an instrument that has the capability to show a measurement. In this unit you are introduced to the principles of comparison measurement, the common tools of comparison measurement and their applications, and to the use of some mechanical dial and angular measuring instruments.

OBJECTIVES

After completing this unit, you should be able to:
1. Define comparison measurement.
2. Identify common comparison measuring tools.
3. Given a measuring situation, select the proper comparison tool for the measuring requirement.
4. Identify common angular measuring tools.
5. Read and record angular measurements using a vernier protractor.

MEASUREMENT BY COMPARISON

All of us, at some time, were probably involved in constructing something in which we used no measuring instruments of any kind. For example, suppose that you had to build some wooden shelves. You have the required lumber available with all boards longer than the shelf spaces. You hold a board to the shelf space and mark the required length for cutting. By this procedure, you have **compared** the length of the board (**the unknown length**) to the shelf space (**the known length or standard**). After cutting the first board to the marked length, it is then used to determine the lengths of the remaining shelves. The board, in itself, has no capacity to show a measurement. However, in this case, it became a measuring instrument.

A great deal of comparison measurement often involves the following steps.

1. A device that has no capacity to show measurement is used to establish and represent an unknown distance.
2. This representation of the unknown is then **transferred** to an instrument that has the capability to show a measurement.

This is commonly known as **transfer measurement**. In the example of cutting shelf boards, the shelf space was transferred to the first board, and then the length of the first board was transferred to the remaining boards.

Transfer of measurements may involve some reduction in reliability. This factor must be kept in mind when using comparison tools requiring that a transfer be made. Remember that an instrument with the capability to show measurement directly is always best.

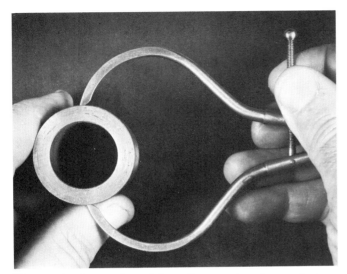

Figure 1. Set one leg of the caliper against the workpiece. *(Machine Tools and Machining Practices)*

Direct reading instruments should be used whenever possible in any situation. Measurements requiring a transfer must be accomplished with proper caution if reliability is to be maintained.

COMMON COMPARISON MEASURING TOOLS AND THEIR APPLICATIONS

Spring Calipers

The spring caliper is a very common comparison measuring tool for rough measurements of inside and outside dimensions. To use a spring caliper, set one jaw on the workpiece (Figure 1). Use this point as a pivot and swing the other caliper leg back and forth over the largest point on the diameter. At the same time, adjust the leg spacing (Figures 2 and 3). When the correct feel is obtained, remove the caliper and compare it to a steel rule to determine the reading (Figure 4). The inside spring caliper can be used in a similar manner (Figure 5). Spring calipers are sometimes used to transfer measurements to micrometers where it is difficult to obtain a reading in other ways. An example of this is measuring a bore when the boring bar cannot be removed, as in line boring.

The use of spring calipers is limited. It has been stated by some in the machining business that they can use a spring caliper to measure to .002 or even .001 of an inch. This is of questionable reliability. In modern machining technology, there is little room for crude measurement practice. The use of the spring caliper is fading, and it has been replaced by measur-

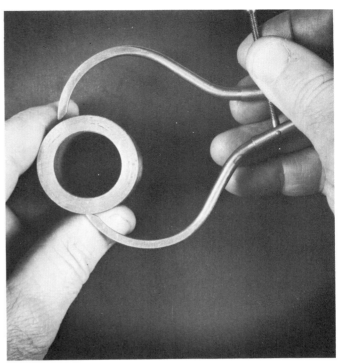

Figure 2. Swing the other leg past center. *(Machine Tools and Machining Practices)*

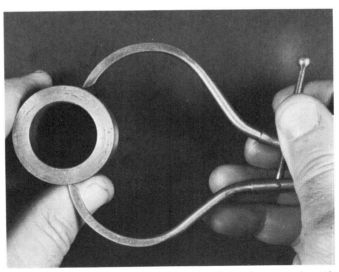

Figure 3. As the leg passes center, adjust the spread until proper feel is obtained. *(Machine Tools and Machining Practices)*

ing instruments of much higher reliability. The spring caliper should be used only for the roughest of measurements.

Telescoping Gage

The telescoping gage is also a very common compari-

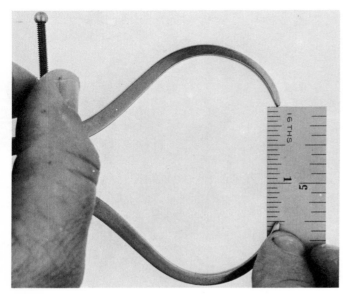

Figure 4. Comparing the spring caliper to a steel rule. *(Machine Tools and Machining Practices)*

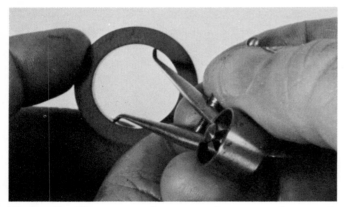

Figure 5. Using an inside spring caliper. *(Machine Tools and Machining Practices)*

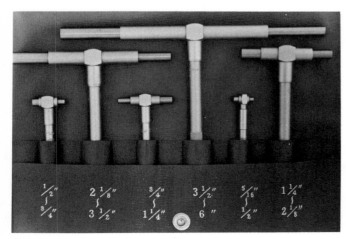

Figure 6. Set of telescoping gages.

son measuring instrument. Telescoping gages are widely used in the machine shop, and they can accomplish a variety of measuring requirements.

Telescoping gages generally come in a set of six gages (Figure 6). The range of the set is usually $\frac{5}{16}$ to 6 in. (8 to 150 mm). The gage consists of two telescoping plungers (some types have only one) with a handle and locking screw. The gage is inserted into a bore or slot, and the plungers are permitted to extend, thus conforming to the size of the feature. The gage is then removed and transferred to a micrometer where the reading is determined. The telescoping gage can be a reliable and versatile tool if proper procedure is used in its application.

Procedure for Using the Telescoping Gage

1. Select the proper gage for the desired measurement range.
2. Insert the gage into the bore to be measured and release the handle lock screw (Figure 7). Rock the gage sideways to insure that you are measuring at the full diameter (Figure 8). This is especially important in large diameter bores.
3. Lightly tighten the locking screw.
4. Use a downward or upward motion, moving the gage through the diameter of the bore. The plungers will be pushed in, thus conforming to the bore diameter (Figure 9). Tighten the locking screw firmly.
5. Remove the gage and measure with an outside micrometer (Figure 10). Place the gage between micrometer spindle and anvil. Try to determine the same feel on the gage with the micrometer as you felt while the gage was in the bore. Excessive pres-

Figure 7. Inserting the telescoping gage into the bore.

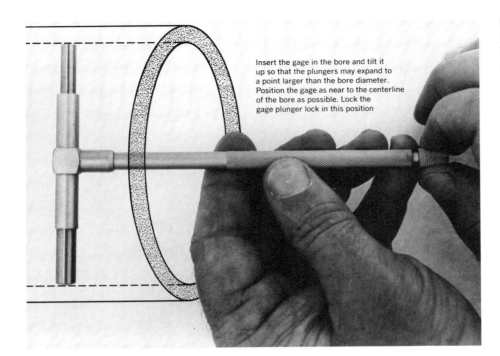

Insert the gage in the bore and tilt it up so that the plungers may expand to a point larger than the bore diameter. Position the gage as near to the centerline of the bore as possible. Lock the gage plunger lock in this position

Figure 8. Release the lock and let the plungers expand larger than the bore. *(Machine Tools and Machining Practices)*

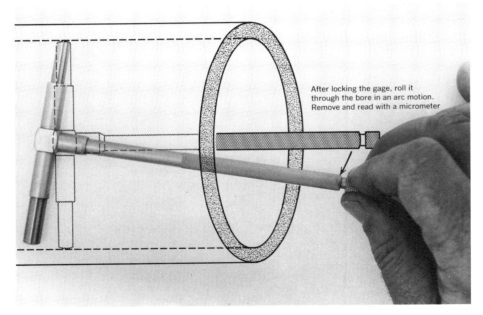

After locking the gage, roll it through the bore in an arc motion. Remove and read with a micrometer

Figure 9. Tighten the lock and roll the gage through the bore. *(Machine Tools and Machining Practices)*

sure with the micrometer will depress the gage plungers and cause an incorrect reading.

6. Take at least two readings or more with the telescoping gage in order to verify reliability. If the readings do not agree, repeat procedure steps 2 to 6.

Small Hole Gages

Small hole gages, like telescoping gages, come in sets with a range of $\frac{1}{8}$ to $\frac{1}{2}$ in. (4 to 12 mm). One type of small hole gage consists of a split ball that is connected to a handle (Figure 11). A tapered rod is drawn between the split ball halves causing them to extend and contact the surface to be measured (Figure 12). The split ball small hole gage has a flattened end so that a shallow hole or slot may be measured. After the gage has been expanded in the feature to be measured, it should be moved back and forth and rotated

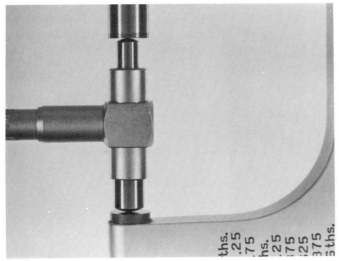

Figure 10. Checking the telescoping gage with an outside micrometer.

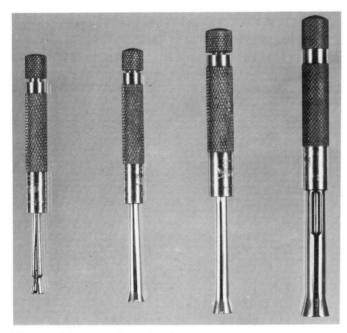

Figure 11. Set of small hole gages.

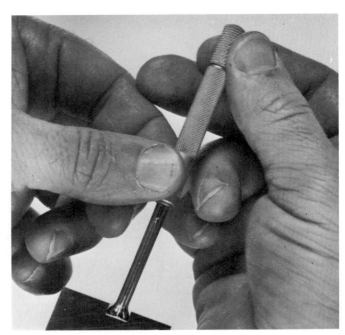

Figure 12. Insert the small hole gage in the slot to be measured. Rotate to locate large diameter of gage.

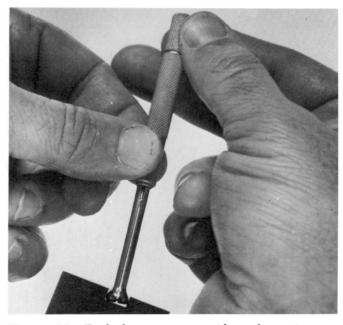

Figure 13. Rock the gage to one side to determine the proper feel.

to determine the proper feel (Figures 13 and 14). The gage is then removed and measured with an outside micrometer (Figure 15).

A second type of small hole gage consists of two small balls that can be moved out to contact the surface to be measured. This type of gage is available in a set ranging from $\frac{1}{16}$ to $\frac{1}{2}$ in. (1.5 to 12 mm) (Figure 16). Once again, the proper feel must be obtained when using this type of small hole gage (Figure 17).

After the gage is set, it is removed and measured with a micrometer.

Another type of small hole gage is the direct reading type (Figure 18). This is not a comparison instru-

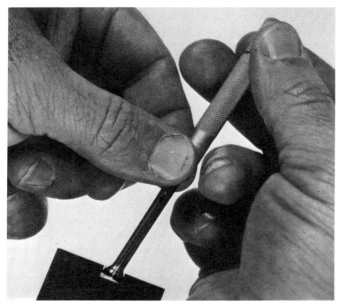

Figure 14. Rock the gage to the other side to determine proper feel.

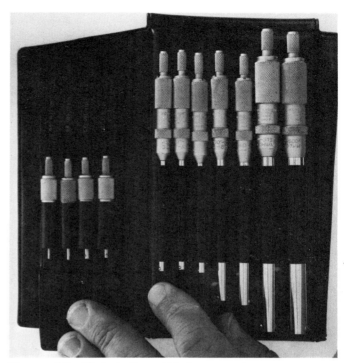

Figure 16. Twin ball small hole gage set.

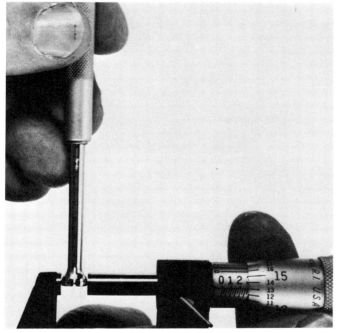

Figure 15. Withdraw the gage and measure with an outside micrometer. Rotate to locate large diameter of gage.

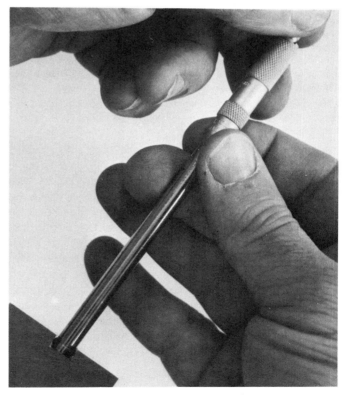

Figure 17. Using the twin ball small hole gage.

ment as it has the capability to display the measurement directly. The gage will discriminate to .001 of an inch and is read from the dial in the handle (Figure 19).

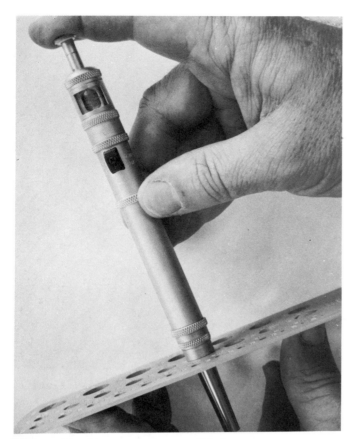

Figure 18. Direct reading small hole gage.

Figure 19. Reading the small hole gage.

Adjustable Parallels

For the purpose of measuring slots, grooves, and keyways, the adjustable parallel may be used. Adjustable parallels are available in sets ranging from about $\frac{3}{8}$ to $1\frac{1}{2}$ in. (10 to 38 mm). They are precision ground for accuracy.

The typical adjustable parallel consists of two parts that slide together on an angle. Adjusting screws are provided so that clearance in the slide may be adjusted or the parallel locked after setting for a measurement. As the halves of the parallel slide, the width increases or decreases depending on direction. The parallel is placed in the groove or slot to be measured and expanded until the parallel edges conform to the width to be measured. The parallel is then locked with a small screwdriver and measured with a micrometer (Figure 20). If possible, an adjustable parallel should be left in place while being measured.

Radius Gages

The typical radius gage set ranges in size from $\frac{1}{32}$ to $\frac{1}{2}$ in. (.8 to 12 mm). Larger radius gages are also available. The gage can be used to check the radii of grooves and external or internal fillets (rounded corners). Radius gages may be separate (Figure 21) or the full set may be contained in a convenient holder (Figure 22).

Thickness Gages

The thickness gage (Figure 23) is often called a feeler gage. It is probably best known for its various automotive applications. However, a machinist may use a thickness gage for such measurements as the thickness of a shim, setting a grinding wheel above a workpiece or determining the height difference of two parts. The thickness gage is not a true comparison measuring instrument, as each leaf is marked as to size. However, it is good practice to check a thickness gage with a micrometer, especially when a number of leaves are

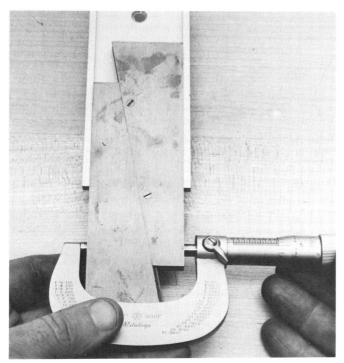

Figure 20. Using adjustable parallels.

Figure 21. Using an individual radius gage. *(Machine Tools and Machining Practices)*

Figure 22. Radius gage set. *(Machine Tools and Machining Practices)*

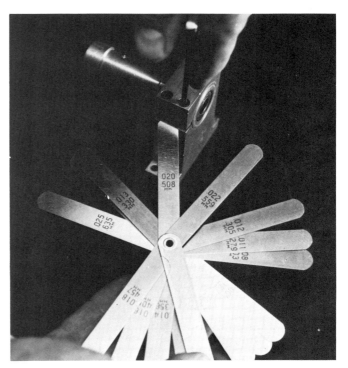

Figure 23. Using a feeler or thickness gage. *(Machine Tools and Machining Practices)*

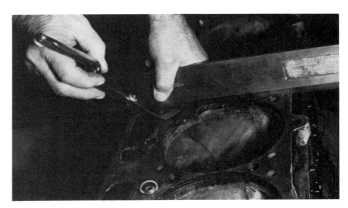

Figure 24. Straightedge and feeler gage being used to check for flatness on an engine block. (Lane Community College)

stacked together. The feeler gage is often used together with a precision straightedge to determine flatness. A common example is in checking an engine head or block for warpage (Figure 24). Another thickness gage, sometimes called a plasti-gage, is used by automotive mechanics to check for crankshaft bearing clearance. The gage is used only once and is placed in the bearing cap which is then tightened in place. When it is removed, the gage has compressed to the amount of clearance and it is then measured with a comparative gage that comes with the plasti-gage.

Planer Gage

The planer gage functions much like an adjustable parallel. Planer gages were originally used to set tool heights on shapers and planers. They can also be used as a comparison measuring tool.

The planer gage may be equipped with a scriber and used in layout. The gage may be set with a micrometer (Figure 25) or in combination with a dial test indicator and gage blocks. In this application, the planer gage is set by using a test indicator set to zero on a gage block (Figure 26). This dimension is then transferred to the planer gage (Figure 27). After the

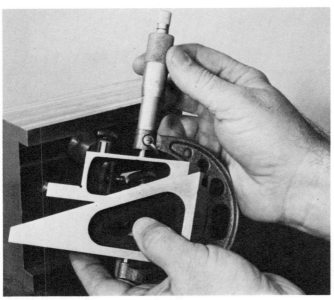

Figure 25. Setting the planer gage with an outside micrometer.

gage has been set, the scriber is attached and the instrument used in a layout application (Figure 28).

Squares

The square is an important and useful tool for the machinist. A square is a comparative measuring instrument in that it compares its own degree of perpendicularity (or squareness) with an unknown degree of perpendicularity on the workpiece. You will use several common types of squares.

Machinist's Combination Square. The combination square (Figure 29) is part of the combination set (Figure 30). The combination set consists of a graduated rule, square head, bevel protractor, and center head. The square head slides on the graduated rule and can be locked at any position (Figure 31). This feature makes the tool useful for layout as the square head can be set according to the rule graduations. The combination square head also has a 45 degree angle along with a spirit level and layout scriber. The combination set is one of the most versatile tools of the machinist.

Indicators

The many types of dial indicators are some of the most valuable and useful tools for the machinist (or mechanic). There are two general types of indicators in general use. These are **dial indicators** (see Figure 32) and **dial test indicators** (see Figure 39). Both types generally take the form of a spring loaded spindle that, when depressed, actuates the hand of an indicating dial. At the initial examination of a dial or test indicator, you will note that the dial face is usually

Figure 26. Setting the dial test indicator to a gage block.

Figure 27. Transferring the measurement to the planer gage.

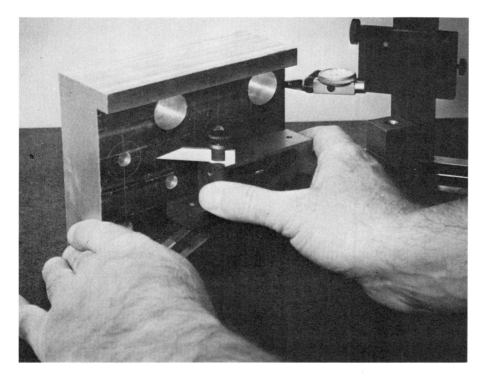

Figure 28. Using the planer gage in layout.

graduated in thousandths of an inch or subdivisions of thousandths. This might lead you to the conclusion that the indicator spindle movement corresponds directly to the amount shown on the indicator face. However, this conclusion is to be arrived at only with the most cautious judgment. **Dial test indicators should not be used to make direct linear measurements.** Reasons for this will be developed in the infor-

mation to follow. Dial indicators can be used to make linear measurements, but only if they are specifically designed to do so and under proper conditions.

Dial Indicators. Dial indicators have discriminations that typically range from .0001 to .001 of an inch. In metric dial indicators, the discriminations typically range from .002 to .01 mm. Indicator ranges

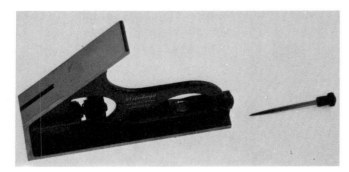

Figure 29. Combination square head with scriber. *(Machine Tools and Machining Practices)*

Figure 30. Machinist's combination set. *(Machine Tools and Machining Practices)*

Figure 31. Using the combination square. *(Machine Tools and Machining Practices)*

Figure 32. Balanced dial indicator.

Figure 34. Dial indicator with .025 in. range and .0001 in. discrimination.

Figure 33. Dial indicator with one inch travel.

(the total reading capacity) of the instrument may commonly range from .003 to 2.000 in., or .2 to 50 mm for metric instruments. On the "balanced" indicator (Figure 32), the face numbering goes both clockwise and counterclockwise from zero. This is convenient for comparator applications where readings above and below zero need to be indicated. The indicator shown has a lever actuated stem. This permits the stem to be retracted away from the workpiece if desired.

The continuous reading indicator (Figure 33) is numbered from zero in one direction. This indicator has a discrimination of .0005 and a total range of one inch. The small center hand counts revolutions of the large hand. Note that the center dial counts each .100 in. of spindle travel. This indicator is also equipped with **tolerance hands** that can be set to mark a desired limit. Many dial indicators are designed for high discrimination and short range (Figure 34). This indicator has a .0001 in. discrimination and a range of .025 in.

The "back plunger" indicator (Figure 35) has the spindle in the back or at right angles to the face. This type of indicator usually has a range of about .200 in. with .001 in. discrimination. It is a very popular

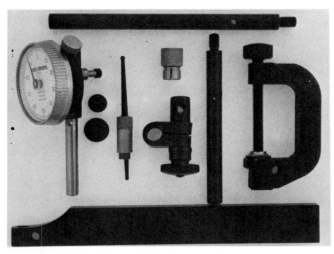

Figure 35. Back plunger indicator with mounting accessories.

Figure 36. Dial indicator tips with holder. (Rank Scherr–Tumico Inc.)

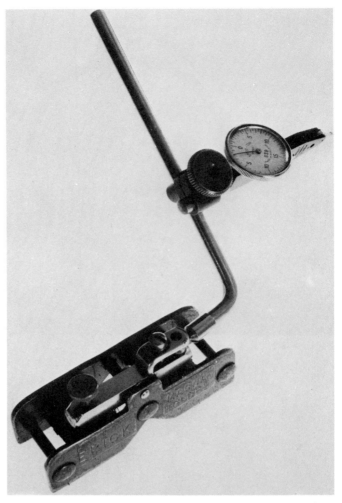

Figure 37. Permanent magnetic indicator base. (*Machine Tools and Machining Practices*)

model for use on a machine tool. The indicator usually comes with a number of mounting accessories.

Indicators are equipped with a **rotating face or bezel.** This feature permits the instrument to be set to zero at any desired place. Many indicators also have a **bezel lock.** Dial indicators may have removable spindle tips, thus permitting use of different shaped tips as required by the specific application (Figure 36).

Care and Use of Indicators. Dial indicators are precision instruments and should be treated accordingly. They **must not be dropped** and should **not be exposed to severe shocks.** Dropping an indicator may bend the spindle and render the instrument useless. Shocks, such as hammering on a workpiece while an indicator is still in contact, may damage the delicate operating mechanism. The spindle should be kept free from dirt and grit. This can cause binding that results in damage and false readings. It is important to **check** indicators **for free travel** before using. When an indicator is not in use, it should be stored carefully with a protective device around the spindle.

One of the problems encountered by indicator users is **indicator mounting.** All indicators must be **mounted solidly** if they are to be reliable. Indicators must be clamped or mounted securely when used on a machine tool. A number of mounting devices are in common use. Some of these have magnetic bases that permit an indicator to be attached at any convenient place on a machine tool. The permanent magnet indicator base (Figure 37) is a useful accessory. This type of indicator base is equipped with an adjust-

Figure 38. Magnetic base indicator holder with on/off magnet. (The L.S. Starrett Company)

ing screw that can be used to set the instrument to zero. Another useful magnetic base has a provision for turning off the magnet by mechanical means (Figure 38). This feature makes for easy locating of the base prior to turning on the magnet. A number of bases making use of flexible link indicator holding arms are also in general use. Often they are not adequately rigid for reliability. In addition to holding an indicator on a magnetic base, it may be clamped to a machine setup by the use of any suitable clamps.

Dial Test Indicators. Dial test indicators frequently have a discrimination of .0005 in. and a range of about .030 in. The test indicator is frequently quite small (Figure 39) so that it can be used to indicate in locations inaccessible to other indicators. The spindle or tip of the test indicator can be swiveled to any desired position. Test indicators are usually equipped with a **movement reversing lever.** This means that the indicator can be actuated by pressure from either side of the tip. The instrument need not be turned around. Test indicators, like dial indicators, have a rotating bezel for zero setting. Dial faces are generally of the balanced design. The same care given to dial indicators should be extended to test indicators.

Figure 39. Dial test indicator.

Potential for Error in Using Dial Indicators. Indicators must be used with appropriate caution if reliable results are to be obtained. The spindle of a dial indicator usually consists of a gear rack that engages a pinion and other gears that drive the indicating hand. In any mechanical device, there is always some clearance between the moving parts. There are also minute errors in the machining of the indicator parts. Because of these, small errors may creep into an indicator reading. This is especially true in long travel indicators. For example, if a one-inch travel indicator with .001 in. discrimination had plus or minus one percent error at full travel, the following condition could exist if the instrument were to be used for a direct measurement.

You wish to determine if a certain part is within the tolerance of .750 ± .003 in. The one-inch travel indicator has the capacity for this; but remember, it is only accurate to plus or minus one percent of full travel. Therefore, .01 × 1.000 in. is equal to ± .010 in. or the total possible error. To calculate the error per thousandth of indicator travel, divide .010 in. by 1000. This is equal to .00001 in., which is the average error per thousandth of indicator travel. This means that at a travel amount of .750 in., the indicator error could be as much as .00001 in. × 750, or ± .0075 in. In a direct measurement of the part, the indicator could read anywhere from .7425 to .7575. As you can see, this is well outside the part tolerance and would hardly be reliable.

The indicator should be used as a comparison

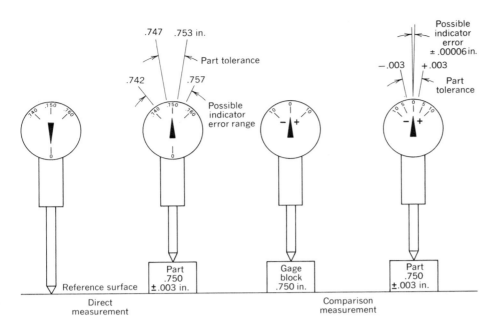

Figure 40. Potential for errors in indicator travel. *(Machine Tools and Machining Practices)*

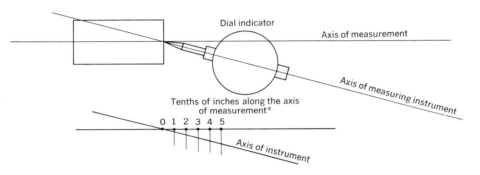

Figure 41. Misalignment of the dial indicator will cause a measuring error. *(Machine Tools and Machining Practices)*

measuring instrument by the following procedure (Figure 40). The indicator is set to zero on a .750 in. gage block. The part to be measured is then placed under the indicator spindle. In this case, the error caused by a large amount of indicator travel is greatly reduced, because the travel is never greater than the greatest deviation of a part from the basic size. The total part tolerance is .006 in. (± .003 in.). Therefore, 6 × .00001 in. error per thousandth is equal to only ± .00006 in. This is well within the part tolerance and, in fact, cannot even be read on a .001 in. discrimination indicator.

Of course, you will not know what the error amounts to on any specific indicator. This can only be determined by a calibration procedure. Furthermore, you would probably not use a long travel indicator in this particular application. A moderate to short travel indicator would be more appropriate. Keep in mind that any indicator may contain some **travel error** and that by using a fraction of that travel, this error can be reduced considerably.

The axis of a linear measurement instrument

must be in line with the axis of measurement. If a dial indicator is misaligned with the axis of measurement, it can be seen that an error in measurement will result (Figure 41). Even if a dial indicator is only being used to check for eccentricity (runout) when adjusting a part in a lathe chuck and the indicator is misaligned to the axis of measurement, a faulty reading will result and correct adjustments will be extremely difficult.

When using dial test indicators, watch for **arc versus chord length errors** (Figure 42) The tip of the test indicator moves through an arc. This distance may be considerably greater than the chord distance of the measurement axis. Dial test indicators should **not** be used to make direct measurements. They should only be applied in comparison applications.

ANGULAR MEASURE

In the metals shop, you will find the need to measure **acute angles, right angles,** and **obtuse angles** (Figure

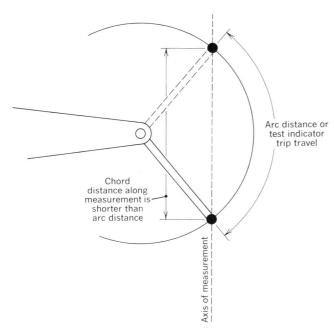

Figure 42. Potential for error in dial test indicator tip movement. *(Machine Tools and Machining Practices)*

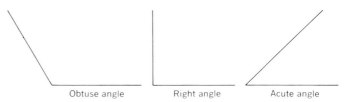

Figure 43. Obtuse, right, and acute angles. *(Machine Tools and Machining Practices)*

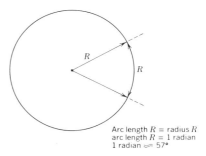

Figure 44. Radian measure. *(Machine Tools and Machining Practices)*

43). Acute angles are less than 90 degrees. Obtuse angles are more than 90 degrees but less than 180 degrees. Ninety degree or right angles are generally measured with squares. However, the amount of angular deviation from perpendicularity may have to be determined. This would require that an angular measuring instrument be used. Straight angles, or those containing 180 degrees, generally fall into the category of straightness or flatness and are measured by other types of instruments.

UNITS OF ANGULAR MEASURE
In the inch system, the unit of angular measure is the **degree.**

Full circle = 360 degrees
1 degree   =   60 minutes of arc ($1° = 60'$)
1 minute   =   60 seconds of arc ($1' = 60''$)

In the metric system, the unit of angular measure is the **radian.** A radian is the length of an arc on the circle circumference that is equal in length to the radius of the circle (Figure 44). Since the circumference of a circle is equal to 2 pi r (radius), there are 2 pi radians in a circle. Converting radians to degrees gives the equivalent:

$$1 \text{ radian} = \frac{360}{2 \text{ pi } r}$$

Assuming a radius of 1 unit:

$$1 \text{ radian} = \frac{360}{2 \text{ pi}}$$
$$= 57° \ 17' \ 44'' \text{ (approximately)}$$

It is unlikely that you will come in contact with much radian measure. All of the common comparison measuring tools you will use read in degrees and fractions of degrees. Metric angles expressed in radian measure can be converted to degrees by the equivalent shown.

REVIEWING ANGLE ARITHMETIC
You may find it necessary to perform angle arithmetic. Use your calculator, if you have one available.

Adding Angles
Angles are added just like any other quantity. One degree contains 60 minutes. One minute contains 60 seconds. Any minute total of 60 or larger must be converted to degrees. Any second total of 60 or larger must be converted to minutes.

EXAMPLE

$3° \ 15' + 7° \ 49' = 10° \ 64'$
Since $64' = 1° \ 4'$, the final result is $11° \ 4'$

EXAMPLE

265° 15′ 52″ + 10° 55′ 17″ + 275° 70′ 69″
Since 69″ = 1′ 9″ and 70′ = 1° 10′, the final result
is 276° 11′ 9″

Subtracting Angles

When subtracting angles where borrowing is necessary, degrees must be converted to minutes and minutes must be converted to seconds.

EXAMPLE

15° − 8° = 7°

EXAMPLE

15° 3′ − 6° 8′ becomes 14° 63′ − 6° 8′ = 8° 55′

EXAMPLE

39° 18′ 13″ − 17° 27′ 52″ becomes
38° 77′ 73″ − 17° 27′ 52″
= 21° 50′ 51″

ANGULAR MEASURING INSTRUMENTS

Plate Protractors

Plate protractors have a discrimination of one degree and are useful in such applications as layout and checking the point angle of a drill (Figure 45).

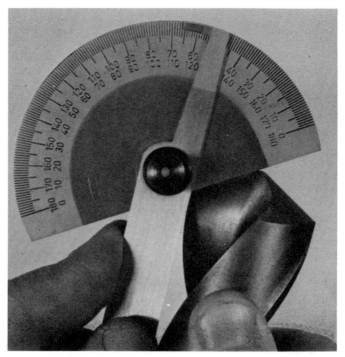

Figure 45. Plate protractor measuring a drill point angle. *(Machine Tools and Machining Practices)*

Figure 46. Using the combination set bevel protractor. (The L.S. Starrett Company)

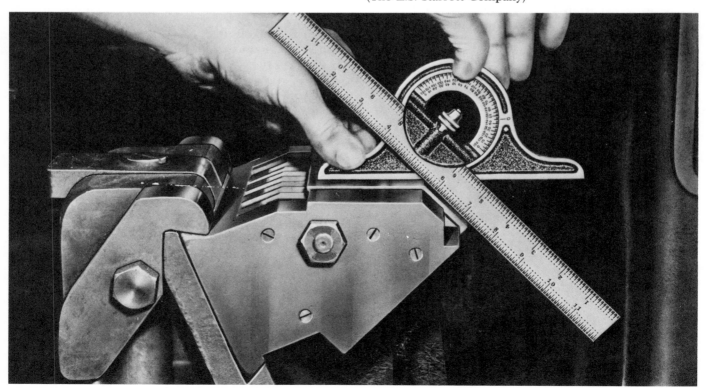

Bevel Protractors

The bevel protractor is part of the machinist's combination set. This protractor can be moved along the rule and locked in any position. The protractor has a flat base permitting it to rest squarely on the workpiece (Figure 46). The combination set protractor has a discrimination of one degree.

Universal Bevel Vernier Protractor

The universal bevel vernier protractor (Figure 47) is equipped with a vernier that permits discrimination to $\frac{1}{12}$ of a degree or 5 minutes of arc.

The instrument can measure an obtuse angle (Figure 48). The acute attachment facilitates the measurement of angles less than 90 degrees (Figure 49). When used in conjunction with a vernier height gage, angle measurements can be made that would be difficult by other means (Figure 50).

Vernier protractors are read like any other instrument employing the vernier. The main scale is divided into whole degrees. These are marked in four

quarters each 0 to 90 degrees. The vernier divides each degree into 12 parts each equal to 5 minutes of arc.

To read the protractor, determine the nearest full degree mark between zero on the main scale and zero on the vernier scale. **Always read the vernier in the same direction as you read the main scale.** De-

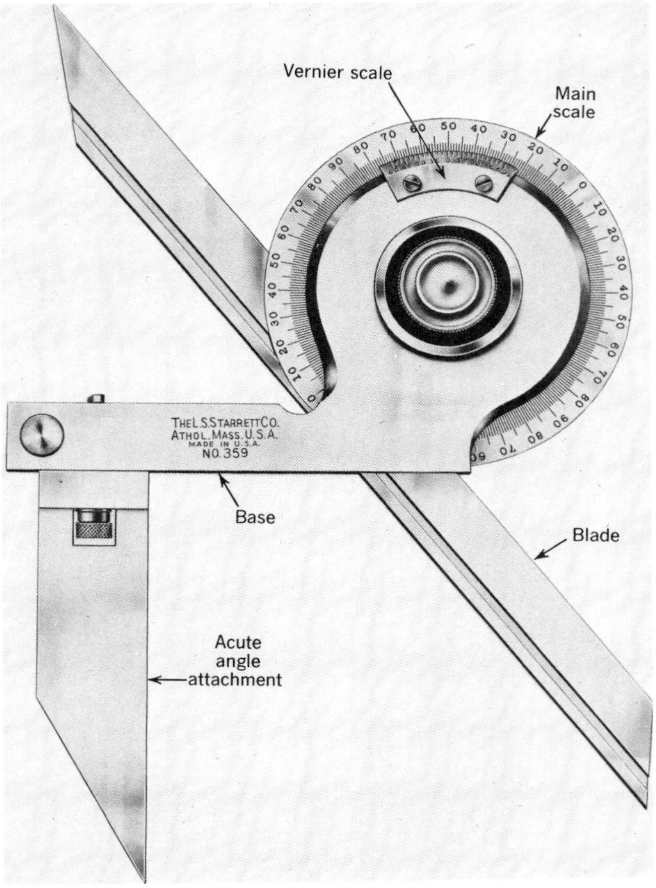

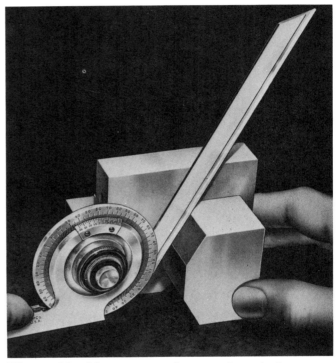

Figure 48. Measuring an obtuse angle with the vernier protractor. (The L.S. Starrett Company)

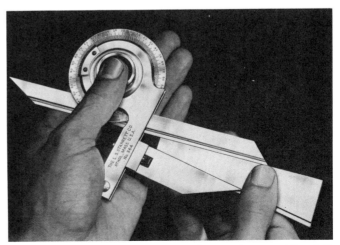

Figure 49. Using the acute angle attachment. (The L.S. Starrett Company)

Figure 47. Parts of the universal bevel vernier protractor. (The L.S. Starrett Company)

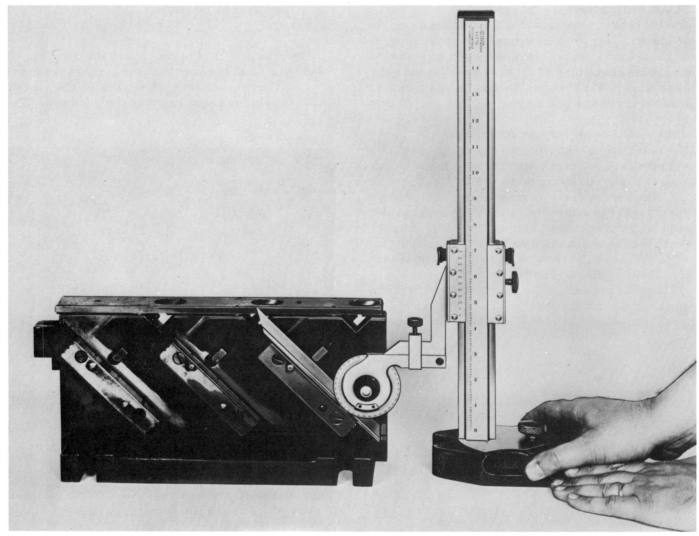

Figure 50. Using the vernier protractor in conjunction with the vernier height gage. (The L.S. Starrett Company)

termine the number of the vernier coincident line. Since each vernier line is equal to 5 minutes, multiply the number of the coincident line by 5. Add this to the main scale reading.

EXAMPLE READING (FIGURE 51)
The protractor shown has a magnifier so that the vernier may be seen more easily.

> Main scale 56°
> Vernier coincident at line 6
> 6 × 5 minutes = 30 minutes
> Total reading is 56° 30′

For convenience, the vernier scale is marked at 0, 30, and 60, indicating minutes.

The vernier bevel protractor can be applied in a variety of angular measuring applications.

Figure 51. Vernier protractor reading of 56 degrees and 30 minutes. *(Machine Tools and Machining Practices)*

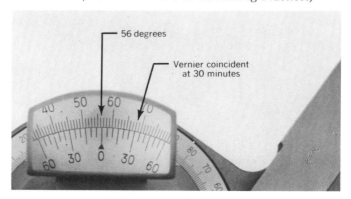

SELF-TEST

1. Define comparison measurement.
2. What can be said of most comparison measuring instruments? Are they the most reliable measuring devices?
3. How can a dial indicator setup cause a measuring error?
4. How can the error in Question 3 be reduced?
5. Match the following measuring situations with the list of comparison measuring tools. Answers may be used more than once.

*A milled slot two inches wide with a tolerance of ±.002 in.	a. Spring caliper
	b. Telescoping gage
	c. Adjustable parallel
*A height transfer measurement.	d. Radius gage
	e. Thickness (feeler) gage
*The shape of a form lathe cutter.	f. Planer gage
	g. Combination square
*Checking an angular setup in a drill press.	h. Dial indicator
	i. Dial test indicator
*The diameter of a 1½ in. hole.	j. Vernier height gage
*Measuring for a shim under a piece of machinery.	

6. Name two angular measuring instruments with one degree of discrimination.

7. What is the discrimination of the universal bevel protractor?
8. Read and record the following vernier protractor readings (Figures 52 to 56).

_____ (Figure 52)   _____ (Figure 53)
_____ (Figure 54)   _____ (Figure 55)
_____ (Figure 56)

Figure 54

Figure 52

Figure 55

Figure 53

Figure 56

SECTION D

SAWING MACHINES

UNIT 1 CUTOFF MACHINES

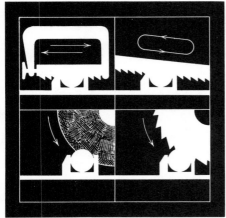

Sawing machines constitute some of the most important machine tools found in the machine shop. Common types of cutoff machines include reciprocating saws, horizontal band saws, universal tilt frame band saws, abrasive saws, and cold saws. This unit will introduce you to safety practices and some of the many uses for these machines.

OBJECTIVES

After completing this unit you should be able to:
1. Identify several types of metal cutting saws used in the shop and name their parts.
2. Explain the uses and advantages of cutoff saws and vertical band machines.
3. Describe correct safety habits when using metal cutting saws.

TYPES OF CUTOFF MACHINES

Reciprocating Saws

The first machine tool that you will probably encounter in the metals shop is a cutoff machine. In the shop, cutoff machines are generally found near the stock supply area. The primary function of the cutoff machine is to reduce mill lengths of bar stock material into lengths suitable for holding in other machine tools. In a large production machine shop where stock is being supplied to many machine tools, the cutoff saw will be constantly busy working on many materials. Material can also be cut with a shear; however, much of the stock used is in the form of a round bar. Mill lengths of round bar are almost always cut by sawing. Shearing is applied to materials in the form of sheets and flat bar; it produces a distorted and rough cut that is not suitable for machine shop use.

The reciprocating saw is often called a power hacksaw. The reciprocating saw is used in many machine shops; however, they are giving way to the band machine. The reciprocating saw is built much like the metal cutting hand hacksaw. Basically, the machine consists of a frame that holds a blade.

Figure 1 shows the basic parts of a reciprocating saw. The blade is tightened with a tensioner nut or screw and the workpiece is held securely in a vise. The saw frame is mounted on a heavy base that usually holds coolant, which is pumped to the saw blade. Reciprocating hacksaw blades are wider and thicker than those used in the hand hacksaw. The reciprocating motion is provided by hydraulics or a crankshaft mechanism.

Reciprocating saws are either the hinge type, as in Figure 1, or the column type. The saw frame on the hinge type pivots around a single point at the

Figure 1. Reciprocating cutoff saw. (Kasto–Racine, Inc.)

rear of the machine. On the column type, both ends of the frame rise vertically. The size of a reciprocating saw is determined by the largest piece of square material that can be cut. Sizes range from about 5 by 5 in. to 24 by 24 in. Large capacity reciprocating saws are often of the column design.

Horizontal Band Cutoff Machine

One disadvantage of the reciprocating saw is that it only cuts in one direction of the stroke. The band machine uses a steel band blade with the teeth on one edge. The horizontal band machine has a high cutting efficiency because the band is cutting at all times with no wasted motion. Band saws are the mainstay of production stock cutoff in the shop (Figure 2).

The major parts of the horizontal band cutoff machine are shown in Figure 3. Some of the parts are similar to those in the reciprocating saw: the blade is here endless—that is, it is a band—and the saw frame pivots on a base. There is a vise to hold the material being sawed and a stock length adjustment for cutting off many pieces of the same length. The blade has a tensioner and the blade guides support it near the material being sawed for maximum rigidity.

A modern band saw may be equipped with a variable speed drive. This permits the most efficient cutting speed to be selected for the material being cut. The feed rate through the material may also be varied. The size of the horizontal band machine is determined by the largest piece of square material that the machine can cut. Large capacity horizontal band saws are designed to handle large dimension workpieces that can weigh as much as 10 tons. With a wide variety of band types available, plus many special workholding devices, the band saw is an extremely valuable and versatile machine tool.

Universal Tilt Frame Cutoff

The universal tilt frame band saw is much like its horizontal counterpart. This machine has the band blade vertical, and the frame can be tilted from side to side (Figure 4). The tilt frame machine is particularly useful for making angle cuts on large structural shapes such as I-beams or pipe.

Abrasive Cutoff Machine

The abrasive cutoff machine (Figure 5) uses a thin circular abrasive wheel for cutting. Abrasive saws are very fast cutting. They can be used to cut a number of nonmetallic materials such as glass, brick, and stone. The major advantages of the abrasive cutoff machine are speed and the ability to cut nonmetals. Each parti-

Figure 2. This large horizontal band cutoff saw is used for production cutting. (The DoAll Company)

Figure 3. Horizontal band cutoff machine. (The DoAll Company)

Figure 4. Tilt frame band saw. (The DoAll Company)

Figure 5. Typical abrasive cutoff machine.

cle of abrasive acts as a small tooth and actually cuts a small bit of material. Abrasive saws are operated at very high speeds. Blade speed can be as high as 10,000 to 15,000 surface feet per minute.

Cold Saw Cutoff Machines

The cold saw uses a circular metal saw with teeth. These machine tools can produce extremely accurate cuts and are useful where length tolerance of the cut material must be held as close as possible. A cold saw blade that is .040 to .080 in. thick can saw material to a tolerance of plus or minus .002 in. Large cold saws are used to cut structural shapes such as angle and flat bar. These are also fast cutting machines.

CUTOFF MACHINE SAFETY

Reciprocating Saws

When you are preparing to use the cutoff saw, be sure that all guards around moving parts are in place before starting the machine. The saw blade must be properly installed with the teeth pointing in the right direction. Check for correct blade tension. Be sure that the width of the workpiece is less than the distance of the saw stroke. The frame will be broken if it hits the workpiece during the stroke. Be sure that the cutting speed and the rate of feed are correct for the material being cut.

When operating a saw with coolant, see that the coolant does not run on the floor during the cutting operation. This can cause an extremely dangerous slippery area around the machine tool.

Horizontal Band Saws

Recent, new regulations require that the blade of the horizontal band machine be fully guarded except at the point of cut (Figure 6). Make sure that the blade tensions are correct on reciprocating and band saws. Check band tensions, especially after installing a new band. New bands may stretch and loosen during their run-in period. Band teeth are sharp. When installing a new band, it should be handled with gloves. This is one of the few places that gloves may be worn around the shop. They must not be worn when operating any machine tool.

Band blades are often stored in double coils. Be careful when unwinding them, as they are under tension. The coils may spring apart and could cause an injury.

Check to see that the band is tracking properly on the wheels and in the blade guides. If a band should break, it could be ejected from the machine and cause an injury.

Make sure that the material being cut is properly secured in the workholding device. If this is a vise, be sure that it is tight. If you are cutting off short pieces of material, the vise jaw must be supported at both ends (Figure 7). It is not good practice to attempt to cut pieces of material that are quite short. The stock cannot be secured properly and may be pulled from the vise by the pressure of the cut (Figure 8). This can cause damage to the machine as well as possible injury to the operator. Stock should extend at least halfway through the vise at all times.

Many cutoff machines have a rollcase that supports long bars of material while they are being cut. The stock should be brought to the saw on a rollcase (Figure 9) or a simple rollstand. The pieces being cut

Figure 6. The horizontal band blade is guarded except in the immediate area of the cut. (*Machine Tools and Machining Practices*)

Figure 7. Support both ends of the vise when cutting short material.

Figure 8. Result of cutting stock that is too short. (*Machine Tools and Machining Practices*)

off can sometimes be several feet long and should be similarly supported. Sharp burrs left from the cutting should be removed immediately with a file. You can acquire a nasty cut by sliding your hand over one of these burrs.

Be careful around a rollcase, since bars of stock can roll, pinching fingers and hands (Figure 10). Also, be careful that heavy pieces of stock do not fall off the stock table or saw and injure feet or toes. Get help when lifting heavy bars of material. This will save your back and possibly your career.

On an abrasive saw, inspect the cutting edge of the blade for cracks and chips (Figure 11). Replace the blade if it is damaged. Always operate an abrasive saw blade at the proper RPM (Figure 12). Overspeeding the blade can cause it to fly apart. If an abrasive saw blade should fail at high speed, pieces of the blade can be thrown out of the machine at extreme velocities. A very serious injury can result if you happen to be in the path of these bulletlike projectiles.

Chips produced by any sawing operation should be brushed away with a suitable brush. Do not use your hands and do not use a rag as it may be caught in a moving part of the machine. A machine cannot

Figure 9. The material is brought into the saw on the rollcase (opposite side) and, when pieces are cut off, they are supported by the stand (this side of the saw). The stand prevents the part from falling to the floor. (The DoAll Company)

Figure 10. Stock on a roller table can pinch fingers and hands.

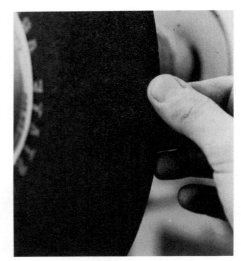

Figure 11. Inspecting the abrasive wheel for chips and cracks.

Figure 12. Abrasive wheel must be operated at the correct RPM.

distinguish between rags and clothing. Roll up your sleeves, keep your shirt tucked in, and if you have long hair, keep it properly secured and out of the way. Remove wristwatches and rings before operating any machine tool.

Safety checklist for sawing machines:

1. Remove wristwatches and rings and wear eye protection.
2. Properly secure the workpiece.
3. See that all guards are in place.
4. Select correct blade and inspect for chips and cracks.

5. Set blade tension correctly.
6. Set feeds and speeds accurately.
7. Adjust blade guides as close as possible to the work.
8. Remember to remove the material stop before making the cut.

SELF-TEST

1. List five types of cutoff machines.
2. What type is used to cut nonmetallic materials?
3. What type is used as the main production cutoff machine in the metals shop?
4. What type uses a back-and-forth motion when cutting?
5. What type uses a circular metal saw with teeth?
6. List three safety considerations on reciprocating and band saws.

7. List three safety considerations on abrasive and cold saws.
8. What hazards exist around a roller stock table?
9. You are going to do extended maintenance on a production cutoff machine. What is the first step that you should take?
10. What can be the result of overspeeding an abrasive saw blade?

UNIT 2 USING RECIPROCATING AND HORIZONTAL BAND CUTOFF MACHINES

The reciprocating saw and the horizontal band saw are the most common cutoff machines that you will encounter. Their primary function is to cut long lengths of material into lengths suitable for other machining operations on other machine tools. The cutoff machine is often the first step in machining a part to its final shape and size. In this unit, you are introduced to saw blades and the applications and operation of these important sawing machines.

OBJECTIVES

After completing this unit, you should be able to:
1. Use saw blade terminology.
2. Describe the conditions that define blade selection.
3. Identify the major parts of the reciprocating and horizontal band cutoff machine.
4. Properly install blades on reciprocating and horizontal band machines.
5. Properly use reciprocating and horizontal band machines in cutoff applications.

CUTTING SPEEDS

An understanding of cutting speeds is one of the most important aspects of machining that you will encounter. Many years of machining experience have shown that certain tool materials are most effective if passed through certain workpiece materials at certain speeds. If a tool material passes through the work too quickly, the heat generated by friction can rapidly dull the tool or cause it to fail completely. Too slow a passage of the tool through a material can result in premature dulling and low productivity.

A cutting speed refers to the amount of workpiece material that passes by a cutting tool in a given amount of time. Cutting speeds are measured in feet per minute. This is abbreviated FPM. In some machining operations, the tool can pass the work. Sawing is an example. The work also may pass the tool, as in the lathe. In both cases, FPM is the same. The shape of the workpiece does not affect the FPM. The circumference of a round part passing a cutting tool is still in FPM. In later units, FPM is discussed in terms of revolutions per minute of a round workpiece.

In sawing, FPM is simply the speed of each saw tooth as it passes through a given length of material in one minute. If one tooth of a band saw passes through one foot of material in one minute, the cutting speed is one foot per minute. This is true of reciprocating saws as well. However, remember that this saw only cuts in one direction of the stroke. Cutting speeds are a critical factor in tool life. Productivity will be low if the sawing machine is stopped most of the time because a dull or damaged blade must be replaced frequently. The additional cost of replacement cutting tools must also be considered. In any machining operation, always keep cutting speeds in mind.

Figure 1. Kerf. *(Machine Tools and Machining Practices)*

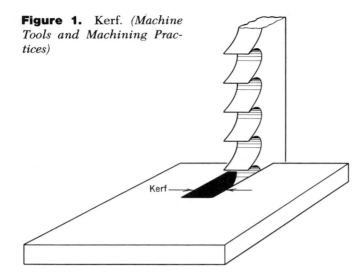

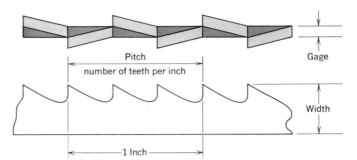

Figure 2. Gage, pitch, and width. *(Machine Tools and Machining Practices)*

On a sawing machine, blade FPM is a function of RPM (revolutions per minute) of the saw drive. That is, the setting of a specific RPM on the saw drive will produce a specific FPM of the blade. Feet per minute is also related to the material being cut. Generally, hard, tough materials have low cutting speeds. Soft materials have higher cutting speeds. In sawing, cutting speeds are affected by the material, size, and cross section of the workpiece.

SAW BLADES

The blade is the cutting tool of the sawing machine. In any sawing operation, at least three teeth on the saw blade must be in contact with the work at all times. This means that thin material requires a blade with more teeth per inch, while thick material can be cut with a blade having fewer teeth per inch. You should be familiar with the terminology of saw blades and saw cuts.

Blade Materials. Saw blades for reciprocating and band saws are made from carbon steels and high speed alloy steels. Blades may also have tungsten carbide tipped teeth or be bi-metal blades.

Blade Kerf. The kerf of a saw cut is the width of the cut as produced by the blade (Figure 1).

Blade Width. The width of a saw blade is the distance from the tip of the tooth to the back of the blade (Figure 2).

Blade Gage. Blade gage is the thickness of the blade (Figure 2). Reciprocating saw blades on large machines can be as thick as .250 in. Common band saw blades are .025 to .035 in. thick.

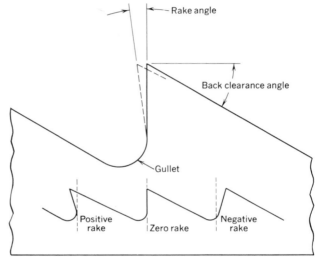

Figure 3. Saw tooth terminology. *(Machine Tools and Machining Practices)*

Blade Pitch. The pitch of a saw blade is the number of teeth per inch (Figure 2). An eight pitch blade has eight teeth per inch (a tooth spacing of $\frac{1}{8}$ in.).

SAW TEETH

You should be familiar with saw tooth terminology (Figure 3).

Tooth Forms. Tooth form is the shape of the saw tooth. Saw tooth forms are either standard, skip, or hook (Figure 4). Standard form gives accurate cuts with a smooth finish. Skip tooth gives additional chip clearance. Hook form provides faster cutting because of the positive rake angle.

Set. The teeth of a saw blade must be offset on each side to provide clearance for the back of the blade. This offset is called set (Figure 5). Set is equal on both sides of the blade. The set dimension is the total dis-

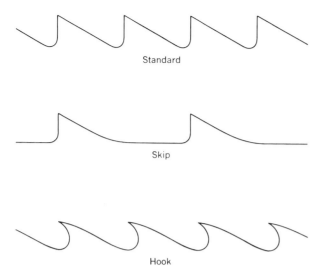

Figure 4. Tooth forms. *(Machine Tools and Machining Practices)*

Figure 6. Job selector on a horizontal band cutoff machine. (The DoAll Company)

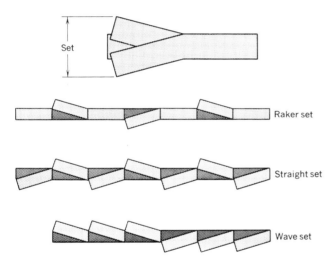

Figure 5. Set and set patterns. *(Machine Tools and Machining Practices)*

tance from the tip of a tooth on one side to the tip of a tooth on the other side.

Set Patterns. Set forms include raker, straight, and wave (Figure 5). Raker and wave are the most common. Raker set is used in general sawing. Wave set is useful where the cross-sectional shape of the workpiece varies.

SELECTING A BLADE
FOR RECIPROCATING AND BAND SAWS

Blade selection will depend upon the material, thickness, and cross-sectional shape of the workpiece. Some

band cutoff machines have a job selector (Figure 6). This will aid you greatly in selecting the proper blade for your sawing requirement. On a machine without a job selector, analyze the job and then select a suitable blade. For example, if you must cut thin tube, a fine pitch blade will be needed so that three teeth are in contact with the work. A particularly soft material may require a zero rake angle tooth form. Sawing through a workpiece with changing cross section may require a blade with wavy set to provide maximum accuracy.

USING CUTTING FLUIDS

Cutting fluids are an extremely important aid to sawing. The heat produced by the cutting action can become so great that the metallurgical structure of the blade teeth can be affected. Cutting fluids will dissipate much of this heat and greatly prolong the life of the blade. Besides their function as a coolant, they also lubricate the blade. Sawing with cutting fluids will produce a smoother finish on the workpiece. One of the most important functions of a cutting fluid is to flush chips out of the cut. This allows the blade to work more efficiently. Common cutting fluids are oils, oils dissolved in water (soluble oils), and synthetic chemical cutting fluids.

OPERATING THE
RECIPROCATING CUTOFF MACHINE

The reciprocating cutoff machine (Figure 7) is often known as the power hacksaw. This machine is an outgrowth of the hand hacksaw. Basically, the machine

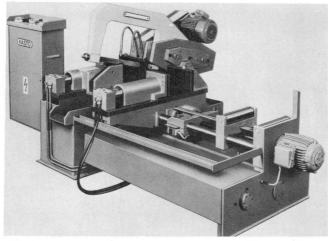

Figure 7. Hinge-type reciprocating cutoff saw. (Kasto–Racine Inc.)

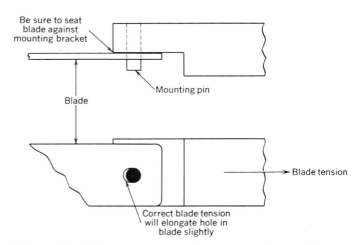

Figure 8. Blade mounting on the reciprocating cutoff saw. (*Machine Tools and Machining Practices*)

consists of a frame supported blade that is operated in a back-and-forth motion. Power hacksaws may be hydraulically driven or driven by a crankshaft mechanism.

Installing the Blade on the Power Hacksaw

Obtain a blade of the correct length and make sure that the teeth are pointed in the direction of the cut. This will be on the back stroke. Make sure that the blade is seated against the mounting plates (Figure 8). Apply the correct tension. The blade may be tightened until a definite ring is heard when the blade is tapped. Do not overtighten the blade, as this may cause the pin holes to break out. If a new blade is installed, the tension should be rechecked after making a few cuts.

Making the Cut

Select the appropriate strokes per minute speed rate for the material being cut. Be sure to secure the workpiece properly. If you are cutting material with a sharp corner, begin the cut on a flat side if possible (Figure 9). When it is absolutely necessary to start a cut on a sharp corner, as on an angle, hold up on the saw frame and gently feed the saw into the work. When a flat is made that is wider than three saw teeth, allow the saw to continue in a normal manner. Before making the cut, go over the safety checklist. Make sure that the length of the workpiece does not exceed the capacity of the stroke. This can break the frame if it should hit the workpiece. Bring the saw gently down until the blade has a chance to start cutting. Apply the proper feed. On reciprocating saws, feed is regulated with a sliding weight or feeding mechanism. If chips produced in the cut are blue, too much

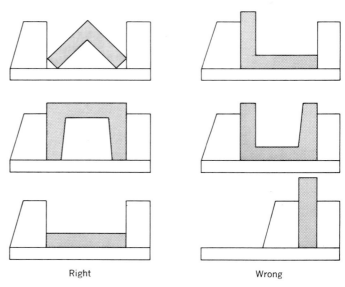

Right   Wrong

Figure 9. Cutting workpieces with sharp corners. (*Machine Tools and Machining Practices*)

feed is being used. The blade will be damaged rapidly. Very fine powder like chips indicate too little pressure. This will dull the blade. If the blade is replaced after starting a cut, turn the workpiece over and begin a new cut (Figure 10). Do not attempt to saw through the old cut. This will damage the new blade. After a new blade has been used for a short time, recheck the tension and adjust if necessary.

OPERATING THE HORIZONTAL BAND CUTOFF MACHINE

The horizontal band cutoff machine (Figure 11) is the

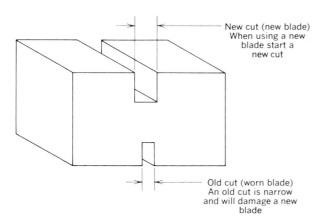

New cut (new blade)
When using a new
blade start a
new cut

Old cut (worn blade)
An old cut is narrow
and will damage a new
blade

Figure 10. If the blade is changed, begin a new cut on the other side of the workpiece. *(Machine Tools and Machining Practices)*

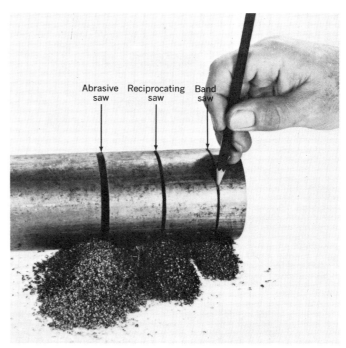

Figure 12. Comparison of kerf widths from band, reciprocating, and abrasive cutoff machines. (The DoAll Company).

Figure 11. Horizontal band cutoff machine. (Lane Community College)

Figure 13. Changing speeds by shifting belts on a horizontal band saw. (The DoAll Company)

most common stock cutoff machine found in the machine shop. This machine tool uses an endless steel band blade with teeth on one edge. Since the blade passes through the work continuously, there is no wasted motion. Cutting efficiency is greatly increased over the reciprocating saw.

The kerf from the band blade is quite narrow as compared to the reciprocating hacksaw or abrasive saw (Figure 12). This is an added advantage in that minimum amounts of material are wasted in the sawing operation.

The size of the horizontal band saw is determined by the largest piece of square material that can be cut. Speeds on the horizontal band machine may be set by manual belt change (Figure 13), or a variable

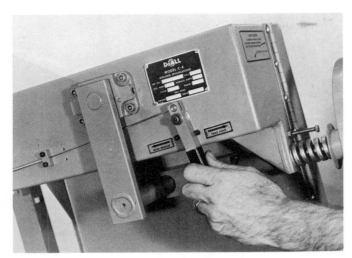

Figure 14. Horizontal band saw head release lever. (The DoAll Company)

Figure 16. Coolant system on the horizontal band saw. (The DoAll Company)

Figure 15. Adjusting the head tension on the horizontal band saw. (The DoAll Company)

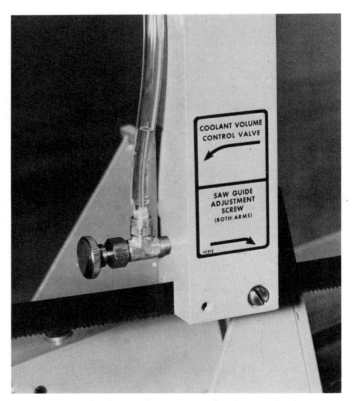

Figure 17. Coolant volume control valve on the horizontal band saw. (The DoAll Company)

speed drive may be used. The variable speed drive permits an infinite selection of band speeds within the capacity of the machine. Cutting speeds can be set precisely. Many horizontal band machines are of the hinge design. The saw head, containing the drive and idler wheels, hinges around a point at the rear of the machine.

The saw head may be raised and locked in the up position while stock is being placed into or removed from the machine (Figure 14). Feeds are accomplished by gravity of the saw head. The feed rate can be regulated by adjusting the spring tension on the saw head (Figure 15). Some sawing machines use

a hydraulic cylinder to regulate the feed rate. The head is held in the up position by the cylinder. A control valve permits oil to flow into the reservoir as the saw head descends. This permits the feed rate to be regulated.

Figure 18. Rotary chip brushes on the band saw blade. (The DoAll Company)

Figure 20. Roller stock table with access gate. (The DoAll Company)

Figure 19. Setting the stock length gage. (The DoAll Company)

Figure 21. Quick setting vise on the horizontal band saw. (The DoAll Company)

flow is controlled by a control valve (Figure 17). Chips can be cleared from the blade by a rotary brush that operates as the blade runs (Figure 18).

Common accessories used on many horizontal band saws include workpiece length measuring equipment (Figure 19) and roller stock tables. The stock table shown has a hinged section that permits the operator to reach the rear of the machine (Figure 20).

Workholding on the Horizontal Band Saw

The vise is the most common workholding fixture. Rapid adjusting vises are very popular (Figure 21). These vises have large capacity and are quickly ad-

Cutting fluid is pumped from a reservoir and flows on the blade at the forward guide. Additional fluid is permitted to flow on the blade at the point of the cut (Figure 16). The saw shown does not have the now required full blade guards installed. Cutting fluid

Figure 22. Band saw vise swiveled for angle cutting. (The DoAll Company)

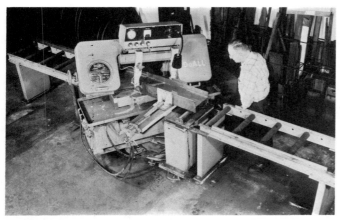

Figure 23. Horizontal cutoff saw frame swiveled for angle cuts. (The DoAll Company)

Figure 24. "Nesting fixture" for sawing multiple workpieces. (The DoAll Company)

justed to the workpiece. After the vise jaws have contacted the workpiece, the vise is locked by operating the lock handle. The vise may be swiveled for miter or angle cuts (Figure 22). On some horizontal band cutoff machines, the entire saw frame swivels for making angle cuts (Figure 23).

The horizontal band saw is often used to cut several pieces of material at once. Stock may be held or nested in a special vise or nesting fixture (Figure 24).

Installing Blades on the Horizontal Band Machine

Blades for the horizontal band saw may be ordered prewelded in the proper length for the machine. Band blades may also be obtained in rolls. The required length is then cut and welded at the sawing machine location.

To install the band, shut off power to the machine and open the wheel guards. Release the tension by turning the tension wheel. Place the blade around the drive and idler wheels. Be sure that the teeth are pointed in the direction of the cut. This will be toward the rear of the machine. See that the blade is tracking properly on the idler and drive wheels (Figure 25). The blade will have to be twisted slightly to fit the guides. Guides should be adjusted so that they have .001 to .002 in. clearance with the blade. Adjust the blade tension using the tension gage (Figure 26) or the manual tension indicator built into the saw (Figure 27). The blade is tightened until the flange on the tension wheel contacts the tension indicator stop.

Figure 25. Make sure that the blade is tracking properly on the band wheel.

Figure 26. Dial band tension indicator gage. (The DoAll Company)

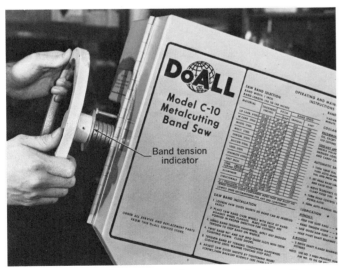

Figure 27. Band tension indicator attached to the machine. (The DoAll Company)

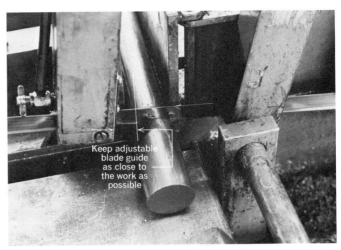

Keep adjustable blade guide as close to the work as possible

Figure 28. Blade guide should be set as close to the work-piece as possible.

Figure 29. Automatic shutoff mechanism.

Making the Cut

Set the proper speed according to the blade type and material to be cut. If the workpiece has sharp corners, it should be positioned in the same manner as in the reciprocating saw (Figure 9). The blade guides must be adjusted so that they are as close to the work as possible (Figure 28). This will insure maximum blade support and maximum accuracy of the cut. Sufficient feed should be used to produce a good chip. Excessive feed can cause blade failure. Too little feed can dull the blade prematurely. Go over the safety checklist for horizontal band saws. Release the head and lower

Figure 30. Stock length stop in position.

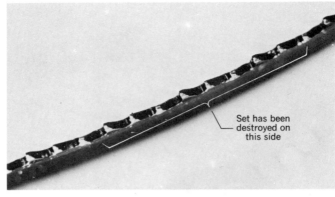

Figure 32. Band saw blade with the set worn on one side.

Figure 31. Stock length stop must be removed before beginning the cut.

Figure 33. Using a band with the set worn on one side will cause the cut to drift toward the side of the blade with the set remaining.

it by hand until the blade starts to cut. Most saws are equipped with an automatic shutoff switch (Figure 29). When the cut is completed, the machine will shut off automatically.

SAWING PROBLEMS ON THE HORIZONTAL BAND MACHINE

You may use the stock stop to gage the length of duplicate pieces of material (Figure 30). It is important to swing the stop clear of the work after the vise has been tightened and before the cut is begun (Figure 31). A cutoff workpiece can bind between the stop and blade. This will destroy the blade set (Figure 32).

A blade with a tooth set worn on one side will drift in the direction of the side that has a set still remaining (Figure 33). This is the principal cause of band breakage. As the saw progresses through the cut, the side drift of the blade will place the machine under great stress. The cut may drift so far as to permit the blade to cut into the vise (Figure 34).

Very thin and quite parallel workpieces can be cut by using a sharp blade in a rigid and accurate sawing machine (Figure 35). Chip removal is important to accurate cutting. If chips are not cleared from the blade prior to it entering the guides, the blade will be scored. This will make it brittle and subject to breakage.

Figure 34. The blade may drift far enough to damage the vise.

Figure 35. Thin and parallel cuts may be made with a sharp blade and rigid sawing machine. *(Machine Tools and Machining Practices)*

SELF-TEST

1. Name the most common saw blade set patterns.
2. Describe the conditions that define blade selection.
3. On a reciprocating saw, what is the direction of the cut?
4. What is set and why is it necessary?
5. What are common tooth forms?
6. What can happen if the stock stop is left in place during the cut?
7. What type of cutoff saw will most likely be found in the machine shop?
8. Of what value are cutting fluids?
9. What can result if chips are not properly removed from the cut?
10. If a blade is replaced after a cut has been started, what must be done with the workpiece?

UNIT 3 VERTICAL BAND MACHINES

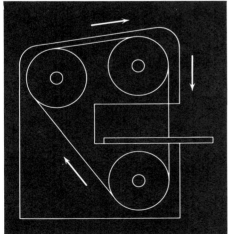

The vertical band machine (Figure 1) is often called the handiest machine tool in the machine shop. Perhaps the reason for this is the wide variety of work that can be accomplished on it. The vertical band machine or vertical band saw is similar in general construction to its horizontal counterpart. Basically, it consists of an endless band blade or other band tool that runs on a driven and idler wheel. The band tool runs vertically at the point of the cut where it passes through a worktable on which the workpiece rests. The workpiece is pushed into the blade and the direction of the cut is guided by hand or mechanical means.

OBJECTIVES

After completing this unit, you should be able to:
1. Identify various types of vertical band machines and describe their different applications.
2. Identify potential hazards on and around vertical band machines.
3. Describe safe vertical band machine cutting procedure.

The vertical band machine, like any other machine tool, is a motor-driven device with a number of moving parts. The machine is designed with sufficient power to perform its required tasks. Anytime that mechanical parts with sufficient power are put into motion, potential hazards exist for the operator. You must remember that any machine tool is only an extension of the operator's intelligence. A machine has no intelligence of its own. Once turned on, it cannot distinguish between cutting material and cutting fingers. The potential hazards of a machine tool must be kept in mind at all times. A lax safety attitude on the part of the operator can only lead to an acci-

dent. Many accidents are caused by doing something that you should not do. Short-cutting proper operating procedure may result in cutting a finger, hand, or arm, or losing an eye.

ADVANTAGES OF BAND MACHINES

Shaping of material with the use of a saw blade or other band tool is often called **band machining.** The reason for this is that the band machine can perform other machining tasks aside from simple sawing. These include band friction sawing, band filing, and band polishing.

In any machining operation, a piece of stock material is cut by various processes to form the final shape and size of the part desired. In most machining operations, all of the unwanted material must be reduced to chips in order to make the final shape and size of the workpiece. With a band saw, only a small portion of the unwanted material must be reduced to chips in order to make the final workpiece shape and size (Figure 2). A piece of stock material can often be shaped to final size by one or two saw cuts. A further advantage is gained in that the band saw cuts a very narrow kerf. A minimum amount of material is wasted. Other machining operations may require that a large amount of material be wasted as chips in order to make the final size and shape of the workpiece.

A second important advantage in band sawing machines is **contouring ability.** Contour band sawing is the ability of the saw to cut intricate curved shapes that would be nearly impossible to machine by other methods (Figure 3). The sawing of intricate shapes can be accomplished by a combination of hand and power feeds. On vertical band machines so equipped, the workpiece is steered by manual operation of the handwheel. The hydraulic table feed varies according to the saw pressure on the workpiece. This greatly facilitates contouring sawing operations.

Figure 1. General purpose vertical band machine. (The DoAll Company)

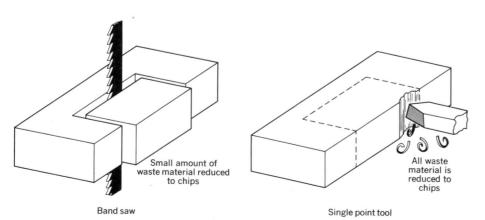

Band saw   Single point tool

Small amount of waste material reduced to chips

All waste material is reduced to chips

Figure 2. Sawing can uncover the workpiece shape in a minimum number of cuts. *(Machine Tools and Machining Practices)*

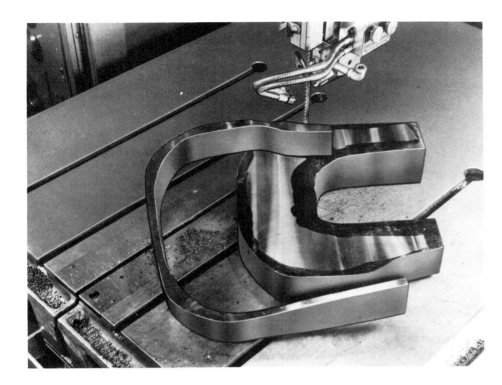

Figure 3. Curved or contour band sawing can produce part shapes that would be difficult to machine by other methods. (The DoAll Company)

Band sawing and band machining have several other advantages. There is no limit to the length, angle, or direction of the cut. Workpieces larger than the band machine can be cut. Since the band tool is fed continuously past the work, cutting efficiency is high. A band tool, whether it be a saw blade, band file, or grinding band, has a large number of cutting points passing the work. In most other machining operations, only one or a fairly low number of cutting points pass the work. With the band tool, wear is distributed over these many cutting points. Tool life is prolonged.

TYPES OF BAND MACHINES

General Purpose Band Machine with Fixed Worktable

The general purpose band machine is found in most machine shops. This machine tool has a nonpower-fed worktable that can be tilted in order to make angle cuts. The table may be tilted 10 degrees left (Figure 4). Tilt on this side is limited by the saw frame. The table may be tilted 45 degrees right (Figure 5). On large machines, table tilt left may be limited to 5 degrees.

The workpiece may be pushed into the blade by hand. Mechanical (Figure 6) or mechanical-hydraulic feeding mechanisms are also used. A band machine may be equipped with a hydraulic tracing attachment. This accessory uses a stylus contacting a template or pattern. The tracing accessory guides the workpiece during the cut (Figure 7).

Band Machines with Power-Fed Worktables

Heavier construction is used on these machine tools. The worktable is moved hydraulically. The operator is relieved of the need to push the workpiece into the cutting blade. The direction of the cut can be guided by a steering mechanism. A roller chain wraps around the workpiece and passes over a sprocket at the back of the worktable. The sprocket is connected to a steering wheel at the front of the worktable. The operator can then guide the workpiece and keep the saw cutting along the proper lines. The workpiece rests on roller bearing stands. These permit the workpiece to turn freely while it is being steered.

High Tool Velocity Band Machines

On the high tool velocity band machine band speeds can range as high as 10 to 15,000 feet per minute (FPM). These machine tools are used in many band machining applications. They are frequently found cutting nonmetal products. These include applications such as trimming plastic laminates (Figure 8) and cutting fiber materials (Figure 9).

Figure 4. Vertical band machine worktable can be tilted 10 degrees left.

Figure 5. Vertical band machine worktable tilted 45 degrees right.

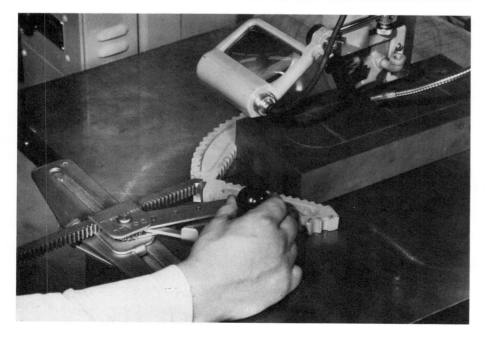

Figure 6. Mechanical work feeding mechanism. (The DoAll Company)

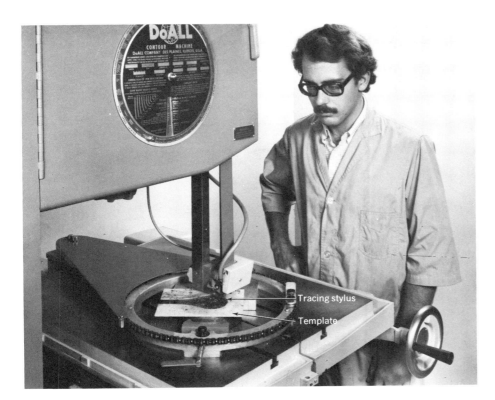

Figure 7. Hydraulic tracing accessory. (The DoAll Company)

Tracing stylus

Template

Figure 8. Trimming plastic laminates on the high tool velocity band machine. (The DoAll Company)

Figure 9. Cutting fiber material on the high tool velocity band machine. (The DoAll Company)

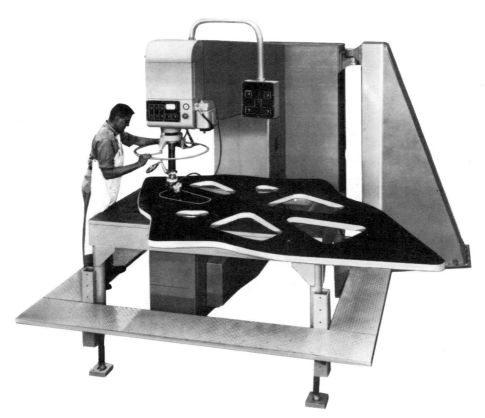

Figure 10. Large capacity vertical band machine. (The DoAll Company)

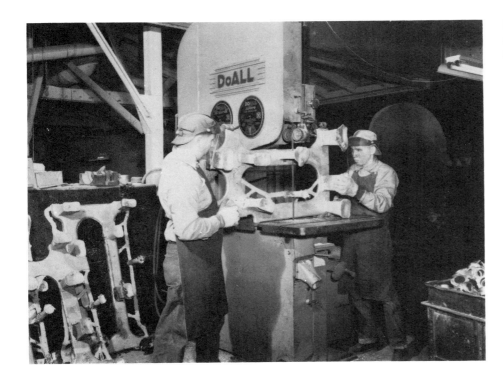

Figure 11. Trimming casting sprues and risers on the vertical band machine. (The DoAll Company)

Large Capacity Band Machines

This type of band machine is used on large workpieces. The entire saw is attached to a swinging column. The workpiece remains stationary and the saw is moved about to accomplish the desired cuts (Figure 10).

APPLICATIONS OF THE VERTICAL BAND MACHINE

Conventional and Contour Sawing

Vertical band machines are used in many conventional sawing applications. They are found in the foundry trimming sprues and risers from castings. The band machine can accommodate a large casting and make widely spaced cuts (Figure 11). Production trimming of castings is easily accomplished with the high tool velocity band machine. Band saws are also useful in ripping operations. In the machine shop, the vertical band machine is used in general purpose, straight line, and contour cutting mainly in sheet and plate stock.

Friction Sawing

Friction sawing can be used to cut materials that would be impossible or very difficult to cut by other means. In friction sawing, the workpiece is heated by friction created between it and the cutting blade. The blade melts its way through the work. Friction sawing can be used to cut hard materials such as files. Tough materials such as stainless steel wire brushes can be trimmed by friction sawing.

Band Filing and Band Polishing

The band file consists of file segments attached to a spring steel band (D) (Figure 12). As the band file passes through the work an interlock closes and keeps the file segment tight (B). The interlock then releases, permitting the file segment to roll around the band wheel. A space is provided for chip clearance between the band and file segment (C). The band file has a locking slot so that the ends can be joined to form a continuous loop (A). Special guides are required for both file and polishing bands. Band files can be used in both internal (Figure 13) and external (Figure 14) filing applications. They are also used in applications such as filing large gear teeth to shape and size (Figure 15).

In band polishing, a continuous abrasive strip is used (Figure 16). The grit of the abrasive can be varied depending on the surface finish desired.

SAFETY CONSIDERATIONS ON VERTICAL BAND MACHINES

Personal Protection

Eye protection is a primary consideration on any ma-

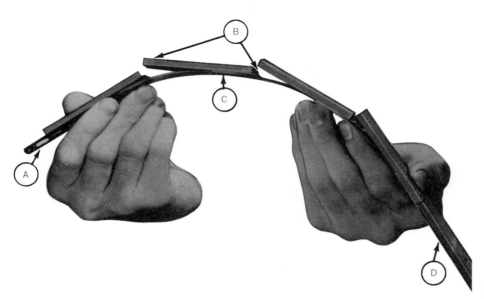

Figure 12. Band file. (The DoAll Company)

Figure 13. Internal band filing. (The DoAll Company)

Figure 14. External band filing. (The DoAll Company)

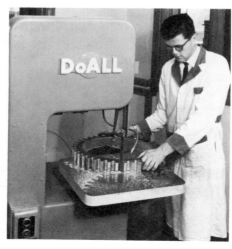

Figure 15. Band filing a large spur gear. (The DoAll Company)

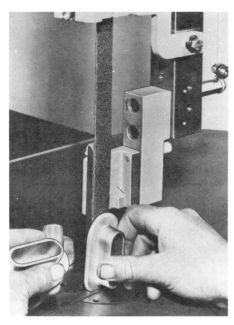

Figure 16. Band polishing. (The DoAll Company)

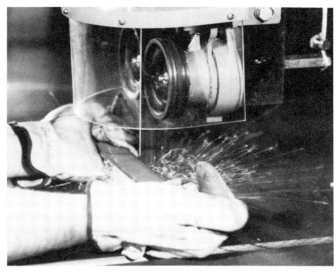

Figure 17. Gloves may be necessary in high speed or friction sawing.

chine tool. Vertical band machines use a hardened steel blade, file, or abrasive band. There is always the chance that metal particles may be ejected from the machine during the cutting operation. Always wear an approved face shield, goggles, or safety glasses. Safety glasses with side shields or a face shield must be worn when friction or high speed sawing.

Gloves should not be worn around any machine tool. An exception to this is for friction or high speed sawing or when handling band blades. Gloves will protect hands from the sharp saw teeth. If you wear gloves during friction or high speed sawing (Figure 17), be extra careful that they do not become entangled in the blade or other moving parts.

Remove your rings and wristwatch before operating any machine tool. These can be extremely hazardous if they become caught in a moving part. Rings are especially dangerous. If they should become caught, a finger can be lost very easily.

Chips produced by sawing operations should be removed with a suitable brush. **Do not use your hands and do not use a rag.** The sharp metal can cut your hands and a rag may be caught in a moving part of the machine. Chips also imbed themselves in rags, creating an additional hazard.

A machine tool cannot distinguish between rags and clothing. Roll up your sleeves, and keep your shirt tucked in. Remove your necktie if you are wearing one. If you have long hair, keep it properly secured out of danger.

Figure 18. Wheel and blade guards on the vertical band machine. (The DoAll Company)

Machine Guarding on Vertical Band Machines

The entire blade must be guarded except at the point of the cut (Figure 18). This is effectively accomplished by enclosing the wheels and blade behind guards that are easily opened for adjustments to the machine.

Figure 19. Guidepost guard.

Figure 20. Left side blade guard when using a short blade over one idler wheel.

Figure 21. Roller blade guide shield.

Wheel and blade guards must be closed at all times during machine operation. The guidepost guard moves up and down with the guidepost (Figure 19). The operator is protected from an exposed blade at this point. For maximum safety, set the guidepost $\frac{1}{8}$ to $\frac{1}{4}$ in. above the workpiece.

Band machines may have one or two idler wheels. On machines with two idler wheels, a short blade running over only one wheel may be used. Under this condition, an additional blade guard at the left side of the wheel is required (Figure 20). This guard is removed when operating over two idler wheels as the blade is then behind the wheel guard.

Roller blade guides are used in friction and high speed sawing. A roller guide shield is used to provide protection for the operator (Figure 21). Depending on the material being cut, the entire cutting area may be enclosed (Figure 22). This would apply to the cutting of hard, brittle materials such as granite and glass. Diamond blades are frequently used in cutting these materials. The clear shield protects the operator while permitting him to view the operation. Cutting fluids are also prevented from spilling on the floor. In any sawing operation making use of cutting fluids, see that

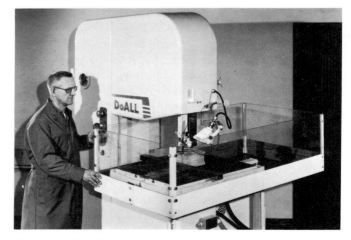

Figure 22. When cutting brittle materials, the entire work area may be guarded. (The DoAll Company)

they do not spill on the floor around the machine. This creates an extremely dangerous situation, not only for you but for others in the shop as well.

Any time that a guard is removed or opened, exposing the operating mechanism of the band machine, the electrical service to the tool should be turned off. Some band machines may be unplugged from the electrical outlet. The circuit breaker supplying the machine circuit can be switched off on machines that are permanently wired. If the machine is to be out of service for an extended period, the circuit breaker should be switched off and tagged with an appropriate warning.

Normally you will not disconnect the electrical service for a routine adjustment such as changing a blade. For total safety, it is good practice to take the time to secure the power to the machine. If a routine adjustment such as changing a belt is being made, you must insure that no other person is likely to turn on the machine while your hands are in contact with belts and pulleys.

Safe Cutting Procedure on the Vertical Band Machine

The primary danger in operating the vertical band machine is accidental contact with the cutting blade. Workpieces are often hand guided. One advantage in sawing machines is that the pressure of the cut tends to hold the workpiece against the saw table. However, hands are often in close proximity to the blade. If you should contact the blade accidentally, an injury is almost sure to occur. You will not have time even to think about withdrawing your fingers before they are cut. Keep this in mind at all times when operating a band saw.

Always use a pusher against the workpiece whenever possible (Figure 23). This will keep your fingers away from the blade. Be careful as you are about to complete a cut. As the blade clears through the work, the pressure that you are applying is suddenly released and your hand or finger could be carried into the blade. As you approach the end of the cut, reduce the feeding pressure as the blade cuts through.

The vertical band machine is generally not used to cut round stock. This can be extremely hazardous and should be done on the horizontal band machine, where round stock can be secured in a vise. Hand-held round stock will turn if it is cut on the vertical band machine. This can cause an injury and may damage the blade as well. If round stock must be cut on the vertical band saw, it must be clamped securely

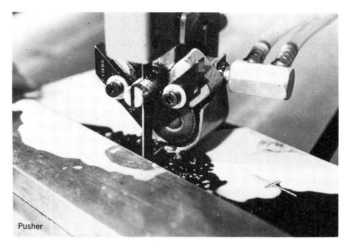

Figure 23. Using a pusher.

in a vise, vee block, or other suitable workholding fixture.

Be sure to select the proper blade for the sawing requirements. Install it properly and apply the correct blade tension. Band tension should be rechecked after a few cuts. New blades will tend to stretch to some degree during their break-in period. Band tension may have to be readjusted.

Safety extends to the machine as well as to the operator. Never abuse any machine tool. They cost a great deal of money and in many cases are purchased with your tax dollars. When cutting on the vertical band machine, bring the workpiece gently into the blade. Use correct speeds, feeds, and blade types. Shift gears only at slow speeds or with the machine stopped. Treat this machine tool with the same respect as you would any expensive precision machine.

SAFETY CHECKLIST FOR VERTICAL BAND MACHINES

1. Have you removed your rings and wristwatch?
2. Are you using the correct blade for the sawing task and is the tension set properly?
3. Are feeds and speeds properly set?
4. Are all guards in place?
5. Is cutting fluid likely to spill on the floor?
6. Do you have eye protection?
7. Is the guidepost with upper blade guide set as close to the work as possible?
8. Are you using a pusher when fingers are close to the blade?

SELF-TEST

1. Name four types of band machines.
2. Which type would be particularly useful in cutting non-metal soft materials like plastics?
3. What other machining operations aside from sawing can be accomplished on the band machine?
4. What is contour sawing?
5. Name two advantages of band machines.
6. What is the biggest danger on the vertical band machine?
7. Describe conditions under which gloves are acceptable.
8. What is a pusher and how is it used?
9. Why should round stock not be cut on the vertical band machine?
10. What is the primary piece of safety equipment for the band machine operator?

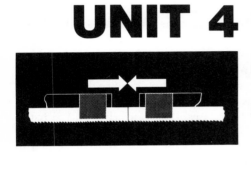

UNIT 4 PREPARING TO USE THE VERTICAL BAND MACHINE

A machine tool can perform at maximum efficiency only if it has been properly maintained, adjusted, and set up. Before the vertical band machine can be used for a sawing or other band machining operation, several important preparations must be made. These include welding saw blades into bands and making several adjustments on the machine tool.

OBJECTIVES

After completing this unit, you should be able to:
1. Weld band saw blades made of carbon steel.
2. Prepare the vertical band machine for operation.

WELDING BAND SAW BLADES

Band saw blades are frequently supplied in rolls. The required length is measured and cut and the ends are welded together to form an endless band. Most band machines are equipped with a band welding attachment. These are frequently attached to the machine tool. They may also be separate pieces of equipment (Figure 1).

The **band welder** is a resistance-type butt welder. They are often called **flash** welders because of the bright flash and shower of sparks created during the welding operation. The metal in the blade material has a certain resistance to the flow of an electric current. This resistance causes the blade metal to heat as the electric current flows during the welding operation. The blade metal is heated to a temperature that permits the ends to be forged together under pressure. When the forging temperature is reached, the ends of the blade are pushed together by mechanical pressure. They fuse, forming a resistance weld. The band weld is then annealed, or softened, and dressed to the correct thickness by grinding.

Welding carbon steel band saw blades is a fairly simple operation and you should master it as soon as possible. Blade welding is frequently done in the machine shop. New blades are always being prepared.

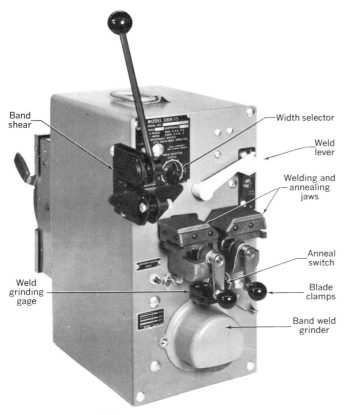

Figure 1. Band blade welder. (The DoAll Company)

Figure 2. Blade shear. The cut should always be started on the edge opposite the teeth.

Figure 3. Placing the blade ends together with the teeth opposite.

Sawing operations, where totally enclosed workpiece features must be cut, require that the blade be inserted through a starting hole in the workpiece and then welded into a band. After the enclosed cut is made, the blade is broken apart and removed. **Note:** High speed steel blades cannot be welded with resistance welders.

PREPARING THE BLADE FOR WELDING

The first step is to cut the required length of blade stock for the band machine that you are using. Blade stock can be cut with snips or with the band shear (Figure 2). Many band machines have a blade shear near the welder. The required length of blade will usually be marked on the saw frame. Blade length, B_L, for two-wheel sawing machines can be calculated by the formula

$$B_L = \pi D + 2L$$

where D is the diameter of the band wheel and L is the distance between band wheel centers. Set the tension adjustment on the idler wheel about midrange so that the blade will fit after welding. Most machine

shops will have a permanent reference mark, probably on the floor, that can be used for gaging blade length.

After cutting the required length of stock, the ends of the blade must be ground so that they are square when positioned in the welder. Place the ends of the blade together so that the teeth are opposed (Figure 3). Grind the blade ends in this position. The grinding wheel on the blade welder may be used for this operation. Blade ends may also be ground on the pedestal grinder, (Figure 4). Grinding the blade ends with the teeth opposed will insure that the ends of the blade are square when the blade is positioned in the welder. Any small error in grinding will be canceled when the teeth are placed in their normal position.

Proper grinding of the blade ends permits correct tooth spacing to be maintained. After the blade has been welded, the tooth spacing across the weld should be the same as any other place on the band. Tooth set should be aligned as well. A certain amount of

blade material is consumed in the welding process. Therefore, the blade must be ground correctly if tooth spacing is to be maintained. The amount consumed by the welding process may vary with different blade welders. You will have to determine this by experimentation. For example, if $\frac{1}{4}$ in. of blade length is consumed in welding, this would amount to about one tooth on a four pitch blade. Therefore, one tooth should be ground from the blade. This represents the amount lost in welding (Figure 5). Be sure to grind only the tooth and not the end of the blade. The number of teeth to grind from a blade will vary according to the pitch and amount of material consumed by a specific welder. The weld should occur at the bottom of the tooth gullet. Exact tooth spacing can be somewhat difficult to obtain. You may have to practice end grinding and welding several pieces of scrap blades until you are familiar with the proper welding and tooth grinding procedure.

The jaws of the blade welder should be clean before attempting any welding. Position the blade ends in the welder jaws (Figure 6). The saw teeth should point toward the back on most welders; but, in any case, the blade should be positioned so that the jaws clamp just in back of the teeth. This prevents scoring of the jaws when welding blades of different widths and also makes a solid electrical contact for performing the weld. A uniform amount of blade should extend from each jaw. The blade ends must contact squarely in the center of the gap between the welder jaws. Be sure that the blade ends are not offset or overlapped. Tighten the blade clamps.

Figure 4. End grinding the blade on the pedestal grinder.

Figure 5. The amount of blade lost in welding. *(Machine Tools and Machining Practices)*

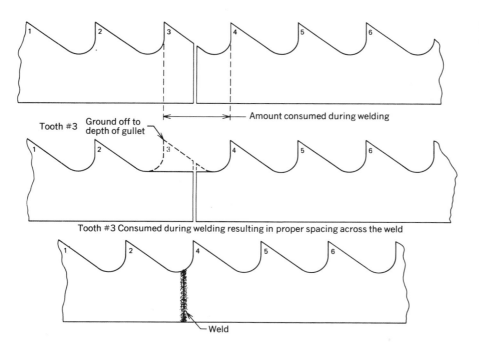

Amount consumed during welding

Tooth #3   Ground off to depth of gullet

Tooth #3 Consumed during welding resulting in proper spacing across the weld

Weld

Figure 6. Placing the blade in the welder.

Figure 7. Welding the blade into a band. (The DoAll Company)

Welding the Blade into an Endless Band

Adjust the welder for the proper width of blade to be welded. **Wear eye protection** and stand to one side of the welder during the welding operation. Depress the weld lever. A flash with a shower of sparks will occur (Figure 7). In this brief operation, the movable jaw of the welder moved toward the stationary jaw. The blade ends were heated to forging temperature by a flow of electric current, and the molten ends of the blades were pushed together, forming a solid joint.

The blade clamps should be loosened before releasing the weld lever. This prevents scoring of the welder jaws by the now welded band. A correctly welded band will have the weld **flash** (flash, in this case, means the upset metal that is pushed out from the weld) evenly distributed across the weld zone (Figure 8). Tooth spacing across the weld should be the same as the rest of the band.

Figure 8. Weld flash should be evenly distributed after welding.

Annealing the Weld

The metal in the weld zone is hard and brittle immediately after welding. For the band to function, the weld must be **annealed** or **softened**. This improves strength qualities of the weld. Place the band in the annealing jaws with the teeth pointed out (Figure 9). This will concentrate annealing heat away from the saw teeth. A small amount of compression should be placed on the movable welder jaw prior to clamping the band. This permits the jaw to move as the annealing heat expands the band.

It is most important not to overheat the weld during the annealing process. Overheating can destroy an otherwise good weld, causing it to become

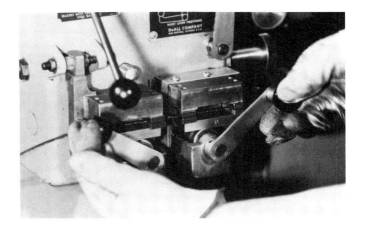

Figure 9. Positioning the band for annealing.

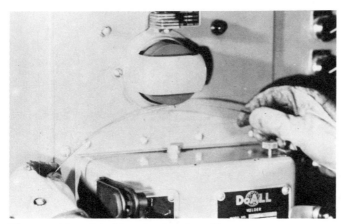

Figure 10. Grinding the band weld.

Figure 11. The saw teeth must not be ground while grinding the band weld.

brittle. The correct annealing temperature is determined by the color of the weld zone during annealing. This should be a dull red color. Depress the anneal switch and watch the band heat. When the dull red color appears, release the anneal switch immediately and let the band begin to cool. As the weld cools, depress the anneal switch briefly several times to slow the cooling rate. Too rapid cooling can result in a band weld that is not properly annealed.

Grinding the Weld

Some machinists prefer to grind the band weld prior to annealing. This permits the annealing color to be seen more easily. More often, the weld is ground after the annealing process. However, it is good practice to anneal the blade weld further after grinding. This will eliminate any hardness induced during the grinding operation. The grinding wheel on the band welder is designed for this operation. The top and bottom of the grinding wheel are exposed so that both sides of the weld can be ground (Figure 10). **Be careful not to grind the teeth** when grinding a band weld. This will destroy the tooth set. Grind the band weld evenly on both sides (Figure 11). The weld should be ground to the same thickness as the rest of the band. If the weld area is ground thinner, the band will be weakened at that point. As you grind, check the band thickness in the gage (Figure 12) to determine proper thickness (Figure 13).

PROBLEMS IN BAND WELDING

Several problems may be encountered in band welding (Figure 14). These include misaligned pitch, blade misalignment, insufficient welding heat, or too much welding heat. You should learn to recognize and avoid these problems. The best way to do this is to obtain

Figure 12. Band weld thickness gage.

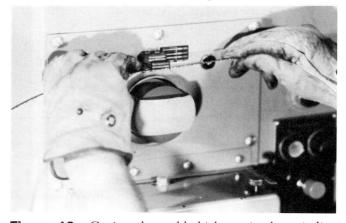

Figure 13. Gaging the weld thickness in the grinding gage.

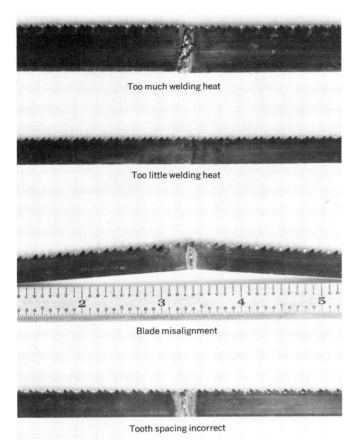

Too much welding heat

Too little welding heat

Blade misalignment

Tooth spacing incorrect

Figure 14. Problems in band welding. *(Machine Tools and Machining Practices)*

Figure 15. Band guides must fully support the band, but must not extend over the saw teeth.

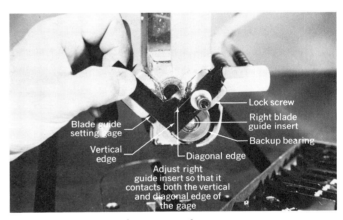

Figure 16. Using the saw guide setting gage.

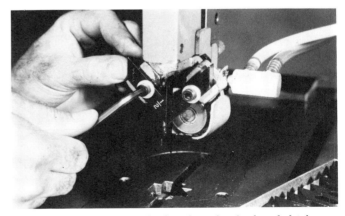

Figure 17. Adjusting the band guides for band thickness.

some scrap blades and practice the welding and grinding operations.

INSTALLING AND ADJUSTING BAND GUIDES ON THE VERTICAL BAND MACHINE

Band guides must be properly installed if the band machine is to cut accurately and if damage to the band is to be prevented. Be sure to use the **correct width guides** for the band (Figure 15). The band must be fully supported except for the teeth. Using wide band guides with a narrow band will destroy tooth set as soon as the machine is started.

Band guides are set with a **guide setting gage.** Install the right-hand band guide and tighten the lock screw just enough to hold the guide insert in place. Place the setting gage in the left guide slot and adjust the position of the right guide insert so that it is in contact with both the vertical and diagonal edges of the gage (Figure 16). Check the **backup bearing** at this time. Clear any chips that might prevent it from turning freely. If the backup bearing cannot turn

freely, it will be scored by the band and damaged permanently.

Install the right-hand guide insert and make the adjustment for band thickness using the same setting gage (Figure 17). The thickness of the band will be

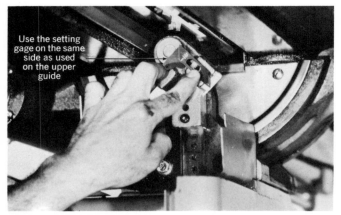

Figure 18. When adjusting the lower guide, use the setting gage on the same side as on the upper guide.

Figure 19. Adjusting roller band guides.

Figure 20. Coolant may be introduced directly ahead of the band. (The DoAll Company)

Figure 21. Mist and flood coolant nozzles.

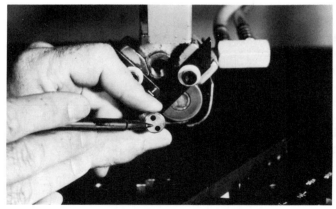

Figure 22. Inlet side of the coolant nozzle.

marked on the tool. Be sure that this is the same as the band that will be used. The lower band guide is adjusted in a like manner. Use the setting gage on the same side as it was used when adjusting the top guides (Figure 18).

Roller band guides are used in high speed sawing applications where band velocities exceed 2000 FPM. They are also used in friction sawing operations. The roller guide should be adjusted so that it has .001 to .002 in. clearance with the band (Figure 19).

ADJUSTING THE COOLANT NOZZLE

A band machine may be equipped with flood or mist coolant. Mist coolant is liquid coolant mixed with air. Certain sawing operations may require only small amounts of coolant. With the mist system, liquid coolant is conserved and is less likely to spill on the floor. When cutting with flood coolant, be sure that the runoff returns to the reservoir and does not spill on the floor. Flood coolant may be introduced directly ahead of the band (Figure 20).

Flood or mist coolant may be introduced through a nozzle in the upper guidepost assembly (Figure 21).

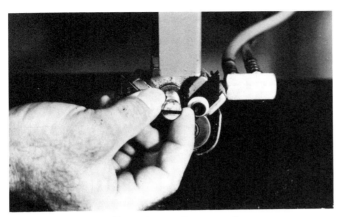

Figure 23. Installing the coolant nozzle.

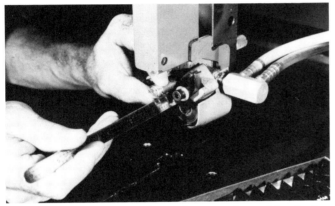

Figure 24. Presetting the coolant nozzle position before installing the band guides.

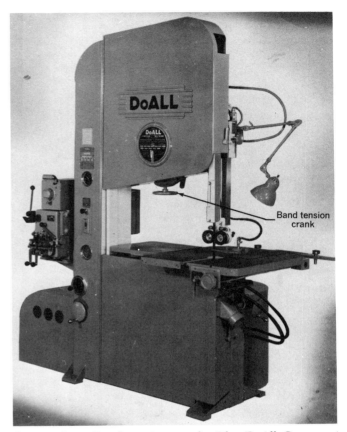

Figure 25. Band tension crank. (The DoAll Company)

Air and liquid are supplied to the inlet side of the nozzle by two hoses (Figure 22). The coolant nozzle must be installed (Figure 23) and preset (Figure 24) prior to installing the band. For mist coolant set the nozzle end $\frac{1}{2}$ in. from the face of the band guide. The setting for flood is $\frac{3}{8}$ in.

INSTALLING THE BAND ON THE VERTICAL BAND MACHINE

Open the upper and lower wheel covers and remove the filler plate for the worktable. It is **safer to handle the band with gloves to protect your hands** from the sharp saw teeth. The hand tension crank is attached to the upper idler wheel (Figure 25). Turn the crank to lower the wheel to a point where the band can be placed around the drive and idler wheels. Be sure to install the band so the teeth point in the direction of the cut. This is always in a **down direction toward the worktable.** If the saw teeth seem to be pointed in the wrong direction, the band may have to be

turned inside out. Place the band around the drive and idler wheels and turn the tension crank so that tension is placed on the band. Be sure that the band slips into the upper and lower guides properly. Replace the filler plate in the worktable.

Adjusting Band Tension

Proper band tension is important to accurate cutting. A high tensile strength band should be used whenever possible. Tensile strength refers to the strength of the band to withstand stretch. The correct band tension is indicated on the **band tension dial** (Figure 26). Adjust the tension for the width of band that you are using. After a new band has been run for a short time, recheck the tension. New bands tend to stretch during their initial running period.

Adjusting Band Tracking

Band tracking refers to the position of the band as it runs on the idler wheel tires. On the vertical band machine, the idler wheel can be tilted to adjust tracking position. The band tracking position should be set so that the back of the band just touches the backup

Figure 26. Band tension dial. (The DoAll Company)

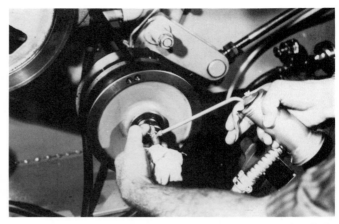

Figure 28. Lubricating the variable speed pulley hub.

Figure 27. Adjusting the band tracking position by tilting the idler wheel.

Figure 29. Band wheel chip brush.

bearing in the guide assembly. Generally, you will not have to adjust band tracking very often. After you have installed a blade, check the tracking position. If it is incorrect, consult your instructor for help in adjusting the tracking position.

The tracking adjustment is made with the motor off and the speed range transmission in neutral. This permits the band to be rolled by hand. Two knobs are located on the idler wheel hub. The outer knob (Figure 27) tilts the wheel. The inner knob is the tilt lock. Loosen the lock knob and adjust the tilt of the idler wheel while rolling the band by hand. When the correct tracking position is reached, lock the inner knob. If the band machine has three idler wheels, adjust band tracking on the top wheel first. Then adjust tracking position on the back wheel.

OTHER ADJUSTMENTS
ON THE VERTICAL BAND MACHINE

The hub of the variable speed pulley should be lubricated weekly (Figure 28). While the drive mechanism guard is open, check the oil level in the speed range transmission. The band machine may be equipped with a chip brush on the band wheel (Figure 29). This should be adjusted frequently. Chips that are transported through the band guides can score the band and make it brittle. The hydraulic oil level should be checked daily on band machines with hydraulic table feeds (Figure 30).

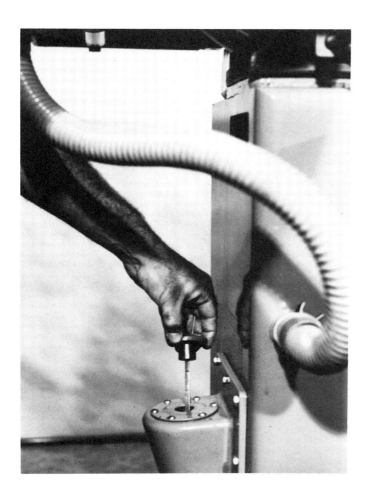

Figure 30. Checking the hydraulic oil level on the vertical band machine.

SELF-TEST

1. Describe the blade end grinding procedure.
2. Describe the band welding procedure.
3. Describe the weld grinding procedure.
4. What is the purpose of the band blade guide?
5. Why is it important to use a band guide of the correct width?
6. What tool can be used to adjust band guides?
7. What is the function of annealing the band weld?
8. Describe the annealing process.
9. What is band tracking?
10. How is band tracking adjusted?
11. Obtain a worn out band blade and cut several suitable lengths. Practice end grinding, welding, annealing, and weld grinding procedures.
12. Practice folding band saw blades into small coils suitable for storage.

UNIT 5 USING THE VERTICAL BAND MACHINE

After a machine tool has been properly adjusted and set up, it can be used to accomplish a machining task. In the preceding unit, you had an opportunity to prepare the vertical band machine for use. In this unit, you will be able to operate this versatile machine tool.

OBJECTIVES

After completing this unit, you should be able to:
1. Use the vertical band machine job selector.
2. Operate the band machine controls.
3. Perform typical sawing operations on the vertical band machine.

SELECTING A BLADE
FOR THE VERTICAL BAND MACHINE

Blade materials include standard carbon steel where the saw teeth are fully hardened but the back of the blade remains soft. The standard carbon steel blade is available in the greatest combination of width, set, pitch, and gage.

The carbon alloy steel blade has hardened teeth and also a hardened back. The harder back permits sufficient flexibility of the blade but, because of increased tensile strength, a higher band tension may be used. Because of this, cutting accuracy is greatly improved. The carbon alloy blade material is well suited to contour sawing.

High speed steel and bimetallic high speed steel blade materials are used in high production and severe sawing applications where blades must have long wearing characteristics. The high speed steel blade can withstand much more heat than the carbon or carbon alloy materials. On the bimetallic blade, the cutting edge is made from one type of high speed steel, while the back is made from another type of high speed steel that has been selected for high flexibility and high tensile strength. High speed and bimetallic high speed blades can cut longer, faster, and more accurately.

Band blade selection will depend on the sawing task. You should review saw blade terminology discussed in Section D, Unit 1, "Cutoff Machines." The first consideration is blade pitch. The pitch of the blade should be such that at least two teeth are in contact with the workpiece. This generally means that fine pitch blades with more teeth per inch will be used in thin materials. Thick material requires coarse pitch blades so that chips will be more effectively cleared from the kerf.

Remember that there are three tooth **sets** that can be used (Figure 1). **Raker** and **wave** set are the most common in the metalworking industries. **Straight** set may be used for cutting thin materials. Wave set is best for accurate cuts through materials with variable cross sections. Raker set may be used for general purpose sawing.

You also have a choice of **tooth forms** (Figure 2). **Precision** or **regular** tooth form is best for accurate cuts where a good finish may be required. **Hook** form

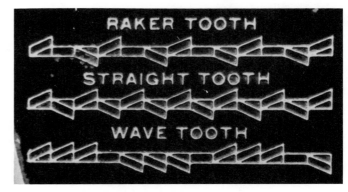

Figure 1. Blade set patterns.

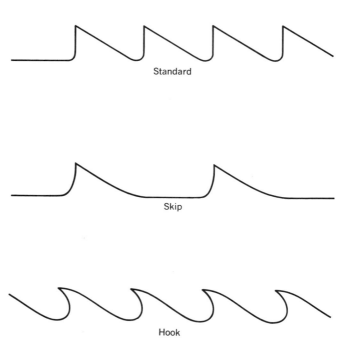

Figure 2. Saw tooth forms. *(Machine Tools and Machining Practices)*

is fast cutting but leaves a rougher finish. **Skip** tooth is useful on deep cuts where additional chip clearance is required.

Several special bands are also used. **Straight, scalloped,** and **wavy edges** are used for cutting nonmetallic substances where saw teeth would tear the material (Figure 3). **Continuous** (Figure 4) and **segmented** (Fig-

Figure 3. Straight, scalloped, and wavy edge bands. (The DoAll Company)

Figure 4. *(left)* Continuous edge diamond band. (The DoAll Company)

Figure 5. *(right)* Segmented diamond edge band. (The DoAll Company)

ure 5) **diamond edged bands** are used for cutting very hard nonmetallic materials.

USING THE JOB SELECTOR ON THE VERTICAL BAND MACHINE

Most vertical band machines are equipped with a **job selector.** This device will be of great aid to you in accomplishing a sawing task. Job selectors are usually attached to the machine tool. They are frequently arranged by material. The material to be cut is located on the rim of the selector. The selector disk is then

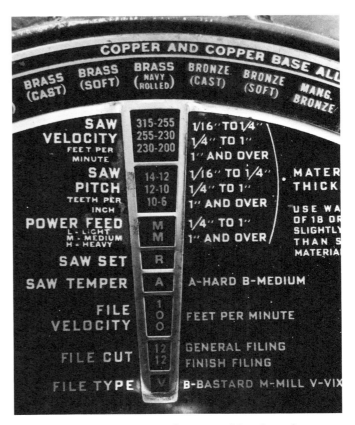

Figure 6. Job selector on the vertical band machine.

turned until the sawing data for the material can be read (Figure 6).

The job selector yields much valuable information. Sawing velocity in feet per minute is the most important. The band must be operated at the correct cutting speed for the material. If it is not, the band may be damaged or productivity will be low. Saw velocity is read at the top of the column and is dependent on material thickness. The job selector also indicates recommended pitch, set, feed, and temper. The job selector will provide information on sawing of nonmetallic materials (Figure 7). Information on band filing can also be determined from the job selector.

SETTING SAW VELOCITY ON THE VERTICAL BAND MACHINE

Most vertical band machines are equipped with a **variable speed drive** that permits a wide selection of band velocities. This is one of the factors that make the band machine such a versatile machine tool. Saw velocities can be selected that permit successful cutting of many materials.

The typical variable speed drive uses a split flange

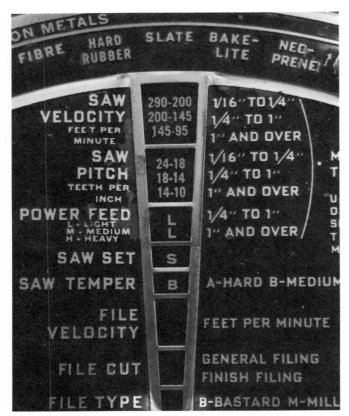

Figure 7. The job selector set for a nonferrous material.

Figure 8. Vertical band machine variable speed drive. (The DoAll Company)

pulley to vary the speed of the drive wheel (Figure 8). As the flanges of the pulley are spread apart by adjusting the speed control, the belt runs deeper in the pulley groove. This is the same as running the drive belt on a smaller diameter pulley. Slower speeds

Figure 9. Band velocity indicator.

are obtained. As the flanges of the pulley are adjusted for less spread, the belt runs toward the outside. This is equivalent to running the belt on a larger diameter pulley. Faster speeds are obtained.

Setting Band Velocity

Band velocity is indicated on the **band velocity indicator** (Figure 9). Remember that band velocity is measured in **feet per minute**. The inner scale indicates band velocity in the low speed range. The outer scale indicates velocity in the high speed range. Band velocity is regulated by adjusting the speed control (Figure 10). Adjust this control only while the motor is running, as this adjustment moves the flanges of the variable speed pulley.

Setting Speed Ranges

Most band machines with a variable speed drive have a **high** and **low speed range**. High or low speed range is selected by operating the **speed range shift lever** (Figure 10). This setting must be made while the band is stopped or is running at the lowest speed in the range. If the machine is set in high range and it is desired to go to low range, turn the band velocity control wheel until the band has slowed to the lowest speed possible. The speed range shift may now be changed to low speed. If the machine is in low speed and it is desired to shift to high range, slow the band to the lowest speed before shifting speed ranges. A speed range shift made while the band is running

Figure 10. Speed range and band velocity controls.

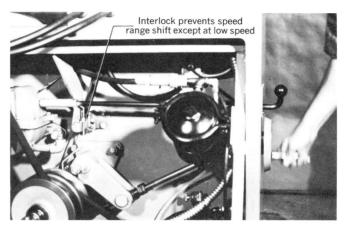

Figure 11. An interlock prevents shifting speed ranges except at low speed.

Figure 12. Adjusting the upper guide post. (The DoAll Company)

at a fast speed may damage the speed range transmission gears. Some band machines are equipped with an interlock to prevent speed range shifts except at low band velocity (Figure 11).

STRAIGHT CUTTING
ON THE VERTICAL BAND MACHINE

Adjust the upper guidepost so that it is as close to the workpiece as possible (Figure 12). This will maximize safety by properly supporting and guarding the band. Accuracy of the cut will also be aided. The guidepost is adjusted by loosening the clamping knob and moving the post up or down according to the workpiece thickness.

Be sure to use a band of the proper pitch for the thickness of the material to be cut. If the band pitch is too fine, the teeth will clog (Figure 13). This can result in stripping and breakage of the saw teeth due to overloading (Figure 14). Cutting productivity will also be reduced. Slow cutting will result from using a fine pitch band on thick material (Figure 15). The correct pitch for thick material (Figure 16) results in much more efficient cutting in the same amount

of time and at the same feeding pressure.

As you begin a cut, feed the workpiece gently into the band. A sudden shock will cause the saw teeth to chip or fracture (Figure 17). This will reduce band life quickly. See that chips are cleared from the band guides. These can score the band (Figure 18), making it brittle and subject to breakage.

Figure 13. Using too fine a pitch blade results in clogged teeth.

Figure 14. Stripped and broken teeth resulting from overloading the saw. (The DoAll Company)

Cutting Fluids

Cutting fluids are an important aid to sawing many materials. They cool and lubricate the band and remove chips from the kerf. Many band machines are equipped with a mist coolant system. Liquid cutting fluids are mixed with air to form a mist. With mist, the advantages of the coolant are realized without

Figure 15. Saw cut with a fine pitch blade in thick material.

Figure 16. Saw cut with correct pitch band for thick material.

Figure 17. Chipped and fractured teeth resulting from shock and vibration. (The DoAll Company)

the need to collect and return large amounts of liquids to a reservoir. If your band machine uses mist coolant, set the liquid flow first and then add air to create a mist (Figure 19). Do not use more coolant than is

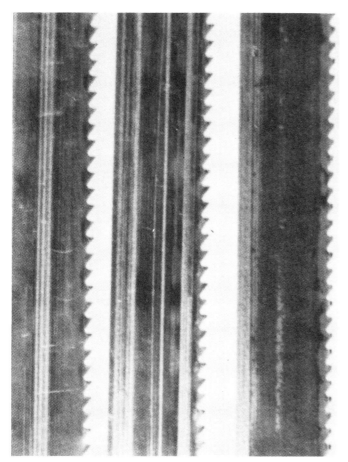

Figure 18. Scored bands can become brittle and lose flexibility. (The DoAll Company)

necessary. Overuse of air may cause a mist fog around the machine. This is both unpleasant and hazardous, as coolant mist should not be inhaled.

CONTOUR CUTTING
ON THE VERTICAL BAND MACHINE

Contour cutting is the ability of the band machine to cut around corners and produce intricate shapes. The ability of the saw to cut a specific radius depends on its **width.** The job selector will provide information on the minimum radius that can be cut with a blade of a given width (Figure 20). As you can see, a narrow band can cut a smaller radius than a wide band.

The set of the saw needs to be adequate for the corresponding band width. It is a good idea to make a test contour cut in a piece of scrap material. This will permit you to determine if the **saw set** is adequate to cut the desired radius. If the saw set is not adequate, you may not be able to keep the saw on the layout line as you complete the radius cut (Figure 21).

If you are cutting a totally enclosed feature, be sure to insert the saw blade through the starting hole in the workpiece before welding it into a band. Also, be sure that the teeth are pointed in the right direction.

Figure 19. Adjusting the coolant and air mix on the vertical band machine.

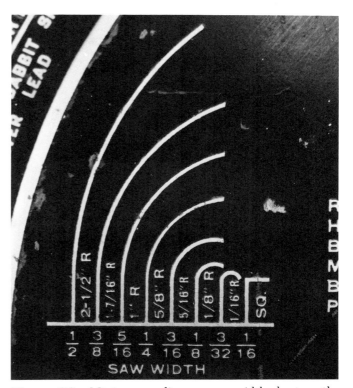

Figure 20. Minimum radius per saw width chart on the job selector.

Figure 21. Band set must be adequate for the band width if the layed out radius is to be cut.

SELF-TEST

1. Name three saw blade sets and describe the applications of each.
2. When might scalloped or wavy edged bands be used?
3. What information is found on the job selector?
4. In what units of measure are band velocities measured?
5. Explain the operating principle of the variable speed pulley.
6. Explain the selection of band speed ranges.
7. What machine safety precaution must be observed when selecting speed ranges?
8. Describe the upper guidepost adjustment.
9. What does band pitch have to do with sawing efficiency?
10. What does band set have to do with contouring?

SECTION E

LAYOUT AND SHOP DRAWINGS

UNIT 1 READING SHOP DRAWINGS
UNIT 2 LAYOUT TOOLS
UNIT 3 LAYOUT PRACTICES

UNIT 1 READING SHOP DRAWINGS

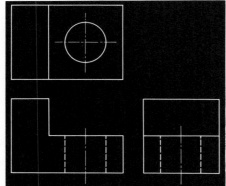

From earliest times, man has communicated his thoughts through drawings. The pictorial representation of an idea is a vital line of communication between the designer and the people who produce the final product. Technological design would be impossible were it not for the several different ways that an idea may be represented by a drawing. The drawing also provides an important testing phase for an idea. Many times an idea may be rejected at the drawing board stage before a large investment is made to equip a manufacturing facility and risk production of an item that does not meet all design requirements.

This does not mean that all design problems can be solved in the drafting room. Almost anything can be represented by a drawing, even to the extent that some designs can be quite impossible to manufacture. It is important that the designer be aware of the problems that confront the machinist. On the other hand, you must fully understand all of the symbols and terminology on the designer's drawing. You must then interpret these terms and symbols in order to transform the ideas of the designer into useful products.

OBJECTIVES

After completing this unit, you should be able to:
1. Identify symbols and terminology on working drawings.
2. Complete a simple orthographic drawing.
3. Read and interpret a typical working drawing for a machinist.

PICTORIAL REPRESENTATIONS

Perspective Drawings

The perspective drawing (Figure 1) is used when it is desired to show an object as it would appear to the eye. Perspective can be either one point, known as parallel perspective, or two point, known as angular perspective. In the perspective view, the lines of the object recede to a single point. The preception of depth is indicated.

Isometric Drawing

An isometric drawing (Figure 2) is also intended to represent an object in three dimensions. However, unlike the perspective, the object lines do not recede but remain parallel. Furthermore, isometric views are drawn about the three isometric axes that are 120 degrees apart.

Oblique Drawing

Object lines in the oblique drawing (Figure 3) also remain parallel. The oblique differs from the isometric in that one axis of the object is parallel to the plane of the drawing.

The perspective view is used mainly by artists and technical illustrators. You will seldom come in contact with a perspective drawing. However, you should be aware of its existence. Isometric and oblique are also not generally used as working drawings for the machinist. However, you may occasionally see them in the machine shop.

Exploded Drawings

The exploded drawing (Figure 4) is a type of pictorial drawing designed to show several parts in their proper location prior to assembly. Although the exploded view is not used as the working drawing for the ma-

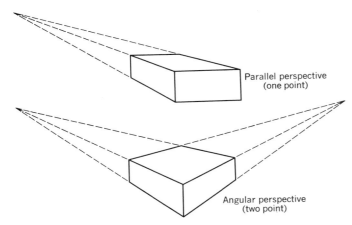

Figure 1. Perspective drawing. *(Machine Tools and Machining Practices)*

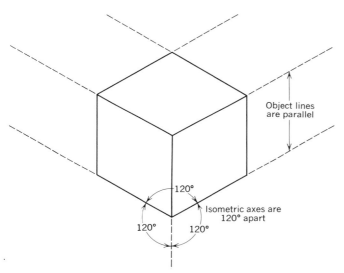

Figure 2. Isometric drawing. *(Machine Tools and Machining Practices)*

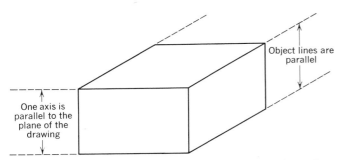

Figure 3. Oblique drawing. *(Machine Tools and Machining Practices)*

chinist, it has an important place in mechanical technology. Exploded views appear extensively in manuals and handbooks that are used for repair and assembly of machines and other mechanisms.

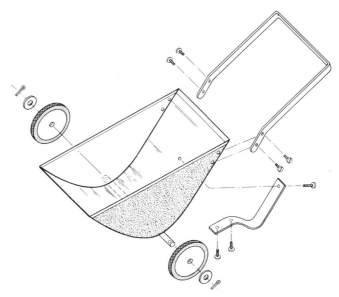

Figure 4. Exploded drawing. *(Machine Tools and Machining Practices)*

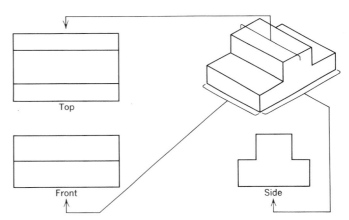

Figure 5. Standard orthographic drawings. *(Machine Tools and Machining Practices)*

ORTHOGRAPHIC DRAWINGS

The Orthographic Projection Drawing

In almost every case, the working drawing for the machinist will be in the form of the **three view** or **orthographic drawing.** The typical orthographic format always shows an object in the three view combination of side, end, and top (Figure 5). In some cases, an object can be completely shown by a combination of only two orthographic views. However, any orthographic drawing must have a minimum of two views in order to show an object completely. The top view is referred to as the **plan** view. The front or side views are referred to as **elevation** views. The terms plan

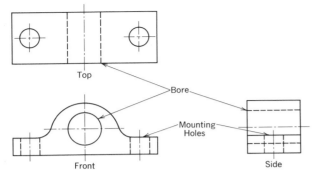

Figure 6. Hidden lines for part features not visible. *(Machine Tools and Machining Practices)*

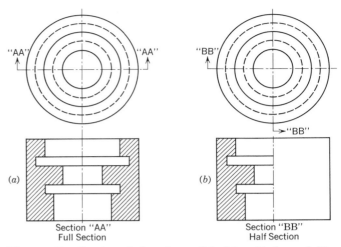

Figure 7. Sectioned drawings. *(Machine Tools and Machining Practices)*

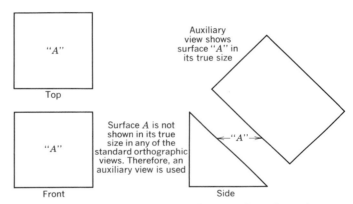

Figure 8. Auxiliary view. *(Machine Tools and Machining Practices)*

and elevation may appear on some drawings, especially those of large complex parts or assemblies.

Hidden Lines for Part Features Not Visible

Features that are not visible are indicated by dotted lines. These are called **hidden lines** as they indicate the locations of part features hidden from view. The plain bearing (Figure 6) is shown in a typical orthographic drawing. The front view is the only one in which the hole through the bearing, or bore, can be observed. In the side and top views, the bore is not visible. Therefore, it is indicated by dotted or hidden lines. The mounting holes through the base are visible only in the top view. They appear as hidden lines in the front and side views.

Sectioned Views

When internal features are complex to the extent that indicating them as hidden lines would be confusing, a **sectioned drawing** may be employed. Two common styles of sections are used. In the **full section** (Figure 7*a*), the object has been cut completely through. In the **half section** (Figure 7*b*), one quarter of the object is removed. The section indicator line shows the plane at which the section is taken. For example (Figure 7*a*), the end view of the object shows the section line marked by the symbol "AA." The section line "BB" (Figure 7*b*) indicates the portion removed in the half section. An object may be sectioned at any plane as long as the section plane is indicated on the drawing.

Auxiliary Views

One of the reasons for adopting the orthographic drawings is to represent an object in its true size and shape. This is not possible with the pictorial drawings discussed earlier. Generally, the orthographic drawing meets this requirement. However, the shape of certain objects is such that their actual size and shape

are not truly represented. An **auxiliary view** may be required (Figure 8). On the object shown in the figure, surface "A" does not appear in its true size in any of the standard orthographic views. Therefore, surface "A" is projected to the auxiliary view, thus revealing its true size.

READING AND INTERPRETING DRAWINGS

Scale

In some cases, an object may be represented by a drawing that is the same size as the object. In other cases, an object may be too large to draw full size, or a very small part may be better represented by a drawing that is larger. Therefore, all drawings are drawn to a specific **scale.** For example, when the draw-

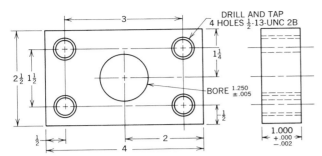

Figure 9. Fractional inch dimensioning. *(Machine Tools and Machining Practices)*

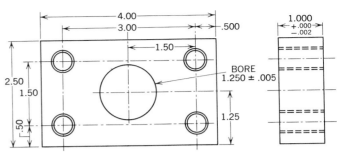

Figure 10. Decimal inch dimensioning. *(Machine Tools and Machining Practices)*

ing is the same size as the object, the scale is said to be full, or 1 = 1. If the drawing is one-half size, the scale is one-half, or $\frac{1}{2}$ = 1. A drawing twice actual size would be double scale, or 2 = 1. The scale used is generally indicated on the drawing.

Dimensioning of Detail and Assembly Drawings

You will primarily come in contact with the **detail drawing**. This is a drawing of an individual part and, in almost all cases, will appear in orthographic form. Depending on the type of work a machinist may be doing, he may also see an **assembly drawing**. The assembly drawing is a drawing of subassemblies or several individual parts assembled into a complete unit. For example, a drawing of a complete automobile engine would be an assembly drawing. In addition, a detail drawing of each engine component would also exist.

A detail drawing contains all of the essential information needed by you in order to make the part. Most important are the **dimensions**. Dimensioning refers generally to the sizes specified for the part and the locations of its features. Furthermore, dimensions reflect many design considerations, such as the fit of mating parts, that will affect the operating characteristics of all machines. Much of the effort that you expend performing the various machining operations will be directed toward controlling the dimensions specified on the drawing.

Several styles of dimensioning appear on drawings. The most common of these is the standard **fractional inch** notation (Figure 9). The outline of the part along with the several holes are dimensioned according to size and location. Generally, the units of the dimensions are not shown. Note that certain dimensions are specified to come within certain ranges. This is known as **tolerance.**

Tolerance refers to an acceptable range of part size or feature location and is generally expressed in the form of a minimum and maximum limit. The bore (Figure 9) is shown to be 1.250 in. ± .005 in. This notation is called a **bilateral tolerance** because the acceptable size range is both above and below the nominal (normal) size of 1.250 in. The bored hole could be any size from 1.245 to 1.255 in diameter. The thickness of the part is specified as 1.000 + .000 and − .002. This tolerance is **unilateral** as all the range is on one side of nominal. Thus, the thickness could range from .998 in. to 1.000 in.

No tolerance is specified for the outside dimensions of the part or the locations of the various features. Since these dimensions are indicated in standard fractional form, the tolerance is taken to be plus or minus $\frac{1}{64}$ of an inch unless otherwise specified on the drawing. This range is known as **standard tolerance** and applies only to dimensions expressed in standard fractional form.

Another system of dimensioning used in certain industries is that of **decimal fraction** notation (Figure 10). In this case, tolerance is determined by the number of places indicated in the decimal notation.

2 places	.00 tolerance is	±.010 in.
3 places	.000 tolerance is	±.005 in.
4 places	.0000 tolerance is	±.0005 in.

Always remember that standard tolerances apply only when no other tolerance is specified on the drawing.

The **coordinate** or **absolute** system of dimensioning (Figure 11) may be found in special applications such as numerically controlled machining. In this system, all dimensions are specified from the same zero point. The figure shows the dimensions expressed in decimal form. Standard fraction notation may also be used. Standard tolerances apply unless otherwise specified.

With the increase in the use of the metric system in recent years, some industries have adopted a system of dual dimensioning of drawings with both metric and inch notation (Figure 12). Dual dimensioning has, in some cases, created a degree of confusion for the machinist. Hence, industry is constantly devising improved methods by which to differentiate metric and inch drawing dimensions. You must use caution when reading a dual-dimensioned drawing to insure that you are conforming your work to the proper system of measurement for your tools. In the figure, metric dimensions appear above the line and inch dimensions appear below the line.

Abbreviations for Machine Operations

Working drawings contain several symbols and abbreviations that convey important information to the machinist (Figure 13). Certain machining operations may

be abbreviated on a drawing. For example, countersinking is a machining operation in which the end of a hole is shaped to accept a flat head screw, and on a drawing countersinking may be abbreviated as C'SINK. The desired angle will also be specified. Table 1 gives some of the abbreviations used on mechanical drawings.

Table 1
Abbreviations on Drawings

Symbol	Definition
BHN	Brinell Hardness Number
B.C.	Bolt circle diameter
C'BORE	Counterbore
C'SINK	Countersink
C to C	Center to center
D, Dia., Diam.	Diameter
F.A.O.	Finish all over
Hardn & Gr	Harden and grind
I.D.	Inside diameter
O.D.	Outside diameter
R, Rad.	Radius
Rc	Rockwell, C scale
S'FACE	Spotface
S.H.C S.	
Soc. Hd. Cp. Scr.	Socket head capscrew
Stl.	Steel
Scr.	Screw
T.I.R.	Total indicator reading
Typ.	Typical

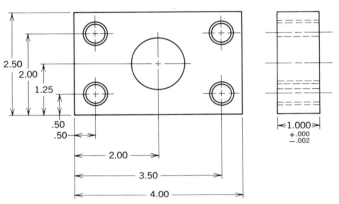

Figure 11. Absolute or coordinate dimensioning. (*Machine Tools and Machining Practices*)

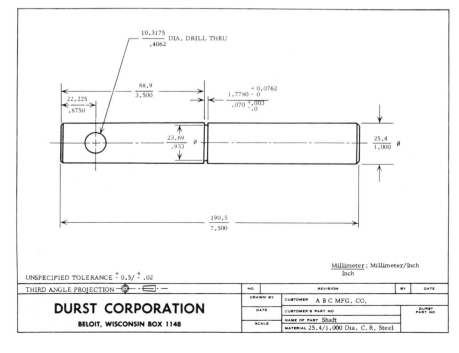

Figure 12. Dual dimensioning: metric and inch. (*Machine Tools and Machining Practices*)

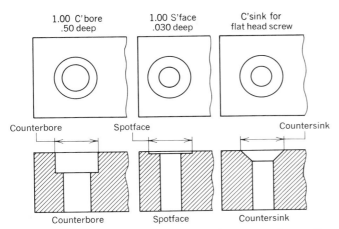

Figure 13. Countersinking, counterboring, and spot-facing symbols. *(Machine Tools and Machining Practices)*

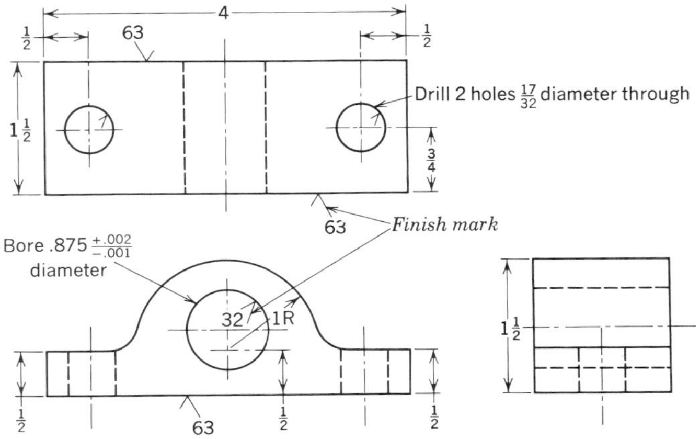

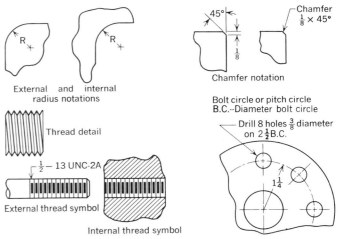

Figure 14. Finish marks. *(Machine Tools and Machining Practices)*

Finish Marks and Symbols

Very often you will perform work on a part that has already been partially shaped. An example of this might be a casting or forging. The finish mark, shaped like a letter "f" and sometimes used to indicate which surfaces are to be machined, is not used much today, being replaced by a "v" symbol. The point of the v touches the work where machining is needed. Surface finish symbols, as shown in Figure 14, tells the machinist what kind of finish is required in terms of microinches. For example, a finish mark notation of 4, 32, or 64 refers to a specific surface finish. A surface roughness comparator gage is sometimes used by the machinist to determine relative smoothness. See Section C, Unit 1, Figure 7 for a surface roughness comparator gage.

Other Common Symbols and Abbreviations

External and internal radii are generally indicated by the abbreviation *R* and the specified size (Figure 15). **Chamfers** may be indicated by size and angle as shown in the figure. **Threads** are generally represented by symbols or they may be drawn in detail. Threads will also have a notation that indicates type, size, and fit. Consider the notation $\frac{1}{2} - 13$ UNC 2A. This thread notation indicates the following:

$\frac{1}{2}$—Major thread diameter
13—Number of threads per inch
UNC—Shape and series of thread
2—Class of fit
A—External thread (internal is denoted B)

A specific bolt circle or pitch circle is often indicated by the abbreviation **B.C.**, meaning **Bolt Circle**

Figure 15. Other symbols and abbreviations. *(Machine Tools and Machining Practices)*

diameter. The size of the diameter is indicated by normal dimensioning or with an abbreviation such as $2\frac{1}{2}$ B.C.

MECHANICAL DRAWING FORMATS

A designer's idea may at first appear as a freehand sketch perhaps in one of the pictorial forms discussed previously. After further discussion and examination, the decision may be made to have a part or an assembly manufactured. This necessitates suitable orthographic drawings that can be supplied to the machine shop. The original drawings produced by the drafting department are not used directly by the machine shop. These original drawings must be carefully pre-

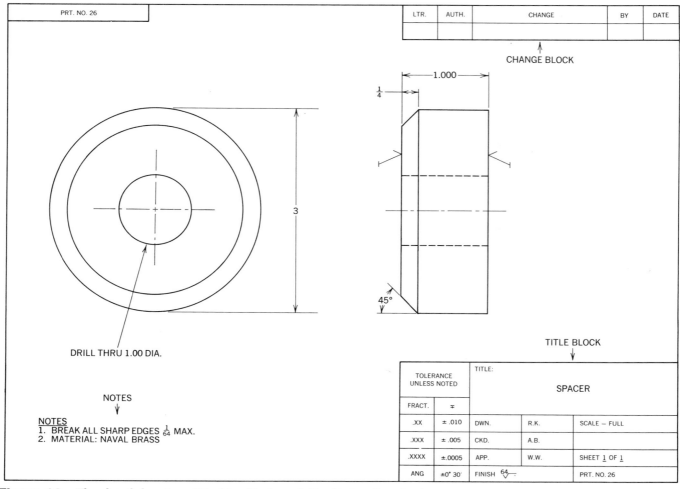

Figure 16. The detail drawing. *(Machine Tools and Machining Practices)*

served, as a great deal of time and money has been invested in them. Were they to be sent directly to the machine shop, they would soon be destroyed by constant handling. Therefore, a copy of the original drawing is made.

Several methods are employed to obtain copies of original drawings. An older method of copying is called blueprinting, where the drawing appears as white lines on a blue background. The process is no longer widely used, but the name has come to mean any of several processes of reproduction. The most commonly used method of reproducing drawings is sometimes called "blue line," where blue lines are on a white background. Any number of prints may be made and distributed to the various departments of a manufacturing facility. For example, assembly prints are needed in the assembly area while prints of individual parts are required at the machine tool stations and in the inspection department.

The typical detail drawing format (Figure 16) contains a suitable title block. In many cases, the name of the firm appears in the title block, as shown in Figure 12. The block also contains the name of the part, specified tolerance, scale, and the initials of the draftsman. A finish mark notation may also appear. The print may also contain a change block. Often designs may be modified after an original drawing is made. Subsequent drawings will reflect any changes, and these will be noted in the change block. A drawing may also contain one or more general notes. The notes contain important information for the machinist. Therefore, you should always find and read any general notes appearing on a print.

A typical assembly drawing format contains essentially the same information as found on the detail drawing plan (Figure 17). However, assembly draw-

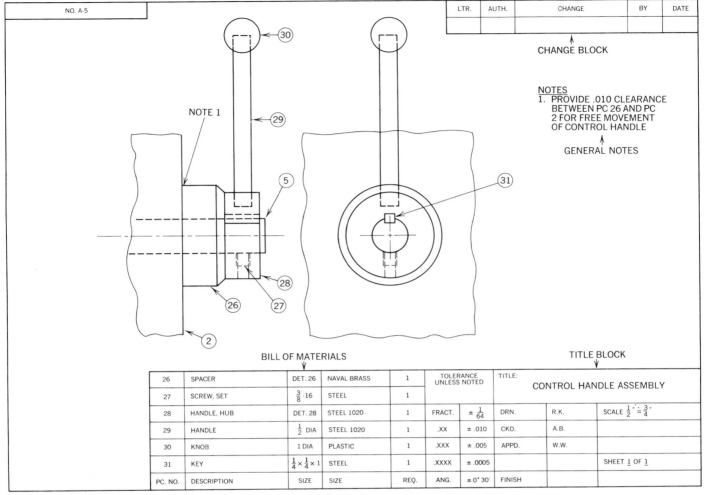

					LTR.	AUTH.	CHANGE	BY	DATE

CHANGE BLOCK

NOTES
1. PROVIDE .010 CLEARANCE BETWEEN PC 26 AND PC 2 FOR FREE MOVEMENT OF CONTROL HANDLE

GENERAL NOTES

BILL OF MATERIALS

TITLE BLOCK

26	SPACER	DET. 26	NAVAL BRASS	1	TOLERANCE UNLESS NOTED		TITLE: CONTROL HANDLE ASSEMBLY		
27	SCREW, SET	$\frac{3}{8}$-16	STEEL	1					
28	HANDLE, HUB	DET. 28	STEEL 1020	1	FRACT.	$\pm \frac{1}{64}$	DRN.	R.K.	SCALE $\frac{1}{2}$"$=\frac{3}{4}$"
29	HANDLE	$\frac{1}{2}$ DIA	STEEL 1020	1	.XX	$\pm$ .010	CKD.	A.B.	
30	KNOB	1 DIA	PLASTIC	1	.XXX	$\pm$ .005	APPD.	W.W.	
31	KEY	$\frac{1}{4} \times \frac{1}{4} \times 1$	STEEL	1	.XXXX	$\pm$ .0005			SHEET 1 OF 1
PC. NO.	DESCRIPTION	SIZE	SIZE	REQ.	ANG.	$\pm 0°30'$	FINISH		

Figure 17. The assembly drawing. *(Machine Tools and Machining Practices)*

ings generally show only those dimensions that pertain to the assembly. Dimensions of the individual parts are found on the detail plans. In addition to the normal information, a bill of materials appears on the assembly plan. This bill contains the part number, description, size, material, and required quantity of each piece in the assembly. Often, the source of a specific item that is not manufactured by the assembler will be specified in the bill of material. An assembly drawing may also include a list of references of detail drawings of the parts in the assembly. Any general notes containing information regarding the assembly will also be included.

SHOP SKETCHING

It often becomes necessary to record pertinent information and dimensions when away from the shop or where drafting equipment is not available. Shop sketching is the method used to record this information for later use when assembling or machining the required parts.

The same standards used for drafting are used for shop drawings; that is, center, section, hidden, dimension, and break lines are all used. It is especially important to use centerlines since most dimensioning is from centerlines or from machined edges (datum or base line). Notes, where needed, should also be included. The most important thing to remember is to get **all** necessary dimensions for later use. In some cases, it may be impractical to return to the location to get an omitted dimension. Since mechanical drawings should not be scaled (measured on the drawing) and the part made only by given dimensions, a crude drawing will suffice to convey the necessary information if all of the needed dimensions are given.

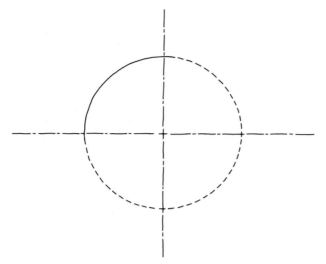

Figure 18. Freehand circles can be made one quadrant at a time.

All that is needed to make a shop drawing is a tablet and pencil and the measuring tools. A small rule will help to make straighter lines but it takes more time to use it. Circles are particularly hard for most people to make freehand.

One method to sketch circles is shown in Figure 18. Centerlines are necessary for all circles as they serve as dimension lines. When one quadrant is drawn at a time, the job is easier.

An example of a large taper shaft to which a new gear must be bored and fitted in a machine shop is shown in Figure 19. The shaft cannot be removed for transport from its remote location. Armed with the correct measuring tools, the machinist goes to the location and makes a sketch. The taper per foot can later be calculated with the given dimensions, and the gear accurately bored to fit.

Shown in Figure 20 is a pipe flange for which a steel plate must be made in the shop for an end cap.

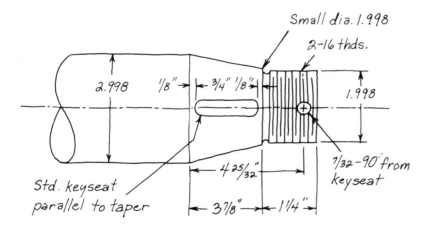

Figure 19. All necessary dimensions and information must be shown in shop sketches.

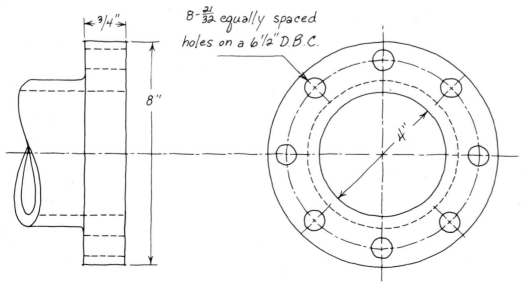

Figure 20. A two-view sketch is sufficient in many cases.

The plate to be made is simply a steel circle with holes drilled in it for bolts, but the holes have to be in the right position, so someone must go to the pipe location and make a sketch.

Two and three view orthographic sketches are often made. Sometimes auxiliary views are necessary. If you are good at sketching, isometric dimensioned drawings are very useful.

SELF-TEST

1. Sketch the object (Figure 21) as it would appear in correct orthographic form.

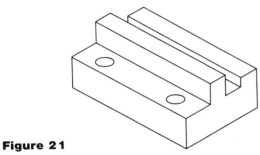

Figure 21

(For Problems 2 to 10 refer to Figure 22.)

2. What is the minimum size of the hole through the clevis head?
3. What length of thread is indicated on the drawing?
4. What is the tolerance of the slot in the clevis head?
5. What radius is specified where the shank and clevis head meet?
6. What is the total length of the part?
7. What is the width of the slot in the clevis head?
8. Name two machining operations specified on the drawing.
9. What is the size and angle of the chamfer on the thread end?
10. What does note 1 mean?

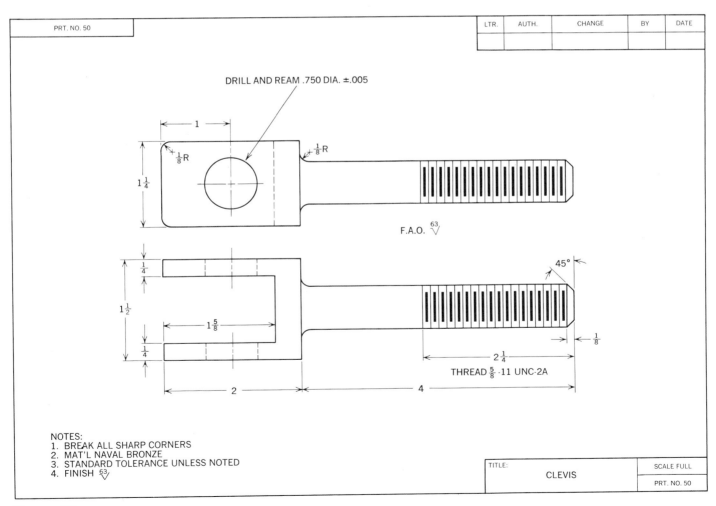

Figure 22

UNIT 2 LAYOUT TOOLS

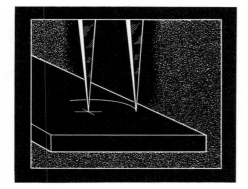

Layout is the process of placing reference marks on the workpiece. These marks may indicate the shape and size of a part or its features. Layout marks often indicate where machining will take place. Before you can cut material for a certain job, you must perform a layout operation. Layout for stock cutoff may involve a simple chalk, pencil, or scribe mark on the material. No matter how simple the layout job may be, you should strive to do it neatly and accurately. In any layout, semiprecision or precision accuracy is the watchword.

Precision layout can be a complex operation making use of sophisticated tools. In the aircraft and shipbuilding industries, reference points, lines, and planes may be layed out using optical and laser instruments. In the machine shop, you will be primarily concerned with layout for stock cutoff, filing and offhand grinding, drilling, milling, and occasionally in connection with lathe work.

OBJECTIVE

After completing this unit, you should be able to:
Identify various layout tools and describe their uses.

LAYOUT CLASSIFICATIONS

The process of layout can be generally classified as **semiprecision** and **precision.** Semiprecision layout is usually done by scale measurement to a tolerance of $\pm\frac{1}{64}$ in. Precision layout is done with tools that discriminate to .001 in. or finer, to a tolerance of $\pm.001$ in. if possible.

TOOLS OF LAYOUT

Layout Dyes

To make layout marks visible on the surface of the workpiece, a **layout dye** is used. Layout dyes are available in several colors. Among these are red, blue, and white. The blue dyes are very common. Depending on the surface color of the workpiece material, different dye colors may make layout marks more visible. Layout dye is available in small containers with a brush applicator, and in spray cans. Layout dye can be removed from the workpiece with acetone. Layout dye should be applied sparingly in an even coat (Figure 1).

Scribers and Dividers

Several types of **scribers** are in common use. The pocket scriber (Figure 2) has a removable tip that can be stored in the handle. This permits the scriber to be carried safely in the pocket. The engineer's scriber (Figure 3) has one straight and one hooked end. The hook permits easier access to the line to be scribed. The machinist's scriber (Figure 4) has only one end with a fixed point. Scribers must be kept sharp. If they become dull, they must be reground or stoned to restore their points. Scriber materials include hardened steel and tungsten carbide.

When scribing against a rule, hold the rule firmly. Tilt the scriber so that the tip marks as close to the rule as possible. This will insure accuracy. An excellent scriber can be made by grinding a shallow angle on a piece of tool steel (Figure 5). This type of scriber is particularly well suited to scribing along a rule. The flat side permits the scriber to mark very close to the rule, thus obtaining maximum accuracy.

Several types of dividers are in common use. The **spring divider** is very common (Figure 6). Spring dividers range in size from 3 to 12 in. The spacing of the divider legs is set by turning the adjusting screw. Dividers are usually set to rules. Engraved rules are best as the divider tips can be set in the engraved rule graduations (Figure 7). Like scribers, divider tips must be kept sharp and at nearly the same length.

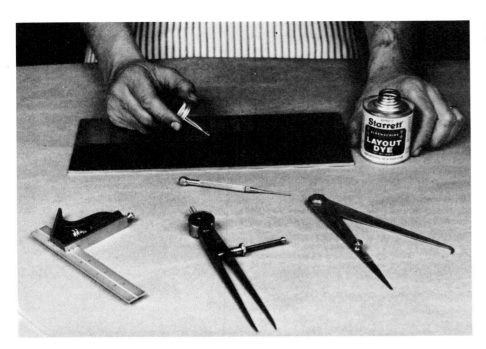

Figure 1. Applying layout dye to the workpiece. (The L.S. Starrett Company)

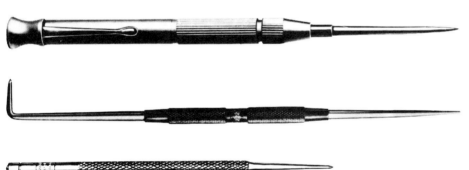

Figure 2. Pocket scriber. (Rank Scherr–Tumico, Inc.)

Figure 3. Engineer's scriber. (Rank Scherr–Tumico, Inc.)

Figure 4. Machinist's scriber. (Rank Scherr–Tumico, Inc.)

Figure 5. Rule scribe made from a high speed toolbit. *(Machine Tools and Machining Practices)*

Hermaphrodite Caliper

The hermaphrodite caliper has one leg similar to a regular divider. The tip is adjustable for length. The other leg has a hooked end that can be placed against the edge of the workpiece (Figure 8). Hermaphrodite calipers can be used to scribe a line parallel to an edge.

The hermaphrodite caliper can also be used to lay out the center of round stock (Figure 9). The hooked leg is placed against the round stock and an arc is marked on the end of the piece. By adjusting the leg spacing, tangent arcs can be layed out. By marking four arcs at 90 degrees, the center of the stock can be established.

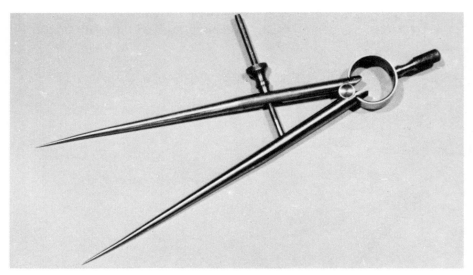

Figure 6. Spring dividers.

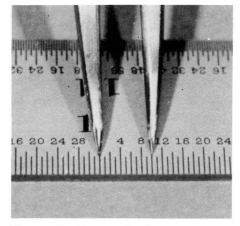

Figure 7. Setting divider points to an engraved rule.

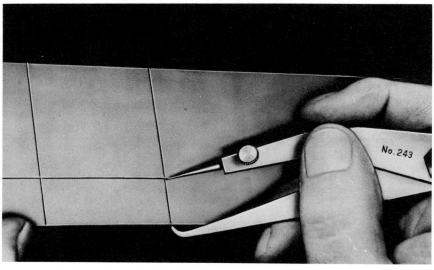

Figure 8. Scribing a line parallel to an edge using a hermaphrodite caliper. (The L.S. Starrett Company)

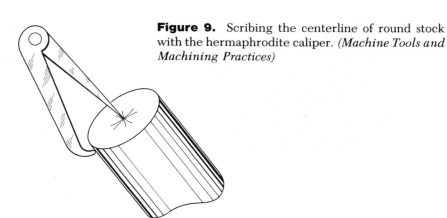

Figure 9. Scribing the centerline of round stock with the hermaphrodite caliper. *(Machine Tools and Machining Practices)*

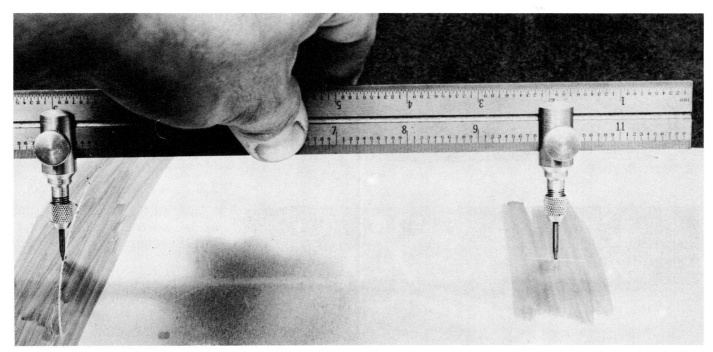

Figure 10. Trammel point attached to a rule. *(Machine Tools and Machining Practices)*

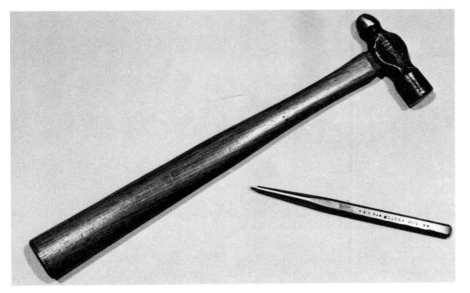

Figure 11. Layout hammer and layout prick punch. *(Machine Tools and Machining Practices)*

Trammel Points

Trammel points are used for scribing circles and arcs when the distance involved exceeds the capacity of the divider. Trammel points are either attached to a bar and set to circle dimensions or they may be clamped directly to a rule where they can be set directly by rule graduations (Figure 10).

Layout Hammers and Punches

Layout hammers are usually light weight machinist's ball peen hammers (Figure 11). A heavy hammer should not be used in layout as it tends to create punch marks that are unnecessarily large. The toolmaker's hammer is also used (Figure 12). This hammer is equipped with magnifier that can be used to help locate a layout punch on a scribe mark (Figure 13).

There is an important difference between a layout punch and a center punch. The layout or prick punch (Figure 12) has an included point angle of 30 degrees. This is the only punch that should be used in layout. The slim point facilitates the locating of the punch on a scribe line. A prick punch mark is

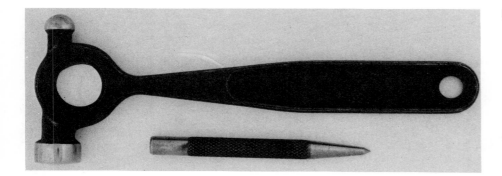

Figure 12. Toolmaker's hammer.

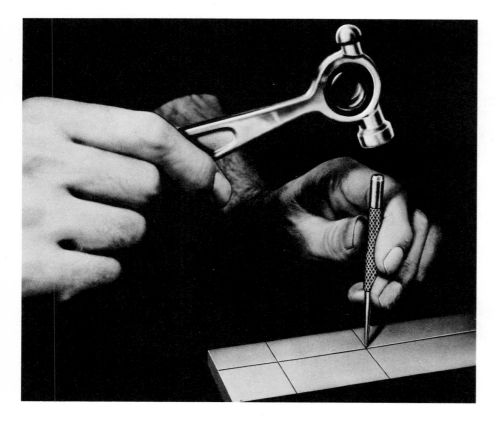

Figure 13. Using a toolmaker's hammer and layout punch. (The L.S. Starrett Company)

only used to preserve the location of a layout mark while doing minimum damage to the workpiece. On some workpieces, depending on the material used and the part application, layout punch marks are not acceptable as they create a defect in the material. A punch mark may affect surface finish or metallurgical properties. Before using a layout punch, you must make sure that it is acceptable. In all cases, layout punch marks should be of minimum depth.

The **center punch** (Figure 14) has an included point angle of 90 degrees and is used to mark the workpiece prior to such machining operations as drilling. A center punch should not be used in place of a layout punch. Likewise, a layout punch should not be used in place of a center punch.

The **automatic center punch** (Figure 15) requires

no hammer. Although called a center punch, its tip is suitably shaped for layout applications (Figure 16). Spring pressure behind the tip provides the required force. The automatic center punch may be adjusted for variable punching force by changing the spring tension. This is accomplished by an adjustment on the handle.

The **optical center punch** (Figure 17) consists of a locator, optical alignment magnifier, and punch. This type of layout punch is extremely useful in locating punch marks precisely on a scribed line or line intersection. The locator is placed over the approximate location and the optical alignment magnifier is inserted (Figure 18). The locator is magnetized so that it will remain in position when used on ferrous metals. The optical alignment magnifier has crossed lines

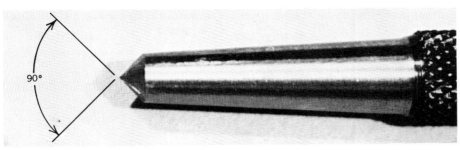

Figure 14. Center punch.

Figure 15. Automatic center punch. (The L.S. Starrett Company)

Figure 16. Using the automatic center punch in layout. (The L.S. Starrett Company)

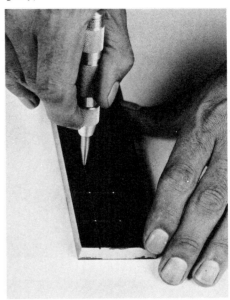

Figure 17. Optical center punch.

Figure 18. Locating the punch holder with the optical alignment magnifier.

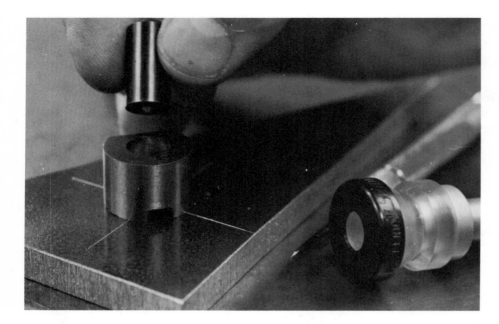

Figure 19. Inserting the punch into the punch holder.

Figure 20. Tapping the punch with a layout hammer.

etched on its lower end. By looking through the magnifier, you can move the locator about until the cross lines are matched to the scribe lines on the workpiece. The magnifier is then removed and the punch is inserted into the locator (Figure 19). The punch is then tapped with a layout hammer (Figure 20).

Centerhead

The centerhead is part of the **machinist's combination set** (Figure 21). Centerheads are used to lay out centerlines on round workpieces (Figure 22). When the centerhead is clamped to the combination set rule,

the edge of the rule is in line with a circle center.

The other parts of the combination set are useful in layout. These include the **rule, square head, and bevel protractor**.

Surface Gage

The surface gage consists of a base, rocker, spindle adjusting screw, and scriber (Figure 23). The spindle of the surface gage pivots on the base and can be moved with the adjusting screw. The scriber can be moved along the spindle and locked at any desired position. The scriber can also swivel in its clamp. A

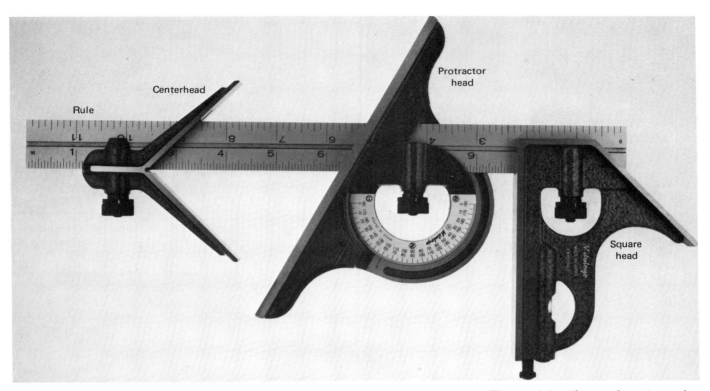

Figure 21. The machinist's combination set is useful for many kinds of layout work. *(Machine Tools and Machining Practices)*

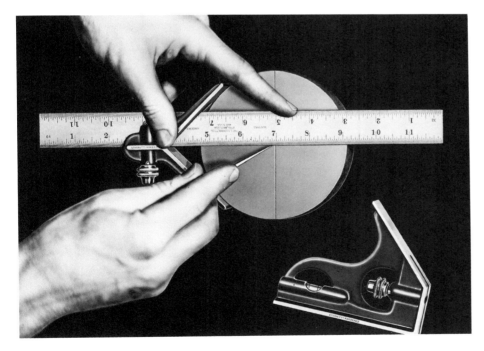

Figure 22. Using the centerhead to lay out a centerline on round stock. (The L.S. Starrett Company)

surface gage may be used as a height transfer tool. The scribe is set to a rule dimension (Figure 24) and then transferred to the workpiece.

The hooked end of the surface gage scriber may be used to mark the centerline of a workpiece. The following procedure should be followed when doing this layout operation. The surface gage is first set as nearly as possible to a height equal to one-half of the part height. The workpiece should be scribed for a short distance at this position. The part should be

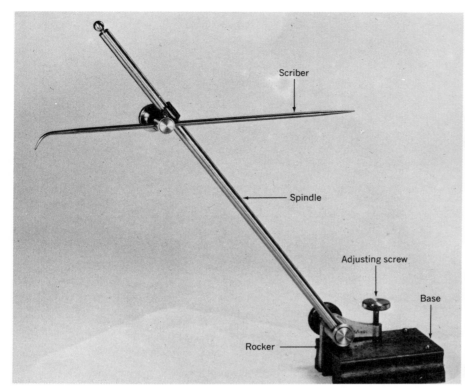

Figure 23.   Parts of the surface gage. (Rank Scherr–Tumico, Inc.)

Figure 25.   Finding the centerline of the workpiece using the surface gage.

Figure 26.   Adjusting the position of the scribe line to center by inverting the workpiece and checking the existing differences in scribe marks.

Figure 24.   Setting a surface gage to a rule.

turned over and scribed again (Figure 25). If a deviation exists, there will be two scribe lines on the workpiece. The surface gage scriber should then be adjusted so that it splits the difference between the two marks (Figure 26). This insures that the scribed line is in the center of the workpiece.

Height Gages

Vernier Height Gage.   The vernier height gage (Figure 27) is the most important tool that you will use in precision layout. The vernier height gage consists of a base, beam, and vernier slide. The scriber is attached to the vernier slide. The capacity of common height gages ranges from 10 to 72 in.

On an inch height gage, the beam is graduated in inches with each inch divided into 10 parts. The $\frac{1}{10}$ in. graduations are further divided into two or four parts depending on the divisions of the vernier. On the 25 division vernier, used on many older height

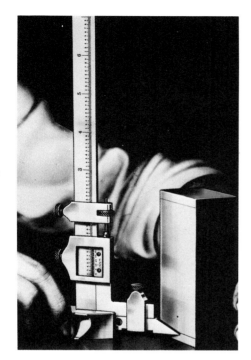

Figure 27. Using the vernier height gage in layout. (The L.S. Starrett Company)

Figure 29. Offset vernier height gage scriber. *(Machine Tools and Machining Practices)*

Figure 28. Straight vernier height gage scriber.

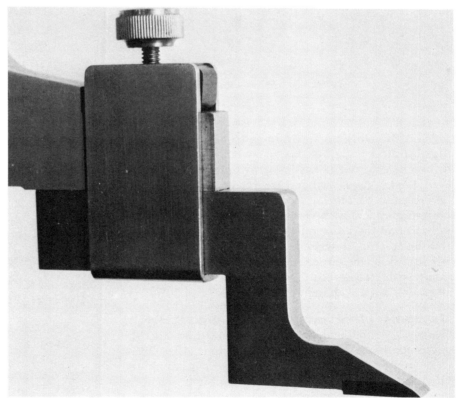

gages, the $\frac{1}{10}$ in. divisions on the beam are graduated into four parts. The vernier permits discrimination to .001 in. Many newer height gages are making use of the 50 division vernier, which permits easier reading. On a height gage with a 50 division vernier, the $\frac{1}{10}$ in. graduations on the beam will be divided into two parts. Discrimination of this height gage is also .001 in. The metric vernier height gage has the beam graduated in millimeters. The vernier contains 50 di-

visions permitting the instrument to discriminate to $\frac{1}{50}$ mm. All vernier height gages are read in the same manner as any instrument employing the principle of the vernier.

The height gage scriber is attached to the vernier slide and can be moved up and down the beam. Scribers are either **straight** (Figure 28) or **offset** (Figure 29). The offset scriber permits direct readings with the height gage. The gage reads zero when the scriber

Figure 30. Universal right angle plate and vee-block used as layout accessories.

rests down on the reference surface. With the straight scriber, the workpiece will have to be raised accordingly, if direct readings are to be obtained. This type of height gage scriber is less convenient.

Height gage scribers are made from tool steel or tungsten carbide. Carbide scribers are subject to chipping and must be treated gently. They do, however, retain their sharpness and scribe very clean narrow lines. Height scribers may be sharpened if they become dull. It is important that any sharpening be done on the slanted surface so that the scriber dimensions will not be changed.

Mechanical Dial and Electronic Digital Height Gages.

Mechanical dial and electronic digital height gages eliminate the need to read a vernier scale. Often these height gages do not have beam graduations. Once set to zero on the reference surface, the total height reading is cumulative on the digital display. This makes beam graduations unnecessary. The electronic digital height gage will discriminate to .0001 in.

LAYOUT ACCESSORIES

Layout accessories are tools that will aid you in accomplishing layout tasks. They are not specifically layout tools as they are used for many other purposes. The layout plate or surface plate used for layout is the

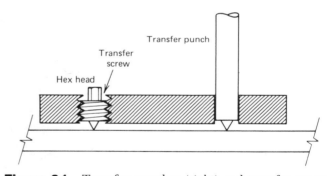

Figure 31. Transfer punches (right) and transfer screws (left) are used to layout a hole pattern from existing holes.

most common accessory as it provides the reference surface from which to work. Other common accessories include vee-blocks and angle plates that hold the workpiece during layout operations (Figure 30).

Transfer punches (Figure 31) come in sets having a range of diameter to fit most drilled holes up to $\frac{1}{2}$ in. They are used for transferring an existing hole pattern from one part to another by making punch marks at the center of the hole. Transfer screws are punches that can be inserted in a threaded hole for transferring a punch mark at the center of the threaded hole to the other part.

SELF-TEST

1. What is semiprecision layout used for in general shop work?
2. Semiprecision layout has a tolerance of $\pm \frac{1}{64}$ in. What tolerance is achieved in precision layout?
3. Why is layout dye used on surfaces to be layed out?
4. Dividers are used to scribe small to medium size circles. What instrument is used to scribe very large circles?
5. What is the function of a prick punch? A center punch?
6. Precision layout punching is done with what instrument?
7. The machinist's combination set is used for various layout jobs. What layout task is the centerhead designed for?
8. The surface gage is sometimes used with a combination square for height transfer on a flat surface. How does a vernier height gage on a surface plate measure height to make layout lines?
9. Precision vee-blocks are used as layout tools. They are accurately made to within ±.0005 in. Name one way they can be used in layout work.
10. Precision right angle layout plates are used in layout and to set up work for precision grinding. What is their purpose?

UNIT 3 LAYOUT PRACTICES

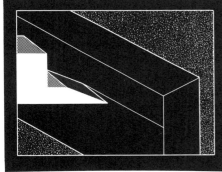

In the last unit, you were introduced to a large number of measuring and layout tools. It is now up to you to put these tools to work in the most productive manner possible. In this unit, you proceed through a typical semiprecision layout task that will familiarize you with basic layout practice.

You were also introduced to one tool used for precision layout: the vernier height gage. Precision layout is generally more reliable and accurate than layout by semiprecision practice. On any job requiring maximum accuracy and reliability, precision layout practice should be used.

The fundamental precision layout tool is the height gage. The vernier height gage is the most common type found in the machine shop. This instrument will discriminate to .001 in. With this ability, a much higher degree of accuracy and reliability is added to a layout task. Whenever possible, you should apply the height gage in all precision layout requirements.

OBJECTIVES

After completing this unit, you should be able to:
1. Prepare the workpiece for layout.
2. Measure for and scribe layout lines on the workpiece, outlining the various features.
3. Locate and establish hole centers, using a layout prick punch and center punch.
4. Lay out a workpiece to a tolerance of $\pm \frac{1}{64}$ in.
5. Identify the major parts of the vernier height gage.
6. Describe applications of the vernier height gage in layout.
7. Read a vernier height gage in both metric and inch dimensions.
8. Accomplish layout using the vernier height gage.

PREPARING THE WORKPIECE FOR LAYOUT

After the material has been cut, all sharp edges should be removed by grinding or filing before placing the stock on the layout table. Place a paper towel under the workpiece to prevent layout dye from spilling on the layout table (Figure 1). Apply a **thin** even coat of layout dye to the workpiece. You will need a drawing of the part in order to do the required layout (Figure 2).

Study the drawing and determine the best way to proceed. The order of steps depends on the layout task. Before some features can be layed out, certain reference lines may have to be established. Measurements for other layout are made from these lines.

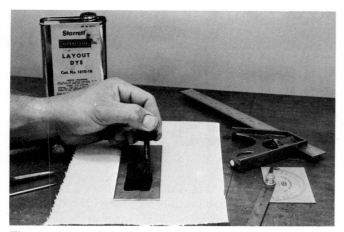

Figure 1. Applying layout dye with workpiece on a paper towel. (Lane Community College)

LAYOUT OF THE DRILL AND HOLE GAGE

If possible, obtain a piece of material the same size as indicated on the drawing. Follow through each step as described in the reading. Refer to the layout drawings to determine where layout is to be done. The pictures will help you to select and use the required tools.

The first operation is to establish the width of the gage. Measure a distance of $1\frac{1}{8}$ in. from one edge of the material. Use the combination square and rule. Set the square at the required dimension and scribe a mark at each end of the stock (Figure 3 and Drawing A).

Remove the square and place the rule carefully on the scribe marks. Hold the rule **firmly** and scribe the line the full length of the material (Figure 4 and Drawing A). Be sure to use a sharp scribe and hold it so that the tip is against the rule. If the scribe is dull, regrind or stone it to restore its point. Scribe a clean visible line. Lay out the 5 in. length from the end of the piece to the angle vertex. Use the combination square and rule (Figure 5 and Drawing B).

Use a plate protractor to lay out the angle. The bevel protractor from the combination set is also a suitable tool for this application. Be sure that the protractor is set to the correct angle. The edge of the protractor blade must be set exactly at the 5 in. mark (Figure 6 and Drawing B). The layout of the 31 degree angle establishes its complement of 59 degrees on the drill gage. This is the correct angle for grinding drills used in general purpose drilling.

The corner radius is $\frac{1}{2}$ in. Establish this dimension

Figure 2. The drill and hole gage. (Lane Community College)

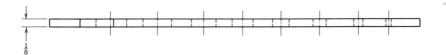

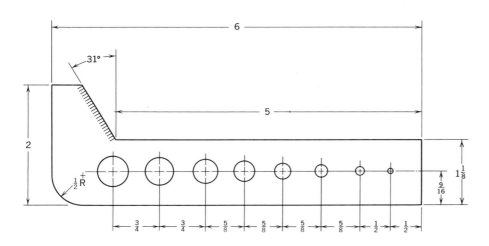

using the square and rule. Two measurements will be required. Measure from the side and from the end to establish the center of the circle (Figure 7 and Drawing C). Prick punch the intersection of the two lines with the 30 degree included point angle layout punch. Tilt the punch so that it can be positioned

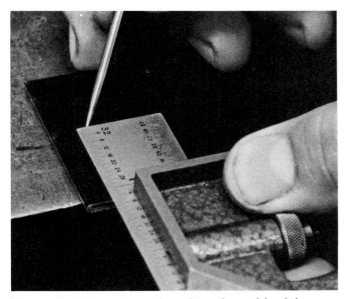

Figure 3. Measuring and marking the width of the gage using the combination square and rule. (Lane Community College)

exactly on the scribe marks (Figure 8). A magnifier will be useful here. Move the punch to its upright position and tap it lightly with the layout hammer (Figure 9 and Drawing C).

Set the dividers to a dimension of $\frac{1}{2}$ in. using the rule. Adjust the divider spacing until you feel the tips drop into the rule engravings (Figure 10). Place one divider tip into the layout punch mark and scribe the corner radius (Figure 11 and Drawing C).

The centerline of the holes is $\frac{9}{16}$ in. from the edge. Use the square and rule to measure this distance. Mark at each end and scribe the line full length (Drawing D). Measure and lay out the center of each hole (Drawing D). Use the layout punch and mark each hole center. After prick punching each hole center, set the dividers to each indicated radius and scribe all hole diameters (Drawing D).

The last step is to center punch each hole center prior to drilling. Use a 90 degree included point angle center punch (Figure 12). Layout of the drill and hole gage is now complete (Figure 13).

THE VERNIER HEIGHT GAGE

The vernier height gage is the most common and important tool used for precision layout. Major parts of the height gage include the base, beam, vernier slide, and scriber (Figure 14). The size of height gages is measured by the maximum height gaging ability

Drawing A. Width line.

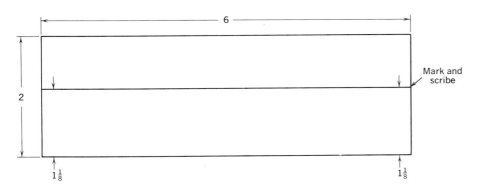

Mark and scribe

6

2

$1\frac{1}{8}$

$1\frac{1}{8}$

Figure 4. Scribing the width line. (Lane Community College)

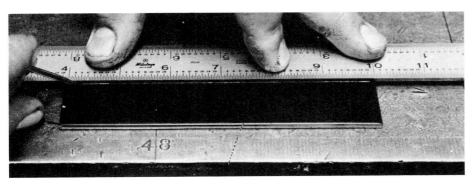

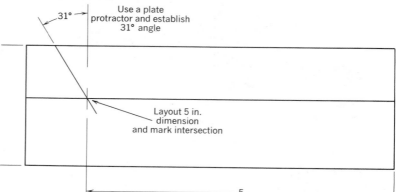

Figure 5. Measuring the 5 in. dimension from the end to the angle vertex. (Lane Community College)

Drawing B. Angle line.

31°

Use a plate protractor and establish 31° angle

Layout 5 in. dimension and mark intersection

5

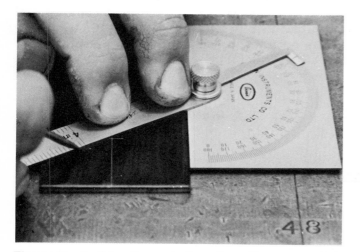

Figure 6. Scribing the angle line. (Lane Community College)

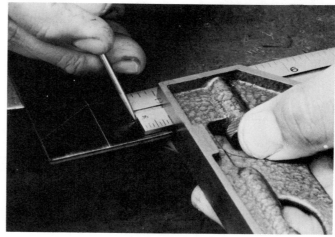

Figure 7. Establishing the center point of the corner radius. (Lane Community College)

Drawing C. Corner radius.

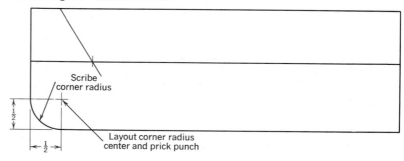

Scribe corner radius

$\frac{1}{2}$

$\frac{1}{2}$

Layout corner radius center and prick punch

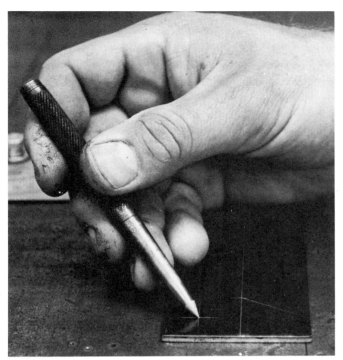

Figure 8. Setting the layout punch on the center point of the corner radius. (Lane Community College)

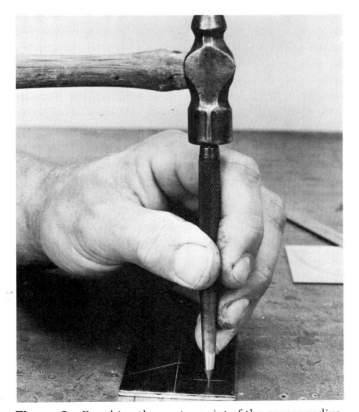

Figure 9. Punching the center point of the corner radius. (Lane Community College)

Figure 10. Setting the dividers to the rule engravings. (Lane Community College)

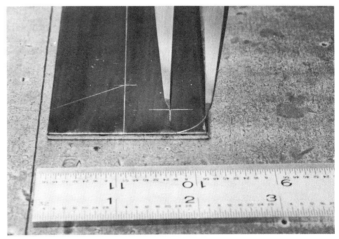

Figure 11. Scribing the corner radius. (Lane Community College)

of the instrument. Height gages range from 10 to 72 in.

The height gage scriber that is used to mark the workpiece is made from hardened steel or tungsten carbide. Many height gage scribers are flat ended; however, rounded end scribers are also used.

READING THE VERNIER HEIGHT GAGE

The vernier height gage is read like any other instrument employing the principle of the vernier. The line on the vernier scale that is coincident with a beam scale graduation must be determined. This value is

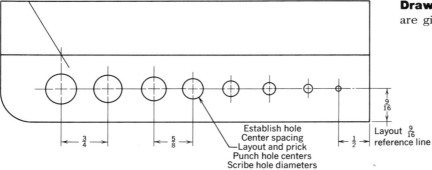

Drawing D. Hole locations. (Note: Hole diameters are given in the worksheet for Section G, Unit 3.)

Establish hole
Center spacing
Layout and prick
Punch hole centers
Scribe hole diameters

Layout $\frac{9}{16}$ reference line

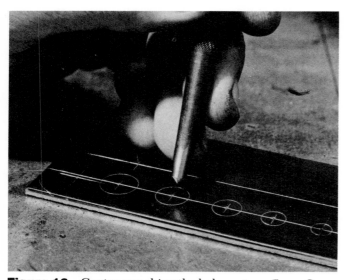

Figure 12. Center punching the hole centers. (Lane Community College)

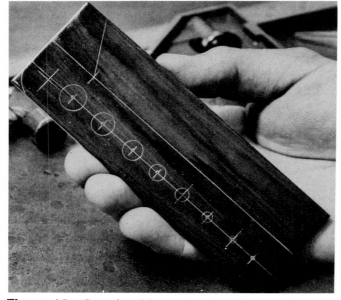

Figure 13. Completed layout for the drill and hole gage. (Lane Community College)

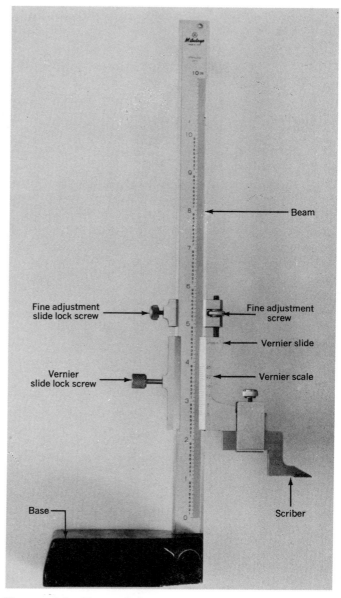

Beam

Fine adjustment screw

Vernier slide

Vernier scale

Fine adjustment slide lock screw

Vernier slide lock screw

Scriber

Base

Figure 14. Parts of the vernier height gage. (*Machine Tools and Machining Practices*)

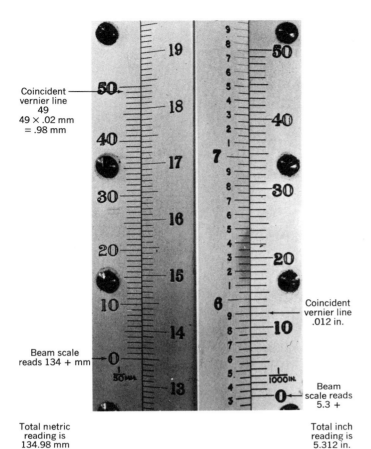

Coincident
vernier line
49
49 × .02 mm
= .98 mm

Beam scale
reads 134 + mm

Coincident
vernier line
.012 in.

Beam
scale reads
5.3 +

Total metric
reading is
134.98 mm

Total inch
reading is
5.312 in.

Figure 15. Reading the 50 division inch/metric vernier height gage.

added to the beam scale reading to make up the total reading. The inch vernier height gage with a 50 division vernier is read as follows (Figure 15, right-hand scale):

Beam reading	5.3
Vernier is coincident at 12	
or .012 in.	.012
Total reading	5.312 in.

On the 50 division inch vernier height gage, the beam scale is graduated in $\frac{1}{10}$ in. graduations. Each $\frac{1}{10}$ in. increment is further divided into two parts. If the zero on the vernier is past the .050 in. mark on the beam, .050 in. must be added to the reading.

The metric vernier scale also has 50 divisions, each equal to $\frac{1}{50}$ or .02 millimeter (Figure 15, left-hand scale):

Beam scale reading	134 mm
Vernier coincident at line 49	
49 × .02 =	.98 mm
Total reading	134.98 mm

On the 25 division inch vernier height gage the $\frac{1}{10}$ in. beam graduations are divided into four parts, each equal to .025 in. Depending on the location of the vernier zero mark, .025, .050, or .075 in. may have to be added to the beam reading. The inch vernier height gage with a 25 division vernier is read as follows (Figure 16):

Beam	5.0
Vernier coincident at 17 or .017	.017
Total reading	5.017 in.

CHECKING THE ZERO REFERENCE ON THE VERNIER HEIGHT GAGE

The height gage scriber must be checked against the reference surface before attempting to make any height measurements of layouts. Clean the surface of the layout table and the base of the gage. Slide the scriber down until it just rests on the reference surface. Check the alignment of the zero mark on the vernier scale with the zero mark on the beam scale. The two marks should coincide exactly (Figure 17). Hold the height gage base firmly against the refer-

Vernier coincident
at .017 in.

Beam scale
reading 5.0 in.

Figure 16. Reading the 25 division inch vernier height gage. (*Machine Tools and Machining Practices*)

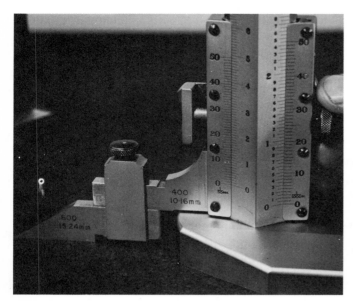

Figure 17. Checking the zero reference.

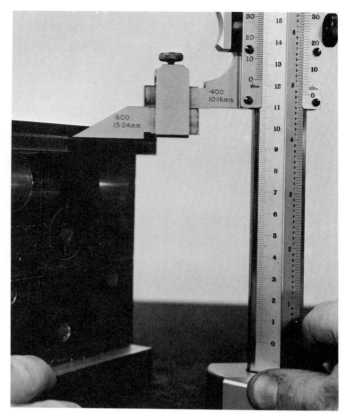

Figure 18. Scribing height lines with the vernier height gage.

ence surface. Be sure that you do not tilt the base of the height gage by sliding the vernier slide past the zero point on the beam scale. If the zero marks on the vernier and beam do not coincide after the scriber has contacted the reference surface, an adjustment of the vernier scale is required.

Some height gages do not have an adjustable vernier scale. A misalignment in the vernier and the beam zero marks may indicate a loose vernier slide, an incorrect scriber dimension, or a beam that is out of perpendicular with the base. Loose vernier slides may be adjusted and scriber dimensions can be corrected. However, if the beam is out of perpendicular with the base, the instrument is unreliable because of angular error. A determination of such a condition can be made by an appropriate calibration process. All height gages, particularly those with nonadjustable verniers, must be treated with the same respect as any precision instrument that you will use.

APPLICATIONS OF THE VERNIER HEIGHT GAGE IN LAYOUT

The primary function of the vernier height gage in layout is to measure and scribe lines of known height on the workpiece (Figure 18). Perpendicular lines may be scribed on the workpiece by the following procedure. The work is first clamped to a right angle plate if necessary and the required lines are scribed in one direction. The height gage should be set at an angle to the work and the corner of the scriber pulled across while keeping the height gage base firmly on the ref-

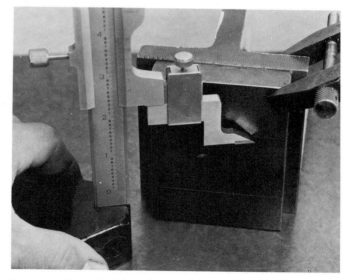

Figure 19. Scribing layout lines with the workpiece clamped to a right angle plate. (*Machine Tools and Machining Practices*)

Figure 20. After turning the workpiece 90 degrees, it can be checked with a square. *(Machine Tools and Machining Practices)*

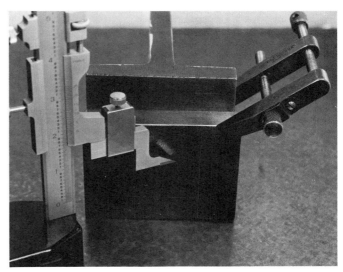

Figure 22. Scribing perpendicular lines. *(Machine Tools and Machining Practices)*

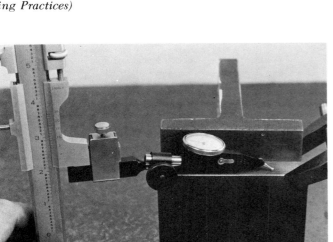

Figure 21. Checking the work using a dial test indicator. *(Machine Tools and Machining Practices)*

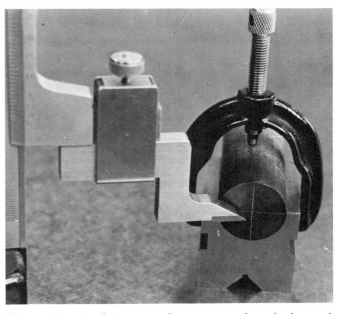

Figure 23. Scribing centerlines on round stock clamped in a vee-block. *(Machine Tools and Machining Practices)*

erence surface (Figure 19). Only enough pressure should be applied with the scriber to remove the layout dye and not actually remove material from the workpiece.

After scribing the required lines in one direction, turn the workpiece by 90 degrees. Setup is quite critical if the scribe marks are to be truly perpendicular. A square (Figure 20) or a dial test indicator may be used (Figure 21) to establish the work at right angles. In both cases the edges of the workpiece must be machined smooth and square. After the clamp has been tightened, the perpendicular lines may be scribed at the required height (Figure 22).

The height gage may be used to lay out center-

lines on round stock (Figure 23). The stock is clamped in a vee-block and the correct dimension to center is determined with the height gage. The height gage is moved to the center of the work, the scriber is attached, and a cross line is made. The stock is rotated 90 degrees and the second line is made.

Parallel bars (Figure 24) are a valuable and useful layout accessory. These bars are made from hardened

Figure 24. Hardened steel parallel bars.

Figure 25. Using parallel bars in layout. (*Machine Tools and Machining Practices*)

steel or granite, and they have extremely accurate dimensional accuracy. Parallel bars are available in many sizes and lengths. In layout with the height gage they can be used to support the workpiece (Figure 25).

Angles may be layed out by placing the workpiece on the sine bar (Figure 26).

LAYOUT OF C-CLAMP BODY
The workpiece should be prepared as in semiprecision layout. Sharp edges must be removed and a thin coat of layout dye applied. You will need a drawing of the part to be layed out (Figure 27). The order of steps will depend on the layout task.

Position One Layouts
In position one (Figure 28) the clamp frame is on edge. In any position, all layouts can be defined as heights above the reference surface. Refer to the drawing on position one layouts and determine all of the layout that can be accomplished there.

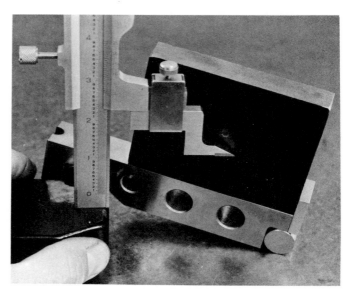

Figure 26. Laying out angle lines using the sine bar. (*Machine Tools and Machining Practices*)

Figure 27. Clamp frame. (Lane Community College)

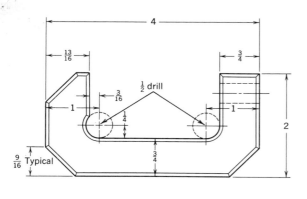
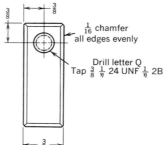

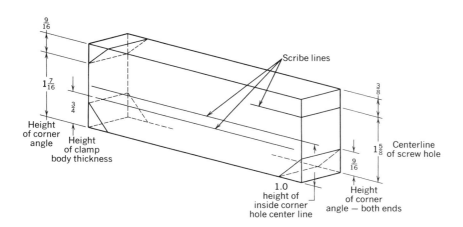

Figure 28. Clamp frame—position one layouts. (Lane Community College)

Figure 30. Attaching the scriber. (Lane Community College)

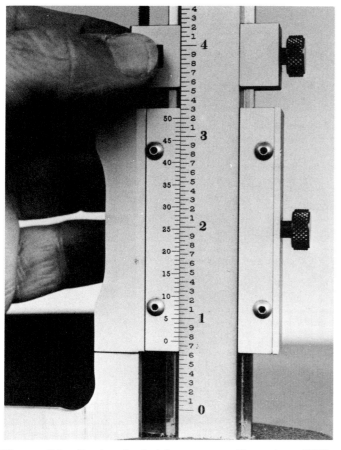

Figure 29. Setting the height gage to a dimension of 750 in. (Lane Community College)

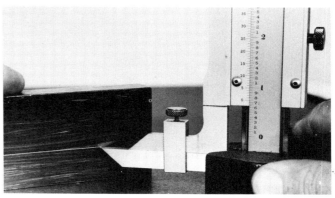

Figure 31. Scribing the height equivalent of the frame thickness. (Lane Community College)

Start by scribing the $\frac{3}{4}$ in. height that defines the width of the clamp frame. Set the height gage to .750 in. (Figure 29). Attach the scriber (Figure 30). Be sure that the scriber is sharp and properly installed for the height gage that you are using. Hold the workpiece and height gage firmly and pull the scriber across the work in a smooth motion (Figure 31). The

height of the clamp screw hole can be layed out at this time. Refer to the part drawing and determine the height of the hole. Set the height gage at 1.625 in. and scribe the line on the end of the workpiece (Figure 32). The line may be projected around on the side of the part. This will facilitate setup in the drill press. Other layouts that can be accomplished

Figure 32. Scribing the height equivalent of the clamp screw hole. (Lane Community College)

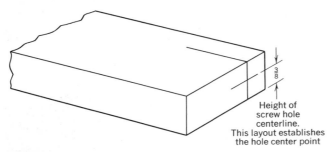

Height of screw hole centerline. This layout establishes the hole center point

$\frac{3}{8}$

Figure 33. Clamp frame—position two layouts. (Lane Community College)

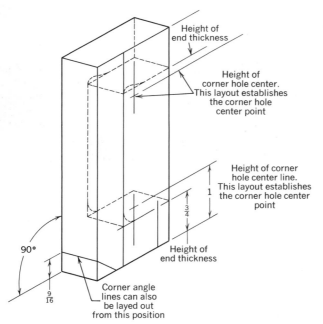

Height of end thickness

Height of corner hole center. This layout establishes the corner hole center point

Height of corner hole center line. This layout establishes the corner hole center point

1

$\frac{3}{4}$

Height of end thickness

90°

$\frac{9}{16}$

Corner angle lines can also be layed out from this position

Figure 34. Clamp frame—position three layouts. (Lane Community College)

Figure 35. Scribing the height equivalent of the end thickness. (Lane Community College)

Figure 36. Completed layout of the clamp frame. (Lane Community College)

at position one include the height equivalent of the inside corner hole centerlines. The starting points of the corner angles on both ends may also be layed out. Refer to the drawing on position one layouts.

Position Two Layouts

In position two, the workpiece is on its side (Figure 33). Check the work with a micrometer to determine its exact thickness. Set the height gage to $\frac{1}{2}$ this amount and scribe the centerline of the clamp screw hole. This layout will also establish the center point of the clamp screw hole. A height gage setting of .375 in. will probably be adequate providing that the stock is .750 in. thick. However, if the thickness varies above or below .750 in. the height gage can be set to ½ of whatever the thickness is. This will insure that the hole is in the center of the workpiece.

Position Three Layouts

In position three, the workpiece is on end clamped to an angle plate (Figure 34). The work must be established perpendicular using a square or dial test indicator. Set the height gage to .750 in. and scribe the height equivalent of the frame end thickness (Figure 35). Other layouts that can be done at position three include the height equivalent of the inside corner hole centerlines. This layout will also locate the center points of the inside corner holes (Figure 36). The height equivalent of the end thickness as well as the ending points of the corner angles can be scribed at position three.

SELF-TEST

1. How should the workpiece be prepared prior to layout?
2. What is the reason for placing the workpiece on a paper towel?
3. Describe the technique of using the layout punch.
4. Describe the use of the combination square and rule in layout.
5. Describe the technique of setting a divider to size using a rule.

Read and record the following 50 division inch/metric height gage readings.

6. (Figure 37) Inch _____ Metric _____
7. (Figure 38) Inch _____ Metric _____
8. (Figure 39) Inch _____ Metric _____

Read and record the following 25 division inch height gage readings.

9. (Figure 40) Inch _____
10. (Figure 41) Inch _____
11. How can the zero reference be adjusted on a height gage?
12. How are perpendicular lines scribed with a height gage?

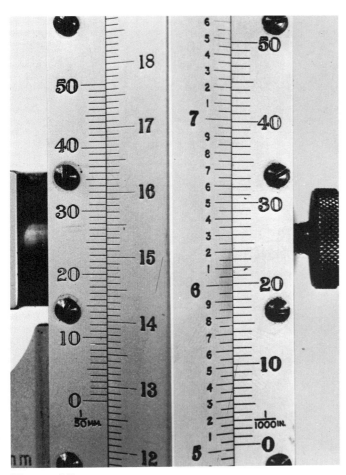

Figure 37

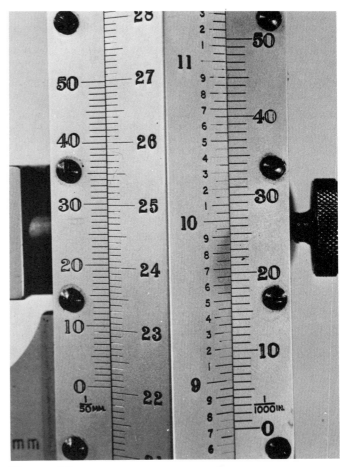

Figure 38

Figure 39

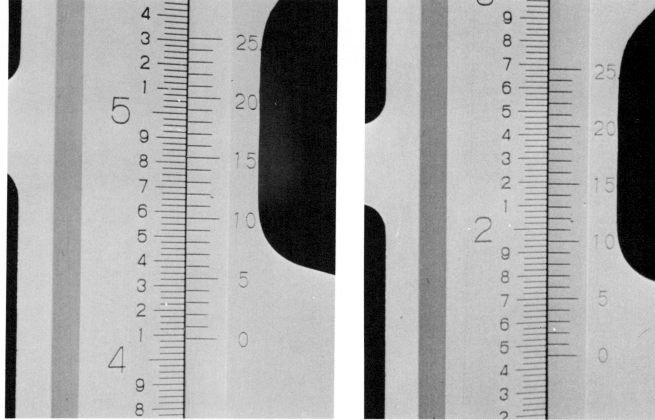

Figure 40   **Figure 41**

SECTION F

EFFECTS OF HEAT TREATMENT AND WELDING ON METALS

UNIT 1   HARDENING, TEMPERING, AND CASE HARDENING PLAIN CARBON STEEL
UNIT 2   ANNEALING, NORMALIZING, AND STRESS RELIEVING
UNIT 3   GAS WELDING REPAIRS AND THEIR EFFECTS
UNIT 4   ARC WELDING AND ITS EFFECT ON METALS

UNIT 1 HARDENING, TEMPERING, AND CASE HARDENING PLAIN CARBON STEEL

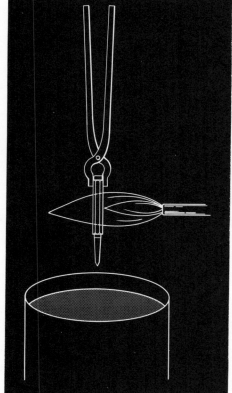

Plain carbon steel has been valued from early times because of certain properties. This soft silver-gray metal could be converted into a superhard substance that would cut glass and many other substances, including itself when soft. Furthermore, its hardness could be controlled. This converting of carbon steel into a steel of useful hardness is done with different heat treatments—two of the most important are hardening and tempering (drawing)—which you will investigate in this unit.

OBJECTIVES

After completing this unit, you should be able to:
1. Correctly harden a piece of tool steel and evaluate your work.
2. Correctly temper the hardened piece of tool steel and evaluate your work.
3. Correctly case harden a piece of soft steel by roll carburizing.

SOLIDIFICATION OF METALS

Metals tend to solidify from the molten state into crystals or grains. As solidification is taking place, the arrangement of the crystalline lattice structure takes on a characteristic pattern. Each unit cell builds on another to form crystalline needle patterns that resemble small pine trees. These structures are called **dendrites** (Figure 1).

The crystalline lattice structures begin to grow first by the formation of seed crystals or nuclei as the metal solidifies. The number of nuclei or grain starts formed determines the fineness or coarseness of the metal grain crystal structure. Slow cooling promotes large grains and fast cooling promotes smaller grains. The grain grows outward to form the dendrite crystal until it meets another dendrite crystal that is also growing. The places where these grains meet are called grain boundaries.

Figure 2 represents the growth of the dendrite from the nucleus to the final grain when metal is solidi-

fying from the melt. The nucleus can be a small impurity particle or a unit cell of the metal. Iron, copper, silver, and other metals are composed of these tiny grain structures that can be seen under a microscope when a specimen is polished and etched. This grain structure can also be seen with the naked eye as small crystals in the rough broken section of a piece of metal (Figure 3).

GRAIN BOUNDARY

As the crystal structures grow in different directions, it can be seen that at the grain boundaries, the atoms are jammed together in a misfit pattern (Figure 4). This strained condition also makes the grain boundaries stronger than the adjacent grain lattice structure at low temperatures (under red heat), but weaker at high temperatures (yellow or white heat). Grain boundaries are only about one or two atoms wide, but their strained condition causes them to etch differ-

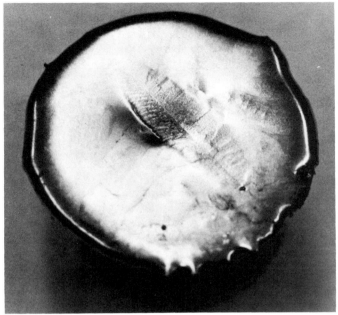

Figure 1. A three-dimensional dendrite crystal on the surface of solidified tin.

Figure 3. Single fracture of steel.

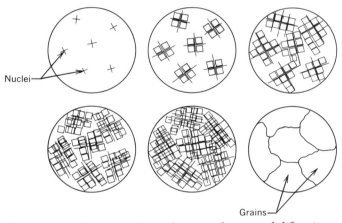

Figure 2. The formation of grains during solidification.

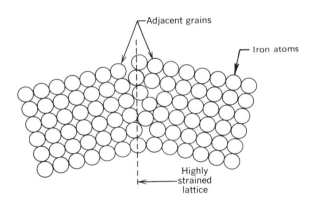

Figure 4. Grain boundary is highly strained.

ently; this means that when a polished surface is etched with acid, grain boundaries may be observed with the aid of a microscope.

CRYSTALLINE CHANGES DURING HEATING

When metals are heated slowly to their melting points, certain changes take place. Most nonferrous metals, such as aluminum, copper, and nickel, do not change their crystalline lattice structure before becoming a liquid. However, this is not the case with iron. (The term iron refers to elemental or pure iron

unless specified as wrought or cast.)

Iron is a special type of metal that does undergo a crystalline change as it is heated to the liquid stage. Iron at room temperature is ferrite (almost pure iron), but, when it is heated to about 1700°F (927°C), the atomic structure changes to austenite. A material that can exist in more than one crystalline lattice structure depending on temperature is called **allotropic.** An allotropic element is able to exist in two or more forms having various properties without a change in chemical composition. For example, carbon exists in three allotropic forms: amorphous (charcoal, soot, coal), graphite, and diamond. Iron also exists in three allotropic forms as shown in Figure 5.

Most metals (except copper when used for elec-

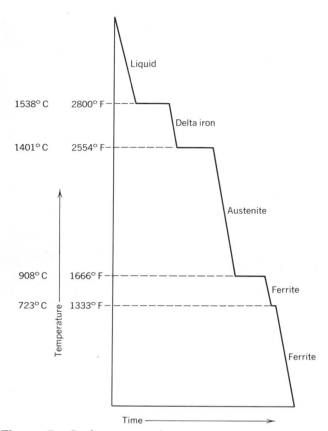

Figure 5. Cooling curve of iron.

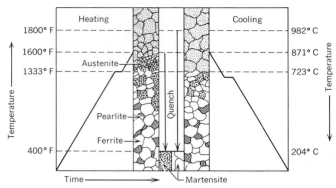

Figure 6. Critical temperature diagram of .83 percent steel showing grain structures in heating and cooling cycles. Center section shows quenching from different temperatures and the resultant grain structure.

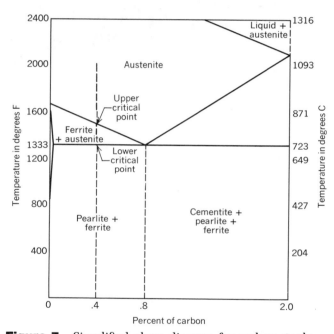

Figure 7. Simplified phase diagram for carbon steels.

tric wire) are not used commercially in their pure states because they are too soft and ductile and have a low tensile strength. When they are alloyed with other elements such as carbon or other metals, they become harder and stronger as well as more useful. A small amount of carbon (1 percent) greatly affects pure iron when alloyed with it. When heat treated, this alloy metal (carbon and iron) becomes a familiar tool steel used for cutting tools, files, and punches. Iron with 2 to 4½ percent carbon content yields cast iron.

In the previous discussion, you learned something about grain structures in metals and crystalline changes during heating and cooling. These transformations from ferrite to austenite on heating and from austenite to ferrite when cooling are called phase changes. The temperatures where one phase changes to another are called **critical points or temperatures** by heat treaters and **transformation temperatures** by metallurgists. For example, the critical points of water are its boiling point (212°F or 100°C) and freezing point (32°F or 0°C). All heat treating is done when

the metal is below its melting point and in a solid state. Figure 6 shows a critical temperature diagram for a carbon steel, illustrating the size and makeup of crystalline grain and the temperatures at which steel changes from ferrite to austenite upon heating and cooling.

The lower critical point in carbon steels is always about 1330°F (721°C), but the upper critical temperature varies as the carbon content changes, as can be seen in Figure 7. The dashed line represents a 0.40 percent carbon steel. At a temperature below 1333°F

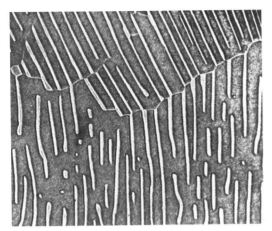

Figure 8. A replica electron micrograph. Structure consists of lamellar pearlite (11,000X). (ASM Committee on Metallography of High-Carbon Steel, "Microstructure of Carbon and Alloy Steels of High Carbon Content," *Atlas of Microstructures of Industrial Alloys,* Volume 7, Eighth Edition, Lyman, T., editor, American Society for Metals, © 1972, p. 48)

(723°C), 0.40 percent carbon steel is composed of pearlite grains (cementite, also called iron carbide, and iron) and ferrite grains (iron). It is in the transformation range between 1333 and 1650°F (723 and 899°C) in which the pearlite and ferrite is transforming to austenite. Above the upper critical point, all of the carbon has been diffused into the austenite.

HARDENING BY QUENCHING

As steel is heated above the lower critical temperature of 1330°F (721°C), the carbon that was in the form of layers of iron carbide in pearlite (Figure 8) begins to dissolve in the iron and form a solid solution called austenite (Figure 9). As this metal is heated through the transformation range to the upper critical point, all of the carbon that was in the pearlite has now been dissolved into the iron, which is now in the austenite phase. When this solution of iron and carbon is suddenly cooled or quenched, a new microstructure is formed. This is called martensite (Figure 10). Martensite is very hard and brittle, having a much higher tensile strength than the steel with a pearlite microstructure. It is quite unstable, however, and must be tempered (drawn) to relieve internal stresses in order to have the ductility and toughness needed to be useful. AISI-C1095, commonly known as water hardening tool (W1) steel, will begin to show hardness when quenched from a temperature just over 1330°F (721°C), but will not harden at all if quenched from a temperature lower than that. This steel will become

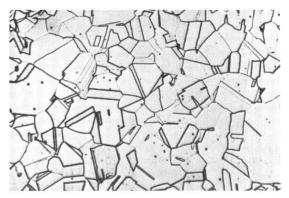

Figure 9. A microstructure of annealed 304 stainless steel that is austenitic at ordinary temperatures (250X). (ASM Committee on Metallography of Wrought Stainless Steels, "Microstructures of Wrought Stainless Steels," *Atlas of Microstructures of Industrial Alloys,* Volume 7, Eighth Edition, Lyman, T., editor, American Society for Metals, © 1972, p. 135)

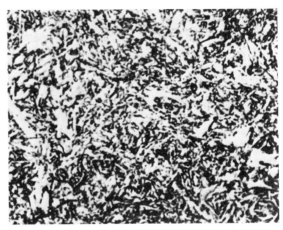

Figure 10. 1095 steel, water quenched from 1500°F (816°C) (1000X). The needlelike structure shows a pattern of fine untempered martensite. (ASM Committee on Metallography of High-Carbon Steel, "Microstructure of Carbon and Alloy Steels of High Carbon Content," *Atlas of Microstructures of Industrial Alloys,* Volume 7, Eighth Edition, Lyman, T., editor, American Society for Metals, © 1972, p. 47)

as hard as it can get when heated to 1450°F (788°C) and quenched in water. This quenching temperature changes as the carbon content changes. It should be 50°F (10°C) above the upper critical temperature for carbon steels (Figure 11). The reason carbon steel should be heated above the upper critical temperature is that the ferrite is not all transformed into austenite below this point, and, when quenched, is retained in the martensitic structure. The retained ferrite causes brittleness even after tempering.

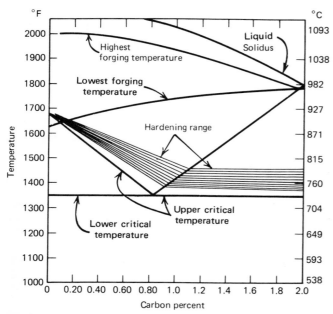

Figure 11. Temperature ranges used for hardening carbon steel. (John E. Neely, *Practical Metallurgy and Materials of Industry,* John Wiley and Sons, Inc., Copyright © 1979, New York.)

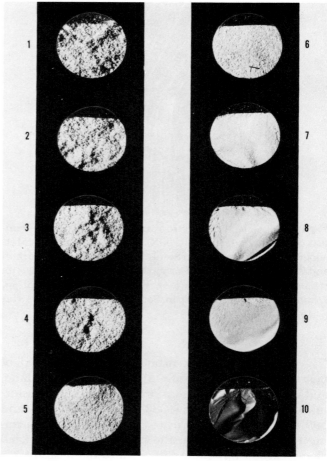

Figure 12. Shepherd grain size fracture standards. The fracture grain size test specimen can be compared visually with this series of ten standards. Number 1 constitutes the coarsest and Number 10 the finest fracture grain size. (Republic Steel Corporation)

Low carbon steels such as AISI-1020, for all practical purposes, will not harden when they are heated and quenched. It is carbon that makes steel hardenable. But there must be enough carbon. Steels with less than 30 points (0.30 percent) of carbon will not harden. Therefore, steels designated 1020, 1115, or 8620 will not harden, but steels designated 4140, 1060, or 6150 will harden when properly heated and quenched. Steels with less than 30 points of carbon may be surface hardened if they have been through a process called "carburizing." Carburizing is a method of injecting carbon into the surface of low carbon steels. Since only the surface of the steel is affected by this process, only the surface will harden. Steels that have been carburized and hardened are sometimes called "case hardened" steels.

Higher carbon, oil, and air hardening alloy steels have a higher hardenability and do not have to be quenched as rapidly as plain carbon steels. Consequently, they are deeper hardening than water hardening types, which must be cooled to about 200°F (93°C) within one or two seconds to become hard. As you can see, it is quite important to know the carbon content and alloying element so that the correct temperature and quenching medium can be used. Fine grained tool steels are much tougher than coarse grained tool steels (Figure 12). If a piece of tool steel is heated above the correct temperature for its specific

carbon content, a phenomenon called grain growth will occur (refer to Figure 6) and a coarse, weak grain structure develops. The grain growth will remain when the part is hardened by quenching and, if used for a tool such as a punch or chisel, the end may simply drop off when the first hammer blow is struck (Figure 13). Tempering (drawing) will not remove the coarse grain structure. If the part has been overheated, simply cooling back to the quenching temperature will not help, as the coarse grain persists well down into the hardening range (see Figure 6). The part should be air cooled below 800°F (427°C) and then reheated to the correct quenching temperature. Plain carbon steels containing 0.83 percent carbon can get as hard (RC67, a little harder than a file) as any plain carbon steel containing more carbon.

Transformation to martensite, the hardened form

Figure 13. This chisel failed in service after being hardened and tempered because it was overheated during the hardening process. (Lane Community College)

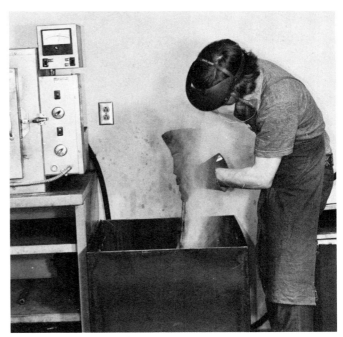

Figure 14. A long part is being quenched vertically. In this case, agitation of the part should be an up-and-down motion. (Lane Community College)

of steel, depends on three factors: (1) mass of the part, (2) severity of the quench, and (3) hardenability of the material. AISI-C1095, or water hardening tool steel (W1), can be quenched in oil, depending on the size of the part, and still become hard. For example, since a piece of small diameter W1 drill rod or thin sheet would have a high cooling rate, oil should be used as a quenching medium. Oil is not as "severe" as water because it conducts heat less rapidly than water and thus prevents quench cracking. Larger sections, however, would not be fully transformed into martensite if they were oil quenched since they have a slower cooling rate, but would instead contain some softer transformation structures. Water quenching should be used in this case, but remember that W1 is shallow hardening and will only harden about $\frac{1}{8}$ in. deep.

If a torch is used to heat a part for quenching, heat colors are used to determine the approximate temperatures. When using a furnace to heat W1 for quenching, the temperature control would be set for 1450°F (788°C). See Table 1. If the part is small, a preheat is not necessary; but if it is thick, it should be heated slowly. If the part is left in a furnace without a controlled atmosphere for any length of time, the metal will form an oxide scale as the carbon leaves the surface. This decarburization of the surface will cause it to remain soft when quenched, while the metal directly under the surface will become hard. This loss of surface carbon can be avoided by using a carburizing compound to replace lost carbon, or by burying the part in cast iron chips. Wearing gloves, a face shield, and a long sleeve shirt for protection, place the part in the furnace using tongs. As a rule, tool steels should be left in the furnace for 1 hr. per inch of cross section. When the part has become the same color as the furnace bricks, remove it by grasping one end with the tongs and immediately plunge

it into the quenching bath (Figure 14). If the part is long, like a chisel or punch, it should be inserted into the quench vertically (straight up-and-down) not at a slant. Quenching at any angle can cause unequal cooling rates and bending of the part. Also, agitate the part in an up-and-down or a figure-8 motion to remove any gases or bubbles that might cause uneven quenching. Brine (salt water) is often used as a quenching medium because it lessens gassing and bubbling.

Oil quenching is used for oil hardening steels (O1). Quenching oils may have to be slightly heated to avoid quench cracking when hardening certain tool steels. A manufacturer's catalog should be consulted to get the correct temperature, time periods, and quenching media.

TEMPERING

Tempering, or drawing, is a process of reheating a steel part that has been previously hardened to transform some of the hard martensite into a softer structure. The higher the tempering temperature used, the less distorted or strained the martensite becomes, and therefore the softer and tougher (less brittle) the piece becomes. Therefore, tempering temperatures are specified according to the strength and ductility

Table 1

Temperatures and Colors for Heating and Tempering Steel

Colors		Fahrenheit	Process
		—	
	White	2500°	
		—	
		2400°	High speed
		—	steel hardening
	Yellow white	2300°	(2250°–2400°F)
		—	
		2200°	
		—	
		2100°	
		—	
	Yellow	2000°	
		—	
		1900°	
		—	
	Orange red	1800°	Alloy steel
		—	hardening
Heat		1700°	(1450°–1950°F)
colors		—	
		1600°	
	Light cherry red	—	
		1500°	
		—	Carbon steel
	Cherry red	1400°	hardening
		—	(1350°–1550°F)
		1300°	
	Dark red	—	
		1200°	
		—	
		1100°	
		—	
	Very dark red	1000°	
		—	
		900°	
		—	
	Black red	800°	High
	in dull light or	—	speed
	darkness	700°	Carbon steel
		—	steel tempering
	Pale blue (590°F)	600°	tempering (350°–
	Violet (545°F)	—	(300°– 1100°F)
Temper	Purple (525°F)	500°	1050°F)
colors		—	
	Yellowish brown (490°F)	—	
	Straw (465°F)	400°	
	Light straw (425°F)	—	
		300°	
		—	
		200°	
		—	
		100°	
		—	
		0°	

Source: *Pacific Quality Steels*, "Stock List and Reference Book," No. 75, Pacific Machinery and Tool Steel Company, 1971.

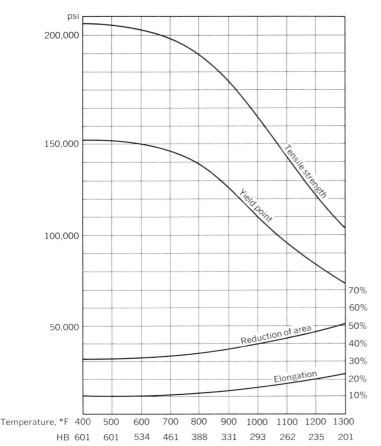

Figure 15. Mechanical properties chart, SAE 1095 steel, water quenched. The bottom two lines refer to tempering temperatures and the resultant hardness in Brinell reading. HB 601 is file hard. (Bethlehem Steel Corporation, Bethlehem, PA)

Figure 16. *(right)* Tempering a punch on a hot plate. *(Machine Tools and Machining Practices)*

desired. Mechanical properties charts, which may be found in steel manufacturers' handbooks and catalogs, give this data for each type of alloy steel. Figure 15 is an example of a mechanical properties chart for water quenched 1095 steel.

A part can be tempered in a furnace or oven by bringing it to the required temperature and holding it there for a length of time, then cooling it in air or water. Some tool steels should be cooled rapidly after tempering to avoid temper brittleness. Small parts are often tempered in liquid baths such as oil, salt, or metals. Specially prepared oils that do not ignite easily can be heated to the tempering temperature. Lead and various salts are used for tempering since they have a low melting temperature.

When there are no facilities to harden and temper a tool with controlled temperatures, tempering by color is done. The oxide color used as a guide in such tempering will form correctly on steel only if it is polished to the bare metal and is free from any oil or fingerprints. An oxy-acetylene torch, steel hot plate, or an electric hot plate can be used. If the part is quite small, a steel plate is heated from the underside, while the part is placed on top. Larger parts such as chisels and punches can be heated on an electric plate (Figure 16) until the needed color shows, then cooled in water. See Table 2 for oxide colors and temperatures.

Table 2
Temper Color Chart

Degrees		Oxide	Suggested Uses for Carbon	
C°	F°	Color	Tool Steels	
220	425	Light straw	Steel cutting tools, files, and paper cutters	Harder
240	462	Dark straw	Punches and dies	
258	490	Gold	Shear blades, hammer faces, center punches, and cold chisels	
260	500	Purple	Axes, wood cutting tools, and striking faces of tools	
282	540	Violet	Spring and screwdrivers	
304	580	Pale blue	Springs	
327	620	Steel gray	Cannot be used for cutting tools	Softer

Source: Warren T. White, John E. Neely, Richard R. Kibbe, and Roland O. Meyer, *Machine Tools and Machining Practices*, John Wiley and Sons, Inc., Copyright © 1977, New York.

When grinding carbon steel tools, if the edge is heated enough to produce a color, you have in effect retempered the edge. If the temperature reached was above that of the original temper, the tool has become softer than it was before you began sharpening it. Table 2 gives the hardnesses of various tools as related to their oxide colors and the temperature at which they form.

Tempering should be done as soon as possible after hardening. The part should not be allowed to cool completely, since untempered it contains very high internal stresses and tends to split or crack. Tempering will relieve the internal stresses. A hardened part left overnight without tempering may develop cracks by itself.

FORGING

Forging temperatures should be well below the solidus or freezing point of steel (refer to Figure 11). If a small tool, such as a chisel made of carbon steel, is being forged (hammered while hot), the temperature should be in the range that produces an orange-red to yellow color (1800 to 1950°F or 982 to 1066°C). If this temperature is exceeded, there is a danger of "burning" the steel and ruining it for any use. If excessive sparking is evident, it is too hot. Carbon steels are somewhat "hot short" above forging temperatures; that is, they tend to split when hammered upon. Excess sulfur also causes hot short at any temperature above a red heat. The part should also be above the lowest forging temperature to avoid the problem of splitting. If a previously hardened tool is to be reshaped by forging, it should first be annealed. Forging temperatures vary with other types of steels. Consult a reference book when forging tool steels other than plain carbon tool steels.

HARDENING A PIECE OF TOOL STEEL

The steps to harden a piece of tool steel in the form of any tool or part, such as a punch or a chisel that has been previously forged, are as follows:

1. Assuming the part is a small tool such as a center punch, it should be placed in a furnace that has already been brought up to the correct temperature.
2. Determine the best place to grip the part with the tongs so that it will not be damaged where it is red hot. Use the properly shaped tongs.
3. Make sure you can grasp the part in the furnace to remove it in the proper orientation to enable you to quench it straight in.

4. Heat the end of the tongs so they will not remove heat from the part.
5. When the piece has become the same color as the furnace bricks, remove it and immediately quench it **completely under** in the bath.
6. Agitate it up and down or in a figure-8 motion.
7. Be sure it has cooled below 200°F (93°C) before you remove it from the quench.
8. Test the piece with an old file in an inconspicuous place. If a Brinell or Rockwell hardness tester is available, check to see if the part is now near 600 BHN or about RC60 hardness.

Note: If no furnace is available, a torch may be used with a temperature chart. See Table 1.

TEMPERING A PART

The steps to temper a part that has just been hardened are as follows.

Furnace Tempering

1. Polish all smooth surfaces of the part with abrasive cloth and remove all oil. A cold furnace should be brought up to the correct temperature.
2. Small parts are then placed in the furnace for about 15 minutes.
3. Remove and cool in air.
4. When this method is used, the striking end of punches, chisels, and other striking tools should be further heated with a torch until they are a blue color on that end. This is done to ensure the safety of the user by preventing the struck end from shattering, since it is softer when tempered to blue.

Torch or Hot Plate Tempering (Color Method)

1. When using a torch or hot plate, make sure that heat is applied to the body of a punch or a part of the tool that can be softer. Allow the heat to travel slowly out to the cutting edge. This way the colors may be observed.
2. See that the striking end of a tool is blue or gray before the proper color arrives at the cutting end.
3. When the proper color, according to Table 2, has arrived, quickly cool the piece in water. **Do not delay** or it will be overtempered.
4. If a Brinell or Rockwell hardness tester is available, recheck the hardness and compare with Figure 15.
5. Leave the temper colors on your project for evaluation.

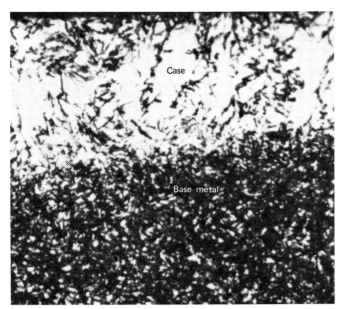

Figure 17. Micrograph of carburized case (250X). (ASM Committee on Metallography of Case Hardening Steel, "Microstructure of Case Hardening Steel," *Atlas of Microstructures of Industrial Alloys*, Volume 7, Eighth Edition, Lyman, T., editor, American Society for Metals, © 1972, p. 62)

CASE HARDENING

Low carbon steels (0.08 to 0.30 percent carbon) do not harden to any great extent even when combined with other alloying elements. Therefore, when a soft, tough core and an extremely hard outside surface are needed, one of several case hardening techniques is used. It should be noted that **surface hardening** is not necessarily the same as **carburizing case hardening.** Surface hardening, which can be done by either flame hardening or induction hardening methods, is performed on surfaces such as gear teeth, lathe ways, and axle shafts. This process depends on sufficient carbon already being present in the material. On the other hand, carburizing case hardening is used where additional carbon needs to be added to the material for it to be hardenable. Low carbon steel mentioned above is an example where additional carbon needs to be added so the metal can be hardened. In this process carbon from an outside source penetrates the surface of the steel by the process of diffusion when it is at a high temperature. Because the diffusion raises the carbon content, it also raises the hardenability of the steel (Figure 17).

SURFACE HARDENING

Flame hardening is mostly used for production hard-ening where long pieces or small parts are passed through a flame and immediately quenched in a spray of water (Figures 18a to 18c). The surface of the part is generally hardened from $\frac{1}{32}$ to $\frac{1}{8}$ in. deep. In this process, since there is no addition or absorption of other elements, the hardenability of the steel determines the depth and hardness of case. Good hardness on the surfaces can be obtained with the core unaffected. There is relative freedom from scaling and pitting. This type of case gives good wearing qualities with a soft ductile core. The induction hardening process is similar to flame hardening except that the heat is generated by electromagnetic induction.

CARBURIZING CASE HARDENING

Carburizing with a case hardening compound can be done by either of two methods. If only a shallow hardened case is needed, roll carburizing may be used. This consists of heating the part to 1650°F (899°C), rolling it in a carburizing compound, reheating, and quenching in water. In roll carburizing, use only a nontoxic compound such as Kasenite® unless special ventilation systems are used. Roll carburizing produces a maximum case of about 0.003 in. When pack carburizing (Figure 19), the part is packed in carburizing compound in a metal box and placed in a furnace long enough to make a case of the required depth. Eight hours at 1700°F (925°C) can produce a case or carbon penetration of $\frac{1}{16}$ in. deep. The part is then removed and quenched in water. To eliminate the possibility of grain growth, the furnace can be shut off and the part allowed to cool slowly in the furnace. The part would then need to be reheated and quenched to harden.

After case hardening, tempering is not usually necessary since the core is still soft and tough. Therefore, unlike a piece hardened completely through that is softened by tempering, the surface of a case hardened piece remains hard, usually RC65 (as hard as a file) or above.

Gas carburizing (another method of introducing carbon into the surface) utilizes a high concentration of a hydrocarbon gas, such as propane or natural gas at high temperatures, to create a carbonizing atmosphere. These gases are usually piped into a furnace called a controlled atmosphere furnace.

Gas nitriding produces an extremely hard case by bringing a ferrous metal into contact with a nitrogenous gas, usually ammonia. Since the heating temperature is relatively low (925 to 1050°F or 496 to 565.5°C), no quench is required for hardening. Little distortion is produced in a hardened part. Steels containing aluminum and molybdenum work best for

Figure 18a. A cam lobe. This gray cast iron cam shaft was sawed halfway through and then broken, showing the flame hardened lobe of white cast iron. *(Practical Metallurgy and Materials of Industry)*

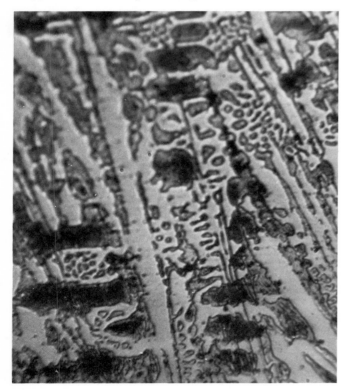

Figure 18b. Micrograph showing the chilled iron area of cam lobe (100X). *(Practical Metallurgy and Materials of Industry)*

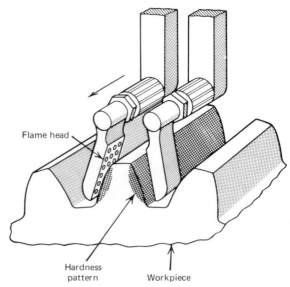

Figure 18c. Hardness pattern developed in sprocket teeth when standard flame tips were used for heating. When space permits this method, hardening one tooth at a time results in low distortion. (ASM Committee on Flame Hardening, "Flame Hardening," *Heat Treating, Cleaning, and Finishing*, Volume 2, Eighth Edition, Lyman, T., editor, American Society for Metals, © 1964, p. 190)

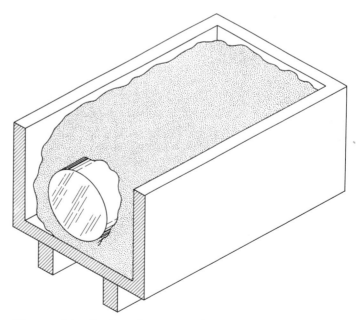

Figure 19. Pack carburizing. The workpiece to be pack carburized should be completely covered with carburizing sand. The metal box should have a close fitting lid. *(Machine Tools and Machining Practices)*

this process. Carbonitriding is a modified gas carburizing process in which steels are held at high temperatures (about 1500°F or 815.5°C) in a gas containing both carbon and nitrogen (ammonia). The higher temperatures that can be used on these metals increase the surface hardenability of the metal over other carburizing or nitriding processes.

Liquid carburizing is an industrial method in which the parts are bathed in carbonate cyanide and chloride salts and held at a temperature between 1500 and 1700°F (815.5 and 927°C). A pot furnace is used for this process. The part is quenched in oil or water after sufficient time is allowed for diffusion into the surface of the metal. Cyanide salts are extremely poisonous and adequate worker protection is essential.

HEAT TREATING SAFETY

When heat treating, always wear a face shield, leather gloves, and long sleeves. There is a definite hazard to the face and eyes when cooling the tool steel by oil quenching, that is, submerging it in oil. The oil, hot from the steel, tends to fly upward, so you should stand to one side of the oil tank and not lean over it.

Always work in pairs during heat treatment. One person can open and close the furnace door, while the other handles the hot part. The heat treated part should be positioned in the furnace so it can be conveniently removed. This will prevent the heat treater from dropping hot parts and help to insure successful heat treatment. Atmospheric furnaces should never be opened until the gas supply is turned off. Failure to do so could result in an explosion.

Very toxic fumes are present when parts are being carburized with compounds containing potassium cyanide. These cyanogen compounds are deadly poisonous and every precaution should be taken when using them. Kasenite®, a trade name for a carburizing compound that is not toxic, is often found in school and machine shops.

CASE HARDENING AND PACK CARBURIZING A PART

The steps to case harden a piece of mild steel using a furnace or heating torch and to pack carburize a piece of mild steel in a furnace are as follows.

Hardening by the Roll Method

1. Heat a part to 1650°F (899°C) and remove from the furnace with tongs.
2. Roll part in carburizing compound.
3. Reheat to 1650°F (899°C).
4. Quench in cool water.

Hardening by the Pack Carburizing Method

1. Place a second part in a steel box containing the carburizing compound.
2. Place in a furnace set at a temperature of 1700°F (927°C). Leave it in the furnace for several hours. If grain growth may be a problem, let the part cool to 800°F (427°C) in the furnace and reheat to 1550 to 1600°F (843 to 871°C) before quenching.
3. Remove the part from the furnace and quench in water.
4. Check the pieces with a file.
5. Grind off a small amount and check again.
6. Determine how deep the case is on the part hardened by the roll method and the part hardened by the pack carburizing method.

SELF-TEST

1. If you heated AISI-C1080 steel to 1200°F (649°C) and quenched it in water, what would be the result?
2. If you heated AISI-C1020 steel to 1500°F (815°C) and quenched it in water, what would happen?
3. List as many problems encountered with water hardening steels as you can think of.
4. Name some advantages of using air and oil hardening tool steels.
5. What is the correct temperature for quenching AISI-C1095 tool steel? For any carbon steel?
6. Why is steel tempered after it is hardened?
7. What factors should you consider when you choose the tempering temperature for a tool?
8. The approximate temperature for tempering a center punch on the point should be _____. The oxide color would be _____.
9. If a cold chisel became blue when the edge was ground on an abrasive wheel, to approximately what temperature was it raised? How would this temperature affect the tool?
10. How soon after hardening should you temper a part?
11. Which method of tempering gives the heat treater the most control of the final product: by color or by furnace?
12. In what ways can decarburization of a part be avoided when it is heated in a furnace?
13. When distortion must be kept to a minimum, which

type of tool steel should be used?

14. What is the advantage of using low carbon steel for parts that are to be case hardened?
15. By which methods of carburizing can a deep case be made?
16. Are parts that are surface hardened always carburized case hardened?

17. Name three methods by which carbon may be diffused into the surface of heated steel.
18. What method of case hardening uses ammonia gas?
19. What can happen to a carbon steel when it is heated to high temperatures in the presence of air (oxygen)?
20. Why should the part or the quenching medium be agitated when you are hardening steel?

UNIT 2 ANNEALING, NORMALIZING, AND STRESS RELIEVING

Since the machinability and welding of metals are so greatly affected by heat treatments, the processes of annealing, normalizing, and stress relieving are important to the welder and machinist. You will learn about these processes in this unit.

OBJECTIVES

After completing this unit, you should be able to:
1. Explain the principles of and differences between the various kinds of annealing processes.
2. Test various steels with annealing, normalizing, and stress relieving heat treatments in order to determine their effect on machinability and welding.

ANNEALING

The heat treatment for iron and steel that is generally called annealing can be divided into several different processes: full anneal, normalizing, spheroidize anneal, stress relief (anneal), and process anneal.

Full Anneal

The full anneal is used to completely soften hardened steel, usually for easier machining of tool steels that have more than 0.8 percent carbon content. Lower carbon steels are full annealed for other purposes. When welding or prior heat treating has been done on a medium to high carbon steel that must be machined, a full anneal is needed. Full annealing is done by heating the part in a furnace to 50°F (10°C) above the upper critical temperature (Figure 1), and then cooling very slowly in the furnace or in an insulating material. By this process, the microstructure is changed from a hard needle-like structure (Figure 2a) to a coarse pearlite and ferrite (Figure 2b), which is soft enough to machine.

Normalizing

Normalizing is somewhat similar to annealing, but it is done for several different purposes. Medium carbon steels are often normalized to give them better machining qualities. Medium (0.3 to 0.6 percent) carbon steel may be "gummy" when machined after a full anneal, but can be made sufficiently soft for machining by normalizing. The finer, but harder, microstructure produced by normalizing gives the piece a better surface finish. The piece is heated to 100°F (38°C) above the upper critical line, and cooled in still air. When

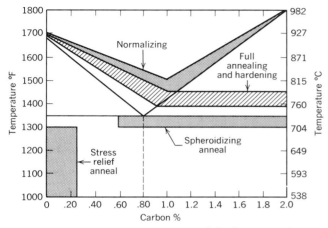

Figure 1. Temperature ranges used for heat treating carbon steels. *(Machine Tools and Machining Practices)*

the carbon content is above or below 0.8 percent, higher temperatures are required. See Figure 1.

Forgings and castings that have unusually large and irregular grain structures are corrected by using a normalizing heat treatment (Figures 3a and 3b). Stresses are removed, but the metal is not as soft as with full annealing. The resultant microstructure is a uniform fine grained pearlite and ferrite, including other microstructures, depending on the alloy and carbon content. Normalizing also is used to prepare steel for other forms of heat treatment such as hardening and tempering. Weldments are sometimes normalized to remove welding stresses that develop in the structure as well as in the weld.

Spheroidizing

Spheroidizing is used to improve the machinability of high carbon steels (0.6 to 1.7 percent carbon). It produces a spherical or globular carbide grain structure in the steel rather than a lamellar (platelike) structure of pearlite (Figure 4). Low carbon steels (0.08 to 0.3 percent carbon) can be spheroidized, but

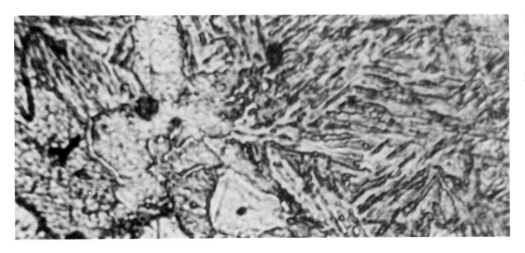

Figure 2a. The structures in the micrograph appear to be needlelike and therefore may be hard. This SAE 1090 steel is probably nonmachinable (500X). *(Practical Metallurgy and Materials of Industry)*

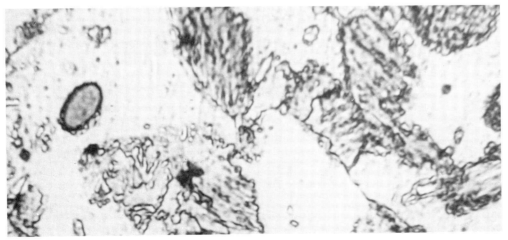

Figure 2b. The same material as shown in Figure 2a has been fully annealed and is now machinable. The martensite has been recrystallized into large grains of pearlite and ferrite (500X). *(Practical Metallurgy and Materials of Industry)*

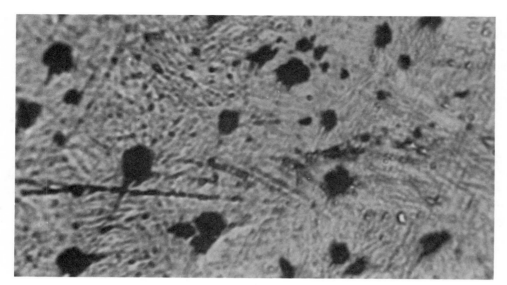

Figure 3a. This carbon steel contains impurities (dark patches) in a matrix of martensite. Normalizing will homogenize this steel and give it a regular grain structure (500X). *(Practical Metallurgy and Materials of Industry)*

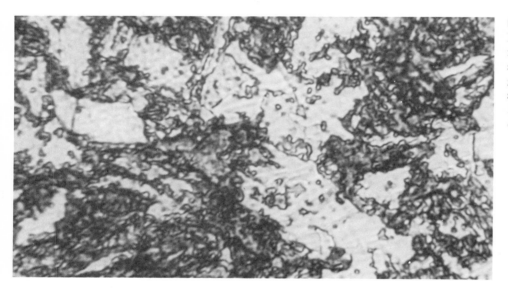

Figure 3b. Same material as in Figure 3a except that it has been normalized. A more homogenous structure can be seen with large grains of pearlite (black) and partial spheroidization (500X). *(Practical Metallurgy and Materials of Industry)*

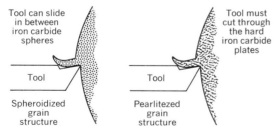

Figure 4. Comparison of cutting action between spheroidized and normal carbon steels.

their machinability gets poorer since they become gummy and soft, causing tool edge buildup and poor finish. The spheroidization temperature is close to 1300°F (704°C). The steel is held at this temperature about four hours. Hard carbides that develop from welding on medium carbon steels, causing them to be brittle, can be changed to the more ductile iron carbide spheroids by the process of spheroidization (Figure 5). Maximum ductility is obtained in steel by this method.

Stress Relief Anneal

Stress relief annealing is a process of reheating low carbon steels to 950°F (510°C). Stresses in the ferrite (mostly pure iron) grains caused by cold working steel such as rolling, pressing, welding, forming, or drawing are relieved by this process. The distorted grains reform or recrystallize into new softer ones (Figures 6 and 7).

The pearlite grains and some other forms of iron carbide remain unaffected by this treatment, unless

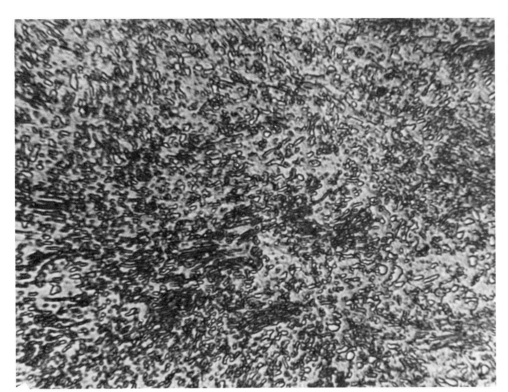

Figure 5. SAE 1090 spheroidized steel. Machining is much easier in this condition than it is when the steel is normalized or full annealed (500X). *(Practical Metallurgy and Materials of Industry)*

Figure 6. Microstructure of flattened grains of 0.10 percent carbon steel, cold rolled (1000X). (ASM Committee on Metallography of Low-Carbon Steel, "Microstructures of Low-Carbon Sheet Steel," *Atlas of Microstructures of Industrial Alloys*, Volume 7, Eighth Edition, Lyman, T., editor, American Society for Metals, © 1972, p. 10)

Figure 7. The same 0.10 percent carbon steel as in Figure 6, but annealed at 1025°F (552°C) (1000X). Ferrite grains are mostly reformed to their original state, but the pearlite grains are still distorted. (ASM Committee on Metallography of Low-Carbon Steel, "Microstructures of Low-Carbon Sheet Steel," *Atlas of Microstructures of Industrial Alloys*, Volume 7, Eighth Edition, Lyman, T., editor, American Society for Metals, © 1972, p. 10)

done at the spheroidizing temperature and held long enough to effect spheroidization. Stress relief is often used on weldments as the lower temperature limits the amount of distortion caused by heating. Full anneal, for example, can cause considerable distortion in steel.

Process Anneal

Process annealing is essentially the same as stress relief annealing. It is done at the same temperatures and with low and medium carbon steels. In the wire and sheet steel industry, the term is used for the annealing or softening that is necessary during the cold rolling or wire drawing processes and for removal of final residual stresses in the material. Wire and other metal products that must be continuously formed and reformed would become too brittle to continue after a certain amount of forming. The anneal, between a series of cold working operations, reforms the grains to the original soft, ductile condition so that cold working can continue. Process anneal is sometimes referred to as bright annealing, and is usually carried out in a closed container with inert gas to prevent oxidation of the surface.

RECOVERY, RECRYSTALLIZATION, AND GRAIN GROWTH

When metals are heated to temperatures less than the recrystallization temperature, a reduction in internal stress takes place. This is done by relieving elastic stresses in the lattice planes and not by reforming the distorted grains. Recovery, an annealing process if used on cold worked metals, is usually not a sufficient stress relief for further extensive cold working (Figure 8), yet it is used for some purposes and is called stress relief anneal. Most often recrystallization is required to reform the distorted grains sufficiently for further cold work.

Recovery is a low temperature effect in which there is little or no visible change in the microstructure. Electrical conductivity is increased and often a decrease in hardness is noted. It is difficult to make a sharp distinction between recovery and recrys-

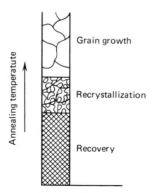

Figure 8. Changes in metal structures that take place during the annealing process. *(Practical Metallurgy and Materials of Industry)*

tallization. Recrystallization releases much larger amounts of energy than does recovery. The flattened, distorted grains are sometimes reformed to some extent during recovery into polygonal grains, while some rearrangement of defects such as dislocations takes place.

Recrystallization not only releases much larger amounts of stored energy, but new grains are formed at the points of stress in the original grains. As temperature increases, there can also be joining of grains to form larger ones. To accomplish this joining of adjacent grains, grain boundaries migrate to new positions, which changes the orientation of the crystal structure. This is called grain growth.

The following factors affect recrystallization.

1. A minimum amount of deformation is necessary for recrystallization to occur.
2. The larger the original grain size, the greater amount of cold deformation is required to give an equal amount of recrystallization with the same temperature and time.
3. Increasing the time of anneal decreases the temperature necessary for recrystallization.
4. The recrystallized grain size depends mostly on the degree of deformation and, to some extent, on the annealing temperature.
5. Continued heating, after recrystallization (reformed grains) is complete, increases the grain size.
6. The higher the cold working temperature, the greater amount of cold work is required to give equivalent deformation.

Body parts and frames on automobiles are made of low carbon steel. These parts are made by rolling, forming, or pressing, which are forms of cold work deformation. It can be seen that any heating above the recrystallization temperature will reform these stressed grains to a larger, softer grain size. Cold worked metal is stronger and more rigid than hot worked metal, so the body and frame are subject to a weakening effect if heat is applied, such as with welding or torch heating. This loss of strength may not be serious in some areas on the body, but at certain high stress areas on the frame, the weakening effect of heat could cause a metal failure.

NONFERROUS METALS

Annealing of most nonferrous metals consists of heating them to the recrystallization temperatures or grain growth range and cooling them to room temperature (Table 1). The rate of cooling usually has no effect on most nonferrous metals such as copper

Table 1
Recrystallization Temperatures of Some Metals

Metal	Recrystallization Temperature °F
99.999% aluminum	175
Aluminum bronze	660
Beryllium copper	900
Cartridge brass	660
99.999% copper	250
Lead	25
99.999% magnesium	150
Magnesium alloys	350
Monel	100
99.999% nickel	700
Low carbon steel	1000
Tin	25
Zinc	50

Source: John E. Neely, *Practical Metallurgy and Materials of Industry*, John Wiley and Sons, Inc., Copyright © 1979, New York.

or brass. Annealing temperatures and procedures are very critical with some metals such as stainless steels and precipitation hardening nonferrous metals. The phenomenon of grain growth, as discussed in Unit 1, is found at higher temperatures. When a large amount of deformation is required in one operation, large grains are sometimes preferred although a surface defect, called orange peel (Figures 9a and 9b), is sometimes seen on formed metals having large grains. In this case, a stress relief anneal could be used; that is, recovery without grain growth.

SELF TEST

1. When might normalizing be necessary?
2. At what approximate temperature should you normalize 0.4 percent carbon steel?
3. What is the spheroidizing temperature of 0.8 percent carbon steel?
4. What is the essential difference between full anneal and stress relieving?
5. When should you use stress relieving?
6. What kind of carbon steels would need to be spheroidized to give them free machining qualities?
7. Explain process annealing.
8. How should the piece be cooled for a normalizing heat treatment?
9. How should the piece be cooled for the full anneal?
10. What happens to machinability in low carbon steels that are spheroidized?

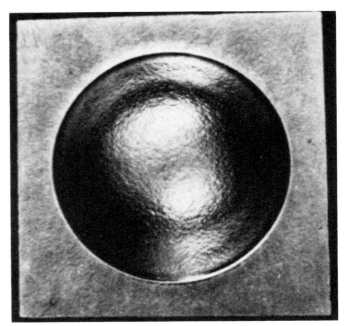

Figure 9a. Alloy 260 (cartridge brass, 70 percent) drawn cup showing rough surface or "orange peel" (actual size). (ASM Committee on Metallography of Copper and Copper Alloys, "Microstructures of Copper and Copper Alloys," *Atlas of Microstructures of Industrial Alloys*, Volume 7, Eighth Edition, Lyman, T., editor, American Society for Metals, © 1972, p. 282)

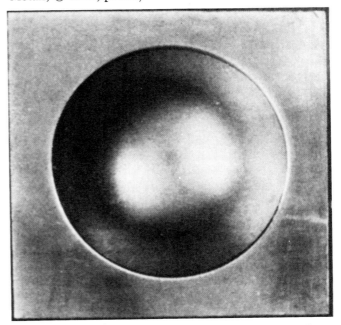

Figure 9b. Alloy 260 (cartridge brass, 70 percent) drawn cup with a smooth surface, no "orange peel" (actual size). (ASM Committee on Metallography of Copper and Copper Alloys, "Microstructure of Copper and Copper Alloys," *Atlas of Microstructures of Industrial Alloys*, Volume 7, Eighth Edition, Lyman, T., editor, American Society for Metals, © 1972, p. 282)

UNIT 3 GAS WELDING REPAIRS AND THEIR EFFECTS

Acetylene and oxygen gases are used to produce a flame for fusion welding because their flame temperatures are considerably higher than those produced with other industrial gases. The approximate temperature range is from 5000 to 6300°F (2760 to 3482°C) at the inner core. Since the melting temperature of common metals is below 3000°F (1649°C) (melting temperature of steel is about 2500°F or 1371°C), most metals can be welded or brazed with the oxy-acetylene process.

OBJECTIVES

After completing this unit, you will be able to:
1. Make a bronze weld.
2. Silver braze a carbide tool.
3. Make an oxy-acetylene torch cut.
4. Explain braze welding, soldering, and torch cutting procedures.

THE GAS WELDING SYSTEM

Oxygen is stored in pressurized tanks, which are perfectly safe as long as they are carefully handled, but can be extremely dangerous if they are dropped so that the valve on the top is broken off. If that happens, the tank can become a missile that will go through brick walls. The cap should therefore **always** be kept on the tank until it is in use, and the tank should always be secured so it cannot tip and fall (Figure 1). **Never** use oil or grease on fittings, especially oxygen fittings. Oxygen under pressure and in contact with oil or grease will explode.

Acetylene tanks use a different method of safely containing gas. Acetylene gas under pressure becomes unstable near 30 PSI and in the free state will explode. The safe working pressure is below 15 PSI. Acetylene, because of its instability, cannot safely be stored under pressure in cylinders. Acetone is a solvent for acetylene and can absorb a considerable volume of gas at about 250 PSI in a tank. The storage cylinder is packed with acetone and an absorbant material (porous calcium silicate) so that the acetylene can be slowly released from the acetone as needed and to keep the acetone from entering the regulators and hoses (Figure 2). An old tank may lose some of its absorbancy

and a small amount of acetone could be released if the tank is used in a horizontal position. This can damage the regulators and create a hazard. Tanks should never be allowed to run to zero pressure and should always be used in the upright position.

The gas welding tanks, regulators, hoses, and torch are shown in Figure 3. The regulators are at-

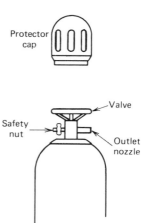

Figure 1. Oxygen tank valve and protector cap. The cap should always be in place when moving or storing the tank.

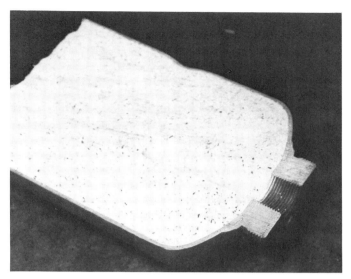

Figure 2. Cutaway view of an acetylene tank showing the porous material within. (Lane Community College)

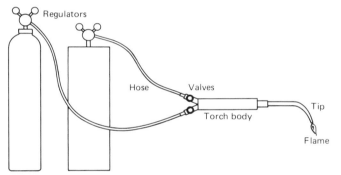

Figure 3. Regulators, hoses, and torch connected to the welding tanks.

tached to the tanks with a threaded fastener. The hoses and torch are also connected with threaded fasteners. These parts cannot be accidentally interchanged (which would be disastrous) because all of the acetylene fittings have left-hand threads and all of the oxygen fittings have right-hand threads.

When turning on the oxy-acetylene system, the regulator diaphragm valve handle (Figure 4) should be turned outward (counterclockwise) so no pressure will enter the hoses. The oxygen tank valve is then cracked open to avoid a sudden surge of pressure on the diaphragm of the valve; it is then completely opened. The valve on the acetylene cylinder is opened just enough to provide a sufficient supply of gas.

When shutting down a system, bleed the hoses and regulators. Turn off the tanks first, then open the acetylene and oxygen valves one at a time on the torch to release pressure on the gages. Turn the diaphragm valve handles counterclockwise.

TYPES OF GAS WELDS

Fusion welding occurs when two edges or surfaces of metal become molten and intermingle with each other or with a molten filler metal (Figure 5). Steel is often fusion welded, using a steel filler rod. This type of fusion welding is very useful on thin stock, such as car bodies, that is difficult to arc weld. Aircraft tubing is oxy-acetylene fusion welded as it is considered more reliable for that application.

Diffusion or braze welding requires the use of a filler metal consisting of brass (an alloy of copper and zinc) or other kinds of bronzes. The filler material has a melting point that is lower than the base metal

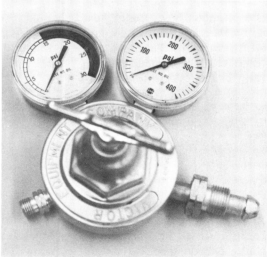

Figure 4. Oxygen (left) and acetylene (right) regulators. (Lane Community College)

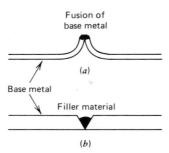

Figure 5. (a) Fusion weld of base metal. (b) Fusion weld using filler rod.

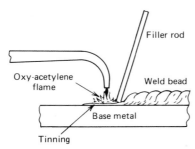

Figure 6. Braze welding requires tinning of the surface prior to filling in the bead.

Table 1

Tip Sizes and Pressures

Metal Thickness (in.)	Tip Size	Drill Size	Oxygen Pressure (P.S.I.G.)		Acetylene Pressure (P.S.I.G.)	
			Min.	Max.	Min.	Max.
up to $\frac{1}{32}$	000	75	3	5	3	5
$\frac{1}{64}-\frac{3}{64}$	00	70	3	5	3	5
$\frac{1}{32}-\frac{5}{64}$	0	65	3	5	3	5
$\frac{3}{64}-\frac{3}{32}$	1	60	3	5	3	5
$\frac{1}{16}-\frac{1}{8}$	2	56	3	5	3	5
$\frac{1}{8}-\frac{3}{16}$	3	53	4	7	3	6
$\frac{3}{16}-\frac{1}{4}$	4	49	5	10	4	7
$\frac{1}{4}-\frac{1}{2}$	5	43	6	12	5	8
$\frac{1}{2}-\frac{3}{4}$	6	36	7	14	6	9
$\frac{3}{4}-1\frac{1}{4}$	7	30	8	16	7	10
$1\frac{1}{4}-2$	8	29	10	19	8	12
$2-2\frac{1}{2}$	9	28	12	22	9	14
$2\frac{1}{2}-3$	10	27	14	24	10	14
$3-3\frac{1}{2}$	11	26	16	26	11	15
$3\frac{1}{2}-4$	12	25	18	28	12	15

Source: *Welding, Cutting & Heating Guide*, Victor Equipment Company, Welding and Cutting Apparatus Division, © 1977, Denton, Texas.

and therefore does not fuse but combines by wetting the surface (capillary attraction) of the base metal and partially combining with the base metal by adhesion (Figure 6).

FLAME ADJUSTMENT

When you light the torch, always use the flint striker provided for that purpose; wear gloves to prevent burning your hand. **Never** use a cigarette lighter or match. Turn on the acetylene first and light it. The acetylene valve should be opened to the point that the flame is burning clean with very little smoke and the flame is not separated from the tip. The next step is to adjust the oxygen to provide the correct flame for the job. There are essentially three kinds of flame: **oxidizing, carburizing,** and **neutral** (Figure 7). See Table 1 for tip sizes and pressures.

The acetylene flame alone mixed with air produces an excess of free carbon that will adhere to any surface it contacts. When a small percentage of oxygen is used, a carbon excess flame that carburizes molten metal is formed. Carburized metal can become hard and brittle when quickly cooled. A carburized weld in low carbon steel will have a boiling action in the molten puddle and will be porous and pitted on the surface. If high carbon steel is welded,

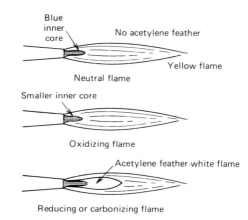

Figure 7. The three types of oxy-acetylene flame are neutral, oxidizing, and carbonizing.

however, an excess carbon flame should be used to replace carbon that is burned away. In contrast, an oxidizing flame is produced by an excess of oxygen. It is characterized by a small, pale blue inner cone and a roaring or loud hissing sound. The oxidizing flame has excess oxygen that oxidizes or burns the weld metal, throwing bright sparks from the weld. Oxidizing flames should not be used for welding steel as it produces a weak weld.

A neutral flame, characterized by a slight hissing sound, has no excess oxygen or carbon and is best for fusion welding of low carbon steel.

BRAZE WELDING

Braze welding is a useful means of joining steel, cast iron, and bronze. Its primary disadvantage for ferrous metal is that it loses its strength at higher temperatures and should not be used for welding exhaust manifolds or stove parts. Also, it does not provide a good color match on ferrous metals.

Brazing is one of the easier processes to learn since only the filler metal becomes molten (Figure 8). A neutral flame is usually used when brazing. A flux is used to clean the base metal so a bond can be made. The flux is usually made to adhere on the hot end of the brazing rod by inserting it into a can of flux from time to time.

Vertical or overhead welds cannot be made by the brazing method, so all welds should be made in a horizontal (level) position. The base metal must be preheated, if it is a large mass, to just under the melting temperature of the filler metal. A flux is applied and the filler metal melted to "wet" the base metal. If this is not done, the weld will not adhere and will peel off (Figure 9). A wide bead should be made on each pass while making sure there is a complete bond with the base metal and previous passes (Figure 10).

Oxy-acetylene welding and brazing operations tend to have the metallurgical effects of stress relieving and softening metals if the welded part is allowed to cool slowly. Grain growth is much greater in the heat affected zone as compared to arc welds.

FURNACE BRAZING AND SILVER SOLDERING

Brazing occurs at a temperature between 800 and 1700°F (426.6 and 926.6°C). The brazing material may be one of several alloys such as copper-zinc or copper-zinc-silver.

Furnace brazing is used in automatic assembly where a ring or ribbon of braze material is heated together with the part until the filler material melts and flows into a relatively tight joint fit by capillary action. This process is also called silver brazing or silver soldering. This process may be used to join parts such as tungsten carbide on a lathe tool. A heating torch can be used in this process as long as the flame is not applied directly on the silver solder.

A tungsten carbide cutting tip may be applied to a tool shank in the following manner.

1. A shank is prepared (milled) to properly fit the

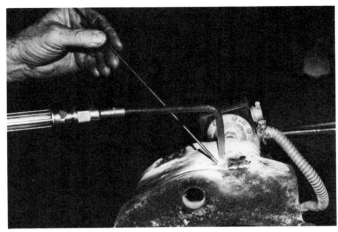

Figure 8. Braze welding. (Lane Community College)

Figure 9. Example of a braze overlay on a steel shaft that was not properly tinned or wetted. When the part was turned on the lathe, the braze weld began to peel off. (Lane Community College)

Figure 10. The diffusion or braze welding bond is not as strong as that of fusion welding so a broader weld area should be used to provide more strength.

carbide bit. It is fastened in a vise so it can be heated from the underside.

2. Silver brazing flux and a ribbon or shim of silver solder are sandwiched between the carbide insert and the shank (Figure 11). Both the shank and carbide insert must be clean and the insert roughed on abrasive to ensure a bond. Also avoid getting any oil as from fingerprints on the silver solder since oil or impurities can cause lack of bonding. The shank is then heated on the underside with a torch until the silver solder melts and flows into all of the joining area. The flame is then removed.

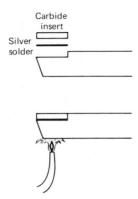

Figure 11. The process of silver soldering a tungsten carbide insert to a tool shank.

Figure 12a. The steel tool shank has been prepared by milling. The carbide insert is being cleaned on abrasive cloth to accept the brazing material. (Lane Community College)

The insert should be held in place with a piece of metal until the solder solidifies since the insert tends to float on the molten metal and may slide out of position.

An alternate method may be used if only silver solder wire or brazing rod is available.

1. Flux is applied to the shank, which is heated from beneath (Figures 12a to 12f) until the silver solder or braze rod melts after being applied to it. The silver solder should flow and adhere to the entire area to be brazed. This is called wetting or "tinning" the base metal.
2. The carbide bit should be cleaned and roughed with abrasive cloth so the solder will adhere to it. The bit is similarly tinned as was the shank. The tinning operation should provide sufficient brazing material to join the surfaces. Apply flux to both tinned surfaces.
3. Place the two tinned surfaces together where the bit is to be located and apply heat from below on the shank. Sometimes heat can be applied to the bit to be sure there is complete melting of the braze filler metal. It is sometimes necessary to apply pressure on the bit with a piece of steel to assure complete seating. Excess solder should be squeezed out from between the surfaces.

Worn carbide throwaway inserts may often be salvaged by this process to make special purpose lathe tools. These tools are then sharpened on silicon carbide or boron wheels. The steel shank should be ground for clearance on an aluminum oxide wheel.

Figure 12b. Carbide insert, which is lightly clamped in a vise, is being heated from beneath and tinned. (Lane Community College)

Figure 12c. Steel shank being tinned. (Lane Community College)

SOLDERING

Soldering is similar to brazing except that it occurs at a temperature below 800°F (426.6°C) and is often

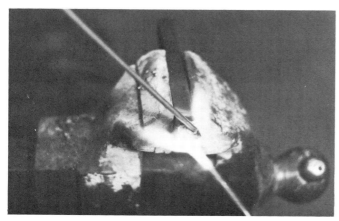

Figure 12d. Carbide insert being sweated to shank. The insert is being held in place with a steel rod. (Lane Community College)

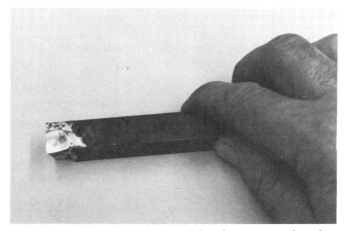

Figure 12e. The finished tool has been ground and is ready for service. (Lane Community College)

Figure 12f. Tool being used in a lathe to make a cut. (Lane Community College)

called soft soldering as compared to hard or silver soldering. Soft solder is an alloy of tin and lead in varying compositions, such as 50:50, 60:40, and 40:60. Soldering is used where high strength is not required or where higher temperatures are not encountered. It is used for electrical wiring, sheet metal, and for connecting tubes in plumbing applications.

Electric soldering coppers or guns are used for soldering wires, but heavy coppers, often heated in a gas oven, or used for sheet metal work. Heating torches are sometimes used in soldering for special applications such as in radiator work.

OXY-ACETYLENE TORCH CUTTING

The metal cutting process that uses the oxy-acetylene torch is widely used in many industrial fields. Ferrous metals can be oxidized by a stream of oxygen if the iron is heated red hot. This is what happens when torch cutting takes place. A preheat flame is used to heat the edge of a piece of steel, for instance, and a jet of oxygen is made to pass over the heated area and across the unheated section. The heated area begins to oxidize (burn) and enough heat is generated by this burning that the unheated metal is also oxidized and converted into an oxide that is blown away. Thus a cut through the steel is made that is about the width of the jet of oxygen. The cutting process is sustained as long as the oxygen jet continues to make contact with the hot metal. This is done by moving the cutting tip slowly along the line to be cut.

THE CUTTING TORCH

In Figure 13 it may be seen that the oxygen and acetylene needle valves are often the same as those used for welding blowpipe. These acetylene and oxygen valves are used to control the adjustment of the preheat flame. Many cutting torches have a separate valve for oxygen adjustment. The high pressure oxygen lever releases the oxygen jet for cutting. The cut-

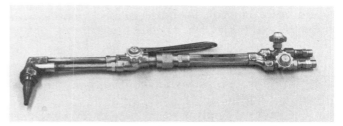

Figure 13. Cutting torch assembly. (Lane Community College)

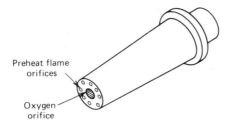

Figure 14. The cutting tip.

Figure 15. Automatic cutting torch in use. (Lane Community College)

Figure 16. Automatic torch that follows a drawing with an electric eye. (El Jay Inc., Eugene, Oregon)

Figure 17. Pipe cutting attachment. (Lane Community College)

Figure 18. Free-hand cutting using a straightedge. (Lane Community College)

ting tip (Figure 14) shows the preheat holes and the oxygen cutting hole.

The cutting torch used on automatic flame-cutting machines (Figure 15) is essentially the same as that used for hand-held cutting. Some machines follow a prepared template and others use an electric eye that follows a drawing (Figure 16). Pipe cutting attachments are used on large steel pipe (Figure 17).

Freehand cutting is usually difficult for the beginner. It is advisable, therefore, when making straight

Table 2
Cutting Pressure

TYPES: 1-101, 3-101, and 5-101 (Oxy-Acetylene)

Metal Thickness (in.)	Tip Size	Cutting Oxygen		Pre-Heat Oxygen		Acetylene		Speed (I.P.M.) (Min./Max.)	KERF (Width)
		Pressure (PSIG)[a] (Min./Max.)	Flow (SCFH) (Min./Max.)	Pressure (PSIG)[b] (Min./Max.)	Flow (SCFH)[b] (Min./Max.)	Pressure (PSIG) (Min./Max.)	Flow (SCFH) (Min./Max.)		
$\frac{1}{8}$	000	20/25	20/25	3/5	3/5	3/5	3/5	28/32	.04
$\frac{1}{4}$	00	20/25	30/35	3/5	4/6	3/5	4/6	27/30	.05
$\frac{3}{8}$	0	25/30	55/60	3/5	5/9	3/5	5/8	24/28	.06
$\frac{1}{2}$	0	30/35	60/65	3/6	7/11	3/5	6/10	20/24	.06
$\frac{3}{4}$	1	30/35	80/85	4/7	9/14	3/5	8/13	17/21	.07
1	2	35/40	140/150	4/9	11/18	3/6	10/16	15/19	.09
$1\frac{1}{2}$	2	40/45	150/160	4/12	13/20	3/7	12/18	13/17	.09
2	3	40/45	210/225	5/14	15/24	4/9	14/22	12/15	.11
$2\frac{1}{2}$	3	45/50	225/240	5/16	18/29	4/10	16/26	10/13	.11
3	4	40/50	270/320	6/17	20/33	5/10	18/30	9/12	.12
4	5	45/55	390/425	7/18	24/37	5/12	22/34	8/11	.15
5	5	50/55	425/450	7/20	29/41	5/13	26/38	7/9	.15
6	6[c]	45/55	500/600	10/22	33/48	7/13	30/44	6/8	.18
8	6[c]	45/55	500/600	10/25	37/55	7/14	34/50	5/6	.19
10	7[c]	45/55	700/850	15/30	44/62	10/15	40/56	4/5	.34
12	8[c]	45/55	900/1000	20/35	53/68	10/15	48/62	3/5	.41

[a] All pressures are measured at the regulator using $25' \times \frac{1}{4}''$ hose through tip size 5 and $25' \times \frac{3}{8}''$ hose for tip size 6 and larger.

[b] Applicable to three-hose machine cutting torches only.

[c] For best results use ST 1600—ST 1900 series torches and $\frac{3}{8}''$ hose when using tip size 6 or larger.

Note: Data compiled using mild steel as test material.

Note: At no time should the withdrawal rate of an individual acetylene cylinder exceed $\frac{1}{7}$ of the cylinder contents per hour. If additional flow capacity is required use an acetylene manifold system of sufficient size to supply the necessary volume.

Source: *Welding, Cutting & Heating Guide*, Victor Equipment Company, Welding and Cutting Apparatus Division, © 1977, Denton, Texas.

Figure 19. Using a circle cutting attachment. (Lane Community College)

cuts, to use a straightedge for a guide (Figure 18), and for cutting circles, to use a circle cutting attachment (Figure 19).

The torch preheat flame should be lit and adjusted in the same way as it is done for welding, using a neutral flame. Press and release the oxygen pressure lever and not the preheat flame; it may be necessary to readjust the preheat flame. Cutting pressures are given in Table 2.

STEPS IN CUTTING

1. Place the plate to be cut on a welding bench so there is clearance beneath the cutting area. Clamp on the straightedge. Wear gloves and goggles.
2. Light the torch. Holding it with both hands for maximum support, start the preheat at the edge of the plate, holding the inner cone about $\frac{1}{16}$ in. above the metal.

3. When a spot has been heated to a bright red, press down on the oxygen lever and move the torch forward along the straightedge. The rate of movement must be even and neither too rapid nor too slow. If it is too rapid, the metal will not burn through and the cut will be ragged. If the rate is too slow, excess slag may build up on the bottom of the kerf. The top edge may melt and roll over making a rounded edge. When the cutting rate is just right, a steady spray of slag flows out the bottom of the cut with no buildup of slag.

PROBLEMS IN OXY-ACETYLENE CUTTING

Clean, accurate, slag-free cuts can be made only through operator skill and care in selecting tips and gas pressures and by maintaining the equipment. The correct size cutting tip should be selected and cleaned with tip cleaners if it has been in use.

If the cut is made too fast, globules of hot metal form at the bottom of the cut and fuse the edges together. Slanting drag lines are evident when the speed is too fast. If the torch is hand-held, care must be taken to maintain an even cutting speed. An irregular kerf and gouges are the result of cutting too slowly and the top edges are melted and rolled over from the preheat flame. Melting of the top edges may also be caused by an excessively large preheat flame or when the flame is held too close to the work. The extra heat can also warp the plate.

For cutting mild steel, a neutral preheat flame should be used. If a carbonizing flame is used, the cut edge will be carburized and the sudden cooling by the plate mass will produce an extremely hard surface that can be very difficult to cut with saws or other cutting tools. However, when cutting medium and high carbon steels, a carbonizing flame is advantageous to avoid decarburization on the cut edge and the cutting action is enhanced.

Slag is usually produced when hand cutting steel because there is less control of the cutting action than with an automatic machine cut. The part sometimes needs to be removed by hammering and the slag chipped off with a slag hammer. Wear safety glasses when doing these operations.

SELF-TEST

1. Propane gas and air will burn at approximately 2000°F (1093°C). This flame is much less expensive to use than that of oxygen and acetylene. Why don't we use propane for welding steel by fusion?
2. The safe working pressure for acetylene gas is 15 PSI. Why is it extremely dangerous to increase the pressure to 30 PSI?
3. Why should the cap be kept on an oxygen tank and the tank secured in a cart or rack when it is not in use?
4. What are the two types of welds made with the oxy-acetylene torch?
5. What is the difference between braze welding and silver brazing?
6. Soldering is very similar to brazing. State one major difference.
7. How does an oxy-acetylene torch cut steel?
8. Besides selecting the tip size for cutting a certain thickness of plate steel, what other two important factors are involved?
9. When globs of metal and slag form on the bottom of a cut and the cutting lines are slanted, is the cutting speed too fast or too slow?
10. An excess acetylene flame (carbonizing) on the preheat can cause what problem on the cut edge?

UNIT 4 ARC WELDING AND ITS EFFECT ON METALS

Arc welding today is one of the principal methods used for joining metals. Before the properties of metals were considered and welding metallurgy was not yet understood, the welding process was not considered to be dependable enough for such purposes as shipbuilding and boilermaking. Prior to World War II, riveting was used to join metals for these purposes. When the welding process was used at that time, disastrous failures sometimes occurred, such as ships breaking in half and pipelines splitting lengthwise for miles. These problems have long since been corrected through the study of welds and their effects.

In this unit you receive some very basic instruction in the arc welding process. It is not intended as a course in welding technology, but only as an overview. You will also be made aware of some of the effects of arc welding on various metals.

OBJECTIVES

After completing this unit, you should be able to:
1. Identify types of arc welding processes and equipment.
2. Identify some arc welding electrodes.
3. Strike an arc and make a simple weld.
4. Identify some causes of failures in metals as a result of improper welding techniques.

ARC WELDING PROCESSES

Arc welding is a process of fusion joining of metals in which a filler metal is usually used and an electric arc provides the heat energy. There are many forms of arc welding. Shielded metal-arc, gas shielded arc (MIG and TIG), and submerged-arc welding are a few of these modern processes (Figure 1). In most of these welding processes, a relatively low voltage, high amperage electric arc is maintained, usually $\frac{1}{16}$ to $\frac{1}{8}$ in. length, between the electrode and the base metal. The electric current may be either AC or DC. When AC is used, an equal amount of heat energy is applied to each pole (the base metal and electrode), but with DC the amount of heat energy at each pole changes with the polarity.

When the term "straight polarity" is used, it means the electrode is negative and the work positive, so more heat energy is applied at the electrode. Penetration of the base metal is limited and the weld metal

Figure 1. Trailer frame being welded with MIG process. (Lane Community College)

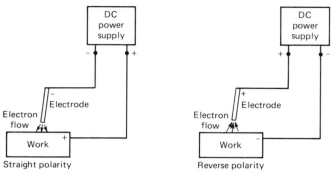

Figure 2. Straight and reverse polarity are both used on welding machines for various applications.

Figure 3. Alternating current (AC) welder. (Lane Community College)

is piled higher than with reverse polarity. When the term "reverse polarity" is used, it means the electrode is positive and more heat is applied to the base metal, causing a greater depth of penetration (Figure 2). Reverse polarity is often used for vertical and overhead welding because the weld metal freezes faster than with straight polarity or when using AC.

WELDING MACHINES
Welding machines may be alternating current (AC) types which are essentially transformers that reduce the line voltage and increase the available amperage to the electrode (Figure 3). Direct current (DC) machines are often motor-generators (Figure 4), but may also be a rectifier type or a combination AC and DC (Figure 5). Heat adjustments are made by changing the amperage.

ELECTRODES
Metal arc welding electrodes (also called "rods" or "stick rods") have a marking system that designates their type and relative tensile strength (Figure 6). The letter "E" is a prefix denoting an arc-welding electrode. The first two digits of four digit numbers and the first three of five digit numbers designate minimum tensile strength in thousand PSI, and the next to last digit designates position. Example:

E601x is an all position electrode, 60,000 PSI minimum tensile strength.

E602x is a flat position electrode, 60,000 PSI minimum tensile strength.

E110xx has 110,000 PSI minimum tensile strength.

The last digit represents a special manufacturer's number designating a special characteristic.

Figure 4. Direct current (DC) welding machine. (Lane Community College)

In general, the selection of the type of electrode is determined by the composition of the base metal and the desired metallurgical characteristics. For example, if you are welding on mild steel (low carbon), use a low tensile electrode such as E6011 or E6013, and if you are welding on high strength tubing or automobile frames, consult the manufacturer for the correct welding process or electrode.

Figure 5. Combination AC and DC welder. (Lane Community College)

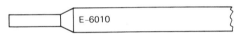

Figure 6. Welding electrode end showing number designation that is usually printed on the electrode.

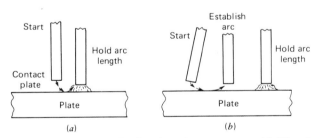

Figure 7. Two methods of striking an arc. (a) The tap method and (b) the scratch method.

Stick electrodes now account for about one-half of the welding electrode market but are rapidly being replaced by welding wire. Coated stick electrodes are somewhat wasteful since a part of the electrode is always discarded and the coating is often damaged in handling or has picked up moisture and so must be thrown out. In the metal arc process, the arc must be started by "striking," which can be done in two ways as shown in Figure 7.

ARC WELDING PROCEDURE

When arc welding, **always** wear a helmet or head shield with the correct lens and protective goggles when chipping slag. Wear heavy leather gloves and suitable clothing that will not melt or burn. Do not wear polyester clothing when welding, heating, or cutting metal. To make a weld with E6010 on a DC machine with reverse polarity, or with E6011 or E6013 on AC machines, use the following steps.

Table 1
Approximate Current Settings for E-6010, E-6011, and E-6013 Electrodes

Dia. of Electrode	Amperes
$\frac{1}{16}$	20–40
$\frac{3}{32}$	30–80
$\frac{1}{8}$	70–120
$\frac{5}{32}$	120–170
$\frac{3}{16}$	140–230
$\frac{7}{32}$	170–275
$\frac{1}{4}$	200–325

1. Set the machine to the correct amperage as shown in Table 1.
2. Place the bare end of the electrode in the holder.
3. Test the arc on a piece of scrap steel. Lower your helmet, having the rod end poised above the work where you will make a weld.
4. Scratch the surface lightly for about an inch and lift the end of the rod slightly, about $\frac{1}{8}$ in. This should start an arc.
5. Practice maintaining the arc to this length by steadily lowering the electrode as it burns off.
6. Move the electrode along the workpiece at a rate that will produce a continuous bead (Figure 8).
7. Raise your helmet, make sure you still have eye protection, and chip off the slag so you can inspect the weld; use a wire brush if necessary. Compare the weld you made with Figure 9.

CRACKING IN THE WELD METAL

Cracking down the center of the weld or microcracking of the weld metal usually takes place at high temperatures (hot cracking) during solidification and is caused by stresses of expansion and contraction in the weld. This is sometimes the result of trying to fill in a gap that is too wide with the weld. High cooling rates tend to increase hot cracking in the weld metal. Slowing down the cooling rate will help to prevent

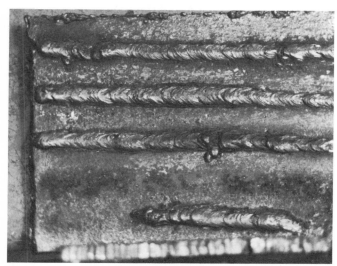

Figure 8. A continuous bead can be made for a practice weld. (Lane Community College)

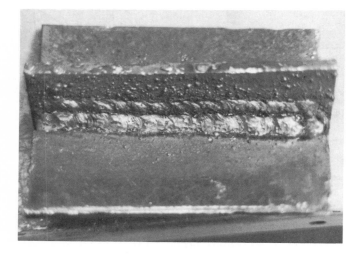

Figure 9. Example of good (top) and bad (bottom) welds. (Lane Community College)

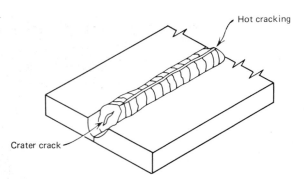

Figure 10. Hot cracking in the weld is often seen as a separation down the center of the weld. *(Practical Metallurgy and Materials of Industry)*

hot cracking (Figure 10). The end or terminal crater crack is one example of hot cracking. This can often be eliminated by back filling of the crater before breaking the arc.

Preheating is also used to slow down the cooling rate in low and high carbon alloy steels. Tables in welding manuals provide correct preheat temperatures for a specific steel and electrode. Some alloy and carbon steels develop coarsened grains if the preheat temperature is too high and the interpass temperatures are consequently above the specified degree. If this occurs, a postheat normalizing procedure is necessary to refine the coarse brittle grain in the heat affected zone (HAZ).

Weld metal cracks may be caused by the following:

1. Wrong choice of filler metal or electrode. For example, a high tensile filler metal on low carbon steel base metal is the wrong application.
2. Welding conditions; too much gap between base metal parts or pickup of embrittling elements such as carbon, nitrogen, or hydrogen.
3. Defects such as porosity, slag or oxide inclusions, and hydrogen gas pockets (hydrogen embrittlement).
4. Too rapid cooling rates (no preheat).
5. Constraint of base metal parts. For example, two heavy plates clamped rigidly so that no movement can take place while welding: all the stresses involved are concentrated in the weld metal.

BASE METAL CRACKING

Cracking in the parent or base metal is a serious problem when welding high carbon or tool steels. These cracks form in the fusion zone or in the heat affected

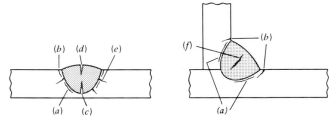

Figure 11. Various forms of cracking in welds: (a) underbead cracking, (b) toe cracks, (c) root cracks, (d) hot cracking, (e) radial cracking, and (f) internal cracks. *(Practical Metallurgy and Materials of Industry)*

zone adjacent to the weld bead (Figure 11). This is called underbead or hard cracking and cannot normally be detected on the surface of the base metal. These cracks are difficult to detect even with magnetic particle inspection or X-ray. Any welding method performed on medium to high carbon steels will result in a hard zone adjacent to the weld if the cooling rate is too fast. The underbead crack follows the contour of the weld bead.

Radial cracks are sometimes a result of welding stress in hard or brittle zones. Hydrogen entrapped in the base metal and weld metal often initiates such underbead cracking. Hydrogen is very soluble in molten steel, but it is forced out of the metal by diffusion if the cooling rate is slow. Hydrogen may be picked up by the use of damp electrodes or by those having a cellulose type of coating. Oil, grease, paint, or water on the base metal will also cause hydrogen embrittlement. When low hydrogen rod that has not picked up any moisture is used on higher carbon steels and preheat is used, underbead cracking is not likely to be a problem.

Toe cracks (Figure 11) are caused by stress at the toe of the weld and across the heat affected zone. Preheating reduces the problem that is usually found in metals with low ductility. A root crack (Figure 11) originates in the root bead. This is caused by a rapid cooling rate of the small weld bead; carbon pickup from oily metal intensifies this problem. Preheat should be used.

SELF-TEST

1. In shielded metal arc welding, what provides the heat energy to fuse the base metal and melt the filler metal?
2. What are the two basic kinds of electric arc welding machines?
3. Metal arc welding rods have a numerical designation. The "E" stands for "arc welding electrode." What do the first two numbers denote?
4. How do you start and hold an arc?
5. What causes welding cracks down the center of the weld metal?
6. How can the cooling rate be slowed?
7. What kind of cracking is sometimes found in the base metal with medium to high carbon content?
8. Why shouldn't you weld over grease or paint?
9. How can toe cracking be reduced?
10. A crack in the root bead (first pass) may be caused from carbon pickup and a rapid cooling rate. How can root bead cracking be reduced?

SECTION G

DRILLING MACHINES

UNIT 1 THE DRILL PRESS

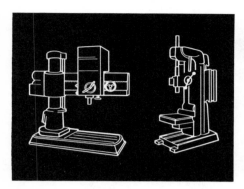

Drilling holes is one of the most basic of machining operations and one that is very frequently done by machinists. Metal cutting requires considerable pressure of feed on the cutting edge. A drill press provides the necessary feed pressure either by hand or power drive. The primary use of the drill press is to drill holes, but it can be used for other operations such as countersinking, counterboring, spot facing, reaming, and tapping, which are processes that modify the drilled hole.

Before operating any machine, a machinist must know the names and the functions of all its parts. In this unit, therefore, you should familiarize yourself with the operating mechanism of several types of drilling machines.

There is a tendency among students to dismiss the dangers involved in drilling since most drilling done in a school shop is performed on small sensitive drill presses with small diameter drills. This tendency, however, only increases the danger to the operator and has turned otherwise harmless situations into serious injuries. Clamps should be used to hold down workpieces, since hazards are always present even with small diameter drills.

The safety habits you develop now will protect you from many injuries in the school shop and all through your career. This unit will alert you to the potential dangers that exist in drilling and help you avoid the hand injuries most likely to occur while working with the drill press.

OBJECTIVES

After completing this unit, you should be able to:
1. List 10 safety hazards involving drilling machines.
2. Describe and explain steps the operator can take to avoid safety hazards.
3. Identify three basic drill press types and explain their differences and primary uses.
4. Identify the major parts of the sensitive drill press.
5. Identify the major parts of the radial arm drill press.

DRILLING SAFETY

One example of a safety hazard on even a small diameter drill is the "grabbing" of the drill when it breaks through the hole. If the operator is holding the workpiece with his hands and the piece suddenly begins to spin, his hand will likely be injured, especially if the workpiece is thin.

Even if the operator were strong enough to hold the workpiece, the drill can break, continue turning, and with its jagged edge become an immediate hazard to the operator's hand. The sharp chips that turn with the drill can also cut the hand that holds the workpiece.

Poor work habits produce many injuries. Chips flying into unprotected eyes, heavy tooling, or parts dropping from the drill press onto toes, slipping on oily floors, and getting hair or clothing caught in a rotating drill are all hazards that can be avoided by safe work habits.

The following are safety rules to be observed around all types of drill presses:

1. Proper dress (Figure 1) does not include loose clothing, gloves, or neckties. Sleeves should be rolled above the elbows, and if hair is long, tie it back or keep it contained in an industrial hairnet. Wristwatches or rings should be taken off

Figure 1. Drill press operator wearing correct attire. (Lane Community College)

Figure 2. Drilling operation on upright drill press with tools properly located on adjacent work table. (Lane Community College)

when operating a drilling machine. Painful injuries can result if any part of you or your attire is caught in a rotating drill.

2. Tools to be used while drilling should never be left lying on the drill press table, but should be placed on an adjacent worktable (Figure 2).

3. Get help when lifting heavy vises or workpieces.

4. Workpieces should always be secured with bolts and strap clamps, C-clamps, or fixtures. A drill press vise should be used when drilling small parts (Figure 3). If a clamp should come loose and a "merry-go-round" results, don't try to stop it from turning with your hands. Turn off the machine quickly; if the drill breaks or comes out, the workpiece may fly off the table.

5. Safety glasses are a necessity and are required by federal safety regulations. The small chips that fly at high speed can penetrate the unprotected eye, though will not usually go through clothing

Figure 3. Properly clamped drill press vise holding work for hard steel drilling.

or skin. Wearing safety glasses will minimize the danger to your eyes. **Wear safety glasses at all times in a machine shop.**

6. Never clean the taper in the spindle when the drill is running, since this practice could result in broken fingers or worse injuries.

7. Always remove the chuck key immediately after using it. A key left in the chuck will be thrown out at high velocity when the machine is turned on. It is a good practice to **never let the chuck key leave your hand when you are using it.** It should not be left in the chuck even for a moment. Some keys are spring loaded so they will automatically be ejected from the chuck when released. Unfortunately, very few of these keys are in use in the industry.

8. Never stop the drill press spindle with your hand after you have turned off the machine. Sharp chips often collect around the chuck or spindle. Do not reach around, near, or behind a revolving drill.

9. When removing taper shank drills with a drift, use a piece of wood under the drills so they will not drop on your toes. This will also protect the drill points.

10. Interrupt the feed occasionally when drilling to break up the chip so it will not be a hazard and will be easier to handle.

11. Use a brush instead of your hands to clean chips off the machine. **Never** use an air jet for removing chips as this will cause the chips to fly at a high velocity and cuts or eye injuries may result. Do not clean up chips or wipe up oil while the machine is running.

12. Keep the floor clean. Immediately wipe up any oil that spills, or the floor will be slippery and unsafe.

13. Remove burrs from a drilled workpiece as soon as possible, since any sharp edges or burrs can cause severe cuts.

14. When you are finished with a drill or other cutting tool, wipe it clean with a shop towel and store it properly.

15. Oily shop towels should be placed in a closed metal container to prevent a cluttered work area and avoid a fire hazard.

16. When moving the head or table on sensitive drill presses, make sure a safety clamp is set just below the table or head on the column; this will prevent the table from suddenly dropping if the column clamp is prematurely released.

Proper drilling and clamping procedures are important to your safety. Reference to these and other safety procedures will be made in other units on drilling.

TYPES OF DRILL PRESSES AND THEIR USES

There are three basic types of drill presses used for general drilling operations: the sensitive drill press, the upright drilling machine, and the radial arm drill press. The sensitive drill press (Figure 4), as the name implies, allows the operator to "feel" the cutting action of the drill as he hand feeds it into the work. These machines are either bench or floor mounted. Since these drill presses are used for light duty applications only, they usually have a maximum drill size of $\frac{1}{2}$ in. diameter. Machine capacity is measured by the diameter of work that can be drilled (Figure 5).

The sensitive drill press has four major parts, not including the motor: the head, column, table, and base. Figure 6 labels the parts of the drill press that you should remember. The spindle rotates within the quill, which does not rotate but carries the spindle up and down. The spindle shaft is driven by a stepped-vee pulley and belt (Figure 7) or by a variable speed drive (Figure 8). **The motor must be running and the spindle turning when changing speeds with a variable speed drive.**

The upright drill press is very similar to the sensi-

Figure 4. A sensitive drill press. These machines are used for light duty application. (Wilton Corporation)

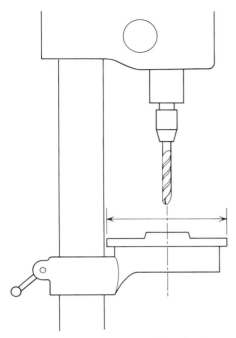

Figure 5. Drill presses are measured by the largest diameter of a circular piece that can be drilled in the center. (*Machine Tools and Machining Practices*)

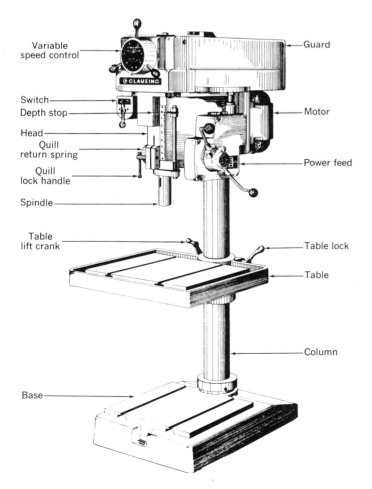

Figure 6. Drill press showing the names of major parts. (Clausing Machine Tools)

Variable speed control — Guard

Switch

Depth stop — Motor

Head

Quill return spring

Quill lock handle — Power feed

Spindle

Table lift crank — Table lock

— Table

— Column

Base

Figure 7. View of a vee-belt drive. Spindle speeds are highest when the belt is in the top steps and lowest at the bottom steps. (Clausing Machine Tools)

Figure 8. View of a variable speed drive. Variable speed selector should only be moved when the motor is running. The exact speed choice is possible for the drill size and material with this drive. (Clausing Machine Tools)

tive drill press, but it is made for much heavier work (Figure 9). The drive is more powerful and many types are *gear driven,* so they are capable of drilling holes to two inches or more in diameter. **The motor must be stopped when changing speeds on a gear drive drill press.** If it doesn't shift into the selected gear, turn the spindle by hand until it meshes. Since power feeds are needed to drill these large size holes, these machines are equipped with power feed mechanisms that can be adjusted by the operator. The operator may either feed manually with a lever or hand wheel or he may engage the power feed. A mechanism is provided to raise and lower the table.

Figure 9. Upright drill press. (Wilton Corporation)

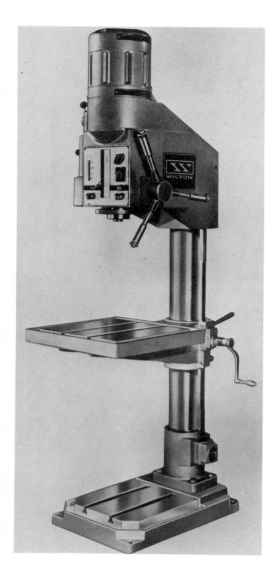

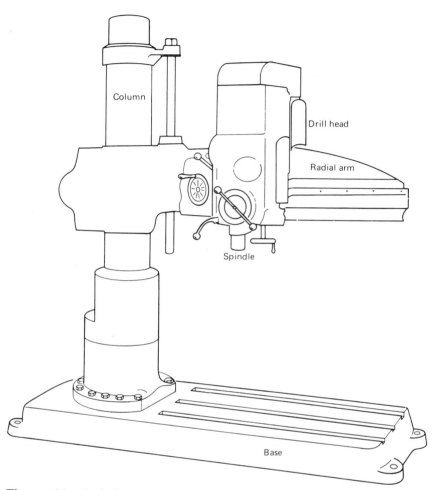

Figure 10. Radial arm drill press with names of major parts.

As Figures 10, 11, and 12 show, the radial arm drill press is the most versatile drilling machine. Its size is determined by the diameter of the column and the length of the arm measured from the center of the spindle to the outer edge of the column. It is useful for operations on large castings that are too heavy to be repositioned by the operator for drilling each hole. The work is clamped to the table or base, and the drill can then be positioned where it is needed by swinging the arm and moving the head along the arm. The arm and head can be raised or lowered on the column and then locked in place. The radial arm drill press is used for drilling small to very large holes and for boring, reaming, counterboring, and countersinking. Like the upright machine, the radial arm press has a power feed mechanism and a hand feed lever.

SELF-TEST

1. What clothing and personal styles could be considered a safety hazard? Explain in your answer.
2. Workpieces that are being drilled can be a danger to the operator. Why is this? How can this be corrected?
3. Why are your eyes in danger when using a drill press? How can you prevent this?
4. Morse tapers in drill press spindles must be clean and free from nicks or burrs. Machinists often use a finger

Figure 11. Small workpiece mounted on the side of the worktable for drilling. (Lane Community College)

Figure 12. Small holes are usually drilled by hand feeding on a radial arm drill. A workpiece is clamped on the tilting table so a hole may be drilled at an angle. (Lane Community College)

to clean the taper and to feel for burrs. Should this be done with the machine running? Explain.

5. The chuck key should not be left in the drill press when you are finished with it. Is it acceptable to leave the chuck key in the chuck just for a moment while you are making your setup? Explain your answer.

6. Is slowing down a drill press spindle with your hand a good work habit to develop? Give the reason for your answer.

7. Large drills (up to $3\frac{1}{2}$ in. diameter) have a taper shank and are removed with a drift. This usually requires the use of both hands, leaving no way to hold the drill from falling. Often the drill is positioned over a hole in the table or, if it is a radial arm drill press, over the floor or workpiece. These drills are sometimes heavy and could cause an injury if dropped on a foot. How would you overcome this problem?

8. Describe a safe method of removing cutting chips from a drilling machine.

9. What are the dangers in having oil on the floor in the work area of a drilling machine?

10. Why should burrs be removed from holes and edges of workpieces be chamfered as soon as possible?

11. Identify these drill press types:

a. List three basic types of drill presses and briefly explain their differences.

b. Describe how the primary uses differ in each of these three drill press types.

12. Sensitive drill press. Fill in the correct letter in the space preceding the name of that part shown on Figure 13.

_____ Spindle
_____ Quill lock handle
_____ Column
_____ Switch
_____ Depth stop
_____ Head
_____ Table
_____ Table lock
_____ Base

_____ Power feed
_____ Motor
_____ Variable speed control
_____ Table lift crank
_____ Quill return spring
_____ Guard

13. Radial drill press. Fill in the correct letters in the space preceding the name of that part shown on Figure 14.

_____ Column
_____ Radial arm
_____ Spindle
_____ Base
_____ Drill head

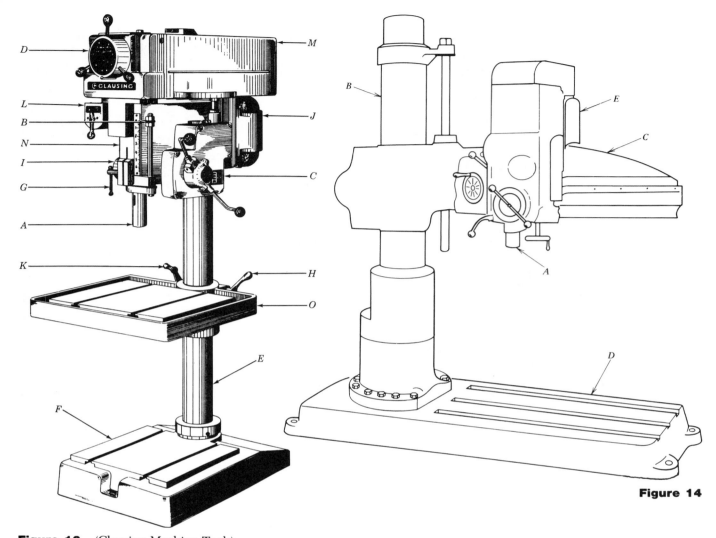

Figure 13. (Clausing Machine Tools)

Figure 14

UNIT 2 DRILLING TOOLS

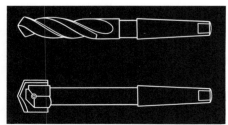

Before you learn to use drills and drilling machines, you will have to know of the great variety of drills and tooling available to the machinist. This unit will acquaint you with these interesting tools as well as to show you how to select the one you should use for a given operation.

OBJECTIVES

After completing this unit, you should be able to:
1. Identify the various features of a twist drill.
2. Identify the series and size of 10 given decimal equivalent drill sizes.

The drill is an end cutting rotary-type tool having one or more cutting lips and one or more flutes for the removal of chips and the passage of coolant. Drilling is the most efficient method of making a hole in metals softer than Rockwell 30. Harder metals can be successfully drilled, however, by using special drills and techniques.

In the past, all drills were made of carbon steel

(a)

(b)

(c)

(d)

(e)

(f)

(g)

Figure 1. Various types of twist drills used in drilling machines. (The DoAll Company): *(a)* High helix drill *(b)* Low helix drill *(c)* Left-hand drill *(d)* Three flute drill *(e)* Taper shank twist drill *(f)* Standard helix jobber drill *(g)* Center or spotting drill.

and would lose their hardness if they became too hot from drilling. Carbon steel drills are still made today for various purposes. Occasionally a machinist will pick one up thinking it is a high speed drill and burn the cutting edge by turning it at an RPM suited to high speed steel. As a rule of thumb, a carbon drill should be turned about half the RPM of a high speed steel drill. Carbon steel drills can be identified by spark testing. (See Section A, Unit 3.) Today, however, many drills are made of high speed steel. High speed steel drills can operate at several hundred degrees Fahrenheit without breaking down and, when cooled, will be as hard as before. Carbide tipped drills are used for special applications such as drilling abrasive materials and very hard steels. Other special drills are made from cast heat-resistant alloys.

Twist Drills

The twist drill is by far the most common type of drill used today. These are made with two or more flutes and cutting lips and in many varieties of design. Figure 1 illustrates several of the most commonly used types of twist drills. The names of parts and features of a twist drill are shown in Figure 2.

The twist drill has either a straight or tapered shank. The taper shank has a Morse taper, a standard taper of about $\frac{5}{8}$ in. per foot, which has more driving power and greater rigidity than the straight shank types. Ordinary straight shank drills are typically held in drill chucks (Figure 3*a*). This is a friction drive, and slipping of the drill shank is a common problem. Straight shank drills with tang drives have a positive drive and are less expensive than tapered shank drills. These are held in special drill chucks with a Morse taper (Figure 3*b*).

Jobbers drills have two flutes, a straight shank design, and a relatively short length-to-diameter ratio that helps to maintain rigidity. These drills are used for drilling in steel, cast iron, and nonferrous metals. Center drills and spotting drills are used for starting holes in workpieces. Oil hole drills are made so that coolant can be pumped through the drill to the cutting

Figure 2. Features of a twist drill. (Bendix Industrial Tools Division)

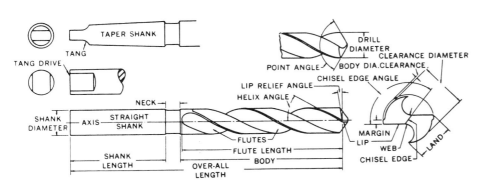

Figure 3a. Drill chucks such as this one are used to hold straight shank drills. *(Machine Tools and Machining Practices)*

Figure 3b. Tang drive drill chuck will fit into a Morse taper spindle. (Illinois/Eclipse, a Division of Illinois Tool Works, Inc.)

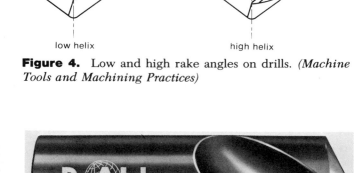

Figure 4. Low and high rake angles on drills. *(Machine Tools and Machining Practices)*

Figure 5. Spotting drill. (The DoAll Company)

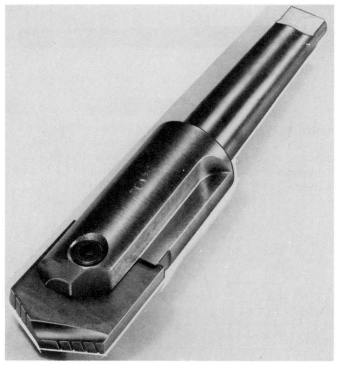

Figure 6. Spade drill clamped in holder. (The DoAll Company)

lips. This not only cools the cutting edges, but also forces out the chips along the flutes. Core drills have from three to six flutes making heavy stock removal possible. They are generally used for roughing holes to a larger diameter or for drilling out cores in castings. Left-handed drills are mostly used on multispindle drilling machines where some spindles are rotated in reverse of normal drill press rotation. The step drill generally has a flat or an angular cutting edge, and can produce a hole with several diameters in one pass with either flat or countersunk shoulders.

Straight fluted drills are used for drilling brass and other soft materials because the zero rake angle eliminates the tendency for the drill to "grab" on breakthrough. For the same reason they are used on thin materials. Low helix drills, sometimes called slow spiral drills (Figure 4), are more rigid than standard helix drills and can stand more torque. Like straight fluted drills, they are less likely to "grab" when emerging from a hole, because of the small rake angle. For this reason the low helix and straight flute drills are used primarily for drilling in brass, bronze, and some other nonferrous metals. Because of the low helix angle, the flutes do not remove chips very well from deep holes, but the large chip space allows maximum drilling efficiency in shallow holes. High helix drills, sometimes called fast spiral drills, are designed to remove chips from deep holes. The large rake angle makes these drills suitable for soft metals such as aluminum and mild steel. Spotting and centering drills (Figure 5) are used to acccurately position holes for further drilling with regular drills. Centering drills

are short and have little or no dead center. These characteristics prevent the drills from wobbling. Lathe center drills are often used as spotting drills.

Spade and Gun Drills

Special drills such as spade and gun drills are used in many manufacturing processes. A spade drill is simply a flat blade with sharpened cutting lips. The spade bit, which is clamped in a holder (Figure 6), is replaceable and can be sharpened many times. Some types provide for coolant flow to the cutting edge through a hole in the holder or shank for the purpose of deep drilling. These drills are made with very large diameters of 12 in. or more (Figure 7) but can also be found as microdrills, smaller than a hair. Twist drills by comparison are rarely found with diameters over $3\frac{1}{2}$ in.

Figure 7. Large hole being drilled with spade drill. This 8 in. diameter hole, 18⅞ in. deep, was spade drilled in solid SAE #4145 steel rolling mill drive coupling housing with a Brinell hardness of 200–240. The machine that did the job was a 6 ft 19 in. Chipmaster radial with a 25 hp. motor. (Giddings & Lewis, Inc.)

Figure 8. Spade drill blades showing various grinds. (The DoAll Company)

Figure 9. Carbide tipped twist drill. (The DoAll Company)

Figure 11. The hard steel drill. (The DoAll Company)

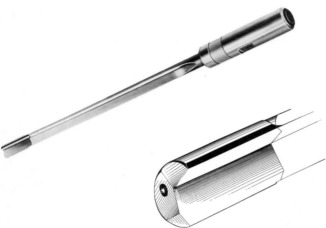

Figure 10. Single flute gun drill with inset of carbide cutting tip. (The DoAll Company)

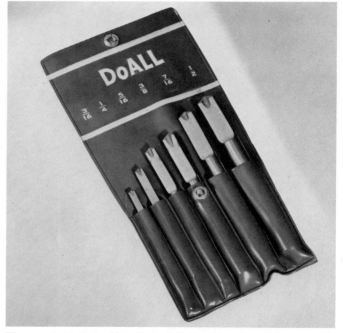

Figure 12. Set of hard steel drills. (The DoAll Company)

Spade drills are usually ground with a flat top rake and with chipbreaker grooves on the end. A chisel edge and thinned web are ground in the dead center (Figure 8).

Some spade drills are made of solid tungsten carbide, usually only in a small diameter. Twist drills with carbide inserts (Figure 9) require a rigid drilling setup. Gun drills (Figure 10) are also carbide tipped and have a single vee-shaped flute in a steel tube through which coolant is pumped under pressure. These drills are used in horizontal machines that feed the drill with a positive guide. Extremely deep precision holes are produced with gun drills.

Another special drill, used for drilling very hard steel, is the Hard Steel Drill (Figure 11). A set of hard steel drills is shown in Figure 12. These drills are cast from a heat-resistant alloy, and the fluted end is ground to a triangular point. These drills work by heating the metal beneath the drill point by friction and then cutting out the softened metal as a chip. (Figures 13*a*–13*d* show this drill in use.)

Figure 13*a*. Hole started in file with a hard steel drill.

Figure 13*c*. Drill removed from file showing chip form.

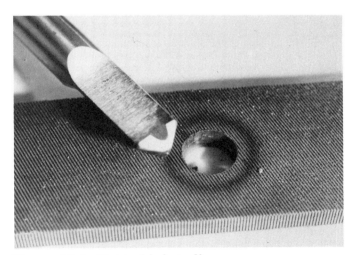

Figure 13*d*. Finished hole in file.

Figure 13*b*. Drilling the hole.

Drill Selection

The type of drill selected for a particular task depends upon several factors. The type of machine being used, rigidity of the workpiece, setup and size of the hole to be drilled are all important. The composition and hardness of the workpiece are especially critical. The job may require a starting drill or one for secondary operations such as counterboring or spot facing, and it might need to be a drill for a deep hole or for a shallow one. If the drilling operation is too large for the size or rigidity of the machine, there will be chatter and the work surface will be rough or distorted.

A machinist also must make the selection of the size of the drill, the most important dimension of which is the diameter. Twist drills are measured across the margins near the drill point (Figure 14). Worn drills measure slightly smaller here. Drills are normally tapered back along the margin so that they will measure a few thousandths of an inch smaller at the shank.

Drilling is basically a roughing operation. This provides a hole that is suitable for many uses such as for bolts or nuts, but holes must be finished by other operations, such as boring or reaming, for applications requiring more precision. Drills almost never make a hole smaller than their measured diameter, but often make a larger hole depending on how they have been sharpened. There are four drill size series: **fractional, number, letter,** and **metric** sizes. The fractional divisions are in $\frac{1}{64}$ in. increments, while the number, letter, and metric series have drill diameters that fall between the fractional inch measures. Together, the four series make up a long-running series in decimal equivalents, as shown in Table 1.

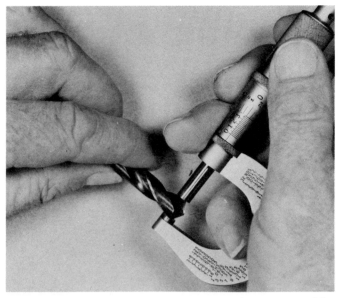

Figure 14. Drill being measured across the margins. *(Machine Tools and Machining Practices)*

Table 1
Decimal Equivalents for Drills

Decimals of an Inch	Inch	Wire Gage	Milli-meter	Decimals of an Inch	Inch	Wire Gage	Milli-meter	Decimals of an Inch	Inch	Wire Gage	Milli-meter
.0135		80		.0330		66		.0550		54	
.0145		79		.0335			.85	.0551			1.4
.0156	$\frac{1}{64}$			.0350		65		.0571			1.45
.0157			.4	.0354			.9	.0591			1.5
.0160		78		.0360		64		.0595		53	
.0180		77		.0370		63		.0610			1.55
.0197			.5	.0374			.95	.0625	$\frac{1}{16}$		
.0200		76		.0380		62		.0630			1.6
.0210		75		.0390		61		.0635		52	
.0217			.55	.0394			1	.0650			1.65
.0225		74		.0400		60		.0669			1.7
.0236			.6	.0410		59		.0670		51	
.0240		73		.0413			1.05	.0689			1.75
.0250		72		.0420		58		.0700		50	
.0256			.65	.0430		57		.0709			1.8
.0260		71		.0433			1.1	.0728			1.85
.0276			.7	.0453			1.15	.0730		49	
.0280		70		.0465		56		.0748			1.9
.0293		69		.0469	$\frac{3}{64}$			.0760		48	
.0295			.75	.0472			1.2	.0768			1.95
.0310		68		.0492			1.25	.0781	$\frac{5}{64}$		
.0313	$\frac{1}{32}$			.0512			1.3	.0785		47	
.0315			.8	.0520		55		.0787			2
.0320		67		.0531			1.35	.0807			2.05

Table 1 (*continued*)

Decimals of an Inch	Inch	Wire Gage	Millimeter
.0810		46	
.0820		45	
.0827			2.1
.0846			2.15
.0860		44	
.0866			2.2
.0886			2.25
.0890		43	
.0906			2.3
.0925			2.35
.0935		42	
.0938	$\frac{3}{32}$		
.0945			2.4
.0960		41	
.0966			2.45
.0980		40	
.0984			2.5
.0995		39	
.1015		38	
.1024			2.6
.1040		37	
.1063			2.7
.1065		36	
.1083			2.75
.1094	$\frac{7}{64}$		
.1100		35	
.1102			2.8
.1110		34	
.1130		33	
.1142			2.9
.1160		32	
.1181			3
.1200		31	
.1220			3.1
.1250	$\frac{1}{8}$		
.1260			3.2
.1280			3.25
.1285		30	
.1299			3.3
.1339			3.4
.1360		29	
.1378			3.5
.1405		28	
.1406	$\frac{9}{64}$		
.1417			3.6
.1440		27	
.1457			3.7
.1470		26	
.1476			3.75
.1495		25	
.1496			3.8
.1520		24	

Decimals of an Inch	Inch	Wire Gage	Millimeter
.1535			3.9
.1540		23	
.1563	$\frac{5}{32}$		
.1570		22	
.1575			4
.1590		21	
.1610		20	
.1614			4.1
.1654			4.2
.1660		19	
.1673			4.25
.1693			4.3
.1695		18	
.1719	$\frac{11}{64}$		
.1730		17	
.1732			4.4
.1770		16	
.1772			4.5
.1800		15	
.1811			4.6
.1820		14	
.1850		13	
.1850			4.7
.1870			4.75
.1875	$\frac{3}{16}$		
.1890			4.8
.1890		12	
.1910		11	
.1929			4.9
.1935		10	
.1960		9	
.1969			5
.1990		8	
.2008			5.1
.2010		7	
.2031	$\frac{13}{64}$		
.2040		6	
.2047			5.2
.2055		5	
.2067			5.25
.2087			5.3
.2090		4	
.2126			5.4
.2130		3	
.2165			5.5
.2188	$\frac{7}{32}$		
.2205			5.6
.2210		2	
.2244			5.7
.2264			5.75
.2280		1	
.2283			5.8

Decimals of an Inch	Inch	Letter Sizes	Millimeter
.2323			5.9
.2340		A	
.2344	$\frac{15}{64}$		
.2362			6
.2380		B	
.2402			6.1
.2420		C	
.2441			6.2
.2460		D	
.2461			6.25
.2480			6.3
.2500	$\frac{1}{4}$	E	
.2520			6.4
.2559			6.5
.2570		F	
.2598			6.6
.2610		G	
.2638			6.7
.2656	$\frac{17}{64}$		
.2657			6.75
.2660		H	
.2677			6.8
.2717			6.9
.2720		I	
.2756			7
.2770		J	
.2795			7.1
.2810		K	
.2812	$\frac{9}{32}$		
.2835			7.2
.2854			7.25
.2874			7.3
.2900		L	
.2913			7.4
.2950		M	
.2953			7.5
.2969	$\frac{19}{64}$		
.2992			7.6
.3020		N	
.3031			7.7
.3051			7.75
.3071			7.8
.3110			7.9
.3125	$\frac{5}{16}$		
.3150			8
.3160		O	
.3189			8.1
.3228			8.2
.3230		P	
.3248			8.25
.3268			8.3
.3281	$\frac{21}{64}$		

Table 1 (*continued*)

Decimals of an Inch	Inch	Letter Sizes	Millimeter
.3307			8.4
.3320		Q	
.3346			8.5
.3386			8.6
.3390		R	
.3425			8.7
.3438	$\frac{11}{32}$		
.3345			8.75
.3465			8.8
.3480		S	
.3504			8.9
.3543			9
.3580		T	
.3583			9.1
.3594	$\frac{23}{64}$		
.3622			9.2
.3642			9.25
.3661			9.3
.3680		U	
.3701			9.4
.3740			9.5
.3750	$\frac{3}{8}$		
.3770		V	
.3780			9.6
.3819			9.7

Decimals of an Inch	Inch	Wire Gage	Millimeter
.3839			9.75
.3858			9.8
.3860		W	
.3898			9.9
.3906	$\frac{25}{64}$		
.3937			10
.3970		X	
.4040		Y	
.4063	$\frac{13}{32}$		
.4130		Z	
.4134			10.5
.4219	$\frac{27}{64}$		
.4331			11
.4375	$\frac{7}{16}$		
.4528			11.5
.4531	$\frac{29}{64}$		
.4688	$\frac{15}{32}$		
.4724			12
.4844	$\frac{31}{64}$		
.4921			12.5
.5000	$\frac{1}{2}$		
.5118			13
.5156	$\frac{33}{64}$		
.5313	$\frac{17}{32}$		
.5315			13.5

Decimals of an Inch	Inch	Millimeter
.5469	$\frac{35}{64}$	
.5512		14
.5625	$\frac{9}{16}$	
.5709		14.5
.5781	$\frac{37}{64}$	
.5906		15
.5938	$\frac{19}{32}$	
.6094	$\frac{39}{64}$	
.6102		15.5
.6250	$\frac{5}{8}$	
.6299		16
.6406	$\frac{41}{64}$	
.6496		16.5
.6563	$\frac{21}{32}$	
.6693		17
.6719	$\frac{43}{64}$	
.6875	$\frac{11}{16}$	
.6890		17.5
.7031	$\frac{45}{64}$	
.7087		18
.7188	$\frac{23}{32}$	
.7283		18.5
.7344	$\frac{47}{64}$	
.7480		19
.7500	$\frac{3}{4}$	
.7656	$\frac{49}{64}$	
.7677		19.5
.7812	$\frac{25}{32}$	
.7874		20
.7969	$\frac{51}{64}$	
.8071		20.5
.8125	$\frac{13}{16}$	
.8268		21
.8281	$\frac{53}{64}$	
.8438	$\frac{27}{32}$	
.8465		21.5
.8594	$\frac{55}{64}$	
.8661		22
.8750	$\frac{7}{8}$	
.8858		22.5
.8906	$\frac{57}{64}$	
.9055		23
.9063	$\frac{29}{32}$	
.9219	$\frac{59}{64}$	
.9252		23.5
.9375	$\frac{15}{16}$	
.9449		24
.9531	$\frac{61}{64}$	
.9646		24.5
.9688	$\frac{31}{32}$	
.9843		25
.9844	$\frac{63}{64}$	6

Source: *Bendix Cutting Tool Handbook,* "Decimal Equivalents—Twist Drill Sizes,"
The Bendix Corporation, Industrial Tools Division, 1972.

Identification of a small drill is simple enough as long as the number or letter remains on the shank. Most shops, however, have several series of drills, and individual drills often become hard to identify since the markings become worn off by the drill chuck. The machinist must then use a decimal equivalent table such as Table 1. The drill in question is first measured by a micrometer, the decimal reading is located in the table, and the equivalent fraction, number, letter, or metric size is found and noted.

Morse taper shanks on drills and Morse tapers in drill press spindles vary in size and are numbered from 1 to 6; for example, the smaller light duty drill press has a number 2 taper. Steel sleeves (Figure 15) have a Morse taper inside and outside with a slot provided at the end of the inside taper to facilitate removal of the drill shank. A **sleeve** is used for enlarging the taper end on a drill to fit a larger spindle taper. Steel **sockets** (Figure 16) function in the reverse manner of sleeves, as they adapt a smaller spindle taper

Figure 15. Morse taper drill sleeve. (The DoAll Company)

Figure 16. Morse taper drill socket. (The DoAll Company)

Figure 17. The drill in use. This tool is one of several sizes used to remove taper shanks, drills, and sleeves.

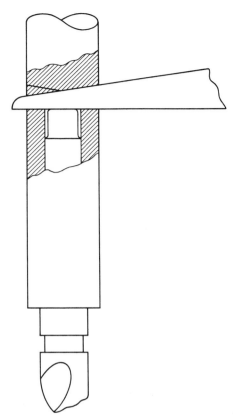

Figure 18. Cutaway of a drill and sleeve showing a drift in place. *(Machine Tools and Machining Practices)*

to a larger drill. The tool used to remove a taper shank drill is called a **drift**. The drift (Figure 17) is made in several sizes and is used to remove drills or sleeves. The drift is placed round side up, flat side against

the drill (Figure 18), and is struck a light blow with a hammer. A block of wood should be placed under the drill to keep it from being damaged and from being a safety hazard.

SELF-TEST

1. Write the correct letters from Figure 19 in the space provided.

_____ Web	_____ Body
_____ Margin	_____ Lip relief angle
_____ Drill point angle	_____ Land
_____ Cutting lip	_____ Chisel edge angle
_____ Flute	_____ Body clearance
_____ Helix angle	_____ Tang
_____ Axis of drill	_____ Taper Shank
_____ Shank length	_____ Straight shank

2. Determine the letter, number, fractional, or metric equivalents of the 10 following decimal measurements of drills.

	Decimal Diameter	Fractional Size	Number Size	Letter Size	Metric Size
a.	.0781				
b.	.1495				
c.	.272				
d.	.159				
e.	.1969				
f.	.323				
g.	.3125				
h.	.4375				
i.	.201				
j.	.1875				

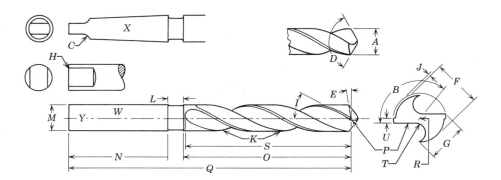

Figure 19. (Bendix Industrial Tools Division)

UNIT 3 HAND GRINDING OF DRILLS ON THE PEDESTAL GRINDER

Hand sharpening of twist drills has been until recent times the only method used for pointing a drill. Of course, various types of sharpening machines are now in use that can give a drill an accurate point. These precision machines are not found in every shop, however, so it is still necessary for a good machinist, mechanic, or metal worker to learn the art of off-hand drill grinding.

OBJECTIVE

After completing this unit, you should be able to:
Properly hand sharpen a twist drill on a pedestal grinder so it will drill a hole not more than .010 in. times diameter oversize.

One of the advantages of hand grinding drills on the pedestal grinder is that special alterations of the drill point such as web thinning and rake modification can be made quickly. The greatest disadvantage to this method of drill sharpening is the possibility of producing inaccurate, oversize holes (Figure 1). If the drill has been sharpened with unequal angles, the lip with the large angle will do most of the cutting (Figure 1a), and will force the opposite margin to cut into the wall of the hole. If the drill has been sharpened with unequal lip lengths, both will cut with equal force, but the drill will wobble and one margin will

cut into the hole wall (Figure 1b). When both conditions exist (Figure 1c), holes drilled may be out of round and oversize. When drilling with the inaccurate points, a great strain is placed on the drill and on the drill press spindle bearings. The frequent use of a drill point gage (Figure 2) during the sharpening process will help to keep the point accurate and avoid such drilling problems.

The web of a twist drill (Figure 3) is thicker near the shank. As the drill is ground shorter a thicker web results near the point. Also the dead center or chisel point of the drill is wider and requires greater

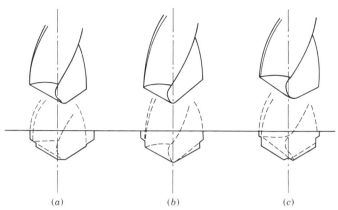

Figure 1. Causes of oversize drilling. (a) Drill lips ground to unequal lengths. (b) Drill lips ground to unequal angles. (c) Unequal angles and lengths.

Figure 2. Using a drill point gage.

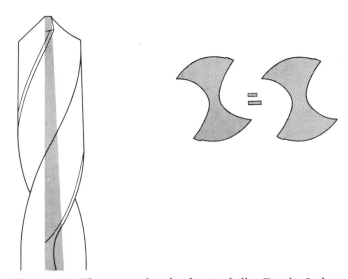

Figure 3. The tapered web of twist drills. (Bendix Industrial Tools Division)

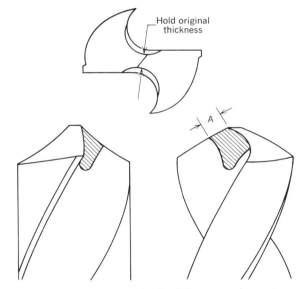

Hold original thickness

A

Figure 4. The usual method of thinning the point on a drill. The web should not be made thinner than it was originally when the drill was new and full length. (Bendix Industrial Tools Division)

55°

Figure 5. Split point design of a drill point. (Bendix Industrial Tools Division)

pressure to force it into the workpiece, thus generating heat. Web thinning (Figure 4) is one method of narrowing the dead center in order to restore the drill to its original efficiency.

Split-point design (Figure 5) is often used for drilling crankshafts and tough alloy steels. The shape of this point is quite critical and too difficult to grind by hand; it should be done on a machine. A sheet metal drill point (Figure 6) may be ground by an experienced hand. The rake angle on a drill can be modified for drilling brass as shown in Figure 7.

The standard drill point angle is an 118 degree included angle, while for drilling hard materials point angles should be from 135 to 150 degrees. A drill point angle from 60 to 90 degrees should be used

Figure 6. Sheet metal drill point.

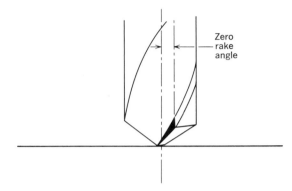

Figure 7. Modification of the rake angle for drilling brass.

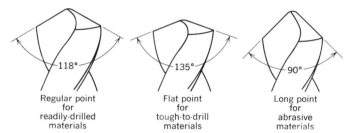

Figure 8. Drill point angles. (Bendix Industrial Tools Division)

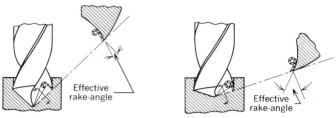

Figure 9. The effective rake angles. (Bendix Industrial Tools Division)

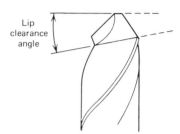

Figure 10. Clearance angles on a drill point.

when drilling soft materials, cast iron, abrasive materials, plastics, and some nonferrous metals (Figure 8). Too great a decrease in the included point angle is not advisable, however, because it will result in an abnormal decrease in effective rake angle (Figure 9). This will increase the required feed pressure and change the chip formation and chip flow in most steels. Clearance angles (Figure 10) should be 8 to 12 degrees for most drilling.

SELF-TEST

Sharpening A Drill

For the following practice in hand grinding a twist drill, you will need a pedestal grinder, a drill point gage, safety goggles or face shield, and a worn or dull drill no smaller than $\frac{3}{8}$ in. in diameter. Check to see if both the roughing and finishing wheels are true (Figure 11). If not, move a wheel dresser across them (Figure 12). If the end of the drill is badly damaged, use the coarse wheel to remove that part. If you overheat the drill, let it cool in air; do not cool high speed steel drills in water. The following method of grinding a drill is suggested.

1. Hold the drill shank with one hand and the drill near the point with the other hand. Rest your fingers that are near the point on the grinder tool rest. Hold the drill lightly at this point so you can manipulate it from the shank end with the other hand (Figure 13).
2. Hold the drill approximately horizontal. The cutting lip (Figure 14) to be ground should also be horizontal. The axis of the drill should be at 59 degrees from the face of the wheel.
3. Using the tool rest and fingers as a pivot, slowly move the shank downward and slightly to the left (Figure 15). The drill must be free to slip forward slightly to keep it against the wheel (Figure 16). Rotate the drill very slightly. It is the most common mistake of the beginner to rotate the drill until the opposite cutting edge has been ground off. Do not rotate small drills at all, only larger ones. As you continue the downward movement

Figure 11. Grinding wheel that is dressed and ready to use.

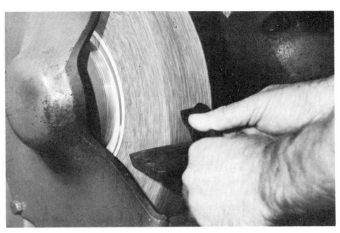

Figure 13. Starting position showing 59 degree angle with the wheel and the cutting lip horizontal. (Lane Community College)

Figure 12. Trueing up a grinding wheel with a wheel dresser. The accepted method of using this type of wheel dresser by many people is to hook the ears behind the tool rest which must first be moved away from the wheel. The method shown here is the one most commonly used in shops, since it is less time consuming and adequately trues the wheel. *(Machine Tools and Machining Practices)*

Figure 14. Drill being held in the same starting position, approximately horizontal. (Lane Community College)

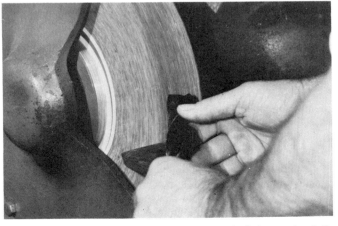

Figure 15. Drill is now moved very slightly to the left with the shank being moved downward. (Lane Community College)

of the shank, crowd the drill into the wheel so that it will grind lightly all the way from the lip to the heel (Figure 17). This should all be one smooth movement. It is very important at this point to allow proper clearance (8 to 12 degrees) at the heel of the drill.

4. Without changing your body position, pull the drill back slightly and rotate 180 degrees so that the opposite lip is now in a level position. Repeat step 3.

Figure 16. Another view of the same position as shown in Figure 15. (Lane Community College)

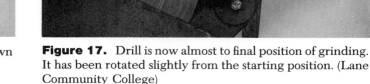

Figure 17. Drill is now almost to final position of grinding. It has been rotated slightly from the starting position. (Lane Community College)

5. Check *both cutting lips* with the drill point gage (Figure 18):
   a. For correct angle.
   b. For equal length.
   c. Check lip clearances visually. These should be between 8 and 12 degrees.
   If errors are found, adjust and regrind until they are correct.
6. When you are completely satisfied that the drill point angles and lip lengths are correct, drill a hole in a scrap metal that has been set aside for this purpose. Consult a drill speed table so you will be able to select the correct RPM. Use cutting oil or coolant.
7. Check the condition of the hole. Did the drill chatter and cause the start of the hole to be misshapen? This could be caused by too much lip clearance. Is the hole oversize more than .005 or .010 in.? Are the lips uneven or the lip angles off or both? Running the drill too slowly for its size will cause a rough hole; too fast will burn the drill. If the hole size is more than .010 in. × diameter over the drill size, resharpen and try again.
8. When you have a correctly sharpened drill, show this and the drilled hole to your instructor for his evaluation.

Figure 18. After the sequence has been completed on both cutting lips, they are checked with a drill point gage for length and angle.

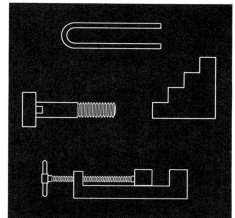

UNIT 4 WORK LOCATING AND HOLDING DEVICES ON DRILLING MACHINES

Workpieces of a great many sizes and shapes are drilled by machinists. In order to hold these parts safely and securely while they are drilled, several types of workholding devices are used. In this unit you will learn how to properly set up these devices for a drilling operation.

OBJECTIVES

After completing this unit, you should be able to:
1. Identify and explain the correct uses for several workholding and locating devices.
2. Set up for drilling holes; align and start a tap using the drill press.

Because of the great forces applied by the machines in drilling, some means must be provided to keep the workpiece from turning with the drill or from climbing up the flutes after the drill breaks through. This is not only necessary for safety's sake but also for workpiece rigidity and good workmanship.

One method of workholding is to use strap clamps (Figure 1) and T-bolts (Figure 2). The clamp must be kept parallel to the table by the use of step blocks (Figure 3) and the T-bolt should be kept as close to the workpiece as possible (Figure 4). Parallels (Figure

5) are placed under the work at the point where the clamp is holding. This provides a space for the drill to break through without making a hole in the table. A thin or narrow workpiece should not be supported too far from the drill, however, since it will spring

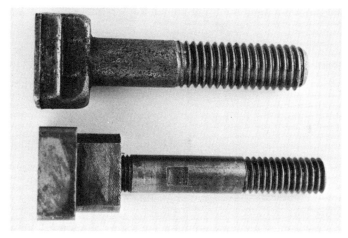

Figure 2. T-bolts.

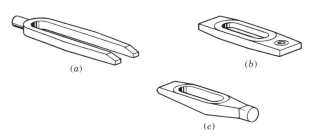

Figure 1. Strap clamps: (a) U-clamp, (b) straight clamp, and (c) finger clamp.

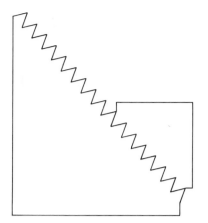

Figure 3. Adjustable step blocks.

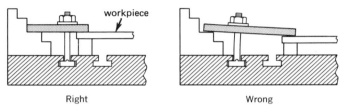

Right   Wrong

Figure 4. Right and wrong setup for strap clamps. The clamp bolt should be as close to the workpiece as possible.

Figure 5. Parallels of various sizes. (Lane Community College)

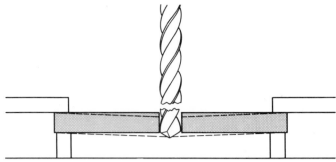

Figure 6. Thin springy material is supported too far from the drill. Drilling pressure forces the workpiece downward until the drill breaks through, relieving the pressure. The work then springs back and the remaining "fin" of material is more than the drill can cut in one revolution. The result is drill breakage.

Figure 7. C-clamp being used on an angle plate to hold work that would be difficult to safely support in other ways. (Lane Community College)

down under the pressure of the drilling. This can cause the drill on breakthrough to suddenly "grab" more material than it can handle. The result is often a broken drill (Figure 6). Thin workpieces or sheet metal should be clamped over a wooden block to avoid this problem. C-clamps of various sizes are used to

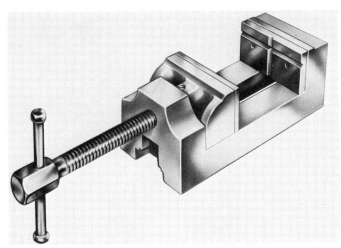

Figure 8. Drill press vise. Small parts are held for drilling and other operations with the drill press vise. (Wilton Corporation)

Figure 9. Part set up in vise with parallels under it. (Lane Community College)

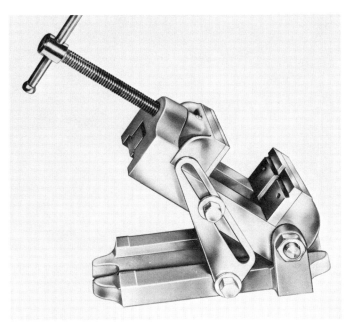

Figure 10. Angle vise. Parts that must be held at an angle to the drill press table while being drilled are held with this vise. (Wilton Corporation)

hold workpieces on drill press tables and on angle plates (Figure 7).

Angle plates facilitate the holding of odd shaped parts for drilling. The angle plate is either bolted or clamped to the table and the work is fastened to the angle plate. For example, a gear or wheel that requires a hole to be drilled into a projecting hub could be clamped to an angle plate.

Drill press vises (Figure 8) are very frequently used for holding small workpieces of regular shape and size with parallel sides. Vises provide the quickest and most efficient setup method for parallel work, but should not be used if the work does not have parallel sides. The workpiece must be supported so the drill will not go into the vise. If precision parallels are used for support, they and the drill can be easily damaged since they are both hardened (Figure 9). For rough drilling, however, cold finished (CF) keystock would be sufficient for supporting the workpiece. Angular vises can pivot a workpiece to a given angle so that angular holes can be drilled (Figure 10). Another method of drilling angular holes is by tilting the drill press table. If there is no angular scale on the vise or table, a protractor head with a level may be used to set up the correct angle for drilling. Angle plates are also sometimes used for drilling angular holes (Figure 11a). The drill press table must be level (Figure 11b).

Figure 11a. View of an adjustable angle plate on a drill press table using a protractor to set up. (Lane Community College)

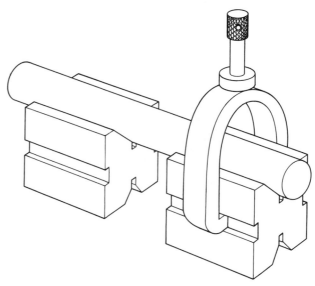

Figure 12. Set of vee-blocks with a vee-block clamp.

Figure 11b. Checking the level of the table. (Lane Community College)

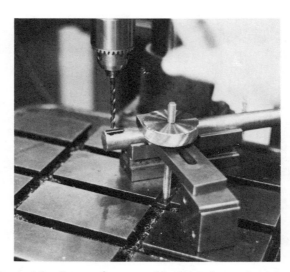

Figure 13. Setup of two vee-blocks and round stock with strap clamp between the vee-blocks. (Lane Community College)

Vee-blocks come in sets of two, often with clamps for holding small size rounds (Figure 12). Larger size round stock is set up with a strap clamp over the vee-blocks (Figure 13). The hole to be cross drilled is first laid out and center punched. The workpiece is lightly clamped in the vee-blocks and the punch mark is centered as shown in Figure 14. The clamps are tightened, and the drill chuck is located precisely over the punch mark by means of a wiggler. A wiggler is a tool that can be put into a drill chuck to locate a punch mark to the exact center of the spindle.

The wiggler is clamped into a drill chuck and the machine is turned on (Figure 15a). Push on the knob near the end of the pointer with a 6 in. rule or other piece of metal until it runs with no wobble (Figure 15b). With the machine still running, bring the pointer down into the punch mark. If the pointer begins to wobble again, the mark is not centered under the spindle and the workpiece will have to be

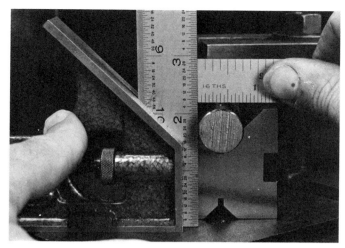

Figure 14. Round stock in vee-blocks. One method of centering layout line or punch mark using a combination square and rule. (Lane Community College)

shifted. When the wiggler enters the punch mark without wobbling, the workpiece is centered.

After the work is centered, use a spotting or center drill to start the hole. Then, for larger holes, use a pilot drill, which is always a little larger than the web of the next drill size used. Pilot drills are not commonly used in industrial applications; only the spotting drill (if used) and the full size drill are used. Use the correct cutting speed and coolant. Chamfer both sides of the finished hole with a countersink or chamfering tool. Round stock can also be cross drilled when held in a vise, using the same technique as with vee-blocks.

A tap may be started straight in the drill press by hand. After tap drilling the workpiece and without removing any clamps, remove the tap drill from the chuck and replace it with a straight shank center. (An alternate method is to clamp the shank of the tap directly in the drill chuck and turn the spindle

Figure 15a. Wiggler set in offset position. (Lane Community College)

Figure 15b. Wiggler centered. (Lane Community College)

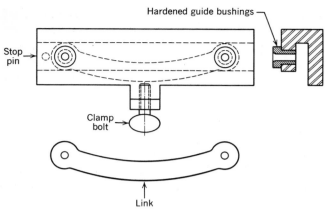

Figure 16. Simple box jig for drilling link. The link is shown below. Hardened guide bushings in the jig are used to limit wear.

by hand two or three turns.) Insert a tap in the work and attach a tap handle. Then put the center into the tap, but **do not turn on the machine.** Apply sulfurized cutting oil and start the tap by turning the tap handle a few turns with one hand while feeding down with the other hand. Release the chuck while the tap is still in the work and finish the job of tapping the hole.

Jigs and fixtures are specially made tools for production work. In general, a fixture references a part to the cutting tool. Jigs guide the cutting tool such as a drill. They both hold and support the part (Figure 16). The use of a jig assures exact positioning of the hole pattern in duplicate and eliminates layout work on every part.

SELF-TEST

1. What is the main purpose for using workholding devices on drilling machines?
2. List the names of all the workholding devices that you can remember.
3. Explain the uses of parallels for drilling setups.
4. Why should the support on a narrow or thin workpiece be as close to the drill as possible?
5. Angle drilling can be accomplished in several ways. Describe two methods. How would this be done if no angular measuring devices were mounted on the equipment?

6. What shape of material is the vee-block best suited to hold for drilling operations? What do you think its most frequent use would be?
7. What is the purpose of using a wiggler?
8. Why would you ever need an angle plate?
9. What is the purpose of starting a tap in the drill press?
10. Do you think jigs and fixtures are used to any great extent in small machine shops? Why?

UNIT 5 OPERATING DRILLING MACHINES

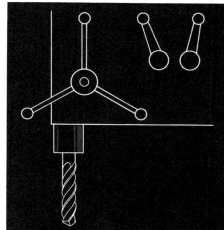

You have already learned many things about drilling machines and tooling. You should now be ready to learn some very important facts about the use of these machines. How fast should the drill run? How much feed should be applied? Which kind of coolant should be used? These and other questions are answered in this unit.

OBJECTIVES

After completing this unit, you should be able to:
1. Determine the correct drilling speeds for five given drill diameters.
2. Determine the correct feed in steel by chip observation.
3. Set up the correct feed on a machine by using a feed table.

After the workpiece is properly clamped and operator safety is assured, the most important considerations for drilling are speeds, feeds, and coolants. Of these, the control and setting of speeds will have the greatest effect on the tool and the work.

CUTTING SPEEDS

Cutting speeds (CS) are normally given for high speed steel cutting tools and are based on surface feet per minute (FPM or SFM). Surface feet per minute means that either the tool moves past the work or the work moves past the tool at a rate based on the number of feet that passes a tool in one minute, whether it be on a flat surface or on the periphery of a cylindrical tool or workpiece. Since machine spindle speeds are given in revolutions per minute (RPM), this can be derived in the following manner:

$$\text{RPM} = \frac{\text{Cutting speed (in feet per minute)} \times 12}{\text{Diameter of cutter (in inches)} \times \pi}$$

If you use 3 to approximate π (3.1416), then the formula becomes

$$\frac{\text{CS} \times 12}{\text{D} \times 3} = \frac{\text{CS} \times 4}{\text{D}}$$

This simplified formula is certainly the most common one used in the machine shop practice and it applies to the full range of machine tool operations, which include the lathe and the milling machine, as well as the drill press. Throughout this text the simplified formula RPM = (CS × 4)/D will be used, where D = the diameter of the drill and CS = an assigned cutting speed for a particular material. For example, for a $\frac{1}{2}$ in. drill in low carbon steel the speed would be

$$\frac{90 \times 4}{\frac{1}{2}} = 720 \text{ RPM}$$

See Table 1 for cutting speeds for some metals.

Table 1
Drilling Speed Table

Material	Cutting Speed (CS)
Low carbon steel	90
Aluminum	300
Cast iron	70
Alloy steel	50
Brass and bronze	120

Cutting speeds/RPM tables for various materials are available in handbooks and as wall charts. Excessive speeds can cause the outer corners and margins of the drill to break down. This will in turn cause

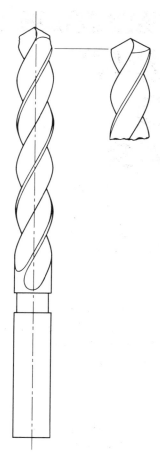

Figure 2. Properly formed chip. (Lane Community College)

Figure 1. Broken down drill corrected by grinding back to full diameter margins and regrinding cutting lips.

the drill to bind in the hole, even if the speed is corrected and more cutting oil is applied. The only cure is to grind the drill back to its full diameter (Figure 1) using methods discussed in Unit 3 of this section.

A blue chip from steel indicates the speed is too high. The tendency with very small drills, however, is to set the RPM of the spindle too slow. This gives the drill a very low cutting speed and very little chip is formed unless the operator forces it with an excessive feed. The result is often a broken drill.

CONTROLLING FEEDS

The feed may be controlled by the "feel" of the cutting action and by observing the chip. A long, stringy chip indicates too much feed. The proper chip in soft steel should be a tightly rolled helix in both flutes (Figure 2). Some materials such as cast iron will produce a granular chip. Drilling machines that have power feeds are arranged to advance the drill a given

amount for each revolution of the spindle. Therefore, .006 in. feed means that the drill advances .006 in. every time the drill makes one full turn. The amount of feed varies according to the drill size and the work material. See Table 2.

Table 2
Drilling Feed Table

Drill Size Diameter (in.)	Feeds per Revolution (in.)
Under $\frac{1}{8}$	.001 to .002
$\frac{1}{8}$ to $\frac{1}{4}$	.002 to .004
$\frac{1}{4}$ to $\frac{1}{2}$	.004 to .007
$\frac{1}{2}$ to 1	.007 to .015
over 1	.015 to .025

Source: Warren T. White, John E. Neely, Richard R. Kibbe, and Roland O. Meyer, *Machine Tools and Machining Practices,* John Wiley and Sons, Inc., Copyright © 1977, New York.

It is a better practice to start with smaller feeds than those given in tables. Materials and setups vary, so it is safer to start low and work up to an optimum feed. You should stop the feed occasionally to break the chip and allow coolant to flow to the cutting edge of the drill.

There is generally no breakthrough problem when using power feed, but when hand feeding, the drill may catch and "grab" while coming through the

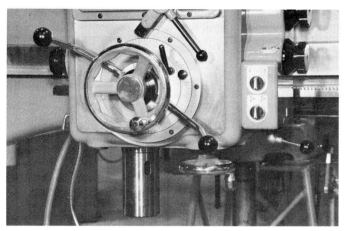

Figure 3. Feed clutch handle. The power feed is engaged by pulling the handles outward. When the power feed is disengaged, the handles may be used to hand feed the drill. (Lane Community College)

Figure 4a. Speed and feed control dials. (Lane Community College)

Figure 4b. Large speed and feed plates on the front of the head of the upright press can be read at a glance. (Giddings & Lewis, Inc.)

Figure 5. Tapping attachment. (Lane Community College)

last $\frac{1}{8}$ in. or so of the hole. Therefore, the operator should let up on the feed handle near this point and ease the drill through the hole. This "grabbing" tendency is especially true of brass and some plastics, but it is also a problem in steels and other materials. Large upright drill presses and radial arm drills have power feed mechanisms with feed clutch handles (Figure 3) that also can be used for hand feeding when the power feed is disengaged. Both feed and speed controls are set by levers or dials (Figure 4a). Speed and feed tables on plates are often found on large drilling machines (Figure 4b).

Tapping with small taps is often done on a sensitive drill press with a tapping attachment (Figure 5) that has an adjustable friction clutch and reverse mechanism that screws the tap out when you raise the spindle. Large size taps are power driven on upright or radial drill presses. These machines provide for spindle reversal (sometimes automatic) to screw the tap back out.

COOLANTS AND CUTTING OILS

A large variety of coolants and cutting oils are used for drilling operations on the drill press. Emulsifying or soluble oils (either mineral or synthetic) mixed in water are used for drilling holes where the main requirement is an inexpensive cooling medium. Operations that tend to create more friction and, hence, need more lubrication to prevent galling (abrasion due to friction), require a cutting oil. Animal or mineral oils with sulfur or chlorine added are often used. Reaming, counterboring, countersinking, and tapping all create friction and require the use of cutting oils, of which the sulfurized type is most used. Cast iron and brass are usually drilled dry, but water soluble

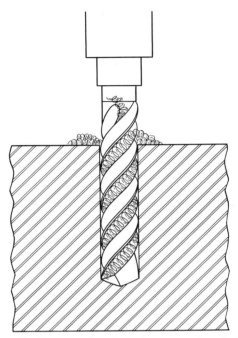

Figure 6. Drill jammed in hole because of packed chips.

oil can be used for both. Aluminum can be drilled with water soluble oil or kerosene for a better finish. Both soluble and cutting oils are used for steel.

DRILLING PROCEDURES

Deep hole drilling requires sufficient drill length and quill stroke to complete the needed depth. A high helix drill helps to remove the chips, but sometimes the chips bind in the flutes of the drill and, if drilling continues, will cause the drill to jam in the hole (Figure 6). A method of avoiding this problem is called "pecking"; that is, when the hole is drilled a short distance, the drill is taken out from the hole allowing the accumulated chips to fly off. The drill is again inserted into the hole, a similar amount is drilled, and the drill is again removed. This pecking is repeated until the required depth is reached.

A depth stop is provided on drilling machines to limit the travel of the quill so that the drill can be made to stop at a predetermined depth (Figure 7). The use of a depth stop makes drilling several holes to the same depth quite easy. Spotfacing and counterboring should also be set up with the depth stop. Blind holes (holes that do not go through the piece) are measured from the edge of the drill margin to the required depth (Figure 8). Once measured, the depth can be set with the stop and drilling can proceed. One of the most important uses of a depth stop, from a maintenance standpoint, is that of setting the depth so that the machine table or drill press vise will not be drilled full of holes.

Holes that must be drilled partly into or across existing holes (Figure 9) may jam or bind a drill unless special precautions are taken. A special drill with a

Figure 7. Using the depth stop. (Lane Community College)

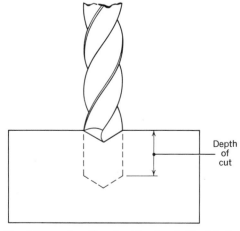

Figure 8. Measuring the depth of a drilled hole.

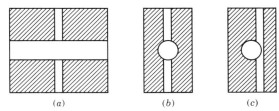

Figure 9. Hole drilled at 90 degree angle into existing hole. Cross drilling is done off center as well as on center. (a) Side view. (b) End view drilled on center. (c) End view drilled off center.

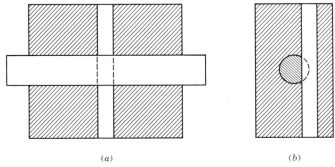

Figure 10. Existing hole is plugged to make the hole more easily drilled. (a) Hole is plugged with the same material that the workpiece consists of. (b) End view showing hole drilled through plug.

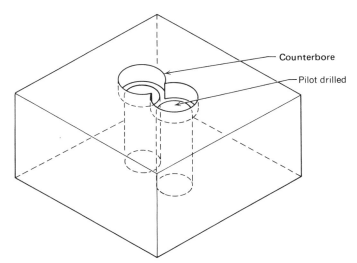

Figure 11. Deep holes that overlap are difficult to drill. The holes are drilled alternately with a smaller pilot drill and a counterbore for final size.

Figure 12. Heavy duty drilling on a radial drill press.

double margin and high helix could be used directly. However, an ordinary jobbers drill will do a satisfactory job if a tight plug made of the same material as the work is first tapped into the cross hole (Figure 10). The hole may then be drilled in a normal manner and the plug removed.

Holes that overlap may be made on the drill press if care is taken, and a set of counterbores with interchangeable pilots is available (Figure 11). First, pilot drill the holes with a size drill that does not overlap. Then counterbore to the proper size with the appropriate pilot on the counterbore.

Heavy duty drilling should be done on an upright or radial drill press (Figure 12). The workpiece should be made very secure since high drilling forces are used with the larger drill sizes. The work should be well clamped or bolted to the work table. The head and column clamp should always be locked when drill-ing is being done on a radial drill press. When several operations are performed in a continuing sequence, quick change drill and toolholders are often used. Coolant is necessary for all heavy duty drilling.

SELF-TEST

1. Name three important things to keep in mind when using a drill press (not including operator safety and clamping work).
2. If RPM = (CS × 4)/D and the cutting speed for low carbon steel is 90, what would the RPMs be for the following drills:
    a. $\frac{1}{4}$ in. diameter _____
    b. 2 in. diameter _____
    c. $\frac{3}{4}$ in. diameter _____
    d. $\frac{3}{8}$ in. diameter _____
    e. $1\frac{1}{2}$ in. diameter _____
3. What are some of the results of excessive drilling speed? What corrective measures can be taken?

4. Explain what can happen to small diameter drills when the cutting speed is too slow.
5. How can an operator tell by observing the chip if the feed is about right?
6. In what way are power feeds designated?
7. Name two differing cutting fluid types.
8. Such operations as counterboring, reaming, and tapping create friction that can cause heat. This can ruin a cutting edge. How can this situation be helped?
9. How can "jamming" of a drill be avoided when drilling deep holes?
10. Name three uses for the depth stop on a drill press.

UNIT 6 COUNTERSINKING AND COUNTERBORING

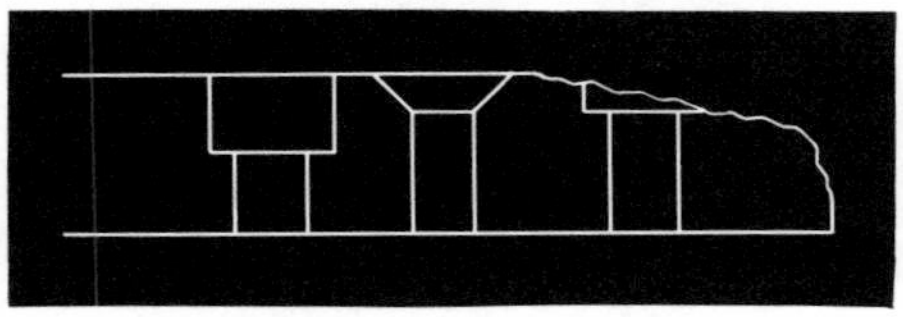

In drill press work it is often necessary to make a recess that will leave a bolt head below the surface of the workpiece. These recesses are made with countersinks or counterbores. When holes are drilled into rough castings or angular surfaces, a flat surface square to these holes is needed, and spot facing is the operation used. This unit will familiarize you with these drill press operations.

OBJECTIVES

After completing this unit, you should be able to:
1. Identify tools for countersinking and counterboring.
2. Select speeds and feeds for countersinking and counterboring.
3. Countersink and counterbore holes.

COUNTERSINKS

A countersink is a tool used to make a conical enlargement of the end of a hole. Figures 1 and 2 illustrate two types of single flute countersinks, both of which are designed to produce smooth surfaces, free from chatter marks. A countersink is used as a chamfering or deburring tool to prepare a hole for reaming or tapping. Unless a hole needs to have a sharp edge, it should be chamfered to protect the end of the hole from nicks and burrs. A chamfer from $\frac{1}{32}$ to $\frac{1}{16}$ in. wide is sufficient for most holes.

Figure 1. (left) Single flute countersink. (The DoAll Company)

Figure 2. (right) Chatterfree countersink. (The DoAll Company)

Figure 3. Center drill or combination drill and counter-sink. (The DoAll Company)

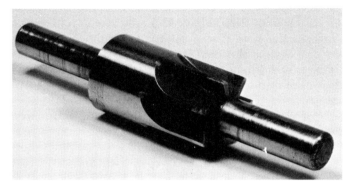

Figure 4. Multiflute counterbore.

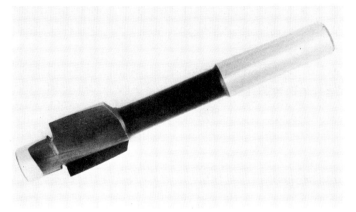

Figure 5. Two flute counterbore.

A hole made to receive a flathead screw or rivet should be countersunk deep enough for the head to be flush with the surface or up to .015 in. below the surface. A flathead fastener should never project above the surface. The included angles on commonly available countersinks are 60, 82, 90, and 100 degrees. Most flathead fasteners used in metalworking have an 82 degree head angle, except for the aircraft industry where the 100 degree angle is prevalent. The cutting speed used when countersinking should always be slow enough to avoid chattering.

A combination drill and countersink with a 60 degree angle (Figure 3) is used to make center holes in workpieces for machining on lathes and grinders.

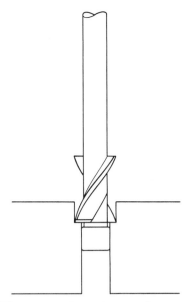

Figure 6. The counterbore is an enlargement of a hole already drilled.

The illustration shows a bell-type center drill that provides an additional angle for a chamfer of the center, protecting it from damage. The combination drill and countersink, known as a center drill, is used for spotting holes when using a drill press or milling machine, since it is extremely rigid and will not bend under pressure.

COUNTERBORES

Counterbores are tools designed to enlarge previously drilled holes, much like countersinks, and are guided into the hole by a pilot to assure the concentricity of the two holes. A multiflute counterbore is shown in Figure 4, and a two flute counterbore is shown in Figure 5. The two flute counterbore has more chip clearance and a larger rake angle than the counterbore in Figure 4 so that the counterbore is freer cutting and better suited for soft and ductile materials. Counterbored holes have flat bottoms, unlike the angled edges of countersunk holes, and are often used to recess a bolt head below the surface of a workpiece. Solid counterbores, such as shown in Figures 4 and 5, are used to cut recesses for socket head capscrews or filister head screws (Figure 6). The diameter of the counterbore is usually $\frac{1}{32}$ in. larger than the head of the bolt.

When a variety of counterbore and pilot sizes is necessary, a set of interchangeable pilot counterbores is available. Figure 7 shows a counterbore in which

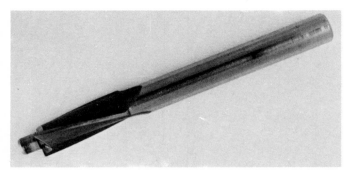

Figure 7. Interchangeable pilot counterbore.

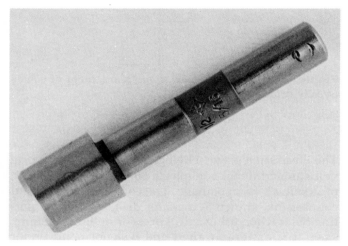

Figure 8. Pilot for interchangeable pilot counterbore.

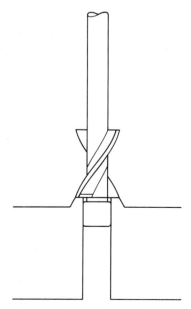

Figure 9. Spot facing on a raised boss.

a number of standard or specially made pilots can be used. A pilot is illustrated in Figure 8.

Counterbores are made with straight or tapered shanks to be used in drill presses, milling machines, or even lathes. When counterboring a recess for a hex head bolt, remember to measure the diameter of the socket wrench so the hole will be large enough to accommodate it. For most counterboring operations the pilot should have from .002 to .005 in. clearance in the hole. If the pilot is too tight in the hole, it may seize and break. If there is too much clearance between the pilot and the hole, the counterbore will be out of round and will have an unsatisfactory surface finish.

It is very important that the pilot be lubricated while counterboring. Usually this lubrication is provided if a cutting oil or soluble oil is used. When cutting dry, which is often the case with brass and cast iron, the hole and pilot should be lubricated with a few drops of lubricating oil.

Counterbores or spot facers are often used to provide a flat bearing surface for nut or bolt heads on rough castings or a raised boss (Figure 9). This operation is called spot facing. Because these rough surfaces may not be at right angles to the pilot hole, great strain is put on the pilot and counterbore, and can cause breakage of either one. To avoid breaking the tool, be very careful when starting the cut, especially when hand feeding. Prevent hogging into the work by tightening the spindle clamps slightly to remove possible backlash.

Recommended power feed rates for counterboring are shown in Table 1. The **feed** rate should be great enough to get under any surface scale quickly, thus preventing rapid dulling of the counterbore. The **speeds** used for counterboring are one-third less than the speeds used for twist drills of corresponding diameters. The choice of speeds and feeds is very much affected by the condition of the equipment, the power available, and the material being counterbored.

Before counterboring a workpiece, it should be

Table 1
Feeds for Counterboring

$\frac{3}{8}$ in. diameter up to .004 in. per revolution
$\frac{5}{8}$ in. diameter up to .005 in. per revolution
$\frac{7}{8}$ in. diameter up to .006 in. per revolution
$1\frac{1}{4}$ in. diameter up to .007 in. per revolution
$1\frac{1}{2}$ in. diameter up to .008 in. per revolution

Source: Warren T. White, John E. Neely, Richard R. Kibbe, and Roland O. Meyer, *Machine Tools and Machining Practices,* John Wiley and Sons, Inc., Copyright © 1977, New York.

securely fastened to the machine table or tightly held in a vise because of the great cutting pressures encountered. Workpieces also should be supported on parallels to allow for the protrusion of the pilot. To obtain several equally deep countersunk or counterbored holes on the drill press or milling machine, the spindle depth stop can be set.

Counterbores can be used on a lathe to rough out a hole before it is finished-bored. This is often more efficient than using a single cutting edge boring bar. It permits the use of a larger diameter boring bar for a more rigid setup.

SELF-TEST

1. When is a countersink used?
2. Why are countersinks made with varying angles?
3. What is a center drill?
4. When is a counterbore used?
5. What relationship exists between pilot size and hole size?
6. Why is lubrication of the pilot important?

7. As a rule, how does cutting speed compare between an equal size counterbore and twist drill?
8. What affects the selection of feed and speed when counterboring?
9. What is spot facing?
10. What important points should be considered when a counterboring setup is made?

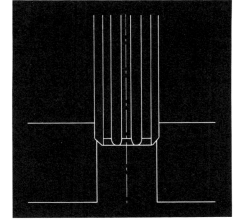

UNIT 7 REAMING IN THE DRILL PRESS

Many engineering requirements involve the production of holes having smooth surfaces, accurate location, and uniform size. In many cases, holes produced by drilling alone do not entirely satisfy these requirements. For this reason, the reamer was developed for enlarging or finishing previously formed holes. This unit will help you properly identify, select, and use machine reamers.

OBJECTIVES

After completing this unit, you should be able to:
1. Identify commonly used machine reamers.
2. Select the correct feeds and speeds for commonly used materials.
3. Determine appropriate amounts of stock allowance.
4. Identify probable solutions to reaming problems.

Reamers are tools used mostly to precision finish holes, but they are also used in the heavy construction industry to enlarge or align existing holes.

COMMON MACHINE REAMERS

Machine reamers have straight or taper shanks, the taper usually is a standard Morse taper. The parts of a machine reamer are shown in Figure 1 and the cutting end of a machine reamer is shown in Figure 2.

Chucking reamers (Figures 3, 4, and 5) are efficient in machine reaming a wide range of materials and are commonly used in drill presses, turret lathes, and screw machines. Helical flute reamers have an extremely smooth cutting action that finishes holes

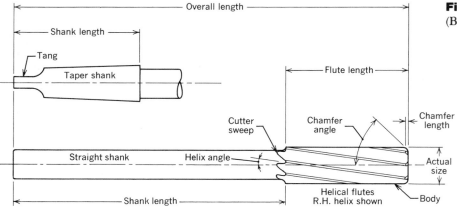

Figure 1. The parts of a machine reamer. (Bendix Industrial Tools Division)

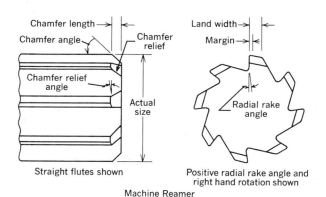

Machine Reamer

Figure 2. The cutting end of a machine reamer. (Bendix Industrial Tools Division)

Figure 3. Straight shank straight flute chucking reamer. (TRW Inc.)

Figure 4. Straight shank helical flute chucking reamer. (TRW Inc.)

Figure 5. Taper shank helical flute chucking reamer. (TRW Inc.)

Figure 6. Taper shank straight flute jobbers reamer. (TRW Inc.)

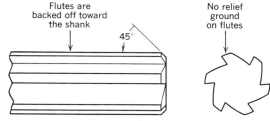

Figure 7. Rose reamer. (*Machine Tools and Machining Practices*)

accurately and precisely. Chucking reamers cut on the chamfer at the end of the flutes. This chamfer is usually at a 45 degree angle.

Jobbers reamers (Figure 6) are used where a longer flute length than chucking reamers is needed. The additional flute length gives added guide to the reamer, especially when reaming deep holes.

The rose reamer (Figure 7) is primarily a roughing reamer used to enlarge holes to within .003 to .005 in. of finish size. The teeth are slightly backed off, which means that the reamer diameter is smaller toward shank end by approximately .001 in./in. of flute length. The lands on these reamers are ground cylindrically without radial relief, and all cutting is done on the end of the reamer. This reamer will remove a considerable amount of material in one cut.

Shell reamers (Figure 8) are finishing reamers. They are more economically produced, especially in larger sizes, than solid reamers because a much smaller amount of tool material is used in making them. Two slots in the shank end of the reamer fit over matching driving lugs on the shell or box reamer (Figure 9). The hole in the shell reamer has a slight taper ($\frac{1}{8}$ in./ft) in it to assure exact alignment with the shell reamer arbor. Shell reamers are made with straight or helical flutes and are commonly produced in sizes from $\frac{3}{4}$ to $2\frac{1}{2}$ in. in diameter. Shell reamer

Figure 8. Shell reamer helical flute. (TRW Inc.)

Figure 9. Taper shank shell reamer arbor. (TRW Inc.)

Figure 10. Morse taper reamer. (TRW Inc.)

Figure 11. Helical taper pin reamer. (TRW Inc.)

Figure 12. Helical flute taper bridge reamer. (TRW Inc.)

Figure 13. Carbide tipped straight flute chucking reamer. (TRW Inc.)

Figure 14. Carbide tipped, helical flute chucking reamer, right-hand helix. (TRW Inc.)

Figure 15. Carbide tipped, helical flute chucking reamer, left-hand helix. (TRW Inc.)

arbors come with matching straight or tapered shanks and are made in designated sizes from numbers 4 to 9.

Morse taper reamers (Figure 10), with straight or helical flutes, are used to finish ream tapered holes in drill sockets, sleeves, and machine tool spindles. Helical taper pin reamers (Figure 11) are especially suitable for machine reaming of taper pin holes. There is no packing of chips in the flutes, which reduces the possibility of breakage. These reamers have a free-cutting action that produces a good finish at high cutting speeds. Taper pin reamers have a taper of $\frac{1}{4}$ in. per foot of length and are manufactured in 18 different sizes ranging from smallest number 8/0 (eight naught) to the largest at number 10.

Taper bridge reamers (Figure 12) are used in structural iron or steel work, bridge work, and ship construction where extreme accuracy is not required. They have long tapered pilot points for easy entry in out-of-line holes often encountered in structural work. Taper bridge reamers are made with straight and helical flutes to ream holes with diameters from $\frac{1}{4}$ to $1\frac{5}{16}$ in.

Carbide tipped chucking reamers (Figure 13) are often used in production setups, particularly where abrasive materials or sand and scale as in castings are encountered. The right-hand helix chucking reamer (Figure 14) is recommended for ductile materials or highly abrasive materials or when machining blind holes. The carbide tipped left-hand helix chucking reamer (Figure 15) will produce good finishes on heat treated steels and other hard materials, but should be used on through holes only. All expansion reamers (Figure 16) after becoming worn can be expanded and resized by grinding. This feature offsets normal wear from abrasive materials and provides for a long tool life. These tools should not be adjusted for reaming size by loosening or tightening the expansion plug, but only by grinding.

Reaming is intended to produce accurate and straight holes of uniform diameter. The required accuracy depends on a high degree of surface finish, tolerance on diameter, roundness, straightness, and absence of bellmouth at the ends of holes. To make an accurate hole it is necessary to use reamers with adequate support for the cutting edges; an adjustable reamer may not be adequate. Machine reamers are often made of either high speed steel or cemented carbide. Reamer cutting action is controlled to a large extent by the cutting speed and feed used.

Figure 16. Carbide tipped expansion reamer. (TRW Inc.)

SPEED

The most efficient cutting speed for machine reaming depends on the type of material being reamed, the amount of stock to be removed, the tool material being used, the finish required, and the rigidity of the setup. A good starting point, when machine reaming, is to use $\frac{1}{2}$ to $\frac{1}{3}$ of the cutting speed used for drilling the same materials. Table 1 may be used as a guide.

Where conditions permit the use of carbide reamers, the speeds may often be increased over those recommended for HSS (high speed steel) reamers. The limiting factor is usually an absence of rigidity in the setup. Any chatter, which is often caused by too high a speed, is likely to chip the cutting edges of a carbide reamer. Always select a speed that is slow enough to eliminate chatter. Close tolerances and fine finishes often require the use of considerably lower speeds than those recommended in Table 1.

FEEDS

Feeds in reaming are usually two to three times greater than those used for drilling. The amount of feed may vary with different materials, but a good starting point would be twice the feed rates given in Table 2 in Unit 5, "Operating Drilling Machines." Too low a feed may "glaze" the hole, which has the result of work hardening the material, causing occasional chatter and excessive wear on the reamer. Too high a feed tends to reduce the accuracy of the hole and the quality of the surface finish. Generally, it is best to use as high a feed as possible to produce the required finish and accuracy.

When a drill press that has only a hand feed is used to ream a hole, the feed rate should be estimated just as it would be for drilling. About twice the feed rate should be used for reaming as would be used for drilling in the same setup when hand feeding.

STOCK ALLOWANCE

The stock removal allowance should be sufficient to assure a good cutting action of the reamer. Too small a stock allowance results in burnishing (a slipping or polishing action), or it wedges the reamer in the hole and causes excessive wear or breakage of the reamer. The condition of the hole before reaming also has an influence on the reaming allowance since a rough hole will need a greater amount of stock removed than an equal size hole with a fairly smooth finish. See Table 2 for commonly used stock allowance for reaming. When materials that work harden readily are reamed, it is especially important to have adequate material for reaming.

Table 1
Reaming Speeds

Aluminum and its alloys	130–200[a]
Brass	130–200
Bronze, high tensile	50– 70
Cast iron	
Soft	70–100
Hard	50– 70
Steel	
Low carbon	50– 70
Medium carbon	40– 50
High carbon	35– 40
Alloy	35– 40
Stainless steel	
AISI 302	15– 30
AISI 403	20– 50
AISI 416	30– 60
AISI 430	30– 50
AISI 443	15– 30

[a] Cutting speeds in surface feet per minute (FPM or SFM) for reaming with an HSS reamer.

Source: Warren T. White, John E. Neely, Richard R. Kibbe, and Roland O. Meyer, *Machine Tools and Machining Practices,* John Wiley and Sons, Inc., Copyright © 1977, New York.

Table 2
Stock Allowance for Reaming

Reamer Size (in.)	Allowance (in.)
$\frac{1}{32}$ to $\frac{1}{8}$	.003 to .006
$\frac{1}{8}$ to $\frac{1}{4}$	.005 to .009
$\frac{1}{4}$ to $\frac{3}{8}$	.007 to .012
$\frac{3}{8}$ to $\frac{1}{2}$	.010 to .015
$\frac{1}{2}$ to $\frac{3}{4}$	$\frac{1}{64}$ or $\frac{1}{32}$
$\frac{3}{4}$ to 1	$\frac{1}{32}$

Table 3
Coolants Used for Reaming

Material	Dry	Soluble Oil	Kerosene	Sulfurized Oil	Mineral Oil
Aluminum		x	x		
Brass	x	x			
Bronze	x	x			x
Cast iron	x				
Steels					
Low carbon		x		x	
Alloy		x		x	
Stainless		x		x	

Source: Warren T. White, John E. Neely, Richard R. Kibbe, and Roland O. Meyer, *Machine Tools and Machining Practices,* John Wiley and Sons, Inc., Copyright © 1977, New York.

CUTTING FLUIDS

To ream a hole to a high degree of surface finish, a cutting fluid is needed. A good cutting fluid will cool the workpiece and tool and will also act as a lubricant between the chip and the tool to reduce friction and heat buildup. Cutting fluids should be applied in sufficient volume to flush the chips away. Table 3 lists some coolants used for reaming different materials.

REAMING PROBLEMS

Chatter is often caused by the lack of rigidity in the machine, workpiece, or the reamer itself. Corrections may be made by reducing the speed, increasing the feed, putting a chamfer on the hole before reaming, using a reamer with a pilot (Figure 17), or reducing the clearance angle on the cutting edge of the reamer. Carbide tipped reamers especially cannot tolerate even a momentary chatter at the start of a hole, as such a vibration is likely to chip the cutting edges.

Oversize holes can be caused by inadequate workpiece support, worn guide bushings, worn or loose spindle bearings, or a bent reamer shank. When reamers gradually start cutting larger holes, it may be because of the work material galling or forming a built-up edge on reamer cutting surfaces (Figure 18). Mild steel and some aluminum alloys are particularly troublesome in this area. Changing to a different coolant may help. Reamers with highly polished flutes, margins, and relief angles, or reamers that have special surface treatment, may also improve the cutting action.

Bell-mouthed holes are caused by misalignment of the reamer with the hole. The use of accurate bushings or pilots may correct bell-mouth, but in many cases the only solution is the use of floating holders. A floating holder will allow movement in some directions while restricting it in others. A poor finish can be improved by decreasing the feed, but this will also increase the wear and shorten the life of the reamer. A worn reamer will never leave a good surface finish as it will score or groove the finish and often produce

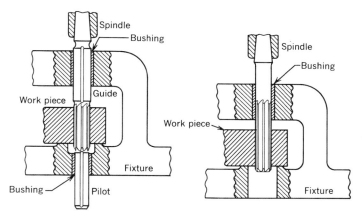

Figure 17. Use of pilots and guided bushings on reamers. Pilots are provided so that the reamer can be held in alignment and can be supported as close as possible while allowing for chip clearance. (Bendix Industrial Tools Division)

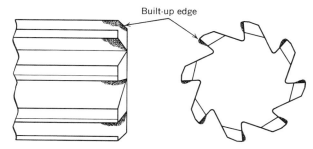

Figure 18. Reamer teeth having built-up edges. *(Machine Tools and Machining Practices)*

a tapered hole.

Too fast a feed will cause a reamer to break. Too large a stock allowance for finish reaming will produce a large volume of chips with heat buildup, and will result in a poor hole finish. Too small a stock allowance will cause the reamer teeth to rub as they cut, not cut freely, which will produce a poor finish and cause rapid reamer wear. Coolant applied in insufficient quantity may also cause rough surface finishes when reaming.

SELF-TEST

1. How is a machine reamer identified?
2. What is the difference between a chucking and a rose reamer?
3. What is a jobbers reamer?
4. Why are shell reamers used?
5. How does the surface finish of a hole affect its accuracy?
6. How does the cutting speed compare between drilling and reaming for the same material?
7. How does the feed rate compare between drilling and reaming?
8. How much reaming allowance will you leave on a $\frac{1}{2}$ in. hole?
9. What is the purpose of using a coolant while reaming?
10. What can be done to overcome chatter?
11. What will cause oversize holes?
12. What causes a bell-mouthed hole?
13. How can poor surface finish be overcome?
14. When are carbide tipped reamers used?
15. Why is vibration harmful to carbide tipped reamers?

SECTION H

TURNING MACHINES

UNIT 1 THE ENGINE LATHE

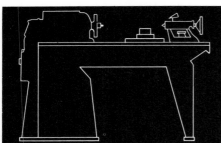

The power drive lathe or engine lathe is truly the father of all machine tools. With suitable attachments, the engine lathe may be used for turning, threading, boring, drilling, reaming, facing, spinning, and grinding, although many of these operations are preferably done on specialized machinery. Sizes range from the smallest jeweler's or precision lathes (Figure 1) to the massive lathes used for machining huge forgings (Figure 2).

Engine lathes (Figure 3) are used by machinists to produce one-of-a-kind parts or a few pieces for a short run production. They are also used for toolmaking, machine repair, and maintenance.

Some lathes have a vertical spindle instead of a horizontal one with a large rotating table on which the work is clamped. These huge machines, called vertical boring mills (Figure 4), are the largest of our machine tools. A 25-foot diameter table is not unusual. Huge turbines, weighing many tons, can be placed on the table and clamped in position to be machined. The machining of such castings would be impractical on a horizontal spindle lathe.

Production lathes that are fully automatic or semiautomatic are generally specialized machines. These include turret lathes (Figure 5), automatic screw machines (Figure 6), tracer lathes (Figures 7a and 7b), and numerically (tape) controlled machines (Figure 8).

In the following units you will be learning how to use the engine lathe, that is, the power driven nonspecialized machine. Modern engine lathes are highly accurate and complex machines capable of performing a great variety of operations. Before attempting to operate a lathe, you should familiarize yourself with its principal parts and their operation.

OBJECTIVE

After completing this unit, you should be able to:
Identify the most important parts of a lathe and their functions.

A lathe is a device in which the work is rotated against a cutting tool. As the cutting tool is moved lengthwise and crosswise to the axis of the workpiece, the shape of the workpiece is generated.

Figure 9 shows a lathe with its most important parts identified. A lathe consists of the following major component groups: headstock, bed, carriage, tailstock, quick-change gearbox, and a base or pedestal. The headstock is fastened on the left side of the bed. It contains the spindle that drives the various workholding devices. The spindle is supported by spindle bearings on each end. If they are sleeve-type bearings, a thrust bearing is also used to take up end play. Tapered roller spindle bearings are often used on modern lathes. Spindle speed changes are also made in the headstock, either with belts or with gears (Figures 10a and 10b). The threading and feeding mechanisms of the lathe are also powered through the headstock.

Most belt driven lathes are equipped with a slow speed range through the use of backgears. Figure 10c shows a backgeared headstock. Slow RPMs are obtained by engaging the backgears with the backgear lever. When higher RPMs are needed, the procedure is reversed by disengaging the backgears. See Section I, Unit 7 for further information on setting speeds on the lathe.

The spindle is hollow, which allows long slender workpieces to pass through. The spindle end facing

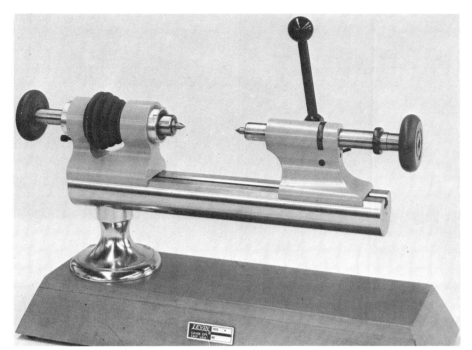

Figure 1. Jeweler's or instrument lathe. (Louis Levin & Son, Inc.)

Figure 2. A massive forging being machined to exact specifications on a lathe. (Bethlehem Steel Corporation)

the tailstock is called the spindle nose. Spindle noses usually are one of three designs: a long taper key drive (Figure 11), a camlock type (Figure 12), or a threaded spindle nose (Figure 13). Lathe chucks and other workholding devices are fastened to and driven by the spindle nose. The hole in the spindle nose typically has a standard Morse taper. The size of this taper varies with the size of the lathe.

The bed (Figure 14) is the foundation and backbone of a lathe. Its rigidity and alignment affect the accuracy of the parts machined on it. Therefore, lathe beds are constructed to withstand the stresses created by heavy machining cuts. The ways are on top of the bed, which usually consist of two inverted vees and two flat bearing surfaces. The ways of the lathes are very accurately machined by grinding or by milling and hand scraping. Many modern lathes have hardened and ground ways. Wear or damage to the ways will affect the accuracy of workpieces machined on them. A gear rack is fastened below the front way of the lathe. Gears that link the carriage handwheel to this rack make the lengthwise movement of the carriage possible by hand-turning the carriage handwheel.

The carriage (Figure 15) is made up of the saddle and the apron. The saddle rides on top of the ways and carries the cross slide and the compound rest. The cross slide is moved perpendicular to the axis of the lathe by manually turning the cross feed screw

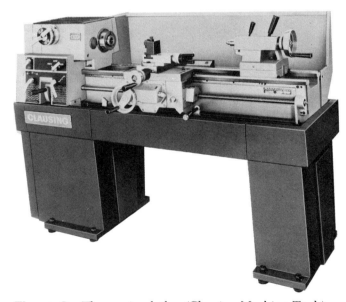

Figure 3. The engine lathe. (Clausing Machine Tools)

handle or by engaging the **cross feed lever** (or clutch knob) for automatic power feed. On some lathes a **feed change lever** (or plunger) on the apron is used to direct power from the feed mechanism to either the longitudinal (lengthwise) travel of the carriage or to the cross slide. In some other lathes, two separate levers or knobs are used to transmit motion to the

Figure 4. Vertical boring mill. (El-Jay Inc., Eugene, Oregon)

Figure 5. Saddle-type turret lathe. (The Warner & Swasey Company)

Figure 6. Small precision automatic screw machine in operation. (Sweetland Archery Products)

Figure 7a. Tracer lathe. (Lane Community College)

Figure 7b. Tracer attachments on an engine lathe. (Clausing Machine Tools)

carriage or cross slide. The direction of movement is controlled by the feed reverse lever that is usually mounted on the headstock. It has the function of changing the direction of the leadscrew rotation and consequently the feed direction. The **compound rest** is mounted on the cross slide and can be swiveled to any angle horizontal with the lathe axis in order to produce bevels and tapers. The compound rest can only be moved manually by turning the compound rest feed screw handle. Cutting tools are fastened on a tool post that is fastened on the compound rest.

The **apron** is the part of the carriage facing the operator. It contains the gears and feed clutches that transmit motion from the feed rod or lead screw to the carriage and cross slide. The thread dial is fastened to the apron, which indicates the exact moment to engage the half-nuts while thread cutting. The **half-nut lever** is used **only** while cutting threads. The entire carriage can be moved along the lathe bed manually by turning the carriage handwheel or under power by engaging the power feed controls on the apron. Once in position, the carriage can be clamped to the

Figure 8. Numerically controlled chucking lathe with turret. (The Monarch Machine Tool Company, Sidney, Ohio)

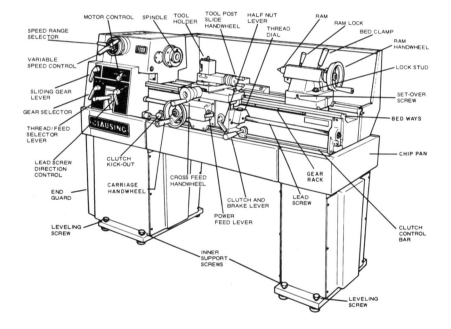

Figure 9. Engine lathe with the parts identified. (Clausing Machine Tools)

bed by tightening the carriage lock screw.

The **tailstock** (Figure 16) is used to support one end of a workpiece for machining or to hold various cutting tools such as drills, reamers, and taps. The tailstock has a sliding spindle that is operated by a handwheel and locked in position with a spindle clamp lever. The spindle is bored to receive a standard Morse taper shank. The tailstock consists of an upper and lower unit and can be adjusted to make tapered workpieces by turning the adjusting screws in the base unit, which offset the upper unit.

The **quick-change gearbox** (Figure 17) is the link

that transmits power between the spindle and the carriage. By using the gear shift levers on the quick-change gearbox, you can select different feeds. Power is transmitted to the carriage through a feed rod, or as on smaller lathes, through the lead screw with a keyseat in it. The index plate on the quick-change gearbox indicates the feed in thousandths of an inch or as threads per inch for the lever positions.

The base of the machine is used to support the lathe and to secure it to the floor. The lathe motor is usually mounted in the base. Figure 18 shows how the lathe is measured.

Figure 10a. Geared headstock for heavy duty lathe. (Lodge & Shipley Company)

Figure 10b. Spindle drive showing gears and shifting mechanism located in the headstock. (Lodge & Shipley Company)

Figure 10c. View of headstock showing back gear disengaged and lock pin engaged for direct belt drive. (Lane Community College)

Figure 11. Long taper key drive spindle nose. (Lane Community College)

Figure 12. Camlock spindle nose.

Figure 13. Threaded spindle nose. (Lane Community College)

Figure 14. Lathe bed. (Clausing Machine Tools)

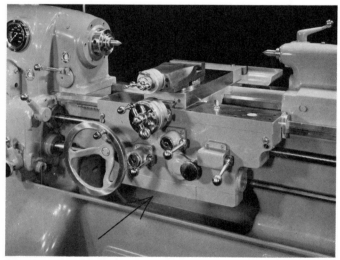

Figure 15. Lathe carriage. The arrow is pointing to the apron. (The Monarch Machine Tool Company, Sidney, Ohio)

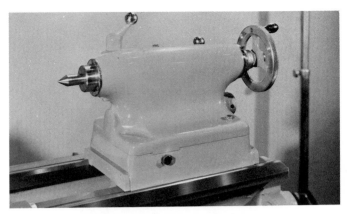

Figure 16. Tailstock. (The Monarch Machine Tool Company, Sidney, Ohio)

Figure 17. Quick-change gearbox showing index plate. (Lane Community College)

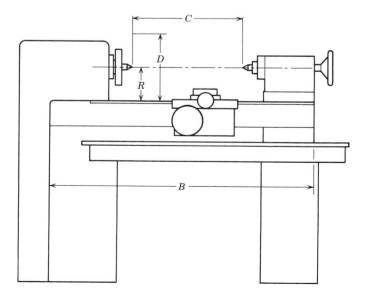

Figure 18. Measuring a lathe for size. C = maximum distance between centers; D = maximum diameter of workpiece over ways—swing of lathe; R = radius, one-half swing; B = length of bed. (*Machine Tools and Machining Practices*)

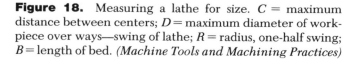

SELF-TEST

At a lathe in the shop, identify the following parts and describe their functions. **Do not** turn on the lathe until you get permission from your instructor.

The Headstock:

1. Spindle
2. Spindle speed changing mechanism
3. Backgears
4. Bull gear lockpin
5. Spindle nose
6. What kind of spindle nose is on your lathe?
7. Feed reverse lever

The Bed:

1. The ways
2. The gear rack

The Carriage:

1. The cross slide
2. The compound rest
3. Saddle

4. Apron
5. Power feed lever
6. Feed change lever
7. Half-nut lever
8. Thread dial
9. Carriage handwheel
10. Carriage lock

The Tailstock:

1. Spindle and spindle clamping lever
2. Tapered spindle hole and the size of its taper
3. The tailstock adjusting screws

The Quick-Change Gearbox:

1. The lead screw
2. Shift the levers to obtain feeds of .005 and .013 in. per revolution. Rotating the lead screw with your fingers aids in shifting these levers.
3. Set the levers to obtain 4 threads/in. and then 12 threads/in.
4. Measure the lathe and record its size.

UNIT 2 TURNING MACHINE SAFETY

Safety programs in industry were virtually unknown and machines were not safeguarded before 1910. Lathes had exposed gears and pulleys. Accidents were very frequent and always considered the fault of careless workers. Since that time many laws have been passed for accident prevention. Guards and other safety devices are required on machines for worker safety, but this is not enough! Without your cooperation in wearing required protective clothing and devices, and in following safe procedure, accidents will still happen.

This unit will make you aware of the many hazards that exist in lathe operations and help you to avoid them. Your responsibility as a student and as a worker is clear: protect yourself and the people around you. Observe the safety rules.

OBJECTIVE

After completing this unit, you should be able to:
Describe and use 10 safety rules or procedures for lathe operations.

Figure 1a. A chuck wrench left in the chuck is a danger to everyone in the shop. (Lane Community College)

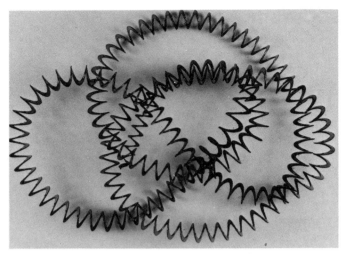

Figure 2. Unbroken lathe chips are sharp and hazardous to the operator. (Lane Community College)

Figure 1b. A safety-conscious lathe operator will remove the chuck wrench when he finishes using it. (Lane Community College)

The lathe can be a safe machine only if the machinist is aware of the hazards involved in its operation. In the machine shop as anywhere, you must always keep your mind on your work in order to avoid accidents. Develop safe work habits in the use of setups, chip breakers, guards, and other protective devices. Standards for safety have been established as guidelines to help you eliminate unsafe practices and procedures on lathes. Some of the hazards are as follows:

1. **Pinch points due to lathe movement.** A finger caught in gears or between the compound rest and

a chuck jaw would be an example. The rule is to keep your hands away from such dangerous positions when the lathe is operating.

2. **Hazards associated with broken or falling components.** Heavy chucks or workpieces can be dangerous when accidentally dropped. Care must be used when handling them. If a threaded spindle is suddenly reversed, the chuck can come off and fly out of the lathe. A chuck wrench left in the chuck can become a missile when the machine is turned on. Always remove the chuck wrench immediately after using it (Figures 1a and 1b).

3. **Hazards resulting from contact with high temperature components.** Burns usually result from handling hot chips (up to 800°F or even more) or a hot workpiece. Gloves may be worn when handling hot chips or workpieces, but never when the machine is running.

4. **Hazards resulting from contact with sharp edges, corners, and projections.** These are perhaps the most common cause of hand injuries in lathe work. Dangerous sharp edges may be found many places: on a long stringy chip, on a tool bit, or on a burred edge of a turned or threaded part. Shields should be used for protection from flying chips and coolant. These shields are usually made of clear plastic and are hinged over the chuck or clamped to the carriage of engine lathes. Stringy chips must not be removed with bare hands; wear heavy gloves and use hook tools or pliers. Always turn off the machine before attempting to remove chips. Chips should be broken and form a figure 9 rather than in a stringy mass or a long wire (Figure 2). Chip breakers on tools and correct feeds will help to

produce safe, easily handled chips. Burred edges must be removed before the workpiece is removed from the lathe. Always remove the tool bit when setting up or removing workpieces from the lathe.

5. **Hazards of workholding devices or driving devices.** When workpieces are clamped, their components often extend beyond the outside diameter of the holding device. Guards, barriers, and warnings such as signs or verbal instructions are all used to make you aware of the hazards. On power chucking devices you should be aware of potential pinch points between workpiece and workholding devices. Make certain sufficient gripping force is exerted by the jaws to safely hold the work. Never run a geared scroll chuck without having something gripped in the jaws. Centrifugal force on the jaws can cause the scroll to unwind and the jaws to come out of the chuck. Keep tools, files, and micrometers off the machine. They may vibrate off into the revolving chuck or workpiece.

6. **Electrical hazards.** Only qualified electricians should install any electric wiring. Enclosures for electrical and power transmission equipment are restricted to authorized personnel only.

7. **Employee's responsiblity.** Setup checks should be made during each shift. Safety glasses and industrial hairnets or hair ties should be worn where required (Figure 3). Jewelry and clothing that will present a hazard must not be worn. An orderly work area must be maintained by the operator. The spindle or workpiece should never be slowed or stopped by hand gripping or by using a pry bar. Always use machine controls to stop or slow it.

8. **Extended workpieces.** Workpieces extending out of the lathe should be supported by a stock tube (Figure 4). If a slender workpiece is allowed to extend beyond the headstock spindle a foot or so without support it can fly outward from centrifugal force. The piece will not only be bent, but it will present a very great danger to anyone standing near.

These and other safety considerations are the responsibility of both employer and employee. Also the manufacturer, reconstructor, and modifier share this responsibility by providing proper safety devices for machines. Additional safety precautions you should take are as follows:

- Never overtighten locking bolts or set screws on toolholders. An overstressed part or bolt may fail suddenly and endanger the operator.
- Always keep the compound rest over its slide unless it is being used for angular cutting.
- Never operate a machine with a guard open or off.
- Keep your hands and fingers away from revolving work or machine parts.

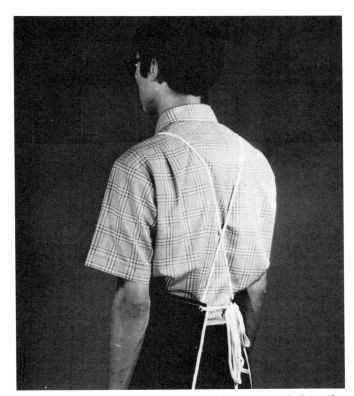

Figure 3. Correct attire for machining on a lathe. The apron is tied in back with a light cord so it will break easily in case the apron catches in the machine.

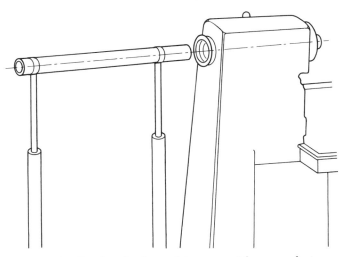

Figure 4. Stock tube is used to support long workpieces that extend out of the headstock of a lathe. (*Machine Tools and Machining Practices*)

Figure 5. Polishing in the lathe with abrasive cloth. (Lane Community College)

Figure 6. Loose hair that is not kept confined or tied back may get caught in machinery with results such as this. (Photo courtesy of John Allan, Jr.)

- When cutting to a shoulder, finish the cut with hand feed.
- When setting up work, rotate the spindle by hand to see if the lathe dog or workpiece clears the carriage or compound rest.
- When setting up work between centers, make sure the dead center (tailstock center) is properly adjusted and the binding lever is tightened.
- Work that extends from the chuck more than five times its diameter should be supported by a steady rest or tailstock.

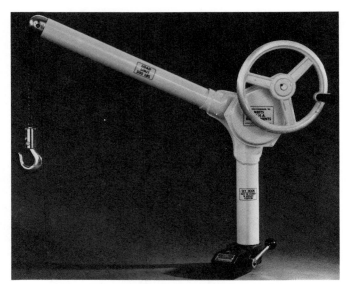

Figure 7a. One way of hoisting materials and equipment into the lathe is a lathe-mounted crane for handling chucks and workpieces. This is adapted for mounting on quick-change toolholders on the compound. (Syclone Products Inc.)

Figure 7b. The skyhook in use bringing a large chuck into place for mounting. (Syclone Products Inc.)

- Hold one end of abrasive cloth strips in each hand when polishing rotating work. Don't let either hand get closer than a few inches from the work (Figure 5).
- Keep rags, brushes, and fingers away from rotating

work, especially when knurling. Roughing cuts tend to quickly drag in and wrap up rags, clothing, neckties, abrasive cloth, and hair (Figure 6).

- Move the carriage back out of the way and cover the tool with a cloth when checking boring work.
- When removing or installing chucks or heavy workpieces, use a board on the ways (a part of the lathe bed) so it can be slid into place. To lift a heavy chuck or workpiece (larger than an 8-in. diameter chuck) get help or use a crane (Figures 7a and 7b). Remove the tool or turn it out of the way during this operation.
- Always clean and oil a machine before you use it, and clean up the machine and work area when you are finished.
- Do not shift gears or try to take measurements while the machine is running and the workpiece is in motion.
- When grinding tool bits, eye protection must be worn, grinder tables or rests must be adjusted within safe limits (about $\frac{1}{16}$ in. from the wheel), and guards must be in place.
- Never use a file without a handle, as the file tang can quickly cut your hand or wrist if the file has been struck by a spinning chuck jaw or lathe dog. Left-hand filing is considered safest in the lathe; that is, the left hand grips the handle while the right hand holds the tip end of the file (Figure 8).
- Never hurry or become distracted from your work. If someone wishes to talk to you, stop your machine until you are finished talking.
- Do not use an air hose to blow away chips. The

Figure 8. Left-hand filing in the lathe. (Lane Community College)

flying chips are hazardous to your eyes as well as the eyes of others. Air pressure will drive small chips and dirt into bearing surfaces and ruin machinery.

- Don't try to fix a machine unless you are authorized and trained to do so.
- Avoid at all times carelessness, horseplay, and acting without thinking, especially when operating a machine. Remember, it is your own attitude that will ultimately either prevent accidents or cause you to suffer pain and loss of working time or worse.

SELF-TEST

1. On whom does the responsibility for worker safety rest?
2. What is a pinch point?
3. How would it be possible for a chuck to come out of a lathe? A chuck wrench?
4. Is it possible for a lathe operator to get burns on his body? How?
5. How can a lathe operator receive cuts on his hands and arms?
6. Because of the nature and use of chucking devices, it is impossible to adequately guard them. How then can the operator avoid danger in this area?
7. The headstock gears are being changed by a mechanic in the lathe next to yours. He accidentally pulls the switch off for your lathe as well as the one he is working on, when he went in the electrical enclosure. What should you do?
8. A student needs to machine several pieces from a $\frac{3}{8}$ in. rod. Since the rod is about 4 ft long and he is reluctant to cut it, he allows it to extend through the spindle and out about 30 in. The machining speed is about 1000 RPM. What do you think will result?
9. Overtightening bolts and set screws can sometimes be more hazardous than undertightening them. Why is this so?
10. When a person knows all the safety rules and procedures and still has accidents, what do you think is wrong?

UNIT 3 TOOLHOLDERS AND TOOL-HOLDING FOR THE LATHE

For lathe work, cutting tools must be supported and fastened securely in the proper position to machine the workpiece. There are many different types of toolholders available to satisfy this need. Anyone working with a lathe should be able to select the best toolholding device for the operation performed.

OBJECTIVES

After completing this unit, you should be able to:
1. Identify standard, quick-change, and turret-type toolholders mounted on a lathe carriage.
2. Identify toolholding for the lathe tailstock.

A cutting tool is supported and held in a lathe by a toolholder that is secured in the tool post of the lathe with a clamp screw. A common tool post found on smaller or older lathes is shown in Figure 1. Tool height adjustments are made by swiveling the rocker in the tool post ring. Making adjustments in this manner changes the effective back rake angle and also the front relief angle of the tool.

Many types of toolholders are used with the standard tool post. A straight shank turning toolholder (Figure 2) is used with high speed tool bits. The tool bit is held in the toolholder at a $16\frac{1}{2}$ degree angle, which provides a positive back rake angle for cutting. Straight shank toolholders are used for general machining on lathes. The type shown in Figure 3 is used with carbide tools.

Offset toolholders (Figures 4 and 5) allow machining close to the chuck or tailstock of a lathe without tool post interference. The left-hand toolholder is intended for use with tools cutting from right to left or toward the headstock of the lathe. The hand of the toolholder can also be determined by looking down at it when it is upside down. A right-hand tool-

Figure 1. Standard-type tool post with ring and rocker. *(Machine Tools and Machining Practices)*

Figure 2. Straight shank toolholder with built-in back rake holding a high speed right-hand tool. (Lane Community College)

Figure 6. Three kinds of cutoff toolholders with cutoff blades. (J. H. Williams Division of TRW, Inc.)

Figure 3. Right-hand toolholder for carbide tool bits without back rake. *(Machine Tools and Machining Practices)*

Figure 4. Left-hand toolholder with right-hand tools. (Lane Community College)

Figure 7. Knuckle-joint knurling tool. (Lane Community College)

Figure 5. Right-hand toolholder with left-hand tool. (Lane Community College)

Figure 8. Triple head knurling tool. (Lane Community College)

holder is bent toward the right in that position.

A toolholder should be selected according to the machining to be done. The setup should be rigid and the toolholder overhang should be kept to a minimum to prevent chattering. A variety of cutoff toolholders (Figure 6) are used to cut off or make grooves in workpieces. Cutoff tools are available in a number of different thicknesses and heights. Knurling tools are made with one pair of rollers (Figure 7) or with three pairs of rollers (Figure 8) that give a choice of three different kinds of knurls.

Another tool used in a standard tool post is the boring bar toolholder (Figure 9). The boring bar tool post (Figure 10) can be used with a number of differ-

ent boring bar sizes. Another advantage of some boring bars is the interchangeability of toolholding end caps. End caps hold the boring tool 90 degrees to the axis of the boring bar or at a 45 or 60 degree angle to it. The heavy duty boring bar holder in Figure

Figure 9. Toolholder for small boring bars. (J. H. Williams Division of TRW, Inc.)

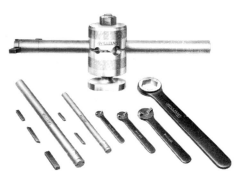

Figure 10. Boring bar tool post and bars with special wrench. (J. H. Williams Division of TRW, Inc.)

Sleeve Bar

Figure 11. Heavy duty boring bar holder. (J. H. Williams Division of TRW, Inc.)

Figure 12. Quick-change tool post, dovetail type. (Aloris Tool Company, Inc.)

Figure 13. Three-sided quick-change tool post. (Lane Community College)

11 is not as rigid as the holder in Figure 10 because it is clamped in the tool post.

A quick-change tool post (Figure 12), so-called because of the speed with which tools can be interchanged, is more versatile than the standard post. The toolholders used on it are accurately held because of the dovetail construction of the post. This accuracy makes for more exact repetition of setups. Tool height adjustments are made with a micrometer adjustment collar and the height alignment will remain constant through repeated tool changes.

A three-sided quick-change tool post (Figure 13) has the added ability to mount a tool on the tailstock side of the tool post. These tool posts are securely clamped to the compound rest. The tool post in Figure 13 uses double vees to locate the toolholders, which are clamped and released from the post by turning the top lever.

Toolholders for the quick-change tool posts include those for turning (Figure 14), threading (Figures 15a and 15b), and holding drills (Figure 16). The drill holder makes it possible to use the carriage power feed when drilling holes instead of the tailstock hand feed. Figure 17 shows a boring bar toolholder in use;

Figure 14. Turning toolholder in use. (Aloris Tool Company, Inc.)

Figure 15a. Threading toolholder, using the top of the blade. (Aloris Tool Company, Inc.)

Figure 15b. Threading is accomplished with the bottom edge of the blade with the lathe spindle in reverse. This assures cutting right-hand threads without hitting the shoulders. (Aloris Tool Company, Inc.)

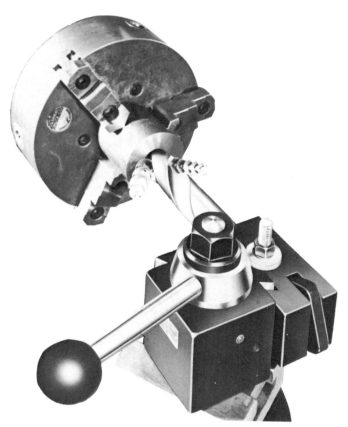

Figure 16. Drill toolholder in the tool post. Mounting the drill in the tool post makes drilling with power feed possible. (Aloris Tool Company, Inc.)

Figure 17. Boring toolholder. This setup provides good boring bar rigidity. (Aloris Tool Company, Inc.)

the boring bar is very rigidly supported.

An advantage of the quick-change tool post toolholders is that cutting tools of various shank thicknesses can be mounted in the toolholders (Figure 18). Shims are sometimes used when the shank is too small for the set screws to reach. Another example of quick-

Figure 18. Toolholders are made with wide or narrow slots to fit tools with various shank thicknesses. (Lane Community College)

Figure 19. Tailstock turret used in quick-change toolholder. (Enco Manufacturing Company)

Figure 20. Quick-change cutoff toolholder. (Enco Manufacturing Company)

Figure 21. Quick-change knurling and facing toolholder. (Enco Manufacturing Company)

Figure 22. Facing cut with a turret-type toolholder. (Enco Manufacturing Company)

change tool post versatility is shown in Figure 19, where a tailstock turret is in use. Figure 20 shows a cutoff tool mounted in a toolholder. Figure 21 is a combination knurling tool and facing toolholder. A four-tool turret toolholder (Figure 22) can be set up

with several different tools such as turning tools, facing tools, threading or boring tools. Often one tool can perform two or more operations, especially if the turret can be indexed in 30-degree intervals. A facing operation (Figure 22), a turning operation (Figure 23), and chamfering of a bored hole (Figure 24) are all performed from this turret. Tool height adjustments are made by placing shims under the tool.

The toolholders studied so far are all intended for use on the carriage of a lathe. Toolholding is also done on the tailstock. Figure 25 shows how the tailstock spindle is used to hold Morse taper shank tools. One of the most common toolholding devices used on a tailstock is the drill chuck (Figure 26). A drill chuck is used for holding straight shank drilling tools.

Figure 23. Turning cut with a turret-type toolholder. (Enco Manufacturing Company)

Figure 24. Chamfering cut with a turret-type toolholder. (Enco Manufacturing Company)

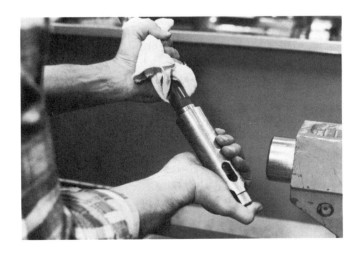

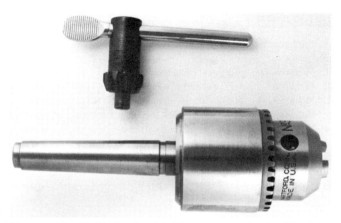

Figure 26. Drill chuck with Morse taper shank. *(Machine Tools and Machining Practices)*

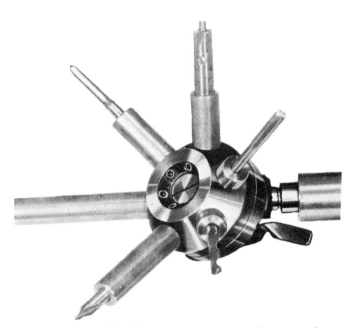

Figure 27. Tailstock turret. (Enco Manufacturing Company)

When a series of operations must be performed and repeated on several workpieces, a tailstock turret (Figure 27) can be used. The illustrated tailstock turret has six tool positions, one of which is used as a workstop. The other positions are for center drilling, drilling, reaming, counterboring, and tapping. Tailstock tools are normally fed by turning the tailstock handwheel.

Figure 25. Taper shank just in front of tailstock spindle hole. (Lane Community College)

SELF-TEST

At a lathe in the shop, identify various toolholders and their functions.

1. What is the purpose of a toolholder?
2. How is a standard left-hand toolholder identified?
3. What is the difference between a standard-type toolholder for high speed steel tools and for carbide tools?
4. Which standard toolholder would be best used for turning close to the chuck?
5. How are tool height adjustments made on a standard toolholder?
6. How are tool height adjustments made on a quickchange toolholder?
7. How are tool height adjustments made on a turret-type toolholder?
8. How does the toolholder overhang affect the turning operation?
9. What is the difference between a standard toolholder and a quickchange toolholder?
10. What kind of tools are used in the lathe tailstock?
11. How are tools fastened in the tailstock?
12. When is a tailstock turret used?

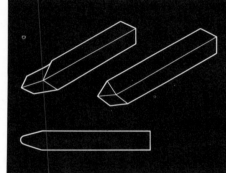

UNIT 4 CUTTING TOOLS FOR THE LATHE

A machinist must fully understand the purpose of cutting tool geometry, since it is the lathe tool that removes the metal from the workpiece. Whether this is done safely, economically, and with quality finishes depends to a large extent upon the shape of the point, the rake and relief angles, and the nose radius of the tool. In this unit, you will learn this tool geometry and also how to grind a lathe tool.

OBJECTIVES

After completing this unit, you should be able to:

1. Explain the purpose of rake and relief angles, chip breakers, and form tools.
2. Grind an acceptable right-hand roughing tool.

On a lathe, metal is removed from a workpiece by turning it against a single point cutting tool. This tool must be very hard and it should not lose its hardness from the heat generated by machining. High speed steel is used for many tools as it fulfills these requirements and is easily shaped by grinding. It should be noted, however, that their use is limited since most production machining today is done with carbide tools. High speed steel tools are required for older lathes that are equipped with only low speed ranges. They are also useful for finishing operations, especially on soft metals.

The most important aspect of a lathe tool is its geometric form: the side and back rake, front and side clearance or relief angles, and chip breakers.

Figure 1 shows the parts and angles of the tool according to a commonly used industrial tool signature. The terms and definitions follow (the angles given are only examples and they could vary according to the application):

Back rake	BR	12°
Side rake	SR	12°
End relief	ER	10°
Side relief	SRF	10°
End cutting edge angle	ECEA	30°
Side cutting edge angle	SCEA	15°
Nose radius	NR	$\frac{1}{32}$ in.

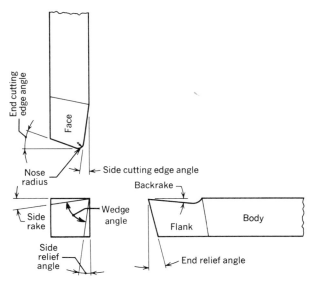

Figure 1. The parts and angles of a tool.

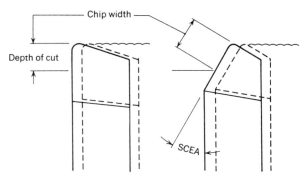

Figure 2. The change in chip width with an increase of the side cutting edge angle. A large SCEA can sometimes cause chatter (vibration of work or tool).

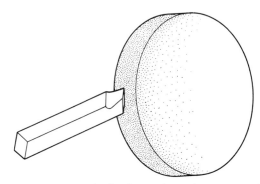

Figure 3. Left-hand and right-hand roughing tools.

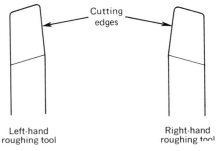

Figure 4. One method of grinding the nose radius on the point of the tool.

Tool Signature

1. The tool shank is that part held by the toolholder.
2. Back rake is very important to smooth chip flow, which is needed to have a uniform chip and a good finish.
3. The side rake directs the chip flow away from the point of cut and provides for a keen cutting edge.
4. The end relief angle prevents the front edge of the tool from rubbing on the work.
5. The side relief angle provides for cutting action by allowing the tool to feed into the work material.
6. The cutting edge angle may vary considerably (from 5 to 32 degrees). For roughing, it should almost be square (5 degrees off 90 degrees), but tools used for squaring shoulders or for other light machining could have angles from 15 to 32 degrees.
7. The side cutting edge angle, which is usually 10 to 20 degrees, directs the cutting forces back into a stronger section of the tool point. It helps to direct the chip flow away from the workpiece. It also affects the thickness of the cut (Figure 2).
8. The nose radius will vary according to the finish required.

Grinding a tool provides both a sharp cutting edge and the shape needed for the cutting operation. When the purpose for the rake and relief angles on a tool are clearly understood, then a tool suitable to the job may be ground. Left-hand tools are shaped just the opposite to right-hand tools (Figure 3). The right-hand tool has the cutting edge on the left side and cuts to the left or toward the headstock. The hand of the lathe tool can be easily determined by looking at the cutting end of the tool from the opposite side of the lathe; the cutting edge is to the right on a right-hand tool.

Tools are given a slight nose radius to strengthen the tip. A larger nose radius will give a better finish (Figure 4), but will also promote chattering (vibration)

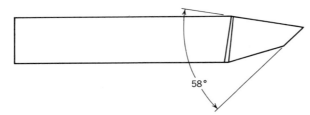

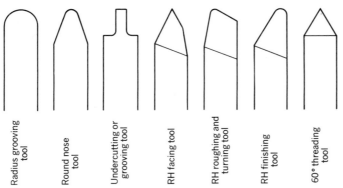

Figure 5. A right-hand facing tool showing point angles. This tool is not suitable for roughing operations because of its acute point angle.

Radius grooving tool | Round nose tool | Undercutting or grooving tool | RH facing tool | RH roughing and turning tool | RH finishing tool | 60° threading tool

Figure 6. Some useful tool shapes most often used. The first tool shapes needed are the three on the right, which are the roughing or general turning tool, finishing tool, and threading tool.

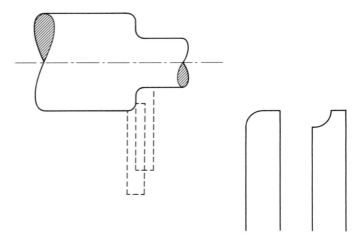

Figure 7. Form tools are used to produce the desired shape in the workpiece. External radius tools, for example, are used to make outside corners round, while fillet radius tools are used on shafts to round the inside corners on shoulders.

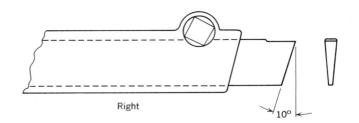

Right   10°

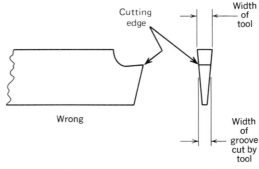

Cutting edge

Width of tool

Wrong

Width of groove cut by tool

Figure 8. Correct and incorrect methods of grinding a cutoff tool for deep parting.

in a nonrigid setup. All lathe tools require some nose radius, however small. A sharp pointed tool is very weak at the point and will usually break off in use, causing a rough finish on the work. A facing tool (Figure 5) for shaft ends and mandrel work has very little nose radius and an included angle of 58 degrees. This facing tool is not used for chucking work, however, as it is a relatively weak tool. A right-hand (RH) or left-hand (LH) roughing or finishing tool is often used for facing in chuck mounted workpieces.

Some useful tool shapes are shown in Figure 6. These are used for general lathe work.

Tools that have special shaped cutting edges are called form tools (Figure 7). These tools are plunged directly into the work, making the full cut in one operation.

Parting or cutoff tools are often used for necking or undercutting, but their main function is cutting off material to the correct length. The correct and incorrect ways to grind a cutoff tool are shown in Figure 8. Note that the width of the cutting edge becomes narrower than the blade as it is ground

lower, which causes the blade to bind in a groove that is deeper than the sharpened end.

However, tools are sometimes specially ground for parting very soft metals or specially shaped grooves (Figure 9). The end is sometimes ground on a slight angle when a series of small hollow pieces is

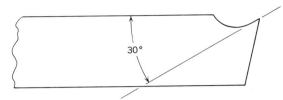

Figure 9. Cutoff tools are sometimes ground with large back rake angles for aluminum and other soft metals.

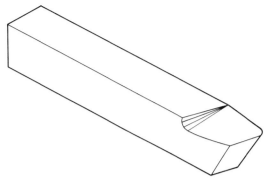

Figure 12. A properly ground right-hand roughing tool.

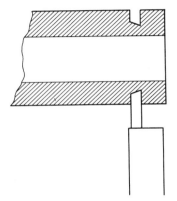

Figure 10. Parting tool ground on an angle to avoid burrs on the cut off pieces.

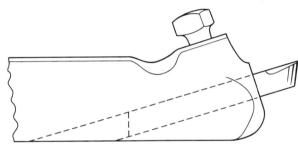

Figure 13. Toolholder with back rake.

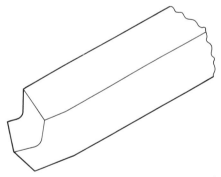

Figure 11. Deformed tool caused by many resharpenings. The chip trap should be ground off and a new point ground on the tool.

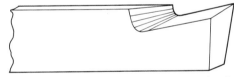

Figure 14. Right-hand roughing tool with back rake.

being cut off (Figure 10). This helps to eliminate the burr on small parts. This procedure is not recommended for deep parting.

Tools that have been ground back for resharpening too many times often form a "chip trap" causing the metal to be torn off or the tool to not cut at all (Figure 11). A good machinist will never allow tools

to get in this condition, but will grind off the useless end and regrind a proper tool shape (Figure 12).

Although many modern lathes have toolholders that hold the tool horizontally, some lathe toolholders have built-in back rake so it is not necessary to grind one into the tool as used in Figure 13. The tool in Figure 14, however, is ground with a back rake and can be used in a toolholder that does not have built-in back rake. Threading tools should have zero rake (Figure 15*a*). If a horizontal toolholder is used for threading, the flat on top is unnecessary as the tool would have a zero rake. The tool should also be checked for the 60 degree angle with a center gage (Figure 15*b*) while grinding. Relief should be ground on each side. A slight flat should be honed on the

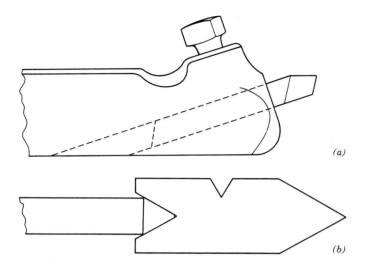

(a)

(b)

Figure 15. (a) Toolholder with back rake showing tool ground for zero rake. (b) Checking a threading tool with center gage.

Figure 16. Side view of back rake angles. The tool point is thinner (and more subject to breakage) on the positive back rake than on the negative rake tool.

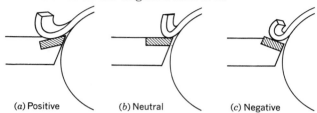

(a) Positive   *(b)* Neutral   *(c)* Negative

Table 1
Angle Degrees for High Speed Steel Tools

Material	End Relief	Side Relief	Side Rake	Back Rake
Aluminum	8 to 10	12 to 14	14 to 16	30 to 35
Brass, free cutting	8 to 10	8 to 10	1 to 3	0
Bronze, free cutting	8 to 10	8 to 10	2 to 4	0
Cast iron, gray	6 to 8	8 to 10	10 to 12	3 to 5
Copper	12 to 14	13 to 14	18 to 20	14 to 16
Nickel and monel	12 to 14	14 to 16	12 to 14	8 to 10
Steels, low carbon	8 to 10	8 to 10	10 to 12	10 to 12
Steels, alloy	7 to 9	7 to 9	8 to 10	6 to 8

Source: Warren T. White, John E. Neely, Richard R. Kibbe, and Roland O. Meyer, *Machine Tools and Machining Practices*, John Wiley and Sons, Inc., Copyright © 1977, New York.

Figure 17. Using a tool gage for checking angles. (a) Checking the side relief angle; (b) checking end relief; and (c) checking the wedge angle for steel.

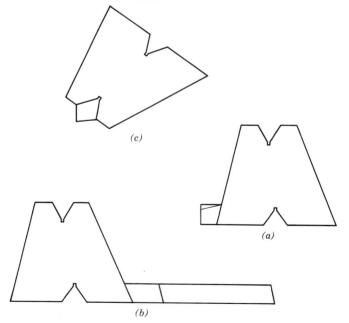

(c)

(a)

(b)

end with an oilstone. (See Unit 12, "Cutting Unified External Threads," for this dimension.)

Tools for brass or plastics should have zero to negative rake to keep the tools from "digging in" (Figure 16). Side rake, back rake, and relief angles are given for tools in Table 1 for machining various metals.

Figure 17 shows the use of a gage for checking angles when grinding a tool. It is designed to check tool angles on any flat surface. The angles are for tools to be used in toolholders having $16\frac{1}{2}$ degree back rake. Tools for straight or horizontal toolholders should have an end relief of 10 degrees; in these cases, change the end relief angle of the gage. A protractor or optical comparator can also be used to check tool angles.

For your safety, it is important to make tools that will produce chips that are not hazardous. Long, unbroken chips are extremely dangerous. Tool geometry, especially side and back rakes, has a considerable effect on chip formation. Smaller side rake angles tend

Figure 18. Chip flow with a plain tool and with chip breaker.

Figure 19. A figure-9 chip is considered the safest kind of chip to produce.

Figure 20. Crowding of the chip is caused by a chip breaker that is ground too deeply. (*Machine Tools and Machining Practices*)

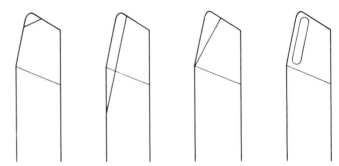

Figure 21. Four common types of chip breakers. (*Machine Tools and Machining Practices*)

to curl the chip more than those with larger angles, and the curled chips are more likely to break up. Coarse feeds for roughing and maximum depth of cut also promote chip breaking. Feeds, speeds, and depth of cut will be further considered in Unit 8, "Turning Between Centers."

Chip breakers are extensively used on both carbide and high speed tools to curl the chip as it flows across the face of the tool. Since the chip is curled back into the work, it can go no further and breaks (Figure 18). A C-shaped chip is often the result, but a figure-9 shaped chip is considered ideal (Figure 19). This chip should drop safely into the chip pan without flying out.

Grinding the chip breaker too deep will form a chip trap that may cause binding of the chip and tool

breakage (Figure 20). The correct depth to grind a chip breaker is approximately $\frac{1}{32}$ in. Chip breakers are typically of the parallel or angular types (Figure 21). More skill is needed to offhand grind a chip breaker on a high speed tool than is required to grind the basic tool angles. Therefore, the basic tool should be ground and an effort made to produce safe chips through the use of correct feeds and depth of cut before a chip breaker is ground.

Care must be exercised while grinding on high speed steel. A glazed wheel can generate heat up to 2000°F (1093°C) at the grinder-tool interface. Do not overheat the tool edge as this will cause small surface cracks that can result in the failure of the tool. Frequent cooling in water will keep the tool cool enough to handle. Do not quench in water, however, if you have overheated it. Let it cool in air.

Since the right-hand roughing tool is the one that is most commonly used by machinists and the first one that you will need, you should begin with it. A piece of keystock the same size as the tool bit should be used for practice until you are able to grind an acceptable tool.

Steps in Grinding a Right-Hand Roughing Lathe Tool

You will need a tool blank, a piece of keystock about 3 in. long, a tool gage, and a toolholder.

A. Grind one acceptable practice right-hand roughing tool. Have your instructor evaluate your progress until it is ground correctly.

B. Grind one acceptable right-hand roughing tool from a high speed tool blank.

C. Wear goggles and make certain the tool rest on the grinder is adjusted properly (about $\frac{1}{16}$ in. from the wheel). True up the wheels with a wheel dresser, if they are grooved, glazed, or out of round.

　　1. Using the roughing wheel, grind the side relief angle and the side cutting angle about 10 degrees by holding the blank and supporting your hand on the tool rest (Figure 22).

　　2. Check the angle with a tool gage (Figure 23). Correct if needed.

　　3. Rough out the end relief angle about 14 degrees and the end cutting edge angle (Figure 24).

　　4. Check the angle with the tool gage (Figure 25). Correct if needed.

　　5. Rough out the side rake. Stay clear of the side cutting edge by $\frac{1}{16}$ in. (Figure 26).

　　6. Check for wedge angle (Figure 27). Correct if needed.

　　7. Now change to the finer grit wheel and very gently finish grind the side and end relief angles. Try to avoid making several facets or grinds on one surface. A side to side oscillation will help to produce a good finish.

　　8. Grind the finish on the side rake as in Figure 26 and bring the ground surface just to the side cutting edge, but avoid going deeper.

　　9. A slight radius on the point of the tool should be ground on the circumference of the wheel (Figure 28) and all the way from the nose to the heel of the tool.

　　10. A medium to fine oilstone is used to remove the burrs from the cutting edge (Figure 29). The finished tool is shown in Figure 30.

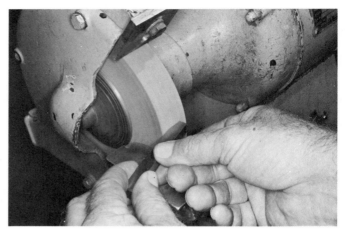

Figure 22. Roughing the side relief angle and the side cutting edge angle. (Lane Community College)

Figure 23. Checking the side cutting edge angle with a tool gage. (Lane Community College)

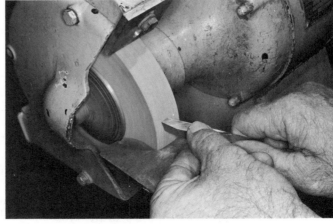

Figure 24. Roughing the end relief angle and the end cutting edge angle. (Lane Community College)

Figure 25. Checking the end relief angle with a tool gage. (Lane Community College)

Figure 28. One method of grinding the nose radius is on the circumference of the wheel. (Lane Community College)

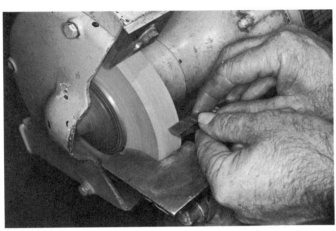

Figure 26. Roughing the side rake. (Lane Community College)

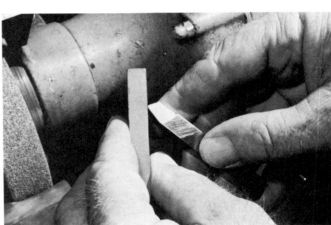

Figure 29. Using an oil stone to remove the burrs from the cutting edge. (Lane Community College)

Figure 27. Checking for wedge angle with a tool gage. (Lane Community College)

Figure 30. The finished tool. (Lane Community College)

SELF-TEST

1. Name the advantages of using high speed steel for tools.
2. Other than hardness and toughness, what is the most important aspect of a lathe tool?
3. How do form tools work?
4. A tool that has been reground too many times on the same place can form a "chip trap." Describe the problems that result from this condition.
5. Why is it not always necessary to grind a back rake into the tool?

6. When should a zero or negative rake be used?
7. Explain the purpose of the side and end relief angles.
8. What is the function of the side and back rakes?
9. How can these angles be checked while grinding?
10. Why should chips be broken up?
11. In what ways can chips be broken?
12. Overheating a high speed tool bit can easily be done by using a glazed wheel that needs dressing or by exerting too much pressure. What does this cause in the tool?

UNIT 5 LATHE SPINDLE TOOLING

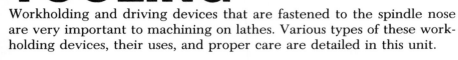

Workholding and driving devices that are fastened to the spindle nose are very important to machining on lathes. Various types of these workholding devices, their uses, and proper care are detailed in this unit.

OBJECTIVES

After completing this unit, you should be able to:
1. Explain the uses and care of independent and universal chucks.
2. Explain the limitations and advantages of collets and describe a collet setup.
3. Explain the use of a face driver or drive center.
4. Explain the uses and differences of drive plates and face plates.

The lathe spindle nose is the carrier of a variety of workholding devices that are fastened to it in several ways. The spindle is hollow and has an internal Morse taper at the nose end, which makes possible the use of taper shank drills or drill chucks (Figure 1). This internal taper is also used to hold live centers, drive centers, or collet assemblies. The outside of the spindle nose can have either a threaded nose (Figure 2), a long taper with key drive (Figure 3), or a camlock (Figure 4).

Threaded spindle noses are mostly used on older lathes. The chuck or face plate is screwed on a coarse, right-hand thread until it is forced against a shoulder on the spindle that aligns it. Two disadvantages of the threaded spindle nose are that the spindle cannot be rotated in reverse against a load and that it is sometimes very difficult to remove a chuck or face plate (Figure 5).

The **long taper key drive spindle nose** relies on the principle that a tapered fit will always repeat its original position. The key gives additional driving power. A large nut having a right-hand thread is turned with a spanner wrench. It draws the chuck into position and holds it there.

Camlock spindle noses use a short taper for alignment. A number of studs arranged in a circle fit into holes in the spindle nose. Each stud has a notch into which a cam is turned to lock it in place.

Figure 1. Section view of the spindle. *(Machine Tools and Machining Practices)*

Figure 2. Threaded spindle nose. (Lane Community College)

Figure 3. Long taper with key drive spindle nose. (Lane Community College)

Figure 4. Camlock spindle nose.

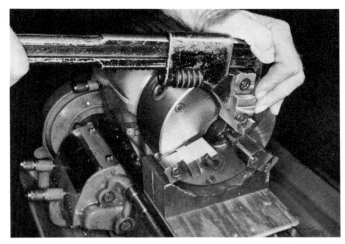

Figure 5. The threaded chuck can be removed by using a large monkey wrench on one of the chuck jaws while the spindle is locked in a low gear. A long steel bar may also be used between the jaws. The knockout bar should never be used to remove a chuck because it is too light and will bend. (Lane Community College)

Figure 6. A spring cleaner is used for cleaning internal threads on chucks. (*Machine Tools and Machining Practices*)

All spindle noses and their mating parts must be carefully cleaned before assembly. Small chips or grit will cause a workholding device to run out of true and be damaged. A spring cleaner (Figure 6) is used on mating threads for threaded spindles. Brushes and cloths are used for cleaning. A thin film of light oil should be applied to threads and mating surfaces.

Independent four-jaw and **universal** three-jaw

Figure 7. Four-jaw independent chuck holding an offset rectangular part.

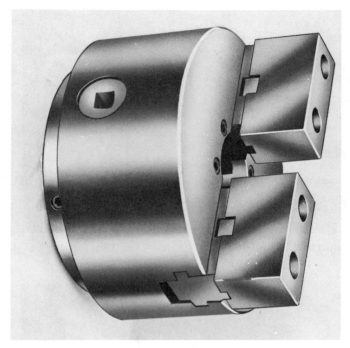

Figure 9. Two-jaw universal chuck. (Hardinge Brothers, Inc.)

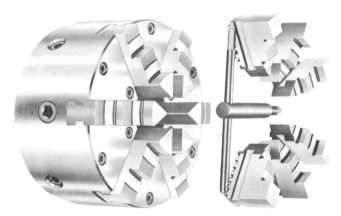

Figure 10. Six-jaw universal chuck. (Buck Tool Company)

Figure 8. Four-jaw chuck in reverse position holding a large diameter workpiece. (Lane Community College)

chucks and, occasionally, drive or face plates are mounted on the spindle nose of engine lathes. Each of the four jaws of the independent chuck moves independently of the others, which makes it possible to set up oddly shaped pieces (Figure 7). The concentric rings on the chuck face help to set the work true before starting the machine. Very precise setups also can be made with the four-jaw chuck by using a dial indicator, especially on round material. Each jaw of the chuck can be removed and reversed to accommodate irregular shapes. Some types are fitted with top jaws that can be reversed after removal of bolts on the jaw. Jaws in the reverse position can grip larger diameter workpieces (Figure 8). The independent chuck will hold work more securely for heavy cutting than will the three-jaw universal chuck.

Figure 11. Exploded view of a universal three-jaw chuck (Adjust-tru) showing the scroll plate and gear drive mechanism. (Buck Tool Company)

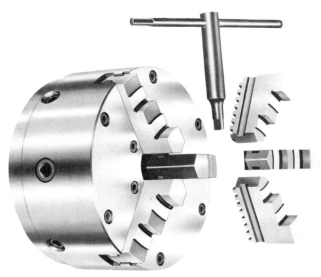

Figure 12. Universal three-jaw chuck (Adjust-tru) with a set of outside jaws. (Buck Tool Company)

Figure 13. Universal chuck with top jaws. (Buck Tool Company)

Universal chucks usually have three jaws, but some are made with two jaws (Figure 9) or six jaws (Figure 10). All the jaws are moved in or out equally in their slides by means of a scroll plate located back of the jaws. The scroll plate has a bevel gear on its reverse side that is driven by a pinion gear. This gear extends to the outside of the chuck body and is turned with the chuck wrench (Figure 11). Universal chucks provide quick and simple chucking and centering of round stock. Uneven or irregularly shaped material will damage these chucks.

The jaws of standard universal chucks will not reverse as with independent chucks, so a separate set of reverse jaws is used (Figure 12) to hold pieces with larger diameters. The chuck and each of its jaws are stamped with identification numbers. Do not interchange any of these parts with another chuck or both will be inaccurate. Also each jaw is stamped 1, 2, or 3 to correspond to the same number stamped by the slot on the chuck. The jaws are removed from the chuck in the order 3, 2, 1 and should be returned in the reverse order, 1, 2, 3.

A universal chuck with **top jaws** (Figure 13), in contrast to the standard type with two sets of jaws, is reversed by removing the bolts in the top jaws and by reversing them. They must be carefully cleaned when this is done. Soft top jaws are frequently used when special gripping problems arise. Since the jaws are machined to fit the shape of the part (Figure 14), they can grip it securely for heavy cuts (Figure 15).

One disadvantage of most universal chucks is that they lose their accuracy when the scroll and jaws wear, and normally there is no compensation for wear other than regrinding the jaws. The three-jaw adjust-

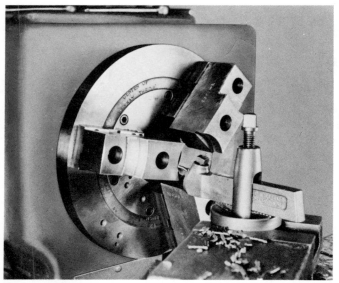

Figure 14. Machining soft jaws to fit an odd-shaped work-piece on a jaw turning fixture. (The Warner & Swasey Company)

Figure 15. Soft jaws have been machined to fit the shape of this cast steel workpiece in order to hold it securely for heavy cuts. (The Warner & Swasey Company)

Figure 16. Universal chuck (Adjust-tru) with special adjustment feature (G) makes it possible to compensate for wear. (Buck Tool Company)

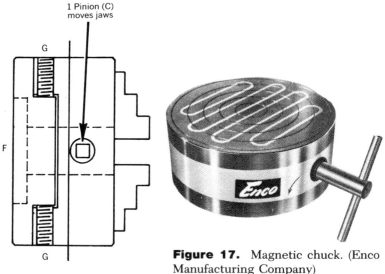

Figure 17. Magnetic chuck. (Enco Manufacturing Company)

able chuck in Figure 16 has a compensating adjustment for wear or misalignment.

Combination universal and independent chucks also provide for quick opening and closing and have the added advantage of independent adjustment on each jaw. These chucks are like the universal type since three or four jaws move in or out equally, but each jaw can be adjusted independently as well.

Magnetic chucks (Figure 17) are sometimes used for making light cuts on ferromagnetic material. They are useful for facing thin material that would be difficult to hold in conventional workholding devices. Magnetic chucks do not hold work very securely and so are not used much in lathe work.

Figure 18. Drive plate for turning between centers. (Lane Community College)

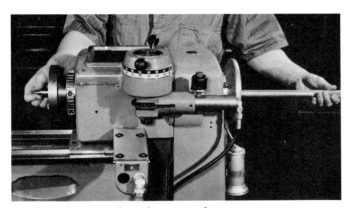

Figure 19. Knockout bar is used to remove centers.

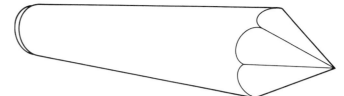

Figure 20. Hardened and serrated drive center is used for very light turning between centers. *(Machine Tools and Machining Practices)*

Figure 21. Face driver is mounted in headstock spindle and work is driven by the drive pins that surround the center. (Copyright © 1976 Sandvik Madison, Inc. All rights reserved. Madison-Kosta is a registered trademark of Sandvik Madison, Inc. All dimensions and specifications subject to change without notice.)

All chucks need frequent cleaning of scrolls and jaws. These should be lightly oiled after cleaning and chucks with grease fittings should be pressure lubricated. Chucks come in all diameters and are made for light, medium, and heavy duty uses.

Drive plates are used together with lathe dogs to drive work mounted between centers (Figure 18). The live center fits directly into the spindle taper and turns with the spindle. A sleeve is sometimes used if the spindle taper is too large in diameter to fit the center. The live center is usually made of soft steel so the point can be machined as needed to keep it running true. Live centers are removed by means of a knockout bar (Figure 19). Soft live centers should not be confused with hardened tailstock centers. A soft center in the tailstock will immediately fuse to

the center hole in the work and damage it.

Often when a machinist wants to machine the entire length of work mounted between centers without the interference of a lathe dog, specially ground and hardened drive centers are used (Figure 20). These are serrated so they will turn the work, but only light cuts can be made. Modern drive centers or face drivers (Figure 21) can also be used to machine a part without interference from a lathe dog. Quite heavy cuts are possible with these drivers, which are used especially for manufacturing purposes.

Face plates are used for mounting workpieces or fixtures. Unlike drive plates that have only slots, face plates have T slots and are more heavily built (Figure 22). Face plates are made of cast iron and so must be operated at relatively slow speeds. If the speed is too high, the face plate could fly apart.

Collet chucks (Figure 23) are very accurate workholding devices and are used in producing small high precision parts. Steel spring collets are available for holding and turning hexagonal, square, and round workpieces. They are made in specific sizes (which are stamped on them) with a range of only a few thousandths of an inch. Rough and inaccurate work-

Figure 22. T-slot face plate. Workpieces are clamped on the plate with T-bolts and strap clamps. (The Monarch Machine Tool Company, Sidney, Ohio)

Figure 23. Side and end views of a spring collet for round work.

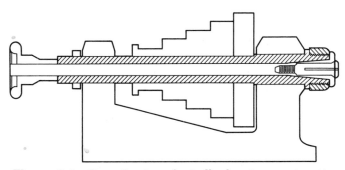

Figure 24. Cross Section of spindle showing construction of draw-in collet chuck attachment.

pieces should not be held in the collet chuck since the gripping surfaces of the chuck would form an angle with the workpiece. The contact area would then be at one point on the jaws instead of along the entire length, and the piece would not be held firmly. If it is not held firmly, workpiece accuracy is impaired and the collet may be damaged. An adapter called a collet sleeve is fitted into the spindle taper and a draw bar is inserted into the spindle at the opposite end (Figure 24). The collet is placed in the adapter, and the draw bar is rotated, which threads the collet into the taper and closes it. **Never tighten a collet without a workpiece in its jaws** because this will damage it. Before collets and adapters are installed, they should be cleaned to insure accuracy.

The rubber flex collet (Figure 25) has a set of tapered steel bars mounted in rubber. It has a much wider range than the spring collet, each collet with a range of about $\frac{1}{8}$ in. A large handwheel is used to open and close the collets instead of a draw bar (Figure 26).

The concentricity that you could expect from each type of workholding device is as follows:

Device	Centering accuracy in inches (indicator reading difference)
Centers	Within .001
Four-jaw chuck	Within .001 (depending on the ability of machinist)
Collets	.0005 to .001
Three-jaw chuck	.001 to .003 (good condition)
	.005 or more (poor condition)

Figure 25. Rubber flex collet.

Figure 26. Collet handwheel attachment for rubber flex collets. (The Monarch Machine Tool Company, Sidney, Ohio)

SELF-TEST

1. Briefly describe the lathe spindle. How does the spindle support chucks and collets?
2. Name the spindle nose types.
3. What is an independent chuck and what is it used for?
4. What is a universal chuck and what is it used for?
5. What chuck types make possible the frequent adjusting of chucks so they will hold stock with minimum runout?
6. Workpieces mounted between centers are driven with lathe dogs. Which type of plate is used on the spindle nose to turn the lathe dog?
7. What is a live center made of? How does it fit in the spindle nose?
8. Describe a drive center and a face driver.
9. On which type of plate are workpieces and fixtures mounted? How is it identified?
10. Name one advantage of using steel spring collets. Name one disadvantage.

UNIT 6 OPERATING THE MACHINE CONTROLS

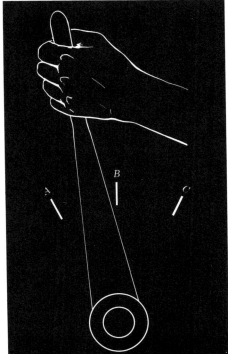

Before using any machine, you must be able to properly use the controls, know what they are for, and how they work. You must also be aware of the potential hazards that exist for you and the machine, if it is mishandled. This unit prepares you to operate lathes.

OBJECTIVES

After completing this unit, you should be able to:

1. Explain drives and shifting procedures for changing speeds on lathes.
2. Describe the use of various feed control levers.
3. Explain the relationship between longitudinal feeds and cross feeds.
4. State the differences in types of cross feed screw micrometer collars.

Most lathes have similar control mechanisms and operating handles for feeds and threading. Some machines, however, have entirely different driving mechanisms as well as different speed controls.

DRIVES

Spindle speed is controlled on some lathes by a belt on a step cone pulley in the headstock (Figure 1a). The speed is changed by turning the belt tension lever

Figure 1a. Speeds are changed on this lathe by moving the belt to various steps on the pulley. (Lane Community College)

Figure 1b. View of headstock showing flat belt drive, the back gear engaged, and the lock pin disengaged. (Lane Community College)

to loosen the belt, moving the belt to the proper step for the desired speed, and then moving the lever to its former position. Several more lower speeds are available by shifting to backgear. To do this, pull out or release the bull gear lockpin to disengage the spindle from the step cone pulley and engage the backgear lever as shown in Figure 1b. It may be necessary to rotate the spindle by hand slightly to bring the gear into mesh. The backgears must **never** be engaged when the spindle is turning with power.

Another drive system uses a variable speed drive (Figure 2) with a high and low range using a backgear. On this drive system, the motor must be running to change the speed on the vari-drive, but the motor must be turned off to shift the backgear lever. Geared head lathes are shifted with levers on the outside of the headstock (Figure 3). Several of these levers are

Figure 2. Variable speed control and speed selector. (Clausing Machine Tools)

Figure 3. Speed change levers and feed selection levers on a geared head lathe. (Lane Community College)

used to set up the various speeds within the range of the machine. The gears will not mesh unless they are perfectly aligned, so it is sometimes necessary to rotate the spindle by hand. **Never try to shift gears with the motor running and the clutch lever engaged.**

FEED CONTROL LEVERS

The carriage is moved along the ways by means of the lead screw when threading, or by a separate feed rod when using feeds. On most small lathes, however, a lead screw-feed rod combination is used. In order to make left-hand threads and reverse the feed, the feed reverse lever is used. This lever reverses the lead screw. It should never be moved when the machine is running.

The quick-change gearbox (Figures 4a and 4b) typically has two or more sliding gear shifter levers. These are used to select feeds or threads per inch. On those lathes also equipped with metric selections, the threads are expressed in pitches (measured in millimeters).

The carriage apron (Figure 5) contains the handwheel for hand feeding and a power feed lever that engages a clutch to a gear drive train in the apron.

Hand feeding should not be used for long cuts as there would be lack of uniformity and a poor finish would result. When using power feed and approaching a shoulder or the chuck jaws, disengage the power feed and hand feed the carriage for the last $\frac{1}{8}$ in. or so. The handwheel is used to quickly bring the tool close to the work before engaging the feed and for rapidly returning to the start of a cut after disengaging the feed. A feed change lever diverts the feed to either the carriage for longitudinal movement or to the cross feed screw to move the cross slide. There is generally some slack or backlash in the cross feed and compound screws. As long as the tool is being fed in one direction against the work load, there is no problem, but if the screw is *slightly* backed off, the readings will be in error. To correct this problem, back off two turns and come back to the desired position.

Cross feeds are geared differently than longitudinal feeds. On most lathes the cross feed is approximately one-third to one-half that of the longitudinal feed, so a facing job (Figure 6) with the quick-change gearbox set at about .012 in. feed would actually only be .004 in. for facing. The cross feed ratio for each lathe is usually found on the quick-change gearbox index plate.

Figure 4a. Quick-change gearbox with index plate. (Lane Community College)

Figure 4b. Exposed quick-change gear mechanism for a large, heavy duty lathe. (Lodge & Shipley Company)

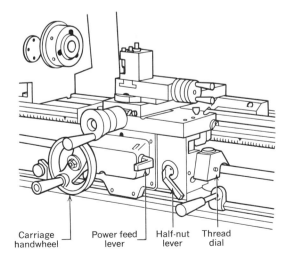

Carriage handwheel   Power feed lever   Half-nut lever   Thread dial

Figure 5. View of carriage apron with names of parts. (Clausing Machine Tools)

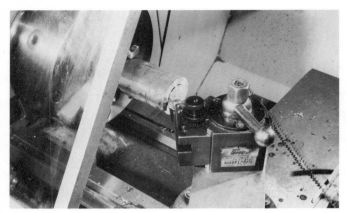

Figure 6. Facing on a lathe. (Lane Community College)

Figure 7. Micrometer collar on the crossfeed screw that is graduated in English units. Each division represents .001 in. (Lane Community College)

Figure 8. Crossfeed and compound screw handles with metric-English conversion collars. (The Monarch Machine Tool Company, Sidney, Ohio)

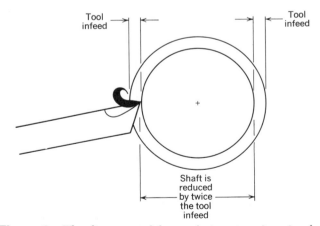

Figure 9. The diameter of the workpiece is reduced twice the amount in which the tool is moved. *(Machine Tools and Machining Practices)*

The **half-nut** or split-nut lever on the carriage engages the thread on the lead screw directly and is used **only** for threading. It cannot be engaged unless the feed change lever is in the neutral position.

Both the cross feed screw handle and the compound rest feed screw handle are fitted with **micrometer collars** (Figure 7). These collars traditionally have been graduated in English units, but metric conversion collars (Figure 8) will help in the transition to the metric system.

Some micrometer collars are graduated to read single depth; that is, the tool moves as much as the reading shows. When turning a cylindrical object such as a shaft, dials that read single depth will remove twice as much from the diameter (Figure 9). For ex-

Figure 10. Clutch rod is actuated by moving the clutch level. This disengages the motor from the spindle. (Lane Community College)

ample, if the cross feed screw is turned in .020 in. and a cut is taken, the diameter will have been reduced by .040 in. Sometimes only the compound dial is calibrated in this way and the crossslide will have graduations on its micrometer dial to compensate for double depth on turning. On this type of lathe, if the cross feed screw is turned in until .020 in. shows on the dial and a cut is taken, the diameter will have been reduced .020 in. The tool would have actually moved into the work only .010 in. This is sometimes called radius and diameter reduction.

To determine which type of graduation you are using, set a fractional amount on the dial (such as .250 in. = $\frac{1}{4}$ in.) and measure on the cross slide with a rule. The actual slide movement you measure with the rule will be either the same as the amount set on the dial, for the single depth collar, or one-half that amount, for the double depth collar.

Some lathes have a brake and clutch rod that is the same length as the lead screw. A clutch lever connected to the carriage apron rides along the clutch rod (Figure 10). The spindle can be started and stopped without turning off the motor by using the clutch lever. Some types also have a spindle brake that quickly stops the spindle when the clutch lever is moved to the stop position. An adjustable automatic clutch kickoff is also a feature of the clutch rod.

When starting a lathe for the first time use the following checkout list:

1. Move carriage and tailstock to the right to clear workholding device.
2. Locate feed clutches and half-nut lever and disengage before starting spindle.
3. Set up to operate at low speeds.
4. Read any machine information panels that may be located on the machine and observe precautions.
5. Note the feed direction; there are no built-in travel limits or warning devices to prevent feeding the carriage into the chuck or against the end of the slides.
6. When you are finished with a lathe, disengage all clutches, clean up chips, and remove any attachments or special setups.

SELF-TEST

1. How can the shift to the low speed range be made on the belt drive lathes that have the step cone pulley?
2. Explain speed shifting procedure on the variable speed drive.
3. In what way can speed changes be made on gearhead lathes?
4. What lever is shifted in order to reverse the lead screw?
5. The sliding gear shifter levers on the quick-change gearbox are used for just two purposes. What are they?
6. When is the proper time to use the carriage handwheel?
7. Why will you not get the same surface finish (tool marks per inch) on the face of a workpiece as you would get on the outside diameter when on the same power feed?
8. The half-nut lever is not used to move the carriage for turning. Name its only use.
9. Micrometer collars are attached on the cross feed handle and compound handle. In what ways are they graduated?
10. How can you know if the lathe you are using is calibrated for single or double depth?

UNIT 7 FACING AND CENTER DRILLING

Facing and center drilling the workpiece are often the first steps taken in a turning project to produce a stepped shaft or a sleeve from solid material. Much lathe work is done in a chuck and requires considerable facing and some center drilling. These important lathe practices will be covered in detail in this unit.

OBJECTIVES

After completing this unit, you should be able to:
1. Correctly set up a workpiece and face the ends.
2. Correctly center drill the ends of a workpiece.
3. Determine the proper feeds and speeds for a workpiece.
4. Explain how to set up to make facing cuts to a given depth and how to measure them.

SETTING UP FOR FACING

Facing is done to obtain a flat surface on the end of cylindrical workpieces or on the face of parts clamped in a chuck or face plate (Figures 1a and 1b). The work most often is held in a three- or four-jaw chuck. If the chuck is to be removed from the lathe spindle, a lathe board must first be placed on the ways. Figure 2 shows a camlock mounted chuck being removed. The correct procedure for installing a chuck on a camlock spindle nose is shown in Figures 3a to 3f. The cams should be tightly snugged (Figure 3f) for one or two revolutions around the spindle.

Setting up work in an independent chuck is simple, but mastering the procedures takes some practice. Round stock can be set up by using a dial indicator (Figure 4). Square or rectangular stock can either be set up with a dial indicator or by using a toolholder turned backwards (Figures 5a and 5b).

Begin the set up by aligning two opposite jaws with the same concentric ring marked in the face of the chuck while the jaws are near the workpiece. This will roughly center the work. Set up the other two jaws with a concentric ring also when they are near the work. Next, bring all of the jaws firmly against

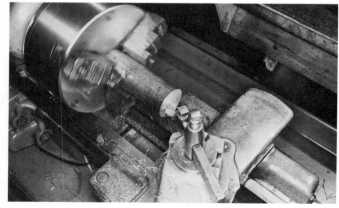

Figure 1a. Facing a workpiece in a chuck. (Lane Community College)

Figure 1b. Facing the end of a shaft. (Lane Community College)

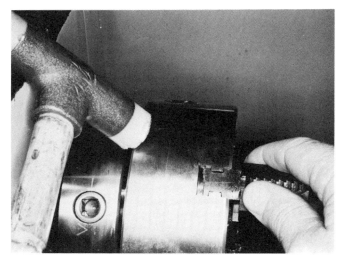

Figure 2. Removing a camlock chuck that is mounted on a lathe spindle.

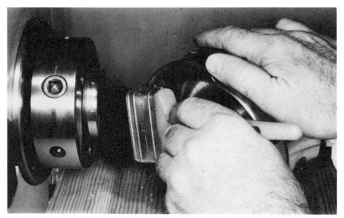

Figure 3a. Chips are cleaned from spindle nose with a brush.

Figure 3b. Cleaning the chips from the chuck with a brush.

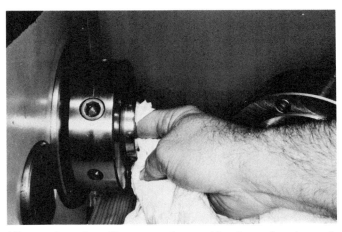

Figure 3c. Spindle nose is thoroughly cleaned with a soft cloth.

Figure 3d. Chuck is thoroughly cleaned with a soft cloth.

Figure 3e. Chuck is mounted on spindle nose.

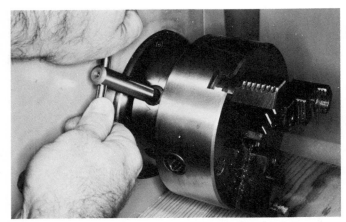

Figure 3f. All cams are turned clockwise until locked securely.

Figure 4. Setting up round stock in an independent chuck with a dial indicator. (Lane Community College)

Figure 5a. Rectangular stock being set up by using a toolholder turned backwards. The micrometer dial is used to center the workpiece. (Lane Community College)

Figure 5b. Adjusting the rectangular stock at 90 degrees from Figure 5a. (Lane Community College)

the work. When using the dial indicator, zero the bezel at the lowest reading. Now rotate the chuck to the opposite jaw with the high reading and tighten it half the amount of the runout. It might be necessary to loosen slightly the jaw on the low side. Always tighten the jaws at the position where the dial indicator contacts the work since any other location will give erroneous readings. When using the back of the toolholder, the micrometer dial on the cross slide will show the difference in runout. Chalk is sometimes used for setting up rough castings and other work too irregular to be measured with a dial indicator. Workpieces can either be chucked normally, internally, or externally (Figures 6a to 6c).

Figure 6a. Normal chucking position. (Lane Community College)

Figure 6c. External chucking position. (Lane Community College)

Figure 6b. Internal chucking position. (Lane Community College)

FACING

The material to be machined usually has been cut off in a power saw and so the piece is not square on the end or cut to the specified length. Facing from the center out (Figure 7) produces a better finish, but it is difficult to cut on a solid face in the center. Facing from the outside (Figure 8) is more convenient since heavier cuts may be taken and it is easier to work to the scribed lines on the circumference of the work. When facing from the center out, a right-hand turning tool in a left-hand toolholder is the best arrangement, but when facing from the outside to the center, a left-hand tool in a right-hand or straight toolholder can be used. Facing or other tool machining should not be done on workpieces extending more than five diameters from the chuck jaws.

The point of the tool should be set to the center of the work (Figure 9). This is done by setting the tool to the tailstock center point or by making a trial cut to the center of the work. If the tool is off center, a small uncut stub will be left. The tool can then be reset to the center of the stub.

The carriage must be locked when taking facing cuts as the cutting pressure can cause the tool and carriage to move away (Figure 10), which would make

Figure 7. Facing from the center to the outside of a workpiece. (Lane Community College)

Figure 8. Facing from the outside toward the center of the workpiece. (Clausing Machine Tools)

Figure 10. Carriage must be locked before taking a facing cut. (Lane Community College)

Figure 11. Facing to length using a hook rule for measuring. (Lane Community College)

Figure 12. The compound set at 90 degrees for facing operations. (Lane Community College)

Figure 9. Setting the tool to the center of the work using the tailstock center. (Lane Community College)

Figure 13. Setting the compound at 30 degrees. (Lane Community College)

slide is moved. For example, if you wanted to remove .015 in. from the workpiece, you would turn in .030 in. on the micrometer dial (assuming it reads single depth). It should be noted that the angle graduations on lathes are not standardized. On some lathes, Figure 12 would show the compound set at 0 degrees, and Figure 13 could show the compound at 60 degrees.

A specially ground tool is used to face the end of a workpiece that is mounted between centers. The right-hand facing tool is shaped to fit in the angle between the center and the face of the workpiece. Half centers (Figure 14) are made to make the job easier, but they should be used only for facing and not for general turning. If the tailstock is moved off center away from the operator, the shaft end will be convex (Figure 15a) and, if it is moved toward the operator, it will be concave (Figure 15b). Both right-hand and left-hand facing tools are used for facing work held on mandrels (Figure 16).

the faced surface curved rather than flat. Finer feeds should be used for finishing than for roughing. Remember, the cross feed is one-half to one-third that of the longitudinal feed. The ratio is usually listed on the index plate of the quick-change gearbox. A roughing feed could be from .005 to .015 in. and a finishing feed from .003 to .005 in. Use of cutting oils will help produce better finishes on finish facing cuts.

Facing to length may be accomplished by trying a cut and measuring with a hook rule (Figure 11) or by facing to a previously made layout line. A more precise method is to use the graduations on the micrometer collar of the compound. The compound is set so its slide is parallel to the ways (Figure 12). The carriage is locked in place and a trial cut is taken with the micrometer collar set on zero index. The workpiece is measured with a micrometer and the desired length is subtracted from the measurement; the remainder is the amount you should remove by facing. If more than .015 to .030 in. (depth left for finish cut) has to be removed, it should be taken off in two or more cuts by moving the compound micrometer dial the desired amount. A short trial cut (about $\frac{1}{8}$ in.) should again be taken on the finish cut and adjustment made if necessary. Roughing cuts should be approximately .060 in. in depth.

Quite often the compound is kept at 30 degrees for threading purposes (Figure 13). It is convenient to know that at this angle, the tool feeds into the face of the work .001 in. for every .002 in. that the

Figure 14. Half centers make facing shaft ends easier.

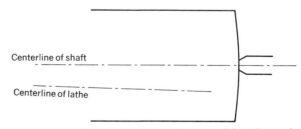

Figure 15a. Convex shaft ends caused by the tailstock being moved off center away from the operator.

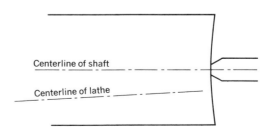

Figure 15b. Concave shaft ends caused by the tailstock being moved toward the operator.

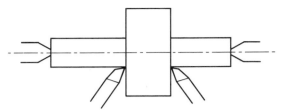

Figure 16. Work that is held between centers on a mandrel can be faced on both sides with right-hand or left-hand facing tools. *(Machine Tools and Machining Practices)*

SPEEDS

Speeds (RPM) for lathe turning a workpiece are determined in essentially the same way as speeds for drilling tools. The only difference is that the diameter of the work is used instead of the diameter of the drill. For facing work, the outside diameter is always used to determine RPM. Thus:

$$RPM = \frac{CS \times 4}{D}, \text{ where}$$

D = diameter of workpiece (where machining is done)

RPM = revolutions per minute

CS = cutting speed (surface feet per minute)

Cutting speeds for various metals are given in Table 1 of Unit 8, "Turning Between Centers."

EXAMPLE 1

The cutting speed for low carbon steel is 90 SFM (surface feet per minute) and the workpiece diameter to be faced is 6 in. Find the correct RPM.

$$RPM = \frac{90 \times 4}{6} = 60$$

EXAMPLE 2

A center drill has a $\frac{1}{8}$ in. drill point. Find the correct RPM to use on low carbon steel (CS 90).

$$RPM = \frac{90 \times 4}{\frac{1}{8}} = \frac{360}{1} \times \frac{8}{1} = 2880$$

These are only approximate speeds and will vary according to the conditions. If chatter marks (vibration marks) appear on the workpiece, the RPM should be reduced. If this does not help, ask your instructor for assistance. For more information on speeds and feeds, see Unit 8, "Turning Between Centers."

CENTER DRILLS AND DRILLING

When work is held and turned between centers, a center hole is required on each end of the workpiece. The center hole must have a 60 degree angle to conform to the center and have a smaller drilled hole to clear the center's point. This center hole is made with a combination drill and countersink, sometimes referred to as a center drill. These drills are available in a range of sizes from $\frac{1}{8}$ to $\frac{3}{4}$ in. body diameter and are classified by numbers from 00 to 8, which are normally stamped on the drill body. For example, a number 3 center drill has a $\frac{1}{4}$ in. body diameter and a $\frac{7}{64}$ in. drill diameter. Full listings can be found in the *Machinery's Handbook.*

Center drills are usually held in a drill chuck in the tailstock, while the workpieces are most often supported and turned in a lathe chuck for center drilling (Figure 17). A workpiece could also be laid out and supported in a vertical position for center drilling in a drill press. This method, however, is not used very often.

Round stock could be clamped in a vee-block and drill press vise (Figure 18) or on an angle plate. Crossed layout lines are scribed on round stock by means of a centerhead (Figure 19), and a punch mark is made where they intersect. Layout for square or rectangular stock to be placed between centers is done simply by scribing two diagonal lines from corner to corner. Figure 20 shows this being done with a height gage.

As a rule, center holes are drilled by rotating the work in a lathe chuck and feeding the center drill into the work by means of the tailstock spindle. Long workpieces, however, are generally faced by chucking

Figure 17. Center drilling a workpiece held in a chuck. (Lane Community College)

Figure 18. Center being drilled in round stock that is held in drill press vise. (Lane Community College)

Figure 19. Layout lines for center drilling are scribed on round stock by using a centerhead. (Lane Community College)

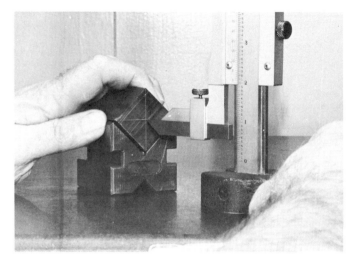

Figure 20. Layout lines for center drilling on square stock are made from corner to corner. (Lane Community College)

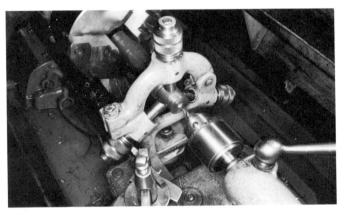

Figure 21. Center drilling long material that is supported in a steady rest. (Lane Community College)

one end and supporting the other in a steady rest (Figure 21). Since the end of stock is never sawed square, it should be center drilled only after spotting a small hole with the lathe tool. A slow feed is needed to protect the small, delicate drill end. Cutting oil should be used and the drill should be backed out frequently to remove chips. The greater the work diameter and the heavier the cut, the larger the center hole should be.

The size of the center hole can be selected by the center drill size and then regulated to some extent by the depth of drilling. You must be careful not to drill too deeply (Figure 22) as this causes the center to contact only the sharp outer edge of the hole, which is a poor bearing surface. It soon becomes loose and

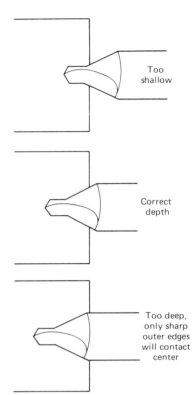

Figure 22. Correct and incorrect depth for center drilling. Remember, the speed of the lathe usually has to be made faster when center drilling to avoid breakage.

Figure 23. Center drill is brought up to work and lightly fed into material.

Figure 24. Center drill is fed into work with a slow even feed.

out of round and causes such machining problems as chatter and roughness. Center drills are often broken from feeding the drill too fast with the lathe speed too slow or with the tailstock off center.

Center drills are often used as starting or spotting drills when a drilling sequence is to be performed (Figures 23 and 24). This keeps the drill from "wandering" off the center and making the hole run eccentric. Spot drilling is done when work is chucked or is supported in a steady rest. Care must be taken that the workpiece is centered properly in the steady rest or the center drill will be broken.

SELF-TEST

1. You have a rectangular workpiece that needs a facing operation plus center drilling, and a universal chuck is mounted on the lathe spindle. What is your procedure to prepare for machining?
2. Should the point of the tool be set above, below, or at the center of the spindle axis when taking a facing cut?
3. If you set the quick-change gearbox to .012 in., would that be considered a roughing feed for facing?

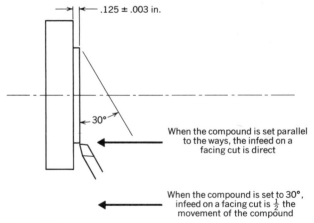

Figure 25

4. An alignment step must be machined on a cover plate .125 in., plus or minus .003 in., in depth (Figure 25). What procedure should be taken to face to this depth? How can you check your final finish cut?

5. What tool is used for facing shaft ends when they are mounted between centers? In what way is this tool different from a turning tool?
6. If the cutting speed of aluminum is 300 SFM and the workpiece diameter is 4 in., what is the RPM? The formula is

$$\text{RPM} = \frac{\text{CS} \times 4}{D}$$

7. Name two reasons for center drilling a workpiece in a lathe.
8. How is laying out and drilling center holes in a drill press accomplished?
9. Name two causes for center drill breakage.
10. What happens when you drill too deeply with a center drill?

UNIT 8 TURNING BETWEEN CENTERS

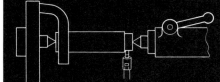

Since for a large percentage of lathe work the workpiece is held between centers or between a chuck and a center, turning between centers is a good way for you to learn the basic principles of lathe operation. The economics of machining time and quality will be detailed in this unit, as heavy roughing cuts are compared to light cuts, and speeds and feeds for turning operations are presented.

OBJECTIVES

After completing this unit, you should be able to:
1. Describe the correct setup procedure for turning between centers.
2. Select correct feeds and speeds for a turning operation.
3. Detail the steps necessary for turning to size predictably.

SETUP FOR TURNING BETWEEN CENTERS

To turn a workpiece between centers, it is supported between the dead center (tailstock center) and the live center in the spindle nose. A lathe dog (Figure 1) clamped to the workpiece is driven by a drive or dog plate (Figure 2) mounted on the spindle nose. Machining with a single point tool can be done anywhere on the workpiece except near or at the location of the lathe dog.

Turning between centers has some disadvantages. A workpiece cannot be cut off with a parting tool while being supported between centers as this will bind and break the parting tool and ruin the workpiece. For drilling, boring, or machining the end of a long shaft, a steady rest is normally used to support the work. But these operations cannot very well be done when the shaft is supported only by centers.

The advantages of turning between centers are many. A shaft between centers can be turned end for end to continue machining without eccentricity if the centers are in line (Figure 3). This is why shafts that are to be subsequently finish-ground between centers must be machined between centers on a lathe. If a partially threaded part is removed from between centers for checking, and everything is left the same on the lathe, the part can be returned to the lathe, and the threading resumed where it was left off.

A considerable amount of straight turning on shafts is done with the work held between a chuck and the tailstock center (Figure 4). The advantages of this method are quick setup and a positive drive. One disadvantage is that eccentricities in the shaft are caused by inaccuracies in the chuck jaws. Another is the tendency for the workpiece to slip endwise into

Figure 1. Lathe dog. *(Machine Tools and Machining Practices)*

Figure 2. Dog plate or drive plate on spindle nose of the lathe. (The Monarch Machine Tool Company, Sidney, Ohio)

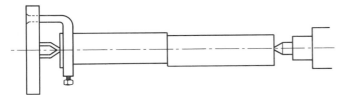

Figure 3. Eccentricity in the center of the part because of the live center being off center. *(Machine Tools and Machining Practices)*

the chuck jaws under a heavy cut, thus allowing the workpiece to loosen or to come out of the tailstock center.

As in other lathe operations, chip formation and handling are important to safety. Coarser feeds,

Figure 4. Work being machined between chuck and tailstock center. Note chip formation. Chip guard has been removed for clarity. (Lane Community College)

Figure 5a. A tangle of wiry chips. These chips can be hazardous to the operator. (Lane Community College)

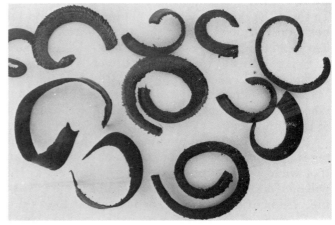

Figure 5b. A better formation of chips. This type of chip will fall into the chip pan and is more easily handled. (Lane Community College)

Figure 6. Inserting the tapered spindle nose sleeve.

Figure 7. Make sure the bushing is firmly seated in the taper. (Lane Community College)

Figure 8. Installing the center.

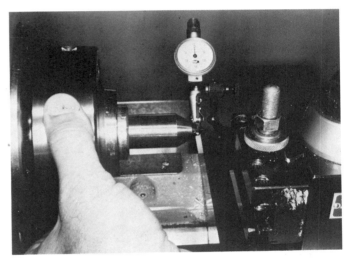

Figure 9. Checking the live center for runout with a dial indicator.

deeper cuts, and smaller rake angles all tend to increase chip curl, which breaks up the chip into small, safe pieces. Fine feeds and shallow cuts, on the other hand, produce a tangle of wiry, sharp hazardous chips (Figures 5a and 5b) even with a chip breaker on the tool. Long strings may come off the tool, suddenly wrap in the work and be drawn back rapidly to the machine. The edges are like saws and can cause very severe cuts.

The center for the headstock spindle is called a live center and it is usually not hardened since its point frequently needs machining to keep it true. Thoroughly clean the inside of the spindle with a soft cloth and wipe off the live center. If the live center is too small for the lathe spindle taper, use a tapered bushing that fits the lathe (Figure 6). Seat the bushing firmly in the taper (Figure 7) and install the center

(Figure 8). Set up a dial indicator on the end of the center (Figure 9) to check for runout. If there is runout, remove the center by using a knockout bar through the spindle. Be sure to catch the center with one hand. Check the outside of the center for nicks or burrs. These can be removed with a file. Check the inside of the spindle taper with your finger for nicks or grit. If nicks are found, *do not* use a file but check with your instructor. After removing nicks, if the center still runs out more than the acceptable tolerances (usually .0001 to .0005 in.) a light cut by tool or grinding can be taken with the compound set at 30 degrees off the center axis of the lathe.

A live center is often machined from a short piece of soft steel mounted in a chuck (Figure 10). It is then

Figure 10. Live center being machined in a four-jaw chuck. The lathe dog on the workpiece is driven by one of the chuck jaws. (Lane Community College)

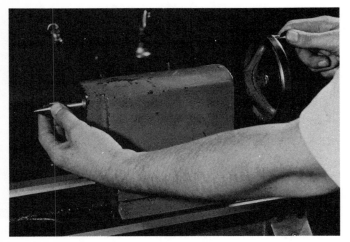

Figure 11. The dead center is hardened to resist wear. It is made of high speed steel or steel with a carbide insert.

Figure 12a. Pipe center used for turning. (The Monarch Machine Tool Company, Sidney, Ohio)

Figure 12b. Antifriction ball bearing center. (The Monarch Machine Tool Company, Sidney, Ohio)

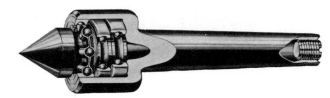

Figure 12c. Cutaway view of a ball bearing tailstock center. (The DoAll Company)

left in place and the workpiece is mounted between it and the tailstock center. A lathe dog with the bent tail against a chuck jaw is used to drive the workpiece. This procedure sometimes saves time on large lathes where changing from the chuck to a drive plate is cumbersome and the amount of work to be done between centers is small.

The tailstock center (Figure 11) is hardened to withstand machining pressures and friction. Clean inside the taper and on the center before installing. Ball bearing, antifriction centers are often used in the tailstock as they will withstand high speed turning without the overheating problems of dead centers.

Pipe centers are used for turning tubular material (Figures 12a to 12c).

To set up a workpiece that has been previously center drilled, slip a lathe dog on one end with the bent tail toward the drive plate. Do not tighten the dog yet. Put antifriction compound into the center hole toward the tailstock and then place the workpiece between centers (Figure 13). The tailstock spindle should not extend out too far as some rigidity in the machine would be lost and chatter or vibration may result. Set the dog in place and avoid any binding of the bent tail (Figure 14). Tighten the dog and then adjust the tailstock so there is no end play, but so

Figure 13. Antifriction compound put in the center hole before setting the workpiece between centers. This step is not necessary when using an antifriction ball bearing center. (Lane Community College)

Figure 14. Lathe dog in position. (Lane Community College)

the bent tail of the dog freely clicks in its slot. Tighten the tailstock binding lever. The heat of machining will expand the workpiece and cause the dead center to heat from friction. If overheated, the center may be ruined and may even be welded into your workpiece. Periodically, or at the end of each heavy cut, you should check the adjustment of the centers and reset if necessary.

When a tool post and toolholder are used, the toolholder must be positioned so it will not turn into the work when heavy cuts are taken (Figures 15a and 15b). The tool and toolholder should not overhang

Figure 15a. Incorrect position of toolholder for roughing. If the toolholder should turn when using heavy feeds, the tool will gouge more deeply into the work. (Lane Community College)

Figure 15b. Correct position of toolholder for roughing. The toolholder will swing away from the cut with excessive feeds. (Lane Community College)

Figure 16. Tools with excessive overhang. Both tool and toolholder extend too far from the tool post for roughing operations. (Lane Community College)

Figure 17. Tool and toolholder in the correct position. (Lane Community College)

Figure 18. Centering a tool by means of a steel rule. (Lane Community College)

Where:

$$RPM = \text{Revolutions per minute}$$
$$D = \text{Diameter of workpiece}$$
$$CS = \text{Cutting speed in surface feet}$$
$$\text{per minute (SFM)}$$

Cutting speeds for various materials are given in Table 1.

EXAMPLE

If the cutting speed is 40 for a certain alloy steel and the workpiece is 2 in. in diameter, find the RPM.

$$RPM = \frac{40 \times 4}{2} = 80$$

After calculating the RPM, use the nearest or next lower speed on the lathe and set the speed.

Feeds are expressed in inches per revolution (IPR) of the spindle. A .010 in. feed will move the carriage and tool .010 in. for one full turn of the headstock spindle. If the spindle speed is changed, the feed ratio still remains the same. Feeds are selected by means of an index chart (Figure 19) found either on the quick-change gearbox or on the side of the headstock housing (Figure 20). The sliding gear levers are shifted to different positions to obtain the feeds indicated on the index plate. The lower decimal numbers on the plate are feeds and the upper numbers are threads per inch.

Feeds and depth of cut should be as much as

too far (Figure 16) for rough turning, but should be kept toward the tool post as far as practicable (Figure 17). Tools should be set on or slightly above the center of the workpiece. The tool may be set to the dead center or to a steel rule on the workpiece (Figure 18).

SPEEDS AND FEEDS FOR TURNING

Since machining time is an important factor in lathe operations, it is necessary for you to fully understand the principles of speeds and feeds in order to make the most economical use of your machine. Speeds are determined for turning between centers by using the same formula as given for facing operations in the last unit.

$$RPM = \frac{CS \times 4}{D}$$

Table 1

Cutting Speeds and Feeds for High Speed Steel Tools

	Low Carbon Steel	High Carbon Steel Annealed	Alloy Steel Normalized	Aluminum Alloys	Cast Iron	Bronze
Roughing speed SFM	90	50	45	200	70	100
Finishing speed SFM	120	65	60	300	80	130
Feed IPR roughing	.010–.020	.010–.020	.010–.020	.015–.030	.010–.020	.010–.020
Feed IPR finishing	.003–.005	.003–.005	.003–.005	.005–.010	.003–.010	.003–.010

Source: Warren T. White, John E. Neely, Richard R. Kibbe, and Roland O. Meyer, *Machine Tools and Machining Practices*, John Wiley and Sons, Inc., Copyright © 1977, New York.

Figure 19. Index chart on the quick-change gearbox. (Lane Community College)

Figure 20. Index chart for feed mechanism on a modern geared head lathe with both metric and inch thread and feed selections. (Lane Community College)

the tool, workpiece, or machine can stand without undue stress. A small 10 or 12 in. swing lathe should handle $\frac{1}{8}$ in. depth of cut in soft steel, but in some cases this may have to be reduced to $\frac{1}{16}$ in. If .100 in. were selected as a trial depth of cut, then the feed could be anywhere from .010 to .020 in. If the machine seems to be overloaded, reduce the feed. Finishing feeds can be from .003 to .005 in. for steel. Feeds smaller than .003 in. often produce poor finishes due to low tool pressure from the small chip. Use a tool with a larger nose radius for finishing.

TURNING TO SIZE

The cut-and-try method of turning a workpiece to size or making a cut and measuring how close you came to the desired result, was used in the past when calipers and rule were used for measuring work di-

ameters. A more modern method of turning to size predictably uses the compound and cross feed micrometer collars and micrometer calipers for measurement. If the micrometer collar on the cross feed screw reads in single depth, it will remove twice the amount from the diameter of the work as the reading shows. A micrometer collar that reads directly or double depth will remove the same amount from the diameter that the reading shows, though the tool will actually move in only half that amount.

After taking one or several roughing cuts (depending on the diameter of the workpiece), .015 to .030 in. should be left for finishing. This can be taken in one cut if the tolerance is large, such as plus or minus .003 in. If the tolerance is small (plus or minus .0005 in.), two finish cuts should be taken, but enough stock must be left for the second cut to make a chip. If insufficient material is left for machining, .001 in.

Figure 21. A trial cut is made to establish a setting of a micrometer dial in relation to the diameter of the workpiece. (Lane Community College)

Figure 22. Measuring the workpiece with a micrometer. (Lane Community College)

for example, the tool will rub and will not cut. Between .005 and .010 in. should be left for the last finish cut.

The position of the tool is set in relation to the micrometer dial reading, and the first of the two finish cuts is made (Figure 21). **The tool is then returned to the start of the cut without moving the cross feed screw.** The diameter of the workpiece is checked with a micrometer (Figure 22) and the remaining amount to be cut is dialed on the cross feed micrometer dial. A short trial cut is taken (about $\frac{1}{8}$ in. long) and the lathe stopped. A final check with a micrometer is made to validate the tool setting, and then the cut is completed. If the lathe makes a slight taper, see the next unit on "Alignment of the Lathe Centers" to correct this problem.

Finishing of machined parts with a file and abrasive cloth should not be necessary if the tools are sharp and honed and if the feeds, speeds, and depth of cut are correct. A machine-finished part looks better than a part finished with a file and abrasive cloth. In the past, filing and polishing the precision surfaces of lathe

workpieces were necessary because of lack of rigidity and repeatability of machines. In the same way, worn lathes are not dependable for close tolerances and so an extra allowance must be made for filing. The amount of surface material left for filing ranges from .0005 to .005 in., depending on the diameter, finish, and finish diameter. If more than the tips of the tool marks are removed with a file and abrasive cloth, a wavy surface will result. For most purposes .002 in. is sufficient material to leave for finishing.

When filing on a lathe, use a low speed, long strokes, and file left handed (Figure 23). For polishing with abrasive cloth, set the lathe for a high speed and move the cloth back and forth across the work. Hold an end of the cloth strip in each hand (Figures 24a and 24b). Abrasive cloth leaves grit on the ways of the lathe, so a thorough cleaning of the ways should be done after polishing.

Machining shoulders to specific lengths can be done in several ways. Using a machinist's rule (Figure 25) to measure workpiece length to a shoulder is a very common but semiprecision method. Preset car-

Figure 23. Filing in the lathe, left-handed. (Lane Community College)

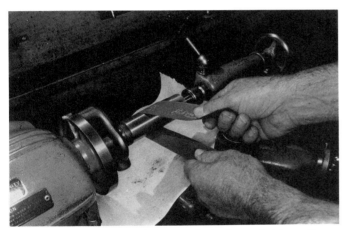

Figure 24a. Using abrasive cloth for polishing. (Lane Community College)

Figure 24b. Using a file for backing abrasive cloth for more uniform polishing. (Lane Community College)

Figure 25. Measuring the workpiece length to a shoulder with a machinist's rule. (Lane Community College)

Figure 26. Micrometer carriage stop set to limit tool travel in order to establish a shoulder. (Lane Community College)

riage stops (Figure 26) can be used to limit carriage movement and establish a shoulder. This method can be very accurate if it is set up correctly. Another very accurate means of machining shoulders is by using special dial indicators with long travel plunger rods.

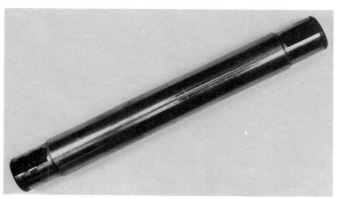

Figure 27a. Tapered mandrel or arbor. *(Machine Tools and Machining Practices)*

Figure 28. Operating principle of a special type of an expanding mandrel. The "Saber-tooth" design provides a uniform gripping action in the bore. (Buck Tool Company)

Figure 27b. Tapered mandrel and workpiece set up between centers. (Clausing Machine Tools)

Figure 29. Expanding mandrel and workpiece set up between centers. (Lane Community College)

These indicators show the longitudinal position of the carriage. Whichever method is used, **the power feed should be turned off one-eighth in. short of the workpiece shoulder** and the tool handfed to the desired length. If the tool should be accidentally fed into an existing shoulder, the feed mechanism may jam and be very difficult to release. A broken tool, toolholder, or lathe part may be the result.

Mandrels, sometimes called lathe arbors, are used to hold work that is turned between centers (Figures 27a and 27b). Tapered mandrels are made in standard sizes and have a taper of only .006 in./ft. A flat is milled on the big end of the mandrel for the lathe dog set screw. High pressure lubricant is applied to the bore of the workpiece and the mandrel is pressed into the workpiece with an arbor press. The assembly is mounted between centers and the workpiece is turned or faced on either side.

Expanding mandrels (Figure 28) have the advantage of providing a uniform gripping surface for the length of the bore (Figure 29). A tapered mandrel grips tighter on one end than the other. Gang mandrels (Figure 30) grip several pieces of similar size, such as discs, to turn their circumference. These are made with collars, thread, and nut for clamping.

Stub mandrels (Figures 31a–31c) are used in chucking operations. These are often quickly machined for a single job and then discarded. Expanding stub mandrels (Figures 32a and 32b) are used when

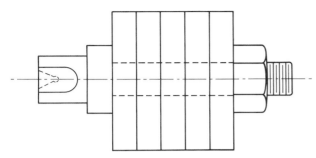

Figure 30. Many similar parts are machined at the same time with a gang mandrel. *(Machine Tools and Machining Practices)*

Figure 31c. Part being machined after assembly on mandrel. (Lane Community College)

Figure 31a. Stub mandrel being machined to size with a slight taper, about .006 in./ft. (Lane Community College)

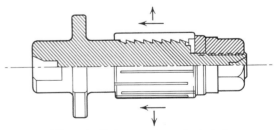

Figure 32a. Special stub mandrel with adjust-tru feature. (Buck Tool Company)

MC standard between centers nut actuated

Figure 31b. Part to be machined being affixed to the mandrel. The mandrel is oiled so that the part can be easily removed. (Lane Community College)

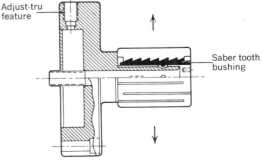

Adjust-tru feature

Saber tooth bushing

MLDR — flange mount — draw bolt actuated stationary sleeve

Figure 32b. Precision adjustment can be maintained on the expanding stub mandrels with the adjust-tru feature on the flange mount. (Buck Tool Company)

Table 2
Coolants and Cutting Oils Used for Turning

Material	Dry	Water Soluble Oil	Synthetic Coolants	Kerosene	Sulfurized Oil	Mineral Oil
Aluminum		x	x	x		
Brass	x	x	x			
Bronze	x	x	x			x
Cast iron	x					
Steel						
Low carbon		x	x		x	
Alloy		x	x		x	
Stainless		x	x		x	

Source: Warren T. White, John E. Neely, Richard R. Kibbe, and Roland O. Meyer, *Machine Tools and Machining Practices,* John Wiley and Sons, Inc., Copyright © 1977, New York.

production of many similar parts is carried out. Threaded stub mandrels are used for machining the outside surfaces of parts that are threaded in the bore.

Coolants are used for heavy duty and production turning. Oil-water emulsions and synthetic coolants are the most commonly used, while sulfurized oils usually are not used for turning operations except for threading. Most job work or single piece work is done dry. Many shop lathes do not have a coolant pump and tank, so, if any coolant or cutting oil is used, it is applied with a pump oil can. Coolants and cutting oils for various materials are given in Table 2.

SELF-TEST

1. Name two advantages and two disadvantages of turning between centers.
2. What other method besides turning between centers is extensively used for turning shafts and long workpieces that are supported by the tailstock center?
3. What factors tend to promote or increase chip curl so that safer chips are formed?
4. Name three kinds of centers used in the tailstock and explain their uses.
5. How is the dead center correctly adjusted?
6. Why should the dead center be frequently adjusted when turning between centers?
7. Why should you avoid excess overhang with the tool and toolholder when roughing?
8. Calculate the RPM for roughing a $1\frac{1}{2}$ in. diameter shaft of machine steel.
9. What would the spacing or distance between tool marks on the workpiece be with a .010 in. feed?
10. How much should the feed rate be for roughing?
11. How much should be left for finishing?
12. Describe the procedure in turning to size predictably.

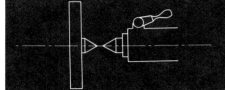

UNIT 9 ALIGNMENT OF THE LATHE CENTERS

As a lathe operator, you must be able to check a workpiece for taper and properly set the tailstock of a lathe. Without these skills, you will lose much time in futile attempts to restore precision turning between centers when the workpiece has an unintentional taper. This unit will show you several ways to align the centers of a lathe.

OBJECTIVES

After completing this unit, you should be able to:
1. Check for taper with a test bar and restore alignment by adjusting the tailstock.
2. Check for taper by taking a cut with a tool and measuring the workpiece and restore alignment by adjusting the tailstock.

The tailstock will normally stay in good alignment on a lathe that is not badly worn. If a lathe has been used for taper turning with the tailstock offset, however, the tailstock may not have been realigned properly (Figure 1). The tailstock also could be slightly out of alignment if an improper method of adjustment was used. It is therefore a good practice to occasionally check the center alignment of the lathe you usually use and to always check the alignment before using a different lathe.

It is often too late to save the workpiece by realigning centers if a taper is discovered while making a finish cut. A check for taper should be made on the workpiece while it is still in the roughing stage. You can do this by taking a light cut for some distance along the workpiece or on each end *without resetting* the cross feed dial. Then check the diameter on each end with a micrometer; the difference between the two readings is the amount of taper in that distance.

Four methods are used for aligning centers on a lathe. In one method, the center points are brought together and visually checked for alignment (Figure 2). This is, of course, not a precision method for checking alignment.

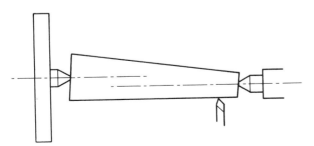

Figure 1. Tailstock out of line causing a tapered workpiece. *(Machine Tools and Machining Practices)*

Figure 2. Checking alignment by matching center points.

Figure 3. Adjusting the tailstock to the witness marks for alignment.

Figure 4. Hexagonal socket set screw that, when turned, moves the tailstock provided the opposite one is loosened.

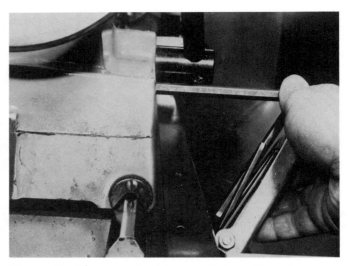

Figure 5. The opposite set screw being adjusted.

Figure 6. Test bar setup between centers with a dial indicator mounted in the tool post.

Another method of aligning centers is by using the tailstock witness marks. Adjusting the tailstock to the witness marks (Figure 3), however, is only an approximate means of eliminating taper. The tailstock is moved by means of a screw or screws. A typical arrangement is shown in Figure 4, where one set screw is released and the opposite one is tightened to move the tailstock on its slide (Figure 5). The tailstock clamp bolt must be released before the tailstock is offset.

Two more accurate means of aligning centers are by using a test bar and by machining and measuring. A test bar is simply a shaft that has true centers (is

not off center) and has no taper. Some test bars are made with two diameters for convenience. When checking alignment with a test bar, no dog is necessary as the bar is not rotated. A dial indicator is mounted, preferably in the tool post, so it will travel with the carriage (Figure 6). Its contact point should be on the center of the test bar.

Begin with the indicator at the headstock end, and set the indicator bezel to zero (Figure 7). Now move the setup to the tailstock end of the test bar (Figure 8), and check the dial indicator reading. If no movement of the needle has occurred, the centers are in line. If the needle has moved clockwise, the tailstock is misaligned toward the operator. This will cause the workpiece to taper with the smaller end near the tailstock. If the needle has moved counterclockwise, the tailstock is away from the operator too far and the workpiece will taper with the smaller end at the headstock.

Figure 7. Indicator is moved to measuring surface at head-stock end and the bezel is set on zero.

Figure 8. The carriage with the dial indicator is moved to the measuring surface near the tailstock. In this case the dial indicator did not move, so the tailstock is on center.

Figure 9. Checking for taper by taking a cut on a work-piece. After the cut is made for the length of the workpiece, a micrometer reading is taken at each end to determine any difference in diameter. (Lane Community College)

Figure 10. Using a dial indicator to check the amount of movement of the tailstock when it is being realigned. (Lane Community College)

In either case, move the tailstock until both diameters have the same reading.

Since usually only a minor adjustment is needed while a job is in progress, cutting and measuring is the most used method of aligning centers on a lathe. It is also the most accurate. This method, unlike the bar test method, usually uses the workpiece while it is in the roughing stage (Figure 9). A light cut is taken along the length of the test piece and both ends are measured with a micrometer. If the diameter at the tailstock end is smaller, the tailstock is toward the operator, and if the diameter at the headstock end is smaller, the tailstock is away from the operator. Set up a dial indicator (Figure 10) and move the tailstock half the difference of the two micrometer readings. Make another light cut and check for taper.

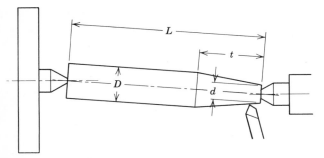

Figure 11. A short taper is made on a longer shaft. *(Machine Tools and Machining Practices)*

When only a short distance of the total length of a shaft (Figure 11) can be turned to check for taper, the amount of tailstock setover may be calculated by the following formula:

$$\text{Offset} = \frac{L \times (D - d)}{2t}$$

Where:

$L =$ total length of shaft
$t =$ length of taper
$D =$ diameter at large end
$d =$ diameter at small end

EXAMPLE

$L = 14$ in.
$t = 4$ in.
$D = 1.495$ in.
$d = 1.490$ in.

$$\text{Offset} = \frac{14 \times (1.495 - 1.490)}{2 \times 4} = .010 \text{ in.}$$

In this case, the tailstock should be moved away from the operator .010 in.

SELF-TEST

1. What is the result on the workpiece when the centers are out of line?
2. What happens to the workpiece when the tailstock is offset toward the operator?
3. Name three methods of aligning the centers.
4. Which measuring instrument is used when using a test bar?
5. By what means is the measuring done when checking taper by taking a cut?

UNIT 10 LATHE OPERATIONS

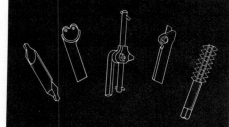

Much of the versatility of the lathe as a machine tool is due to the variety of tools and workholding devices used. This equipment makes possible the many special operations that you will begin to do in this unit.

OBJECTIVES

After completing this unit, you should be able to:

1. Explain the procedures for drilling, boring, reaming, knurling, recessing, parting, and tapping in the lathe.
2. Set up to drill, ream, bore, and tap on the lathe and complete each of these operations.
3. Set up for knurling, recessing, die threading, and parting on the lathe and complete each of these operations.

DRILLING

The lathe operations of boring, tapping, and reaming usually begin with spotting and drilling a hole. The workpiece, often a solid material that requires a bore, is mounted in a chuck, collet, or face plate, while the drill is typically mounted in the tailstock spindle that has a Morse taper. If there is a slot in the tailstock spindle, the drill tang must be aligned with it when inserting the drill.

Drill chucks with Morse taper shanks are used to hold straight shank drills and center drills (Figure 1a). Center drills are used for spotting or making a start for drilling (Figure 1b). When drilling with large size drills, a pilot drill should be used first. If a drill wobbles when started, place the heel of a toolholder against it near the point to steady the drill while it is starting in the hole. Taper shank drills (Figure 2) are inserted directly into the tailstock spindle. The friction of the taper is usually all that is needed to keep the drill from turning while a hole is being drilled (Figures 3 and 4), but when using larger drills, the friction is not enough. A lathe dog is sometimes clamped to the drill just above the flutes (Figure 5) with the bent tail resting on the compound. Hole depth can be measured with a rule or by means of the graduations on top of the tailstock spindle. The alignment of the tailstock with the lathe center line should be checked before drilling or reaming.

Drilled holes are not accurate enough for many applications, such as for gear or pulley bores, which should not be made over the nominal size more than .001 to .002 in. Drilling typically produces holes that are oversize and run eccentric to the center axis of

Figure 1a. Mounting a straight shank drill in a drill chuck in the tailstock spindle. (Lane Community College)

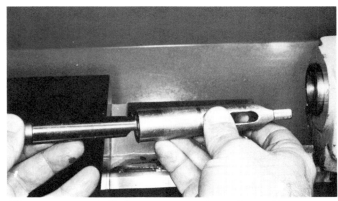

Figure 2. A drill sleeve is placed on the drill so that it will fit the taper in the tailstock spindle.

Figure 1b. Center drilling is the first step prior to drilling, reaming, or boring.

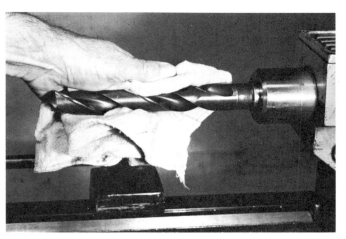

Figure 3. The drill is then firmly seated in the tailstock spindle. (Lane Community College)

Figure 4. The feed pressure when drilling is usually sufficient to keep the drill seated in the tailstock spindle, thus keeping the drill from turning.

Figure 5. A lathe dog is used when the drill has a tendency to turn in the tailstock spindle. (Lane Community College)

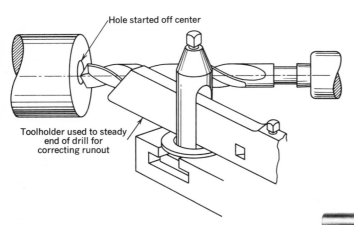

Hole started off center

Toolholder used to steady end of drill for correcting runout

Figure 6. An exaggerated view of the runout and eccentricity that is typical of drilled holes. (*Machine Tools and Machining Practices*)

the lathe (Figure 6). This is not true in the case of some manufacturing types such as gun drilling. However, truer axial alignment of holes is possible when the work is turned and the drill remains stationary, as in a lathe operation, in comparison to operations where the drill is turned and the work is stationary, as on a drill press. Drilling also produces holes with rough finishes, which along with size errors can be corrected by boring or reaming. The hole must first be drilled slightly smaller than the finish diameter in order to leave material for finishing by either of these methods.

BORING

Boring is the process of enlarging and truing an existing or drilled hole. A drilled hole for boring can be from $\frac{1}{32}$ to $\frac{1}{16}$ in. undersize, depending on the situation. Speeds and feeds for boring are determined in the same way as they are for external turning. Boring to size predictably is also done in the same way as in external turning except that the cross feed screw is turned counterclockwise to move the tool into the work.

An inside spring caliper and a rule are sometimes useful for rough measurement. Vernier calipers are also used by machinists for internal measuring, though the telescoping gage and outside micrometer are most commonly used for the precision measurement of small bores. Inside micrometers are used for larger bores. Other means of measurement are an inside spring caliper used with an outside micrometer, and on large bores an inside micrometer used with an outside caliper. Precision bore gages are used where many bores are checked for similar size, such as for acceptable tolerance.

A boring bar is clamped in a holder mounted on the carriage compound. Several types of boring bars and holders are used. Boring bars designed for small holes ($\frac{1}{2}$ in. and smaller) are usually the forged type (Figure 7). The forged end is sharpened by grinding. When the bar gets ground too far back, it must be reshaped or discarded. Boring bars for holes with diameters over $\frac{1}{2}$ in. (Figure 8) use high speed tool inserts, which are typically hand ground in the form of a left-hand turning tool. These tools can be removed from the bar for resharpening when needed. The cutting tool can be held at various angles to obtain differ-

Figure 7. A small forged boring bar made of high speed steel. (*Machine Tools and Machining Practices*)

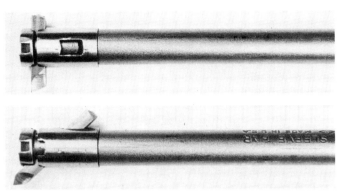

Figure 8. Two boring bars with inserted tools set at different angles. *(Machine Tools and Machining Practices)*

Figure 9. A boring bar with carbide insert. When one edge is dull, a new one is selected. (Kennametal Inc., Latrobe, Pa.)

Figure 10. A boring bar setup with a large overhang for making a deep bore. It is difficult to avoid chatter with this arrangement.

ent results, which makes the boring bar useful for many applications. Standard bars generally come with a tool angle of 30, 45, or 90 degrees. Some boring bars are made for carbide inserts (Figure 9).

Chatter is the rattle or vibration between a workpiece and a tool because of the lack of rigid support for the tool. Chatter is a great problem in boring operations since the bar must extend away from the support of the compound (Figure 10). For this reason boring bars should be kept back into their holders

Figure 11. Tuned boring bars contain dampening slugs of heavy material that can be adjusted by applying pressure with a screw. (Kennametal Inc., Latrobe, Pa.)

Figure 12. Boring tools must have sufficient side relief and side rake to be efficient cutting tools. Back rake is not normally used. (Lane Community College)

as far as practicable. Tuned boring bars can be adjusted so that their vibration is dampened (Figure 11). Stiffness of boring bars is increased by making them of solid tungsten carbide. If chatter occurs when boring, one or more of the following may help to eliminate the vibration of the boring tool.

1. Shorten the boring bar overhang, if possible.
2. Make sure the tool is on center.
3. Reduce the spindle speed.
4. Use a boring bar as large in diameter as possible without it binding in the hole.
5. Reduce the nose radius on the tool.
6. Apply cutting oil to the bore.

Boring bars sometimes spring away from the cut and cause bell-mouth, a slight taper at the front edge of a bore. One or two extra (free) cuts taken without moving the cross feed will usually eliminate this problem.

A large variety of boring bar holders are used. Some types are designed for small, forged bars while others of more rigid construction are used for larger, heavier work.

Boring tools are made with side relief and end relief, but usually with zero back rake (Figure 12). Insufficient end relief will allow the heel of the tool to rub on the workpiece (Figure 13). The end relief should be between 10 and 20 degrees. The machinist must use his judgment when grinding the end relief because the larger the bore, the less end relief is required (Figure 14). If the end of the tool is relieved too much, the cutting edge will be weak and break down.

Figure 13. A tool with insufficient end relief will rub on the heel of the tool and will not cut. (Lane Community College)

Figure 15a. The point of the boring tool must be positioned on the centerline of the workpiece. (Lane Community College)

Figure 14. End relief angle varies depending on the diameter of the workpiece bore. These views are looking outward from inside the chuck. (Lane Community College)

Figure 15b. If the boring tool is too low, the heel of the tool will rub and the tool will not cut, even if the tool has the correct relief angle. (Lane Community College)

Figure 15c. If the tool is too high, the back rake becomes excessively negative and the tool point is likely to be broken off. A poor quality finish is the result of this position. (Lane Community College)

Figure 16. Allowance must be made so the chips can clear the space between the bar and the surface being machined. This setup has insufficient chip clearance (where the arrow is pointing). (Lane Community College)

Figure 17. Bar and tool arrangement for through boring. (Lane Community College)

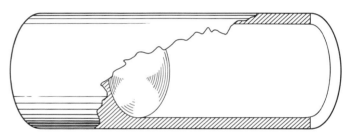

Figure 18. A blind hole machined flat in the bottom. *(Machine Tools and Machining Practices)*

The point of the cutting tool should be positioned exactly on the centerline of the workpiece (Figures 15a to 15c). There must be a space to allow the chips to pass between the bar and the surface being machined, or the chips will wedge and bind on the back side of the bar, forcing the cutting tool deeper into the work (Figure 16).

"Through boring" is the boring of a workpiece from one end to the other or all the way through it. For through boring, the tool is held in a bar that is perpendicular to the axis of the workpiece. A slight side cutting edge angle is often used for through boring (Figure 17). Back facing is sometimes done to true up a surface on the back side of a through bore. This is done with a straight ground right-hand tool also held in a bar perpendicular to the workpiece. The amount of facing that can be done in this way is limited to the movement of the bar in the bore.

A blind hole is one that does not go all the way through the part to be machined (Figure 18). Machining the bottom or end of a blind hole to a flat is easier

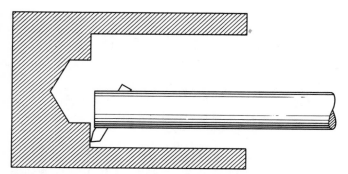

Figure 19. A bar with an angled tool used to square the bottom of a hole with a drilled center. *(Machine Tools and Machining Practices)*

Figure 20. Workpiece clamped on face plate has been located, drilled, and bored.

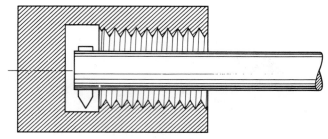

Figure 21. Ample thread relief is necessary when making internal threads. *(Machine Tools and Machining Practices)*

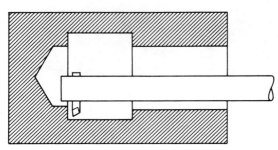

Figure 22. The hole is drilled deeper than necessary to allow room for the boring bar. *(Machine Tools and Machining Practices)*

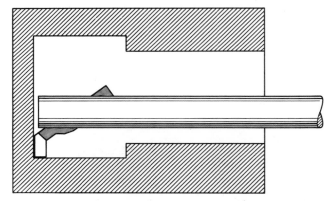

Figure 23. A special tool is needed when the thread relief must be next to a flat bottom. *(Machine Tools and Machining Practices)*

when the drilled center does not need to be cleaned up. A bar with the tool set at an angle, usually 30 or 45 degrees, is used to square the bottom of a hole with a drilled center (Figure 19).

Most boring is performed on workpieces mounted in a chuck. But it is also done in the end of workpieces supported by a steady rest. Boring and other operations are infrequently done on workpieces set up on a face plate (Figure 20).

A "thread relief" is a bore that is a little larger at the blind end of a hole. The purpose of a thread relief is to allow a threading tool to disengage the work at the end of a pass (Figure 21). When the work will allow it, a hole can be drilled deeper than necessary. This will give the end of the boring bar enough space so that the tool can reach into the area to be

relieved and still be held at a 90 degree angle (Figure 22). When the work will not allow for the deeper drilling, a special tool must be ground (Figure 23).

Grooves in bores are made by feeding a form tool (Figure 24) straight into the work. Snap ring, O-ring, and oil grooves are made in this way. Cutting oil should always be used in these operations.

Counterboring in a lathe is the process of enlarging a bore for definite length (Figure 25). The shoulder

Figure 24. A tool that is ground to the exact width of the desired groove can be moved directly into the work to the correct depth. (Lane Community College)

Figure 25. A square shoulder is made with a counterboring tool. (Lane Community College)

that is produced in the end of the counterbore is usually made square (90 degrees) to the lathe axis. Boring and counterboring are also done on long workpieces that are supported in a steady rest. All boring work should have the edges and corners broken or chamfered.

REAMING

Reaming is done in the lathe to quickly and accurately finish drilled or bored holes to size. Machine reamers, like drills, are held in the tailstock spindle of the lathe. Floating reamer holders are sometimes used to assure alignment of the reamer, since the reamer follows the eccentricity of drilled holes. This helps eliminate bell-mouth bores that result from reamer wobble, but does not eliminate the hole eccentricity. Only boring will remove the bore runout.

Roughing reamers (rose reamers) are often used in drilled or cored holes followed by machine or finish reamers. When drilled or cored holes have excessive eccentricity, they are bored .010 to .015 in. undersize and machine reamed. If a greater degree of accuracy

Figure 26. Hand reaming in the lathe. (Lane Community College)

is required, the hole is bored to within .003 to .005 in. of finish size and hand reamed (Figure 26). For hand reaming, the machine is shut off and the hand reamer is turned with a tap wrench. Types of machine reamers are shown in Section G, Unit 7, "Reaming in the Drill Press." Hand reamers are shown in Section B, Unit 6, "Hand Reamers."

Cutting oils used in reaming are similar to those used for drilling holes. (See Table 3, "Coolants Used for Reaming," in Section G, Unit 7, "Reaming in the Drill Press.") Cutting speeds are dependent on machine and workpiece material finish requirements. A rule of thumb for reaming speeds is to use one-half the speed used for drilling. (See Table 1, "Reaming Speed," in Section G, Unit 7, "Reaming in the Drill Press.")

Feeds for reaming are about twice that used for drilling. The cutting edge should not rub without cutting as it causes glazing, work hardening, and dulling of the reamer.

A simple machine reaming sequence would be as follows:

1. Assuming that the hole has been drilled $\frac{1}{64}$ in. undersize, a taper shank machine reamer is seated in the taper by hand pressure (Figure 27).

2. Cutting oil is applied to the hole and the reamer is started into the hole by turning the tailstock handwheel (Figure 28). The reamer should be backed out with the machine running forward. Never reverse the machine when reaming.

3. The reamer is removed from the tailstock spindle and cleaned with a cloth (Figure 29). The reamer is then returned to the storage rack.

4. The lathe is cleaned with a brush (Figure 30).

Figure 27. The reamer must be seated in the taper.

Figure 28. Starting the reamer in the hole. Kerosene is being used as a cutting fluid since the material is aluminum.

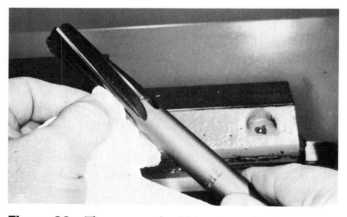

Figure 29. The reamer should be cleaned and put away after using it.

Figure 30. The chips are brushed into the chip pan.

TAPPING

The tapping of work mounted in a chuck is a quick and accurate means of producing internal threads. Tapping in the lathe is similar to tapping in the drill press, but it is generally reserved for small size holes, as tapping is the only way they can be internally threaded. Large internal threads are made in the lathe with a single point threading tool. A large tap requires considerable force or torque to turn, more than can be provided by hand turning. A tap that is aligned by the dead center will make a straight tapped hole that is in line with the lathe axis.

A plug tap (Figure 31) or spiral point tap (Figure 32) may be used for tapping through holes. When tapping blind holes, a plug tap could be followed by a bottoming tap (Figure 33), if threads are needed to the bottom of the hole. A good practice is to drill a blind hole deeper than the required depth of threads.

Two approaches may be taken for hand tapping. Power is not used in either case. One method is to turn the tap by means of a tap wrench or adjustable wrench with the spindle engaged in a low gear so it will not turn (Figure 34). The other method is to disengage the spindle and turn the chuck by hand while the tap wrench handle rests on the compound (Figure 35). In both cases the tailstock is clamped to the ways, and the dead center is kept in the center of the tap by slowly turning the tailstock handwheel. The tailstock on small lathes need not be clamped to the ways for small taps, but held firmly with one hand. Cutting oil should be used and the tap backed off every one or two turns to break chips unless it is a spiral point tap, sometimes called a gun tap.

The correct tap drill size should be obtained from

Figure 31. Plug tap. (TRW Inc., 1980)

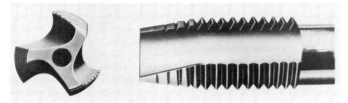

Figure 32. Spiral point tap or gun tap. (TRW Inc., 1980)

Figure 33. Bottoming tap. (TRW Inc., 1980)

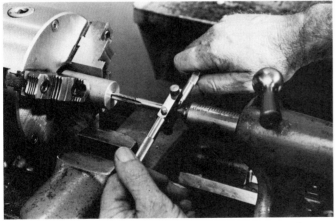

Figure 34. Hand tapping in the lathe by turning the tap wrench. (Lane Community College)

Figure 35. Tapping by turning the chuck. (Lane Community College)

Figure 36. Starting die on rod to be threaded in the lathe. The tailstock spindle (without a center) is used to start the die squarely onto the work. (Lane Community College)

a tap drill chart. Drills tend to drill slightly oversize, and tapping the oversize hole can produce poor internal thread with only a small percentage of thread cut. Make sure the drill produces a correctly sized hole by drilling first with a slightly smaller drill, then use the tap drill as a reamer.

Tapping can be done on the lathe with power, but it is recommended that it be done only if the spindle rotation can be reversed, if a spiral point tap

is used, and if the hole is clear through the work. The tailstock is left free to move on the ways. Insert the tap in a drill chuck in the tailstock and set the lathe on a low speed. Use cutting oil and slide the tailstock so the tap engages the work. Reverse the tap and remove it from the work every $\frac{3}{8}$ to $\frac{1}{2}$ in. When reversing, apply light hand pressure on the tailstock to move it to the right until the tap is all of the way out.

External threads cut with a die should only be used for nonprecision purposes, since the die may wobble and the pitch (the distance from a point on one thread to the same point on the next) may not be uniform. The rod to be threaded extends a short distance from the chuck and a die and diestock are started on the end (Figure 36). Cutting oil is used. The handle is rested against the compound. The chuck

Figure 37. A button die holder used to guide the die onto the work. This ensures a more uniform thread than by using a die stock. (Lane Community College)

Figure 38a. The undercutting tool is brought to the workpiece and the micrometer dial is zeroed. Cutting oil is applied to the work. (Lane Community College)

may be turned by hand, but if power is used, the machine is set for low speed and reversed every $\frac{3}{8}$ to $\frac{1}{2}$ in. to clear the chips. Finish the last $\frac{1}{4}$ in. by hand if approaching a shoulder. Reverse the lathe to remove the die. An alternate method is to use a button die holder to guide the die onto the work (Figure 37). Thread cutting only takes place when the holder is gripped by the operator, a method that provides more control over the cutting action. This method provides a much truer, more aligned thread than does the method shown in Figure 36.

RECESSING, GROOVING, AND PARTING
Recessing and grooving on external diameters (Figures 38a to 38c) is done to provide grooves for thread relief, snap rings, and O-rings. Special tools (Figure 39) are ground for both external and internal grooves and recesses. Parting tools are sometimes used for external grooving and thread relief.

Parting or cutoff tools (Figure 40) are designed to withstand high cutting forces, but if chips are not sufficiently cleared or cutting oil is not used, these tools can quickly jam and break. Parting tools must be set on center and square with the work (Figure 41). Lathe tools are often specially ground as parting tools for small or delicate parting jobs (Figure 42). Diagonally ground parting tools leave no burr.

Parting alloy steels and other metals is sometimes difficult, and step parting (Figure 43) may help in these cases. When deep parting difficult material, extend the cutting tool from the holder a short distance

Figure 38b. The tool is fed to the single depth of the thread or the required depth of the groove. If a wider groove is necessary, the tool is moved over and a second cut is taken as shown. (Lane Community College)

and part to that depth. Then back off the cross feed and extend the tool a bit farther; part to that depth. Repeat the process until the center is reached. Sulfurized cutting oil works best for parting unless the lathe is equipped with a coolant pump and a steady flow of soluble oil is available. Parting tools are made in either straight or offset types. A right-hand offset cutoff tool is necessary when parting very near the chuck.

All parting and grooving tools have a tendency

Figure 38c. The finished groove. (Lane Community College)

Figure 41. Parting tools must be set to the center of the work. (Lane Community College)

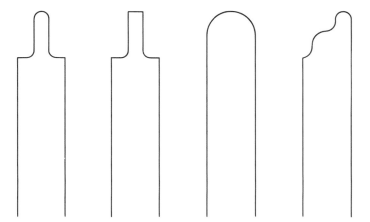

Figure 39. Recessing or grooving tools for internal and external use. *(Machine Tools and Machining Practices)*

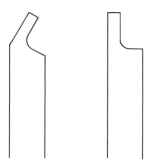

Figure 42. Special parting tools that have been ground from lathe tools for small or delicate parting jobs. *(Machine Tools and Machining Practices)*

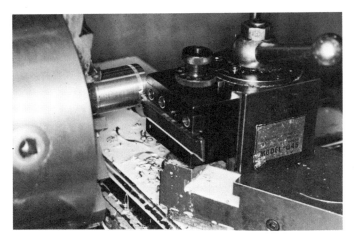

Figure 40. Parting tool making a cut. (Lane Community College)

Figure 43. Step parting. (Lane Community College)

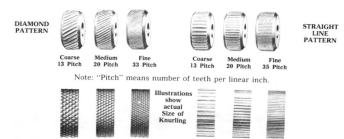

Figure 44. Set of straight knurls and diagonal knurls. (J. H. Williams Division of TRW Inc.)

Figure 45a. Knuckle-joint knurling toolholder. (*Machine Tools and Machining Practices*)

Figure 45b. Revolving head knurling toolholder. (*Machine Tools and Machining Practices*)

to chatter; therefore any setup must be as rigid as possible. A low speed should be used for parting; if the tool chatters, reduce the speed. Work should not extend very far from the chuck when parting or grooving, and no parting should be done in the middle of a workpiece or at the end near the dead center. A feed that is too light can cause a chatter, but a feed that is too heavy can jam the tool. The tool should always be making a chip. Hand feeding the tool is best at first.

KNURLING

A knurl is a raised impression on the surface of a workpiece produced by two hardened rolls, and is usually of two patterns, diamond or straight (Figure 44). The diamond pattern is formed by a right-hand and a left-hand helix mounted in a self-centering head. The straight pattern is formed by two straight rolls. These common knurl patterns can be either fine, medium, or coarse.

Diamond knurling is used to improve the appearance of a part and to provide a good gripping surface for levers and tool handles. Straight knurling is used to increase the size of a part for press fits in light duty applications. A disadvantage to this use of knurls is that the fit has less contact area than a standard fit.

Three basic types of knurling toolholders are used: the knuckle-joint holder (Figure 45a), the revolving head holder (Figure 45b), and the straddle holder (Figure 45c). The straddle holder allows small diameters to be knurled with less distortion. This principle is used for knurling on production machines.

Knurling works best on workpieces mounted between centers. When held in a chuck and supported by a center, the workpiece tends to crawl back into the chuck and out of the supporting center with the high pressure of the knurl. This is especially true when the knurl is started at the tailstock end and the feed

Figure 45c. Straddle knurling toolholder. (Ralmike's Tool-A-Rama)

is toward the chuck. Long slender pieces push away from the knurl and will stay bent if the knurl is left in the work after the lathe is stopped.

Knurls do not cut, but displace the metal with high pressures. Lubrication is more important than cooling, so a lard oil or lubricating oil is satisfactory. Low speeds (about the same as for threading) and a feed of about .015 to .030 in. are used for knurling.

Figure 46. Knurls are centered on the workpiece. (Lane Community College)

Figure 47. Angling the toolholder 5 degrees often helps establish the diamond pattern. (Lane Community College)

Figure 48. The knurl is started approximately half depth. Notice the diamond pattern is not fully developed. (Lane Community College)

Figure 49. Double impression on the left is the result of the rolls not tracking evenly. (Lane Community College)

The knurls should be centered vertically on the workpiece (Figure 46) and the knurl toolholder should be square with the work, unless the knurl pattern is difficult to establish, as it often is in tough materials. In that case, the toolholder should be angled about 5 degrees to the work so the knurl can penetrate deeper (Figure 47).

A knurl should be started in soft metal about half depth and the pattern checked. An even diamond pattern should develop (Figure 48). But, if one roll is dull or placed too high or too low, a double impression will develop (Figure 49) because the rolls are not tracking evenly. If this happens, move the knurls

Figure 50. More than one pass is usually required to bring the knurl to full depth. (Lane Community College)

Figure 51. A knurling tool that cuts a knurl rather than forming it by pressure. (*Machine Tools and Machining Practices*)

Figure 52. A knurl being cut showing formation of chip. (Lane Community College)

to a new position along the workpiece, readjust up or down, and try again. The knurls should be cleaned with a wire brush between passes.

Material that hardens as it is worked, such as high carbon or spring steel, should be knurled in one pass if at all possible, but in not more than two passes. Even in ordinary steel, the surface will work harden after a diamond pattern has developed to points. It is best to stop knurling just before the points are sharp (Figure 50). Metal flaking off the knurled surface is evidence that work hardening has occurred. Avoid knurling too deeply as it produces an inferior knurled finish.

Knurls are also produced with a type of cutting tool (Figure 51) that is similar in appearance to a knurl-

ing tool. The serrated rolls form a chip on this edge (Figure 52). Material that is difficult to knurl by pressure rolling, such as tubing and work hardening metals, can be knurled by this cutting tool. Sulfurized cutting oil should be used when knurling steel with this kind of knurling tool.

SELF-TEST

1. Why are drilled holes not used for bores in machine parts such as pulleys, gears, and bearing fits?
2. Describe the procedure used to produce drilled holes on workpieces in the lathe with minimum oversize and runout.
3. What is the chief advantage of boring over reaming in the lathe?
4. List five ways to eliminate chatter in a boring bar.
5. Explain the differences between through boring, counterboring, and boring blind holes.
6. By what means are grooves and thread relief made in a bore?
7. Reamers will follow an eccentric drilled hole, thus producing a bell-mouth bore with runout. What device can

be used to help eliminate bell-mouth? Does it help remove the runout?

8. Machine reamers produce a better finish than is obtained by boring. How can you get an even better finish with a reamer?

9. Cutting speeds for reaming are (twice, half) that used for drilling; feeds used for reaming are (twice, half) that used for drilling.

10. Are large internal threads produced with a tap or a boring tool? Explain the reason for your answer.

11. How can you avoid drilling oversize with a tap drill?

12. Standard plug or bottoming taps can be used when hand tapping in the lathe. If power is used, what kind of tap works best?

13. Why would threads cut with a hand die in a lathe not be acceptable for using on a feed screw with a microme-

ter collar?

14. By what means are thread relief on external grooves produced?

15. If cutting oil is not used on parting tools or chips do not clear out of the groove because of a heavy feed, what is generally the immediate result?

16. How can you avoid chatter when cutting off stock with a parting tool?

17. State three reasons for knurling.

18. Ordinary knurls do not cut. In what way do they make the diamond or straight pattern on the workpiece?

19. If a knurl is producing a double impression, what can you do to make it develop a diamond pattern?

20. How can you avoid producing a knurled surface on which the metal is flaking off?

UNIT 11 SIXTY DEGREE THREAD INFORMATION AND CALCULATIONS

$P = \frac{1}{n}$ $d = \frac{.613}{n}$

To cut threads, a good lathe operator must know more than how to set up the lathe. He or she must know the thread form, class of fit, and thread calculation. This unit prepares you for the actual cutting of threads, which you will do in the next unit.

OBJECTIVES

After completing this unit, you should be able to:

1. Describe the several 60 degree thread forms, noting their similarities and differences.
2. Calculate thread depth, infeeds, and minor diameters of threads.

THE SHARP VEE THREAD FORM

Various screw thread forms are used for fastening and for moving or translating parts against loads. The most widely used of these forms are the 60 degree thread

types. These are mostly used for fasteners. An early form of the 60 degree thread is the sharp V (Figure 1). The sides of the thread form a 60 degree angle with each other. Theoretically, the sides and the base

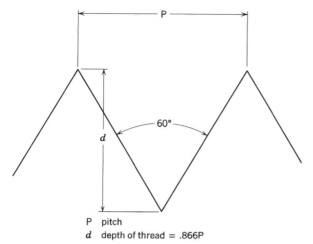

P pitch
d depth of thread = .866P

Figure 1. The 60 degree sharp V thread. *(Machine Tools and Machining Practices)*

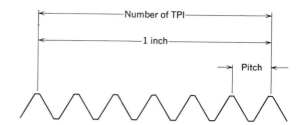

Figure 2. The difference between threads per inch and pitch. *(Machine Tools and Machining Practices)*

between two thread roots would form an equilateral triangle, but in practice this is not the case; it is necessary to make a slight flat on top of the thread in order to deburr it. Also, the tool will always round off and leave a slight flat at the thread root. The greatest drawback to this thread form is that it is so easily damaged while handling. The sharp V thread will fit closer and seal better than most threads, but is seldom used today. The depth *(d)* for the sharp V thread is calculated as follows:

$$d = \text{pitch} \times \cos 30$$
$$= \text{pitch} \times .866, \text{ or}$$
$$= \frac{.866}{\text{number of threads per inch}}$$

The relationship between pitch and threads per inch should be noted (Figure 2). Pitch is the distance between a point on one screw thread and the corresponding point on the next thread, measured parallel to the thread axis. Threads per inch means the number of threads in one inch. The pitch *(P)* may be derived by dividing the number of threads per inch (TPI) into one.

$$P = \frac{1}{\text{TPI}}$$

EXAMPLE
Find the pitch of a $\frac{1}{2}$ in. diameter-20 TPI machine screw thread.

$$P = \frac{1}{20} = .050 \text{ in.}$$

Pitch is checked on a screw thread with a screw pitch gage (Figures 3a and 3b). General dimensions and symbols for screw threads are shown in Figure 4.

Figure 3a. Screw pitch gage for inch threads. *(Machine Tools and Machining Practices)*

Figure 3b. Screw pitch gage for metric threads. *(Machine Tools and Machining Practices)*

UNIFIED AND AMERICAN NATIONAL FORMS

The American National form (Figure 5), formerly United States Standard, was used for many years for screws, bolts, and other products. These National form threads are in either the national fine (NF) or the national coarse (NC) series. Now, the Unified Series Thread Form is used, which is only slightly different from American National Form, and has a coarse series (UNC) and a fine series (UNF). Other screw threads are listed in machinist's handbooks. See Appendix Table 11.

Taps and dies are marked with letter symbols to designate the series of the threads they form. For

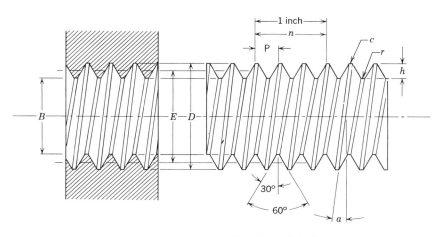

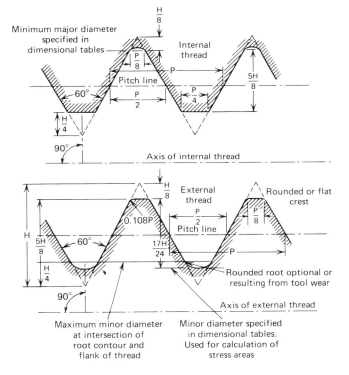

Figure 4. General dimensions of screw threads. *(Machine Tools and Machining Practices)*

D Major Diameter
E Pitch Diameter
B Minor Diameter
n Number of threads per inch (TPI)
P Pitch

a Helix (or Lead) Angle
c Crest of Thread
r Root of Thread
h Basic Thread Height (or depth)

Figure 6. Unified screw threads. (Reprinted from ASA B1.1—1960, Unified Screw Threads, with permission of the publisher, The American Society of Mechanical Engineers; data, John Neely).

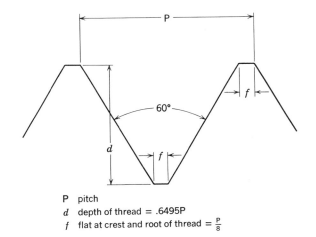

P pitch
d depth of thread = .6495P
f flat at crest and root of thread = $\frac{P}{8}$

Figure 5. The American National form thread. *(Machine Tools and Machining Practices)*

(H = height of sharp V-thread = 0.86603 x pitch)

example, the symbol for American Standard Taper pipe thread is NPT, for Unified coarse thread it is UNC, and for Unified fine thread the symbol is UNF.

Thread Depth

The American Standard for Unified threads (Figure 6) is very similar to the American National Standard with certain modifications. The thread forms are practically the same and the basic 60 degree angle is the same. The depth of an external American National thread is .6495 × pitch, and the depth of the Unified thread is .6134 × pitch. The constant for American National thread depth may be rounded off from .6495 to .65, and the constant for Unified thread depth may

Unified Thread Form Data:

$$\text{Pitch} = \frac{1}{\text{Number of threads per inch}}$$

Depth, external
 thread   = 0.613 × pitch
Depth, internal
 thread   = 0.541 × pitch
Flat at crest,
 external thread = 0.125 × pitch
Flat at crest,
 internal thread = 0.25 × pitch
Flat at root,
 internal thread = 0.125 × pitch
 (tool flat)
Flat at root,
 external thread = 0.125 × pitch
 (permissible tool flat
 on new tool)

be rounded to .613. The thread depth and the root flat of the American National thread is fixed or definite, but these factors are variable within limits for Unified threads. A rounded root for Unified threads is desirable whether from tool wear or by design. A rounded crest is also desirable but not required. The constants .613 for thread depth and .125 for the flat on the end of the tool were selected for calculations on Unified threads in this unit.

Thread Fit Classes and Thread Designations

Unified and American National Standard form threads are interchangeable. An NC bolt will fit an UNC nut. The principal difference between the two systems is that of tolerances. The Unified system, a modified version of the old system, allows for more tolerances of fit. Thread fit classes 1, 2, 3, 4, and 5 were used with the American National Standard; one being a very loose fit, two a free fit, three a close fit, four a snug fit, and five a jam or interference fit. The Unified system expanded this number system to include a letter, so the threads could be identified as class 1A, 1B, 2A, 2B, and so on. "A" indicates an external thread and "B" an internal thread. Because of this expansion in the Unified system, tolerances are now possible on external threads and are 30 percent greater on internal threads. These changes make easier the manufacturer's job of controlling tolerances to insure the interchangeability of threaded parts. See the *Machinery's Handbook* for tables of Unified thread limits. Limits are the maximum and minimum allowable dimensions of a part, in this case, internal and external threads.

Threads are designated by the nominal bolt size or major diameter, the threads per inch, the letter series, the thread tolerance, and the thread direction. Thus, $1\frac{1}{4}$ in.–12 UNF-2BLH would indicate a $1\frac{1}{4}$ in. Unified nut with 12 threads per inch, a class 2 thread fit, and a left-hand helix.

Unified screw thread systems are the American Standard for fastening types of screw threads. Manufacturing processes where V threads are produced are based on the Unified system. Many job and maintenance machine shops, on the other hand, still use the American National thread system when chasing a thread with a single point tool on an engine lathe.

Tool Flats and Infeeds for Thread Cutting

The flat on the crest of the thread on both the Unified and American National systems is $P/8$ or $P \times .125$. The root flat (flat on the end of the external threading tool) is calculated $P/8$ or $P \times .125$ for the American National system, but it varies in the Unified system. However, for purposes of convenience in cutting threads on a lathe, the same tool flat $(P/8)$ may also

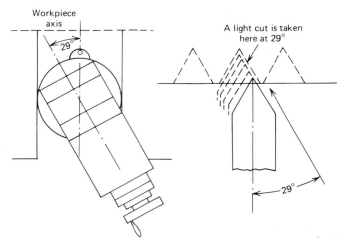

Figure 7. The compound at 29 degrees for cutting 60 degree threads. The reason for setting the compound at 29 degrees instead of 30 degrees for 60 degree threads is to provide a finish on the trailing side of the thread.

be used for the Unified form thread.

To cut 60 degree form threads, the tool is fed into the work with the compound (Figure 7), which is set at 29 degrees. The infeed depth along the flank of the thread at 30 degrees is greater than the depth at 90 degrees from the work axis. This depth may be calculated for American National threads by dividing the number of threads per inch *(n)* into .75,

$$\text{Infeed} = \frac{.75}{n} \quad \text{or} \quad P \times .75$$

Thus, for a thread with 10 threads per inch (.100 in. pitch):

$$\text{Infeed} = \frac{.75}{10} = .075 \text{ in.}$$

or

$$\text{Infeed} = .75 \times .100 = .075 \text{ in.}$$

For external Unified threads the infeed at 29 degrees may be calculated by the formula:

$$\text{Infeed} = \frac{.708}{n} \quad \text{or} \quad .708P$$

Thus, for a thread with 10 threads per inch (.100 in. pitch):

$$\text{Infeed} = \frac{.708}{10} = .0708 \text{ in.}$$

or

$$\text{Infeed} = .708 \times .100 = .0708 \text{ in.}$$

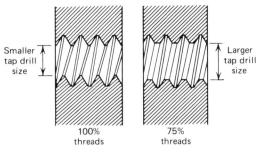

Smaller tap drill size

Larger tap drill size

100% threads

75% threads

Figure 8. The difference between 100 percent and 75 percent threads is the minor diameter (tap drill size). Most tap-drill charts give drill sizes that will produce 75 percent threads.

PITCH DIAMETER, HELIX ANGLE, AND PERCENT OF THREADS

The making of external and internal threads that are interchangeable depends upon the selection of thread fit classes. The clearances and tolerances for thread fits are derived from the pitch diameter. The pitch diameter on a straight thread is the diameter of an imaginary cylinder that passes through the thread profiles at a point where the width of the groove and thread are equal. The mating surfaces are the flanks of the thread.

The percent of thread has little to do with fit, but refers to the actual minor diameter of the internal thread. The typical nut for machine screws has 75 percent threads, which are easier to tap than 100 percent threads and retain sufficient strength for most thread applications (Figure 8).

The helix angle of a screw thread (Figure 9) is larger for greater lead threads than for smaller leads; and the larger the diameter of the workpiece, the smaller the helix angle for the same lead. Helix angles should be taken into account when grinding tools for threading. The relief plus the helix angle must be ground on the leading or cutting edge of the tool (Figure 10). A protractor may be used to check this angle.

Helix angles may be determined by the following formula:

$$\frac{\text{Tangent of}}{\text{helix angle}} = \frac{\text{lead of thread}}{\text{circumference of screw}}$$
$$= \frac{\text{lead of thread}}{\pi D}$$

where $\pi = 3.1416$, $D =$ the major diameter of the screw. (Also note that pitch and lead are the same for single lead screws.) Helix angles are given for Unified and other thread series in handbooks such as the *Machinery's Handbook*.

A taper thread is made on the internal or external

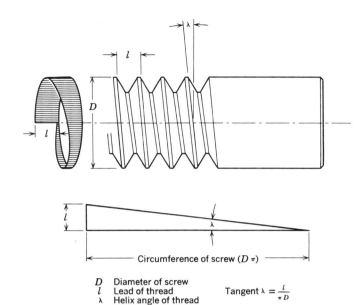

D Diameter of screw
l Lead of thread
λ Helix angle of thread

Tangent $\lambda = \frac{l}{\pi D}$

Figure 9. Screw thread helix angle, $D =$ diameter of screw; $l =$ lead of thread, and $\lambda =$ helix angle of thread. Tangent $\lambda = 1/\pi D$. *(Machine Tools and Machining Practices)*

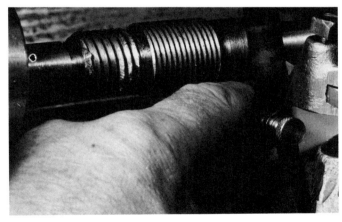

Figure 10. Checking the relief and helix angle on the threading tool with a protractor. (Lane Community College)

surface of a cone. An example of a 60 degree taper thread is the American National Standard pipe thread (Figure 11). A line bisecting the 60 degree thread is perpendicular to the axis of the workpiece. On a taper thread the pitch diameter at a given position on the thread axis is the diameter of the pitch cone at that position.

The British Standard Whitworth thread (Figure 12) has rounded crests and roots and has an included angle of 55 degrees. This thread form has been largely replaced by the Unified and metric thread forms.

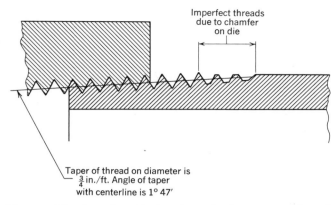

Figure 11. American National standard taper pipe thread. *(Machine Tools and Machining Practices)*

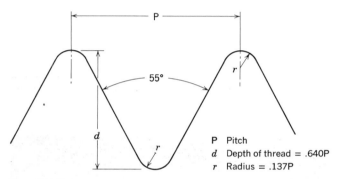

Figure 12. Whitworth thread. *P* = pitch; *d* = depth of thread, .640*P*; *r* = radius, .137*P*. *(Machine Tools and Machining Practices)*

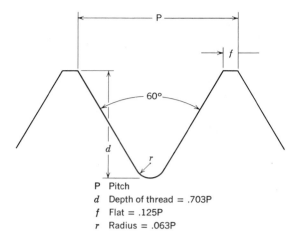

Figure 13. The SI metric thread form. *P* = pitch; *d* = depth of thread, .703*P*; *f* = flat, .125*P*; *r* = radius, .063*P*. *(Machine Tools and Machining Practices)*

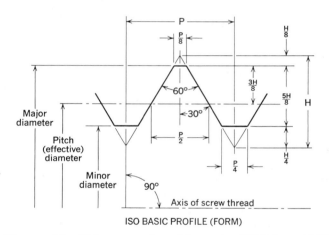

ISO BASIC PROFILE (FORM)

Figure 14. ISO metric thread form. (© TRW Inc., 1980)

METRIC THREAD FORMS

Several metric thread systems such as the SAE standard spark plug threads and the British Standard for spark plugs are in use today. The Système Internationale (SI) thread form (Figure 13), adapted in 1898, is similar to the American National Standard. Metric bolt sizes differ slightly from one European country to the next. The British Standard for ISO (International Organization for Standardization) metric screw threads was set up to standardize metric thread forms. The basic form of the ISO metric thread (Figure 14) is similar to the Unified thread form. These and other metric thread systems are listed in the *Machinery's Handbook*. See Table 10 in the Appendix. A common practice to find the tap drill sizes for metric threads is simply to subtract the pitch from the major diameter.

A new metric standard was endorsed by the Industrial Fasteners Institute (IFI) January 31, 1974. It is called the IFI-500 Trial Standard. For further information on the new thread system, refer to *Machinery's Handbook*.

BASIC METHODS FOR CHECKING SCREW THREADS

The simplest method for checking a thread is to try the mating part for it. The fit is determined solely by feel with no measurement involved. While a loose, medium or close fit may be determined by this method, the threads cannot be depended upon for interchangeability with others of the same size and pitch.

More precise methods of checking threads depend on the Go/No Go principle. The thread plug

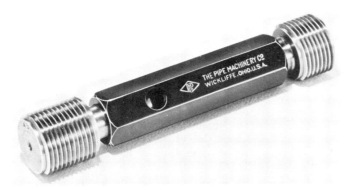

Figure 15. Thread plug gage. (PMC Industries)

Figure 17. Thread roll snap gage. (PMC Industries)

Figure 16. Thread ring gage (PMC Industries)

gage (Figure 15) is used for checking internal threads. These gages are available in various sizes, which are stamped on the handle. An external screw thread is on each end of the handle. The longer threaded end is called the "Go" gage, while the shorter end is called the "No Go" gage. The No Go end is made to a slightly larger dimension than the pitch diameter for the class of fit that the gage tests. To test an internal thread, both the Go and No Go gages should be tried in the hole. If the part is within the range or tolerance of

the gage, the Go end should turn in flush to the bottom of the internal thread, but the No Go end should just start into the hole and become snug with no more than three turns. The gage should never be forced into the hole.

Thread ring gages are used to check the accuracy of external threads (Figure 16). The outside of the ring gage is knurled and the No Go gage can be easily identified by a groove on the knurled surface. When these gages are used, the Go ring gage should enter the thread fully. The No Go gage should not exceed more than $1\frac{1}{2}$ turns on the thread being checked.

Thread roll snap gages are used to check the accuracy of external screw threads (Figure 17). These common measuring tools are easier and faster to use than thread micrometers or ring gages. The part size is compared to a preset dimension on the roll gage. The first set of rolls are the Go and the second the No Go rolls.

Thread roll snap gages, ring gages, and plug gages are used in production manufacturing where quick gaging methods are needed. These gaging methods depend on the operator's "feel" and the level of precision is only as good as the accuracy of the gage. The thread sizes are not measurable in any definite way.

The most common method of checking thread fits in maintenance machine shops is to use the mating part or a standard nut. When this is done, the mating thread should be tried on as the thread nears completion and the fit made as close as possible, yet free to turn with the fingers all the way on the threaded part.

SELF-TEST

1. Name one disadvantage of the sharp V thread.
2. Explain the difference between the threads per inch and the pitch of the thread.
3. Name two similarities and two differences between American National and Unified threads.
4. What is a major reason for thread allowances and classes of fits?
5. What does $\frac{1}{2}$–20 UNC—2A describe?
6. The root flat for Unified threads and for American National threads is found by .125P. What should the flat

on the end of the threading tools be for a $\frac{1}{2}$–20 thread?
7. How far should the compound set at 29 degrees move to cut a $\frac{1}{2}$–20 Unified thread? The formula is .708/n. How far should the compound move to cut a $\frac{1}{2}$–20 American National thread? The formula is .75/n.
8. Explain the difference between the fit of threads and the percent of thread.
9. Name two metric thread standard systems.
10. Explain a simplified method of finding tap drill sizes for metric threads.

UNIT 12 CUTTING UNIFIED EXTERNAL THREADS

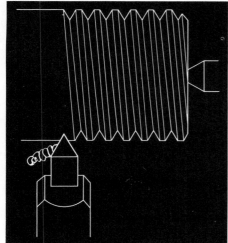

A lathe operator is frequently called upon to cut threads of various forms on the engine lathe. The threads most commonly made are the V form, American National, or Unified. This unit will show you how to make these threads on a lathe. You will need much practice to gain confidence in your ability to make external Unified threads on any workpiece.

OBJECTIVES

After completing this unit, you should be able to:
1. Detail the steps and procedures necessary to cut a Unified thread to the correct depth.
2. Set up a lathe for threading and cut several different thread pitches and diameters.

HOW THREADING IS DONE ON A LATHE

Thread cutting on a lathe with a single point tool is done by taking a series of cuts in the same helix of the thread. This is sometimes called chasing a thread. A direct ratio exists between the headstock spindle rotation, the lead screw rotation, and the number of threads on the lead screw. This ratio can be altered by the quick-change gearbox to make a variety of threads. When the half-nuts are clamped on the thread of the lead screw, the carriage will move a given distance for each revolution of the spindle. This distance is the lead of the thread.

If the infeed of a thread is made with the cross slide (Figure 1), equal size chips will be formed on both cutting edges of the tool. This causes higher tool pressures that can result in tool breakdown, and sometimes causes tearing of the threads because of insufficient chip clearance. A more accepted practice is to feed in with the compound, which is set at 29 degrees (Figure 2) toward the right of the operator, for cutting right-hand threads. This assures a cleaner cutting action than with 30 degrees with most of the chip taken from the leading edge and a scraping cut from the following edge of the tool. (See Figure 7 in Unit 11 of this section.)

Figure 1. An equal chip is formed on each side of the threading tool when the infeed is made with the cross slide. (Lane Community College)

Figure 2. A chip is formed on the leading edge of the tool when the infeed is made with the compound set at 29 degrees to the right of the operator to make right-hand threads. (Lane Community College)

SETTING UP FOR THREADING

Begin setup by obtaining or grinding a tool for cutting Unified threads of the required thread pitch. The only difference in tools for various pitches is the flat on the end of the tool. For Unified threads this is .125 *P*, as discussed in the last unit. If the toolholder you are using has no back rake, no grinding on the top of the tool is necessary. If the toolholder does have back rake, the tool must be ground to provide zero rake (Figure 3).

A center gage (Figure 4) may be used to check

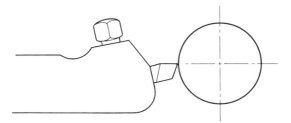

Figure 3. The tool must have zero rake and be set on the center of the work in order to produce the correct form. (*Machine Tools and Machining Practices*)

Figure 4. Checking the tool angle with a center gage. (*Machine Tools and Machining Practices*)

the tool angle. An adequate allowance for the helix angle on the leading edge will assure sufficient side relief.

The part to be threaded is set up between centers, in a chuck, or in a collet (Figures 5*a* to 5*c*). The tool is clamped in the holder and set on the centerline of the workpiece (Figures 6*a* to 6*c*). A center gage is used to align the tool to the workpiece (Figure 7). The toolholder is clamped tightly after the tool is properly aligned.

Setting Dials on the Compound and Cross Feed

The point of the tool is brought into contact with the work by moving the cross feed handle, and the micrometer collar is set on the zero mark (Figure 8). The compound micrometer collar should also be set on zero (Figure 9), but first be sure all slack or backlash is removed by turning the compound feed handle clockwise.

Setting Apron Controls

On some lathes a feed change lever, which selects either cross or longitudinal feeds, must be moved to

Figure 5a. Lathe is set up for a threading project by inserting collet in collet sleeve.

Figure 5b. A stub mandrel is inserted in the collet and the collet is tightened.

Figure 5c. The sleeve to be threaded externally is mounted on the stub mandrel.

Figure 6a. The threading tool is placed in the holder and lightly clamped.

Figure 6b. The tool is adjusted to the dead center for height. A tool that is set too high or too low will not produce a true 60 degree angle in the cut thread.

a neutral position for threading. This action locks out the feed mechanism so that no mechanical interference is possible. All lathes have some interlock mechanism to prevent interference when the half-nut lever is used. The half-nut lever causes two halves of a nut to clamp over the lead screw. The carriage will move the distance of the lead of the thread on the lead screw for each revolution of the lead screw.

Threading dials operate off the lead screw and continue to turn when the lead screw is rotating and the carriage is not moving. When the half-nut lever

Figure 6c. An alternate method of adjusting the tool for height is to use the steel rule. It will be in a vertical position when the tool is on center.

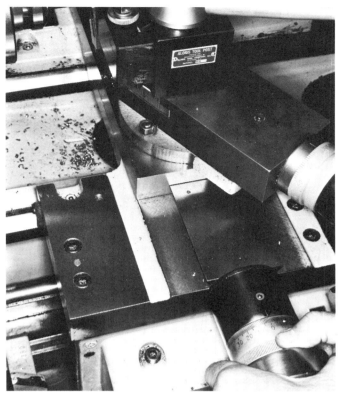

Figure 8. After the tool is brought into contact with the work, the cross feed micrometer collar is set to the zero index.

Figure 7. The tool is properly aligned by using a center gage. The toolholder is adjusted until the tool is aligned. The toolholder is then tightened.

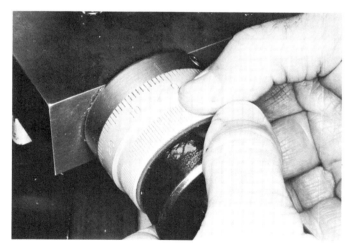

Figure 9. The operator will then set the compound micrometer collar to the zero index.

is engaged, the threading dial stops turning and the carriage moves. The marks on the dial indicate when it is safe to engage the half-nut lever. If the half-nuts are engaged at the wrong place, the threading tool will not track in the same groove as before but may cut into the center of the thread and ruin it. With any even number of threads such as 4, 6, 12, and 20, the half-nut may be engaged at *any line*. Odd

Figure 10. The threads per inch selection is made on the quick-change gearbox.

Figure 11. The half-nut lever is engaged at the correct line or numbered line depending upon whether the thread is odd, even, or a fractional numbered thread.

Figure 12. A light scratch cut is taken for the purpose of checking the pitch.

numbered threads such as 5, 7, 13 may be engaged at any *numbered line*. With fractional threads it is safest to engage the half-nut at the same line every time.

The Quick-Change Gearbox
The settings for the gear shift levers on the quick-change gearbox are selected according to the threads per inch desired (Figure 10). If the lathe has an interchangeable stud gear, be sure the correct one is in place.

Spindle Speeds
Spindle speeds for thread cutting are approximately one-fourth turning speeds. The speed should be slow enough so you will have complete control of the thread cutting operation.

CUTTING THE THREAD
The following is the procedure for cutting right-hand threads:

1. Move the tool off the end of the work ready to start the cut. Reset the cross feed micrometer dial to zero.
2. Feed in .002 in. on the compound dial.

3. Turn on the lathe and engage the half-nut lever (Figure 11).
4. Take a scratch cut without using cutting oil (Figure 12). Stop the lathe at the end of the cut and back out the tool using the cross feed. Disengage the half-nut. Return the carriage to the starting position. This method of ending the pass can only be done when there is an undercut. See "Methods of Terminating Threads" at the end of this unit.
5. Check the thread pitch with a screw pitch gage or a rule (Figure 13). If the pitch is wrong, it can still be corrected.
6. Apply cutting oil to the work (Figure 14).
7. Feed the compound in .005 in. and reset the cross

Figure 13. The pitch of the thread is being checked with a screw pitch gage.

Figure 14. Cutting fluid is applied before taking the first cut.

Figure 16. The finish cut is taken with infeed of .001 to .002 in.

feed dial to zero. Make the second cut (Figure 15).

8. Continue this process until the tool is within .010 in. of the finish depth (Figure 16).

9. Brush the threads to remove the chips. Check the thread fit with a ring gage (Figure 17a), mating part or standard nut (Figure 17b), or comparison thread micrometer (Figure 17c). The work may be removed from between centers and returned without disturbing the threading setup, provided that the tail of the dog is returned to the same slot.

10. Continue to take cuts of .001 to .002 in. (as shown in Figure 16) and check the fit between each cut. Thread the nut with your fingers; it should go on easily but without end play. A class 2 fit is desirable for most purposes.

11. Chamfer the end of the thread to protect it from damage.

Figure 15. The second cut is taken after feeding in the compound .005 in.

Figure 17a. The thread is checked with a ring gage.

Figure 17c. A thread comparison micrometer may be used to check the threads against a known standard such as a precision thread plug gage.

Figure 17b. A standard nut is often used to check a thread.

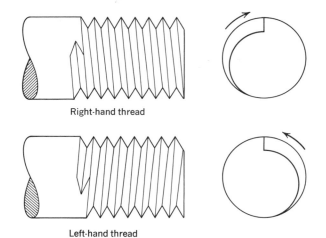

Right-hand thread

Left-hand thread

Figure 18a. The difference between right-hand and left-hand threads as seen from the side and end. (*Machine Tools and Machining Practices*)

Left-Hand Threads

The procedure for cutting left-hand threads (Figure 18a) is the same as that used for cutting right-hand threads with two exceptions. The compound is set at 29 degrees to the left of the operator (Figure 18b) and the lead screw rotation is reversed so the cut is made from the left to the right. The feed reverse lever is moved to reverse the lead screw. Sufficient undercut must be made for a starting place for the tool. Also, sufficient relief must be provided on the right side of the tool.

Methods of Terminating Threads

Undercuts are often used for terminating threads. They should be made the single depth of the threads plus .005 in. The undercut should have a radius to lessen the possibility of fatigue failure resulting from stress concentration in the sharp corners.

Machinists sometimes simply remove the tool quickly at the end of the thread while disengaging the half-nuts. If a machinist misjudges and waits too long, the point of the threading tool will be broken off. A dial indicator is sometimes used to locate the

Figure 18b. Compound set for cutting a left-hand thread. (Lane Community College)

Figure 20. By placing the blade of the threading tool in the upper position, this tool can be made to thread on the bottom side. The lathe is reversed and the thread is cut from left to right making the job easier when threading next to a shoulder. (Aloris Tool Company, Inc.)

Figure 19. A threading tool that is used for threading to a shoulder. (Lane Community College)

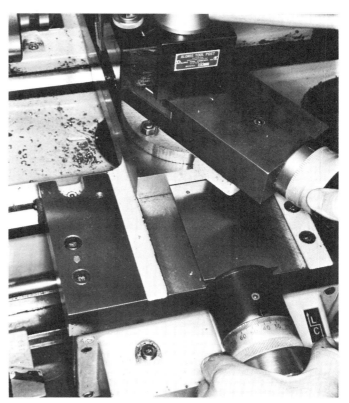

Figure 21. Repositioning the tool.

exact position for removing the tool. When this tool withdrawal method is used, an undercut is not necessary.

Terminating threads close to a shoulder requires specially ground tools (Figure 19). Sometimes it is convenient to turn the tool upside down and reverse the lathe when cutting right-hand threads to a shoulder. Some commercial threading tools are made for this purpose (Figure 20).

Picking Up a Thread

It sometimes becomes necessary to reset the tool

when its position against the work has been changed during a threading operation. This position change may be caused by removing the threading tool for grinding, by the work slipping in the chuck or lathe dog, or by the tool moving from the pressure of the cut.

To reposition the tool the following steps may be taken:

1. Check the tool position with reference to the work by using a center gage. If necessary, realign the tool.
2. With the tool backed away from the threads, en-

gage the half-nuts with the machine running. Turn off the machine with the half-nut still engaged and the tool located over the partially cut threads.
3. Position the tool in its original location in the threads by moving both the cross feed and compound handles (Figure 21).
4. Set the micrometer dial to zero on the cross feed collar and set the dial on the compound to the last setting used.
5. Back off the cross feed and disengage the half-nuts. Resume threading where you left off. This method is also useful when repairing damaged threads on a part. (See Section I, Unit 4.)

SELF-TEST

1. By what method are threads cut or chased with a single point tool in a lathe? How can a given helix or lead be produced?
2. The better practice is to feed the tool in with the compound set at 29 degrees rather than with the cross slide when cutting threads. Why is this so?
3. By what means should a threading tool be checked for the 60 degree angle?
4. How can the number of threads per inch be checked?
5. How is the tool aligned with the work?
6. Is the carriage moved along the ways by means of gears

when the half-nut lever is engaged? Explain.
7. Explain which positions on the threading dial are used for engaging the half-nuts for even, odd, and fractional numbered threads.
8. How fast should the spindle be turning for threading?
9. What is the procedure for cutting left-hand threads?
10. If for some reason it becomes necessary for you to temporarily remove the tool or the entire threading setup before a thread is completed, what procedure is needed when you are ready to finish the thread?

UNIT 13 CUTTING UNIFIED INTERNAL THREADS

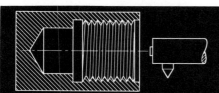

While small internal threads are tapped, larger sizes from one inch and up are often cut in a lathe. The problems and calculations involved with cutting internal threads differ in some ways from those of cutting external threads. This unit will help you understand these differences.

OBJECTIVES

After completing this unit, you should be able to:
1. Calculate the dimensions for a given internal Unified thread.
2. Determine the procedure for cutting a 1–8 UNC internal thread to fit a plug gage.

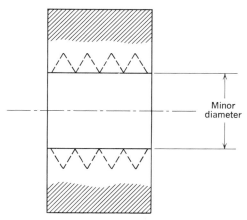

Figure 1. Diagram of internal thread showing minor diameter. *(Machine Tools and Machining Practices)*

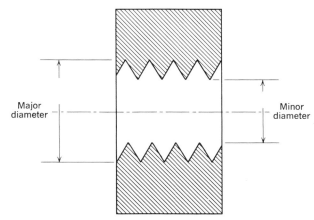

Figure 2. View of internal threads showing major and minor diameters. *(Machine Tools and Machining Practices)*

Many of the same rules used for external threading apply to internal threading: the tool must be shaped to the exact form of the thread, and the tool must be set on the center of the workpiece. When cutting an internal thread with a single point tool, the inside diameter of the workpiece should be the *minor diameter* of the internal thread (Figure 1). On the other hand, if the thread is made by tapping in the lathe, the inside diameter of the workpiece can be varied to obtain the desired percent of thread. The purpose of making an internal thread less than 100 percent is for easier tapping, especially when tough materials, such as stainless steel, are used.

TAPPING THREADS ON THE LATHE
Full depth or 100 percent threads are very difficult to tap in soft metals and impossible in tough materials. Tests have proven that above 60 percent of thread very little additional strength is gained. Lower percentages provide less flank surface for wear, however. Most commercial internal threads in steel are about 75 percent. Tap drill charts are generally based on 75 percent thread calculations for American National threads. The larger the tap drill, the lower the percent of thread. These figures are correct, however, only if the drilled hole is not oversize.

The probable percent of thread may be 5 to 10 percent lower than the calculated percent since drills usually make an oversize hole. In practice, a drill slightly under the tap drill size should first be used so the tap drill will make a more accurate hole.

The major and minor diameters of the internal thread are shown in Figure 2. For tapped internal threads, the minor diameter is varied according to the percent of thread desired. The percent of thread for Unified threads is calculated: the basic major diameter minus the actual minor diameter, divided by two times the basic thread height, expressed as a percentage. Thus,

Percent of thread
$$= \frac{\text{major } D - \text{actual minor } D}{2 \times \text{basic thread height}} \times 100$$

If you should need to select a tap drill size, or determine the bore size for tapping an internal thread, use this formula in a different form. Solving for the actual minor diameter, which should be the inside diameter of the hole, the formula becomes:

Actual minor D
$$= \text{major } D - 2 \times \text{basic thread height} \\ \times \text{Percent of thread}$$

EXAMPLE
Determine the drill size for tapping an 80 percent 1 in.–8 UNC internal thread.

$$\text{Actual minor } D = 1 - \left(\frac{.541}{8} \times 2 \times .8\right) = .892$$
$$\text{Tap drill size} = .892$$

A simple formula for calculating the drill size for an approximately 75 percent thread is:

$$\text{Major diameter} - \text{Pitch} = \text{Tap drill}$$

or,

$$\text{Major diameter} - \frac{1}{\text{TPI}} = \text{Tap drill}$$

Figure 3*a*. For right-hand internal threads, the compound is swiveled to the left. (Lane Community College)

Figure 3*b*. The compound rest is swiveled to the right for left-hand internal threads. (Lane Community College)

MAKING INTERNAL THREADS WITH A SINGLE POINT TOOL

A bolt or external thread is made with 100 percent threads, and the outside diameter is always close to its nominal size within tolerances. There is a clearance of a few thousandths of an inch between the flank of the internal thread or nut and the flank of the bolt, which is provided by making the major diameter of the internal thread .002 to .005 in. oversize.

The advantages of making internal threads with a single point tool are that large threads of various forms can be made and that the threads are concentric to the axis of the work. The threads may not be concentric when they are tapped. There are some difficulties encountered when making internal threads. The tool is often hidden from view and tool spring must be taken into account.

Figure 4. Aligning the threading tool with a center gage. (Lane Community College)

PROCEDURE

The hole to be threaded is first drilled to $\frac{1}{16}$ in. diameter less than the minor diameter. Then a boring bar is set up, and the hole is bored to the minor diameter of the thread. If the thread is to go completely through the work, no recess is necessary, but if threading is done in a blind hole, a recess must be made. The compound rest should be swiveled 29 degrees to the left of the operator for cutting right-hand threads (Figures 3*a* and 3*b*). A threading tool is clamped in the bar and aligned by means of a center gage (Figure 4).

The compound micrometer collar is moved to the zero index after the slack has been removed by turning the screw outwards or counterclockwise. The tool is brought to the work with the cross slide handle and its collar is set on zero. Threading may now proceed in the same manner as it is done with external threads. The compound is advanced outwards a few thousandths of an inch, a scratch cut is made and the thread pitch is checked with a screw pitch gage.

The cross slide is backed out of the cut and reset to zero before the next pass. Cutting oil is used. The compound is advanced a few thousandths (.001 to .010 in.). The exact amount of infeed depends on how rigid the boring bar and holder are and how deep the cut has progressed. Too much infeed will cause the bar to spring away and produce a bell-mouth internal thread.

If a slender boring bar is necessary or there is more than usual overhang, lighter cuts must be used to avoid chatter. The bar may spring away from the cut causing the major diameter to be less than the calculated amount, or that amount fed in on the compound. If several free cuts are taken through the thread with the same setting on the compound, this problem can often be corrected.

The single depth of the Unified internal thread (Figure 5) equals $P \times .541$. The minor diameter is found by subtracting the double depth of the thread from the major diameter.

Thus, if

$$D = \text{Major diameter}$$
$$d = \text{Minor diameter}$$
$$P = \text{Thread pitch}$$
$$P \times .541 = \text{Single depth}$$

the formula is

$$d = D - (P \times .541 \times 2)$$

EXAMPLE

A $1\frac{1}{2}$–6 UNC nut must be bored and threaded to fit a stud. Find the dimension of the bore.

$$P = \tfrac{1}{6} = .1666$$
$$d = 1.500 - (.1666 \times .541 \times 2)$$
$$d = 1.500 - .180$$
$$d = 1.320$$

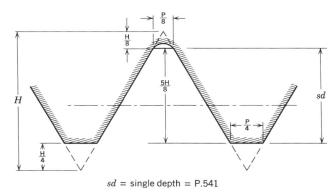

$sd = \text{single depth} = P.541$

Figure 5. Single depth of the Unified internal thread. *(Machine Tools and Machining Practices)*

Thus, the bore should be made 1.320 in.

The infeed on the compound is calculated in the same way as with external threads: except for Unified internal threads, use the formula:

$$\text{Infeed} = P \times .625$$

for the depth of cut with the compound set at 29 degrees. Using the pitch from the previous example,

$$\text{Infeed} = .1666 \times .625 = .104 \text{ in.}$$

Often it is necessary to realign an internal threading tool with the thread when the tool has been moved for sharpening or when the setup has moved during the cut. The tool is realigned in the same way as it is done for external threads: by engaging the half-nut and positioning the tool in a convenient place over the threads, then moving both the compound and cross slides to adjust the tool position.

SELF-TEST

1. When internal threads are made with a tool, what should the bore size be?
2. In what way is percent of thread obtained? Why is this done?
3. What percent of thread are tap drill charts usually based on?
4. Drills often make an oversize hole that lowers the percent of thread that a tap will cut. How can a more precise hole be drilled?
5. Your specifications call for a 60 percent thread in a tough stainless steel casting for a $\frac{1}{2}$–13 UNC tapped thread. If the Tap drill size = Major $D - 2 \times$ basic thread height $\times$ percent of thread what would your tap drill size be?
6. Name two advantages of making internal threads with a single point tool on the lathe.
7. When making internal right-hand threads, which direction should the compound be swiveled?
8. After a scratch cut is made, what would the most convenient method be to measure the pitch of the internal thread?
9. What does deflection or spring of the boring bar cause when cutting internal threads?
10. Using $P \times .541$ as a constant for Unified single depth internal threads, what would the minor diameter be for a 1–8 thread?

UNIT 14 TAPER TURNING, TAPER BORING, AND FORMING

Tapers are very useful machine elements that are used for many purposes. The machinist should be able to quickly calculate a specific taper and to set up a machine to produce it. The machinist should also be able to accurately measure tapers and determine proper fits. This unit will help you understand the various methods and principles involved in making a taper.

OBJECTIVE

After completing this unit, you should be able to:
Describe different types of tapers and the methods used to produce and measure them.

USES OF TAPERS

Tapers are used on machines because of their capacity to align and hold machine parts and to realign when they are repeatedly assembled and disassembled. This repeatability assures that tools such as centers in lathes, taper shank drills in drill presses, and arbors in milling machines will run in perfect alignment when placed in the machine. When a taper is slight, such as a Morse taper that is about $\frac{5}{8}$ in. taper/ft, it

is called a self-holding taper since it is held in and driven by friction (Figure 1). A steep taper, such as a quick-release taper of $3\frac{1}{2}$ in./ft and used on most milling machines, must be held in place with a draw bolt (Figure 2).

A taper may be defined as a uniform increase in diameter on a workpiece for a given length measured parallel to the axis. Internal or external tapers are expressed in taper per foot (TPF), taper per inch

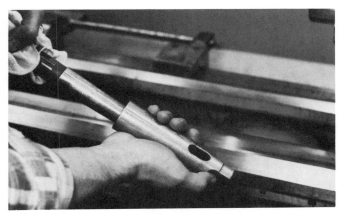

Figure 1. The Morse taper shank on this drill keeps the drill from turning when the hole is being drilled. (Lane Community College)

Figure 2. The milling machine taper is driven by lugs and held in by a draw bolt. (Lane Community College)

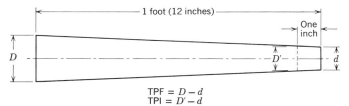

Figure 3. The difference between taper per foot (TPF) and taper per inch (TPI). *(Machine Tools and Machining Practices)*

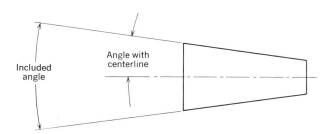

Figure 4. Included angles and angles with centerline. *(Machine Tools and Machining Practices)*

(TPI), or in degrees. The TPF or TPI refers to the difference in diameters in the length of one foot or one inch, respectively (Figure 3). This difference is measured in inches. Angles of taper, on the other hand, may refer to the included angles or the angles with the centerline (Figure 4).

Some machine parts that are measured in taper per foot are mandrels (.006 in./ft), taper pins and reamers ($\frac{1}{4}$ in./ft), the Jarno taper series (.600 in./ft), the Brown and Sharpe taper series ($\frac{1}{2}$ in./ft), and the Morse taper series (about $\frac{5}{8}$ in./ft). Morse tapers include eight sizes that range from size 0 to size 7. Tapers and dimensions vary slightly from size to size in both the Brown and Sharpe and the Morse series. For instance, a No. 2 Morse taper has .5944 in./ft taper and a No. 4 has .6233 in./ft taper. See Table 1 for more information on Morse tapers.

METHODS OF MAKING A TAPER

There are four methods of turning a taper on a lathe. They are the compound slide method, the offset tailstock method, the taper attachment method, and the use of a form tool. Each method has its advantages and disadvantages, so the kind of taper needed on a workpiece should be the deciding factor in the selection of the method that will be used.

The Compound Slide Method

Both internal and external short steep tapers can be turned on a lathe by hand feeding the compound slide

(Figure 5). The swivel base of the compound is divided in degrees. When the compound slide is in line with the ways of the lathe, the 0 degree line will align with the index line on the cross slide (Figure 6). When the compound is swiveled off the index, which is parallel to the centerline of the lathe, a direct reading may be taken for the half angle or angle to centerline of the machined part (Figure 7). When a taper is machined off the lathe centerline, its included angle will be twice the angle that is set on the compound. Not all lathes are indexed in this manner.

When the compound slide is aligned with the axis of the cross slide and swiveled off the index in either direction, an angle is directly read off the cross slide centerline (Figure 8). Since the lathe centerline is 90

Figure 5. Making a taper using the compound slide. (Lane Community College)

Figure 6. Alignment of the compound parallel with the ways. (Lane Community College)

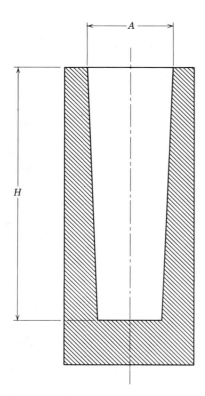
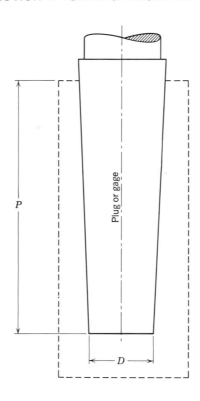

Table 1

Morse Tapers Information

Number of Taper	Taper per Foot	Taper per Inch	P Standard Plug Depth	D Diameter of Plug at Small End	A Diameter at End of Socket	H Depth of Hole
0	.6246	.0520	2	.252	.356	$2\frac{1}{32}$
1	.5986	.0499	$2\frac{1}{8}$	.396	.475	$2\frac{3}{16}$
2	.5994	.0500	$2\frac{9}{16}$	.572	.700	$2\frac{5}{8}$
3	.6023	.0502	$3\frac{3}{16}$	.778	.938	$3\frac{1}{4}$
4	.6232	.0519	$4\frac{1}{16}$	1.020	1.231	$4\frac{1}{8}$
5	.6315	.0526	$5\frac{3}{16}$	1.475	1.748	$5\frac{1}{4}$
6	.6256	.0521	$7\frac{1}{4}$	2.116	2.494	$7\frac{3}{8}$
7	.6240	.0520	10	2.750	3.270	$10\frac{1}{8}$

Source: Warren T. White, John E. Neely, Richard R. Kibbe, and Roland O. Meyer, *Machine Tools and Machining Practices,* John Wiley and Sons, Inc., Copyright © 1977, New York.

degrees from the cross slide centerline, the reading on the lathe centerline index is the complementary angle. So, if the compound is set off the axis of the cross slide $14\frac{1}{2}$ degrees, the lathe centerline index reading is $90 - 14\frac{1}{2} = 75\frac{1}{2}$ degrees, as seen in Figure 8.

Tapers of any angle may be cut by this method, but the length is limited to the stroke of the compound slide. Since tapers are often given in TPF, it is sometimes convenient to consult a TPF to angle conversion table, as in Table 2. A more complete table may be

found in the *Machinery's Handbook.*

If a more precise conversion is desired, the following formula may be used to find the included angle.

Divide the taper in inches per foot by 24; find the angle that corresponds to the quotient in a table of tangents and double this angle. If the angle with centerline is desired, do not double the angle.

EXAMPLE
What angle is equivalent to a taper of $3\frac{1}{2}$ in./ft?

Figure 7. An angle may be set off the axis of the lathe from this index. (Lane Community College)

Figure 8. The compound set 14½ degrees off the axis of the cross slide. (Lane Community College)

Table 2
Tapers and Corresponding Angles

Taper per Foot	Included Angle		Angle with Centerline		Taper per Inch
	Degrees	Minutes	Degrees	Minutes	
$\frac{1}{8}$	0	36	0	18	.0104
$\frac{3}{16}$	0	54	0	27	.0156
$\frac{1}{4}$	1	12	0	36	.0208
$\frac{5}{16}$	1	30	0	45	.0260
$\frac{3}{8}$	1	47	0	53	.0313
$\frac{7}{16}$	2	5	1	2	.0365
$\frac{1}{2}$	2	23	1	11	.0417
$\frac{9}{16}$	2	42	1	21	.0469
$\frac{5}{8}$	3	00	1	30	.0521
$\frac{11}{16}$	3	18	1	39	.0573
$\frac{3}{4}$	3	35	1	48	.0625
$\frac{13}{16}$	3	52	1	56	.0677
$\frac{7}{8}$	4	12	2	6	.0729
$\frac{15}{16}$	4	28	2	14	.0781
1	4	45	2	23	.0833
$1\frac{1}{4}$	5	58	2	59	.1042
$1\frac{1}{2}$	7	8	3	34	.1250
$1\frac{3}{4}$	8	20	4	10	.1458
2	9	32	4	46	.1667
$2\frac{1}{2}$	11	54	5	57	.2083
3	14	16	7	8	.2500
$3\frac{1}{2}$	16	36	8	18	.2917
4	18	56	9	28	.3333
$4\frac{1}{2}$	21	14	10	37	.3750
5	23	32	11	46	.4167
6	28	4	14	2	.5000

Source: Warren T. White, John E. Neely, Richard R. Kibbe, and Roland O. Meyer, *Machine Tools and Machining Practices*, John Wiley and Sons, Inc., Copyright © 1977, New York.

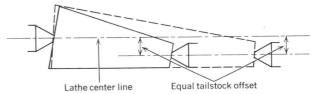

Figure 9. When tapers are of different lengths, the TPF is not the same with the same offset. *(Machine Tools and Machining Practices)*

$$\frac{3.5}{24} = .14583$$

The angle of this tangent is 8 degrees 18 minutes, and the included angle is twice this, or 16 degrees 36 minutes.

The Offset Tailstock Method
Long, slight tapers may be produced on shafts and external parts between centers. Internal tapers cannot be made by this method. Power feed is used so good finishes are obtainable. The taper per foot or taper per inch must be known so the amount of offset for the tailstock can be calculated. Since tapers are of different lengths, they would not be the same TPI or TPF for the same offset (Figure 9). When the taper per inch is known, the offset calculation is as follows: Where

$$TPI = \text{taper per inch}$$
$$L = \text{length of workpiece}$$
$$\text{Offset} = \frac{TPI \times L}{2}$$

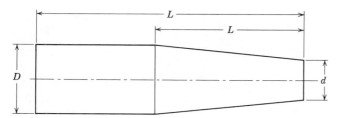

Figure 10. Long workpiece with a short taper. *(Machine Tools and Machining Practices)*

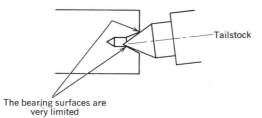

Figure 11. The contact area between the center hole and the center is small. *(Machine Tools and Machining Practices)*

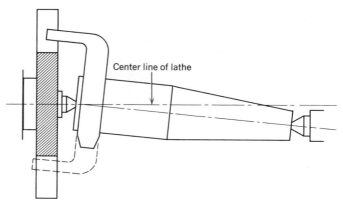

Figure 12. The bent tail of the lathe dog should have adequate clearance. *(Machine Tools and Machining Practices)*

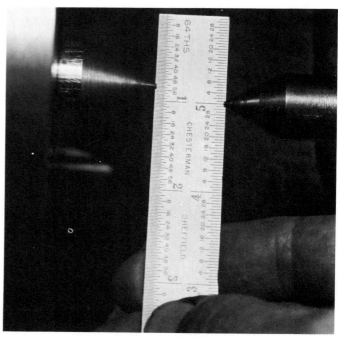

Figure 13. Measuring the offset on the tailstock by use of the centers and the rule. (Lane Community College)

When the taper per foot is known, use the following formula:

$$\text{Offset} = \frac{\text{TPF} \times L}{24}$$

If the workpiece has a short taper in any part of its length (Figure 10) and the TPI or TPF is not given use the following formula:

$$\text{Offset} = \frac{L \times (D-d)}{2 \times L_1}$$

Where

D = diameter at large end of taper
d = diameter at small end of taper
L = total length of workpiece
L_1 = length of taper

When you set up for turning a taper between centers, remember that the contact area between the center and the center hole is limited (Figure 11). Frequent lubrication of the centers may be necessary.

You should also note the path of the lathe dog bent tail in the drive slot (Figure 12). Check to see that there is adequate clearance.

To measure the offset on the tailstock, use either the centers and a rule (Figure 13) or the witness mark and a rule (Figure 14); both methods are adequate for some purposes. A more precise measurement is possible with a dial indicator as shown in Figure 15.

The indicator is set on the tailstock spindle while the centers are still aligned. A slight loading of the indicator is advised, since the first .010 or .020 in. movement of the indicator may be inaccurate or the mechanism loose due to wear, causing fluctuating readings. The bezel is set at zero; the tailstock is loosened and moved toward the operator the calculated amount. Clamp the tailstock to the way. If the indicator reading changes, loosen the clamp and readjust.

Figure 14. Measuring the offset with the witness mark and a rule. (Lane Community College)

Figure 15. Using the dial indicator to measure the offset. (Lane Community College)

Figure 16a. The toolholder is brought to the tailstock spindle using a paper strip as a feeler gage. The micrometer dial is set to zero. (Lane Community College)

Another accurate method for offsetting the tailstock is to use the cross slide (Figures 16a to 16c). With the centers aligned, bring the reverse end of the toolholder in contact with the tailstock spindle. A paper strip may be used as a feeler gage. Set the micrometer dial to zero. Back off the cross slide the calculated amount plus a full turn to remove backlash; then turn back in to the calculated amount. Move the tailstock until it contacts the paper strip held at the end of the toolholder.

Figure 16b. The cross slide is backed off the desired amount plus one full turn. (Lane Community College)

Figure 16c. The toolholder is brought forward to the desired setting and the tailstock is moved over until it contacts the paper strip. (Lane Community College)

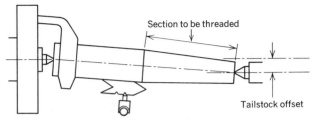

Figure 17. Adjusting the threading tool for cutting tapered threads. The tool is set square to the centerline of the work rather than the taper. *(Machine Tools and Machining Practices)*

Figure 18. The plain taper attachment. (Lane Community College)

When cutting tapered threads such as pipe threads, the tool should be square with the centerline of the workpiece, not the taper (Figure 17). When you have finished making tapers by the offset tailstock method, realign the centers to .001 in. or less in 12 in. When more than one part must be turned by this method, all parts must have identical lengths and center hole depths if the tapers are to be the same.

The Taper Attachment Method

The taper attachment features a slide independent to the ways that can be angled and will move the cross slide according to the angle set. Slight to fairly steep tapers ($3\frac{1}{2}$ in./ft) may be made, but length is limited to the stroke of the taper attachment. Centers may remain in line without distortion of the center holes. Work may be held in a chuck and both external and internal tapers may be made, often with the same setting for mating parts. Power feed is used. Taper attachments are graduated in taper per foot (TPF) or in degrees.

There are two types of taper attachments, the plain taper attachment (Figure 18) and the telescop-ing taper attachment (Figure 19). The cross feed binding screw must be removed to free the nut when the plain type is set up. The depth of cut must then be made by using the compound feed screw handle. The cross feed may be used for depth of cut when using the telescoping taper attachment since the cross feed binding screw is not disengaged with this type.

When a workpiece is to be duplicated or an internal taper is to be made for an existing external taper, it is often convenient to set up the taper attachment by using a dial indicator (Figure 20). The contact point of the dial indicator must be on the center of the

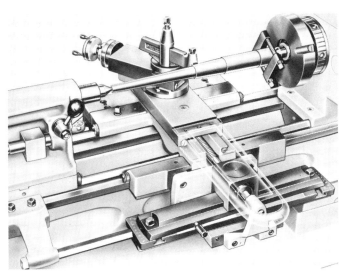

Figure 19. The telescopic taper attachment. (Clausing Machine Tools)

Figure 20. Adjusting the taper attachment to a given taper with a dial indicator. (Lane Community College)

workpiece. The workpiece is first set up in a chuck or between centers so there is no runout when it is rotated. With the lathe spindle stopped, the indicator is moved from one end of the taper to the other. The taper attachment is adjusted until the indicator does not change reading when moved.

The angle, the taper per foot, or the taper per inch must be known to set up the taper attachment to cut specific tapers. If none of these are known, proceed as follows.

If the end diameters (*D* and *d*) and the length of taper (*L*) are given in inches:

$$\text{Taper per foot} = \frac{D - d}{L} \times 12$$

If the taper per foot is given, but you want to know the amount of taper in inches for a given length, use the following formula.

$$\text{Amount of taper} = \frac{\text{TPF}}{12} \times \text{given length of tapered part}$$

When the TPF is known, to find TPI divide the TPF by 12. When the TPI is known, to find TPF multiply the TPI by 12.

To set up the taper attachment (refer to Figure 21) proceed as follows:

1. Clean and oil the slide bar (*a*).
2. Set up the workpiece and the cutting tool on center. Bring the tool near the workpiece and to the center of the taper.
3. Remove the cross feed binding screw (*b*) that

binds the cross feed screw nut to the cross slide. *Do not remove* this screw if you are using a telescoping taper attachment. The screw is removed *only* on the plain type. Put a temporary plug in the hole to keep chips out.

4. Loosen the lock screws (*c*) on both ends of the slide bar and adjust to the required degree of taper.
5. Tighten the lock screws.
6. Tighten the binding lever (*d*) on the slotted cross slide extension at the sliding block, *plain type only*.
7. Lock the clamp bracket (*e*) to the lathe bed.
8. Move the carriage to the right so that the tool is from ½ to ¾ in. past the start position. This should be done on every pass to remove any backlash in the taper attachment.
9. Feed the tool in for the depth of the first cut with the cross slide unless you are using a plain-type attachment. Use the compound slide for the plain type.
10. Take a trial cut and check for diameter. Continue the roughing cut.
11. Check the taper for fit and readjust the taper attachment, if necessary.
12. Take a light cut, about .010 in. and check the taper again. If it is correct, complete the roughing and final finish cuts.

Internal tapers (Figure 22) are best made with the taper attachment. They are set up in the same manner as prescribed for external tapers.

Other Methods of Making Tapers

A tool may be set with a protractor to a given angle

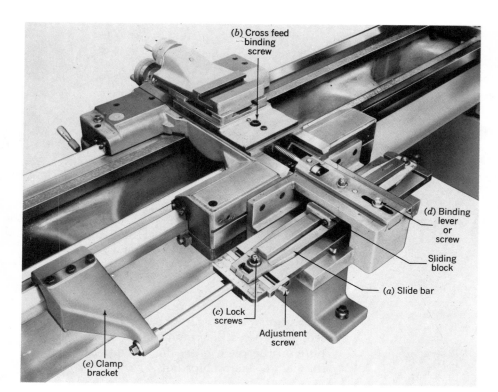

Figure 21. The parts of the taper attachment. (The Lodge & Shipley Company)

(b) Cross feed binding screw

(d) Binding lever or screw

Sliding block

(a) Slide bar

(c) Lock screws

Adjustment screw

(e) Clamp bracket

Hole   Screw

Figure 22. Internal taper being made with a plain taper attachment. Note that the cross feed nut locking screw has been removed and the hole has not been plugged. This hole *must* be plugged and the screw stored in a safe place. (Lane Community College)

(Figures 23a and 23b) and a single plunge cut may be made to produce a taper. This method is often used for chamfering a workpiece to an angle such as the chamfer used for hexagonal bolt heads and nuts. Tapered form tools sometimes are used to make V-

shaped grooves. Only very short tapers can be made with form tools.

Tapered reamers are sometimes used to produce a specific taper such as a Morse taper. A roughing reamer is first used, followed by a finishing reamer. Finishing Morse taper reamers are often used to true up a badly nicked and scarred internal Morse taper.

METHODS OF MEASURING TAPERS

The most convenient and simple way of checking tapers is to use the taper plug gage (Figure 24) for internal tapers and the taper ring gage (Figure 25) for external tapers. Some taper gages have Go and No Go limit marks on them (Figure 26).

To check an internal taper, a chalk or prussian blue mark is first made along the length of the taper plug gage (Figures 27a and 27b). The gage is then inserted into the internal taper and turned slightly. When the gage is taken out, the chalk mark will be partly rubbed off where contact was made. Adjustment of the taper should be made until the chalk mark is rubbed off along its full length of contact, indicating a good fit. An external taper is marked with chalk to be checked in the same way with a taper ring gage (Figures 28a to 28c).

The taper per inch may be checked with a micrometer by scribing two marks one inch apart on the taper and measuring the diameters (Figures 29a

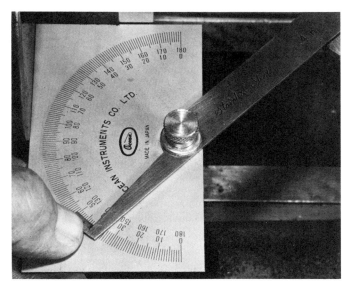

Figure 23a. Tool is set up with protractor to make an accurate chamfer or taper. (Lane Community College)

Figure 24. Taper plug gage. (Lane Community College)

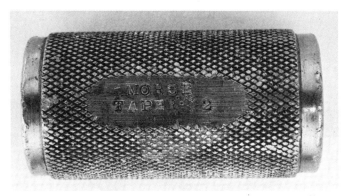

Figure 25. Taper ring gage. (Lane Community College)

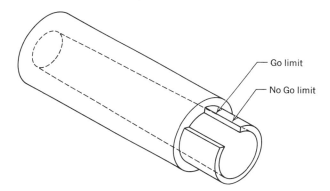

Figure 26. Go/No Go taper ring gage. (*Machine Tools and Machining Practices*)

Figure 23b. Making the chamfer with a tool. (Lane Community College)

and 29b) at these marks. The difference is the taper per inch. A more precise way of making this measurement is shown in Figures 30a and 30b. A surface plate is used with precision parallels and drill rods. The tapered workpiece would have to be removed from the lathe if this method is used, however.

Perhaps an even more precise method of measuring a taper is with the sine bar and gage blocks on the surface plate (Figure 31). When this is done, it is important to keep the centerline of the taper paral-

Figure 27a. Chalk mark is made along a taper plug gage prior to checking an internal taper. (Lane Community College)

Figure 27b. The taper has been tested and the chalk mark has been rubbed off evenly, indicating a good fit. (Lane Community College)

Figure 28a. The external taper is marked with chalk or prussian blue before being checked with a taper ring gage. (Lane Community College)

Figure 28b. The ring gage is placed on the taper snugly and is rotated slightly. (Lane Community College)

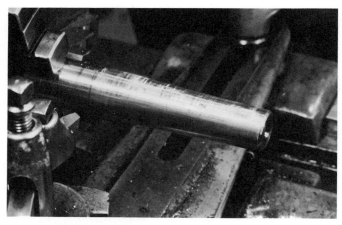

Figure 28c. The ring gage is removed and the chalk mark is rubbed off evenly for the entire length of the ring gage, which indicates a good fit. (Lane Community College)

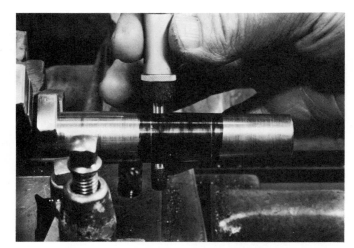

Figure 29a. Measuring the taper per inch (TPI) with a micrometer. The larger diameter is measured on the line with the edge of the spindle and the anvil of the micrometer contacting the line. (Lane Community College)

lel to the sine bar and to read the indicator at the highest point.

Tapers are also measured with taper micrometers. The outside taper micrometer (Figure 32) and inside taper micrometer (Figure 33) measure the taper directly and without removing the part from the lathe.

Figure 29b. The second measurement is taken on the smaller diameter at the edge of the line in the same manner. (Lane Community College)

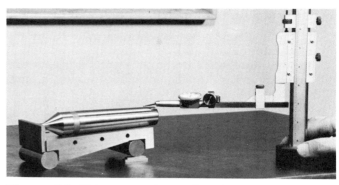

Figure 31. Using a sine bar and gage blocks with a dial indicator to measure a taper. (Lane Community College)

Figure 30a. Checking the taper on a surface plate with precision parallels, drill rod, and micrometer. The first set of parallels is used so that the point of measurement is accessible to the micrometer. (Lane Community College)

Figure 32. Outside taper micrometer. (Taper Micrometer Corporation)

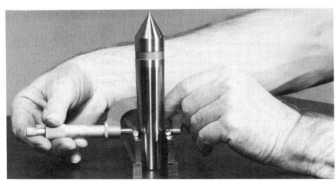

Figure 30b. When one inch wide parallels are in place, a second measurement is taken. The difference is the taper per inch. (Lane Community College)

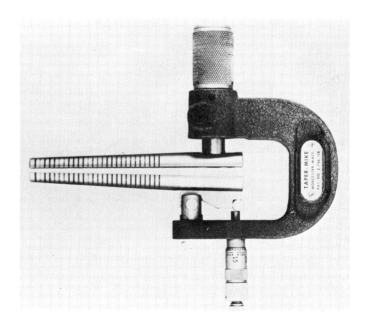

Figure 33. Inside taper micrometer. (Taper Micrometer Corporation)

SELF-TEST

1. State the difference in use between steep tapers and slight tapers.
2. In what three ways are tapers expressed (measured)?
3. Briefly describe the four methods of turning a taper in the lathe.
4. When a taper is produced by the compound slide method, is the reading in degrees on the compound swivel base the same as the angle of the finished workpiece? Explain.
5. If the swivel base is set to a 35 degree angle at the cross slide centerline index, what would the reading be at the lathe centerline index?
6. Calculate the offset for the taper shown in Figure 34. The formula is

$$\text{Offset} = \frac{L \times (D - d)}{2 \times L_1}$$

7. Name four methods of measuring the offset on the tailstock for making a taper.

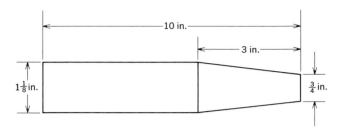

Figure 34

8. What are the two types of taper attachments and what are their advantages over other means of making a taper?
9. What is the most practical and convenient way to check internal and external tapers when they are in the lathe? Name four methods of measuring tapers.
10. Describe the kinds of tapers that may be made by using a form tool or the side of a tool.

UNIT 15 LATHE ACCESSORIES

Many lathe operations would not be possible without the use of the steady and follower rests. These valuable attachments make internal and external machining operations on long workpieces possible on a lathe. Other useful lathe accessories such as milling attachments are available for use where a lathe is the only machine tool.

OBJECTIVES

After completing this unit, you should be able to:
1. Identify the parts and explain the uses of the steady rest.
2. Explain the correct uses of the follower rest.
3. Correctly set up a steady rest on a straight shaft.
4. Correctly set up a follower rest on a prepared shaft.
5. Explain the uses of the milling attachment.
6. Wind a spring on the lathe.
7. Describe methods of making spheres in a lathe.
8. Explain eccentricity and throw and how eccentrics are made.

THE STEADY REST

On a lathe, long shafts tend to vibrate when cuts are made, leaving chatter marks. Even light finish cuts will often produce chatter when the shaft is long and slender. To help eliminate these problems, use a steady rest to support workpieces that extend from a chuck more than four or five diameters of the workpiece for turning, facing, drilling, and boring operations.

The steady rest (Figure 1) is made of a cast iron

Figure 1. The parts of the steady rest. (Lane Community College)

Figure 2. A long, slender workpiece is supported by a steady rest near the center to limit vibration or chatter. (Lane Community College)

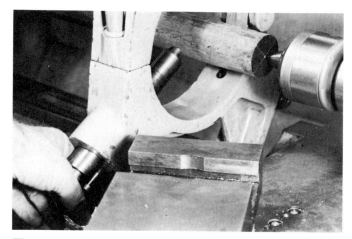

Figure 3. Adjusting the steady rest jaws to a centered shaft. (Lane Community College)

or steel frame that is hinged so it will open to accommodate workpieces. It has three or more adjustable jaws that are tipped with bronze, plastic, or ball bearing rollers. The base of the frame is machined to fit the ways of the lathe and it is clamped to the bed by means of a bolt and crossbar.

A steady rest is also used to support long workpieces for various other machining operations such as threading, grooving, and knurling (Figure 2). Heavy cuts can be made by using one or more steady rests along a shaft.

Adjusting the Steady Rest

Workpieces should be mounted and centered in a chuck whether a tailstock center is used or not. If the shaft has centers and finished surfaces that turn concentric (have no runout) with the lathe centerline, setup of the steady rest is simple. The steady rest is slid to a convenient location on the shaft, which is supported in the dead center and chuck, and the base is clamped to the bed. The two lower jaws are brought up to the shaft finger tight only (Figure 3). A good high pressure lubricant is applied to the shaft and the top half of the steady rest is closed and clamped. The upper jaw is brought to the shaft finger tight, and then all three lock screws are tightened. Some clearance is necessary on the upper jaw to avoid scoring of the shaft. As the shaft warms or heats up from friction during machining, readjustment of the upper jaw is necessary.

A finished workpiece can be scored if any hardness or grit is present on the jaws. To protect finishes, brass or copper strips or abrasive cloth is often placed between the jaws and the workpiece; with the abrasive cloth, the abrasive side is placed outward against the jaws.

When there is no center in a finished shaft, that is of uniform diameter, the setup procedure is as follows.

1. Position the steady rest near the end of the shaft with the other end lightly chucked in a three- or four-jaw chuck.

2. Scribe two cross center lines with a center head on the end of the shaft and prick punch (Figures 4*a* and 4*b*).

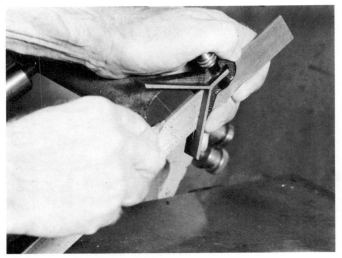

Figure 4a. Laying out the center of a shaft. (Lane Community College)

Figure 5. Turning a concentric bearing surface on rough stock for the steady rest jaws. (Lane Community College)

Figure 4b. Aligning the shaft center with the dead center. (Lane Community College)

Figure 6. Using a cat head for supporting a square piece. Drilling and boring in the end of a heavy square bar requires the use of an external cat head. (Lane Community College)

3. Bring up the dead center near to the punch mark.

4. Adjust the lower jaws of the steady rest so the dead center is aligned to the center of the shaft.

5. Tighten the chuck. If it is a four-jaw chuck, check for runout with a dial indicator.

6. The steady rest may now be moved to any location along the shaft.

Stepped shafts may be set up by using a similar procedure, but the steady rest must remain on the diameter on which it is set up.

Using the Steady Rest

A frequent misconception among students is that the

steady rest may be set up properly by using a dial indicator near the steady rest on a rotating shaft. This procedure would never work since the indicator would show no offset or runout, no matter where the jaws were moved.

Steady rest jaws should never be used on rough surfaces. When a forging, casting, or hot rolled bar must be placed in a steady rest, a concentric bearing with a good finish must be turned (Figure 5). Thick walled tubing or other materials that tend to be out of round also should have bearing surfaces machined on them. The usual practice is to remove no more in diameter than necessary to clean up the bearing spot.

When the piece to be set up is very irregular, such as a square or hexagonal part, a cat head is used

Figure 7a. Using a centered cat head to provide a center when the end of the shaft or tube cannot be centered conveniently. (Lane Community College)

Figure 7b. The cat head is adjusted over the irregular end of the shaft. (Lane Community College)

Figure 8. Tubing being set up with a cat head using a dial indicator to true the inside diameter. (Lane Community College)

Figure 9. A follower rest is used to turn this long shaft. (Lane Community College)

(Figure 6). The piece is placed in the cat head and the cat head is mounted in the steady rest while the other end of the workpiece is centered in the chuck. The workpiece is made to run true near the steady rest by adjusting screws on the cat head. In most cases the workpiece is given a center to provide more support for turning operations. A centered cat head (Figures 7a and 7b) is sometimes used when a permanent center is not required in the workpiece. Internal cat heads (Figure 8) are used for truing to the inside diameter of tubing that has an irregular wall thickness, so that a steady rest bearing spot can be machined on the outside diameter. These also have adjustment screws.

THE FOLLOWER REST
Long, slender shafts tend to spring away from the tool, vary in diameter, chatter, and often climb the cutting tool. To avoid these problems when machining

a slender shaft along its entire length, a follower rest (Figure 9) is often used. Follower rests are bolted to the carriage and follow along with the tool. Most follower rests have two jaws placed to back up the work opposite to the tool thrust. Some types are made with different size bushings to fit the work.

Figure 10. Adjusting the follower rest. (Lane Community College)

Figure 12. Both steady and follower rests being used. (Lane Community College)

Figure 11. Long, slender Acme threaded screw being machined with the aid of a follower rest. (Lane Community College)

Figure 13. Internal grinding on the lathe with a tool post grinder. Truing chuck jaws. (Lane Community College)

Using the Follower Rest

The workpiece should be one to two inches longer than the job requires to allow room for the follower rest jaws. The end is turned to smaller than the finish size. The tool is adjusted ahead of the jaws about one and one-half inches and a trial cut of two or three inches is made with the jaws backed off. Then the lower jaw is adjusted finger tight (Figure 10) followed by the upper jaw. Both locking screws are tightened. A cutting oil should be used to lubricate the jaws.

The follower rest is often used when cutting

threads on long, slender shafts, especially when cutting square or Acme threads (Figure 11). Burrs should be removed between passes to prevent them cutting into the jaws. Jaws with rolls are sometimes used for this purpose. On quite long shafts, sometimes both a steady rest and follower rest are used (Figure 12).

TOOL POST GRINDERS

Cylindrical grinders and centerless grinders have largely replaced the tool post grinder for finishing parts to accurate dimensions. Tool post grinding on the lathe, however, is still required for some special applications. The grinder may be set up for internal grinding to true up chuck jaws (Figure 13). Normal wear on chuck jaws cause them to become bell-mouthed and to have some runout. The chuck jaws

Figure 14. External tool post grinding on the lathe. A diameter is being ground to precise dimensions. (Clausing Machine Tools)

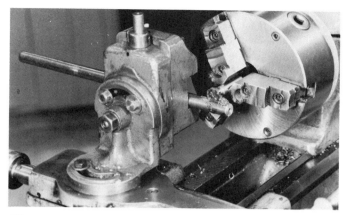

Figure 15. A woodruff keyseat being cut in a shaft that is held in a milling attachment. (Lane Community College)

Figure 16. The finished woodruff key seat. (Lane Community College)

are tightened against a slotted ring to keep pressure on them while they are being ground.

Tool post grinding is sometimes done (Figure 14) in job and maintenance shops when welded or metal sprayed surfaces are too hard to machine with a tool. This grinding may be either internal or external. Straight wheels or cup wheels may be used depending on the job requirement. The lathe ways must be protected by a suitable covering, as grinding grit will severely damage the ways. The operator should always stay out of the line of a rotating wheel.

USING THE LATHE FOR MILLING

Milling attachments can be used on many small lathes to make them more versatile. They are especially useful for doing small milling jobs in a small shop where milling machines are not available. A woodruff key seat may quickly be cut (Figures 15 and 16) in a shaft held in a milling attachment. Flats may be made and other milling operations are possible by using an endmill. Straight keyways and small milling jobs are also possible, but other applications are limited.

RADIUS TURNING

A convex or concave radius may be roughly produced by offhand turning and by filing in the lathe, while frequently checking with a radius gage or template. Small radii are produced with form tools. A more precise method of producing a convex or concave radius or a ball is by using a radii-cutter attachment (Figures

17 and 18). Large spheres may be produced by using a power driven radii-cutter (Figure 19). Radii and complex curves are produced with lathe tracer attachments (Figure 20) or on numerically controlled lathes.

SPRING WINDING ON THE LATHE

Although small springs may be wound by hand, larger springs with heavier wire must be wound with a machine or on a lathe (Figure 21). The spring wire is wrapped on an arbor that is slightly smaller than the desired inside diameter of the finished spring. The exact size may either be found by trial and error or by referring to the *Machinery's Handbook*. A hole slightly larger than the spring wire should be drilled through the arbor (Figure 22) as a starting point for the spring.

A friction device held in the tool post, such as

Figure 17. The radii-cutter forming a 2 in. groove radius on the outer portion of a bending die. (The DoAll Company)

Figure 18. The radii-cutter cutting a convex internal radius. (The DoAll Company)

Figure 19. Radii-cutter provides motor drive in place of manually operated handle for diameters larger than 5 in. (The DoAll Company)

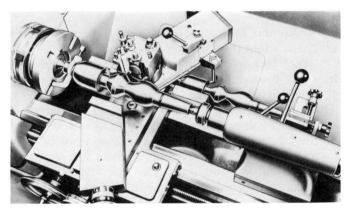

Figure 20. Contour turning with the tracing attachment. The stylus follows the template and reproduces the same pattern at the point of the tool. (Clausing Machine Tools)

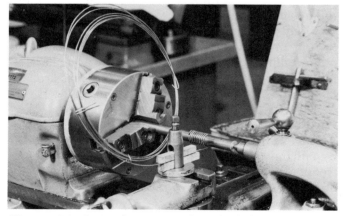

Figure 21. Spring being wound on the lathe. (Lane Community College)

the two hardwood blocks shown in Figure 21, controls the winding tension on the wire. The amount of tension influences the final size of the spring since it causes the spring to expand more or less when it is released from the arbor. Speeds are set at a low RPM.

The lathe is set in the same way as for threading according to the desired pitch of the spring; one or two wraps are usually made before engaging the half-nuts. After disengaging the half-nuts, allow one or two wraps to finish the spring. These extra turns at the ends are needed for either compression or tension-type springs. Spring wire can be snipped with wire cutters or cut by notching on a grinder and breaking. Spring ends are finished on a grinder. Some kinds

Figure 22. A hole drilled in the arbor is the starting point for the spring. (Lane Community College)

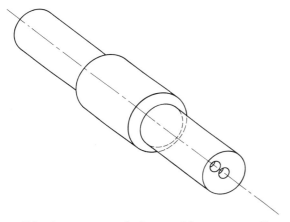

Figure 23. An eccentric shaft is used for many small mechanisms and for converting rotary motion into linear motion. (*Machine Tools and Machining Practices*)

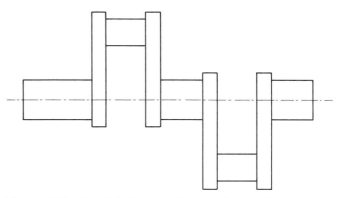

Figure 24. Crankshafts are often used in engines to convert reciprocating motion into rotary motion. (*Machine Tools and Machining Practices*)

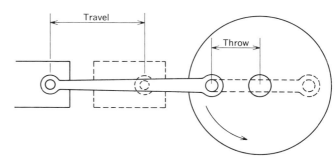

Figure 25. The travel of the reciprocating part is always twice the throw of the crankshaft or eccentric. (*Machine Tools and Machining Practices*)

of spring wires used are music wire, stainless steel wire, and brass wire.

ECCENTRIC TURNING

Eccentric shafts (Figure 23) are used for converting rotary motion (circular motion around an axis) into reciprocating motion (back-and-forth linear motion). Crankshafts (Figure 24) are often used to convert reciprocating motion into rotary motion. The throw of an eccentric is the amount the center holes of the eccentric have been set off from the normal axis of the work. The total linear movement or travel of the reciprocating part is always twice the throw of the crankshaft or eccentric (Figure 25).

A short eccentric may be turned in a four-jaw chuck (Figure 26) by offsetting one set of jaws the amount of the throw.

Figure 26. Short eccentrics may easily be machined in a four-jaw chuck. (Lane Community College)

SELF-TEST

1. When should a steady rest be used?
2. In what ways can a steady rest be useful?
3. How is the steady rest set up on a straight finished shaft when it has centers in the ends?
4. What precaution can be taken to prevent scoring of a finished shaft?
5. How can a steady rest be set up when there is no center hole in the shaft?
6. Is it possible to correctly set up a steady rest by using a dial indicator on the rotating shaft in order to watch for runout?
7. Should a steady rest be used on a rough surface? Explain.
8. How can a steady rest be used on an irregular surface such as square or hex stock?
9. When a long, slender shaft needs to be turned or threaded for its entire length, which lathe attachment could be used?
10. The jaws of the follower rest are usually one or two inches to the right of the tool on a setup. If the workpiece happens to be smaller than the dead center or tailstock spindle, how would it be possible to bring the tool to the end of the work to start a cut without interference by the follower rest jaws?
11. Name two uses for the tool post grinder.
12. What two precautions should be taken when using the tool post grinder?
13. When should lathe milling attachments be used?
14. What can be done with a lathe milling attachment?
15. Name four ways of producing radii.
16. What is the best means of producing a true sphere on the lathe?
17. Describe the lathe setup for spring winding.
18. What effect does the tension on the wire have on the final spring?
19. What are eccentric shafts used for?
20. Name two basic methods of producing an eccentric.

SECTION I

MECHANICAL PROCESSES

UNIT 1 MECHANICAL SYSTEMS

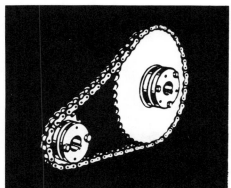

The earliest power sources were humans, animals, wind power, and water power. In order to utilize these prime energy sources for specific operations such as milling grain into flour or processing mineral ores, power transmission systems had to be developed (Figure 1). Later, when power sources such as steam, electrical, and internal combustion engines were developed, many more complex and dependable power transmission devices and systems were developed.

If an auto mechanic were to take an engine completely apart, only six or less basic mechanical elements could be found: the lever, the inclined plane, the wedge, the screw, the wheel and axle, and the pulley. These are called simple machines and one or more of them can be found in all mechanical devices. Power transmission and mechanical drives utilize these simple mechanical systems to provide powerful and sometimes complex machines that are used to manufacture the products we need and provide us with many forms of transportation. Machine components such as bearings, gears, pulleys, and shafting have been standardized to facilitate manufacture and replacement of parts. The purpose of this unit is to acquaint you with these various machine elements and mechanical systems as they are used in industrial applications.

OBJECTIVES

After completing this unit, you should be able to:
1. Identify various mechanical and hydraulic drive systems and their components.
2. Calculate drive ratios and speeds.
3. Identify sleeve, ball and roller bearings, their limitations and uses.

Although there are hundreds of mechanisms designed to transfer and alter power from one point to another, there are only about ten basic mechanical devices used. These are belts and pulleys, chains and sprockets, gears, shafting and bearings, clutches, wire rope and sheaves, friction drives, levers and linkages, cams, screws, and cranks. Of course, hydraulic, air, steam, and electrical systems could also be considered as means of power transmission.

BELTS AND PULLEYS

Flat Belts

Flat belts and pulleys were one of the most common methods of transferring power between two shafts before World War II. These have been largely replaced by V-belts and pulleys (Figure 2). An important

principle used in flat belt pulleys is called "crowning" (Figure 3). Since the belt has a tendency to climb to the highest point on the pulley, a slight "crown" in the middle centered the belt while a high side with a flange ensured a steel band position against the flange. Band saw pulleys utilize this principle to keep the saw band in place, and belt conveyor systems use center crowned pulleys. Belts tend to climb over flanges.

V-Belts

V-belts are so named because their cross section is in the shape of the letter "V" (Figure 4). They have the advantage of being able to deliver much more horsepower than flat belts in comparison to their belt and pulley sizes. (Pulleys are called sheaves when used for V-belts.) V-belts can also be single or multiple strand (Figure 5). In general, V-belts are more efficient

Figure 1. In the eighteenth century, ingenious mechanisms were used to transfer motion from a waterwheel for processing ores. (Dover Publications, Inc.)

Figure 4. Cross section of a V-belt. (Reliance Electric Company)

Figure 2. Multiple V-belts and pulleys are used to deliver a great amount of horsepower in this reduction drive. (Reliance Electric Company)

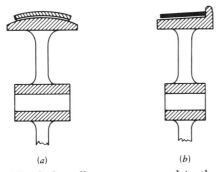

(a)   *(b)*

Figure 3. Flat belt pulleys are crowned in the center to keep the belt in place as in *(a)*. Steel saw bands are sometimes kept in place by a combination crown and flange as in *(b)*.

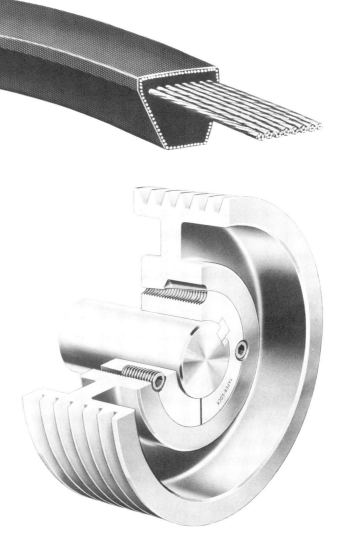

Figure 5. A multiple strand V-belt pulley. (Reliance Electric Company)

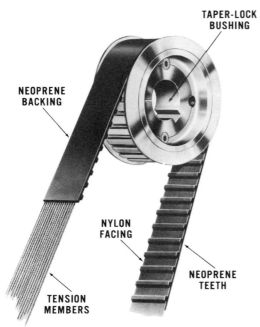

Figure 6. Synchronous belt and sprocket. (Reliance Electric Company)

Figure 7. Lumber being graded as it moves along on a conveyor chain. (Weyerhaeuser Company)

in the high speed ranges (500 to 5000 RPM, depending on size).

SYNCHRONOUS BELT DRIVES

These toothed belts and sprockets (Figure 6), sometimes called cog belt drives, have no slippage, which is often a problem in flat and V-belt drives. The advantages of synchronous belt drives are wide speed and load range, less heat generated, more power delivery in less space, and constant drive ratio, to name a few. They are used by many automotive engine manufacturers for timing belts and for many other power transmission applications.

CHAINS AND SPROCKETS

Many kinds of industrial chain and sprocket systems are in use, such as various types of conveyor chains used in mills and heavy chain drives for low speeds (Figure 7). Link chain (either formed, cast, or forged links) and cast iron sprockets are used on farm machinery and bucket conveyors. Many forms of heavy duty conveyor chains are used in mills. The most versatile and universally used chain, however, is roller chain with machine-cut steel sprockets. There are several types of roller chain, but all utilize the method of rollers on a roller link connected by a pin link (Figures 8a and 8b).

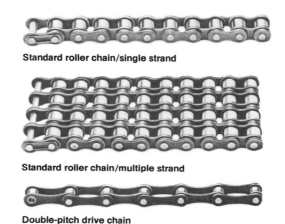

Standard roller chain/single strand

Standard roller chain/multiple strand

Double-pitch drive chain

Figure 8a. Various types of roller chain used by industry. (Reprinted with permission of FMC Corporation)

Figure 8b. Parts of roller chain links. These are hardened and tempered for high strength. (Reliance Electric Company)

Slip-fit connecting links

Spring clip type is standard for all single and multiple strand standard roller chains from .250″ to .625″ pitch.

Cotter type is standard for all single strand standard roller chains from .750″ to 2.000″ pitch, and for all double pitch chains.

Press-fit connecting links

Cotter type is standard for all single strand standard roller chains of 2.250″ pitch and larger, and for all multiple strand standard roller chains of .750″ pitch and larger.

Offset links

Offset links are standard for all single and multiple strand standard roller chains except .250″ pitch, and for all double pitch chains.

Offset assemblies

Offset assemblies are standard for .250″ pitch standard roller chain and are furnished on request for all other standard roller chains.

Roller links

Roller links, standard roller chain components, can be furnished individually. For multiple strand chains, one roller link is required for each strand.

Figure 9. Various types of special links that are used for connecting the ends and adjusting the length of the roller chain. (Reprinted with permission of FMC Corporation)

Figure 10. Roller chain pin extractor or chain breaker removes roller chain pins when various lengths of chain are needed. (Lane Community College)

Type A

Type B

Type C

Figure 11. Three common styles of roller chain sprockets. Type A is a plate sprocket; Type B is made with a hub on one side; and Type C has a hub on both sides. (Reprinted with permission of FMC Corporation)

Roller chains and sprocket drives are among the most efficient (least power loss) of any systems of power transmission. This is the reason they are used on bicycles and motorcycles. There is much more power loss in belts and pulleys and even more in hydrostatic drive systems. Roller chains, like gears, have high efficiency if they are kept lubricated in a gear box; if they run dry, their efficiency drops considerably. The efficiency and working life of roller chains can be extended by running them in oil. High reduction ratios are possible with roller chains as compared to belt systems.

Special links are made for roller chains to connect ends and to shorten or lengthen one-half link for fit up (Figure 9). Roller chain comes in lengths (usually 10 feet) that can be extended with coupler links or shortened by using a pin extractor (Figure 10). If a pin extractor is not available, the two pins in a pin link can be removed by grinding the heads flush with the link plate. Then, holding the chain in a vise and using a pin punch, drive out the two pins.

Roller chain sprockets are normally available in three styles. Type A is a plate sprocket which is adaptable to machining or as a bolt-on sprocket; Types B and C with hubs are usually bored and keyseated to fit a shaft (Figure 11).

Figure 12. Taper-Lock[R] gears, pulleys, and sprockets are manufactured with reusable, interchangeable tapered bushings that tighten on the shaft as they are installed. Bushings are made with many standard bore sizes. (Reliance Electric Company)

Figure 14. Spur gear and gear rack. (Lane Community College)

Figure 13. Silent chain. (Reprinted with permission of FMC Corporation)

Figure 15. Helical gears. (Lane Community College)

Another very commonly used sprocket is one with the patented Taper-Lock® bushing that grips the shaft, assuring true rotation and eliminating much of the wear problems associated with the standard bored, keyseated, and setscrewed Type B sprockets (Figure 12).

"Silent chain" is used with special sprockets for such purposes as timing chains on automobile engines (Figure 13). As its name implies, it is a very smooth running, quiet drive system.

Formulas and tables for RPM, horsepower, center distances, and adjustments are available in many industrial power transmission catalogs (which are usually free for the asking) and also in the *Machinery's Handbook*. Also included are sheave and belt sizes and sprocket and chain dimensions and information.

GEAR SYSTEMS

Gears provide positive, no-slip power transmission and are used to increase or decrease the turning effort or speed in machine assemblies. When two gears are running together, the one with the larger number of teeth is called the gear and the one with the smaller number of teeth is called the pinion. Gears are generally used when shaft center distances are short, to provide a constant speed ratio between shafts, or to transmit high torques.

Spur Gears

Gears fall into several categories: the first is gears that connect parallel shafts. The best known of these gears is the spur gear. Spur gears have a cylindrical form

Figure 17. Helical internal gears. (Lane Community College)

Figure 16. Herringbone gear. (Lane Community College)

with straight teeth cut into the periphery (Figure 14). When teeth are cut on a straight bar, a gear rack is made. A gear rack converts rotary gear motion into a linear movement. When the teeth on a cylindrical gear are at an angle to the gear axis, it is a helical gear (Figure 15). On a helical gear, several teeth are in mesh simultaneously with the mating gear, which provides a smoother operation than with spur gears. Because of the angle of the teeth, both radial and thrust loads are imposed on the gear support bearings. To offset this thrust effect, double helical gears, also called herringbone gears, (Figure 16) are sometimes used. They consist of a right-hand and a left-hand helix. The hand of a helical gear is determined by facing the gear teeth; if the teeth slant **downwards** to the right it is a right-hand helix.

Common reduction ratios for gears are 1:1 to 5:1 for spur gears, 1:1 to 10:1 for helical gears, and 1:1 to 20:1 for herringbone gears. These same three gear types are also manufactured as internal gears (Figure 17). An internal gear has a greater tooth strength than that of an equivalent external gear. An internal gear rotates in the same direction as its mating pinion. Internal gears permit close spacing of parallel shafts. Internal gears mesh with external pinions.

Bevel Gears

The second category concerns gears that connect

Figure 18. Bevel gears. (Lane Community College)

shafts at any angle, providing the shaft's axes would intersect if extended. Gears used to transmit power between intersecting shafts are often bevel gears (Figure 18). Bevel gears are conical gears and may have straight or spiral teeth. Spiral bevel gears are smoother running and usually will transmit more power than straight bevel gears because more than one pair of teeth is in contact at all times.

Figure 19. Miter gears. (Lane Community College)

Figure 20. A crossed helical gear drive. (Lane Community College)

Mating bevel gears with an equal number of teeth producing a 1:1 ratio are called miter gears (Figure 19). When the angle of the two shafts is other than 90 degrees, angular bevel gears are used. Bevel gears are used to provide ratios from 1:1 to 8:1. Face gears have teeth cut on the end face of a gear. The teeth on a face gear can be straight or helical. Ratios for face gears range from 3:1 to 8:1.

Helical Gears

The third category is for nonintersecting, nonparallel shafts. Helical gears of the same hand, cut with a 45 degree helix angle, will mesh when their two shafts are at 90 degree or right angles from each other. By changing the helix angle of the gears, the angle of the shafts in relation to each other can be changed. This type of gear arrangement is called a cross helical gear drive and is used for ratios of 1:1 to 100:1 (Figure 20).

Figure 21a. Worm and worm gear. (Lane Community College)

Worm Gears

Worm and worm gears (Figure 21a) are used for power transmission and speed reduction on shafts that are at 90 degrees to each other. Worm gear drives operate smoothly and quietly and give reduction ratios of 3:1 to over 100:1. In a worm gear set, the worm acts as the pinion, driving the worm gear. Two basic kinds of worm shapes are made: the single-enveloping worm gear set with a cylindrical worm, and the double-enveloping worm gear set where the worm is hourglass shaped (Figure 21b). The worm of the double-enveloping worm gear set is wrapped around the worm gear and gives a much larger load carrying capacity to the worm gear drive. Worms are made with single thread or lead, double lead thread, triple lead thread, or other multiple lead threads. The number of leads in a worm is determined by counting the number of thread starts at the end of the worm.

Figure 21b. A double-enveloping worm gear.

Figure 22. Hypoid gears. (Lane Community College)

A single thread worm is similar to a single lead thread when the ratio of a worm gear set is determined. A worm gear is made to run with the worm of a given number of leads. A double thread worm gear must be run with a double lead worm and cannot be interchanged with a single lead worm.

Another commonly used gear form is hypoid gears (Figure 22). Hypoid gears are similar in form to spiral bevel gears except that the pinion axis is offset from the gear axis. Hypoid gear ratios range from 1:1 to 10:1.

Gear Materials

Gear materials fall into three groups: ferrous, nonferrous, and nonmetallic. In the ferrous materials, steel and cast iron are often used. Steel gears, when hardened, carry the greatest load relative to their size. Steel gears can be hardened and tempered to exacting specifications. The composition of the steel can be changed. When the carbon content is increased, the wear resistance also increases. Lowering the carbon content gives better machinability. Cast iron is low in cost and is easily cast into any desired shape. Cast iron machines easily and has good wear resistance. Cast iron gears run relatively quietly and have about three-quarters of the load carrying capacity of an equal size steel gear. One drawback of cast iron is its low impact strength, which prohibits its use where severe shockloads occur. Other ferrous gear materials are ductile iron, malleable iron, and sintered metals.

Nonferrous gear materials are used where corrosion resistance, light weight, and low cost production are desired. Bronze gears are very tough and wear

resistant. Gear bronzes make very good castings and have a high machinability. Lightweight gears are often made from aluminum alloys. When these alloys are anodized, a hard surface layer increases their wear resistance.

Low cost gears can be produced by die casting. Most die cast gears are completely finished when ejected from the mold with the exception of the removal of the flash on one side. Die casting materials are zinc base alloys, aluminum base alloys, magnesium base alloys, and copper base alloys.

Nonmetallic gears are used primarily because of their quiet operation at high speed. Some of these materials are layers of canvas impregnated with phenolic resins; they are then heated and compressed to form materials such as formica or micarta. Other materials include thermoplastics, such as nylon. These nonmetallic materials exhibit excellent wear resistance. Some of these materials need very little lubrication. When plastic gears are used in gear trains, excessive temperature changes must be avoided to control damaging of dimensional changes. Good mating materials with nonmetallic gears are hardened steel and cast iron.

In many instances, gear sets are made up with different gear materials. Many worm drives use a bronze worm gear with a hardened steel worm. Cast iron gears work well with steel gears. To equalize the wear in gear sets, the pinion is made harder than the gear. Even wear in gear sets can be obtained when a gear ratio is used that allows for a "hunting tooth." For example, a gear set ratio of approximately 4:1 is needed. This is possible by using an 80 tooth gear in mesh with a 20 tooth gear. In this gear arrangement, the same tooth of the pinion will mesh with the same tooth of the gear in every revolution. If an 81 tooth gear were used, the teeth of the pinion will not equally divide into it, and each tooth of one gear will mesh with each of the mating teeth one after the other, distributing wear evenly over all teeth.

Gear Drive Systems

Transmission and differential gear trains are familiar examples of the use of gearing. The great advantages of gear systems are long life, quietness, and their ability to withstand high torque loads while being compact and lightweight compared to equal torque drive systems. Gear drive systems are used on practically every kind of industrial machinery. Farm machinery and some industrial machines often use exposed gearing (not in a gearbox) with only safety guards around them, while drive systems such as found in machine shop machines are usually enclosed (Figure 23). Add-on gear reduction drive systems are widely used in

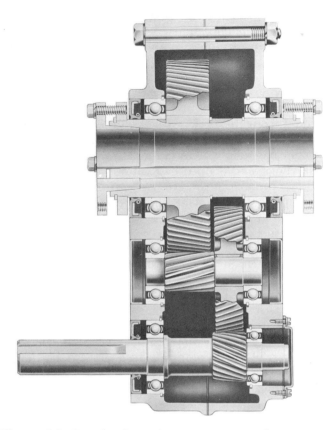

Figure 23. Speed reducer showing a compound gear train. Gears and bearings operate in an oil bath, giving them a long working life. (Reliance Electric Company)

heavy industry for slow moving machinery such as conveyor systems (Figure 24).

HYDRAULIC POWER TRANSMISSION

The use of fluids (both air and oil are classed as fluids) to power machinery or actuate mechanisms is familiar to everyone. We often see earth moving machinery on construction projects that use hydraulic systems that actuate powerful cylinders to do work. Just as mechanical systems transfer power from one place to another and change its speed or force, hydraulic systems transfer power but in a more flexible manner. Instead of belts or shafts, the power is transmitted through a fluid in a steel tube or flexible line so the force can be applied at just about any location on the machine.

Hydraulic pumps, rotary motors, and valves can be combined to form variable speed drives for almost an unlimited range of applications. These drive systems are called "hydrostatic drives" since they use a positive displacement (no slippage) pump and motor in a closed loop system. This is in contrast to the "hy-

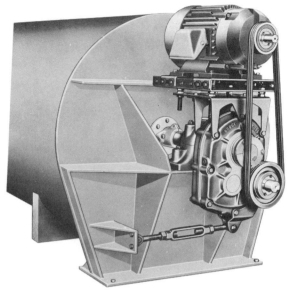

Figure 24. A gear reduction unit saves space. (Reliance Electric Company)

drodynamic drives" found in automobile automatic transmissions that do have slippage between the driving and driven member.

Hydrostatic drives provide full torque from zero to full load speed and can be stalled without damage. High horsepower can be utilized in relation to a small package size and weight. A disadvantage is the heat build-up due to friction in lines and components. This energy loss is responsible for the slightly lower efficiency that is typical of many hydraulic systems as compared to equivalent mechanical drives.

MECHANICAL DRIVE COMPONENTS

Gears, chains, and belts are used to transfer power from one rotating member (usually a shaft) to another. These rotating members must be securely fastened to the gears, pulleys, and sprockets that they drive. Shafts often require connecting devices to couple them together or to another drive system. The most commonly used of these components are clutches, couplings, keys, splines, shafts (straight and flexible), taper and straight pins, and universal joints.

Clutches

The friction clutch is the most common type and is used in the automobile and for some industrial applications (Figure 25). Two basic types of friction clutches are disc and conical. They can be actuated mechanically (with a lever), with oil or air pressure, or by an electric solenoid. Clutches are used to connect and disconnect drive systems.

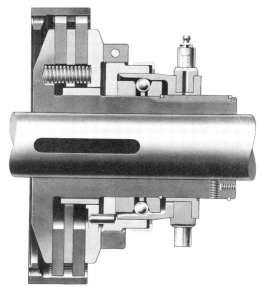

Figure 25. Cross section of a typical friction clutch. (Reliance Electric Company)

Figure 28. Chain coupler showing the method of assembly, using a two-strand chain. (Reliance Electric Company)

Figure 26. Steel rigid coupling. (Reliance Electric Company)

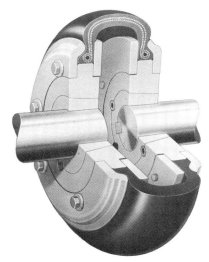

Figure 29. Cutaway view of a flexible Para-Flex^R coupling. This coupling can withstand high shock loads. (Reliance Electric Company)

Figure 27. Ribbed rigid coupling. (Reliance Electric Company)

Couplings

The simplest method of connecting two in-line shafts or rotating members is with a rigid coupling (Figures 26 and 27). With this type of coupling, the driving

member and driven member must be in line both radially and axially (not offset or at an angle). Since this kind of precision alignment is not always possible, several types of flexible couplings have been designed. One of the most common of these is the chain coupler (Figure 28). Chain and gear type couplers are very efficient on constant load applications. Where shock loads are a problem, some kind of rubberized, flexible coupling should be used (Figure 29). The coupler size depends on the horsepower, torque, and shock load. Manufacturer's catalogs contain this information.

Keys

Various types of keys are used as a positive means for transmitting torque between a shaft and hub. A

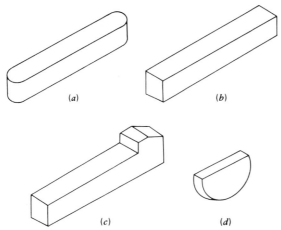

Figure 30. Various types of keys used to connect shafts to hubs. (a) Round end, straight key; (b) straight key; (c) tapered gib head key; and (d) woodruff key.

Table 1
Standard Keyseats/Keyways

Shaft Size	W Width	D Depth Regular	D Depth Shallow	R Cutter Runout
$\frac{1}{8}$ to $\frac{9}{16}$	$\frac{1}{8}$	$\frac{1}{16}$	. . .	$\frac{9}{16}$
$\frac{5}{8}$ to $\frac{7}{8}$	$\frac{3}{16}$	$\frac{3}{32}$	. . .	$\frac{11}{16}$
$\frac{15}{16}$ to $1\frac{1}{4}$	$\frac{1}{4}$	$\frac{1}{8}$	. . .	$\frac{13}{16}$
$1\frac{5}{16}$ to $1\frac{3}{8}$	$\frac{5}{16}$	$\frac{5}{32}$	. . .	$\frac{15}{16}$
$1\frac{7}{16}$ to $1\frac{3}{4}$	$\frac{3}{8}$	$\frac{3}{16}$	. . .	$1\frac{1}{16}$
$1\frac{13}{16}$ to $2\frac{1}{4}$	$\frac{1}{2}$	$\frac{1}{4}$	$\frac{1}{8}$	$1\frac{1}{16}$
$2\frac{5}{16}$ to $2\frac{3}{4}$	$\frac{5}{8}$	$\frac{5}{16}$	$\frac{3}{16}$	$1\frac{5}{16}$
$2\frac{13}{16}$ to $3\frac{1}{4}$	$\frac{3}{4}$	$\frac{3}{8}$	$\frac{3}{16}$	$1\frac{9}{16}$
$3\frac{5}{16}$ to $3\frac{3}{4}$	$\frac{7}{8}$	$\frac{7}{16}$	$\frac{1}{4}$	$1\frac{11}{16}$
$3\frac{13}{16}$ to $4\frac{1}{2}$	1	$\frac{1}{2}$	$\frac{1}{4}$	$1\frac{3}{4}$
$4\frac{9}{16}$ to $5\frac{1}{2}$	$1\frac{1}{4}$	$\frac{5}{8}$	$\frac{1}{4}$	$1\frac{15}{16}$
$5\frac{9}{16}$ to $6\frac{1}{2}$	$1\frac{1}{2}$	$\frac{3}{4}$	$\frac{1}{4}$	$2\frac{1}{8}$
$6\frac{9}{16}$ to $7\frac{1}{2}$	$1\frac{3}{4}$	$\frac{3}{4}$	$\frac{1}{4}$	$2\frac{1}{8}$
$7\frac{9}{16}$ to 9	2	$\frac{3}{4}$	$\frac{3}{8}$	$2\frac{1}{8}$
$9\frac{1}{16}$ to 11	$2\frac{1}{2}$	$\frac{7}{8}$	$\frac{3}{8}$	$2\frac{5}{16}$
$11\frac{1}{16}$ to 13	3	1	$\frac{3}{8}$	$2\frac{7}{16}$

Source: Reliance Electric Company, *D78 Dodge Engineering Catalog,* © 1977, page 98–7.

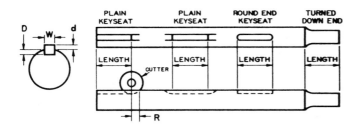

keyseat as defined in the *Machinery's Handbook* is "an axially located rectangular groove in a shaft or hub." The common types of keys used are square and rectangular plain keys, plain and gib head taper keys, and woodruff keys (Figure 30). Plain, standard keyseat sizes and depths are given in Table 1. Figure 31 shows the method of measurement for standard and round end keyseats.

Splines
Another positive drive method that is used when keys are not able to handle the load is the spline. The two common types of splines used in machinery are the square and the involute forms (Figure 32). An involute spline on a hardened shaft (as in automobile transmissions) can transmit a much higher load for its size than a square spline on a nonhardened shaft.

Shafts
A shaft is a round bar that is used to transmit rotational motion. Shafts can take many forms, from standard bar stock to special manufactured shafts that can have splines, keyways, gears, and flanges. Flexible shafts are used to transmit rotational motion in locations other than a straight line. They are used for speedometers in automobiles and to turn hand-held cutters and grinders.

A shaft is usually called an axle when a wheel rotates on it. Short shafts are usually supported by two bearings, while long, slender shafts may have several bearings. Long shafts that are used to drive a number of machines are called line shafts; they are generally obsolete as a means of power transmission and distribution since individual electric motors are used to drive machines today.

Pins
Another common method of securing hubs, handwheels, and other members to a shaft for light drives is by using pins—small diameter straight or tapered rods that are driven in a drilled and reamed hole through the shaft and driven member (Figure 33). Straight pins are kept in place by a tight fit, or more often by peening over each end to make a slight head. Shear pins are straight pins that are designed to fail by shearing before a more expensive member fails. Machinery that is subject to infrequent shock loads is fitted with shear pins as a safety device. An alternate

Figure 31. The method of measuring standard keyseats. Depth *(D)* is always ½ *W* on regular (square) keys and is measured at the shoulder of the keyseat, not on the centerline. (Reliance Electric Company)

Figure 32. Two common types of splines: (a) involute and (b) square. (Lane Community College)

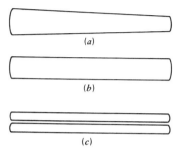

(a)

(b)

(c)

Figure 33. Three kinds of pins that are used to secure hubs on shafts: (a) taper pin, (b) straight pin, and (c) roll pin.

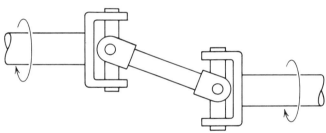

Figure 34. Light duty universal joint. Correct installation of two universal joints on a common shaft to avoid vibration.

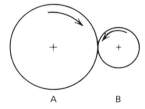

A   B

Figure 35. Friction drive between two discs.

Unit 6, "Hand Reamers.") These pins are kept in place by the friction of the taper and, of course, can only be driven out in one direction.

Pins should be driven out by starting them with a drift punch (on the small end of taper pins) and continuing to drive them out with a pin punch. If a taper or straight pin (that is not hard) fails to come out, they can be drilled out with a slightly smaller drill size than the pin; the remaining shell will usually come out easily. Roll pins are hard and cannot be drilled; they can only be driven out.

Universal Joints

The purpose of universal joints is to connect two rotating members (usually shafts) that are not in alignment. Simple joints, as shown in Figure 34, are used for light drives and sometimes on agricultural equipment. Automotive universal joints, in contrast, are able to withstand high torque loads. Since a single universal joint on misaligned shafts produces a harmonic motion (like an ellipse) where its speed varies during each revolution, two joints are required, each cancelling out the other to produce an even rotational speed. This requires each joint to be in a specific angular location to the other; for instance, the two yokes on a propeller (drive) shaft should be positioned on the same plane.

DRIVE RATIOS AND SPEEDS

A ratio is simply the proportion of one thing to another. The ratio of a shaft turning 800 RPM to one turning at 400 RPM is 2 to 1 (2:1).

EXAMPLE

Two circular discs are running together (driven by friction), disc A and disc B (Figure 35). If disc B were turning at 600 RPM at the ratio of 3 to 1, how fast would disc A turn?

ANSWER

Disc A would turn 200 RPM, one-third of the speed of disc B.

type of straight pin that requires no peening to retain it is actually a rolled up steel spring called a roll pin. It should be remembered that a hole drilled in a shaft weakens it by reducing its cross section.

Taper pins have a taper of $\frac{1}{4}$ in. per foot, and come in a series of sizes as do the taper reamers that are used to fit the holes for the pins. (See Section B,

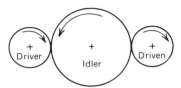

Figure 36. Gear train with an idler.

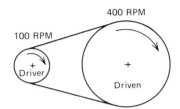

Figure 37. Belt drive.

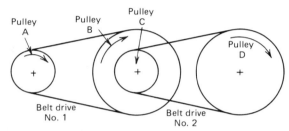

Figure 38. Compound belt drive.

The member that drives the other is the "driver" and the other one is the "driven." Disc B in Figure 35 is the driver and disc A is the driven disc. Disc B makes three turns for every one turn of A. This example is similar to simple gear drives, A and B rotate in opposite directions. Ratios may be determined by gear or pulley diameters or circumferences, speeds, or number of teeth (on gears and sprockets).

In gear trains, only driver and driven gears contribute to the ratio. Idler gears as in Figure 36 have no effect on ratio or speed, no matter what their size. Their function is to change the direction of rotation and to transfer rotation between two gears.

In belt and chain drives, the driver and driven member rotate in the same direction (Figure 37). The ratio or speeds of the belt drive in Figure 37 has nothing to do with the belt; the diameters of the pulleys determine the ratio. The ratio is found by dividing 400 by 100, which equals 4 to 1, or 4:1.

A compound belt drive can be seen in Figure 38, showing the location of pulleys, A to D. It is called a compound drive because it has more than one drive,

each with its own ratio. Pulley B and pulley C are on the same shaft so both have the same RPM. In a compound drive, the total ratio is the product of all drive ratios. That is, gear ratios are multiplied to get the total ratio in a system. Therefore, if belt drive Number 1 is 3:1 and belt drive Number 2 is 4:1 in Figure 38, the total ratio is 12:1. If pulley A were turning at 1800 RPM, pulley B (at 3:1) would be turning at 600 RPM and pulley D would turn at 150 RPM (1800 divided by 12).

Gear, belt, and chain drives may be combined in a compound drive system, but these principles of ratios and speeds will still be the same for even mixed or combined drives, providing the factors remain the same. That is, do not mix diameters and gear teeth to calculate ratios.

BEARINGS

Sleeve or plain bearings use a sliding contact between the two bearing surfaces that is more or less separated by a lubricating film. In contrast, roller and ball bearings use a rolling contact that has much less friction than sleeve bearings. Ball and roller bearings are therefore called antifriction bearings (although there is some rolling friction) and sleeve bearings are called friction bearings (Figures 39*a* to 39*c*).

Plain bearings can be made of many materials, metallic and nonmetallic. Automobile engine bearings were once made of cast babbitt (usually tin based, containing tin, antimony, and copper) metal and then bored to size. In modern engines, insert bearings are made of such materials as lead-tin babbitt and aluminum-copper alloys that are sintered as fine powders onto a steel backing. Sometimes many layers of metals are used. Before antifriction bearings were introduced, many machinery bearings were made of babbitt metal that was poured into a removable cast iron housing and then scraped or bored to fit the shaft. Some large electric motor bearings are still being made this way (Figure 40). Plain bearings are also made of such materials as bronze, aluminum, silver (for aircraft engines), cast iron, plastics (nylon, teflon, phenolic), porous (sintered) metals, and carbon-graphite. Even wood has been used for bearings, especially on farm machinery and for propellor shafts on ships in which a tough, greasy wood, called lignum vitae, is used since it works well under water as a bearing surface.

Ball and roller bearings are available for industrial uses in single and double row, light, medium, and heavy duty sizes. These include simple bearings with inner and outer races that fit directly in the machine;

Figure 39a. A porous sintered bronze sleeve bearing. (Reliance Electric Company)

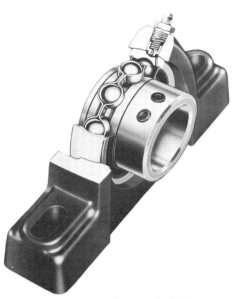

Figure 39b. Single race ball bearing pillow block. (Reliance Electric Company)

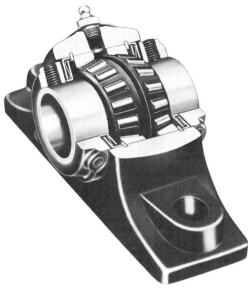

Figure 39c. Tapered roller bearing pillow block. (Reliance Electric Company)

Figure 40. This large electric motor babbitt bearing has just been poured and is ready for scraping and oil groove cutting before reassembly.

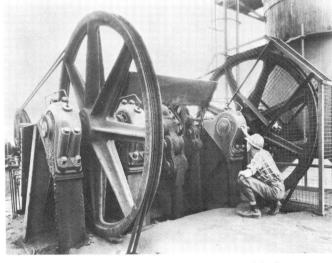

Figure 41. These large seven-inch pillow blocks support the shaft on this skip hoist for a blast furnace for an eastern steel mill. (Reliance Electric Company)

for example, the ball bearings in an automotive transmission. A number of shapes in bearing housings are used on machinery to adapt to various structures. The ball or roller bearings in these housings are often self-aligning; that is, they adjust to slight misalignments that would put a severe strain on a bearing not having this feature.

Pillow block bearing housings are made to bolt onto a flat surface (Figure 41). Flange bearings are made to bolt to vertical or horizontal frames and supports. They are available in several designs (Figure 42). Figure 43 shows one of many types of take-up bearing mounts that are commonly used for adjusting conveyor belts or chains.

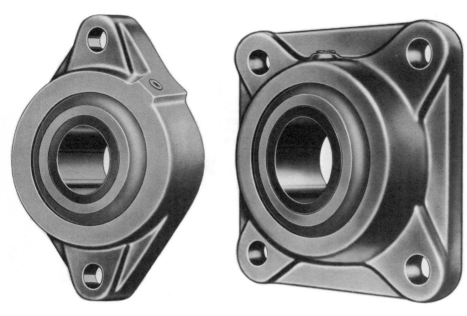

Figure 42. Two common designs of flange bearings (2- and 4-bolt). (Reliance Electric Company)

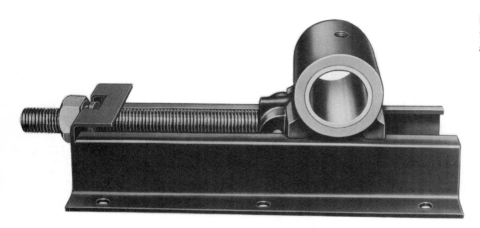

Figure 43. Take-up bearing mount used for adjusting conveyor chains. (Reliance Electric Company)

SELF-TEST

1. What is the reason that flat belt drives have largely been replaced with V-belts?
2. Gear and chain drives have the advantage of no slip drive from very low to high speeds. For what range is a V-belt drive best suited?
3. The chain drive most used for power transmission is roller chain. Name two advantages of roller chains over belt drives.
4. Name three ways in which gears are superior for power transmission.
5. Why are double helix gears (herringbone) sometimes used?
6. Which kind of gears are used to transmit power between two intersecting shafts at a 90 degree angle? Between two nonintersecting shafts?
7. Name five different materials used for making gears.
8. Hydraulic, variable speed drives provide no-slip, full torque from zero to full speed and deliver high horsepower in a small package. What are these drives called?
9. What is the difference between a clutch and a coupling?
10. What is the difference between a keyseat and a spline?
11. Name two types of punches used for removing taper pins.
12. What is the purpose of universal joints?
13. A pulley 12 in. in diameter is driven clockwise by a 3 in. diameter pulley with a belt. (a) What direction does the driver turn? (b) What is the ratio between the two pulleys? (c) If the driver is turning 1800 RPM, what is the RPM of the driver?
14. Explain the difference between sleeve bearings and ball or roller bearings in terms of the kind of friction involved.
15. Pillow block and flange mount are names of what product? What are they used for?

UNIT 2 CAUSES OF MECHANICAL FAILURES

One of the most common problems found in industrial equipment and other machinery is mechanical breakdown due to wear, overload, or metal fatigue. Operational failures sometimes cause a loss of production, loss of work time for employees, and, occasionally, hazardous conditions. Many service problems can be prevented by proper maintenance, selection of materials, and design.

With the exception of gold, silver, and some other noble metals, all metals are subject to the deterioration caused by corrosion. Iron, for example, tends to revert back to its natural state of iron oxide. Other metals revert to sulfides and oxides or carbonates. Buildings, ships, machines, and automobiles are all subject to attack by the environment. The corrosion that results often renders them useless and they have to be scrapped. Billions of dollars a year are lost as a result of corrosion. Corrosion can also cause dangerous conditions to prevail such as on bridges, where the supporting structures have been eaten away, or aircraft in which an insidious corrosion, called intergranular corrosion, can weaken the structural members of the aircraft and cause a sudden failure.

Those who work with metals need to have a knowledge of the principles involved in these service failures in order to be able to use the correct preventative measures. This unit will help you identify and correct many of these service problems.

OBJECTIVE

After completing this unit, you should be able to:
Explain the causes of several industrial problems that lead to failures and list corrective measures for them.

CAUSES OF FAILURES IN METALS

Metal failure occurs only in an extremely low percentage of the millions of tons of steel fabricated every year. Those that do occur fall into the following groups.

Improper design
  Sharp corners in high stress areas
  Insufficient safety stress factor
  Incorrect material selection
Operational failures
  Overload
  Wear
  Corrosion and stress corrosion

Metal fatigue
Brittle failures
Failures caused by thermal treatments
  Forging
  Hardening and tempering
  Welding
Crazing or surface cracks caused by the heat of grinding

OVERLOAD

Overload failures are usually attributed to faulty design, to extra loads applied, or to an unforeseen ma-

Figure 1. This CR shaft was stressed beyond its yield point by an overload. This could have been prevented by using a tougher alloy shafting. (John E. Neely, *Practical Metallurgy and Materials of Industry,* John Wiley and Sons, Inc., Copyright © 1979, New York)

chine movement. Shock loads or loads applied above the design limit are quite often the cause of the breakdown of machinery.

Although mechanical engineers always plan for a high safety factor in designs (for example, the 10 to 1 safety factor above the yield strength is used in fasteners), the operators of machinery often tend to use machines above their design limit. Of course, this kind of overstress is due to operator error. Inadequate design can sometimes play a part in overload failures. Improper material selection in the design of the part or improper heat treatment can cause some failures when overload is a factor.

Often a machinist or welder will select a metal bar or piece for a job based upon its ultimate tensile strength rather than its yield point. Tensile strength is the breaking point of a material. The yield point is that amount of stress applied to a metal member that begins to deform it so that when the stress is relieved, the part will not return to its original shape or condition. In effect this is a design error and can ultimately cause overload and breakdown (Figure 1). The strength of any material selected for a job should be based on its yield strength plus an adequate safety factor. When part of a mechanism or structure is stressed beyond its yield point, it has been damaged or ruined for most purposes. For example, an engine crankshaft would be ruined if it were permanently deformed or bent a small fraction of an inch, while a fender, though badly damaged, would continue to function as a fender. The tensile and yield strengths of commonly used materials may be found in the *Machinery's Handbook.*

WEAR

Excessive wear can also be caused by continuous overload, but wear is ordinarily a slow process that is related to the friction between two surfaces (Figure 2). Rapid wear can often be attributed to lack of lubrication or the improper selection of material for the wear

Figure 2. The teeth on this mild steel roller chain sprocket have worn almost to the point of complete failure. Wear life would be increased considerably if an alloy steel were used for the sprocket. *(Practical Metallurgy and Materials of Industry)*

surface. Some wear is to be expected, however, and could be called normal wear.

Wear is one of the most frequent causes of failure. We find normal wear in machine tooling, such as carbide and high speed tools, that have to be replaced or resharpened. Parts of automobiles ultimately wear until an overhaul is required (Figure 3). Machines are regularly inspected for worn parts, which when found are replaced; this is called preventative maintenance. Often normal wear cannot be prevented; it is simply accepted, but it can be kept to a minimum by the proper use of lubricants. Rapid wear can occur if the load distribution is concentrated in a small area be-

Figure 3. This type of failure can be hazardous. The spindle bearing seized up due to lack of lubrication and began to twist the spindle. Had the operator of the automobile not stopped very quickly, the wheel would have come off. *(Practical Metallurgy and Materials of Industry)*

cause of the part design or shape. This can be altered by redesign to offer more wear surface. Speeds that are too high can increase friction considerably and cause rapid wear.

Metallic wear is a surface phenomenon caused by the displacement and detachment of surface particles. All surfaces subjected to either rolling or sliding contact show some wear. In some severe cases, the wear surface can be cold welded to the other surface.

In fact, some metals are pressure welded together in machines, taking advantage of their tendency to be cold welded. This happens when tiny projections of metal make a direct contact on the other surface and produce friction and heat, causing them to be welded to the opposite surface if the material is soft. Metal is torn off if the material is brittle. Insufficient lubrication is usually the cause of this problem (Figure 4).

High pressure lubricants are often used while pressing two parts together in order to prevent this sort of welding (often called galling). Two steel parts such as a steel shaft and a steel bore in a gear or sprocket, if pressed together dry, will virtually always seize or weld and cause the two parts to be ruined for further use (Figure 5). In general, soft metals, when forced together, have a greater tendency to cold weld than harder metals. Two extremely hard metals even when dry will have very little tendency to weld together. For this reason, hardened steel bushings and hardened pins are often used in earth moving machinery to avoid wear. Some soft metals when used together for bearing surfaces (aluminum to aluminum) have a very great tendency to weld or seize. Among these metals are aluminum, copper, and austenitic stainless steels.

Cast iron, when sliding on cast iron, as is found in machine tools on the ways of lathes or milling machine tables, has less tendency than most metals to

Figure 4. The babbitted surface of this tractor engine bearing insert has partially melted and torn off. This failure was not due to normal wear but to lack of lubrication. *(Practical Metallurgy and Materials of Industry)*

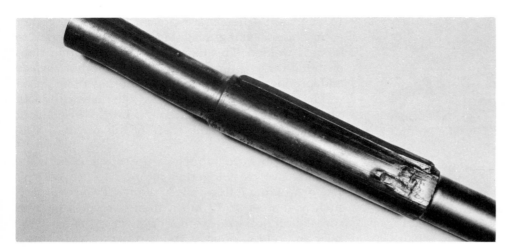

Figure 5. This shaft had just been made by a machinist and was forced into an interference fit bore for a press fit. No lubrication was used, and it immediately seized and welded to the bore, which was also ruined. *(Practical Metallurgy and Materials of Industry)*

Figure 6. The bronze bushing in this arm has seen severe use, is badly worn, and will be replaced, while the steel shaft that turns in the arm shows relatively little wear. *(Practical Metallurgy and Materials of Industry)*

seize because the metal contains graphite flakes that provide some lubrication, although additional lubrication is still necessary.

As a general rule, however, it is not good practice to use the same metal for two bearing surfaces that are in contact. However, if a soft steel pin is used in a soft steel link or arm, it should have a sufficiently loose fit to avoid seizing. In this application it is better practice to use a bronze bushing or other bearing material in the hole and a steel pin because the steel pin is harder than the bronze so that when a heavy load is applied, the small projections of bronze are flattened instead of torn out. Also, the bronze will wear more than the steel and usually only the bushing will need replacing when a repair is needed (Figure 6).

Some metals have a tendency to work harden and, although they would gall or seize in their soft condition as they begin sliding together, they begin to harden on the surface and minimize the tendency to cold weld. An example of this is in the austenitic manganese steels used in rock crusher machinery. When these have work hardened sufficiently, they do not tend to cold weld to their own surfaces.

In **abrasive wear** small particles are torn off the surface of the metal creating friction. Friction involving abrasive wear is sometimes used or even required in a mechanism such as on the brakes of an automobile. The materials are designed to minimize wear with the greatest amount of friction in this case. Where friction is not desired, a lubricant is normally used to provide a barrier between the two surfaces.

Erosive wear is often found in areas that are subjected to a flow of particles or gases that impinge on the metal at high velocities. Sandblasting, which is sometimes used to clean parts, utilizes this principle.

Corrosive wear takes place as a result of an acid, caustic, or other corrosive medium in contact with metal parts. When lubricants become contaminated with corrosive materials, pitting can occur in such areas as machine bearings.

Surface fatigue is often found on roller or ball bearing races or sleeve bearings where excessive side thrust has been applied to the bearing. It is seen as fine checks (cracks) or spalling (small pieces falling out on the surface). See Section I, Unit 3.

PROTECTION AGAINST WEAR

Various methods are used to limit the amount of wear in the part. One of the most commonly used methods is simply to harden the part. Also the part can be surface hardened by **diffusion** of a material, such as carbon or chrome, into the surface of the part. Parts can also be **metallized, hard faced,** or **heat treated.** Other methods of limiting wear are electroplating (especially the use of hard industrial chromium) and anodizing of aluminum. Chromium plate can either be hard or porous. The porous type can hold oil to provide a better lubrication film. Some internal combustion engines are chromium plated in the cylinders and piston rings (Figures 7a and 7b). Some nickel plate is used and also rhodium, which is very hard and has high heat resistance.

The oxide coating that is formed by anodizing on certain metals such as magnesium, zinc, aluminum, and their alloys is very hard and wear resistant. These oxides are porous enough to form a base for paint or stain to give it further resistance to corrosion.

Some of the types of diffusion are **carburizing, carbo-nitriding, cyaniding, nitriding, chromizing,** and **siliconizing.** Chromizing consists of the introduction of chromium into the surface layers of the base metal. This is sometimes done by the use of chromium powder and lead baths in which the part is immersed at a relatively high temperature. This, of course, produces a stainless steel on the surface of low carbon steel or an iron base metal, but it may also be applied to nonferrous materials such as tungsten, molybdenum, cobalt, or nickel to improve corrosion and wear resistance.

The fusion of silicon, which is called **irighizing,** consists of impregnating an iron base material with silicon. This also greatly increases wear resistance. **Hard facing** is put on a metal by the use of several types of welding operations, and it is simply a hard type of metal alloy such as an alloy of cobalt and tungsten or tungsten carbide that produces an extremely hard surface that is very wear resistant. Metal spraying is used for the purpose of making hard wear resistant surfaces and for repairing worn surfaces (see "Welding Repairs" in Unit 3). Metalizing is usually done by either feeding a metal powder or a metal wire at a controlled rate through a tool that provides a heat source; the molten particles of metal are blown onto the surface of the base metal at a high velocity. In this process, which is not the same as welding, the liquid metal particles simply flatten out on the base metal and make a mechanical bond with the base metal and the previously deposited material instead of a metallurgical bond since the cooling is rapid and an oxide film forms over the particle, preventing fu-

Figure 7a. These aircraft engine cylinders must withstand high temperatures and wear. The inside of the cylinder wall is porous chromium plated. *(Practical Metallurgy and Materials of Industry)*

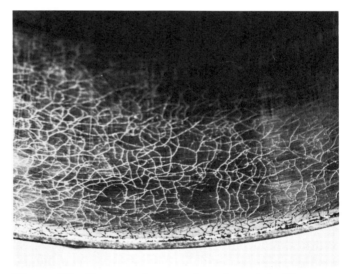

Figure 7b. Close up of porous chromium plate. The many grooves or channels will hold the lubricant. This plate is applied in small droplets, which, when ground off, produce this effect. *(Practical Metallurgy and Materials of Industry)*

sion of the metal particles. Thus there is only a loose metallic or oxide bond between the particles. This determines to a great extent how strong or how porous the deposited material becomes. Some manufacturers of sprayweld equipment claim a partial metallurgical bond takes place in their process.

Figure 8. Shafts sometimes fail by fatigue as shown here. The cracks were initiated from a badly worn keyway and slowly moved toward the center until the load was greater than the tensile strength of the remaining area.

FATIGUE

Failures caused by fatigue are found in many of the materials of industry. Some plastics and most metals are subject to fatigue in varying degrees. Fatigue can be caused by a crack that is initiated by a notch, bend, or scratch that continues to grow gradually as a result of stress reversals on the part. The crack growth continues until the cross sectional area of the part is reduced sufficiently to weaken the part to the point of brittle failure (Figure 8). Even welding spatter on a sensitive surface such as a steel spring can initiate fatigue failure (Figure 9). Fatigue is greatly influenced by the kind of material, grain structure, and the kind of loading. Some metals such as alloy steels are more sensitive to sharp changes in section (notch sensitive) than others, tending to fail crosswise by fatigue. Figure 10 shows a forged shaft with slag inclusions that run the length of the material. Wrought iron also contains inclusions that give the material high transverse (crosswise) fatigue strength. These metals have a lowered resistance to high torque values, however, and tend to split lengthwise (Figure 11).

A study of individual fatigue problems based on the service conditions and by direct observation of

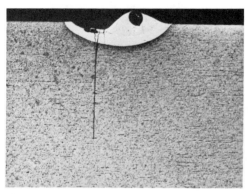

Figure 9. Fatigue failure in a leaf spring nucleated by a weld spatter. The smooth "oyster shell" character is typical of fatigue cracks. Note the nature of brittle failure of the spring in service. The polished section through another weld spatter clearly illustrates the formation of fatigue failure (10X). (Republic Steel Corporation)

Figure 10. Wrought iron forging, showing inclusions that run the length of the material.

Figure 11. Splined shaft that has split along its lengthwise fibers caused by overload.

Figure 12. Classic example of fatigue in a shaft. The three distinct areas can be seen here.

the failure can often lead to a conclusion that explains the cause or causes and suggests some corrective measures. The loading of the part can be high or low for its size; it can have high or low speed stress reversals caused by misalignment. Vibration or cyclic bending is often a cause of fatigue, the frequency and intensity being a factor. Occasional overloads are also instrumental in initiating failure.

Fatigue failures are usually characterized by three distinct surfaces (Figure 12).

1. A smooth surface with wave marks such as seen on clam shells. This area represents a slow progression of the initial crack.
2. A similar but rougher surface showing coarse wave marks progressing toward the center.
3. A crystalline area showing the final brittle and sudden failure of the part. It is this last portion that prompts the erroneous conjecture that the part has "crystallized."

Continued use and many stress reversals continued the propagation of the crack until only a small section was holding. When the small, final section suddenly failed, the natural, crystalline structure of the steel was revealed. Any metal, when suddenly broken, will show a crystalline section.

Welding on highly stressed parts often initiates fatigue in the part. A weld, due to contraction during cooling, produces a tensile stress in the base metal. This stress concentration can start a crack that continues to grow until the part fails. Shafting, since it often produces stress reversals because of rotation, is particularly sensitive to stress concentrations, such as those produced by a weld. It is therefore poor practice to weld anything on the side of the shafting; even a tack weld to hold a key in place can cause a failure (Figure 13). Welding an arm or hub to a shaft is very likely to cause an ultimate failure of the shaft by fatigue.

Figure 13. Stress concentrations caused by welds on this shaft precipitated the fatigue failure.

Figure 14. Bent connecting rod and cap. Note the necked down screws indicating that they were overtorqued.

Figure 15. End of one of the capscrews that held the cap on the connecting rod. A typical fatigue pattern is evident.

The failure of a tractor engine connecting rod may be seen in Figure 14. The bolts in the bearing caps suddenly failed and the crankshaft pushed the connecting rod through the side of the engine block. The capscrews that fasten the connecting rod bearing cap had evidently been overtorqued by the mechanic to the point that they had "necked down" to a smaller

diameter. This put a much higher stress on the screws that started a fatigue fracture (Figure 15).

Fatigue failures on connecting rod capscrews can also be caused by undertorquing. A loose fastener is even more likely to have a high cyclic stress than an overtight one. See Section B, Unit 3 for correct torque values on fasteners.

CORROSION IN METALS

The most common form of corrosion is a deterioration of metals by an electrochemical action that is generally slow and continuous. Another form is high temperature scaling and the formation of oxides on metals called oxidation corrosion. The oxide of iron formed at high temperatures is black, often called mill scale. Corrosion in metals is the result of their desire to unite with oxygen in the atmosphere or in other environments to return to a more stable compound, usually called ore. Iron ore, for example, is in some cases simply iron rust. Corrosion may be classified by the two different processes by which it can take place: **direct oxidation corrosion,** which usually happens at high temperatures, and **galvanic corrosion,** which takes place at normal temperatures in the presence of moisture or an electrolyte.

Direct Oxidation Corrosion

Oxidation at high temperatures is often seen in the scaling that takes place when a piece of metal is left in a furnace for a length of time. The black scale is actually a form of iron oxide, called magnetite (Fe_3O_4) (Figure 16). This oxide coating is also called mill scale

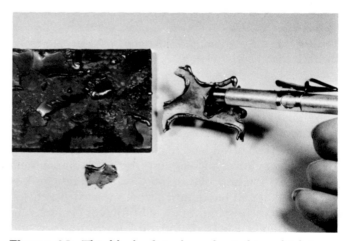

Figure 16. This block of steel was heated to a high temperature in a furnace in the presence of air. It is covered with a loose black scale, which is magnetite-iron oxide (Fe_3O_4). Magnetite, as the name implies, is magnetic.

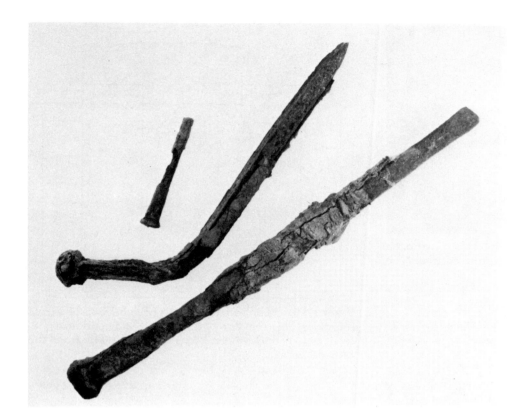

Figure 17. These large iron spikes were found on an ocean beach. They are almost completely changed into iron oxide rust.

because it is formed on heated ingots or slabs that are rolled in steel mills. The red hot steel is constantly scaling since it is in contact with the oxygen in the atmosphere. In time, if the high temperature is maintained, all of the iron will be converted to scale.

Galvanic Corrosion

Galvanic corrosion is essentially an electrochemical process that causes a deterioration of metals by a very slow but persistent action. In this process, part or all of the metal becomes transformed from the metallic state to the ionic state and often forms a chemical compound if an electrolyte is present (see below). On the surface of some metals such as copper or aluminum, the corrosion product sometimes exists as a thin film that resists further corrosion. In other metals such as iron, the film of oxide that forms is so porous that it does not resist further corrosive action, and corrosion continues until the whole piece has been converted to the oxide (Figure 17).

Electrolytes

An electrolyte is any solution that conducts electric current and contains negative or positive ions. An ion is simply an atom that has either gained or lost electrons. One that has gained electrons is a negative ion

and one that loses electrons is a positive ion. Metals generally lose electrons and nonmetals gain them. Corrosion requires the presence of an electrolyte to allow metal ions to go into solution. The electrolyte may be fresh or salt water, acid or alkaline solutions of any concentration. Even a fingerprint on a clean metal can form an electrolyte and produce corrosion. There must be a completed electric circuit and a flow of direct current before any galvanic action can take place. There also must be two electrodes, an anode and a cathode, and they must be electrically connected. (See note at the end of the unit.) The anode and cathode may be of two different kinds of metals or they may be located on two different areas of the same piece of metal. The connection between the anode and the cathode may be made by the metal itself or by a metallic connection such as a bolt or rivet.

If a piece of metal is immersed in hydrochloric acid, hydrogen bubbles will collect rapidly and be released, indicating some corrosion is taking place on the surface of the metal. Some metals corrode very quickly when placed in acids. In this case, there are anodes and cathodes and the deterioration of the metal occurs at the anodes. Very tiny, well defined cathode and anode areas are formed all over the piece

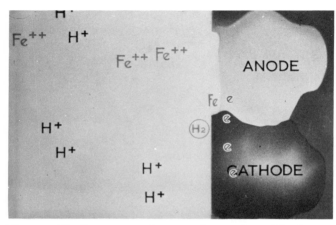

Figure 18. Formation of ions at the anode and hydrogen at the cathode. This is the process of rusting on iron. (The International Nickel Company, Inc.)

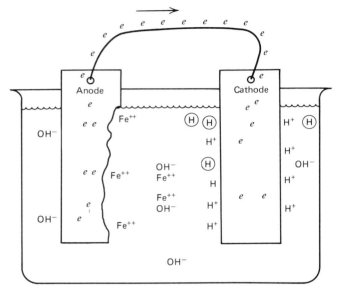

Figure 19. When iron ions (Fe^{++}) become detached from the anode, electrons flow to the cathode and neutralize positively charged hydrogen ions, releasing them from the cathode surface.

of metal. However, they may shift, causing a very uniform corrosion to take place (Figure 18).

How Corrosion Takes Place

When corrosion occurs, positively charged atoms of metal are released or detached from the solid surface and enter into solution as metallic ions while the corresponding negative charges in the form of electrons are left behind in the metal. The detached positive ions bear one or more positive charges. In the corrosion of iron, each iron atom releases two electrons and then becomes an iron ion carrying two positive charges. Two electrons (e) must then pass through a conductor to the cathode area (Figure 19). (Without this electron flow, no metal ions can be detached at all from the anode.) The electrons reach the surface of the cathode material and neutralize positively charged hydrogen ions (H^+) that have become attached to the cathode surface. Two of these ions will now become neutral atoms and are released generally in the form of hydrogen gas Ⓗ. This release of the positively charged hydrogen ions leaves an accumulation and a concentration of OH negative ions (OH^-) that increases the alkalinity at the cathode. When this process is taking place, it can be observed that hydrogen bubbles are forming at the cathode only. When cathodes and anodes are formed on a single piece of metal, their particular locations are determined, for example, by the lack of homogeneity in the metal, surface imperfections, stresses, inclusions in the metal, or anything that can form a crevice, such as a washer, a pile of sand, or a lapping of the material.

The Galvanic Series

Anodes and cathodes can form at various places on the surface of a metal, but on the surface of dissimilar metals, the corrosion rates produced are often greater. Some metals have a greater tendency to corrode than others because they are more active chemically and tend to become anodes, while others are less active and are cathodic. Gold, for example, is very inactive and is cathodic and has about the lowest possible corrosion potential. The galvanic series of metals and alloys shows this difference in the activity of metals in sea water (Table 1). The galvanic series (also called the electromotive series) is so named because of the direct electric current produced (galvanism) when two dissimilar metals are immersed in an electrolyte. At the more active end, we find the metals that most readily become anodic are magnesium and zinc. Except for those metals at the extreme opposite ends (gold and magnesium), any given metal may be an anode or cathode when it is coupled with another metal in the series.

In the galvanic series the metals become increasingly active toward the anode end. For example, if potassium is placed in water, a violent reaction takes place, but gold or platinum in strong nitric acid will not even show evidence of attack. The rate at which galvanic action and corrosion takes place depends on the degree of difference in electrical potential and the resultant current flow. If zinc is coupled with copper, a large current will flow, but if brass is coupled

Table 1

Galvanic Series of Metals and Alloys in Sea Water

Anode End (Highest Corrosion)

Magnesium	Copper
Beryllium	Bronze
Aluminum	Copper-nickel alloys
Cadmium	Nickel
Aluminum alloys	Tin
Uranium	Lead
Manganese	Hydrogen
Zinc	Inconel
Plain carbon steel	Silver
Alloy steels	Stainless steel (passive)
Cast iron	Monel
Cobalt	Titanium
Stainless steel (active)	Platinum
Brass	Gold

Cathode End
(Least Corrosion)

Source: John E. Neely, *Practical Metallurgy and Materials of Industry*, John Wiley and Sons, Inc., Copyright © 1979, New York.

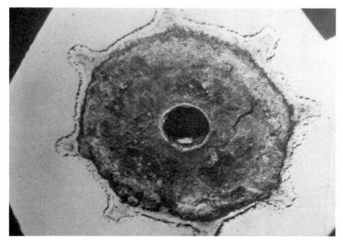

Figure 20. Galvanic corrosion of magnesium that surrounds a steel core. (The International Nickel Company, Inc.)

with copper, a very small current will flow and less reaction will be seen. A galvanic corrosion of magnesium that is in contact with a steel core can be seen in Figure 20.

Dezinctification

Dezinctification is a form of corrosion that often takes place in brass, cast iron, and other alloys. In dezinctification, the removal of zinc from brass continues throughout the alloy until the entire part has failed. Corrosion cells are set up between the zinc and the copper in the brass. The zinc remains in the electrolyte solution and the copper remains by plating on the brass. The final structure is very porous and brittle since it is made up of copper oxide and copper.

Cast iron can be completely corroded in that the iron goes into solution as ferrous ions, leaving mostly the graphite. This final condition of the iron is a very weak, soft, brittle, and porous material.

Concentration Cells

Concentration cells are created on a single piece of metal on which a deposit of sand or any material can produce a low oxygen area and a high oxygen area (Figure 21). Lapped metal on riveted or bolted joints is especially susceptible to this form of corrosion. Droplets of water can produce concentration cells (Figure 22). Cathodes are formed at the areas of high oxygen concentration, and anodes at areas of low oxygen concentration when corrosion occurs. This can be seen by observing a drop of salt water placed on a freshly polished steel surface. In only an hour or

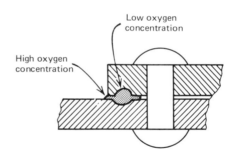

Figure 21. Concentration cell. The crevice in this concentration cell hinders the diffusion of oxygen and causes high and low oxygen areas. The low oxygen area is anodic. When the solution surrounding a metal contains more metal ions at one point than another, metal goes into solution where the ion concentration is low.

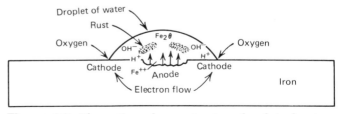

Figure 22. The action of corrosion in a droplet of water. Pitting corrosion is the result of this type of action.

two, a ring of rust will form inside the drop (anode) while the outer edges (cathode) remain clear.

Stainless steel is protected from the environment with an invisible but effective film that can be strengthened by immersing the stainless steel in concentrated nitric acid. If stainless steel, however, becomes oxygen starved in any particular area by a washer or a pile of sand, the passivity of the stainless steel can break down in these places. The area of the stainless steel in the immediate vicinity of the washer or pile of sand that is freely exposed to the dissolved oxygen acts as a cathode. Corrosion will take place under these deposits or crevices where oxygen cannot penetrate because they have become anodic. Therefore, when stainless steel is used around sea water or other corrosive media, it is important to avoid such crevices or nitches where oxygen cannot penetrate. In these cases, stainless steel can corrode just like ordinary steel.

Erosion Corrosion

Corrosion can also take the form of erosion in which the protective film, usually an oxide film, is removed by a rapidly moving atmosphere or fluids (Figure 23). Depolarization can also take place, for example, on the propellers of ships because of the movement through the water, which is the electrolyte. This causes an increased corrosion rate of the anodic steel ship's hull. Impellers or pumps are often corroded by this form of erosion corrosion in which metal ions are rapidly removed at the periphery of the impeller but are concentrated near the center where the velocity is lower.

Intergranular Corrosion

Another form of corrosion is intergranular corrosion. This takes place internally. Often the grain boundaries form anodes and the grains themselves form cathodes, causing a complete deterioration of the metal in which it simply crumbles when it fails (Figure 24). This often occurs in stainless steels in which chromium carbides precipitate into the grain boundaries. This lowers the chromium content adjacent to the grain boundaries, thus creating a galvanic cell. Most types of austenitic stainless steel when held at temperatures of 1500°F (815°C) and cooled slowly are subject to this kind of intergranular corrosion (Figure 25).

Pitting Corrosion

Differences in environment can cause a high concentration of oxygen ions. This is called cell concentration corrosion. Pitting corrosion is localized and results in small holes on the surface of a metal caused by a concentration cell at that point.

Figure 23. Butane gas impinges on the internal passage of this conversion part in a carburetor on a gasoline engine and has badly eroded the metal. This is an example of erosion corrosion.

Figure 24. The very large grains are outlined in this zinc based die casting because of intergranular corrosion. The metal has deteriorated at the grain boundaries, which have become anodes.

Figure 25. Section through type 310 stainless steel hot rolled plate in the as-welded condition, showing intergranular microcracking through the multiple-pass weld metal and the base metal (¾ size). (ASM Committee of Metallography of Wrought Stainless Steels, "Microstructures of Wrought Stainless Steels," *Atlas of Microstructures of Industrial Alloys*, Volume 7, 8th Edition, Lyman, T., Editor, American Society for Metals, 1972, page 132)

Stress Corrosion

Another form of corrosion is called stress corrosion. When high stresses are formed on metals in a corrosive environment, cracking can also be accelerated in the form of corrosion fatigue failure (Figure 26). It is a very localized phenomenon and results in cracking type of failures (Figure 27).

PROTECTION AGAINST CORROSION

Cathodic protection is often used to protect buried steel pipelines and steel hulls of ships. This is done by using zinc and magnesium sacrificial anodes that are bolted to the ship's hull or buried in the ground at intervals and electrically connected to the metal to be protected. In the case of the ship, the bronze propeller acts as a cathode, the steel hull as an anode, and the sea water as an electrolyte. Severe corrosion can occur on the hull as a result of galvanic action. The sacrificial anodes are very near the anodic end of the galvanic series and have a large potential difference between both the steel hull of the ship and the bronze propeller (Figures 28a and 28b). Both the hull and propeller become cathodic and consequently do not deteriorate. The magnesium anodes are replaced from time to time.

Selection of materials is of foremost importance. Even though a material may be normally resistant to corrosion, it may fail in a particular environment or if coupled with a more cathodic metal.

Coatings are extensively used to prevent corrosion. They are classified as follows:

1. Anodic coatings (anodizing)
2. Cathodic coatings (electroplating, chrome, copper, nickel)
3. Organic and inorganic coatings
4. Inhibitive coatings (red lead, zinc chromate)
5. Inhibitors (placed in the electrolyte)

Metal coatings may be listed as follows:

1. Hot dip process (galvanizing)
2. Metal spraying
3. Metal cementation (diffusion of atoms into the surface of a metal part)
   (a) Sherardizing metal parts heated with zinc powder to 600 to 850°F (315 to 454°C)
   (b) Chromizing
   (c) Irighizing
4. Metal cladding (rolling a "sandwich" of metal with the outer layers of a corrosion resistant metal). For example, clad aluminum sheet on which a thin outer layer of pure aluminum covers the strong aluminum alloy that is not as resistant to corrosion.

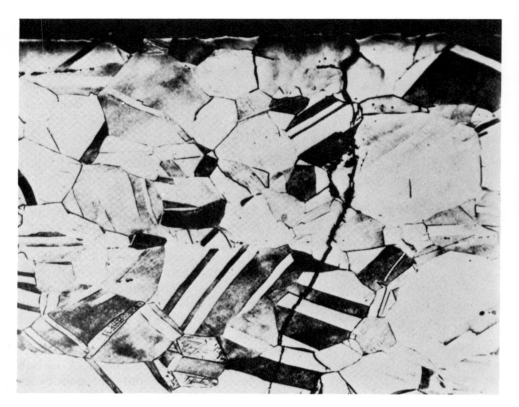

Figure 26. Typical transgranular corrosion-fatigue crack in alloy 260 (cartridge brass 70 percent). Note the lack of branching in the inner, or fatigue section of the crack. (ASM Committee on Metallography of Copper and Copper Alloys, "Microstructures of Copper and Copper Alloys," *Atlas of Microstructures of Industrial Alloys,* Volume 7, 8th Edition, Lyman, T., Editor, American Society for Metals, 1972, page 282)

Figure 27. Typical intergranular stress-corrosion cracks in alloy 260 (cartridge brass 70 percent) tube that was drawn, annealed, and cold reduced 5 percent. The cracks show some branching. (ASM Committee on Metallography of Copper and Copper Alloys, "Microstructures of Copper and Copper Alloys," *Atlas of Microstructures of Industrial Alloys,* Volume 7, 8th Edition, Lyman, T., Editor, American Society for Metals, 1972, page 282)

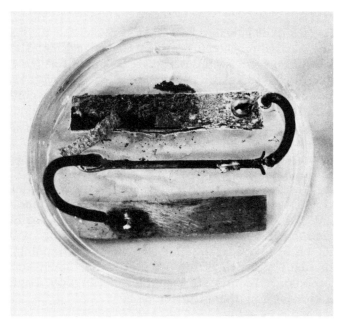

Figure 28a. In this experiment, a brass strip (bottom) is soldered to a nail (center) and a magnesium strip (top) bolted to the nail. The magnesium has started to deteriorate in the sodium chloride electrolyte. *(Practical Metallurgy and Materials of Industry)*

Figure 28b. The sacrificial anode has deteriorated and separated from its connection. The iron nail has not corroded since it was cathodic.

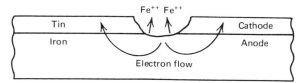

Figure 29. The action of corrosion on tin plate.

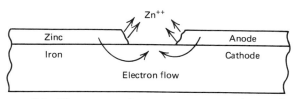

Figure 30. The action of corrosion on zinc plate (galvanized iron).

Anodizing is the process of thickening and toughening the oxide film. On metals such as aluminum, the thicker layer of oxides increases resistance to further corrosion. Anodized films provide not only a hard wear surface, but also a somewhat porous base for paint or other coatings.

The familiar electroplating process is similar to galvanic corrosion in that the metal ions are detached from the anode and deposited on the cathode to provide protection from corrosion. In most cases, the noble metals that are cathodic in the galvanic series are used for plating materials. Zinc plated steel, however, is used for roofing material and many other products, and tin plate is used extensively for food containers. If there is a break in the plating material, corrosion will begin at that point if the plating is cathodic to the base metal. If a break occurs in tin plate on iron, the tin is more cathodic than iron and therefore accelerates the corrosion of the iron at the break (Figure 29).

In air-free conditions, however, as when food is sealed in "tin" cans, the tin plate is generally anodic to steel. Zinc is more anodic than iron and, if there is a break in the zinc plate on iron, the zinc will be a sacrificial metal in this case and will corrode instead of the iron (Figure 30). Large patches of zinc may corrode away on galvanized sheet iron before denuded areas become sufficiently wide to allow anodic areas to form on the steel. Only then will it begin to rust.

Chemical inhibitors used in the corrosive medium make it inert, that is, unable to transfer metal ions from the anode to the cathode. Some types are used as a protective coating on the surface of the metal. Sodium phosphate in water can be used to produce a passive (inactive) ferrous oxide (Fe_2O_3) film on steel.

Organic coatings such as paint, tar, grease, and varnishes prevent corrosion by keeping the corrosive atmosphere from contacting the surface. Inorganic coatings such as vitreous enamels or even mill scale also create a barrier, but one drawback is their brittleness. Corrosive attack can take place where the coating is chipped or broken.

Metal cementation or diffusion of a material into the surface of metal is done by applying heat. Carburizing of low carbon steel is one example of metal cementation. Zinc powder can be diffused into the surface of steel by heating both powder and steel together to a few hundred degrees Fahrenheit. This is contrasted to the hot dip zinc process or electroplating in that here the powder penetrates to some extent into the surface of the steel. Silicon, aluminum, chromium, and many other elements are diffused or cemented into steel and other metals.

NOTE

The study of galvanic corrosion as presented in this unit is based on the current flow theory in which the flow of current is considered to be opposite to the electron flow and in which the anode deteriorates because of corrosion as the electrons leave it. In contrast, the electron theory as used in the study of electronics shows the cathode with a negative charge as the emitter of electrons and it deteriorates (as in a vacuum tube) while the anode with its positive charge (as the plate in a vacuum tube) does not.

Positive and negative signs are not used in order to avoid confusion about polarity. The terms anode and cathode as related to the current theory are in common use in industrial circles and probably will not be changed to the electron theory terminology for many years, if at all. The terms anodizing and cathodic protection are examples of common usage of current flow theory terminology.

SELF-TEST

1. Name four major causes of operational failures in machinery.
2. Why can an overload ruin a precision mechanism for further use?
3. Name three causes of rapid, excessive wear between two moving surfaces.
4. Are the best wear surfaces made of two similar or dissimilar metals? Name one exception to this rule.
5. The tendency for a metal shaft to develop fatigue cracking can be lessened in several ways. Name three of these.
6. How can welding initiate a fatigue crack in a stressed area such as on a spring or bumper of a car?
7. When metals are heated to a high temperature and an oxide scale forms on them, it is called direct oxidation corrosion. What kind of corrosion takes place at normal temperatures in the presence of an electrolyte (usually water)?
8. Can two separate pieces of metal, one an anode and one a cathode, immersed in an electrolyte develop any corrosion if they are not connected together in any way?
9. Which of the two coupled metals, zinc or nickel, would form the anode in an electrolyte and begin to deteriorate?
10. How can stainless steel rust like ordinary iron?
11. Name five methods used to prevent corrosion.

UNIT 3 MECHANICAL COMPONENT REPAIRS

Workers in all mechanical fields occasionally are faced with the responsibility of making a decision whether to repair, continue to use a worn or damaged mechanical component, or obtain a new replacement part. The choice involves identification of the problem, economy (cost), availability, operational life, safety, and a time factor. Of course, the mechanic should know if the part can be repaired and how it would be done should that be the decision. This unit will help you to make an intelligent decision when you are confronted with these mechanical problems.

OBJECTIVES

After completing this unit, you should be able to:
1. Identify and list types of failures found in mechanical devices.
2. Identify methods of repair and their correct applications.
3. Make decisions involving repair versus replacement of parts.
4. Explain the correct method of disassembling, testing, and reassembling machinery.

METHODS OF REPAIR

Worn components may be repaired by the use of several methods: welding, sprayweld buildup, applying industrial adhesives, boring and sleeving, or by remachining and replacing the mating part with one that fits the new dimension; for example, regrinding crankshaft bearings and replacing the insert bearings with slightly smaller ones.

Mechanical repairs may be made by replacing the worn or damaged component with a new one. This requires the techniques used by mechanics of troubleshooting, disassembly, inspection, and reassembly. Most automotive mechanical repairs are done in this fashion. In other machinery, it is often necessary to make a decision whether to repair or replace parts and whether a repair is practical in any particular case.

Perhaps wear is the most common reason a mechanical part must be replaced or repaired. Wear is most often found in moving parts such as gears and sprockets, bearing journals (areas on shafts that turn in sleeve bearings), roller and ball bearings and their supportive surfaces in housings and on shafts. Wear is only a serious problem when it is found in a critical area where it will affect the proper function of a ma-

chine or cause a part to be weakened to the point at which it will eventually fail from continuing wear. Essentially, wear is the eroding or tearing away of bits of metal on the surface in contact. When wear occurs, metal may be replaced by various means and the part remachined. However, these problems can be minimized by methods of wear prevention such as proper lubrication and the use of the correct material (Figure 1).

Corrosion and metal fatigue are two serious problems found in mechanical devices and industrial operations. Cracking due to metal fatigue often cannot or should not be repaired, while repairs on a corroded part can often be made by metal replacement and machining techniques.

Corrosion and stress corrosion are related to the atmosphere surrounding the part and to its resistance to corrosion. This subject was discussed in Section I, Unit 2, "Causes of Mechanical Failures."

Metal fatigue is a very common phenomenon associated with metals that are subjected to cyclic loading and vibration. This subject was also discussed in Section I, Unit 2. Roughness, notches, sharp fillets (corners), and stress concentration from welds are the most common causes of fatigue failures.

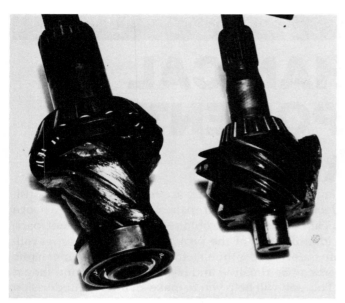

Figure 1. The hypoid gear on the right, which is from a truck, shows considerable wear, but not resulting in failure, as happened to the pinion at the left. The kind of wear shown on the right is largely unavoidable where extremely heavy loading must take place as in truck differentials. The complete failure as on the left could have been caused by loss of lubricants. The solution is the replacement of parts. *(Practical Metallurgy and Materials of Industry)*

Figure 2a. Weld on a cast iron engine housing. *(Practical Metallurgy and Materials of Industry)*

Figure 2b. Cracks between cylinders in an engine block are welded and remachined before the sleeves are put back in place. *(Practical Metallurgy and Materials of Industry)*

Casting Cracks and Porosity Repairs

Overloading equipment is most often the cause of casting cracks. Misalignment during assembly can cause cracking and internal damage, such as broken gears in a gearbox, which can create excessive forces that can greatly damage or ruin a casting. Porosity in castings is not usually a strength problem, although the casting is weakened in the porous area. The problem is usually leakage or seeping of lubricant, cooling water, or other fluids. Cracking and porosity can sometimes be found in all metal castings, but most commonly in cast iron, aluminum, and steel. Industrial sealants are available that can be used to seal porous areas of castings. Most of these sealants will withstand moderate heating without failing, such as that found in automobile engines. Porosity can sometimes be sealed by welding techniques, but this is a very difficult procedure that often creates more problems than it solves; however, cracks are often welded.

Some mechanics seal porous castings in engine blocks with soft solder that will not melt at the operating temperature of the engine. However, when an engine or a piece of equipment is under warranty, the mechanic **must** follow standard procedure recommended by the factory. This may involve the replacement of a porous casting.

Welding Repairs

Welding is often used to reattach broken pieces of castings, forgings, and even machined parts (Figures

Figure 3a. The threaded end of this small engine crankshaft has been broken off by overtorquing the nut. (Lane Community College)

Figure 3c. After the stub was welded, the crankshaft was mounted in a small lathe using a steadyrest and the end will be turned to size. (Lane Community College)

Figure 3b. A small stub that is a larger diameter than the threaded end has been prepared for welding on the crankshaft. (Lane Community College)

Figure 3d. Threads were cut on the stubbed end and the crankshaft is now ready to be installed in the engine. (Lane Community College)

2a and 2b). Various metals such as aluminum, cast iron, brass, and steel require different welding techniques and present different problems. Steel is perhaps the least difficult metal to weld, but cast iron, though difficult to weld, is the easiest metal for maintaining the original alignment of the part since there is very little distortion at the break. When any part or casting has been machined, this original alignment must be maintained, usually within plus or minus .001 in. or less. Obviously, it is not normally possible to maintain this kind of alignment when welding a previously machined part. The most common method used to restore this alignment is called **remachining**. If an end of a shaft has broken off or it simply needs to be extended, a piece of metal is "stubbed on," that is, welded on the end. After it cools, the shaft is set

up in a lathe and the end is machined in alignment with the machined surfaces (Figures 3a to 3d).

Outer race bearing surfaces are usually not in perfect alignment in housings that have been cracked and welded (Figure 4). They can be realigned by machining in several ways: by welding buildup, spray-weld and reboring to size, or by boring for a sleeve that is fitted to the bearings. Of course, in such housings, flat machined surfaces that are sealed by gaskets are usually warped and must be milled or ground to restore flatness (Figure 5). It is obvious that welding and machining operations are closely related; the welder and machinist should each know the abilities and limitations of the other. For instance, if the welder uses the wrong welding electrode on cast iron, it may be hard to machine. Worn surfaces on metal parts

Figure 4. The worn bore that supports a bearing on this differential drop-in unit has been built-up welded and is being remachined in a lathe. (Lane Community College)

Figure 5. The damaged sealing surface on this aluminum casting has been built up with weld and is now being remachined in a vertical milling machine. (Lane Community College)

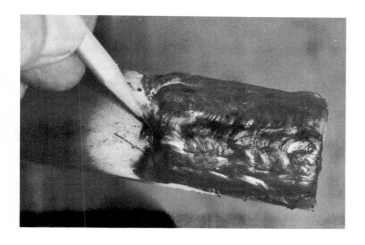

Figure 7. Crawler tractor idler rolls are being built up with wireweld on this automatic machine. Horizontal passes are made on the two diameters and circular welds are made on the flanges to give maximum strength and wear resistance. (Lane Community College)

are commonly restored by either fusion or diffusion (brazing) welding or metal spray welding. When arc welding with a stick rod for buildup, the termination of the weld passes, especially on shafts, can cause stress concentration and subsequent early failure from metal fatigue (Figure 6). A better way is to use wire welding with the shaft in a fixture that slowly rotates it, advancing the weld bead like a thread. This helical buildup weld is not as subject to stress concentration and fatigue as are the lengthwise passes of stick welding. Figure 7 shows an automatic wire welding machine where the welds are made lengthwise on the cylinder and spiral on the flanges to minimize stress concentration.

Many kinds of wear on machinery are repaired with welding processes. Splines become worn along

Figure 6. Horizontal weld passes for build up such as on this shaft tend to create stress concentration at the termination of the welds, causing early failure. (Lane Community College)

Figure 8a. This splined shaft drives a hydraulic pump on a tractor. The load is in one direction of rotation and therefore the spline is worn on one side only. (Lane Community College)

Figure 8b. Before the welding passes were made on this shaft, it was spark tested for carbon content and had approximately .40 percent carbon content, which would cause severe hardening and very low machinability. The shaft was therefore preheated and postheated, making it machinable. (Lane Community College)

one side usually and, if the shaft has a low carbon content so it will not harden, the splines can be welded and remachined (Figures 8a to 8c). Broken gear teeth on steel gears are easily welded and remachined. Cast iron gears are more difficult as a strong bond is not always possible, either by braze weld or with Eni electrodes. Studding (Figure 9) is often helpful for cast gears.

Metal spray buildup is superior in terms of having less stress concentration so there is less danger of fatigue failure later. However, one drawback is that an undercut must be made to allow room for the metal

Figure 8c. The welded spline has been turned on a lathe and recut on a horizontal milling machine. It is now ready for service. (Lane Community College)

Figure 9. Reinforcing a cast iron weld can be done by studding.

spray (Figures 10a to 10c). The undercut reduces the effective stress area of the part, weakening it to some extent. The sprayweld material is not very strong and has a limited bonding strength so it does not contribute much strength to the part after it is machined (Figures 11a and 11b). This is usually not a problem since most shafts are many times stronger than necessary for the applied load.

One of the best uses for metal spray is for worn ball or roller bearing surfaces on shafts and in bores. When a bearing race becomes loose on a shaft, for example, the bearing race begins to slip on the shaft, creating local heating that often ruins the bearing, causing it to seize; the final result can be extensive damage to the shaft or housing (Figure 12). When the loose fit exceeds a few thousandths of an inch, a metal buildup is necessary; if the fit is closer, mechanics sometimes use an industrial adhesive. If the adhesive is applied according to manufacturer's instructions, good results are obtained (Figure 13).

Knurling can be used to restore some worn parts. For example, valve guides and pistons in automobile engines. In these cases, the knurl, which only provides

Figure 10a. The undercut that is made ranges from 0.015 to 0.020 in. deep where the buildup is required. Grooves are often made to insure bonding.

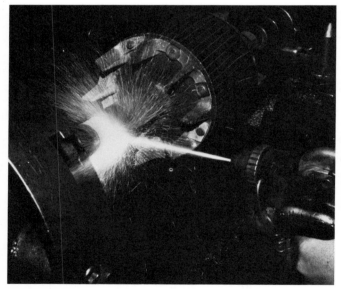

Figure 10b. Metal being sprayed on prepared surface of shaft. Often a light undercoat of molybdenum is applied first to create a bond that has a physio-chemical nature. Steel spray has only an adhesive bond, not a metallic bond as in welding.

Figure 10c. When sufficient material has been applied, the surface may be machined to the correct diameter after it has cooled.

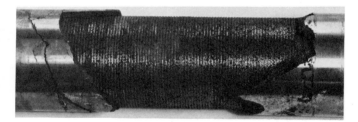

Figure 11a. An example of a very poor spraywelding application that peeled off after being in service for a short time. (Lane Community College)

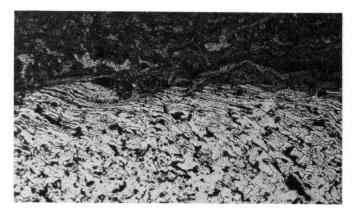

Figure 11b. This microphotograph reveals that the bond between the sprayweld metal and the base metal is very good. This build up was correctly applied and would not tend to peel off as in Figure 11a. The distortion at the surface of the base metal was caused by machining (100X). (Lane Community College)

Figure 12. A loose fit on this shaft caused the inner ball bearing race to slip, damaging the shaft. Excessive slipping of the race caused heating and subsequent damage to the bearing. (Lane Community College)

Figure 13. Mounting a bearing with an adhesive compound. This adhesive material will keep the bearing race from slipping in the bore. (Locktite Corporation)

a partial contact of metal parts, can hold more oil and even reduce wear. However, knurling should **never** be used as a means to build up a worn shaft, especially where heavy loads are applied, because the few high points of contact would very soon wear off, leaving the parts worse off than before.

Machine Shop Repairs

When welding repairs are not practical and the part is not manufactured or is not available, a machine shop can often duplicate the original part. Maintenance or job machine shops are found in every industrial area. They usually have their own welding facilities and can machine new parts from standard bar or round stock in cold rolled low carbon steel, alloy steels, and nonferrous metals such as aluminum. A few of these shops may have heat treating (hardening, tempering, annealing) facilities; however, usually only tool and die shops have heat treating equipment and expertise. A job shop can usually drill holes, make various shafts with keyseats or splines and threads, bore chain sprockets, gears, and housings. Many are equipped to make gears and sprockets and do many other operations related to metal cutting.

Automotive machine shops are a specialized branch of the machining field. These shops remachine automotive parts usually in engines (Figures 14a to 14f). Crankshafts are ground to a slightly smaller size and smaller insert bearings are then used to restore an accurate fit. Engine heads and other parts are surface ground to restore flatness. Cylinders are rebored. If an engine block has severely damaged cylinder walls and it is not equipped with replaceable sleeves, the cylinder wall can be bored completely out and a sleeve installed, restoring it to a new condition.

Figure 14a. This damaged crankshaft journal cannot be used as it is but needs to be reground to a slightly smaller diameter. (Lane Community College)

Figure 14b. Crankshaft journal being reground in a crankshaft grinder. (Huffman's Auto Grinding)

Figure 14c. Cylinders in an engine block being rebored with a boring bar. This is usually done when worn engines are rebuilt. (Huffman's Auto Grinding)

Figure 14d. Cylinder bores being checked for size by the machinist. (Huffman's Auto Grinding)

Figure 14e. A final check is made by inserting the piston into the bore along with a feeler gage. (Huffman's Auto Grinding)

Figure 14f. Cylinder heads are sometimes resurfaced on a vertical spindle grinding machine such as this one. (Huffman's Auto Grinding)

MAKING THE DECISION TO REPAIR OR REPLACE

The mechanic is often called upon to make the difficult decision whether to scrap a part and purchase a new one or to have the part repaired. With small, relatively inexpensive parts, this is no problem since, in most cases, repair is not feasible as small parts manufactured by mass production methods and readily available cost far less than one produced in a machine shop (Figure 15). It is the large costly mechanical parts that can often be salvaged at far less cost than the price of a new one (Figure 16).

Welds should usually never be made in high stress parts such as automobile rear axles. The stress concentration of the weld, coupled with high stresses and load reversals, would quickly lead to failure by metal fatigue. However, in some special applications such as some shop built dune buggies, for example, the axles are shortened by cutting out a section and welding the two parts together in a fixture to insure alignment. These welds are generally successful because of the correct electrode and welding technique used, but also because the weld is made in the largest diameter of the axle where less stress would be on the weld.

There are some parts that should not be considered for repair by weld buildup and remachining: flame hardened gears, parts in a gearbox (such as a transmission), and case hardened shafts are among them. Sometimes, however, very large shafts in tractor and truck transmissions are built up at bearing areas by sprayweld to salvage them for continued use (Figure 17). These are extremely expensive items when newly purchased and the risk of failure and down time is more acceptable. When parts fail in a gearbox, they can cause extensive damage to all other parts, including the case. Large parts that are not in

Figure 15. These small parts can be purchased from a parts store at far less cost than new ones can be made in a machine shop. (Lane Community College)

Figure 16. This large machine part can be salvaged by rebuilding and remachining techniques at less cost than a new one.

Figure 17. This shaft from a tractor transmission is worn a few thousandths of an inch on a bearing area. It will be restored by using metal spray techniques and remachining.

a gearbox can almost always be repaired by weld and remachine techniques at a substantial savings. This, of course, depends on the capability of the local machine shop.

Another consideration the mechanic must use when deciding to have a part repaired is the time factor. If the savings of repairing the part is offset because of lost time when a more expensive part can be taken immediately off the shelf, the decision, of course, should be to use the readily available part. On the other hand, if the new part must be sent from the factory or warehouse with a two or three day wait and an assembly line or a needed machine is down, the obvious decision is to have the old part repaired or a new one made in a machine shop. The wait may be only a few hours.

New parts can be manufactured in the local machine shop and, of course, are only practical in terms of cost if the original part is not readily available and is priced near or above the cost of making a new one in a shop. One would be willing to pay several times the cost of a new part if the down time was exceedingly costly in the machine that needed the part.

In many cases, rebuilt or repaired parts will outlast the original. Many problems exist, however, that can bring about an early failure of repaired parts. The greatest of these problem areas is the stress concentration of welds and subsequent fatigue failure. **Often a simple preheat or postheat treatment can eliminate this problem.** Tool marks, grooves, and sharp fillets can also contribute to early failure; yet, new equipment is often faulty in these areas.

DISASSEMBLY, TESTING, AND REASSEMBLY

Automotive and heavy equipment mechanics, millwrights, and machinists are often required to disassemble complicated mechanisms for troubleshooting and replacing worn parts. If adequate drawings and procedures are available, reassembly is usually no problem (Figures 18a to 18d). Even if drawings are available, the parts should be laid out in the order that they are removed to facilitate reassembly. Parts should be marked in the position and orientation (the direction it faces) in which they were removed. Gears should be match marked in case they must be synchronized in a certain order. Bolted flanges should be match marked with a number stamp or center punch so the same holes will be matched when they are assembled (Figure 19). In complicated assemblies, a sketch should be made of each subassembly that is removed, including location, dimensions, and orien-

Figure 18a. This drive on a vertical spindle milling machine became inoperative and was removed from the machine for repairs. It is being disassembled. (Lane Community College)

Figure 18b. When this bevel gear was exposed, it became immediately apparent that the gear teeth were stripped. (Lane Community College)

Figure 18c. The one-piece gear and shaft is removed as a subassembly. A new one must be purchased and the bearings replaced. (Lane Community College)

Figure 18d. It is obvious that the mating bevel gear as shown here will be damaged as well and must be removed for inspection. (Lane Community College)

Figure 19. All flanges that are to be removed must be match marked for later reassembly. (Lane Community College)

tation of each small part removed from the subassembly (Figure 20).

When disassembling threaded parts, especially those that rotate, make certain that you are not actually tightening a left-hand thread, assuming it is a right-hand thread. If the end of the thread can be seen, you can quickly determine if it is right or left hand. If the thread is not visible, try to determine if it would be logical to use a left-hand thread to keep the part from unscrewing in use. For example, the right-hand arbor of a pedestal grinder uses a right-hand thread (since the nut would tend to tighten when the wheel was in use), and a left-hand thread on the

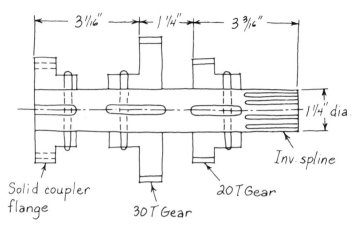

Figure 20. Mechanic's sketch showing position and location of parts on a subassembly shaft. Such a sketch makes it possible for the mechanic to reassemble parts exactly as they were originally.

Figure 21. This mating bevel gear that was seen in the housing in Figure 18d has been removed and cleaned for close inspection. The gear is damaged and must be replaced as well as the ball bearing. (Lane Community College)

left arbor of the grinder. When a threaded fastener cannot be removed with an optimum amount of torque applied to it in the counterclockwise rotation of a right-hand thread, it may be a left-hand thread. If you suspect it is a left-hand thread, apply the same torque in the clockwise rotation to remove it.

When the mechanism has been dismantled to the necessary extent, each part should be cleaned in solvent and inspected for wear, cracks, or corrosion (Figure 21). Ball or roller bearings should be closely inspected for any signs of breakdown. Small, relatively inexpensive bearings (perhaps 3 in. in diameter or smaller) should be discarded if they do not have a new appearance. Any imperfections on a ball or roller or on the races is usually an indication that the bearing is beginning to break down and may fail before the

mechanism is again dismantled for repair or maintenance. Large expensive bearings, on the other hand, should be studied by an expert who can determine the need for replacement and the approximate hours of running life remaining in the bearing.

BALL AND ROLLER BEARING SERVICE FAILURES

The series of bearing photographs below will help you to determine the causes of bearing breakdown. It is a general rule that flexible seals (usually made of neoprene rubber) should always be replaced because they get hard and brittle in use and are usually damaged when they are removed.

Most bearing failures can be attributed to one or more of the following causes:

1. Defective bearing seats on shafts and in housings.
2. Misalignment.
3. Faulty mounting practice.
4. Incorrect shaft and housing fits.
5. Inadequate lubrication.
6. Ineffective sealing.
7. Vibration while the bearing is not rotating.
8. The passage of electric current through the bearing.
9. Excessive lubrication.

However, even when properly applied and maintained, the bearing will still be subjected to one cause of failure: fatigue of the bearing material. Fatigue is the result of shear stresses cyclically applied immediately below the load carrying surfaces and is observed as spalling away of surface metal as seen in Figure 22.

Figure 23 shows the condition resulting when a bearing outer ring is not fully supported. The impression made on the bearing O.D. by a turning chip left in the housing when the bearing was installed is seen in the left-hand view. This outer ring was subsequently supported by the chip alone with the result that the entire load was borne by a small portion of the roller path. The heavy specific load imposed on that part of the ring immediately over the turning chip produced the premature spalling seen in the right-hand illustration.

When the contact between a bearing and its seat is not close, relative movement results. Small movements between the bearing and its seat produce a condition called fretting corrosion (Figure 24). Fretting corrosion can also be found in applications where machining of the seats is accurate but where, because of service conditions, the seats deform under load.

Figure 22. Advanced spalling on inner ball bearing race. (SKF Industries, Inc.)

Figure 23. Fatigue from chip in housing bore. (SKF Industries, Inc.)

Figure 24. Wear due to fretting corrosion. (SKF Industries, Inc.)

Figure 25. Fretting caused by yield in the shaft journal. (SKF Industries, Inc.)

Shaft bearing seats as well as housing bores can yield and produce fretting corrosion. Figure 25 is an illustration of damage by movement on a shaft. The fretting corrosion covers a large portion of the surface of both the inner ring bore and the shaft bearing surface. The axial crack through the inner ring started from surface damage caused by the fretting.

Bearing damage is also caused by bearing seats that are concave, convex, or tapered. On such a seat, a bearing ring cannot make contact throughout its width. The ring therefore deflects under loads and fatigue cracks commonly appear axially along the raceway. Cracks caused by faulty contact between a ring and its housing are seen in Figure 26.

Impact damage during handling or mounting results in brinelled depressions that become the start of premature fatigue. An example of this is shown in Figure 27 where the spacing of flaked areas correspond to the distance between the balls. The bearing has obviously suffered impact and, if it is installed, the fault should be apparent by the noise or vibration during operation.

If a bearing is subjected to loads greater than those calculated to arrive at the life expectancy, premature fatigue results. Unexpected loads can arise from faulty mounting practice. Examples of unanticipated loads would be found in mounting the front

Figure 26. Cracks caused by faulty housing fit. (SKF Industries, Inc.)

Figure 27. Fatigue caused by impact damage during handling or mounting. (SKF Industries, Inc.)

Figure 28. Spalling from excessive thrust. (SKF Industries, Inc.)

wheel of an automobile if the locknut were not backed off after applying the specified torque to seat the bearing. Another example is in a bearing that should be free in its housing but, because of pinching or cocking, cannot move with thermal expansions and an abnormal thrust is induced on the bearing (Figure 28). Excessive fits also result in bearing damage by internally preloading the bearing or inducing dangerously high stresses around the inner ring (Figure 29).

Figure 29. Axial cracks caused by an excessive interference fit. (SKF Industries, Inc.)

Figure 30. Spalling caused by inadequate lubrication. (SKF Industries, Inc.)

An **insufficient** quantity of lubricant at medium to high speeds generates a temperature rise and usually a whistling sound. An **excessive** amount of lubricant also produces a sharp temperature rise due to churning of the lubricant in all but exceptionally slow speed bearings. Conditions inducing abnormally high temperatures can render a normally adequate lubricant inadequate under the temperature conditions. For correct lubrication, consult your local bearing dealer or the manufacturer.

When lubrication is inadequate, surface damage will result. This damage will progress rapidly to failures that are often difficult to differentiate from a primary fatigue failure. Spalling will occur, and often will destroy the evidence of inadequate lubrication. However, if caught soon enough, one can find indications that will pinpoint the real cause of the short bearing life. Figure 30 shows an advanced stage of spalling caused by inadequate lubrication.

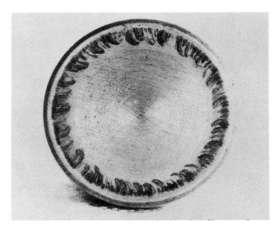

Figure 31. Smearing on spherical roller end. (SKF Industries, Inc.)

Figure 32. Smearing on cylindrical rollers caused by ineffective lubrication. (SKF Industries, Inc.)

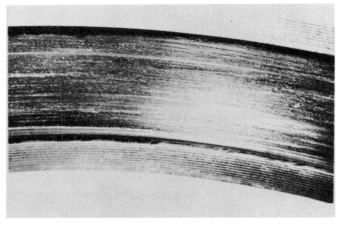

Figure 33. Smearing on outer raceway. (SKF Industries, Inc.)

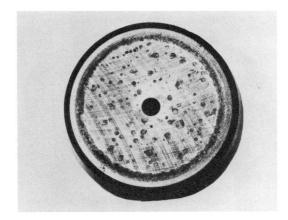

Figure 34. Rust on end of roller caused by moisture in lubricant. (SKF Industries, Inc.)

Another form of surface damage is called smearing. It appears when two surfaces slide and the lubricant cannot prevent adhesion of the surfaces. Minute pieces of one surface are torn away and rewelded to either surface. Examples are shown in Figures 31 through 33.

Both abrasive material and corrosive agents such as water and acid can cause premature bearing failure. Figure 34 shows how moisture in the lubricant can rust the end of a roller bearing.

REASSEMBLY

When the mechanism is reassembled, care must be taken that parts are replaced in the exact order that they were removed. If the proper sequence is not used, often the entire job must be done over. It is very important to avoid damage during reassembly.

Do not use excessive force in either assembly or disassembly. Cleanliness of parts should be maintained and they should be lubricated with light oil (in most cases) when they are returned to their proper location.

When the job of assembly has been completed and all adjustments are made, the mechanism should be tested for operation. Parts should be moved or rotated by hand so that any resistance or unusual tightness can be detected. In some cases the device can be tested in an operational mode as in the case of hydraulic pumps on a test stand. When this is done, any lack of proper function or any spurious noises can be checked out before returning the mechanism to service.

SELF-TEST

1. Besides proper lubrication, name three methods of preventing wear.
2. What is usually the greatest problem resulting from casting porosity?
3. Name three methods of repair for worn mechanical parts.
4. When broken pieces of castings or other mechanical parts that have precision machined surfaces on them are welded together, how can the original alignment of machine surfaces be maintained?
5. What is the major service problem subsequent to weld buildup on worn parts and is not a problem with metal spray buildup?
6. Name one drawback of metal spray buildup.
7. A millwright is called to repair a machine on which one man works on each of three shifts and is now idle (he is sweeping and cleaning up around his area); he earns $12 per hour. The parts he makes supply an assembly line of 30 workers, but there are enough parts ahead for eight hours of work. The cause of the breakdown is a seized bearing (which can be purchased locally) and a completely damaged shaft that must be replaced. After a few telephone calls, a shaft is located in a warehouse 1,000 miles distant. Its cost is $35. It can be shipped by air which should get it to the millwright in about eight hours. It is a simple shaft and a reputable local machine shop can make a new one in two or three hours, but the cost will be about $200. What do you think the millwright's decision should be?
8. Precision hardened parts are generally not repaired by weld buildup and remachining because welding heat softens adjacent areas and any heat treatment to restore hardness will warp critical surfaces. Welds on case hardened parts can cause pieces to spall (break off the surface). Why then should parts in a gearbox not normally be repaired by welding?
9. In disassembling a machine, why should you match mark gears and flanges?
10. When disassembling a gearbox for inspection and repair, when should you replace bearings and seals?

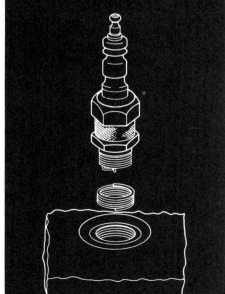

UNIT 4 RETHREADING AND RESTORING DAMAGED THREADS

One of the most frequent causes of delay on a mechanical repair job is damaged threads on an expensive or hard-to-replace part. The damage might be caused by corrosion, wear, or mechanic's carelessness. However, in most cases, damaged threads can be restored to an acceptable degree of usefulness. This unit will prepare you to restore threads, using various methods and tools.

OBJECTIVES

After completing this unit, you should be able to:
1. Identify the various methods of restoring threads.
2. Use hand rethreading tools for external threads.
3. Use a small engine lathe to restore damaged threads.
4. Restore damaged internal threads, using several methods.

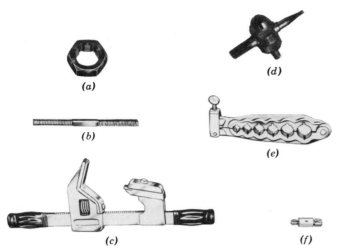

Figure 1. Several tools that are used for rethreading: *(a)* hexagon nut die, *(b)* thread file, *(c)* universal thread restorer, *(d)* valve stem thread restorer, *(e)* split-type thread restorer, and *(f)* internal thread chaser. (Snap-On Tools Corporation)

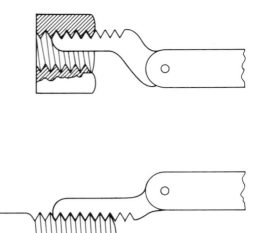

Figure 2. Using a screw pitch gage to determine the pitch (threads per inch) of internal and external threads.

Various tools may be used for stripped threads, each of them having a special application (Figure 1). In the following discussion of the five kinds of damaged threads, one or more of these tools will be recommended. Thread restorers are made of hardened and tempered tool steel to provide cutting edges and wear resistance. They should never be used on heated metal or forced over a hardened thread. Each cutter or die has its own thread pitch (numbers of threads per inch) stamped on it and it *must* be used on a thread of the same pitch (Figure 2). In the case of a nut die, the pitch diameter (or bolt diameter) must be the same as that stamped on the die. Be careful that you do not use inch measure rethreading dies on a metric bolt; they are often very close to the same diameter and pitch, but not close enough to avoid ruining the thread altogether.

Thread restorers may be divided into two categories: those designed for external threads and those for internal threads. Thread restorers often come in kits having various sizes of threading dies (Figure 3).

DAMAGED EXTERNAL THREADS

External threads that are damaged fall into five different types:

1. Stripped threads from overtorquing or heavy loading.
2. Galled threads from welding or seizing.
3. Corroded (rusted) threads.

Figure 3. External thread restoring kit. (Lane Community College)

4. Distorted threads from hammering upon the thread end of a shaft or by overtorquing the nut or screw.
5. Cross-threading damage from starting the thread at a tilted angle.

Stripped Threads

In automotive manufacture and some other machinery, fasteners are sometimes made with hardened bolts and soft nuts, or with hardened nuts and soft bolts (as in auto spring U-bolts and hi-nuts). Where this is the case, when threads are stripped, only the softer part will be damaged. An example is found in aluminum engine heads where steel spark plugs are screwed into the aluminum threads. If Anti-Seize®

Figure 4. Anti-Sieze® or its equivalent should be applied to threaded assemblies to avoid corrosion, especially when dissimilar metals are contacting each other. (Lane Community College)

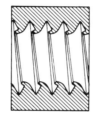 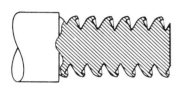

Figure 5. Internal and external stripped threads.

Figure 6. Hexagonal rethreading die. These dies are used for restoring existing threads that are damaged. They should never be used for cutting new threads. (Snap-On Tools Corporation)

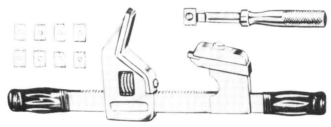

Figure 7. Universal external and internal thread chasers. Interchangeable chasers of many thread pitches are included. Metric chasers are also available. (Snap-On Tools Corporation)

(Figure 4) was not used when the parts were assembled, corrosion may take place in the threads and, when the plug is removed, the internal aluminum thread will tear and come out with the plug. Stripped threads are almost always caused by overtorquing and are usually found in fine series threads. If both parts are about the same hardness, such as front axle spindle threads, or wheel studs and nuts, internal and external threads will be about equally damaged. In coarse series threads, the stud or bolt usually breaks before the thread will strip. If screws that are subjected to vibration become loose, the threads may actually wear out and will resemble stripped threads.

Usually, on stripped threads, the thread crest is flattened and turned over and thread material is displaced, not cut off (Figure 5). A solid, hexagonal, rethreading die (Figure 6) can be used on a part that has not been hardened. These dies should never be used on a hardened bolt. If it is hard to make an indentation on the bolt with a center punch, or if a file does not bite in easily but tends to slide over it, the bolt is too hard and will ruin the die. The cutting action of the die reforms and resizes the thread so

it is almost at the original condition. Some of the deformed thread is forced back into the original thread form by tool pressure, and it is not all cut away. Rethreading dies are often used on axle spindles and engine studs.

The universal external thread restorer is perhaps the most versatile of these thread repair tools (Figure 7). Interchangeable chasers with different thread pitches may be selected for the job. Since the diameter of the tool can be freely adjusted, the tool can be used on nonstandard threads such as those found on pipes, tubes, and some axles. Even if the tube has been slightly expanded, which is often the case, this tool can be used to gradually reduce the diameter of the tube while cutting and restoring the thread by adjusting the tool slightly smaller for each cutting revolution. The nut should be tried frequently for a correct fit. Cutting oil should be used when much stock is removed.

Thread files are square in cross section and have four different thread pitches on each end. They are made for both inch and metric pitches for all generally used coarse and fine pitches. They are useful for light

Figure 8. Thread file being used to clean up a dented thread on an axle. (Lane Community College)

Figure 9. A knife file is especially useful when large threads are slightly damaged in one place. (Snap-On Tools Corporation)

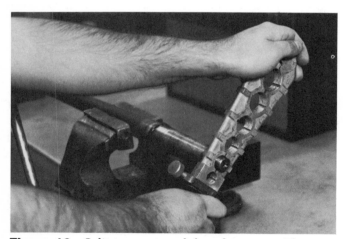

Figure 10. Split-type external thread restorers. This type is most often used on axles that have been damaged by hammering. (Lane Community College)

thread cleanup jobs as shown in Figure 8, or they may be used on a small lathe while the threaded part is being rotated slowly. The knife file (Figure 9) also is frequently used for light cleanup jobs on damaged threads.

Split-type external thread restorers have several thread diameter sizes in standard thread pitches that can only be used on those standard threads (Figure

Figure 11. Badly damaged thread caused by hammering on the end of this axle which also damaged the center hole. The axle is set up in a steady rest in a lathe to correct the problem. (Lane Community College)

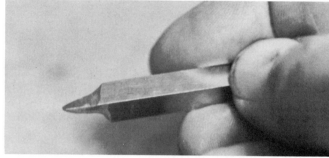

Figure 12. This special tool has been ground with a narrow point so that it will fit into a center hole and cut on one side. (Lane Community College)

10). Their advantage is that they can be clamped on the inside, undamaged thread and rotated outward to repair a thread that has been expanded by hammering.

RESTORING EXPANDED THREADS WITH A SMALL ENGINE LATHE

Threaded ends of shafts sometimes become swelled because of hammer blows or by presswork when hub or bearing removal is difficult. This swelling precludes the use of nut dies or even the universal thread restorer in many cases. The only practical method of restoring these threads is to use a lathe. To use this method, you must already have completed the section on lathes in this book. The following steps may be used for restoring a damaged or expanded thread:

1. Inspect the workpiece center to see if it has been damaged by hammering (Figure 11). If it has been damaged, steps 2 to 6 will need to be completed; if not, skip to step 7.

Figure 13*a*. Center hole being recut. (Lane Community College)

Figure 13*b*. Completed job of restoring the center. (Lane Community College)

Figure 14. Checking the thread pitch. (Lane Community College)

Figure 15. Adjusting the threading tool. (Lane Community College)

2. Set up the workpiece in a chuck and dead center.

3. Set up a steady rest on a machined area of the work near the tailstock end. Use high pressure lubricant on the steady rest jaws. Remove the dead center.

4. Mount a special tool (Figure 12) in the tool post for truing the center hole. (If there is no center hole at all, simply drill one.)

5. Set the tool edge at the correct angle (30 degrees) and shave one side of the center hole while the lathe is set on a very low speed; use a drop or two of cutting oil (Figures 13*a* and 13*b*).

6. Remove the steady rest setup and adjust the dead center to the corrected center hole.

7. Determine the number of threads per inch (or pitch if it is metric) and set the lathe gearbox for that number. Set a low RPM and prepare the lathe for threading (Figure 14).

8. Set up a threading tool (Figure 15).

9. With the tool away from the work, engage the half-nuts, noting the correct line on the thread dial. Turn the motor on and allow the tool to move across the threads, past the damaged area to an undamaged area. Turn off the motor but **do not disengage the half-nuts.**

10. Now adjust the compound (which should be set at 30 degrees for threading) and the cross slide together until the tool just fits the thread (Figure

Figure 16. Adjusting the tool to the threads. (Lane Community College)

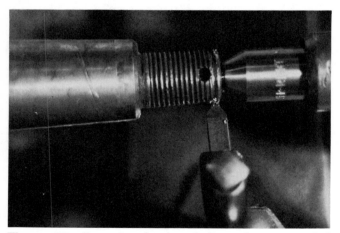

Figure 17. Restoring the thread. (Lane Community College)

Figure 18. Testing the thread with a nut. (Lane Community College)

Figure 19. Tapping a thread by hand. (Lane Community College)

16). Zero the cross slide dial and the compound dial.

11. Back off the tool one turn of the cross slide and return to the starting position, either by reversing the lathe motor or by disengaging the half nuts and resetting them at the starting place using the threading dial. Run the lathe until the tool is over the damaged area of the thread.

12. Move the tool in with the cross slide until it just touches the highest place on the damaged thread and note the reading on the dial. Back the tool out and reposition at the noted reading at the starting place. You are now ready to take a skim cut.

13. Take the first cut and note if the tool is tracking in the original thread. It will probably not track very well on the damaged area.

14. Continue to take successive cuts until the cross-slide dial reads zero (your original setting and the correct diameter of the thread) (Figure 17). Use cutting oil. Test the thread frequently with a nut or mating part if possible (Figure 18).

RESTORING DAMAGED INTERNAL THREADS

Internal threads, which were originally made with a tap and have become damaged, can often be restored with a tap. Care must be exercised when using a tap that it is aligned square to the work (Figure 19). This is especially important if the internal thread has been cross threaded.

A thread chaser is simply an external thread on a plug that has been gashed on the end to produce a cutting edge that is then hardened and tempered. Internal thread chasers (Figure 20) are perhaps more

Figure 20. Internal thread chaser. (Orchard Auto Parts Co., Eugene, Oregon)

Figure 21. Rethreading tap set. (Snap-On Tools Corporation)

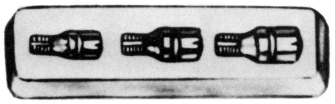

Figure 22. Inverted flare tube fitting chaser set. (Snap-On Tools Corporation)

convenient to use and they often come in kits (Figure 21). They are also available for special purpose threads such as for flare tubing fittings and spark plug threads (Figure 22). Spark plug thread restoring kits are available that include a counterbore that is used to make

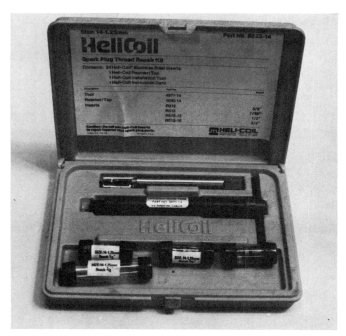

Figure 23. Spark plug thread restoring kit. (Lane Community College)

the next larger spark plug size thread when the threads are stripped beyond repair (Figure 23).

REPLACING UNRESTORABLE INTERNAL THREADS

There are three basic methods of making new threads in a hole when the old threads are damaged beyond repair.

1. Drilling and retapping to a larger size thread.
2. Replacing the thread with a steel coil insert.
3. Drilling and tapping to a larger size, inserting a threaded plug, and drilling and retapping the plug to the original thread size.

When the wall thickness of the casting or threaded part permits, and other considerations allow it, the ruined thread can simply be drilled with a tap drill for the next larger size thread (Figure 24). A larger diameter capscrew or stud can then replace the old one. The mating part will usually be too small for the new stud, so it will also need to be drilled out to the body size of the new stud. When using this method, make sure that the tap drill for the new thread is large enough to remove all of the old threads. If it is necessary to drill through into a cavity where the chips could cause damage, such as when repairing spark plug holes, the part should be removed. In some

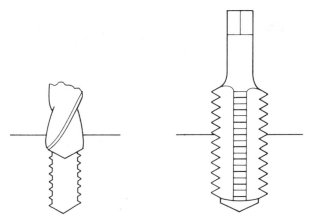

Figure 24. Steps in drilling out a damaged thread and making a larger one. The tap drill for the larger tap should be of sufficient size to just remove the old damaged threads.

cases that are not so critical as an engine head, heavy grease can be applied to the drill and tap to collect and hold most of the chips until the tool is removed.

Using a Steel Coil Thread Replacement Kit

Heli-Coil® inserts are precision formed screw thread coils of 18-8 stainless steel wire having shaped cross section. They are inserted in a special pitch and diameter size internal thread, which is larger than the restored thread size. The tap used to make the internal threads is not a standard thread or size; it is provided in coil insert kits along with tools, inserts, and tap drills. These inserts are widely accepted for many uses, such as replacement threads for studs, capscrews, and spark plugs. They are also used for threads that are tough, wear and corrosion resistant in soft metals and many other materials.

Procedures for insert applications should be followed in the steps given with the particular kit being used or from a Heli-Coil® catalog. There are four basic steps for most applications, which are shown in Figure 25.

Insert kits come in all sizes from the small, single thread size packet (Figure 26) to the larger kit having most frequently used thread sizes, which is widely

Figure 25. The basic steps for installing inserts: (a) Damaged thread. (b) Drilling out the hole with a tap drill. (c) Tapping the hole with a special tap. (d) Steel coil insert with insertion tool. (e) Coil being installed. (f) Completed replacement thread. (Lane Community College)

(a)

(b)

(c)

(d)

(e)

(f)

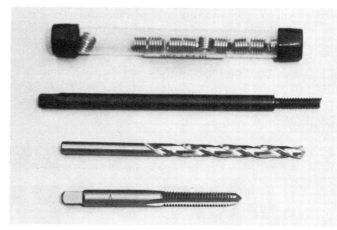

Figure 26. Single size thread insert kit. (Lane Community College)

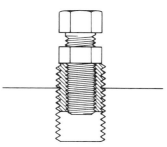

Figure 27. Inserted plug method to repair stripped threads.

used in the U.S. Air Force, Army, and Navy repair and maintenance activities. Strip-feed inserts are used in manufacturing where coil inserts can be quickly inserted.

The Drill and Plug Method

Another method of replacing damaged threads to the original size is to drill out and retap the thread, as in Figure 24, and then insert a plug. A somewhat larger thread is made in the workpiece by drilling and tapping and a section of threaded rod is screwed in tightly. It is then hacksawed off almost flush, and filed flat and even with the workpiece surface. The center of the threaded rod is marked and center punched. It is often helpful to stake the thread (deform the metal) with a punch so the plug cannot screw out, or peen it slightly to expand it in the hole before filing it flat. It is then center punched, drilled, and tapped to the original thread size and pitch.

Special self-tapping plugs are available, which are drilled and tapped to the original screw size. With this method you simply drill a hole to the required size, then turn the threaded plug into the hole by using a capscrew and jam nut. When the plug is completely seated, the jam nut is loosened and the capscrew removed (Figure 27).

SELF-TEST

1. What are the three major causes of damaged threads?
2. Name at least three kinds of damaged threads.
3. Are stripped threads most often found in coarse thread or fine thread series bolts? Soft or hard metals?
4. Why should rethreading dies not be used on hardened bolts?
5. What methods can be used to restore a thread having a nonstandard pitch and diameter?

6. If a mechanic has hammered on a thread end of a shaft, expanding it too much to restore with a hand tool, how can the shaft be repaired?
7. How can damaged internal threads be restored?
8. Name three methods of replacing damaged threads in a hole.

9. Name one limitation that would prohibit using an enlarged thread or a plug when an internal thread is damaged beyond repair.
10. Heli-Coil® thread inserts are widely used to restore internal screw threads and spark plug threads. Can you name another common use of these inserts?

UNIT 5 REMOVING BROKEN STUDS, TAPS, AND SCREW EXTRACTORS

Much working time is lost by mechanics and machinery maintenance persons because of broken fasteners. Stud bolts and capscrews (Figure 1) are the most common fasteners in machinery. They sometimes are weakened by metal fatigue, corrosion, or prolonged heat so that they tend to break off when torque is applied to remove them. When this happens, special procedures and tools are needed. This unit covers these methods and will help you choose the right one for the job. You will also learn how to deal with the difficult problem of broken screw extractors and taps.

OBJECTIVES

After completing this unit, you should be able to:
1. Make the right selection of removal tools and procedures for a specific problem.
2. Remove a broken stud by any of several methods.
3. Remove a broken tap or screw extractor.

REMOVING BROKEN THREADED FASTENERS

Studs are usually removed with a special tool (Figure 2) or by using two jam nuts (Figure 3), and capscrews are removed with a box wrench or a socket. A broken capscrew or stud can often be avoided in the first place by knowing how much torque can be applied to a particular size fastener without breaking it and not exceeding this limit. If the fastener will not turn with a safe torque limit, other methods must then be used. Of course, some fasteners will fail with very little torque applied because of metal fatigue or corrosion.

When ordinary methods and tools fail to remove threaded fasteners, or if they break off, the methods that may be used in the order of the severity of the problem are as follows:

1. Applying penetrating oil.

Figure 1. The difference between capscrews and stud bolts is that capscrews always have a head and stud bolts require nuts for fastening. (Lane Community College)

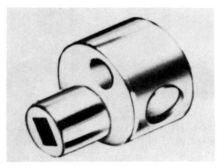

Figure 2. One of the several kinds of tools used for removing studs. (Snap-On Tools Corporation)

Figure 3. A box wrench is used with two jam nuts to remove a stud. (Lane Community College)

2. Removing the screw with a punch.
3. Applying heat and using a locking plier.
4. Using screw extractors.
5. Drilling out broken studs or screws.
6. Using welding techniques.

Applying Penetrating Oil

When a stud bolt or screw has been in service for a long time, the threads are likely to be corroded. If

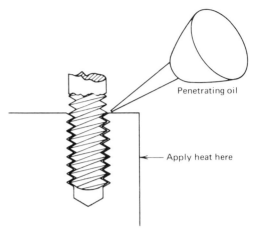

Figure 4. Penetrating oil has the ability to seep into narrow crevices by capillary action. In severe cases, longer periods of time are required. Heating can accelerate this action.

they have been in a "dry" environment (not soaked with oil) or in an area subjected to heating, there will be metal-to-metal contact between the mating threads with nothing to prevent galling or seizing of the threads when they are turned. For this reason, penetrating oil should always be used before attempting to remove any screw that has been in service for more than a month or two in a "dry" environment.

Several brands of penetrating oil are available for loosening stubborn threads. One or two applications and allowing a soaking time (overnight is best) is often sufficient to loosen most stubborn threads (Figure 4).

Removing the Screw with a Punch

The first attempt in removing a broken screw should be made with a small sharp pointed punch or chisel to try turning the broken screw in a counterclockwise direction (if it is a right-hand thread) (Figure 5). Penetrating oil may be used also. Quite often a broken screw is loose and will turn out freely by this method. If you are not getting results, try another method.

Applying Heat and Using a Locking Plier

Fasteners are often broken off as a result of over-torquing by the mechanic or by overload on a machine part. They sometimes fail by metal fatigue, which can be seen as a smooth, ripple pattern on the broken end of the screw. The removal job is easiest if the break is well above the surface, in which case a locking plier may be used as a removal tool. An extended broken stud may also be removed by filing a flat on

Figure 5. Quite often a broken screw or stud is loose enough to be easily removed, but is broken off flush with the surface. In this case, the screw can often be rotated with the point of a punch to remove it. (Lane Community College)

Figure 6. When the stud is broken off above the surface, a locking plier can often be used as a removal tool. (Lane Community College)

each side and using a wrench or by sawing a slot on the top with a hacksaw and turning the screw out with a screwdriver. The process of stud removal can sometimes be speeded up by applying heat around the bolt area. **Caution:** Be careful not to use a flame near a source of gasoline or other flammable substance and only raise the temperature to about two or three hundred degrees.

The process may be repeated several times, applying more penetrating oil (after the flame has been removed). The heating and cooling tend to separate the threads and allow the penetrating oil to flow the entire length of the threads. Torque should again be applied and, if the screw fails to break free, other measures must be taken. Tapping the stud or screw head with a hammer will sometimes break the threads loose. Also a slight rocking back and forth may start a thread turning. A locking plier (Vise-Grip®) is a useful tool for this operation (Figure 6). When the thread begins to release, do not apply much force as the threads may seize or the stud break. Continue to use the back and forth motion while applying penetrating oil. If this method fails, then other methods of removal may be used, such as drilling out the stud or capscrew and rethreading, or the use of the Heli-Coil® method when the threads are damaged beyond repair. A screw extractor (Easy-Out®) would be useless in this case where torque applied with a wrench would not turn the frozen threads. It is obvious the lesser torque that could be applied with a screw extractor would be ineffective, so a different method should be used. In many cases, when the break occurs at the surface or slightly under the surface, any of

several other methods may be applied. They should be used in the order of severity of the problem.

Screw Extractor Method

When a screw extractor is used, a hole must be drilled in or through the broken end of the stud. Some mechanics prefer to drill this hole with a left-hand rotating drill. This technique tends to turn the screw in a counterclockwise rotation, which will screw the thread outward. If the screw is thus removed during the drilling operation, then of course, the screw extractor operation is not necessary.

Care must be exercised in drilling a hole for the screw extractor. Read the manufacturer's instructions before using these tools. This involves the selection of the proper drill size and drill guide, if needed. Should the fastener be broken off below the surface, a drill guide will be needed; it is usually included in extractor kits (Figures 7a and 7b). Placing the hole in the exact center of the screw is of primary importance, not only for efficient removal with the extractor, but in case it cannot be removed with an extractor, it can then be drilled out with the tap drill size of the screw thread and the remainder removed with a tap. (See "Drilling Out Broken Studs and Screws" below.) The hole should be drilled at least three-fourths of the screw length or completely through if possible, taking care not to drill too deeply into adjacent parts and damage them. If the broken screw is slightly above the surface, prepare it by filing it flat, then center punch in the center. Drill first with a small pilot drill before drilling the size for the extractor. Tap or twist the screw extractor into the hole and apply pressure with a tap wrench in a counterclockwise motion (Figure 8). (If you must use an end wrench, be careful not to apply much side pressure.) Do not apply more torque to the extractor than you would normally apply to a screw of that size using a

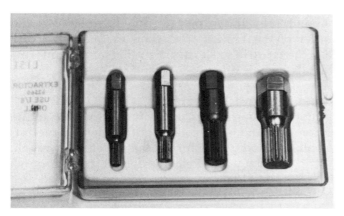

Figure 7a. A screw contractor kit. (Lane Community College)

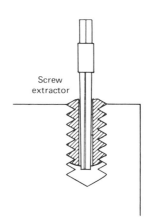

Figure 8. Using a screw extractor to remove broken stud.

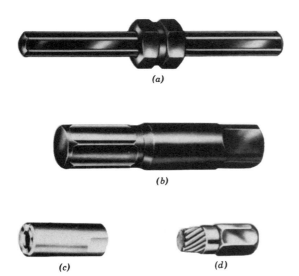

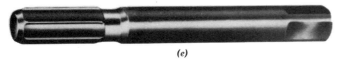

Figure 7b. Screw extractors and components: *(a)* spline-type screw extractor with hex nut for turning; *(b)* and *(e)* spline extractors with squared ends for turning; (c) drill guide used for centering a drill on a stud below the surface; (d) multispline extractor with shoulder, often used to remove damaged socket head setscrews. (Snap-On Tools Corporation)

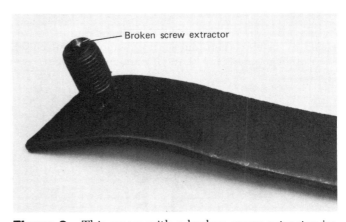

Figure 9. This screw with a broken screw extractor imbedded in it was successfully removed by the strap method of welding. (Lane Community College)

wrench. If the screw extractor should break, it can pose one of the most difficult removal problems a mechanic may face, especially if it is in a blind (one ended) hole. If the screw will not come out with reasonable force, remove the extractor and try another method. Broken screw extractors can sometimes be removed with the strap method of welding (Figure 9).

Extractors may also be used to remove broken pipe ends that are threaded into fittings. Threaded tubing and pipe, when broken off, leave a thin threaded shell in the fitting. If other methods fail, the shell may be chiseled out with a diamond point chisel and the shell peeled toward the center. The loosened shell can then be removed easily (Figure 10).

Drilling Out Broken Studs and Screws

Probably the most important step in drilling out broken studs is to start the drill exactly in the center so that when the tap size drill goes through, no threads are removed. The tap drill should go completely through the piece of broken stud, leaving only a helix of threads that can be pried out or removed by running a plug tap through the internal threads (Figures

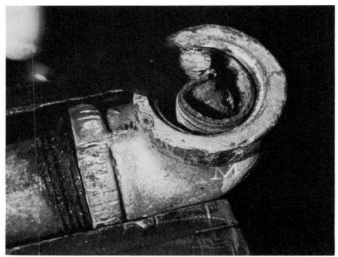

Figure 10. Broken pipe threads can sometimes be removed by peeling them out with a diamond point chisel when other methods fail. (Lane Community College)

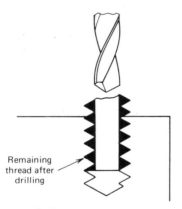

Remaining thread after drilling

Figure 11a. Tap drilled thread showing the thread helix that remains in the internal thread.

Figure 11b. Removing the thread helix by peeling it out with a plier. (Lane Community College)

11a and 11b). If the pilot drill is not exactly at the center of the broken stud, then use a smaller drill than the tap drill size to avoid cutting into the internal threads on one side.

Using Welding Techniques

In more stubborn cases, a nut may be welded to the broken end and, after cooling and applying penetrating oil, the stud can be turned out with a box wrench. A strip of metal with a hole (the diameter of the screw) drilled in one end can be welded on and used as a lever to remove the screw (Figures 12a to 12c). The nut or strap method of welding the stud or screw may be used even when the screw is broken off flush with the surface of the part. When this method is used, care must be taken that the weld is kept away from the casting or part, thus a smaller size nut or drilled hole should be used when the screw is broken off flush. Use the "tap" method of starting the arc and weld around the edge first, then remove the slag and fill in at the center. Do not cool with water; allow the part to cool in air. The heat of welding often helps to break the frozen threads loose. Apply penetrating oil after it cools and use the back and forth motion to loosen the threads; sometimes tapping the screw with a hammer helps. Do not use excessive force as that will break the weld.

Some studs, such as on an exhaust manifold flange, are so severely seized that removal is next to impossible. In these cases, it is best simply to drill out the stud and restore the threads with a tap. However, some mechanics claim that they can successfully remove these manifold studs with a locking plier by applying heat to the flange around the stud area until it is red hot.

REPLACING DAMAGED INTERNAL THREADS

In some difficult cases when removing broken studs or taps, the thread becomes enlarged or damaged beyond repair. If the situation permits, the thread can simply be drilled to the next larger tap drill size and tapped to a larger stud or capscrew size. If an enlarged thread size is not acceptable, a steel coil insert repair can be made. See Section I, Unit 4 for this and other methods of replacing damaged internal threads.

REMOVING FROZEN NUTS

Nuts are sometimes so badly rusted that they cannot be removed with a wrench even when using penetrating oil. Nuts can be removed in several ways.

1. Applying heat.

Figure 12a. A steel strap that has a hole drilled into it (the diameter of the screw) is welded on the screw. After cooling, penetrating oil is applied. (Lane Community College)

Figure 12c. The stud is removed by rocking back and forth until it loosens. Penetrating oil is frequently applied during the removal of the screw. (Lane Community College)

Figure 12b. Since this screw thread would not loosen in the first operation, heat is now applied carefully to avoid cracking the cast iron. (Lane Community College)

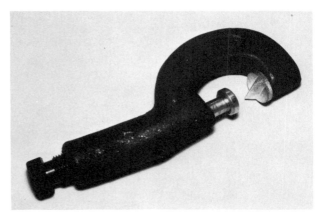

Figure 13. Nut splitter. (Orchard Auto Parts Co., Eugene, Oregon)

2. Using a nut splitter.
3. Using a chisel.
4. Cutting them off with an oxy-acetylene torch.

Applying Heat
In this method, heat is applied to the nut with a heating torch in order to expand it sufficiently to turn it off with a wrench. Alternate applications of heat and cooling sequences, then applying penetrating oil sometimes works.

Using a Nut Splitter
When it becomes evident that the nut cannot be removed with a wrench, much time can be saved by using a nut splitter (Figure 13). This is a simple tool in which a chisel is forced into the side of the nut by turning a screw with a wrench.

Using a Chisel
When a nut splitter is not available, a chisel can be used to expand the nut by cutting into opposite sides (Figure 14). As with a nut splitter, it is possible to cut the nut completely in two by using this method although it is much more laborious.

Cutting the Nut with a Torch
An oxy-acetylene cutting torch is sometimes used to

Figure 14. Using a chisel to split a nut. (Lane Community College)

Figure 15. Using a torch to remove a nut. (Lane Community College)

remove frozen nuts when time is of importance or the nut is in an inaccessible location for other tools. The nut is simply cut along each side, leaving two pieces (Figure 15). An alternate method is to cut off both bolt and nut. **Caution:** Be sure there is no oil or other flammable substance in the area where you are cutting with a torch.

REMOVING BROKEN TAPS AND SCREW EXTRACTORS

Several methods may be used to remove broken taps and screw extractors; in general, these fall into the following categories:

1. Using a tap extractor.
2. Drilling out the tap.
3. Breaking up the tap.
4. Punching out the tap.

Figure 16. Tap extractor. (Lane Community College)

5. Use of EDM to remove broken taps and screw extractors.

Removing Broken Taps with Tap Extractors

One of the most challenging problems in metal working is the removal of broken taps. Of course, the objective, when using taps, is to avoid breakage in the first place. This can be done by not using dull taps, using the correct tap drill size, using the right kind of cutting oil, not putting side stress on the tap (by using an end wrench, for example), and by using only the amount of torque necessary to cut the threads and sensing an increase in torque such as when the tap is stopped by the bottom of a blind hole or by jammed chips. When an increase in torque is noted, you should stop turning the tap wrench and back the tap out of the hole.

Since taps are very hard, they are also quite brittle and often break into several pieces that are jammed in the hole. When taps are broken off near the surface and are relatively free to turn in the hole, they can be removed with a tap extractor (Figure 16). In most cases, however, taps are severely jammed in the hole and cannot be screwed out.

There are basically two types of steel used to make taps. They are high speed steel and high carbon tool steel. You can tell the difference by spark testing the broken off piece. See Section A, Unit 3 for identifying steels.

Drilling Out the Tap

High speed steel cannot be readily softened by heating but carbon steel can. If the tap and adjacent area can be heated to red heat without harming the part, it is possible to anneal (soften) carbon steel taps so they can be removed by drilling. An oxyacetylene heating torch may be used to bring the tap to a cherry red and holding that temperature for two or three

Figure 17. Breaking up and removing a tap. (Lane Community College)

Figure 18a. EDM machine. (Elox, Division Colt Industries)

Figure 18b. Part in an EDM machine. (Heath Tool & Die)

Figure 18c. Die at left is made with the electrode shown in Figure 18b. The four splines at the right were made in a finished die. (Heath Tool & Die)

minutes. The tap is then allowed to cool slowly by backing off the flame. The slower the rate of cooling, the softer the tap will be. Most screw extractors (Easy-Out®) are made of high carbon tool steel and will also respond to annealing and drilling out. As in the case of broken taps, screw extractors should be checked by spark testing for high speed or carbon steel.

Breaking Up the Tap

High speed taps cannot be softened by this method of annealing. High speed steel can only be annealed in a furnace over a long period of time under controlled conditions. However, high speed steel can be made brittle by heating it to a bright red heat and then quenching it in water. One method used to remove broken high speed taps is to break them up into small chunks that can be taken out piece by piece (Figure 17). The previously mentioned heat treatment makes it easier to break up the tap. Always wear eye and face protection when using this method; this includes anyone in the vicinity since these small chunks of tap will fly at a high velocity.

Punching Out the Tap

If the tap is in a through hole, it can sometimes be punched out from the bottom side of the hole. This measure will tear out the threads where the flutes of the tap are located. Thread restoration methods, such as Heli-Coil®, will need to be used in this case.

Removing Taps and Extractors with EDM

The best method of tap and extractor removal is the use of Electrical Discharge Machining (EDM) (Figures 18a to 18c). Electro-discharge machining is a relatively new metal removal process in which the metal is removed through the action of electrical discharges of short duration and high current density between

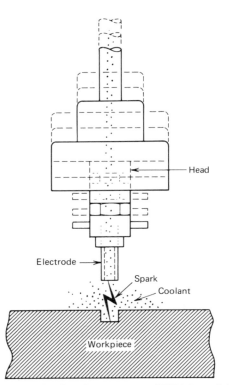

Figure 19. Principle of operation of EDM machine. The Elox Electron Disintegrator works on the principle of a vibrating head with an electric power supply creating a series of intermittent arcs that disintegrate even the hardest metals. The eroded minute bits of metal are continuously flushed away by coolant. (Elox, Division Colt Industries)

the electrode and the work in the presence of a dielectric medium (Figure 19). The electrode, which can be made in almost any shape, advances slowly through the metal to be cut by an eroding process. The greatest advantage of this machining method over others is that it can easily cut through very hard, low machinable materials such as tungsten carbide, high speed steel, fully hardened carbon tool steel, and other difficult materials.

These machines consist of a power supply unit that supplies the automatic feeding control of the electrode while providing a pulsing (in some machines) current for eroding the workpiece, a machine table, a dielectric tank with a fluid bath, and a servo unit that feeds the electrode into the work. Some electrode materials are copper, graphite, brass, copper, and tungsten carbide.

Electro-discharge machining is sometimes used for tap removal when the part is small enough to be placed in the EDM machine. The electrode should

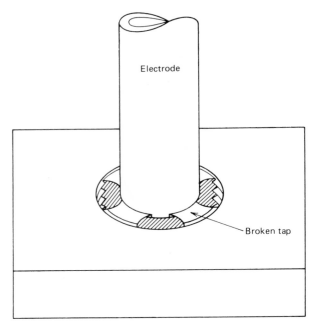

Figure 20. Removal of tap by EDM.

Table 1

Time Required to Remove Broken Taps and Drills

			Set Up (sec.)	Drilling (sec.)	Cleaning (sec.)	Total (min.)
A	8-32	$\frac{1}{2}$″ deep	60	35	120	3.4
Taps	10-32	$\frac{3}{4}$″ deep	60	60	120	4
	$\frac{1}{4}$-20	$\frac{3}{4}$″ deep	60	72	120	4.3
	$\frac{1}{2}$-13	$\frac{3}{4}$″ deep	60	140	120	5.2
B	$\frac{3}{16}$″	1″ deep	60	90	120	4.5
Drills		2″ deep	60	210	120	4.8
		3″ deep	60	360	120	4

Source: *Elox Electron Disintegrator*, Elox, Division Colt Industries Corporation, © 1977.

be a rod of sufficient diameter to remove the core or web of the tap, leaving the flutes in the thread. The flutes can then be easily removed when the electrode is withdrawn (Figure 20). EDM is not a rapid process of metal removal; a small machine may remove from one to ten cubic inches per hour. A medium sized tap, $\frac{1}{2}$ to $\frac{3}{4}$ in., may contain one cubic inch of material in the web. See Table 1.

Portable EDM units (Figure 21) are sometimes used right on the job to remove broken taps. These machines are usually too expensive to be used in small shops for the few taps that are broken.

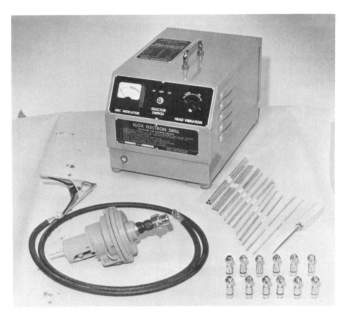

Figure 21. Portable EDM machine. The head may be mounted in a drill press to support it. Water coolant may be used with this machine for removing taps. Electrodes and collets are shown in the foreground. (Elox, Division Colt Industries)

SELF-TEST

1. When you have applied a reasonable amount of force attempting to loosen a stud or capscrew and it will not turn, what should you do next?
2. When a screw extractor is being used to remove a broken stud, can you apply as much torque to it as you would to the original screw to loosen it?
3. How can welding be used as a tool to remove broken studs?
4. If threads are so severely seized that removal of the stud is impossible, what other method can be used to remove the broken stud and restore the threads?
5. When a broken stud that is flush with the surface is discovered, how should you *first* attempt to remove it?
6. Why do some mechanics use a left-hand drill when drilling a broken stud for a screw extractor?
7. Why should the screw extractor drill be exactly on center?
8. Name four methods of removing broken taps.
9. What two types of steel are taps and screw extractors made of?
10. How can you tell the difference between high speed steel and high carbon tool steel?

UNIT 6 NONDESTRUC-TIVE TESTING

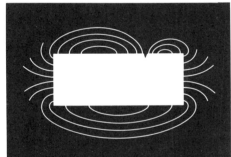

Nondestructive testing methods are among the most useful tools of modern industry. Inspection and testing of each part are sometimes necessary for critical aircraft parts. This is regularly done with the testing systems explained in this unit.

OBJECTIVES

After completing this unit, you should be able to:
1. Name the several methods of nondestructive testing and explain specific uses and operation of each.
2. Use testing equipment for inspecting test pieces.

Nondestructive testing, as the name implies, in no way impairs the specimen that it tests. Usually these tests are not a direct measurement of mechanical properties, such as tensile strength or hardness, but are made for the purpose of locating defects or flaws.

When machines or parts have large safety factors built into them, there is little need for nondestructive testing. However, many products used in aircraft and space technology require a high level of reliability. This is achieved by inspection at the time of manufac-

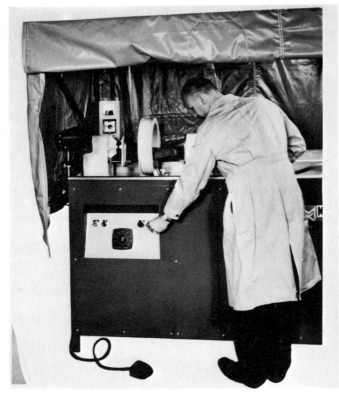

Figure 1. Magnaflux-type "H-700" series machine. A versatile shop unit that handles all magnetic particle test needs up to 144 in. long and 20 in. in diameter. (Magnaflux Corporation)

Figure 2. Front axle kingpin for a truck as it appeared (left) under visual inspection, apparently safe for service; (center) with dangerous cracks revealed by Magnaflux; and (right) with the same cracks revealed by fluorescent Magnaglo, under black light. (Magnaflux Corporation)

ture and, in some cases, continued testing during the service life of the part. The most common types of nondestructive tests are magnetic-particle inspection, fluorescent-penetrant inspection, ultrasonic inspection, radiography (X-ray and gamma ray), and eddy current inspection.

MAGNETIC-PARTICLE INSPECTION

There are several methods of magnetic-particle inspection (Magnaflux) used to detect various kinds of flaws in ferromagnetic metals such as iron and steel (Figure 1). A magnetized workpiece is sprinkled with dry iron powder or submerged in a liquid in which the particles are suspended. The Magnaflux Corporation has developed a method called *Magnaglo* in which fluorescent-magnetic particles are suspended in solution. The solution is flowed over the magnetized workpiece, which is then viewed under a black light. Cracks that are only a few millionths of an inch wide can be found by this method (Figure 2).

The wet method is useful for inspection in the manufacture of parts and for maintenance inspection.

The dry method is generally used for inspection of welds, large forgings, castings, and other parts having rough surfaces. The basic principle of magnetic testing is shown in Figure 3. A new magnetic pole is formed at a crack, which causes the magnetic powder to concentrate at that point. When a part is magnetized lengthwise, as shown in Figure 4, transverse (crosswise) cracks can be detected. This is done by energizing a coil around the bar. If an electric current is passed through the bar, however, a circular magnetic field results and the defects can be found that are lengthwise to the part (Figure 5). In many cases the part must be demagnetized after testing, so this feature is built into the machine. This system of inspection is limited to the magnetic metals such as iron and steel.

FLUORESCENT-PENETRANT INSPECTION

Invisible cracks, porosity, and other defects are also found in the nonferrous metals such as aluminum, bronze, tungsten carbide, and the nonmetals such as glass, plastics, and ceramics. Fluorescent-penetrant in-

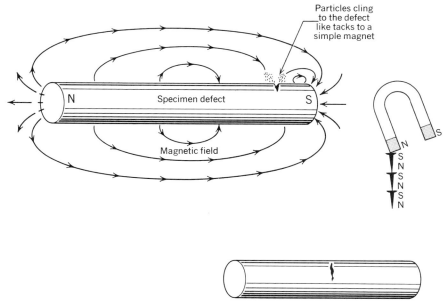

Figure 3. By inducing a magnetic field within the part to be tested, and applying a coating of magnetic particles, surface cracks are made visible; the cracks in effect form new magnetic poles. Particles cling to the defect like tacks to a simple magnet. (Magnaflux Corporation)

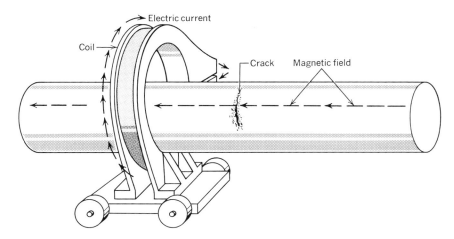

Figure 4. Longitudinal method of magnetization. (Magnaflux Corporation)

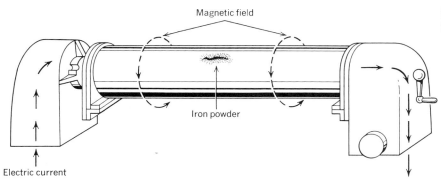

Figure 5. Circular method of magnetization. (Magnaflux Corporation)

Figure 6. Freshly reground carbide tipped tools as they appear: (left) to normal visual inspection, perfectly good; (center) test results with Zyglo penetrant; and (right) with supersensitive Zyglo Pentrex. All cracks, however small, are found. (Magnaflux Corporation)

spection is widely used for testing and inspection on these materials. *Zyglo*, the Magnaflux Corporation copyrighted name for this method, is similar to Magnaglo in the use of black light to make the defects glow with fluorescense (Figure 6). Once applied, Zyglo penetrant is drawn into every defect no matter how fine or deep, but it must be given the time to do so. The length of time depends upon the material and the type of defect you are testing for. The surface film is rinsed off with water, developer is applied to draw out the penetrant, and the part is inspected for defects under black light where cracks and other flaws will fluoresce brilliantly. Zyglo systems are available from hand portable test kits to huge automated systems.

DYE PENETRANTS

A similar method of nondestructive testing using dye penetrants is used for visual inspection, but without the black light and fluorescent penetrant. As with Zyglo, it may be used on almost any dense material. This method, called *Spotcheck* by the Magnaflux Corporation, works as follows:

1. The defect is penetrated by a dye penetrant.
2. The excess penetrant is removed.
3. The developer is applied.
4. Inspection shows a bright colored indication marking the defect.

Spotcheck (Figure 7) is available in sealed pressure spray cans or for brush or spray gun application. Its advantages are portability for remote uses (Figure 8) or for rapid inspection of small sections in the shop, low cost, and ease of application.

ULTRASONIC INSPECTION

The pulse-echo system and the through-transmission

Figure 7. Spotcheck is used to locate fatigue crack in punch press frame. (Magnaflux Corporation)

Figure 8. Spotcheck is used to find dangerous crack in aircraft wheel. (Magnaflux Corporation)

system are two methods of ultrasonic inspection (Figure 9) used to check for flaws in metal parts. Both systems use ultrasonic sound waves (millions of cycles per second) in their testing.

In the pulse-echo system, a pulse generator produces short bursts of sound that activate a transducer (a device that converts mechanical energy into electrical energy) fastened to the metal being tested. The signal pulse is seen as a pattern or "pip" on an oscillo-

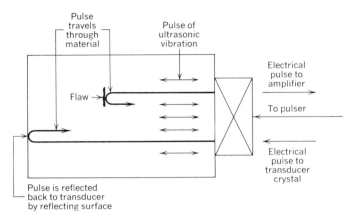

Figure 9. How ultrasonic sound waves are used to locate flaws in material. (Machine Tools and Machining Practices)

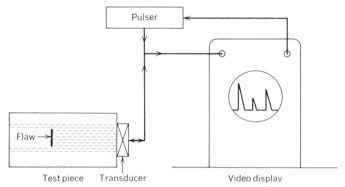

Figure 10. Ultrasonic pulse echo system. (Machine Tools and Machining Practices)

operator when similar parts are being checked.

Ultrasonic inspection is used in production inspection of testing equipment and other parts. Internal defects, cracks, porosity, laminations, thickness, and weld bonds are tested in metal and hard nonmetallic structures (Figure 12). Some testing equipment is relatively light and portable (Figure 13), and some require the part to be immersed in a liquid. Test results are immediate and accurate.

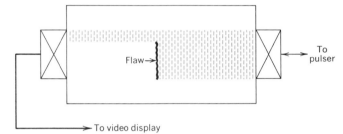

Figure 11. Through-ultrasonic transmission. (Machine Tools and Machining Practices)

Figure 12. Portable ultrasonic machine instruments can be used for routine manual inspection under field conditions. This is very useful for structural weld testing and corrosion surveys. (Magnaflux Corporation)

scope screen when the sound wave enters the test piece (Figure 10). Part of the sound wave is reflected back when it reaches the other side of the material and shows on the oscilloscope as a second "pip." If there is a flaw in the specimen, a smaller "pip" will be seen between the other two. Since the distance between the "pips" on the oscilloscope screen represent elapsed time of the reflected pulse, the distance to a flaw can be accurately measured.

The through-transmission method uses a transducer on each side or end of the test piece (Figure 11). The signal pulse enters the material at one transducer, travels through the material to be received by the other transducer, and is translated into another signal shown on an oscilloscope. If no flaws are present, a clear, strong signal will be seen on the oscilloscope, but if the material contains any flaws, a weaker or distorted signal will be seen. The reduced signal can be set up to actuate a bell or light to alert the

Figure 13. The ultrasonic method results in swift inspection of critical weldments with dependability. (Magnaflux Corporation)

Figure 14. X-ray method of testing. A specialist makes a test of pilot run parts to evaluate manufacturing techniques and procedures. (Magnaflux Corporation)

RADIOGRAPH TESTING

Radiographic testing utilizes the ability of X-rays (Figures 14 and 15) or gamma rays (Figures 16 and 17) emitted from radioactive materials to pass through solids. The test results are determined from a radiograph, which is a film exposed to radiation that has gone through the test materials. Shadows on the radiograph reveal defects since the radiation will pass through a void, crack, or area of lower density at a greater intensity and will appear darker on the negative. This testing method is widely used for forgings, castings, welded vessels including pipelines, and corrosion analysis. Since there is a radiation hazard, only trained technicians should use the equipment. The test is quite sensitive and provides a permanent record.

A radiation detector may be used to determine material thickness since the radiation that passes through the test material decreases as the thickness increases. A moving, continuous strip of material can then be constantly monitored for thickness without any physical contact of the radiation source or detector.

Eddy current inspection can only be used to test electrically conducting materials. An alternating cur-

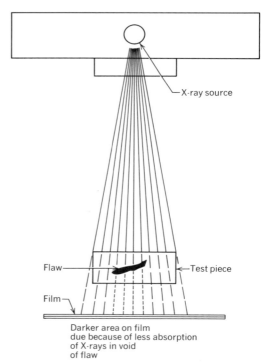

Darker area on film due because of less absorption of X-rays in void of flaw

Figure 15. X-ray inspection. (Machine Tools and Machining Practices)

Figure 16. An experienced technician is preparing to make gamma ray inspection of a section of pipe. (Magnaflux Corporation)

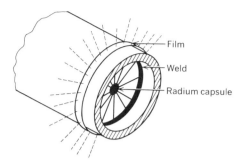

Figure 17. Gamma ray inspection. (Machine Tools and Machining Practices)

Figure 18. New Magnatest ED-100 for nondestructive comparison of magnetic materials. Used to separate metallurgical properties as alloy variations, heat treat condition, hardness differences, internal stress, and tensile strength. (Magnaflux Corporation)

rent in a coil produces a corresponding magnetic field. Eddy currents, which flow opposite the main current, are produced in the test material if the coil is placed near or around it. The eddy currents then produce a magnetic field, which is converted into voltage that is read on a meter or oscilloscope (Figure 18). This method is used for detecting seams or variations in thickness and for sorting metals in various compositions and heat treatments. Physical differences in mass, dimension, and shape can also be detected.

SELF-TEST

1. What is the purpose of nondestructive testing?
2. Name a limiting factor of magnetic-particle inspection.
3. How must the part be magnetized to form new magnetic poles on a crosswise crack? On a lengthwise crack?
4. Describe how fluorescent penetrant works to reveal defects. On what materials can this method be used?
5. What are some of the advantages in using a dye penetrant method (without black light)?
6. Name the two basic systems of ultrasonic testing. Briefly explain their differences.
7. What can be determined in a specimen with ultrasonic inspection systems?
8. How can X-rays help us to detect flaws in solid steel or other metals?
9. In what way does gamma ray inspection differ from that of X-ray inspection?
10. Explain some of the uses for eddy current inspection. To what metals is it limited?

UNIT 7 MANUFACTURE OF MECHANICAL PARTS

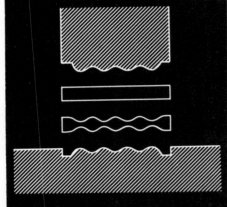

Mass production had its beginnings in such men as Eli Whitney who produced rifles that for the first time in history had interchangeable parts. This required the use of jigs, fixtures, and repeatability of machines. Tolerances were developed where dimensions were required to fall within certain limits of accuracy, thus ensuring a fit of mating parts even though different people made them on different machines. Formerly, when one part was made at a time, fitting it to others and the many hours required to produce a part made it extremely expensive. Today parts are often produced on automatic machines that further reduce the cost per piece. The great wealth of our modern societies is due for the most part to our ability to mass produce goods such as machinery, textiles, farm produce, and thousands of other items. This unit will acquaint you with many of the methods of manufacture used to produce mechanical parts and help you make decisions in selecting replacement parts and how to keep them from being damaged in handling.

OBJECTIVES

After completing this unit, you should be able to:
1. Identify methods of manufacture used to produce several mechanical parts.
2. Explain methods used to avoid damage to a number of parts.

MATERIALS

Most of the mechanical parts that are made of steel have their beginnings in the form of bar stock—either flat, square, or round—and also in flat stock—sheet or plate. These shapes lend themselves to further manufacture in machines. Most bar stock is either hot rolled (HR) or cold rolled (CR) and contains various amounts of carbon and alloying elements. Plain carbon steel contains mostly carbon (C) as an alloy that can give the steel the property of hardenability. However, low carbon steel that contains .10 to .20 percent C does not harden appreciably and is most used in manufacture. It is sometimes called "mild steel" and most structural forms such as angles, I beams, and pipe sections are made of it. There are also many kinds of tough alloys of steel that are used in the manufacture of parts, as well as many nonferrous metals and alloys too numerous to mention here. The study of metals and alloys is called metallurgy; further information on this subject may be found in a good metallurgy text or reference book. Some nonmetallic materials are also used in mechanisms such as phenolic gears, nylon bushings, and industrial sealants and gaskets.

METHODS OF MANUFACTURE

Metals are shaped into useful items by several general methods; they are:

1. Forming (drawing, swaging, bending, rolling)
2. Forging [hammering, pressing (hot or cold), drop forging]
3. Machining (removing unwanted metal with a cutting tool under power)
4. Hand operations (sawing, filing, chiseling, polishing, and lapping)
5. Metal spinning

Figure 1a. Vertical press brakes. (Carothers Company, Eugene, Oregon)

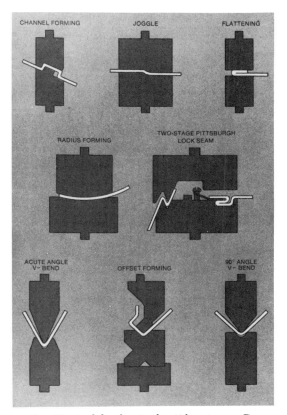

Figure 2. Typical brake tools. (Aluminum Company of America)

Figure 1b. Short 90 degree flange being formed on sheet metal. (Carothers Company)

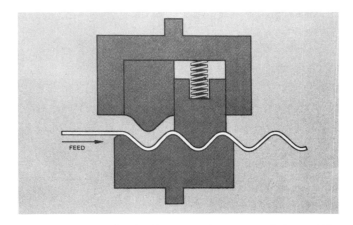

Figure 3. Press-brake forming of corrugated sheet. (Aluminum Company of America)

6. Casting (green sand molding, die cast, and investment casting)
7. Weldments, standard bar and plate stock welded together to make a useful part.

FORMING

Probably one of the most common operations in forming sheet metal is bending. The usual methods are press brake bending and roll bending. Additional or complementary operations are flanging, beading, curling, crimping, and lock seaming. Vertical press brakes (Figures 1a and 1b) are used for many forming operations. Figure 2 shows some typical brake tools for bending sheet metal shapes. A common product, corrugated sheet metal, is manufactured on press brakes (Figure 3).

Roll bending is used to form curves, cylinders,

and conical shapes in sheet metal (Figure 4). Roll bending bars and shapes can be done on a set of rolls (Figure 5), but tubing is commonly bent by draw bending (Figure 6), compression bending (Figure 7), or press bending (Figure 8).

Contour roll forming is a bending process that transforms the metal strips into a desired shape of constant cross section with any length. The strip is formed by a series of opposing rolls that progressively bend the metal (Figure 9). Continuous gutters for

Figure 4. Roll bending is used to form cylindrical parts. (Aluminum Company of America)

Figure 5. Vertical roll-bending machine with hydraulically actuated toggle mechanism. (Aluminum Company of America)

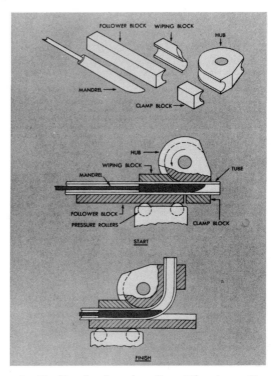

Figure 6. Tooling for draw bending. (Aluminum Company of America)

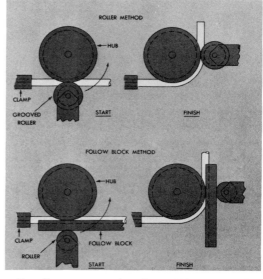

Figure 7. Compression bending. (Aluminum Company of America)

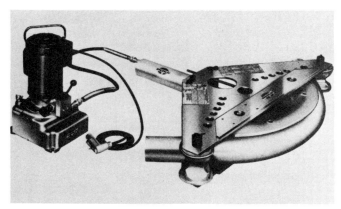

Figure 8. Hydraulic ram bender with electrically driven pump. (Aluminum Company of America)

Figure 9. (Left) Forming machine, and (right) products. (Aluminum Company of America)

Figure 10. Blanked and drawn parts. (Aluminum Company of America)

Figure 11. Blanking operation. (Aluminum Company of America)

roofs are a common application for this type of forming.

Sheet metal drawing is essentially converting flat sheet metal into a shape (Figure 10). Sometimes in thicker metal, the blanks are first made in a blanking die (Figure 11). Kitchenware such as pots and pans are made in this way. Deep drawing usually takes several operations. Many automotive parts are made by stamping and drawing; among these are wheels, clutch housings, rocker arm covers, body parts, and bumpers.

Metal stamping is also used to raise projecting figures and designs on a surface. This is called embossing. Usually embossing is a simple operation in which the outline is made in the female die, pressing on a rubber die. Coining is a similar method that raises designs on one or both sides of a piece. Medals and

coins are made by this process.

Other methods of forming, such as stretch forming, high energy rate (explosive) forming, and electromagnetic forming, produce similar results to mechanical methods, and they are superior for certain materials. However, for the purposes of identification of methods of manufacture, basic forming techniques are sufficient here.

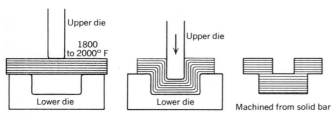

Figure 12. Grain flow in a solid bar as it is being forged compared to a machined solid bar. (Machine Tools and Machining Practices)

It should usually be obvious to the mechanic if a part has been made by a forming process. Several observations can be made.

1. Is it of a relatively uniform thickness (indicating sheet metal material)?
2. Does it show cut edges (from blanking)?
3. Does it show an imprint of a die and drawing lines or marks?

This, of course, is in contrast to forgings or castings that are of varied thickness and usually show a seam where the "flash" was removed where the dies came together. Metal spinning is a method that produces shapes similar to the drawing process, but the difference can be seen in the tool marks around the periphery of the spun part. (See "Metal Spinning" below.) Formed metals are usually ductile (can be bent) and tough. They do not usually break when deformed but may remain distorted or out of shape. The strength of formed metals is reduced dramatically when heat is applied to them since the metal tends to recrystallize into a softer condition. The strength of formed metals is greater in the direction of forming or rolling (anistropy).

FORGINGS

Forging is basically a process by which steel is heated and then forced into a specific shape by pressing or hammering. The process of forging is used when a strong, tough product is needed, such as tools. Wrenches, sockets, and hammers are usually drop-forged if they are good tools. Forging retains the grain flow in such a way that it makes a stronger part than one of equivalent shape machined from bar stock or stamped from sheet (Figure 12). In contrast, very large parts, such as huge generator shafts, are hot forged before they are machined to finish size (Figure 13).

Upset forging is used to enlarge the diameter or change the shape of a metal piece, using extremely high forces. Pinion gear blanks, valve stems, bolts,

Figure 13. Between the jaws of a 7500 ton hydraulic forging press, what was once a large glowing ingot is now taking on the desired shape and strength. (American Iron & Steel Institute)

and nails are made by this process. Usually heated steel is used because it is a tough metal, but many steel parts are now made by the cold upset process (Figure 14).

Cold forging is practically a coining operation in many cases, but by frequent annealing and by using high forces considerable change in shape is possible. Aluminum can be hot or cold forged or extruded (pushed through a die) to form many shapes such as tubes, rods, and molding. Forgings usually have a "flash" caused by extruded metal between the die halves that has been ground off. Also, embossed (raised) letters or numbers may often be found, as on wrench handles. Since forged parts are superior for applications, such as automotive front spindles, no attempt should ever be made to replace these parts with a shop-made duplicate from bar stock (Figure 15).

MACHINING

By far, the greatest amount of machining done is on production machines that are arranged and set up

Figure 14. This alternator rotor part is an example of cold upset forging. Clockwise from top left, the progression of forging can be seen. (Lane Community College)

Figure 15. This front axle spindle from a tractor was shop made from bar stock instead of forged. The result was an early fatigue failure. (Lane Community College)

Figure 16a. One of a group of screw machines being set up for the next run. (Sweetland Archery Products)

Figure 16b. Tools are indexed on a turret to perform several operations and quickly produce arrow points. (Sweetland Archery Products)

Figure 16c. A collection bin for parts made on the automatic screw machine. A new part drops every few seconds. (Sweetland Archery Products)

to make precision parts many times faster than a single machinist could produce them in a machine shop. These machines can be either semi-automatic or fully automatic. Turret lathes are usually semi-automatic and require the constant attention of a machine operator. Automatic screw machines, on the other hand, can rapidly produce small parts with one operator for many machines (Figures 16a to 16d). Multiple spindle machines are used normally for mass producing somewhat larger parts such as pinion gear blanks. Each spindle rotates and advances the work into the tool, then retracts and indexes to the next step (Figure 17).

Figure 16d. Typical aluminum parts turned on screw machines. (Aluminum Company of America)

Figure 17. The multiple spindle machine is another type of high production manufacture. (Sweetland Archery Products)

In many cases, a machined part that is manufactured on high production machines can be produced with just as high quality in a job shop (one-of-a-kind) situation, but it will take much longer to produce— perhaps ten times or more man and machine hours.

Figure 18. Part being made in a machine shop. (Lane Community College)

This is why a part should be purchased from a dealer as a manufactured item, if one is available, in preference to having it made locally. Tool room shops are generally involved in making prototypes of new machines or parts; tool and die shops make tooling for manufacturing; job shops make single items or short production runs; and maintenance machine shops repair existing machinery. Each has its purpose and should be used accordingly.

An example of the use of a job machine shop would be a need for a replacement shaft for an obsolete machine or perhaps a part for a vintage automobile (Figure 18). The part made in the job shop will cost more than the original part, but there is no other recourse.

HAND OPERATIONS

In any kind of metal cutting operation, certain finishing processes are required including deburring, cleaning, testing, and assembly. All of these are normally hand operations that are necessary to produce a finished part (Figure 19). If there is a great deal of hand work required to produce a part, the unit cost goes up accordingly. Much of this hand work has been replaced in recent years by bulk finishing methods, such as tumbling, ball burnishing, shot peening, and by mechanical systems designed to finish mass produced pieces (Figure 20).

METAL SPINNING

Perhaps metal spinning should be considered a forming operation, but since it is done on a lathe, it is in a class by itself. Metal spinning is a method of forming

Figure 19. Hand operation—deburring. (Lane Community College)

Figure 20. Polishing and buffing flatiron aluminum sole plates. (Aluminum Company of America)

Figure 21. Alcoa 6061 missile nose cone being spun on Hydrospin machine. (Aluminum Company of America)

flat metal discs over a form that is rotating in a lathe or similar machine. Metal spinning was originally a hand operation, used to form such things as musical instruments. A hand-held wooden stick would be used to push the metal over the spinning form. Shear spinning, as it is sometimes called when thicker plate is formed, is commonly used for aerospace products (Figure 21). Steel may be either hot or cold spun by hydroforming (Figures 22a and 22b). Metal spun products are symmetrical in shape and bear circumferential tool marks that give them a unique appearance. Photographic floodlight reflectors are examples of metal spinning.

CASTING

Several methods of casting molten metals are used to produce industrial products. Briefly, they are sand casting, investment casting (lost wax process), permanent mold, and die casting. Sand casting is used to form the higher melting point metals such as iron and steel. Machine bases, housings, and internal combustion engine blocks are made of cast iron and are sand cast (Figure 23). Pieces that require more tensile strength are made of cast steel. Small, precision parts are manufactured by the process of investment casting (Figure 24). Odd shaped parts that could not be cast in a conventional mold can easily be made by this process (Figure 25). Permanent molds are used for short run production (Figure 26). Their major advantage is that instead of a new sand mold being made for each pouring, the mold halves are simply closed; a disadvantage is that they have a limited life of only a few thousand castings.

Perhaps the greatest number of small parts used today in home appliances, office equipment, and automobiles are made by the process of die casting (Figure 27). Unlike most other forms of casting molten metal

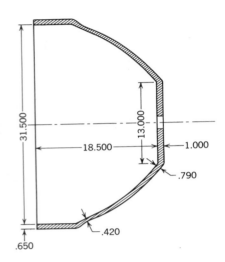

Figure 22b. The principle of the Floturn process. (Floturn Inc., Division of the Lodge & Shipley Co.)

Figure 22a. Floturn lathe. The photograph shows the completion of the first two operations. Starting with the flat blank, the workpiece is shear formed (Floturn) to the shape shown on the mandrel. In the process the 1 in. thickness of the blank is reduced to .420 in. The second operation, although performed on a Floturn machine, is more properly spinning than shear forming. In this operation, the workpiece is brought to the finished shape as seen in the background. (Floturn Inc., Division of the Lodge & Shipley Co.)

Figure 23. Rough casting of a machine table before machining. (Lane Community College)

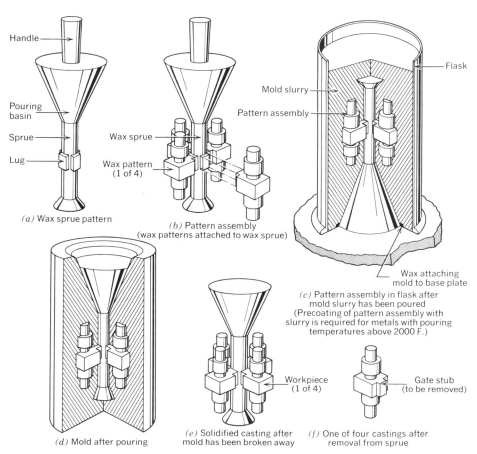

Handle
Pouring basin
Sprue
Lug

(a) Wax sprue pattern

Wax sprue
Wax pattern (1 of 4)

(b) Pattern assembly (wax patterns attached to wax sprue)

Mold slurry
Pattern assembly
Flask

Wax attaching mold to base plate

(c) Pattern assembly in flask after mold slurry has been poured (Precoating of pattern assembly with slurry is required for metals with pouring temperatures above 2000 F.)

(d) Mold after pouring

Workpiece (1 of 4)

(e) Solidified casting after mold has been broken away

Gate stub (to be removed)

(f) One of four castings after removal from sprue

Figure 24. Steps in the production of a casting by the solid investment molding process, using a wax pattern. (ASM Committee on Production of Investment Castings, "Investment Casting," *Forging and Casting,* Volume 5, Eighth Edition, Lyman, T., Editor, American Society for Metals, 1970, p. 239)

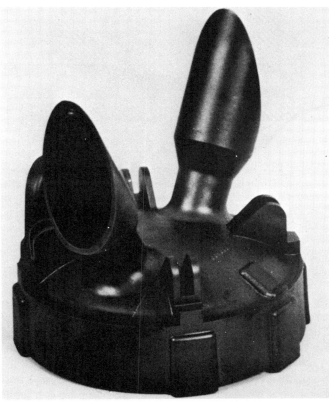

in which the metal is poured, that is, using gravity to fill the mold, die casting forces the metal into the closed mold halves with extremely high injection pressures from 1000 to 100,000 PSI (Figure 28).

Several of the metal alloys used for die casting are zinc base, tin base, lead base, aluminum base, copper base (brass and bronze), and magnesium base. Zinc base alloys are perhaps most commonly found in automotive parts. They are made by alloying zinc with aluminum and copper.

It is not normally economically feasible to have any cast part duplicated in a local foundry because patterns would have to be made before a mold could be prepared. Single die cast parts would be out of the question to duplicate because die cast molds cost many thousands of dollars to make and consequently a great many castings must be made to make the operation profitable. Die cast metals are usually extremely

Figure 25. An experimental rocket part cast in Maraging steel by the investment casting process. (Hitchiner Manufacturing Company, Inc.)

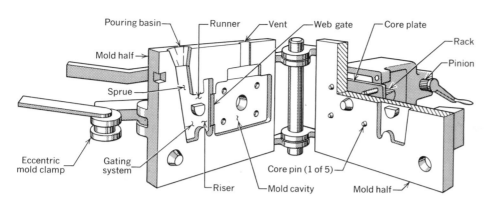

Figure 26. Book-type manually operated mold casting machine, used principally with molds having shallow cavities. (ASM Committee on Production of Permanent Mold Casting, "Permanent Mold Casting," *Forging and Casting*, Volume 5, Eighth Edition, Lyman, T., Editor, American Society for Metals, 1970, p. 266)

Figure 27. Die cast aluminum outboard motor parts. (Aluminum Company of America)

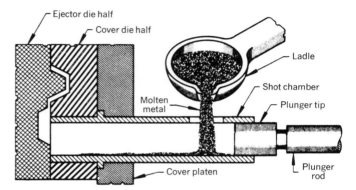

Position 1 Pouring

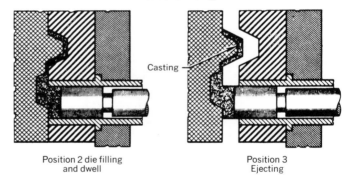

Position 2 die filling and dwell   Position 3 Ejecting

Figure 28. Operating cycle of a horizontal cold-chamber die casting machine. (ASM Committee on Die Casting, "Die Casting," *Forging and Casting*, Volume 5, Eighth Edition, Lyman, T., Editor, American Society for Metals, 1970, p. 286)

difficult to weld or solder when a piece is broken off. Normally, when a die cast part is broken, it should be discarded and a new one purchased.

WELDMENTS

The welding process has opened up a new method of making mechanical parts such as machine frames, gear and sprocket blanks, and hydraulic cylinders. These were all formerly made by the casting process that had some limitations and was often more costly. Furthermore, steel bar and plate stock (which is read-

ily available) when welded makes an extremely tough product that lends itself well to machining (Figure 29). Almost any mechanical device can be produced by the fabrication of metal by welding and by subsequent machining. Prototype of new machines or inventions are almost always made first by this method because it is readily available to anyone who can weld

Figure 30. Power steering hydraulic valve parts that were ruined by mishandling. (Lane Community College)

Figure 31. Hydraulic valve body showing damage at the lip in which the rotor was misaligned and forced in by hammering. (Lane Community College)

Figure 29. Student-made heavy duty drill press that was fabricated from steel (a weldment) and was subsequently machined in various ways to produce a precision tool. (Lane Community College)

and machine metal. High production methods such as die casting can then later be used when the device has been tested and proven.

CARE IN HANDLING MECHANICAL PARTS

The main reason that modern machinery, autos, tractors, and space vehicles are so dependable and normally trouble free is that each part is made to a set of specifications that requires the part to be within allowable tolerances. In many cases, the part would simply not function if that tolerance is disturbed or altered by the mechanic. An example of this is the power steering valve in Figure 30. You can see by the marks on the body of the valve that the mechanic had it squeezed in a vise with no protective pad. The body will therefore not fit in its housing without hammering it in. Figure 31 is a close-up view of the edge of the bore into which the valve rotor fits. Hydraulic valves depend on a very small clearance for sealing that is less than .001 in. The rotor must be installed squarely with the bore and, if it is misaligned even slightly, it will stick. The cause of the damaged bore is the mechanic's **carelessness** since it was not fitted

in properly but was hammered on instead. The rotor now will not fit in the bore at all and is ruined for use. The valve standing on end in Figure 30 suffered the same fate, but the mechanic succeeded in hammering it in all the way. A hammer dent can be seen on the end of the valve. This valve is also ruined since it cannot be moved now except with a hammer. If a delicate part is dropped, it should be inspected for nicks. Any imperfection, dirt, or grit can seize up a close fitting part. Cleanliness in handling is extremely important when assembling machinery. Bearings should be installed with the correct tools. Using a punch is poor practice since the bearing tends to become misaligned and then jams in the bore. There is also the danger of damaging the bearing seal.

As a rule, flexible seals are replaced when repairing equipment because they become dry and are subject to cracking and failure (Figure 32). The proper installation tool must be used to avoid damage to the

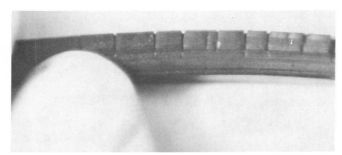

Figure 32. Flexible seals such as this tend to develop cracks as they age and should not be reused. (Lane Community College)

Figure 33. When proper installation tools are not used, flexible seals often suffer this kind of damage, which limits their usefulness. (Lane Community College)

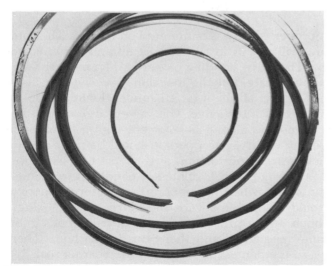

Figure 34. A group of seals with various cross sections (square and O-ring) that were ruined by shearing during assembly. This is usually caused by sharp edges when the seal is installed. (Lane Community College)

Figure 35. Too much sealant was applied to this gasket before it was installed. (Lane Community College)

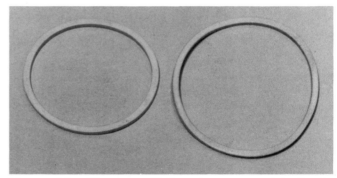

Figure 36. These two Teflon rings were originally the same size. The one on the right was stretched and remained in that shape. (Lane Community College)

Figure 37. Cast iron rings such as this one should not be bent out of shape since they tend to remain that way. (Lane Community College)

seal (Figure 33). Seals are exceptionally subject to damage from mishandling. A common cause of leakage in a newly repaired device is the shearing damage to ring seals during assembly (Figure 34). Gaskets should always be replaced since they lose their resiliency in use.

If a sealant is used with a gasket, be sure not to apply too much as it will extrude into the case and sometimes gum up the mechanism (Figure 35). Teflon rings should never be stretched since, unlike rubber, they will remain in that shape (Figure 36). Metal (cast iron) rings should not be distorted or bent sideways during installation since they will remain that way and lose some of their sealing ability (Figure 37).

SELF-TEST

1. From which general shapes of steel are most mechanical parts manufactured?
2. Name six methods of manufacture.
3. What differences between formed metals and castings or forgings can help you in identifying the method of manufacture?
4. What is the effect of heat, such as from welding, on formed metals?
5. Why is it unwise to replace a forged part in a high stress situation with one machined from bar stock?
6. Parts made on a turret lathe or an automatic screw machine cost far less than if they were made on an engine lathe in a machine shop. Why is this?
7. Name two methods of casting used to produce automotive parts such as engine blocks, door handles, and carburetors.
8. A production-cast item cannot be reproduced in a local foundry at less cost for a single part. Why?
9. What is a weldment?
10. When assembling close fitting mechanical parts, what three things must a mechanic watch for to avoid difficulties?

SECTION J

MILLING MACHINES

UNIT 1   HORIZONTAL MILLING MACHINES
UNIT 2   USING THE HORIZONTAL MILLING MACHINE
UNIT 3   VERTICAL MILLING MACHINES
UNIT 4   USING THE VERTICAL MILLING MACHINE

UNIT 1 HORIZONTAL MILLING MACHINES

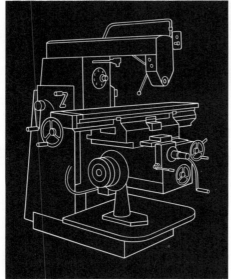

The horizontal milling machine serves the unique function of removing material from a workpiece with a rotating circular cutter. This metal-cutting system, unlike turning machines, makes possible the machining of flat surfaces, slots, grooves, gears, splines, and many other forms in metal and other materials. In addition, it provides a fairly high metal removal rate as compared to other machine tools, while holding a relatively high level of precision. There are two basic types of horizontal milling machines found in the shop: plain and universal. The table of the plain horizontal machine (Figure 1) does not swivel as does the universal milling machine (Figure 2). The main difference between these machines is that the universal machine has an additional housing that swivels on the saddle and supports the table. This allows the table to be swiveled 45 degrees in either direction in a horizontal plane. The universal milling machine is specially designed to machine helical slots or grooves as in twist drills and milling cutters. Other than these special applications, a universal mill and a plain milling machine can perform the same operations.

OBJECTIVES

After completing this unit, you should be able to:
1. Identify the important parts of the horizontal milling machine.
2. Identify various mounting devices used to drive milling cutters.
3. Identify by name and explain the application for a number of milling cutters.

Machine tool components and parts are identified by name. Operators of machine tools should be familiar with these names and with the location of these parts. The major components of a horizontal milling machine are the column, knee, saddle, table, spindle, and overarm. Figure 2 shows a horizontal milling machine with the major parts identified.

The column is the main part of the milling machine. Its face is machined to provide an accurate guide for the up and down travel of the knee. The column also contains the main drive motor and the spindle. The spindle is hollow, and holds and drives the various cutting tools, chucks, and arbors. The end facing the operator is called the spindle nose; it has a tapered hole with a standard milling machine taper, which is $3\frac{1}{2}$ in. per ft.

The overarm is mounted on top of the column and supports the arbor through an arbor support. The overarm slides in and out and can be clamped securely in any position. The knee can be moved vertically on the face of the column. The knee supports the saddle and the saddle provides the sliding surface for the table. The saddle can move toward and away from the column to give crosswise movement of the machine table. The table provides the surface on which the workpieces are fastened. T-slots are machined along the length of the top surface of the table to align and hold fixtures and workpieces. Hand wheels or hand cranks are used to manually position the table. Micrometer collars make possible positioning movements as small as .001 in.

Power feed levers control automatic feeds in

Figure 1. Small, plain, automatic knee and column milling machine. Note that this machine has no provision for swiveling the table. (Cincinnati Milacron)

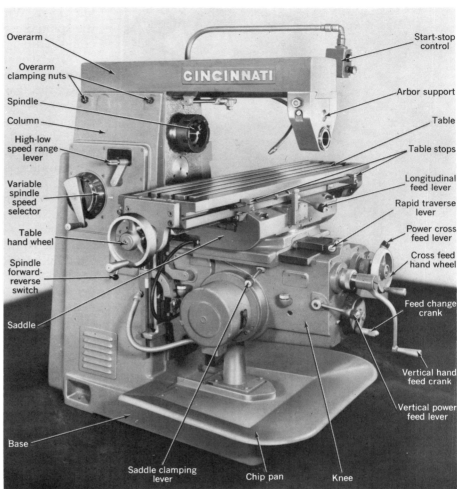

Figure 2. Universal horizontal milling machine. (Cincinnati Milacron)

three axes; the feedrate is adjusted by the feed change crank (Figure 3). On many milling machines, the power feed only operates when the spindle is turning. On others, however, the feed continues after the spindle is shut off if the motor switch is still on. This is an unsafe situation since the work can be forced into a cutter that is not turning, causing extensive damage. Two safety stops are provided on each slide movement to limit travel and thus prevent accidental damage to the feed mechanisms by providing automatic kickout of the power feed. In addition, two adjustable trip dogs for each axis allow the operator to preset specific power feed kickout travel distances. Rapid positioning of the table is accomplished with the rapid traverse lever (Figure 4). The direction of the rapid advance is dependent on the position of the respective feed lever. When using rapid traverse, it is extremely important to see that the moving part is clear of obstructions to avoid damage to the machine and tooling.

Locking devices on the table, saddle, and knee

Figure 3. Feed change crank. (Cincinnati Milacron)

are used to prevent unwanted movements in any or all of these axes. The locking devices should be released only in the axis in which power feed is used. The spindle can rotate clockwise or counterclockwise,

Figure 4. Rapid traverse lever. The table feed lever (top of picture) is in the left movement position. (Cincinnati Milacron)

Figure 5. Speed change levers. (Cincinnati Milacron)

low range. When the lever is in this neutral position, the spindle can easily be rotated by hand during machine setup.

The size of a horizontal milling machine is usually given as the range of movement possible and the power rating of the main drive motor of the machine. An example would be a milling machine with a 28 in. longitudinal travel, 10 in. cross travel, and 16 in. vertical travel with a 5 horsepower main drive motor. As the physical capacity of a machine increases, more power is also available at the spindle through a larger motor.

Before any machine tool is operated, it should be lubricated. A good starting point is to wipe all sliding surfaces clean and apply a coat of a good way lubricant to them. Way lubricant is a specially formulated oil for sliding surfaces. Dirt, chips, and dust will act like a lapping compound between sliding members and cause excessive machine wear. Most machine tools have a lubrication chart that outlines the correct lubricants and lubrication procedures. When no lubrication chart is available, check all oil sight gages for the correct oil level and refill if necessary. Too much oil causes leakage. Lubrication should be performed progressively, starting at the top of the machine and working down. Machine points that are hand oiled should only receive a small amount of oil at any one time, but this should be repeated at regular intervals, at least daily. If the machine is equipped with a pump oiling system, a few pumps on the handle before starting the machine is sufficient in most cases.

Before operating a milling machine, you should be familiar with all control levers. Do not use force to engage or disengage controls or levers. Check that all operating levers are in the neutral position before the machine is turned on. Make sure that the locking levers are loosened on all moving slides. On a **variable speed drive** milling machine, change spindle speeds only while the spindle motor is running. On **geared** models, the spindle has to be stopped. Stop the spindle motor in either case when shifting from one range to another. All power feed levers should be in their neutral position before feed changes are made. Spindle rotation should be reversed only after the machine has come to a complete standstill.

depending on the position of the spindle forward-reverse switch. Spindle speeds are changed to a high or low range with the speed range lever (Figure 5). The variable spindle speed selector makes any spindle speed possible between the minimum and maximum RPM available in each speed range. The speed range lever has a neutral position between the high and

SPINDLES AND ARBORS

Spindles

The spindle of the milling machine holds and drives milling cutters. Cutters can be mounted directly on the spindle nose, as with face milling cutters, or by means of arbors and adaptors that have tapered shanks that fit into the tapered hole or socket in the spindle nose.

Self-releasing tapers have a large included angle generally over 15 degrees. This steep taper permits easy and quick removal of arbors from the spindle nose. Most manufacturers have adapted the standard national milling machine taper. This taper is $3\frac{1}{2}$ in. per ft (IPF), or about $16\frac{1}{2}$ degrees. National milling machine tapers are available in four standard sizes, numbered 30, 40, 50, and 60. Self-releasing taper type shanks must be locked in the spindle socket with a draw-in bolt. Positive drive is obtained through two keys in the spindle nose that engage in keyways in the flange of arbors and adaptors. For most purposes these keys should **never** be removed.

Arbors

Two common arbor styles are shown in Figure 6. Style

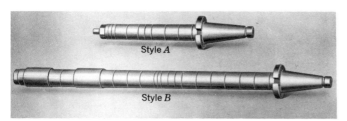

Figure 6. Arbors, styles A and B. (Cincinnati Milacron)

A arbor has a cylindrical pilot on the end opposite the shank. The pilot is used to support the free end of the arbor. Style A arbors are used mostly on small milling machines. They are also used on larger machines when a style B arbor support cannot be used because of a small diameter cutter or interference between the arbor support and the workpiece.

Style B arbors are supported by one or more bearing collars and arbor supports. Style B arbors are used to obtain rigid setups in heavy duty milling operations.

Style C arbors are also known as shell end mill arbors or as stub arbors (Figure 7). Shell end milling cutters are face milling cutters up to 6 in. in diameter. Because of their relatively small diameter, these cutters cannot be counterbored large enough to be mounted directly on the spindle nose, as face mills are, but they are mounted on shell end mill arbors. Since face mills (Figure 8) are mounted directly on the spindle nose, they can be much larger than shell end mills and make heavier cuts, removing more metal.

The method by which arbors are mounted in the horizontal milling machine is illustrated in Figure 9. The draw-in bolt is screwed **by hand** into the arbor as far as it will go, then the arbor is pulled into the spindle nose by tightening the draw-in bar lock nut. Note that the cutters are mounted close to the spindle and the first bearing support is close to the cutter.

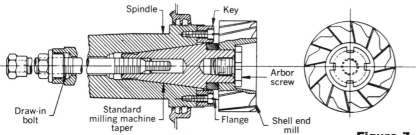

Figure 7. Style C arbor; shell end mill arbor. (Cincinnati Milacron)

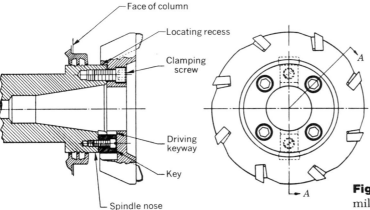

Figure 8. A face mill mounted on the spindle nose of a milling machine. (Cincinnati Milacron)

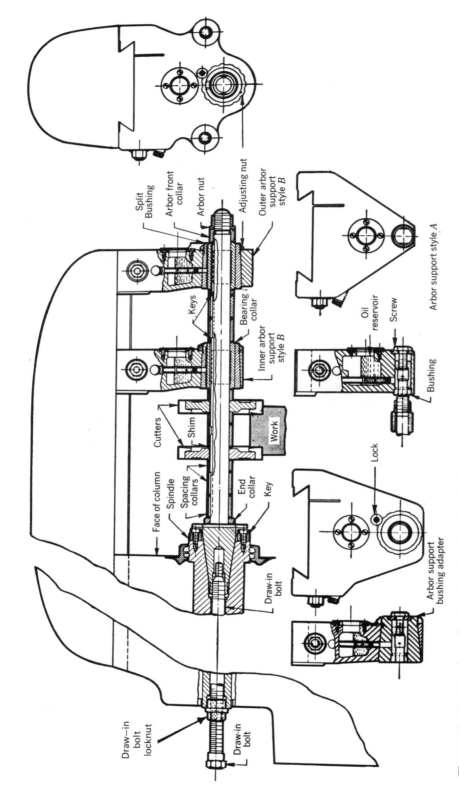

Figure 9. Section through arbor showing location of arbor collars, keys, bearing collars, and various arbor supports. (Cincinnati Milacron)

Cutters should always be mounted as near the bearing supports as possible to avoid chatter and to keep from springing the arbor.

Hardened collars are used to position cutters on a bar. Collars are manufactured to very close tolerances with their ends or faces being parallel and also square to the hole. It is very important that the collars and other parts fitting on the arbor are handled carefully to avoid damaging the collar faces. Any nicks, chips, or dirt between the collar faces will misalign the cutter or deflect the arbor and cause cutter runout. The arbor nut should be tightened or loosened only with the arbor support in place. Without the arbor support, the arbor can easily be sprung and permanently bent. Drive keys are used in a keyway in the arbor to drive the cutter.

Steps in Removing and Mounting an Arbor

1. To remove an arbor, loosen the lock nut on the draw-in bolt one full turn (Figure 10a).
2. Tap the end of the draw-in bolt with a lead hammer. This releases the arbor shank from the spindle socket (Figure 10b).
3. Some arbors are heavy; you may need someone to hold the arbor while you unscrew the draw-in bolt from the rear of the machine (Figure 10c).
4. To install an arbor, the reverse process is used. Before inserting a tapered shank into the spindle socket, clean all mating parts and check for nicks and burrs, which should be removed with a honing stone (Figure 10d).
5. When the arbor is in place in the spindle, screw

Figure 10a. Loosening the lock nut on the draw-in bolt. (Lane Community College)

Figure 10c. Have another person hold heavy arbors when you are loosening them. (Lane Community College)

Figure 10b. Tapping the end of the draw-in bolt with a lead hammer. (Lane Community College)

Figure 10d. Cleaning the external and internal taper prior to installation of the arbor. (Lane Community College)

Figure 10e. Installing the arbor and tightening the draw-in bolt lock nut. (Lane Community College)

Figure 10f. Arbor storage. (Lane Community College)

Figure 11a. Parts should be placed on a clean area. (Lane Community College)

the draw-in bolt in **by hand** as far as it will go, then tighten the lock nut (Figure 10e).

Arbors should be stored in an upright position. Long arbors laying on their sides, if not properly supported, may bend (Figure 10f).

MILLING CUTTERS

Steps in Mounting a Cutter

1. When changing cutters on an arbor, place the cutter and spacers on a smooth and clean area on the worktable to avoid damage to their accurate bearing or contact surfaces (Figure 11a).

2. The cutter, spacing collars, and bearing collars should be a smooth sliding fit on the arbor. Nicks should be stoned off. Clean all parts (Figure 11b).

3. If possible, use an arbor length that does not give much arbor overhang beyond the outer arbor support. Arbor overhang may cause vibration and chatter. Mount cutters as close to the column as the work permits. Cutters are sharp; handle carefully with shop towels. Cutters may be assembled on the arbor when mounted in the machine or at the workbench (Figure 11c), but tighten the arbor nut only by hand.

4. Tighten the arbor nut with a wrench that fits accurately **after** the arbor support is in place. Do not use a hammer to tighten the arbor nut. Overtightening will spring or bend the arbor. Never mount a dull cutter on an arbor. Dull cutters can produce poor results on the workpiece and may be ruined beyond repair if they are used in this condition (Figure 11d).

Arbor-Driven Milling Cutters

Milling cutters are the cutting tools of the milling

Figure 11c. Assembling the cutter on the arbor. (Lane Community College)

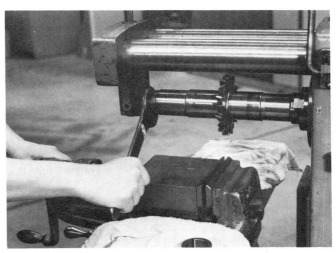

Figure 11d. Tightening the arbor nut. (Lane Community College)

Figure 11b. (Top) Stoning off nicks on the contact face of the spacers. (Bottom) Stoning off nicks on the arbor. (Lane Community College)

machines. They are made in many different shapes and sizes. Each of these various cutters is designed for a specific application. A milling machine operator should be capable of selecting a cutting tool for the required application. One should be able to identify milling cutters by sight as well as to know their capa-

bilities and limitations. For the most part, milling cutters are made from solid high speed steel, but large cutters are often made with inserted teeth of high speed steel. Today, inserted cemented carbide cutters are more often used. Milling cutters can be divided into profile sharpened cutters and form relieved cutters. Profile sharpened cutters are resharpened by grinding a narrow land (Figure 12) back of the cutting edges. An example of this type is the plain milling cutter. Form relieved cutters are resharpened by grinding the face of the tooth parallel to the axis of the cutter. This does not alter the form or shape of the cutter. An example of a form relieved cutter is the concave or convex milling cutter.

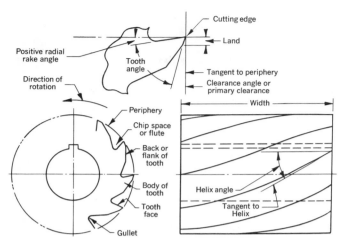

Figure 12. Nomenclature of plain milling cutter. (Cincinnati Milacron)

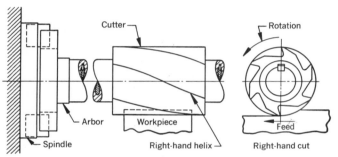

Figure 13. Plain milling cutter with right-hand helix and right-hand cut. (Cincinnati Milacron)

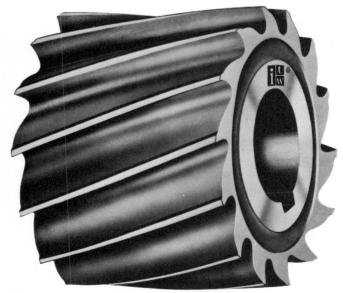

Figure 14. Light duty plain milling cutter. (Illinois/Eclipse, Division Illinois Tool Works Inc.)

Milling cutters are manufactured for either right-hand or left-hand helix. The hand of a milling cutter is determined by looking at the side of a cutter from the end of the table (Figure 13). If the flutes slope down and to the right, it is a right-hand helix; a left-hand helix slopes down and to the left. Cutters may be either right-hand or left-hand cut.

Plain Milling Cutters

Plain milling cutters are designed for milling plain surfaces where the width of the work is narrower than the cutter (Figure 14). Plain milling cutters less than $\frac{3}{4}$ in. wide have straight teeth. On straight tooth cutters, the cutting edge will cut along its entire length at the same time. Cutting pressure increases until the chip is completed. At this time the sudden change in tooth load causes a shock that is transmitted through the drive and often leaves chatter marks or an unsatisfactory surface finish. Light duty milling cutters have a large number of teeth, which limits their use to light or finishing cut because of insufficient chip space for heavy cutting.

Heavy duty plain mills (Figure 15) have fewer and much coarser teeth than light duty mills, giving them stronger teeth with ample chip clearance. These

Figure 15. Heavy duty plain milling cutter. (Illinois/Eclipse, Division Illinois Tool Works Inc.)

Figure 16. Helical plain milling cutter. (Illinois/Eclipse, Division Illinois Tool Works Inc.)

Figure 17. Side milling cutter. (Lane Community College)

Figure 18. Stagger tooth milling cutter. (Illinois/Eclipse, Division Illinois Tool Works Inc.)

are therefore used when heavy cutting with a high removal rate is desired. The helix angle of heavy duty mills is about 45 degrees. The helical form enables each tooth to take a cut gradually, which reduces shock and lowers the tendency to chatter. Plain milling cutters are also called slab mills. Plain milling cutters with a helix angle over 45 degrees are known as helical mills (Figure 16). These milling cutters produce a smooth finish when used for light cuts or on intermittent surfaces. Plain milling cutters do not have side cutting teeth and should not be used to mill shoulders or steps on workpieces.

Side Milling Cutters

Side milling cutters are used to machine steps or grooves. These cutters are made from $\frac{1}{4}$ to 1 in. in width. Figure 17 shows a straight tooth side milling cutter. To cut deep slots or grooves, a staggered tooth side milling cutter (Figure 18) is preferred because the alternate right-hand and left-hand helical teeth reduce chatter and give more chip space for higher speeds and feeds than is possible with straight tooth side milling cutters. To cut slots over 1 in. wide, two or more side milling cutters may be mounted on the arbor simultaneously. Shims between the hubs of the side mills can be used to get any precise width cutter combination or to bring the cutter again to the origi-

nal width after sharpening.

Half side milling cutters are designed for heavy duty milling where only one side of the cutter is used (Figure 19). For straddle milling, a right-hand and a left-hand cutter combination is used.

Plain metal slitting saws are designed for slotting and cutoff operations (Figure 20). Their sides are slightly relieved or "dished" to prevent binding in a slot. Their use is limited to a relatively shallow depth of cut. These saws are made in widths from $\frac{1}{32}$ to $\frac{5}{16}$ in.

To cut deep slots or when many teeth are in contact with the work, a side tooth metal slitting saw (Figure 21) will perform better than a plain metal slitting saw. These saws are made from $\frac{1}{16}$ to $\frac{3}{16}$ in. wide.

Extra deep cuts can be made with a staggered tooth metal slitting saw (Figure 22). Staggered tooth saws have greater chip carrying capacity than other saw types. All metal slitting saws have a slight clearance ground on the sides toward the hole to prevent

Figure 19. Half side milling cutter. (Illinois/Eclipse, Division Illinois Tool Works Inc.)

Figure 20. Plain metal slitting saw. (Illinois/Eclipse, Division Illinois Tool Works Inc.)

binding in the slot and scoring of the walls of the slot. Stagger tooth saws are made from $\frac{3}{16}$ to $\frac{5}{16}$ in. wide.

Angular milling cutters are used for angular milling such as cutting of dovetails, V-notches, and serrations. Single angle cutters (Figure 23) form an included angle or 45 or 60 degrees, with one side of the angle at 90 degrees to the axis of the cutter.

Double angle milling cutters (Figure 24) usually have an included angle of 45, 60, or 90 degrees. Angles other than those mentioned are special milling cutters.

Convex milling cutters (Figure 25) produce concave bottom grooves, or they can be used to make a radius in an inside corner. Concave milling cutters (Figure 26) make convex surfaces. Corner rounding milling cutters (Figure 27) make rounded corners. The cutters illustrated in Figures 25 to 28 are form relieved cutters.

Involute gear cutters (Figure 28) are commonly available in a set of eight cutters for a given pitch,

depending on the number of teeth for which the cutter is to be used. The ranges for the individual cutters are given in Table 1. These eight cutters are designed so that their forms are correct for the lowest number of teeth in each range. If an accurate tooth form near the upper end of a range is required, a special cutter is needed.

Table 1

Gear Cutters

Number of Cutter	Range of Teeth
1	135 to rack
2	55 to 134
3	35 to 54
4	26 to 34
5	21 to 25
6	17 to 20
7	14 to 16
8	12 and 13

Source: Warren T. White, John E. Neely, Richard R. Kibbe, and Roland O. Meyer, *Machine Tools and Machining Practices*, John Wiley and Sons, Inc., Copyright © 1977, New York.

Figure 21. Side tooth metal slitting saw. (Illinois/Eclipse, Division Illinois Tool Works Inc.)

Figure 22. Staggered tooth metal slitting saw. (Illinois/Eclipse, Division Illinois Tool Works Inc.)

Figure 23. Single angle milling cutter. (Illinois/Eclipse, Division Illinois Tool Works Inc.)

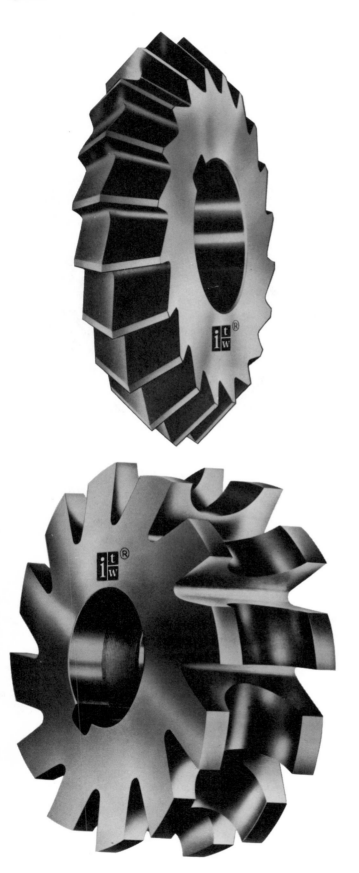

Figure 24. Double angle milling cutter. (Illinois/Eclipse, Division Illinois Tool Works Inc.)

Figure 25. Convex milling cutter. (Illinois/Eclipse, Division Illinois Tool Works Inc.)

Figure 26. Concave milling cutter. (Illinois/Eclipse, Division Illinois Tool Works Inc.)

Figure 27. Corner rounding milling cutter. (Illinois/Eclipse, Division Illinois Tool Works Inc.)

Figure 28. Involute gear cutter. (Lane Community College)

SELF-TEST

1. What are the two basic kinds of milling cutters with reference to their tooth shape?
2. What is the difference between a light duty and a heavy duty plain milling cutter?
3. Why are plain milling cutters not used to mill steps or grooves?
4. What kind of cutter is used to mill grooves?
5. How does the cutting action of a straight tooth side milling cutter differ from a stagger tooth side mill?
6. Give an example of an application of half side milling cutters.
7. When are metal slitting saws used?
8. Give two examples of form relieved milling cutters.
9. When are angular milling cutters used?
10. What kind of cutters are mounted directly on the spindle nose?
11. Milling machine spindle sockets have two classes of taper. What are they?
12. What is the amount of taper on a national milling machine taper?
13. Why is it important to carefully clean the socket, shank, and arbor spacers prior to mounting them on a milling machine?
14. When is a style A arbor used?
15. Where should the arbor supports be in relation to the cutter?
16. What is a style C arbor?
17. Why should the arbor support be in place before the arbor nut is tightened or loosened?

UNIT 2 USING THE HORIZONTAL MILLING MACHINE

When a student or apprentice begins to use the horizontal milling machine, there are several things to consider before turning on the power and starting a cut. The first consideration is that of personal safety, and the second is care for the machine so it will not be damaged. It also involves the selection and alignment of a workholding device, milling cutter, speed, feed, and depth of cut.

OBJECTIVES

After completing this unit, you should be able to:
1. Observe safety precautions on a horizontal milling machine.
2. Align workholding devices.
3. Mill flat surfaces, grooves, and slots.
4. Mill surfaces square to each other.

HORIZONTAL MILLING MACHINE SAFETY

As mentioned earlier in Section A, loose fitting clothing may catch in rotating machinery and pull the operator into it. Rings, bracelets, and watches can be very dangerous because they may catch in a cutter and tear off a finger or pull your hand or arm into the cutter with obvious consequences. Long or loose hair can be caught in a cutter or even be wrapped around a rotating smooth shaft, resulting in a quick, painful scalping. Persons with long hair should wear a cap or hairnet in the machine shop.

A milling machine should not be operated while wearing gloves because of the danger of getting caught in the machine. When gloves are needed to handle sharp-edged materials, the machine should be stopped. Eye protection should be worn at all times in machine shops. Eye injuries can be caused by flying chips, tool breakage, or cutting fluid sprays. Keep your fingers away from the moving parts such as the cutter, gears, spindle, or arbor. Do not reach over a rotating spindle or cutter.

Safe operation of a machine tool requires that you think before you do something. Before starting up a machine, know the location and operation of its controls. Operate all controls on the machine yourself; do not have another person start or stop the machine for you. While operating a milling machine, observe the cutting action at all times so that you can stop the machine immediately when you see or hear something unfamiliar. **Always stay within reach of the controls while the machine is running.** An unexpected emergency may require quick action on your part. **Never** leave a running machine unattended.

Before operating the rapid traverse control on a milling machine, loosen the locking devices on the machine axis to be moved. Check that the hand wheels or hand cranks are disengaged, or they will spin and injure anyone near them when the rapid traverse is engaged. The rapid traverse control will move any machine axis that has its feed lever engaged. Do not try to position a workpiece too close to the cutter with rapid traverse, but approach the final two inches by using the hand wheels or hand cranks.

A person concentrating on a machining operation should not be approached quietly from behind, since it may annoy and alarm the person thus causing a ruined workpiece or injury. Do not lean on a running machine; moving parts can hurt you. Signs posted on a machine indicating a dangerous condition or a repair in progress should only be removed by the person making the repair or by a supervisor.

Measurements should only be taken on a milling machine after the cutter has stopped rotating and after the chips have been cleared away. Milling machine chips are dangerously sharp and often hot and contaminated with cutting fluids. They should not be handled with bare hands. Chips should be removed with a brush. Compressed air should not be used to clean off chips from a machine because it will make small missiles out of chips that can injure a person at quite a distance away. A blast of air will also force small chips into the ways and sliding surfaces of the milling machine where they will cause scoring and rapid premature wear. Cleaning chips and cutting fluids from the machine or workpiece should only be done after the cutter has stopped turning. Keep the area around the machine clean of chips, oil spills, cutting fluids, and other obstructions to prevent the operator from slipping or stumbling. Wash your hands and arms thoroughly after being splashed with cutting fluid to prevent getting dermatitis, a skin disease.

Injuries can be caused by improper setups or the use of wrong tools. Use the correct size wrench when loosening or tightening nuts or bolts, preferably a box wrench or a socket wrench. An oversized wrench will round off the corners on bolts and nuts and prevent sufficient tightening or loosening; a slipping wrench can cause mashed fingers or other injuries to the hands or arms. Milling machine cutters have very sharp cutting edges; handling cutters carefully and with a cloth will avoid cuts on the hands.

SETTING SPEEDS AND FEEDS ON THE HORIZONTAL MILLING MACHINE

Speeds

To get maximum use from a milling cutter and to avoid damaging it, it is important that it is operated at the correct cutting speed. Cutting speed is expressed in surface feet per minute, which varies for different work materials and cutting tool materials. The cutting speed of a milling cutter is the distance the cutting edge of a cutter tooth travels in 1 minute. Table 1 is a table of cutting speeds for a number of commonly used materials.

The cutting speeds given in Table 1 are only intended to be starting points. These speeds represent experience in instructional settings where cutter and machine conditions are often less than ideal. Different hardnesses within each materials group account for the wide range of cutting speeds. Generally, speeds should be lower for hard materials, abrasive materials, and deep cuts. Speeds are higher for soft materials, better finishes, light cuts, delicate workpieces, and light setups. Cutting edges will dull rapidly if the cutting speed is too high for the material being machined. Excess speed can destroy an expensive cutter in a

Table 1
Cutting Speeds for Milling

| Material | Cutting Speed SFM | |
	High Speed Steel Cutter	Carbide Cutter
Free machining steel	100–150	400–600
Low carbon steel	60–90	300–550
Medium carbon steel	50–80	225–400
High carbon steel	40–70	150–250
Medium alloy steel	40–70	150–350
Stainless steel	30–80	100–300
Gray cast iron	50–80	250–350
Bronze	65–130	200–400
Aluminum	300–800	1000–2000

Source: Warren T. White, John E. Neely, Richard R. Kibbe, and Roland O. Meyer, *Machine Tools and Machining Practices*, John Wiley and Sons, Inc., Copyright © 1977, New York.

few seconds by burning the cutting edges beyond saving by resharpening. If the cutting speed is too slow, cutting action will be inefficient, causing a relatively slow removal rate; also, a built-up edge will form on the cutting edge which increases tool pressure that often results in tool breakage.

It is a good practice to use the lower cutting speed value as it is given in Table 1 to begin with, and then, if the setup allows, increase it toward the higher speed until chatter develops or the coolant begins to vaporize excessively, indicating heat build up. To use these cutting speed values on a milling machine, they have to be expressed in RPM. The formula used to convert cutting speed into RPM is:

$$RPM = \frac{CS \times 4}{D}$$

CS = cutting speed found in Table 1
4 = constant
D = diameter of cutter in inches

EXAMPLE:
Calculate the RPM for a 3 in. diameter high speed steel cutter to be used on cast iron.

$$RPM = \frac{CS \times 4}{D}$$

$$RPM = \frac{50 \times 4}{3} = \frac{200}{3} = 67$$

Tool materials other than high speed steels, such as cast alloys or cemented carbides, can be used at higher cutting speeds because they retain a sharp cutting edge at elevated temperatures. As a rule, if we consider high speed steel cutting speeds as 100 percent, then the cutting speed for cast alloys can be 150 percent and cemented carbides can have a cutting speed rated between 200 to 600 percent. Determining the speeds and feeds accurately for milling is complicated by the fact that the cutting edge does not remain in the work continuously and that the chip being made varies in thickness during the cutting period. The type of milling being done (slab, face, or end milling) also makes a difference, as does the way that the heat is transferred from the cutting edge.

Feeds

Only secondary to speed in efficient milling machine operation is the feed. It is expressed as a feedrate and given in inches per minute (IPM). Most milling machines have two separate drive motors: one to power the spindle and one to power the feed mechanism. Independent changes in feedrate and spindle speed are made possible by having two motors.

The feedrate is the product of the feed per tooth times the number of teeth on the cutter times the RPM of the spindle. You have already determined how to calculate the RPM of a cutter; by counting the number of teeth in a cutter and knowing the feed per tooth, you can determine the feedrate. Table 2 is a chart of commonly used feeds per tooth (FPT). As you can see in Table 2, there is only a slight difference in the feed per tooth allowance between high speed steel cuttters and carbide cutters.

EXAMPLE

Calculate the feedrate for a 3 in. diameter six tooth helical mill that is cutting free machining steel, first for a high speed steel tool and then for a carbide tool. The formula is:

$$FPT \times N \times RPM = \text{feedrate in inches per minute (IPM)}$$
$$FPT = \text{feed per tooth}$$
$$N = \text{number of teeth on cutter}$$
$$RPM = \text{revolutions per minute of cutter}$$

Table 2
Feed in Inches Per Tooth (Instructional Setting)

Type of Cutter	Aluminum		Bronze		Cast Iron		Free Machining Steel		Alloy Steel	
	HSS	Carbide	HSS	Carbide	HSS	Carbide	HSS	Carbide	HSS	Carbide
Face mills	.007 to .022	.007 to .020	.005 to .014	.004 to .012	.004 to .016	.006 to .020	.003 to .012	.004 to .016	.002 to .008	.003 to .014
Helical mills	.006 to .018	.006 to .016	.003 to .011	.003 to .010	.004 to .013	.004 to .016	.002 to .010	.003 to .013	.002 to .007	.003 to .001
Side cutting mills	.004 to .013	.004 to .012	.003 to .008	.003 to .007	.002 to .009	.003 to .012	.002 to .007	.003 to .009	.001 to .005	.002 to .008
End mills	.003 to .011	.003 to .010	.003 to .007	.002 to .006	.002 to .008	.003 to .010	.001 to .006	.002 to .008	.001 to .004	.002 to .007
Form relieved cutters	.002 to .007	.002 to .006	.001 to .004	.001 to .004	.001 to .005	.002 to .006	.001 to .004	.002 to .005	.001 to .003	.001 to .004
Circular saws	.002 to .005	.002 to .005	.001 to .003	.001 to .003	.001 to .004	.002 to .006	.001 to .003	.001 to .004	.005 to .002	.001 to .004

Source: Warren T. White, John E. Neely, Richard R. Kibbe, and Roland O. Meyer, *Machine Tools and Machining Practices*, John Wiley and Sons, Inc., Copyright © 1977, New York.

Note: The feedrate should be adjusted according to the capability of the setup. Lower feedrates should be used at first, then adjusting to higher feedrates as the situation permits.

For a starting point, calculate the RPM (refer to Table 1):

$$\text{RPM} = \frac{\text{CS} \times 4}{D} = \frac{100 \times 4}{3} = \frac{400}{3} = 134 \text{ RPM}$$

Table 2 gives an FPT of .002 in. The cutter has six teeth. The feedrate is

$$.002 \times 6 \times 134 = 1.608 \text{ IPM}$$

for the high speed steel cutter.

EXAMPLE
For a carbide cutter the RPM is

$$\text{RPM} = \frac{\text{CS} \times 4}{D} = \frac{400 \times 4}{3} = \frac{1600}{3} = 534 \text{ RPM}$$

The FPT is .003 in. The cutter has six teeth. The feedrate is

$$.003 \times 6 \times 534 = 9.612 \text{ IPM}$$

for the carbide cutter.

To calculate the starting feedrate, use the lower figure from the feed per tooth table and, if conditions permit, increase the feedrate from there. The most economical cutting takes place when the greatest amount (expressed in cubic inches of metal per minute) is removed and a long tool life is obtained. The tool life is longest when a relatively low speed and high feed rate is used. Try to avoid feedrates of less than .001 in. per tooth, because this will cause rapid dulling of the cutter. An exception to this limit is the use of small diameter end mills on harder materials. The depth and width of cut also affect the feedrate. Wide and deep cuts require a smaller feedrate than do shallow, narrow cuts. Roughing cuts are made to remove material rapidly. The depth of cut may be $\frac{1}{8}$ in. or more, depending on the rigidity of the machine, the setup, and the horsepower available. Finishing cuts are made to produce precise dimensions and acceptable surface finishes. The depth of cut for a finishing cut should be between .015 and .030 in. A depth of cut of .005 in. or less will cause the cutter to rub instead of cut, which also results in excessive cutting edge wear.

Cutting fluids should be used when machining most metals with high speed steel cutters on a milling machine. A cutting fluid cools the tool and the workpiece. The lubricating action of the coolant reduces friction between the tool face and chip. Cutting fluids prevent rust and corrosion and, if applied in sufficient quantity, will flush away chips. Because of these characteristics and the use of higher cutting speeds, increased production is made possible through the use of good cutting fluids.

Most milling with carbide cutters is done dry unless a large constant flow of cutting fluid can be directed at the cutting edge. An interrupted coolant flow on a carbide tool causes thermal cracking and results in subsequent chipping of the tool.

WORKHOLDING DEVICES ON THE MILLING MACHINE

The Machine Vise

Probably the most common method of workholding on a milling machine is a vise. Vises are simple to use and can quickly be adjusted to the size of the workpiece. A vise should be used to hold work with parallel sides if it is within the size limits of the vise, because it is the quickest and most economical workholding method. The plain vise (Figure 1) is bolted to the machine table. Alignment with the table is provided by two slots at right angles to each other on the underside of the vise. These slots are fitted with removable keys that align the vise with the table T-slots either lengthwise or crosswise. A plain vise can be converted to a swivel vise (Figure 2) by mounting it on a swivel plate. The swivel plate is graduated in degrees. This allows the upper section to be swiveled to any angle in the horizontal plane. When swivel bases are added to a plain vise, the versatility increases, but rigidity is lessened.

For work involving compound angles, a universal vise (Figure 3) is used. This vise can be swiveled 90 degrees in the vertical plane and 360 degrees in the horizontal plane. The most rigid setup is the one

Figure 1. Plain vise. (Cincinnati Milacron)

Figure 2. Swivel vise. (Cincinnati Milacron)

Figure 3. Universal vise. (Cincinnati Milacron)

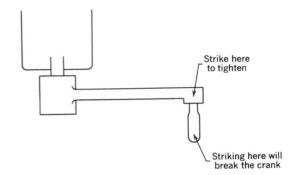

Strike here
to tighten

Striking here will
break the crank

Figure 4. Tightening a vise. (Cincinnati Milacron)

Figure 5. Cutting pressure against solid jaw. (Lane Community College)

where the workpiece is clamped close to the table surface.

Air or hydraulically operated vises are often used in production work; but in general toolroom work, vises are opened and closed by cranks or levers. To hold workpieces securely without slipping under high cutting forces, a vise must be tightened by striking the crank with a lead hammer (Figure 4).

Whenever possible, position the vise so that the cutting pressure will be against the solid jaw (Figure 5). Often references are made to the "solid jaw" of a vise (for example, alignment is made on the solid

jaw). The solid jaw will not move or change when the vise is tightened, although the movable jaw will align itself to some degree with the work. The vise should never be aligned with a dial indicator on the movable jaw since it can move on its guide a few thousandths of an inch.

A good machine vise is an accurate and dependable workholding device. When milling only the top of a workpiece, it is not necessary that the vise be square to the column or parallel to the table travel. However, when the job requires that the outside surface is parallel to a step or groove in the workpiece, the vise must be precisely aligned and positioned on the table.

The base of the vise should be located with keys (Figure 6) that fit snugly into the T-slots on the milling machine table. This normally positions the solid jaw of the vise parallel with or square to the face of the column. Before mounting a vise or other fixture on a machine table, inspect the base carefully for small chips and nicks and remove any that you find. When the base is clean, fasten the vise to the table.

Figure 6. Fixture alignment keys. (Lane Community College)

Figure 7. Aligning the universal milling machine. (Lane Community College)

Preparing a machine tool prior to machining is called setting up the machine. Before a setup can be made, the machine should be cleaned, especially all sliding surfaces such as the ways and the machine table. After wiping the table clean, use your hand to feel for nicks or burrs. If you find any, use a honing stone to remove them. Workpieces must be fastened securely for the machining operation. They can be held in a vise, clamped to an angle plate, or clamped directly to the table. Odd shaped workpieces may be held in a fixture designed for that purpose. On a universal milling machine it is good practice to check the alignment of the table before mounting a vise or fixture on it.

Table Alignment on a Universal Milling Machine

1. Clean the face of the column and the machine table.
2. Fasten a dial indicator to the table with a magnetic base or other mounting device (Figure 7).
3. Preload the indicator to approximately one-half revolution of its dial and set the bezel to zero.
4. Move the table longitudinally with the handwheel to indicate across the column.
5. If the indicator hand moves, loosen the locking bolts on the swivel table and adjust the table one-half of the indicated difference.
6. Tighten the locking bolts and indicate across the column again; make another adjustment if needed.

Never indicate the table with the indicator mounted

Figure 8. Aligning the vise parallel to the table. (Lane Community College)

on the column, as this would always show alignment even though the table was off.

Aligning a Vise Parallel with the Table Travel

A number of different methods of aligning a vise on a table are shown as follows.

1. Fasten a dial indicator with a magnetic base to the arbor (Figure 8), set the dial indicator to the solid vise jaw, and preload the indicator contact point to one-half revolution of the dial. Set the bezel to zero.

Figure 9. Aligning the vise square to table travel. (Lane Community College)

2. Move the table so that the indicator slides along the solid jaw. Record any indicator movement.
3. Loosen the holddown bolts and lightly retighten. Lightly tap the vise with a lead or soft faced hammer to move the vise one-half the distance of the indicator movement.
4. Indicate the solid jaw again to check the alignment. **Always** take another indicator reading after securely tightening the clamping bolts. Often the final tightening will move a vise, fixture, or workpiece.

Aligning a Vise at a Right Angle to the Table Travel

1. Fasten a dial indicator with a magnetic base to the arbor (Figure 9) and preload the indicator.
2. Move the table with the cross feed hand wheel and indicate the solid jaw.
3. Loosen the vise holddown bolts and make any necessary correction.
4. Indicate the solid jaw again to check the alignment. **Always** take another indicator reading after securely tightening the clamping bolts. Often the final tightening will move a vise, fixture, or workpiece.

If no indicator is available to align a vise on a table, a combination square may be used as shown in Figure 10. The beam of the square is slid along the machined surface of the column until contact is made with the solid jaw of the vise. Two strips of paper used as feeler gages help in locating the contact point. A soft headed

Figure 10. Using a square to align a vise on the table. (Lane Community College)

Figure 11. Using a protractor to align a vise on the table. (Lane Community College)

hammer or lead hammer is used to tap the vise into position.

Aligning a Vise at an Angle Other Than 90 Degrees to the Table Travel

Occasionally a vise needs to be mounted on the table at an angle other than square to the table travel. This can be done with a protractor (Figure 11). Paper strips are used as feeler gages, the angular setting being

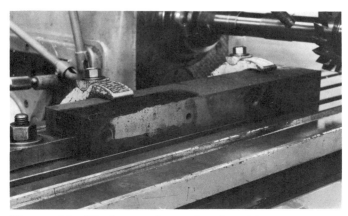

Figure 12. Workpiece clamped to the table. (Lane Community College)

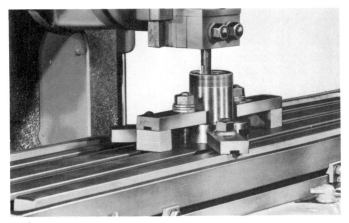

Figure 13. Work clamped to the table with T-slot bolts and clamps. (Cincinnati Milacron)

Figure 14. Shafts are often clamped in T-slots when keyseats are being cut. (Lane Community College)

correct when both strips contact the protractor blade and the vise jaw at the same time. This is not a precise method of setting an angle because of the limitations in setting an angle accurately with a protractor, maintaining the level of the protractor blade, and accurately testing the "drag" on the paper strips.

Workholding on the Machine Table

Before machining a workpiece to size on a milling machine, several important decisions need to be made. One consideration is how to hold the workpiece while it is being machined. Large workpieces can be clamped directly to the table (Figure 12). Workpieces tend to move on the table from the cutting pressure against them. This movement can be prevented by clamping a stop block on the table and placing the workpiece against it.

T-slots, which run lengthwise along the top of the table, are accurately machined and parallel to the sides of the table. These T-slots are used to retain the clamping bolts. Workpieces can also be aligned against snug fitting parallels that are set into the T-slot; the workpiece is pushed against these parallels while the work is being clamped. Figure 13 shows the workpiece clamped to the table with T-slot bolts and clamps. The bolts are placed close to the workpiece and the block supporting the outer end of the clamp is the same height as the shoulder being clamped. When the bolt is closer to the work than to the clamp support block, maximum leverage is obtained. The support block should never be lower than the work being clamped.

When workpieces with finished or soft surfaces are clamped, care must be taken to protect those surfaces from damage by clamping. A shim should be placed between the work surface and the clamp. Before placing rough castings or weldments on a machine table, protect the table surface with a shim. This shim can be paper, sheet metal, or even plywood, depending on the accuracy of the machining to be performed.

A workpiece should have a support directly underneath the location where a clamp exerts pressure. Clamping an unsupported workpiece may cause it to bend or spring, and it will bend back after clamping pressure is released. If the workpiece material is brittle, clamping pressure may break it.

Round stock may be milled while clamped to the table in the T-slot groove for alignment (Figure 14). Keyways are often milled on long shafting in this manner, the overhang being supported on a stand or with

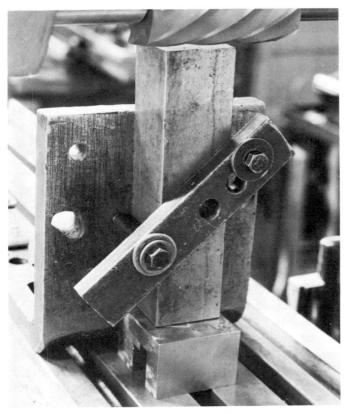

Figure 15. Use of angle plate to mill ends of workpiece. (Lane Community College)

Figure 16. Setup of a workpiece to machine an end square. (Lane Community College)

a hoist. Keyways on short shafts are sometimes milled while they are clamped in a vise. Flats and square ends may be milled on shafts by rotating them 90 degrees after each cut.

Angle plates are sometimes used for a rigid setup when the end of a piece is machined (Figure 15). The vise can also be used to set up relatively short pieces for milling the ends square (Figure 16). When work must be clamped off-center in a vise, a spacer must be used (Figure 17).

A dividing head is often used for machining parts that require accurate spacing on a circle. Figure 18 shows a gear being cut from a blank that is mounted on a mandrel supported by centers on the head and the footstock and driven by a dog. Dividing heads are also used for milling splines, squares, and hexagons on ends of shafts.

Setting Up the Machine

Many workpieces can be held securely in a machine vise (Figure 19). If the workpiece is high enough, seat it on the bottom of the vise. If it is not, use parallels to raise it. It is the friction between the vise jaws

Figure 17. Work clamped off center needs a spacer. (Lane Community College)

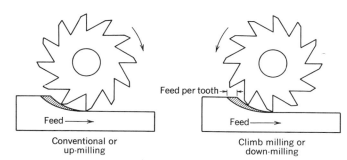

Figure 20. Conventional and climb milling. (Machine Tools and Machining Practices)

Figure 18. Dividing head and foot stock. (Cincinnati Milacron)

Figure 19. Workpiece held in a vise for slab milling. (Lane Community College)

and the workpiece that holds the workpiece and the more contact area there is, the better it is gripped. After the cutter for the job has been selected, the speed and feed can be calculated and set on the machine. The depth of cut depends largely on the amount of material that is to be removed and the rigidity of the setup.

Conventional and Climb Milling

Figure 20 shows one cutter operated in a conventional milling mode and the other cutter in a climb milling operation. In conventional milling modes (sometimes called up-milling), the workpiece is forced against the cutter with the teeth of the cutter trying to lift the workpiece up, especially at the beginning of a cut. In climbing (or down-milling), the cutter tends to hold the workpiece down.

Climb milling should only be performed on machines equipped with an antibacklash device. Backlash, which is slack between the table drive screw and nut assembly, would let the cutter pull the workpiece under it, break the cutter, and ruin the workpiece. **Always** use the conventional or up-milling mode unless the machine you are using is equipped for climb milling.

Setting Up the Arbor and Cutter

Prior to assembly of the spacing collars and cutter on the arbor, all pieces should be cleaned. Keys should always be used to drive the cutter. Do not depend on the friction between the spacers and cutter. The drive keys should extend into the spacing collars on both sides of the cutter. Tighten and loosen the arbor lock nut only when the arbor support is in place, and do not use a hammer on the wrench. Select the proper cutter for the job.

Milling a Slot

The diameter of the cutter to be used on a job depends on the depth of the slot or step. As a rule, the smallest diameter cutter that will do the job should be used, as long as sufficient clearance remains between arbor and work and between arbor support and vise (Figure 21). Use only a sharp cutter to minimize cutting pressures and to get a good surface finish. Resharpen a cutter when it becomes slightly dull. A slightly dull cutter can be resharpened easily and quickly.

Milling cutters with side cutting teeth are used when grooves and steps have to be machined on a workpiece. The size and kind of cutter to be used depends largely on the operation to be performed.

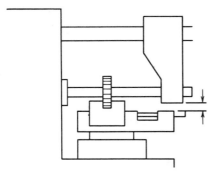

Figure 21. Check for clearance. (Machine Tools and Machining Practices)

Figure 22. Full side (staggered tooth) milling cutter machining a groove. (Lane Community College)

Full side mills with cutting teeth on both sides are used when slots or grooves are cut (Figure 22).

Cutters usually make a slot that is slightly wider than the nominal width of the cutter. Cutters that wobble because of dirt or chips between the arbor spacers will cut slots that are considerably oversize. A slot will become wider when more than one cut is made through it. If a slot needs to be .375 in. wide, a $\frac{3}{8}$ in. cutter probably cannot be used because it may cut a slot .3755 to .376 in. wide. A trial cut in a piece of scrap metal will tell the exact slot width. It may be necessary to use a $\frac{5}{16}$ in. wide cutter and make two or more cuts.

The width of a slot from a given cutter is also affected by the amount of feed used. A very slow feed will tend to allow the cutter to cut more clearance for itself and make a wider groove. A fast feed crowds the cutter, which means a narrower slot. A cutter tends to cut a wider slot in soft material than it will in harder material. The width of slots are measured with venier calipers or with adjustable parallels. If it is a keyway, the key itself can be used as a gage to test the slot width.

Squaring a Block

Slab milling (Figure 23) cutters are used in a setup where a uniform, flat surface is needed and where square or rectangular workpieces are milled with sides parallel and square. Shims are often needed in a vise to achieve true squareness since most milling machine vises become slightly out of square from heavy use.

A good machinist will mark the workpiece with layout lines before fastening it in a vise. The layout should be an exact outline of the part to be machined. The reason for making the layout prior to machining is that reference surfaces are often removed by machining. After the layout has been made, make diagonal lines on the portion to be machined away. This helps in identifying on which side of the layout line the cut is to be made (Figure 24). The cutter can be accurately positioned on the workpiece with the hand feed cranks and the micrometer dials. When the finished outside surface of a workpiece must not be marred or scratched by a revolving cutter, a paper strip is held between the workpiece and the cutter (Figure 25). The power is turned off and the spindle is rotated by hand. Make sure the paper strip is long

Figure 23. Location of shim to square up work. (Lane Community College)

Figure 24. Work laid out for milling. (Lane Community College)

Figure 26. Using a paper strip to set the depth of cut. (Lane Community College)

Figure 25. Setting up a cutter by using a paper strip. (Lane Community College)

Figure 27. Positioning a cutter using a steel rule. (Lane Community College)

enough so your hands are not near the cutter. Carefully move the table toward the revolving cutter. When the cutter pulls the paper strip from your fingers, the cutter is about .002 in. from the workpiece. At this time set the cross feed dial to zero. Lower the knee until the cutter clears the top of the workpiece. Then, by using the cross feed handwheel, posi-

tion the work where the cut is to be made. The same method will work in positioning a cutter above a workpiece and establishing the zero position for the depth of cut without actually touching the workpiece with the cutter (Figure 26).

A quicker, but not as accurate, method is illustrated in Figure 27. A steel rule is used to position a side mill a given distance from the outside edge of a workpiece. The end of the rule is held firmly against the side cutting edge of a tooth. The distance is indicated by the edge of the workpiece. The micrometer dials of the cross feed and knee controls should be zeroed when the cutter contacts the side and top of the workpiece. When these zero positions are established, additional cuts can easily be made by positioning from these points with the micrometer dials.

After the first cut is made, the distance of the side of the step or groove to the outside of the workpiece should be measured. A measurement made on both ends shows if the cut is parallel to the sides of the workpiece. If the step is not parallel, the vise needs to be aligned or, on a universal milling machine, the table may need aligning.

Making a Step

Good milling practice is to take a roughing cut and then a finish cut. Better surface finishes and higher dimensional accuracy are achieved when roughing and finishing cuts are made. The depth of the roughing cut is often limited by the horsepower of the machine or the rigidity of the setup. A good starting point for roughing is .100 to .200 in. deep. The finishing cut should be .015 to .030 in. deep. Depth of cut less than .015 in. deep should be avoided, because a milling cutter, especially in conventional or up-milling, has a strong rubbing action before the cutter actually starts cutting. This rubbing action causes a cutter to dull rapidly. Assuming that a cut .100 in. deep has to be taken, the following steps outline the procedure to be used.

1. Loosen the knee locking clamp and the cross slide lock.
2. Turn on the spindle and check its rotation.
3. Position the table so that the workpiece is under the cutter.
4. Raise the knee slowly by turning the vertical hand feed crank until the cutter just touches the workpiece. (If the cutter cuts a groove, you have gone too far and should try again on a different place on the workpiece.)
5. Set the micrometer dial on the knee feedscrew to zero.
6. Lower the knee by approximately one-half revolution of the hand feed crank. (If the knee is not lowered, the cutter will leave tool marks on the workpiece in the following operation.)
7. Move the table lengthwise until the cutter is clear of the workpiece. Move the cutter to that side

of the workpiece, which will result in making a conventional milling cut.

8. Raise the knee past the zero mark to the 100 mark on the micrometer dial.
9. Tighten the knee lock and the cross slide lock. **Always** tighten all locking clamps prior to starting a machining operation, except the one that would restrict table movement while cutting. This aids in making a rigid chatterfree setup.
10. The machine is now ready for the cut. Turn on the coolant. Move the table slowly into the revolving cutter until the full depth of cut is obtained before engaging the power feed.
11. When the cut is completed, disengage the power feed, stop the spindle rotation, and turn off the coolant before returning the table to its starting position. If the revolving cutter is returned over the newly machined surface, it will leave cutter marks and mar the finish.
12. After brushing off the chips and wiping the workpiece clean, the workpiece should be measured while it is still fastened in the machine. If the workpiece is parallel at this time, additional cuts can be made if more material needs to be removed.

Cutting a Keyseat on a Shaft

Keyseats are commonly milled on horizontal milling machines with plain or staggered tooth cutters. Standard key sizes for shafts are given in Table 1 in Section I, Unit 1. Keyseats are cut on the end of a shaft or at locations along the shaft. When a cut is started on the shaft, away from the end, the table lock must be tightened while making the plunge cut to the depth of the keyseat. The table lock is then loosened and the table feed is engaged. The shaft is clamped in a vise if it is short. If it is a long shaft, it is often clamped in a T-slot groove. The cutter is centered over the shaft by touching its side with the cutter with a paper strip for a feeler gage (Figures 28a and 28b). The micrometer collar is set to zero and the cutter is moved a distance equal to the radius of the shaft plus half the width of the cutter. The work is brought to the rotating cutter and a spot is cut equal to the width of the cutter so the correct depth can be made from the side of the slot. The collar on the knee feedscrew is zeroed and the cutter relocated away from the end of the shaft. The correct depth is set and the cut is started (Figure 29).

Coolants and Cutting Fluids

Most milling machines are equipped with a coolant tank and pump (Figure 30). Usually some kind of solu-

Figure 28a. Cutter positioned to shaft with paper strip for feeler gage. (Lane Community College)

Figure 29. Starting the cut. (Lane Community College)

Figure 28b. Cutter positioned over shaft. A spot is made equal to the width of the cutter and the index zeroed for vertical travel. The shaft has been lowered for viewing. (Lane Community College)

Figure 30. Coolant cools the cutter and flushes away chips. (Lane Community College)

ble oil-water mix is used as a cutting fluid. Unless the soluble oil contains a bactericide, the coolant often becomes rancid and can cause skin rashes. There

are many synthetic cutting fluids available today that eliminate this problem. Smaller machines are often equipped with a coolant spray device that is quite effective. The operator should try to avoid breathing the spray by standing on the opposite side of the nozzle. Sulfurized (black) cutting oil from a pump should not be used at all on a milling machine since the soluble oil coolant would be contaminated by it.

SELF-TEST

1. Name at least three hazards to a person around or near a revolving cutter on a milling machine.
2. When cleaning chips off the machine and when taking measurements, should the cutter be rotating? Should compressed air be used for cleaning?
3. A cut is to be made on low carbon steel with a 4 in. diameter high speed cutter with 16 teeth. A cutting speed of 90 FPM is selected from Table 1. What is the correct RPM?
4. A feed per tooth of .003 in. is chosen from Table 2

for selecting the table feed of the cutter in Question 3. What is the correct table feed?

5. What is the most common workholding device used on the milling machine?
6. The milling machine vise should be aligned to what part?
7. Name two ways that round stock may be set up for milling a keyseat.
8. On older milling machines or on those not equipped

with an anti-backlash device, should a conventional or climb milling mode be used?
9. How can backlash in a machine cause the cutter to be broken and the workpiece ruined?
10. Coolant is always used to flood the cutter and workpiece with high speed cutters and sometimes with carbide cutters. When is it best not to use coolant with carbide cutters?

UNIT 3 VERTICAL MILLING MACHINES

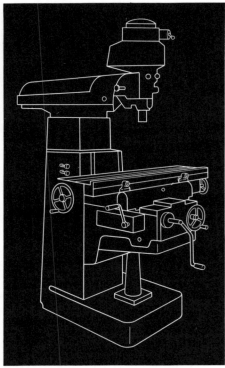

The vertical milling machine is a relatively recent development compared to the horizontal milling machine. The basic difference is that the spindle of the vertical milling machine is normally in a vertical position and cutting tools are mounted at the end of the spindle. There is a similarity to the drill press in that the quill moves the spindle up and down; but there the similarity ends. The spindle and quill on the vertical milling machine are heavy duty and can usually be tilted or rotated in one or two axes in order to make angular cuts. The spindle has a power feed and the table is also equipped with power feed on some machines. These unique characteristics make the vertical milling machine one of the most versatile machines found in the shop. The knee and column principle has been adapted from the horizontal milling machine, which allows the milling table to be raised and lowered in relation to the spindle.

OBJECTIVES

After completing this unit, you should be able to:
1. Identify safe vertical milling machine practices.
2. Identify the important parts of a vertical milling machine.
3. Identify and select from commonly used vertical milling machine cutting tools.

Aside from the safety instructions given earlier in this text—including proper dress, eye protection, jewelry removal, and alertness—there are a few safety instructions that pertain to the vertical milling machine. All machine guards should be in place prior to starting a machine. Observe other machines in operation around you to make sure they are guarded properly.

Report any unsafe or missing guards to your supervisor. If you observe any unsafe practices by other machine operators, report them to the person in charge. A safe workplace involves everyone.

Safety also involves keeping a clean machine and keeping the area surrounding it clean. Any oil or coolant spills on the floor should be wiped up immediately to avoid slipping and falling. Chips should be swept up with a brush or broom and deposited in chip or trash containers. Do not handle chips with your bare hands or you will get cut. Dirty and oily rags should be kept in closed containers and should not accumulate in piles on the floor. Do not use an air hose to clean a machine. Flying chips can hurt you or those around you. When you lift heavy workpieces or machine attachments on or off the machine, ask someone to help you. When you are lifting anything, use proper lifting methods.

Be careful when handling tools and sharp edged workpieces to avoid getting cut. Use a rag to protect your hand. Workpieces should be rigidly supported and tightly clamped to withstand the usually high cutting forces encountered in machining. When a workpiece comes loose while machining, it is usually ruined and often the cutter is ruined. The operator can also be hurt by flying particles from the cutter or workpiece.

The cutting tools should be securely fastened in the machine spindle to prevent any movement during the cutting operation. Cutting tools need to be operated at the correct revolutions per minute (RPM) and feedrate for any given material. Excessive speeds and feeds can break the cutting tools. On vertical milling machines, care has to be exercised when swiveling the workhead to make angular cuts. After loosening the clamping bolts that hold the workhead to the ram, retighten them lightly to create a slight drag. There should be enough friction between the workhead and ram that the head only swivels when pressure is applied to it. If the clamping bolts are completely loosened, the weight of the heavy spindle motor will flip the workhead upside down or until it hits the table, possibly injuring the operator's hand or a workpiece.

Measurements are frequently made during machining operations. Do not make any measurements until the spindle has come to a complete stop.

THE VERTICAL SPINDLE MILLING MACHINE

Identification

Milling machine parts and components are named to make their identification easier. Knowing the names is useful in locating trouble spots or in operating the machine controls. Figure 1 identifies many of the important parts of a light duty vertical milling machine. The column is the backbone of the machine; it rests on the base. The front or face of the column is accurately machined to provide a guide for the vertical travel of the knee. The top part of the column

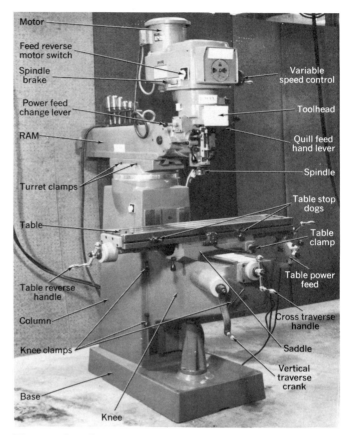

Figure 1. The important parts of a vertical milling machine. (Lane Community College)

is machined to provide a swivel on which the ram can be rotated. The knee supports and guides the saddle. The saddle provides cross travel for the machine and is the support and guide for the table. The table provides lengthwise travel for the machine and supports the workpiece or workholding devices. The ram can be adjusted toward or away from the column to increase the working capacity of the milling machine. The toolhead on this machine is attached to the end of the ram and can be swiveled on some milling machines in one or two planes. These six assemblies are the major components of the ram style vertical milling machine.

Some of these components have controls or parts that are important for you to know. For instance, the toolhead (Figure 2) contains the motor, which powers the spindle. Speed changes are made with a variable speed drive. On variable drives, the spindle has to be revolving while speed changes are made. When changing speeds to the high or low speed range, the spindle must be stopped. The same is true for V-belt or gear-driven speed changes on other milling ma-

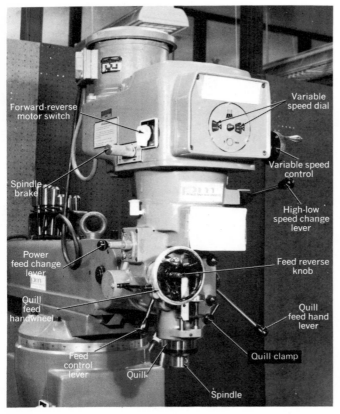

Figure 2. The toolhead. (Lane Community College)

Figure 3. Quill stop. (Lane Community College)

chines. The spindle is contained in a quill. The quill can be extended from and retracted into the toolhead by a quill feed hand lever or hand wheel. The quill feed hand lever is used to rapidly position the quill or for drilling holes. The quill feed hand wheel gives a controlled slow manual feed as needed when boring holes.

Power feed to the quill is obtained by engaging the feed control lever. Different power feeds are available through the power feed change lever. The power feed is automatically disengaged when the quill dog contacts the adjustable micrometer depth stop (Figure 3). When feeding upward, the power feed disengages when the quill reaches its upper limit. The micrometer dial allows depth stop adjustments in .001 in. increments. The quill clamp is used to lock the quill in the head to get maximum rigidity when milling. The spindle brake or spindle lock is needed to keep the spindle from rotating when installing or removing tools from it. The toolhead is swiveled on the ram by loosening the clamping nuts on the toolhead and then turning the swivel adjustment until the desired angle is obtained.

The ram is adjusted toward or away from the column by the ram positioning pinion. The ram also

Figure 4. Clamping devices. (Lane Community College)

can be swiveled on the column after the turret clamps are loosened.

The table can be moved manually with the table traverse hand wheel. Table movement toward or away from the column is accomplished with the cross traverse hand wheel. Raising and lowering the knee is done with the vertical traverse crank. Each of these three axes of travel can be adjusted in .001 in. incre-

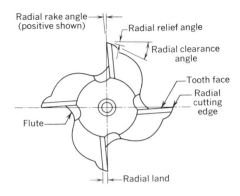

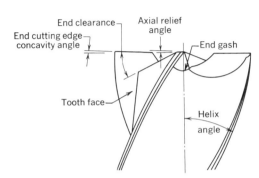

Figure 5. End mill nomenclature. (National Twist Drill & Tool Div., Lear Siegler, Inc.)

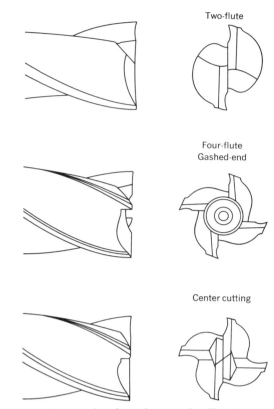

Figure 6. Types of end teeth on end mills. (National Twist Drill & Tool Div., Lear Siegler, Inc.)

Figure 7. Single end helical teeth end mill. (The Weldon Tool Company)

Figure 8. Two flute, double end, helical teeth end mill. (The Weldon Tool Company)

ments with micrometer dials. The table, saddle, and knee can be locked securely in position with the clamping levers shown in Figure 4. During machining, all table axes except the moving one should be locked. This will increase the rigidity of the setup.

Before operating any of the machine controls, the machine should be cleaned and lubricated. Lubrication is needed on all moving parts. Follow the machine manufacturer's recommendation as to the kind of lubricant required.

Cutting Tools for the Vertical Milling Machine
The most frequently used tool on a vertical milling machine is the end mill. End mills are made in either a right-hand or a left-hand cut. Identification is made by viewing the cutter from the cutting end. A right-hand cutter rotates counterclockwise. The helix of the flutes can also be left or right hand; a right-hand helix flute angles downward to the right when viewed from the side. Figure 5 is an illustration of the cutting end of a four flute end mill, which is an example of a right-hand cut, right-hand helix end mill.

The end teeth of an end mill can vary, depending on the cutting to be performed (Figure 6). Two flute end mills are center cutting, which means they can make their own starting hole. This is called plunge cutting. Four flute end mills may have either center cutting teeth or a gashed or center drilled end. End mills with center drilled or gashed ends cannot be used to plunge cut their own starting holes. These end mills only cut with the teeth on their periphery. End mills can be single end (Figure 7) or double end (Figure 8). Double end type end mills are usually more economical because of the savings in tool material in their production.

Figure 9. Straight tooth, single end end mill. (The Weldon Tool Company)

Figure 10. Forty-five degree helix angle aluminum-cutting end mill. (The Weldon Tool Company)

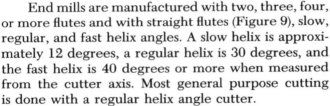

Figure 11. Two flute, carbide tipped end mill. (Brown & Sharpe Mfg. Co.)

Figure 12. Inserted blade end mill. *(Machine Tools and Machining Practices)*

Figure 13. Roughing mill. (Illinois/Eclipse, Division Illinois Tool Works Inc.)

Figure 14. Three flute, tapered end mill. (The Weldon Tool Company)

End mills are manufactured with two, three, four, or more flutes and with straight flutes (Figure 9), slow, regular, and fast helix angles. A slow helix is approximately 12 degrees, a regular helix is 30 degrees, and the fast helix is 40 degrees or more when measured from the cutter axis. Most general purpose cutting is done with a regular helix angle cutter.

Aluminum is efficiently machined with a fast helix end mill and highly polished cutting faces to minimize chip adherence (Figure 10). The end mills illustrated so far are made from high speed steel. High speed steel end mills are available in a great variety of styles, shapes, and sizes as stock items. High speed steel end mills are relatively low in cost when compared with carbide tipped or solid carbide end mills. But to machine highly abrasive or hard and tough materials, or in production milling, carbide tools should be considered.

Carbide tools may have carbide cutting tips brazed to a steel shank (Figure 11). This two flute carbide tipped end mill is designed to cut steel. It has a negative axial rake angle and a slow left-hand helix. Another kind of carbide tipped end mill uses throwaway inserts (Figure 12). Each of the carbide inserts has three cutting edges; when all three cutting edges are dull, the insert is replaced with a new one. No sharpening is required.

When large amounts of material need to be removed, a roughing end mill (Figure 13) should be used. These end mills are also called hogging end mills and have a wavy thread form cut on their periphery. These waves form many individual cutting edges. The tip of each wave contacts the work and produces one short compact chip. Each succeeding wave tip is offset from the next one, which results in a relatively smooth surface finish. During the cutting operation, a number of teeth are in contact with the work. This reduces the possibility of vibration or chatter.

Tapered end mills (Figure 14) are used in mold making, die work, and pattern making, where precise tapered surfaces need to be made. Tapered end mills have included tapers ranging from 1 degree to over 10 degrees.

Ball-end end mills (Figure 15) have two or more flutes and form an inside radius or fillet between surfaces. Ball-end end mills are used in tracer milling and in die sinking operations. Round bottom grooves can also be machined with them. Precise convex radii can be machined on a milling machine with corner rounding end mills (Figure 16). Dovetails are machined with single angle milling cutters (Figure 17). The two commonly available angles are 45 degrees and 60 degrees. T-slots in machine tables and workholding devices are machined with T-slot cutters (Fig-

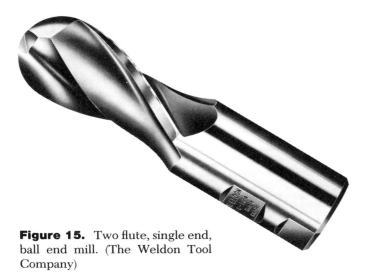

Figure 15. Two flute, single end, ball end mill. (The Weldon Tool Company)

Figure 16. Corner rounding milling cutter. (Illinois/ Eclipse, Division Illinois Tool Works Inc.)

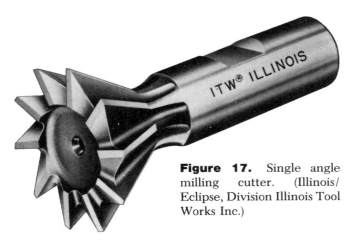

Figure 17. Single angle milling cutter. (Illinois/ Eclipse, Division Illinois Tool Works Inc.)

Figure 18. T-slot milling cutter. (The Weldon Tool Company)

Figure 19. Woodruff keyslot milling cutter. (Illinois/ Eclipse, Division Illinois Tool Works Inc.)

Figure 20. Shell end mill. (Illinois/Eclipse, Division Illinois Tool Works Inc.)

ure 18). T-slot cutters are made in sizes to fit standard T-nuts.

Woodruff key slots are cut into shafts to retain a woodruff key as a driving and connecting member between shafts and pulleys or gears. Woodruff key slot cutters (Figure 19) come in many standard sizes. When larger flat surfaces need to be machined, a shell end mill (Figure 20) can be used. Shell end mills are more economical to produce because a smaller amount of the expensive tool material is needed to make a shell mill then for a solid shank end mill of the same size. Shell mills are also made with carbide inserts (Figure 21). The ease with which new sharp cutting edges can be installed makes this a very practical, efficient cutting tool. Shell end mills are mounted

Figure 21. Shell mill with carbide inserts. (Lane Community College)

Figure 23. Fly cutter. (Lane Community College)

Figure 22. Shell mill arbor. (The Weldon Tool Company)

and driven by shell mill arbors (Figure 22). All the tools and toolholders will perform satisfactorily only if they are cleaned and inspected for nicks and burrs and corrections made before mounting them in the spindle.

The fly cutter (Figure 23) is an inexpensive face-milling tool. High speed or carbide tool bits can be used if they are sharpened to have the correct clearance and rake angles.

All of these cutting tools must be mounted on and driven by the machine spindle. End mills or other straight shank tools can be held in collets. The most rigid type of collet is the solid collet (Figure 24). This collet has been precision ground with a hole that is concentric to the spindle and the exact size of the tool shank. Driving power to the tool is transmitted by one or two set screws engaging in flats on the tool shank.

Another type of frequently used collet is the split collet (Figure 25). The shank of the tool is held by the friction created when the tapered part of the collet is pulled into the taper of the spindle nose. When a heavy side thrust is created by a deep cut, a large

Figure 24. Solid collet. (Lane Community College)

feedrate, or a dull tool, helical flute end mills have a tendency to be pulled out from the collet. With the solid collet, the set screws prevent any end movement of the cutting tool.

To speed up frequent tool changes, a quick-change toolholder (Figure 26) is used. Different tools can be mounted in their own holder, preset to a specific length, and interchanged with a partial turn of a clamping ring.

Figure 25. Split collet. (Lane Community College)

Figure 26. Quick-change adapter and tool holders. (Lane Community College)

SELF-TEST

1. How are chips handled?
2. When is eye protection worn?
3. Why do machine guards need to be in place before operating a machine?
4. Why is a clean work area important?
5. What are danger points when handling tools?
6. How is a vertical milling machine head swiveled safely?
7. How are safe measurements made on a machine?
8. How is a right-hand cut end mill identified?
9. What is characteristic of end mills that can be used for plunge cutting?
10. What is the main difference between a general purpose end mill and one designed to cut aluminum?
11. When are carbide tipped end mills chosen over high speed steel end mills?
12. Name the six major components of a vertical milling machine.

UNIT 4 USING THE VERTICAL MILLING MACHINE

Using the vertical milling machine involves the selection of cutting tools and workholding devices as well as cutting speed, feed, depth of cut, and coolant. Machine adjustments such as the alignment of the head and ram positions may also need to be made. The vertical milling machine is one of the most versatile machines found in a machine shop. This unit will illustrate some of the many possible vertical milling machine operations.

OBJECTIVES

After completing this unit, you should be able to:

1. Select cutting speeds and calculate RPM for end mills.
2. Select and calculate feedrates for end mills.
3. Use end mills to machine grooves and cavities.
4. Align vises and toolheads, locate the edges of workpieces, and find the centers of holes.
5. Identify and select vertical milling machine setups for a variety of different machining operations.

SPEED

One of the major requirements in efficient cutting with end mills and other cutting tools is the intelligent selection of the correct cutting speed. The cutting speed of a cutting tool is influenced by the tool material, condition of the machine, rigidity of the setup, and use of coolant. The most commonly used tool materials are high speed steel and cemented carbide. After the cutting speed has been selected for a job, the cutting operation should be observed carefully so that speed adjustments can be made before a job is ruined. Table 1 gives starting values for some commonly used materials. As a rule, lower speeds are used for hard materials, tough materials, abrasive materials, heavy cuts, minimum tool wear, and maximum tool life. Higher speeds are used when machining softer materials, for better surface finishes, with smaller diameter cutters, for light cuts, for delicate workpieces, and nonrigid setups. When you calculate an RPM to use on a job, use the lower cutting speed value to start. The formula for this is:

$$RPM = \frac{CS \times 4}{D}$$

D is the diameter of the cutting tool in inches.

FEED

Another very important factor in efficient machining is the feed. The feed in milling is calculated by starting with a desired feed per tooth. The feed per tooth determines the chip thickness. The chip thickness affects the tool life of a cutter. Very thin chips dull the cutting edges very rapidly. Very thick chips produce high pressures on the cutting edges and cause a built-up edge (BUE) that produces rough finishes and leads to tool breakage. Commonly used feed per tooth values for end mills are given in Table 2. Usually the feed per tooth is the same for HSS and carbide end mills.

The values in Table 2 are only intended as starting

Table 1
Starting Values for Some Commonly Used Materials

| Work Material | Tool Material | |
	High Speed Steel	Cemented Carbide
Aluminum	300–800	1000–2000
Brass	200–400	500–800
Bronze	65–130	200–400
Cast iron	50–80	250–350
Low carbon steel	60–100	300–600
Medium carbon steel	50–80	225–400
High carbon steel	40–70	150–250
Medium alloy steel	40–70	150–350
Stainless steel	30–80	100–300

Source: Warren T. White, John E. Neely, Richard R. Kibbe, and Roland O. Meyer, *Machine Tools and Machining Practices*, John Wiley and Sons, Inc., Copyright © 1977, New York.

points and may need to be adjusted up or down, depending on the machining conditions of the job at hand. The highest feed per tooth without BUE will usually give the longest tool life between sharpenings. Excessive feeds will cause tool breakage or the chipping of the cutting edges. When feed per tooth for a cutter is selected, the feedrate can be calculated. The feedrate on a milling machine is expressed in inches per minute (IPM) times the number of teeth on the cutter (*n*). The formula for feedrate is:

$$Feedrate = f \times RPM \times n$$

EXAMPLE

Using the values given in Tables 1 and 2, calculate the RPM and feedrate for a $\frac{1}{2}$ in. diameter HSS two flute end mill that is cutting aluminum.

$$RPM = \frac{CS \times 4}{D} = \frac{300 \times 4}{\frac{1}{2}} = \frac{1200}{0.5} = 2400$$

$$Feedrate = f \times RPM \times n = .005 \times 2400 \times 2 = 24 \text{ IPM}$$

Table 2

Feeds for End Mills (Feed per Tooth in Inches)

Cutter Diameter	Aluminum	Brass	Bronze	Cast Iron	Low Carbon Steel	High Carbon Steel	Medium Alloy Steel	Stainless Steel
$\frac{1}{8}$	.002	.001	.0005	.0005	.0005	.0005	.0005	.0005
$\frac{1}{4}$	.002	.002	.001	.001	.001	.001	.0005	.001
$\frac{3}{8}$	.003	.003	.002	.002	.002	.002	.001	.002
$\frac{1}{2}$	.005	.003	.003	.0025	.003	.002	.001	.002
$\frac{3}{4}$	.006	.004	.003	.003	.004	.003	.002	.003
1	.007	.005	.004	.0035	.005	.003	.003	.004
$1\frac{1}{2}$	.008	.005	.005	.004	.006	.004	.003	.004
2	.009	.006	.005	.005	.007	.004	.003	.005

Source: Warren T. White, John E. Neely, Richard R. Kibbe, and Roland O. Meyer, *Machine Tools and Machining Practices*, John Wiley and Sons, Inc., Copyright © 1977, New York.

DEPTH OF CUT

The third factor to be considered in using end mills is the depth of cut. The depth of cut is limited by the amount of material that needs to be removed from the workpiece, by the power available at the machine spindle, and by the rigidity of the workpiece, tool, and setup. As a rule, the depth of cut in mild steel for an end mill should not exceed one-half of the diameter of the tool. But if deeper cuts need to be made, the feedrate needs to be reduced to prevent tool breakage. In softer metals such as aluminum, the depth of cut can be increased considerably, especially if the correct end mill is used for the material being cut. The end mill must be sharp and should run concentric in the end mill holder, and it should be mounted with no more tool overhang than necessary to do the job.

A problem that occasionally arises when using end mills to machine grooves or slots is a slot with nonperpendicular (leaning) sides. The causes for these out-of-square grooves are worn spindles, loose spindle bearings, dull end mills, and excessive feedrates. The leaning slot is produced by an end mill that is deflected by high cutting forces (Figure 1). To correct the problem, reduce the feedrate, use end mills with only a short projection from the spindle, and use end mills with straight or low helix angle flutes.

CUTTING FLUIDS

Cutting fluids should be used when high speed steel cutters are used. The cutting fluid dissipates the heat generated while cutting. It reduces the heat by acting as a lubricant between the tool and chip. Higher cutting speeds can be used with cutting fluids. A stream

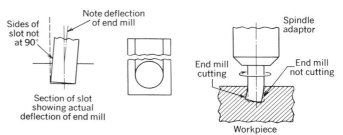

Figure 1. The causes of a leaning slot in end milling. *(Machine Tools and Machining Practices)*

of coolant also washes the chips away. Water base coolants have very good cooling qualities and oil base coolants produce very good surface finishes. Most milling with carbide cutters is done dry unless a large enough flow of coolant at the cutting edge can be maintained to keep the cutting edge from being intermittently heated and cooled, which usually results in thermal cracking and premature tool failure.

Some materials, such as cast iron, brass, and plastics, are commonly machined dry. A stream of compressed air can be used to cool tools and to keep the cutting area clear of chips, but precautions have to be taken to prevent flying chips from injuring anyone.

WORKHOLDING METHODS

The two most commonly used methods of holding workpieces for machining on the vertical milling machine are clamping the work to the machine table and using a machine vise. When a workpiece is fastened to the machine table, it must be aligned with the axis of the table. Milling machine tables are accu-

Figure 2. Work aligned by locating against stops in T-slots. (Lane Community College)

Figure 4. Aligning a workpiece with the aid of a dial indicator. (Lane Community College)

Figure 3. Measuring the distance from the edge of the table to the workpiece. (Lane Community College)

rately machined and the table travels parallel to its outside surfaces and also parallel to its T-slots. Workpieces can be aligned by placing them against stops that fit snugly into the T-slots (Figure 2), or by measuring the distance from the edge of the table to the workpiece in a few places (Figure 3). More accurate alignments can be made when a dial indicator is used to indicate the edge of a workpiece (Figure 4). When a vise is used to hold the workpiece, the solid jaw of the vise should be indicated to assure its alignment

with the axis of the table travel. See Unit 2 in this section, "Using the Horizontal Milling Machine," for explicit details on aligning machine vises.

ALIGNING THE TOOLHEAD

For precise machining operations, the toolhead needs to be aligned squarely to the top surface of the machine table. This is normally done when others have been using the machine before you, or you have a critical machining operation to perform. The following is the recommended procedure for aligning the toolhead.

1. Fasten a dial indicator onto the machine spindle (Figure 5). The dial indicator should sweep a circle slightly smaller than the width of the table.
2. Lower the quill until the indicator contact point is deflected .015 to .020 in. Lock the quill in this position.
3. Tighten the knee clamping bolts. If this is neglected, the knee will sag on the front, causing an erroneous reading.
4. Next, set the indicator bezel to read zero.
5. Loosen the head clamping bolts one at a time and retighten each one to create a slight drag. This slight drag makes fine adjustments easier.
6. Rotate the spindle by hand until the indicator is to the left of the spindle in the center of the table, and note the indicator reading.

Figure 5. Aligning the toolhead square to the table with a dial indicator. (Lane Community College)

Figure 6. Offset edge finder. (Lane Community College)

Figure 7. Work approaches the tip of the offset edge finder. (Lane Community College)

7. Rotate the spindle 180 degrees so that the indicator is to the right of the spindle, and note the indicator reading at that place. Be careful that the indicator contact does not catch and hang up when crossing the T-slots.

8. Turn the head-tilting-screw to correct the error between the right-hand reading and the left-hand reading. Tilt the head so that one-half the difference between these two readings is noted on the indicator.

9. Check and compare the indicator reading at the left side of the table. If both readings are the same, tighten the head clamping bolts. If the readings differ, repeat Step 8.

10. After the head clamping bolts are tight, make another comparison on both sides. Often the tightening of the bolts changes the head location and additional adjustments need to be made.

11. The next step is to align the head crosswise to the table (if the machine also tilts on that axis). The procedure is the same as for the lengthwise alignment. A final check should be made to be sure all clamping bolts are tight.

ALIGNING THE SPINDLE CENTERLINE TO THE WORK

When the workpiece edges are aligned parallel with the table travel and the toolhead is aligned square with the table top, it becomes necessary to locate the spindle centerline with the edges of the workpiece. A commonly used tool to locate edges on a milling machine is an offset edge finder (Figure 6). An offset edge finder consists of a shank and a tip that is held against the shank by an internal spring. The shank is usually $\frac{1}{2}$ in. in diameter and the tip is either .200 or .500 in. in diameter. The edge finder is usually mounted in a collet when in use. With the spindle revolving at 600 to 800 RPM, the tip is moved off center so that it wobbles. An edge or locating point of the workpiece is then moved slowly toward the wobbling tip of the edge finder until it just touches (Figure 7). Continue to advance the work more slowly than before, gradually reducing the eccentric run-out of the tip until it seems to turn without any wobble. Now, the spindle axis is exactly (that is, within one or two ten-thousandths of an inch) one-half of the tip diameter from the edge of the workpiece. If the tip diameter is .200 in., then the centerline of the spindle is .100 in. away from the workpiece edge. Set the micrometer dial of the now adjusted machine axis .100 in. from zero, taking care that the blacklash is removed in the direction you intend to move the work. You should practice the approach to the workpiece a few more times with the edge finder while observing the micrometer dial position until you feel secure in locating an edge with the edge finder. Also

Figure 8. Indicator used to locate the edge of a workpiece. (Lane Community College)

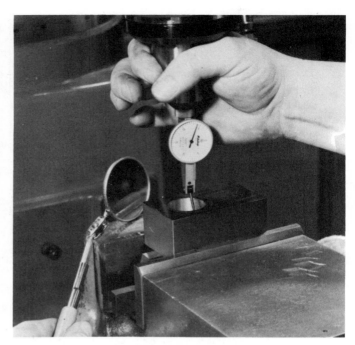

Figure 10. Dial indicator locating the center of a hole. (Lane Community College)

Figure 9. Indicator against parallel to locate the edge of a workpiece. (Lane Community College)

practice the edge finding process for the other machine axis.

If an edge finder is not available in your shop, an edge can be located with the aid of a dial indicator in the following way. The dial indicator is mounted in the spindle. Rotate the spindle by hand and set the indicator contact point as close to the spindle cen-

terline as possible. Lower the spindle so that the indicator contact point touches the workpiece edge and registers a .010 to .020 in. deflection (Figure 8). A slight rotating movement of the spindle forward and backward is used to find the lowest reading on the dial indicator. Set the dial indicator to register zero. Raise the spindle so that the indicator contact point is $\frac{1}{2}$ in. above the workpiece to clear it, and then turn the spindle 180 degrees from the original zeroed position. Hold a precision parallel against the edge of the workpiece so it extends above the workpiece. Lower the spindle until the indicator contact point is against the parallel (Figure 9). Read the indicator value; it is easier if you use a mirror to read the indicator when it faces away from you. Now turn the table hand wheel and move the table to where the indicator pointer is halfway between the present reading against the parallel and the zero on the indicator dial. Now set the dial on zero and check the position of the spindle, as in Figure 8. Continue to make corrections until both readings are the same.

To pick up the center of an existing hole, the indicator is mounted in the spindle and swiveled so that the contact point touches the side of the hole (Figure 10). The spindle should be rough centered, first in one table axis and then in the other. The spindle is then rotated by hand and the table adjustments are made until the same reading is obtained throughout a complete circle.

SETTING UP THE MACHINE

1. Clamp the workpiece securely on the table, vise, or fixture.

2. Mount the cutter in the spindle and set the correct RPM.

3. Turn on the power to the spindle and check its rotation. If it is the wrong direction for the cutter, use the reverse rotation.

4. Loosen the cross slide lock and the knee locking lever.

5. Bring the quill up into its housing and tighten the quill locking clamp. The quill should not be extended from its housing for rough milling work unless necessary, since it is less rigid when extended. Make sure the feed mechanism is disengaged.

6. Position the table so the cutter is directly over the workpiece.

7. Raise the knee slowly by turning the vertical hand feed crank until the cutter just touches the workpiece (in a place where metal will be removed, so a finished surface will not be marred). If you cut below the surface, you have gone too far and should try again in another place.

8. Set the micrometer dial on the knee feed screw to zero.

9. Move the cutter clear of the work (unless you intend to plunge cut) and feed in the depth of cut on the micrometer dial. Remember, this should be no more than one-half the diameter of the end mill in steel.

10. Tighten the knee and cross slide locking levers. Always tighten **all** locks on a machine except the one that is being used in order to have a rigid setup and avoid chatter.

11. Turn on the coolant.

12. Set the table feed mechanism. If it is to be hand fed, observe the chip form, always producing a chip and backing off when no cutting is being done to avoid rubbing, which dulls the cutter. Avoid heavy feeds by hand since the erratic feed-rate of hand feeding can cause momentary feed-rates that are much too high, resulting in tool breakage.

13. When the cut is completed, turn off the coolant and stop the spindle rotation before returning the table to the original position.

14. Brush off the chips and wipe the workpiece clean.

15. Make the necessary measurements before removing the workpiece from the machine. Additional cuts may be required or finish cuts may be needed.

VERTICAL MILLING MACHINE OPERATIONS

Many vertical milling machine operations such as the milling of steps are performed with end mills (Figure 11). Two surfaces can be machined in one setup, both square to each other. The ends of workpieces can be machined square and to a given length by using the peripheral teeth of an end mill (Figure 12). Center cutting end mills make their own starting hole when used to mill a pocket or cavity (Figure 13).

Prior to making any milling cuts, the outline of the cavity should be accurately laid out on the workpiece for a guide or reference line. Only when finish cuts are made should these layout lines disappear. Good milling practice is to rough out the cavity to within .030 in. of finish size before making any finishing cuts.

When you are milling a cavity, the direction of the feed should be against the rotation of the cutter (Figure 14a). This assures positive control over the distance the cutter travels and prevents the workpiece from being pulled into the cutter because of backlash. When you reverse the direction of table travel, you will have to compensate for the blacklash in the table feed mechanism. When conventional milling in a pocket, as in Figure 14a, an undercut in the

Figure 11. Using an end mill to mill steps. (Lane Community College)

Figure 12. Using an end mill to square stock. (Lane Community College)

Figure 13. Using an end mill to machine a pocket. (Lane Community College)

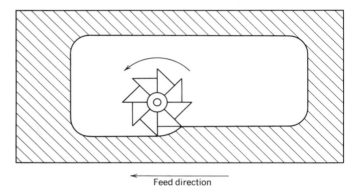

Figure 14a. Feed direction is against cutter rotation when conventional milling procedures are used. (*Machine Tools and Machining Practices*)

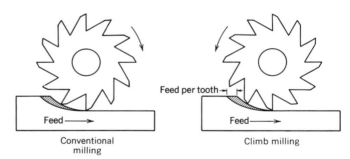

Figure 14b. Conventional and climb milling. (*Machine Tools and Machining Practices*)

corner is often made, especially when making heavy cuts. One way to overcome this problem is to take a light finishing cut by climb milling. However, the machine you are using must have an antibacklash device on the feed screws in order to use climb milling techniques. Climb milling without an antibacklash device will usually break cutters and damage the workpiece. Conventional and climb milling are illustrated in Figure 14b. During any milling operation, all table movements should be locked except the one that is moving in order to obtain the most rigid setup possible. Spiral (helical) fluted end mills may work their way out of a split collet when deep, heavy cuts are made or when the end mill gets dull. As a precaution,

to warn you that this is happening, you can make a mark with a felt tip pen on the revolving end mill shank where it meets the collet face. Observing this mark during the cut will give you an early indication if the end mill is changing its position in the collet.

One often performed operation with end mills is the cutting of keyways in shafts. It is very important that such a keyway be centrally located in the shaft. A very accurate method of doing this is by positioning the cutter with the help of the machine dials. After

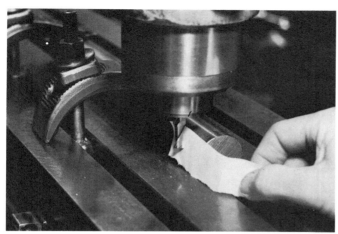

Figure 15. Setting an end mill to the side of a shaft with the aid of a paper feeler. (Lane Community College)

Figure 16a. Cutter centered on work and lowered to make a circular mark. The cutter was raised in order to show the spot. (Lane Community College)

Figure 16b. After plunging the cutter to the correct depth, table feed is engaged and the keyseat is cut. (Lane Community College)

clamping the shaft in a vise, or possibly to the table, the quill is lowered so that the cutter is along side the shaft but not touching it. Then, with the spindle motor off and the spindle rotated by hand, the table is moved until a paper feeler strip is pulled from your hand (Figure 15). At this point, the cutter is approximately .002 in. from the shaft. Zero the cross slide micrometer dial compensating for the .002 in. Raise the quill so that the cutter clears the workpiece. Move the cross slide a distance equal to half the shaft diameter plus half the cutter diameter. This will locate the cutter centrally over the shaft. Lock the cross slide table at this position. Raise the quill to its top position and lock it there. Move the table lengthwise to position the cutter where the keyseat is to begin. Start the spindle motor and raise the knee until the cutter makes a circular mark equal to the cutter diameter (Figure 16a). Zero the vertical travel micrometer dial. Turn on the coolant. Slowly move the knee upward a distance equal to half the cutter diameter plus an additional .005 in. Lock the knee. Cut the keyseat the required length (Figure 16b).

To machine a T-slot or a dovetail into a workpiece, two operations are performed. First a slot is cut with a regular end mill, and then a T-slot cutter or a single angle milling cutter is used to finish the contour (Figures 17 and 18). Angular cuts on workpieces can be made by tilting the workpiece in a vise with the aid of a protractor (Figure 19) and its built-in spirit level. The angle cut in Figure 19 can be made with an end mill as shown in Figure 20. The workhead may also be tilted to produce an angle as shown in

Figure 17. Milling a slot and then the dovetail. (Lane Community College)

Figure 18. First a slot is milled and then the T-slot cutter makes the T-slot. (Lane Community College)

Figure 19. Setting up a workpiece for an angular cut with a protractor. (Lane Community College)

Figure 20. Machining an angle with an end mill. (Lane Community College)

Figure 21. Cutting an angle by tilting the workhead and using the end teeth of an end mill. (Lane Community College)

Figure 22. Cutting an angle by tilting the workhead and using the peripheral teeth of an end mill. (Lane Community College)

Figure 21, but the length of cut is limited to the diameter of the cutter in this setup. To finish this angle, the tool must be raised, the table moved, and the cutter lowered to the line. However, in the tilted-head setup in Figure 22, an angle can be milled equal to the length of flutes on the cutter. A quite practical method of milling an angle can be seen in Figure 23, using a shell mill and making a series of cuts along the work to depth.

Accurate holes can be drilled at any angle the head can be swiveled to. These holes can be drilled

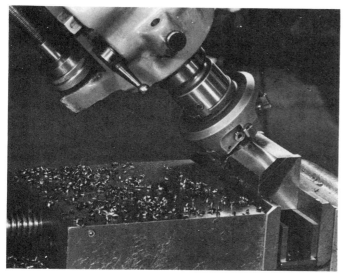

Figure 23. Using a shell mill to machine an angle. (Lane Community College)

Figure 25. Tapping in a vertical milling machine. (Cincinnati Milacron)

Figure 24. Drilling of accurately located holes. (Cincinnati Milacron)

Figure 26. Boring with an offset boring head. (Cincinnati Milacron)

by using the sensitive quill feed lever or the power feed mechanism (Figure 24) or, in the case of vertical holes, the knee can be raised.

Holes can be machine tapped by using the sensitive quill feed lever and the instant spindle reversal knob (Figure 25). When an offset boring head is mounted in the spindle, precisely located and accurately dimensioned holes can be bored (Figure 26). Circular slots can be milled when a rotary table is

Figure 27. Using a rotary table to mill a circular slot. (Cincinnati Milacron)

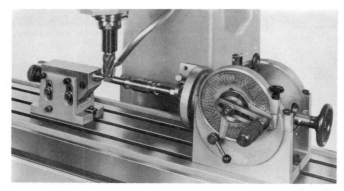

Figure 28. A dividing head in use. (Cincinnati Milacron)

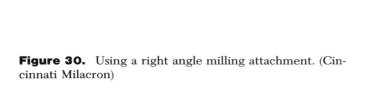

Figure 30. Using a right angle milling attachment. (Cincinnati Milacron)

Figure 29. A shaping head used to cut a square corner hole. (Lane Community College)

used (Figure 27). Precise indexing can be performed when a dividing head is mounted on the milling machine. Figure 28 shows a square being milled on the end of a shaft. On many vertical milling machines a shaping attachment is mounted on the rear of the ram. This shaping attachment can be brought over the machine table by swiveling the ram 180 degrees.

Shaping attachments are used to machine irregular shapes on or in workpieces such as the milled cavity being shaped to a square hole, as shown in Figure 29. When a right angle milling attachment (Figure 30) is mounted on the spindle, it is possible to machine hard-to-get-at cavities that are often at difficult angles on workpieces.

SELF-TEST

1. Name four factors that influence cutting speed of milling cutters.
2. How is tool breakage caused by thick chips and high tool pressures?
3. The depth of cut can be varied according to the situation. As a rule of thumb, what should the depth be in mild steel for end milling?
4. With what tool material should coolant always be used with a few exceptions, such as cast iron or brass?
5. When setting up on a machine in which the previous operator used a vise and left it in place, should you assume the vise and toolhead are in perfect alignment?
6. When accurately setting up work in a vertical milling machine, the spindle centerline must usually be first aligned to the edge of the workpiece. What instrument is used for this purpose?
7. How can a cutter be aligned with the edge of a workpiece without any precision tools?
8. When milling a cavity or drilling precisely located holes where the direction of travel is reversed, what machine characteristic can cause considerable error and how can this problem be corrected?
9. How can angular cuts be taken on a vertical mill?
10. Name four attachments that can be used on vertical milling.

SECTION K

SURFACE GRINDERS

UNIT 1   SELECTION AND IDENTIFICATION OF GRINDING WHEELS
UNIT 2   GRINDING WHEEL SAFETY
UNIT 3   USING THE SURFACE GRINDER
UNIT 4   PROBLEMS AND SOLUTIONS IN SURFACE GRINDING

UNIT 1 SELECTION AND IDENTIFICATION OF GRINDING WHEELS

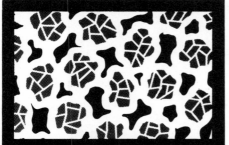

A grinding wheel is both a many-toothed cutting tool and a toolholder. Selecting a grinding wheel is somewhat more complicated than selecting a lathe tool or a milling cutter because there are more factors to be considered. These factors, including size, shape, and composition of the wheel, are expressed in a system of symbols (numbers and letters) that you must be able to interpret and apply. Whether you are selecting a wheel for a surface grinder or a pedestal grinder, the selection process is the same if you expect to get the right wheel for the job.

OBJECTIVES

After completing this unit, you should be able to:
1. List the five principal abrasives with their general areas of best use.
2. List the four principal bonds with the types of applications where they are most used.
3. Identify by type number and name, from unmarked sketches or from actual wheels, the four most commonly used shapes of grinding wheels.
4. Interpret wheel shape and size markings together with the five basic symbols of a wheel specification into a description of the grinding wheel.
5. Given several standard, common grinding jobs, recommend the kind of abrasive, approximate grit size and grade, and bond.

IDENTIFYING FACTORS

Six or seven factors must be considered in selecting a grinding wheel for a particular job. All these factors—wheel shape, wheel size, kind of abrasive, abrasive grain size, hardness, grain spacing, and bond—are expressed in symbols consisting of numbers and letters, most of which are easy to understand and interpret. For example, the size and shape of the wheel are usually determined by the type and size of the grinder. The abrasive, of which there are five major kinds, is determined primarily by the material being ground. The abrasives are not equally efficient on all materials. The size of the abrasive particles selected (the largest are perhaps $\frac{1}{8}$ to $\frac{1}{4}$ in. long and the smallest less than .001 in. long) depends on the amount of stock to be removed and the finish desired.

Selecting the grade or hardness of the wheel, which is a measure of the force required to pull out the abrasive grains, is typically a choice between one of two or, at most, three grades. Grain spacing or structure is most often standard, according to the grain or grit size and the grade of the wheel. Finally, there is the bond that holds the wheel together. Although there are four common bonds, the choice is usually made clear from the job.

Color can also be useful for identification, and it is used to some extent. One of the best abrasives for grinding tools is white when it is manufactured, and a more commonly used abrasive is gray or brown in color. Some wheels are green, pink, or black. Among a given number of grinding wheels, there are wheels of different diameters, thicknesses, or hole

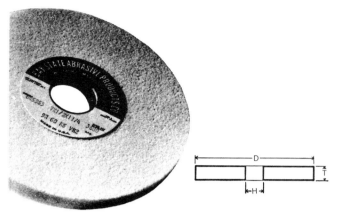

Figure 1. Straight or Type 1 wheel, whose grinding face is the periphery. This wheel usually comes with the grinding face at right angles to the sides, in what is sometimes called an "A" face. (Bay State Abrasives, Dresser Industries, Inc.)

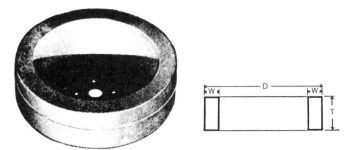

Figure 2. Cylinder or Type 2 wheel, whose grinding face is the rim or wall end of the wheel. Has three dimensions: diameter, thickness, and wall thickness. (Bay State Abrasives, Dresser Industries, Inc.)

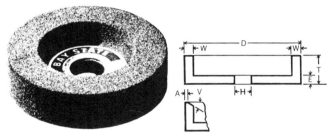

Figure 3. Straight cup or Type 6 wheel, whose grinding face is the flat rim or wall end of the cup. (Bay State Abrasives, Dresser Industries, Inc.)

sizes. Different wheels are made of different sizes of abrasive grain, or with different proportions of grain and the bond that holds the grain together in the wheel.

A code of numbers and letters has been developed by the grinding wheel makers that provides the needed information about a given wheel in just a few letters and numbers. Within a given group of symbols, the order of listing is important.

Grinding wheels are designed for grinding either on the periphery (outside diameter), which is a curved surface, or on the flat side, but rarely on both. It is not a safe practice to grind on the side of a wheel designed for peripheral grinding. However, there are some exceptions to this rule. The shape of the wheel determines the type of grinding performed.

WHEEL SHAPES

The shapes of grinding wheels are designated according to a system published in an American National Standard, *Specifications for Shapes and Sizes of Grinding Wheels*, whose number is ANSI B74.2-1974. The various shapes have been given numbers ranging from 1 to 28, but only five are important for you now. These are described below.

Type 1 (Figure 1) is a peripheral grinding wheel, a straight wheel with three dimensions: diameter, thickness, and hole, in that order. A typical wheel for cylindrical grinding is 20 in. (diameter) × 3 in. (thickness) × 5 in. (hole). Probably most wheels are of this type.

A *Type 2* or cylinder wheel (Figure 2) is a side

grinding wheel, to be mounted for grinding on the side instead of on the periphery. This also has three dimensions; for example, 14 in. (diameter) × 5 in. (thickness) × 1½ in. (wall). This, of course, might also be called a 14 × 5 × 11 (14 in. D minus 2 times 1½—the two wall thicknesses), but the wall thickness is more important than the hole size. Hence the change.

A *Type 6* or straight cup wheel (Figure 3) is a side grinding wheel with one side flat and the opposite side deeply recessed. It has four essential dimensions: the diameter, thickness, hole size (for mounting), and wall.

A *Type 11* or flaring cup wheel (Figure 4) is a side grinding wheel that resembles a Type 6, except that the walls flare out from the back to the diameter and are thinner at the grinding face than at the back. This introduces a couple of new dimensions: the diameter at the back, called the "J" dimension, and the recess diameter at the back, the "K" dimension. This is mentioned only to emphasize that the "D" dimension, the diameter, is always the largest diameter of any wheel.

The *Type 12* dish wheel (Figure 5a) is essentially a very shallow Type 11 wheel, mostly for side grinding. The big difference is the dish wheel has a sec-

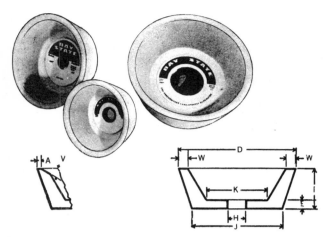

Figure 4. Flaring cup or Type 11 wheel, whose grinding face is also the flat rim or wall of the cup. Note that the wall of the cup is tapered. (Bay State Abrasives, Dresser Industries, Inc.)

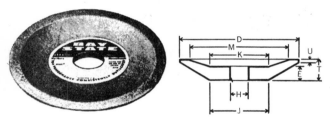

Figure 5a. Dish or Type 12 wheel, similar to Type 11, but a narrow, straight peripheral grinding face in addition to the wall grinding face. Only wheel of those shown that is considered safe for both peripheral and wall or rim grinding. (Bay State Abrasives, Dresser Industries, Inc.)

Figure 5b. An assortment of mounted wheels most often used for deburring and other odd jobs. (Bay State Abrasives, Dresser Industries, Inc.)

ondary grinding face on the periphery, the "U" dimension, so that it is an exception to the rule of grinding **only** on the side or the periphery.

This factor of grinding on the side or the periphery of a wheel is important because it affects the grade of the wheel to be chosen. The larger the area, the softer the wheel should be. In peripheral grinding the contact is always between the arc of the wheel and either a flat (in surface grinding) or another arc (in cylindrical grinding). This makes for small areas of contact and somewhat harder wheels. On the other hand, if the flat side of the wheel is grinding the flat surface of a workpiece, then the contact area is larger and the wheel can be still softer.

It is important to understand that any grinding machine grinds either a flat surface or a round or cylindrical surface. The first group of grinders is collectively called surface grinders; the second group is called cylindrical grinders, whether the workpiece is held between centers or not, and whether the grinding is external or internal. Various forms can be cut into the grinding faces of peripheral grinding wheels, and these can then be ground into either flat or cylindrical surfaces.

Mounted wheels like the ones in Figure 5b can be used in a variety of ways around a shop. Often they are used in portable grinders for jobs like deburring or breaking the edges of workpieces where the tolerances are not too critical. They are also used in internal grinding.

STANDARD MARKING SYSTEM

The description of a grinding wheel's composition is contained in a group of symbols known as the *standard marking system*. That is, the basic symbols for the various elements are standard, but they are usually amplified by individual manufacturer's symbols, so that it does not follow that two wheels with the same basic markings, but made by two different suppliers, would act the same. However, it is a useful tool for anyone concerned with grinding wheels.

There are five basic symbols. The first is a letter indicating the kind of abrasive in the wheel, called the **abrasive type.** The second is a number to indicate the approximate size of the abrasive; this is commonly called **grit size.** In the third position, a letter symbol indicates the **grade** or relative hardness of the wheel. The fourth, **structure,** is a number describing the spacing between abrasive grains. The fifth is a letter indicating the **bond,** the material that holds the grains together as a wheel. Thus, a basic toolroom wheel specification (Figure 6) might be

A60-J8V

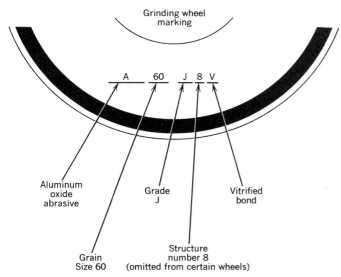

Figure 6. Sketch to illustrate the wheel specification that will be used as an illustration in following pages.

But, since most wheel makers have, for example, a number of different aluminum oxide or other types of abrasives and a number of different vitrified or other bonds, the symbol sometimes appears cluttered:

9A80-K7V22

This means that the wheel maker is using a particular kind of aluminum oxide (A) abrasive, which is indicated by the "9," and a particular vitrified (V) bond, which is indicated by the "22."

The wheel markings for diamond or cubic boron nitride wheels are a little different and are not standard enough for a simple explanation.

FIRST SYMBOL—
TYPE OF ABRASIVE (A60-J8V)

The symbol in the first position, as suggested above, denotes the type of abrasive. Basically, there are five:

A Aluminum oxide.

C Silicon carbide.

D Natural diamond.

MD or SD Manufactured diamond (sometimes called synthetic diamond).

B Cubic boron nitride.

The first two are "cents per pound" abrasives, cheap enough so that it is practical to make whole wheels of the abrasive. Both are made in electric furnaces that hold literally tons of material, and both are crushed and graded by size for grinding wheels and other uses.

Diamond, both natural and manufactured, and cubic boron nitride are expensive enough that most wheels are made of a layer of abrasive around a core of other material. However, they have made a place for themselves because they will grind materials that no other abrasive will touch and because they stay sharp and last so long that they are actually less expensive per piece of parts ground. Natural or mined diamond that is definitely diamond, but is of less than gem quality, is crushed and sized. Both manufactured diamond and cubic boron nitride are made by a combination of high heat (in the range of 3000°F) and tremendous pressure (1 million or more PSI). This heat, incidentally, is somewhat less than that for other manufactured abrasives, which require something in the range of 4000°F for fusing or crystallization.

Each of these abrasives generally has an area in which it excels, but there is no one abrasive that is first choice for all applications. Aluminum oxide is best for grinding most steels, but on the very hard tool steel alloys it is outclassed by cubic boron nitride. However, aluminum oxide (Al_2O_3) may have close to 75 percent of the market, because it is used in foundries for grinding castings and steel mills for billet grinding, and in other high volume applications. Silicon carbide, which does poorly on most steels, is excellent for grinding nonferrous metals and nonmetallic materials. Diamond abrasive excels on cemented carbides, although green silicon carbide is occasionally used. Green silicon carbide was the recommended abrasive for cemented carbides until the introduction of diamonds. Cubic boron nitride (CBN), finally, is superior for grinding high speed steels. CBN is a hard, sharp, cool cutting, long wearing abrasive.

There is a lot still unknown about why certain abrasives grind well on some materials and not on others, and this makes for some interesting speculation. All abrasives are harder than the materials ground. But their relative hardness is apparently only one of the factors in effectiveness. Aluminum oxide is three to perhaps five times harder than most steels, but it grinds them easily. Silicon carbide is harder than aluminum oxide, but not at all effective on steel; on the other hand, it grinds glass and other nonmetallics that are as hard or harder than many steels. Diamond, which is much harder than cubic boron nitride and many times harder than the hardest steel, does not grind steels well, either. The best theory is that there are chemical reactions between certain abrasives and certain materials that make some abrasives ineffective on some metals and other materials.

The basic symbol in the first position, A, C, D,

or B, is often preceded and sometimes followed by a manufacturer's symbol that indicates which abrasive within the group is meant, for example, 9A, 38A, AA.

SECOND SYMBOL—GRIT SIZE (A60-J8V)

The symbol in the second position of the standard marking represents the size of the abrasive grain, usually called grit size. This is a number ranging from 4 to 8 on the coarse side to 500 or higher on the fine side. The number is derived from the approximate number of openings per linear inch in the final screen used to size the grain; the larger the number, the smaller the abrasive grain. Any standard grit size contains grain of smaller and larger sizes, whose amounts are strictly regulated by the federal government, because it would be very expensive to reduce the mix to just the size indicated.

While there is no real agreement as to what is coarse or fine, for general purposes anything from 46 to 100 might be considered medium, with everything 36 and lower considered as coarse and anything 120 and higher considered as fine. Selection of grit in any shop depends on the kind of work it is doing. Thus, where the job is to remove as much metal as fast as possible, 46 or 60 grit size would be considered very fine. On the other hand, in a shop specializing in fine finishes and close tolerances, 240 might be considered coarse.

A final point is in order about grit size. Most standard symbols end in a "0," particularly the three digit sizes 100 and finer. However, every abrasive grain manufacturer also makes some combinations that are not standard for special uses, and these usually end in a "1" or other low digit. Thus, a 240 grit is the finest that is sized by screening. Finer grits are sized by other means. On the other hand, 241 grit is a coarse 24 grit in a "1" combination.

Coarse grain is used for fast stock removal and for soft, ductile materials. Fine grain is used to obtain good finishes and for hard or brittle materials. Some materials are hard enough that fine grain removes as much stock as coarse, and neither removes very much. In a general machine shop or toolroom, most of the wheels used will be between 46 and 100 grit.

THIRD POSITION—GRADE OR "HARDNESS" (A60-J8V)

In the third position is a letter of the alphabet called grade. The later the letter, the "harder" the grade. Thus, a wheel graded F or G would be considered "soft," and one graded R to Z would be very "hard." Actually, these descriptions are put into quotation

Weak holding power

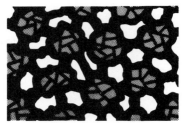
Medium holding power

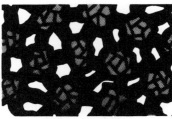
Strong holding power

Figure 7. Three sketches illustrating (from top down) a soft, medium, and hard wheel. This is the "grade" of the wheel. The white areas are voids with nothing but air, the black lines are the bond, and the others are the abrasive grain. The harder the wheel, the greater the proportion of bond and, usually, the smaller the voids. (Bay State Abrasives, Dresser Industries, Inc.)

marks because the words are the best we have, but are still not quite accurate. What is being measured is the hold that the bond has on the abrasive grain (Figure 7), and the greater the proportion of bond to grain, the stronger the hold and the harder the wheel. Precision grinding wheels tend to be on the soft side, because it is necessary to have the grains pull out as they become dull; otherwise, the wheel glazes and its grinding face becomes shiny, but the abrasive is dull. On high speed, high pressure applications like foundry snagging, the pressures to pull out the grains are much greater, and a harder wheel is needed to hold in the grains until they have lost their sharpness. Ideally, a wheel should be self-sharpening. The bond should hold each grain only long enough for it (the grain) to become dull. In practice, this is difficult to achieve.

One thing, however, should be mentioned. Grade is a much less measurable thing than type or size of abrasive or, as you will see later, than bond. Grade depends on the formula for the mix used in the wheels, but it must be checked after the wheel is finished.

FOURTH POSITION—STRUCTURE (A60-J8V)

Following the grade letter, in the fourth position of the symbol, is a number from 1 (dense) to 15 (open); this number describes the spacing of the abrasive grain in the wheel (Figure 8). The use of structure is to provide chip clearance, so that the chips of ground material have some place to go and will be flung out of the wheel by centrifugal force or washed out by the coolant. If the chips remain in the wheel, then the wheel becomes loaded (Figure 9), stops cutting and starts rubbing; it then has to be resharpened, or dressed, which is the trade term.

Structure is also a result of grain size and proportion of bond, similar to grade. Quite often, large grain size wheels tend to have open structure, while smaller sized abrasive grain is often associated with dense structure. On the open side, 11 to 12 and up, the openness is aided by the inclusion of something in the mix like ground up walnut shells, which will burn out as the wheel is fired and leave definite open spacing.

However, for many grit size and grade combinations, a "best" or standard structure has been worked out through experience and research, and so the structure number may be omitted.

FIFTH POSITION—BOND (A60-J8V)

The fifth position of the wheel marking is a letter indicating the bond used in the wheel. This is always a letter, as follows: vitrified—V; resinoid—B (originally the bakelite bond); rubber—R (rubber was used well before resinoid); and shellac—E (originally the "elastic" bond, and also preceded by the now obsolete silicate bond). These are really general bond groups; each wheel maker uses extra symbols to indicate, for instance, which vitrified bond he has used in a particular wheel, and there is no standardization in these extra symbols.

The bond used has important influence on both the manufacturing process and on the final use of the wheel. Vitrified-bonded wheels are fired at temperatures between 2000 and 2500°F; for that reason, no steel inserts can be used. If such inserts are needed, they must be cemented in afterward. The others, also grouped together as organic bonds, are all baked at

Dense spacing

Medium spacing

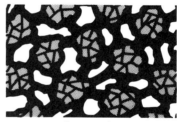
Open spacing

Figure 8. Three similar sketches showing structure. From the top down, dense, medium, and open structure or grain spacing. The proportions of bond, grain, and voids in all three sketches are about the same. (Bay State Abrasives, Dresser Industries, Inc.)

around 400°F, and inserts may be molded in without problems.

Vitrified, resinoid, and shellac wheels are all pressed in molds after mixing. Rubber bonded wheels, on the other hand, are mixed in a process similar to that of making dough for cookies; they are then reduced to thickness by passing the mass or grain-impregnated rubber between precisely spaced rolls. For that reason, it is possible to make much thinner wheels with rubber bond than can be made with any other bond. Thin grinding wheels with either resinoid or rubber bonds can be used in cutoff operations such as abrasive saws, but the rubber wheels can be made thinner than the resinoid bonded wheels.

In general, vitrified wheels are used for precision grinding; resinoid wheels are used for rough grinding with high wheel speeds and heavy stock removal; and rubber or shellac wheels, are used for more specialized applications. The first two bonds monopolize over

Figure 9. The wheel at the left is called "loaded" with small bits of metal imbedded in its grinding face. It is probably too dense in structure or perhaps has too fine an abrasive grain. Same wheel at the right has been dressed to remove all the loading. (Desmond-Stephan Mfg. Co.)

90 percent of the market.

Bond also determines maximum safe wheel speed. Vitrified bonded wheels, with a few minor exceptions, are limited to about 6500 SFPM or a little over a mile a minute. The others run much faster, with some resinoid wheels getting up to 16,000 SFPM or more. Of course, all grinding wheels must be properly guarded to protect the operator in the unlikely case of wheel breakage. Grinding machines in general, however, are safe machines. Many will not operate unless the guards are properly in place and closed.

FACTORS IN GRINDING WHEEL SELECTION

It is generally considered that there are seven or eight factors to consider in the choice of a grinding wheel. Some people group together the elements of the amount of stock removal and of the finish required. Dividing them seems to be more logical.

These factors have also been divided into two

groups—three concerned with the workpiece, which thus change frequently, and five concerned with the grinder, which are more constant.

Variable Factors

In the first group are things such as composition, hardness of the material being ground, amount of stock removal, and finish.

The composition of the material generally determines the abrasive to be used. For steel and most steel alloys, use aluminum oxide. For the very hard, high speed steels, use cubic boron nitride. For cemented carbides, use diamond. Whether this is natural or manufactured is too specialized a question to be discussed here, but the trend appears to be toward the use of manufactured diamond. For cast iron, nonferrous metals, and nonmetallics, use silicon carbide. On some steels, aluminum oxide may be used for roughing and silicon carbide for finishing. Between aluminum oxide and cubic boron nitride, because of the greatly increased cost, the latter is generally used only when it is superior by a wide margin, or if a large percentage of work done on the machine is in the very hard steel alloys like T15.

Material hardness is of concern in grit size and grade. Generally, for soft, ductile materials, the grit is coarser and the grade is harder; for hard materials, finer grit and softer grades are the rule. Of course, it is understood that for most machine shop grinding these "coarse" grits are mostly in the range of 36 to 60, and the finer grits are perhaps in the range of 80 to 120. Likewise, the "soft" wheels are probably something in the range of F, G, H, and perhaps I, and the "hard" wheels are in the range of J, K, or L, and maybe M or N. Too coarse a grit might leave scratches that would be difficult to remove later. And sometimes on very hard materials, coarse grit removes no more stock than fine grit, so you use fine grit. Too soft a wheel will wear too fast to be practical and economical. Too hard a wheel will glaze and not cut.

If **stock removal** is the only objective, then you can use a very coarse (30 and coarser) resinoid bonded wheel. However, in machine shop grinding, you are probably in the 36 to 60 grit sizes mentioned above, and definitely in vitrified bonds.

For **finishing** on production jobs, the wheel may be rubber, shellac, or resinoid bonded. But resinoid bonds are often softened by coolants and therefore rarely used. And for finishing, fine grit sizes are usually preferred; however, as you will learn later when you study wheel dressing (sharpening or renewing the grinding face of a wheel), it is possible to dress a wheel so that a comparatively coarse grit like 54 or 64 will finish a surface as smoothly as 100 pr 120 grit.

Fixed Factors

For any given machine the following five factors are likely to remain constant.

Horsepower of the machine, of course, is a fixed consideration. Grinding wheel manufacturers and grinding machine builders are constantly pressing for higher horsepower, because that gives the machine and the wheels the capacity to do more work, but that is a factor, usually, only in the original purchase of the machine. In wheel selection, this affects only grade. The general rule is the higher the horsepower, the harder the wheel that can be used.

The severity of the grinding also remains pretty constant on any given machine. This affects the choice of a particular kind of abrasive within a general group. Thus, you would probably use a regular or intermediate aluminum oxide on most jobs. But, in the toolroom, where pressures are low, you would probably want an easily fractured abrasive, white aluminum oxide.

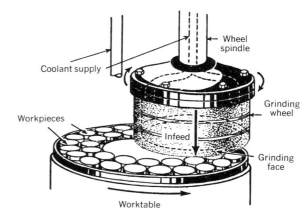

Figure 10. With the flat wall or rim of the wheel grinding a flat surface, as shown here, the wheel must be soft in grade and can have somewhat coarser abrasive grain. The area of contact between wheel and work is large. (Bay State Abrasives, Dresser Industries, Inc.)

On the other end of the scale, for very severe operations like foundry snagging, you need the toughest abrasive you can get, probably an alloy of aluminum oxide and zirconium oxide. For most machine shop grinding, you will probably look first at white wheels.

The **area of grind contact** is also important, but, again, it remains constant for a given machine. The rule is finer grit sizes and harder wheels for small areas of contact; and coarser grit sizes and softer wheels for larger areas of contact. All of this, of course, is within the grit size range of about 36 to 120, or 150 or 180, and within a grade range from E or F to L or N.

It is easy to understand that on a side grinding wheel, where a flat abrasive surface is grinding a flat surface, the contact area is large and the wheels are fairly coarse grained and soft (Figure 10). However, on peripheral grinding wheels, it is a different story. The smallest area is in ball grinding, where the contact area is a point, the point where the arc of the grinding wheel meets the sphere or ball. Thus ball grinding wheels are very hard and very fine grained, for instance, 400 grit, Z grade. In cylindrical grinding the contact area is the line across the thickness of the wheel, usually where the arc of the wheel meets the arc of the workpiece (Figure 11). Here grit sizes of 54, 60, or 80, with grades of K, L, or M, are common. A still larger area is in surface grinding with peripheral grinding where the line of contact is slightly wider because the wheel is cutting into a flat surface (Figure 12). And a combination like 46 I or 46 J is not unusual. An internal grinding wheel where the OD of the wheel grinds the ID of the workpiece may have just

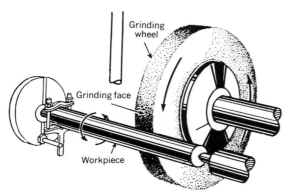

Figure 11. In center-type cylindrical grinding, as shown here, the arc of the grinding face meets the arc of the cylindrical workpiece, making the area of contact a line. This requires a "harder" wheel than in Figure 10. (Bay State Abrasives, Dresser Industries, Inc.)

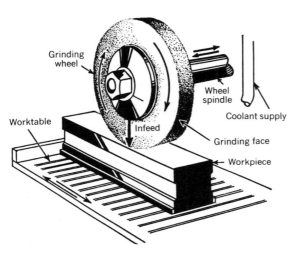

Figure 12. The contact area between the arc of the grinding face and the flat surface of the workpiece in surface grinding makes a somewhat wider line of contact than in cylindrical grinding. (Bay State Abrasives, Dresser Industries, Inc.)

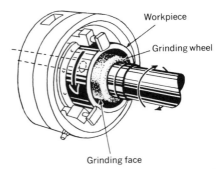

Figure 13. The contact area of the OD grinding face of the wheel and the ID surface of the workpiece creates a still larger area of contact and requires a somewhat "softer" grinding wheel than the two previous examples. (Bay State Abrasives, Dresser Industries, Inc.)

Figure 14. The blotter on the wheel, besides serving as a buffer between the flange and the rough abrasive wheel, provides information as to the dimensions and the composition of the wheel, plus its safe speed in RPMs. This wheel is 7 in. in diameter x ¼ in. thick x 1¼ in. hole. It is a white aluminum oxide wheel, 100 grit, I grade, 8 structure, vitrified 52 bond. It can be run safely at up to 3600 RPMs. (Bay State Abrasives, Dresser Industries, Inc.)

a shade more area of grinding contact (Figure 13). And then when you get to side grinding wheels (cylinder Type 2), cup wheels, and segmental wheels, which are flats grinding flats, you get grit sizes and grades like 30 J or 46 J (see Figure 10). Of course, you have to realize that there may be other factors important enough to override contact area. For example, for grinding copper, you might use a grit size and grade like 14 J, in which the softness of the metal is probably the key factor.

Wheel speed is a factor that can be dealt with quickly. You must always stay within the safe speeds,

which are shown on the blotter or label on every wheel of any size (Figure 14). Vitrified wheels generally have a maximum safe speed of 6500 SFPM or a little more; organic wheels (resinoid, rubber, or shellac) go up to 16,000 SFPM or sometimes higher, but these speeds are generally set by the machine designer, and they are safe speeds for the recommended wheels.

Wet or dry grinding is a factor only in that using a coolant will usually permit the use of about one grade harder wheel than would be used for dry grinding, without as much concern about burning the workpieces. Burning is a discoloration of the workpiece surface caused by overheating. The most common cause is usually the use of a wheel that is too hard.

In any shop, however, unless you are really starting from scratch, there will be some information on what wheels have been used and how they have worked. If a factor seems to need changing, it will probably be grit size or grade. You must remember that the shop probably handles a range of work, and that it does not pay to switch wheels all the time. But change only one element, either grade or grit size, at a time.

SELF-TEST

1. In the course of a week's grinding you might come up with some of each of the following to grind: bronze valve bodies, steel fittings, tungsten carbide tool inserts, and high speed steel tools. If you could pick the ideal abrasive for each metal, what would you use? List four abrasives. If you were limited to three, which one of the four could be eliminated most easily?

2. Straight (Type 1) and cylinder (Type 2) wheels both have three dimensions: diameter, thickness, and a third. What is the third dimension for each and why is it stated that way?

3. Five shapes of grinding wheels are described in this unit. Four are for side grinding and two are for peripheral grinding. List the wheels in the two groups either by name or shape number.

4. Tungsten used in the points of automobile engines is very expensive, which makes it necessary to use the thinnest abrasive cutting wheels possible, $6 \times .008 \times 1$ in. What bond would be used and why?

5. Area of contact between wheel and workpiece is probably the most important factor in picking a wheel grade. Five different sets of grinding conditions are discussed, ranging from flat surfacing with a cylinder wheel to ball grinding. List the five in order by wheel grade, starting with the hardest.

6. Here are two wheel specifications, both for straight (Type 1) wheels: (a) A14-Z3 B, and (b) C14-J6 V. Describe the composition of each wheel in a sentence or two, and suggest the material to be ground by each.

7. Here are two more specifications: (a) C36-K8V and (b) C24-H9V, one for peripheral grinding and one for side grinding. From these specifications, tell which is which.

8. Here is an actual wheel specification: 32A46-H8VBE. Describe the wheel's composition, stating at least the abrasive used, the size of the abrasive, the grade, structure, and bond.

9. A wheel specification for cylindrical grinding of a hard steel fitting with a straight wheel is: A54-L5 V. If you were grinding a flat piece of the same steel with a straight wheel, what elements of the specification might change? Which way? For flat grinding of the same material with a segmental or a cup wheel, what further changes might be made?

10. Write one or two sentences about each of the following to show what elements of a wheel specification it affects.
    a. Material to be ground.
    b. Hardness of the material.
    c. Amount of stock to be removed.
    d. Kind of finish required.

UNIT 2 GRINDING WHEEL SAFETY

An old saying among grinding wheel manufacturers is: A grinding wheel doesn't break, but it can be broken. Grinding machine builders go to great lengths to design guards to hold wheel fragments in case of breakage without cutting down on the ease of operation or the productivity of the grinder. They have succeeded to the extent that the modern grinder, as a machine tool, is a pretty safe piece of equipment. A modern grinder, treated with due care and respect, is a safe machine, but it does require a knowledge and the practice of safe methods of operation.

OBJECTIVES

After completing this unit, you should be able to:

1. List the steps in and, if possible, demonstrate checking a grinding wheel for soundness.
2. List and, if possible, demonstrate the preliminary steps in mounting a wheel and starting a machine safely.
3. List at least six things you should do or not do when grinding, aside from those covered above.
4. Demonstrate safe practices in handling and mounting grinding wheels.
5. Demonstrate safe personal practices around grinding machines.
6. Given a wheel diameter, calculate in revolutions per minute its maximum safe speed (6500 SFPM or less). Convert RPMs to SFPMs for given wheel diameters.

THE GRINDING MACHINE AND SAFETY

Here is the situation on practically all machine shop grinders. In the machine, spinning at 5000 FPM or more, is a vitrified abrasive wheel made of the same material as dishes. The wheel is susceptible to shocks or bumps. It can easily be cracked or broken. If that happens, even though the machine has been designed with a safety guard that will contain most of the pieces (Figures 1 and 2), there is a possibility of broken pieces from the wheel flying around the shop. That could be, at the least, unpleasant and even dangerous. Fortunately, there is not much chance that it will happen; but it could.

Other grinding wheels with organic bonds can be operated safely at speeds over 15,000 FPM, but these are mostly for rough grinding. The wheels are built for it, but the principles are the same. **Every grinding wheel, wherever used, has a safe maximum speed, and this should never be exceeded.**

Any shop grinding machine, properly handled, is a safe machine. It has been designed that way. It should be maintained to be safe. It should be respected for its possibility of causing injury, even though that possibility is low.

Much of the image of grinding wheels and their breakage comes from portable grinders, which are often not well maintained and are sometimes operated by unskilled and careless people.

DETERMINING WHEEL SPEED

Grinding wheels are always marked with a maximum safe speed, but, because of the importance of wheel speed in grinding wheel safety, it is important to know how it is calculated. This quantity is expressed in surface feet per minute (SFPM), which is the distance a given spot on a wheel travels in a minute. It is calculated by multiplying the diameter (in inches) by 3.1416, dividing the result by 12 to convert to feet, and multiplying that result by the number of revolu-

Figure 1. Safety guard on a surface grinder. Note that the guard is somewhat squared off and covers well over half the wheel. (The DoAll Company)

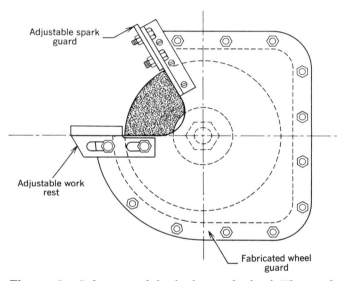

Adjustable spark guard

Adjustable work rest

Fabricated wheel guard

Figure 2. Safety guard for high speed wheel. The work is hand-held against the exposed peripheral grinding face of the wheel on top of the work rest. The squared corners tend to retain fragments in case of wheel breakage. (*Machine Tools and Machining Practices*)

tions per minute (RPM) of the wheel. Thus, a 10 in. diameter wheel traveling at 2400 RPM would be rated at approximately 6283 SFPM, under the safe speed of most vitrified wheels of 6500 SFPM.

To find the safe speed in RPMs of a 10 in. diameter wheel, the formula becomes

$$\frac{SFPM}{D \times 3.1416 \div 12}$$

or,

$$\frac{6500}{10 \text{ in.} \times 3.1416 \div 12} = 2483 \text{ RPM}$$

Most machine shop-type flat surface or cylindrical

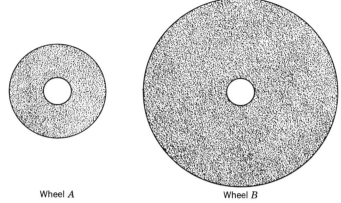

Wheel *A*   Wheel *B*

Figure 3. Wheel speed in SFPM, at a given RPM, increases as the wheel diameter increases. If the diameter of wheel *B* is twice that of wheel *A*, then the SFPM of wheel *B* is twice that of wheel *A*, at the same RPM.

grinders are preset to operate at a safe speed for the largest grinding wheel that the machine is designed to hold. As long as the machine is not tampered with and no one tries to mount a larger wheel on the machine than it is designed for, there should be no problem.

It should be clear that with a given spindle speed (RPM) the speed in SFPM increases as the wheel diameter is increased (Figure 3), and it decreases with a decrease in wheel diameter. Maximum safe speed may be expressed in either way, but on the wheel blotter it is usually expressed in RPM.

THE RING TEST

The primary method of determining whether or not a wheel is cracked is to give it the ring test. A crack may or may not be visible; of course, if there is a visible crack you need go no further and should discard the wheel immediately.

The test is simple. All that is required is to hold the wheel on your finger if it is small enough (Figure 4) or rest it on a clean, hard floor if it is too big to be held (Figure 5) and preferably strike it about 45 degrees either side of the vertical centerline with a wooden mallet or a similar tool. If it is sound, it should give forth a clear ringing sound. If it is cracked, it will sound dead. The sound of a vitrified wheel is clearer than the sound of any other, but there is always a different sound between a solid wheel and a cracked one.

There are a few permissible variations. Some operators prefer to hold the wheel on a stick or a metal pin instead of a finger. Some shops prefer to suspend

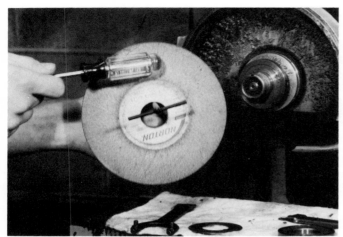

Figure 4. Making a ring test on a small wheel. (Lane Community College)

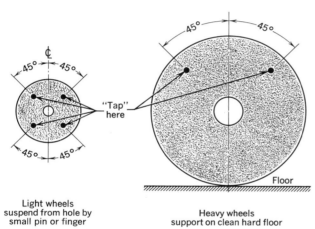

Light wheels
suspend from hole by
small pin or finger

Heavy wheels
support on clean hard floor

Figure 5. Sketch shows where to tap wheels for ring test, 45 degrees off centerline and 1 to 2 in. in from the periphery. After first tapping, rotate wheel about 45 degrees and repeat the test. (Bay State Abrasives, Dresser Industries, Inc.)

Figure 6. Wheel being lifted by a sling for ring testing. (The DoAll Company)

large wheels by a sling (chain or cable covered by a rubber hose to protect the wheel) instead of resting them on the floor (Figure 6).

The important point is that the test be done when each lot of wheels is received from the supplier, just before a wheel is mounted on a grinder, and again each time before it is remounted. If a wheel sounds cracked, or even questionable, then discard it or set it aside to be checked by the supplier, and get another wheel that, of course, must also be ring tested.

WHEEL STORAGE AT THE MACHINE

It should be clear that the wheels ought to be stacked

carefully and separated from each other by corrugated cardboard or another buffer (Figure 7). Tools and other materials, particularly metals, should never be stacked on top of the wheels in transit or at the machine. In short, *handle with care.*

General wheel storage is a responsibility of shop management, but at the machine this is usually the operator's concern. Operator carelessness can ruin an otherwise good wheel storage plan. On the other hand, care by the operator can do much to help out a substandard shop plan.

If the shop provides you with a proper storage area, all you have to do is to use it and keep it clean. If the shop does not provide such an area, you may need some ingenuity. Here are some suggestions.

1. Keep on hand only the wheels you really need. Do not allow extra wheels to accumulate around your machine. They can become cracked or broken and may get in your way. If you do not need a wheel, take it back to the general storage.

2. Store wheels above floor level, either on a table or under it, on pegs in the wall (Figure 8), or in a cabinet. Be especially careful to protect wheels

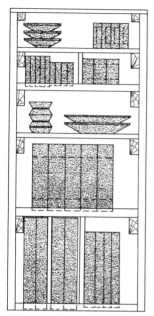
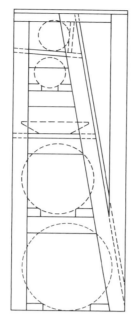

Figure 7. A recommended design for the safe storage of grinding wheels. Corrugated paper cushions should be placed between grinding wheels that are stored flat. The storage racks should be designed to handle the type and sizes of wheels used. *(Machine Tools and Machining Practices)*

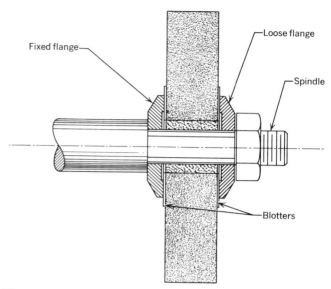

Figure 9. Typical set of flanges with flat rims and hollow centers with blotters separating the wheel and the flanges. Tightening the nut too much could spring the flanges and perhaps even crack the wheel. *(Machine Tools and Machining Practices)*

Figure 8. Storing extra wheels at the machine on pegs is often convenient and practical. The main requirement is to keep the wheels separate or protected, and off the floor. (The DoAll Company)

Figure 10. Well-designed guard for a bench grinder. Note that the side of the guard, like the one on the surface grinder, is easily removed for access to the wheel. (Lane Community College)

from each other and from metal tools or parts. Use corrugated cardboard, cloth, or even newspaper to keep wheels apart. *Keep wheels off the floor.*

MACHINE SAFETY REQUIREMENTS

Mounting flanges must be clean and flat, and equal in size, at least one-third the wheel diameter (Figure 9). Between each flange and the wheel there should be a blotter, a circular piece of compressible paper to cushion the flange from the rough wheel and distribute the flange pressure. The side of the flange facing the wheel has a flat rim, but the rest is hollowed out so that there is no pressure at the hole, which is the weakest part of the wheel.

The wheel guard must be in place. Depending on the machine design, the safety guard covers half or more of the wheel (Figure 10). On older model grinders the guard tends to be circular in shape and fits closely around the wheel. However, the latest models have square shaped safety guards, sometimes called cavernous, on the theory that pieces of any wheel that breaks will be retained in the corners instead of being slung out against the machine or possibly the operator.

MOUNTING THE WHEEL SAFELY, STARTING THE GRINDER

Any time you mount a wheel on the grinder or start the machine, good safety practice requires a set routine like the following.

1. Ring test the wheel as shown in Figure 4, then check the safe wheel speed printed on the blotter with the spindle speed of the machine. The spindle speed must never exceed the safe wheel speed, which is established by the wheel manufacturer after considerable research.

2. The wheel should fit snugly on the spindle or mounting flange. Never try to force a wheel onto the spindle or enlarge the hole in the wheel. If it is too tight or too loose, get another wheel.

3. Be sure the blotters on the wheel are larger in diameter than the flanges, and be sure the flanges are flat, clean, and smooth. Smooth up any nicks or burrs on the flanges with a small abrasive stone. Do not overtighten the mounting nut. It just needs to be snug.

4. Always stand to one side when starting the grinder; that is, stand out of line with the wheel.

5. Before starting to grind, let the wheel run at operating speed for about a minute with the guard in place.

Steps 4 and 5 are essentially a double check on the ring test. If anything is going to happen to a wheel, it usually happens very quickly after it starts to spin. Jog the switch when starting a new wheel.

OPERATOR RESPONSIBILITIES

In grinding, there are a number of operator duties that have long been a part of company safety policy

Figure 11. Approved safety glasses are required for all grinding. Today, as a matter of fact, they are usually required for anyone, even visitors, in the shop area. (Sellstrom Manufacturing Co., Palatine, Ill.)

and state safety regulations, and that, with the passage of the OSHA regulations, also became a matter of federal regulations. These are mostly designed for your protection, and not, as some operators have thought, to make the job more awkward or less productive.

1. Wear approved safety glasses or other face protection when grinding (Figure 11). This is the first and most important rule.

2. Do not wear rings, a wristwatch, gloves, long sleeves, or anything that might catch in a moving machine.

3. Grind on the side of the wheel *only* if the wheel is designed for this purpose, such as Type 2 cylinder wheels, Types 6 and 11 cup wheels, and Type 12 dish wheels (Figure 12).

4. If you are grinding with coolant, turn off the coolant a minute or so before you stop the wheel. This prevents coolant from collecting in the bottom half of the wheel while it is stopped and throwing it out of balance.

5. Never jam work into the grinding wheel. This applies particularly in off-hand grinding on a bench grinder.

Figure 12. Straight wheels are designed for grinding on the periphery. Never grind on the side of one of these, because it is not considered safe. On the other hand, cylinder wheels, cup wheels (both straight and flaring), and seg- ments or segmental wheels (shown here without the holder) are all designed and safe for side grinding. (Bay State Abrasives, Dresser Industries, Inc.)

SELF-TEST

1. A 12 in. vitrified wheel is available for mounting on a 2000 RPM grinder. Can this wheel be used safely on the grinder? Show the calculations to prove your answer.

2. List at least three important requirements for a ring test.

3. There are at least three times when a wheel should be ring tested. What are they, and why is the test needed each time?

4. Handling wheels carefully is most important. List at least four specific precautions to be taken in handling grinding wheels.

5. Wheel flanges are not always considered as important as they really are. Put down at least four points to be considered in their selection and care.

6. There are five or six essential steps to be followed in starting up a grinder. List them in order.

7. Providing a safe place to work is an employer's responsibility, but an operator also has safety responsibilities. List at least four things that an operator should or should not do to operate a grinder safely.

8. Explain the relationship of wheel diameter and spindle speed (RPM) to wheel speed in SFPM.

9. At the factory, grinding wheels are tested running free at 150 percent of safe operating speed. Why, then, isn't it all right to operate the wheels at the same higher speed?

10. Safe grinding practice requires that you **not** wear certain things (clothing and similar items) around a grinding machine. Name at least three.

UNIT 3 USING THE SURFACE GRINDER

The small horizontal spindle, reciprocating table surface grinder is considered to be basic to most general purpose machine shops. With accessories, it is extremely versatile and can do a remarkable variety of precision work. It is important that you learn the nomenclature and functions of this type of grinder and that you are aware of the many accessories that can be obtained to make this machine very useful in a wide variety of situations.

OBJECTIVES

After completing this unit, you should be able to:
1. Name the components of the horizontal spindle surface grinder.
2. Define the functions of the various component parts of this grinder.
3. Name and describe the functions of at least four accessory devices used to increase the versatility of the surface grinder.

The small horizontal spindle, reciprocating table surface grinder (Figure 1) must be considered the basic grinder. There is at least one in nearly every machine shop, toolroom, or school shop. Operating one on straight flat work can be relatively simple. In fact, it has been said that three-fourths or more of all grinding work could be done on a small 6 by 12 in. capacity surface grinder.

In a sense, of course, all grinding is done on a surface, but a surface grinder is one for producing flat surfaces (or, with these horizontal spindle machines, formed surfaces) as long as the hills and valleys of the form run parallel with the path of the wheel as the magnetic chuck moves the work under it. Figure 2 shows a simple form. Vertical spindle machines, however, are usually limited to flat surfaces, because the grinding is done on the flat of the wheel with the abrasive "scratches" making an overlapping, circular path.

It is also possible on the horizontal spindle, reciprocating table grinder, with proper fixturing, to grind surfaces that are not parallel but are either flat or formed. Finally, with accessories, it is possible to do almost any kind of grinding: center-type, cylindrical, centerless, and internal. This is usually incidental work and small workpieces. But these show that the small, hand-operated surface grinder is a very versatile machine.

The best place to start learning the part names of a hand-operated horizontal spindle, reciprocating table grinder is the wheel-work interface, the point where the action is. In this unit, you should understand that the term "grinder" means this type of machine.

The cutting element is the periphery, or OD, of the grinding wheel. The wheel moves up and down under the control of handwheel A, as shown in Figure 1. The doubleheaded arrow A shows the direction of this movement. The wheel, together with its spindle, motor drive, and other necessary attachments, is often referred to as the wheelhead.

MAGNETIC CHUCKS

The work usually is held on a flat magnetic chuck clamped to the table that is supported on saddle ways. The magnetic chuck, the most common accessory for this grinder, holds iron and steel firmly enough to be ground. Using blocking of iron or steel, it will also hold nonmagnetic metals like aluminum, brass, or bronze. Only very rarely is it necessary to fasten a workpiece directly to the machine table.

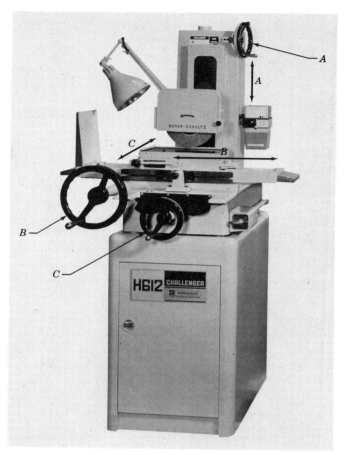

Figure 2. Grinding a simple rounded form. Note that the wheel is dressed to the reverse of the form in the finished workpiece. *(Machine Tools and Machining Practices)*

Figure 1. Surface grinder with direction and control of movements indicated by arrows. Wheel *A* controls downfeed *A*. Large wheel *B* controls table traverse *B*. Wheel *C* controls crossfeed *C*. (Boyar–Schultz Corporation, An Esterline Company, Broadview, Ill.)

Figure 3. Common type of magnetic chuck for reciprocating surface grinder. The guards at the back and left side are usually adjustable and help keep work from sliding off the chuck. (The DoAll Company)

In industrial shops the electromagnetic chuck is the basic workholder for the surface grinder. Only large machines have the electromagnetic type, but the permanent magnetic chuck is often used on small machines. There is no basic difference in principle (Figure 3). Permanent magnetic chucks are made up of a series of alternating plates composed of powerful alnico magnets and a nonmagnetic material. Some improved types use ceramic magnets alternating with stainless steel plates. They exert a more concentrated holding force and can be used for milling as well as grinding. The electromagnetic chuck does have one advantage over the permanent magnet type, however, in that the electric control for the electromagnet usually has a built-in de-mag (demagnetization) mode that will de-mag the workpiece after grinding. One disadvantage of the electromagnetic chuck is the pos-

sibility of a power failure or a shorted cord. The result would be a damaged workpiece and wheel breakage.

So, for many workpieces, all that is needed is to clean off the chuck surface and the workpiece, place

Figure 4. Magnetic sine chuck needed for grinding non-parallel surfaces. It is, of course, adjustable. (Hitachi Magna-Lock Corp.)

Figure 5. Operator in position at grinder. On many small surface grinders the crossfeed and traverse are hydraulic. (Lane Community College)

the part, turn on the magnet, check for firm workholding, and start grinding (assuming that the wheel on the grinder is suited to the workpiece). On all such pieces, it is imperative that the surface to be ground is parallel with the surface resting on the magnetic chuck. The surface of the chuck and the "grinding line" of the wheel must always be parallel.

However, surfaces that are not parallel can be ground. As a simple illustration, consider a beveled edge to be ground around all four sides of the top of a rectangular block. You have two choices. One is to use a fixture that can be set to the angle you want, which is a magnetic sine chuck, as shown in Figure 4. The other is to dress the angle you want on the wheel with an attachment called a sine dresser and then grind in the usual fashion. The choice is up to the operator.

PROCEDURES

The chuck moves from left to right and back again (traverse), as shown in Figure 1, arrow *B*. On many grinders, traverse and crossfeed increments are controlled hydraulically with the table moving back and forth between preset stops. When the hydraulic control is actuated, the traverse wheel is automatically disengaged. This movement is controlled by large handwheel *B*. The chuck also moves toward and away from the operator (crossfeed), as shown by arrow *C*, with the motion controlled by handwheel *C*. The grinding wheel, incidentally, always runs clockwise.

On a completely hand-operated machine, the operator stands in front of the machine with his left hand on the traverse wheel and his right hand on the crossfeed wheel (if he is right-handed), swinging the table and the magnetic chuck back and forth with his left hand and cross feeding with his right hand at the end of each pass across the workpiece (Figure 5). The wheel clears the workpiece at both ends of the pass, and the crossfeed is always less than the width of the wheel, so that there is an overlap. At the end of each complete pass across the entire surface to be ground, he feeds the wheel down with the downfeed wheel on the column.

The traverse handwheel is not marked, because all that is necessary is to clear both ends of the surface. However, both of the other wheels have very accurately engraved markings, so that downfeed and crossfeed can be accurately measured. The downfeed handwheel (Figure 6) has 250 marks around it, and turning the wheel from one mark to the next lowers or raises the wheel .0001 in. This is one ten-thousandth of an inch, known in the shop usually as a "tenth." The crossfeed handwheel has 100 marks, and moving the wheel from one mark to the next moves the workpiece toward or away from the operator .001 in.

With the 250 marks on the downfeed handwheel and .0001 in. movement per mark, a complete revolu-

Figure 6. Closeup of downfeed handwheel. Moving from one mark to the next lowers or raises the grinding wheel .0001 in. (The DoAll Company)

Figure 7. Same control wheel with slip ring set to zero. Now it is simpler for operator to down feed the grinding wheel as he grinds. (The DoAll Company)

tion of the handwheel moves the grinding wheel .025 in. (250 × .0001 in.). A complete revolution of the crossfeed wheel, which has 100 marks, moves the table and the chuck .100 in. (100 × .001 in.).

The other feature needing mention is the zeroing slip ring on the downfeed handwheel or the crossfeed handwheel. You cannot predict in advance just where on the scale either wheel will be when the grinding wheel first contacts the workpiece, and where the control wheels ought to be when the surface is ground. These can be figured out mathematically, but with considerable chance of error. However, with the zeroing slip ring, the starting point on the scale on each wheel is simply set at zero, locked in place (Figure 7), and then ground until the required amount of stock has been removed. Allowance must be made for wheel wear, especially where considerable stock is removed by grinding. On most small surface grinders, downfeed for roughing should be no more than .002 in. per pass and no more than .0005 in. per pass for finishing operations.

On a hand-operated grinder a skilled operator develops a rhythm as he traverses the workpiece back and forth under the grinding wheel, cross feeding

at the end of each pass and downfeeding when the whole surface has been covered. This is a knack you develop with experience. Using machines with hydraulic traverse and crossfeed takes out the need for physical coordination on the part of the operator, but the skill necessary to make good choices on the traverse speed and on the amount of crossfeed for each pass is very important to first-class grinding results. Generally, combinations of large crossfeed movements on the order of one-half the wheel width and relatively small amounts of downfeed are preferred because wheel wear is distributed better this way.

Actually, with the wheelhead, the magnetic chuck, and the three control wheels, you have the basic parts of a surface grinder. You also need a dressing device, which may be just a holder with the diamond mounted at the proper angle that can be mounted on the magnetic chuck or a built-in dresser (Figure 8). But everything else on the machine is either for support (for instance, the table, saddle ways, base, and column that holds the wheelhead) or an accessory that makes it possible to do something you could not do otherwise or that makes the job easier to do.

TRUEING THE WHEEL

Chances are that you will most often be trueing the wheel so that the peripheral surface is concentric with

Figure 8. Built-in wheel dresser. Lever traverses dresser across wheel. (The DoAll Company)

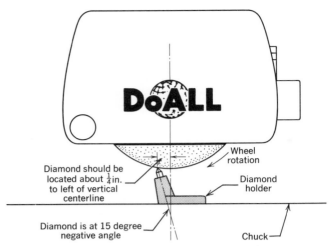

Figure 9. This is one of several ways of mounting a dresser on a surface grinder. The dresser with its diamond is simply spotted on the clean magnetic chuck. Note, however, that the diamond is slanted at a 15 degree angle and slightly past the vertical centerline of the wheel, as the wheel turns. (The DoAll Company)

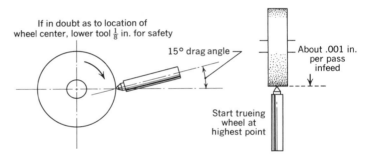

Figure 10. This illustrates the idea of wheel dressing instead of any specific setup. Dressing is rarely done free hand. Note that the diamond is always a little past the centerline and on an angle. (The Desmond–Stephan Manufacturing Co., Urbana, Ohio)

the center of the machine spindle and parallel to the centerline of the spindle, and with a single point diamond dresser. The dresser may be mounted above the wheel (Figure 8), on the magnetic chuck that is the workholder of the machine (Figure 9), or to the side (Figure 10). In any of these positions, it is possible to move the dresser back and forth across the face of the wheel. This is called traversing the dresser.

The diamond is mounted at an angle of 15 degrees, so that it contacts the wheel just after the low or high point of the wheel, or just below the centerline, depending on the location of the dresser (Figures 8, 9, and 10). Thus the wheel is cutting toward the point of the diamond. Remember that dressing with a diamond is always a two-way operation; that is, the diamond is kept sharp while it sharpens the wheel grinding face. Often there is an arrow or other indicator on the dresser to indicate its position. The diamond always points in the direction of the wheel's rotation.

Any new wheel, or a wheel that has just been reflanged, must first be trued. In trueing, the wheel and the dresser are brought together so that the dresser is touching the high point of the wheel. Otherwise, the traversing of the diamond might cause it

to dig too deeply into the wheel, which could ruin the diamond. On the other hand, if you start at the high point, which is the point furthest from the center, the cross cut or traverse is short at first and gradually becomes longer until finally you are dressing the entire width of the wheel.

Infeed of the diamond into the wheel should be light, about .001 in. per pass; if dressing is being done dry, there should be frequent pauses, after every three or four passes, to allow the diamond to cool off. A hot diamond can be shattered if a drop of water or other liquid hits it. Turn the diamond frequently. This helps to keep it sharp.

Trueing the wheel is accomplished, then, by moving the dresser back and forth across the grinding face of the wheel while the wheel is rotating at operating speed. It is preferably done wet. If trueing is done wet, continuous coolant must be assured.

The speed of traverse is probably the remaining point of concern. Generally, the faster the dresser traverses across the wheel's grinding face, the sharper the dress will be and the better suited the wheel will be for rough grinding. Slower traverse means that the diamond does more cutting on the abrasive, dulling it a little bit, so that the wheel is better suited to finishing than to roughing work. If the diamond is traversed back over a newly dressed wheel without infeed, the wheel will be dulled, causing it to burn or glaze the workpiece. Sometimes, however, when a wheel is intended for finishing, it is good to take a few passes across the wheel without any infeed. The point is that with a little experience you can tell the degree of sharpness that is needed in the wheel face and dress the wheel accordingly, depending on what you want to do with it afterward. As grain becomes dulled, it tends to polish more and cut less, and size becomes less of a factor in the action of the grain.

Generally, a coarse grain wheel is recommended for cutting and material removal and fine grain is recommended for finishing. This is true, provided both are in the same degree of sharpness. It can now be said that by proper dressing, it is possible to make a finish with comparatively coarse grain like that of a much finer grain. For example, a 46 grit wheel, dressed to dull the grain, could give the same finish as a 120 grit wheel. The reverse is not true, however.

GRINDING FLUIDS

Grinding produces very high temperatures. Temperatures at the interface (the small area where the abrasive grains are actually cutting metal) are reliably estimated to be over 2000°F (1093°C), and that is enough to warp even fairly thick workpieces. Nor is it safe to assume that just because there is a lot of grinding fluid (sometimes called coolant in the shop) flowing around the grinding area, the interface is cooled. With the grinding wheel rotating at its usual 5000 SFPM plus rate, it creates enough of a fan effect to blow coolant away from the contact area. This creates a condition sometimes referred to as "grinding dry with water." In other words, in spite of the amount of coolant close by, the actual cutting area may be dry and hot. However, the coolant can do a good job of cooling the contact area if it is properly applied. Most grinding fluids are water-based; that is, a mixture of soluble oil and water.

Figure 11. This is a very common method of flood coolant application. For the photograph, the volume of coolant has been reduced. (Lane Community College)

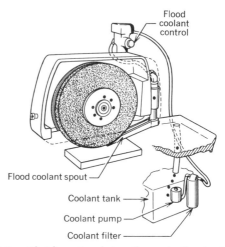

Figure 12. Fluid recirculates through the tank, piping, nozzle, and drains in flood grinding system. (The DoAll Company)

Methods of Coolant Application

Most grinding fluids are applied in what is called flooding. A stream of coolant under pressure is directed from a pipe, sometimes shaped but often just round, in the general direction of the grinding area (Figure 11). The fluid collects beneath the grinding area, is piped back to a tank where it is allowed to settle, and is cleaned; once more it is pumped back around the wheel (Figure 12). There is always a little waste, and periodically either more coolant concentrate or water is added to the solution. Flooding is an effective method of applying coolant; provided the solution stays within the effective range, neither too rich nor too lean, it works very well. However, the fluid cannot just be in the vicinity of the grinding area; it must be right *in* that area if it is to do its job. For example, on a small surface grinder processing small parts, the nozzle can almost be pointed at the grinding wheel-workpiece contact area.

ACCESSORIES

The list of accessories for a hand and hydraulically operated toolroom-type surface grinder may be quite extensive. As mentioned before, with the proper ac-

cessories, almost any type of grinding can be done on one of these little grinders within maximum size ranges. Finally, accessories are a major point of difference between toolroom and production machines. Toolroom grinders must handle a variety of work, so accessories are needed. Production machines are used mostly for one purpose; if that purpose changes, the machine is rebuilt or modified for its new use or it is removed from service.

ATTACHMENTS

For practical purposes it probably makes very little difference whether something is called an attachment or an accessory. Both make it possible to do something with the machine that could not otherwise be done, or at least could not be done so easily or quickly.

Center-Type Cylindrical Attachment

This is basically a workholder with a headstock and a tailstock (Figure 13). It is mounted crosswise on the locked table so that the crossfeed of the grinder makes the wheel traverse end to end on the workpiece. If a flat is needed on an essentially cylindrical part, then all that is necessary is to stop the rotation of the work, reciprocate the table just a little, and cross feed as for any flat surface.

High-Speed Attachment

For incidental internal grinding, this attachment (Figure 14), driven by a belt from the grinder spindle, provides the high speed that is needed to make the small mounted wheels run at the high RPMs needed to make them grind efficiently. Of course, it is essential, usually, to provide an attachment for mounting the workpiece also.

Figure 13. Center-type cylindrical attachment mounted on surface grinder. Attachment can be tilted for grinding a taper, as shown here, or set level for grinding a straight cylinder. (Harig Manufacturing Corporation)

Figure 14. High speed spindle adds capability for internal grinding. (Whitnon Spindle, Division of Mite Corporation)

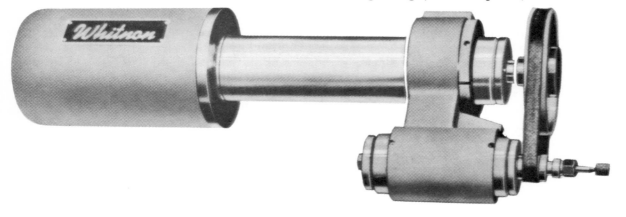

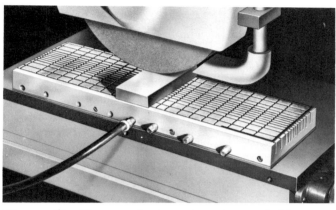

Figure 15. Vacuum chucks such as this one hold practically anything and are considered good for thin work. (Hitachi Magna-Lock Corp., Big Rapids, Mich.)

Vacuum Chuck

This replacement for an electromagnetic chuck (Figure 15) holds the work by exhausting the air from under it. Thus, it makes no difference whether the workpiece is magnetic or not. It is also recommended by some experts for holding pieces as thin as only a few thousandths of an inch.

SURFACE GRINDING A VEE BLOCK

Demonstrating any skill is, of course, the final test of whether you have learned it. The workpieces selected for this unit on using the surface grinder are two matching vee blocks of SAE 4140 or a similar alloy steel. A series of grinding steps are given in the text to detail the process for making these vee blocks. Some of these steps will include grinding operations on both blocks in the same setup at the same time. The specifications for surface grinding are often much closer than those required on other machine tools. These grinding machines, however, are capable of holding tolerances within one-tenth of a thousandth of an inch.

Selecting the Wheel

A SAE 4140 steel in the Rockwell C hardness range of 48 to 52 is regarded as being not too difficult to grind. It can be ground satisfactorily with a number of wheel specifications that are likely to be on hand in practically any grinding shop. Wheels for machine shop use have to be able to grind a wide range of materials, because the number of parts to be ground in any single lot is not likely to be large enough to warrant trials to find *exactly* the best wheel specification. Here, however, are some guidelines in the selection.

The abrasive will be aluminum oxide because this is steel, and the bond will be vitrified because this is precision grinding. Given the requirements for this particular workpiece, below are the recommendations of several specialists in the grinding field giving a range of four possible selections of wheels.

1. 9A46-H8V.
2. 9A60-K8V.
3. 32A46-I8V.
4. DA46-J9V.

All of these may be regarded as general purpose specifications for a part of this configuration, material, and hardness.

Looking at these recommendations in order, the abrasive recommendation of the first two is a white aluminum oxide, which is the most friable (brittle). The third and fourth are for a mixture of white and regular (gray) aluminum oxide, which is slightly tougher and perhaps wears a little longer. In wet grinding, as this is, probably the tougher abrasive would be preferred. However, in dry grinding, there would probably be no question about the use of white abrasive; it does not tend to burn the work as much as a tougher abrasive.

A grit size of 46 would be indicated for grinding efficiency, but 60 would provide a slightly better finish. Wheel grade provides the widest range (H to K), which could be interpreted to mean that any grade within this range would do the job. However, the wheel with the finest grit (60) also has the hardest grade (K). Such a wheel would wear a little longer than the others. Grit size and abrasive types are both likely to be fairly constant from various manufacturers.

The range of structure is only from 8 to 9, which is hardly significant. Grade and structure, however, result from manufacturing procedures, and hence are at least comparable from one wheel supplier to another.

Vitrified bond is indicated. Vitrified bonds are somewhat varied, although all wheel manufacturers have one or two general purpose or standard bonds for machine shop work.

Grinding the Vee-Blocks

The part selected for this precision grinding project is illustrated in Figure 16. Figure 17 is a part drawing with the necessary dimensions and other data. These particular workpieces are examples of student projects sometimes used for surface grinding training. The purpose of this project is to show that you can surface grind a part flat and square to common tolerances and surface finish.

Figure 16. Finished, hardened, and ground precision vee block. *(Machine Tools and Machining Practices)*

Figure 17. Dimensions and information for grinding the vee block. *(Machine Tools and Machining Practices)*

STEPS IN GRINDING

1. *Select the wheel.* The previous information would help you do this.

2. *Clean the spindle (Figure 18).* Use a soft cloth to remove any grit or dirt from the spindle. Note that the chuck is protected by a cloth to prevent nicks or burrs from tools laid out on it.

3. *Ring test the wheel (Figure 19).* As indicated in Unit 2, this is a safety precaution any time a wheel is mounted.

4. *Mount the wheel (Figure 20).* The wheel should fit snugly on the spindle, and the outside flange must be of the same size as the inner flange, as shown in Figure 19. This wheel has blotters attached, but if they were damaged or there were none attached, it would be necessary to get some new blotters. Flanges should also be checked occasionally for burrs or nicks and flatness. The flange is held by a nut.

5. *Tighten the nut (Figure 21).* The nut should be tightened snugly. The blotters will take up a little extra force, but if either flange is warped or otherwise out of flat, it is possible to crack a wheel. Overtightening can also crack a wheel.

6. *Replace the safety guard (Figure 22).* This is a necessary precaution for safety.

7. *Place the diamond for dressing the wheel (Figure 23).* The wheel, as noted by the arrow on the

---

SLOT $\frac{3}{16}$ x $\frac{3}{16}$ [4.7 x 4.7]

$\frac{1}{4}$ [6.3]

SEE NOTE

SLOT $\frac{1}{8}$ x $\frac{1}{8}$ [3.1 x 3.1]

$1\frac{1}{2}$ [38.10]

$\frac{1}{4}$ [6.3]

90°

90°

$\frac{9}{16}$ [14.28] $\frac{23}{32}$ [18.25]

$\frac{11}{16}$ [17.462]

2 [50.80]

Y. I. H.

$\frac{1}{2}$ [12.70]

2 [50.80]

$\frac{1}{16}$ [1.5]

45°

TYP. CORNER

$2\frac{1}{4}$ [57.15]

NOTE: RECESS ON ONE SIDE $\frac{1}{16}$ [1.58] DEEP x $\frac{3}{8}$ [9.52] wide

INCH [MILLIMETER]

ALL SIDES SQUARE TO EACH OTHER TO WITHIN LESS THAN .0005″ WITH A CYLINDRICAL SQUARE. DIM. TOLERANCES ± $\frac{1}{64}$ OR .0005″. HARDEN AND TEMPER TO RC-48-52

1	2	SAE 4140	$2\frac{1}{4}$ x $2\frac{1}{4}$ x $4\frac{3}{4}$
DET.	REQD.	MATL.	STK. SIZE
V — BLOCK			SCALE: $1\frac{1}{2}$ = 1

Figure 18. Cleaning the wheel spindle with a soft cloth. (Lane Community College)

Figure 21. Tightening the nut with spanner wrenches. (Lane Community College)

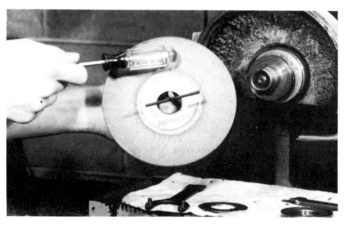

Figure 19. Ring testing the wheel. (Lane Community College)

Figure 22. The safety guard being replaced. (Lane Community College)

Figure 20. Mounting the wheel, flange, and nut. (Lane Community College)

Figure 23. Diamond dresser in the correct position. The camera angle does not show the location of the diamond clearly. (Lane Community College)

Figure 24. Dressing the wheel using coolant. (Lane Community College)

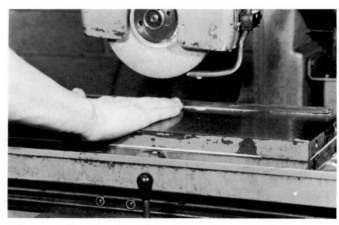

Figure 26. Checking for nicks or burrs on the magnetic chuck. (Lane Community College)

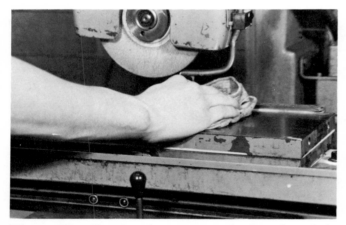

Figure 25. Cleaning the magnetic chuck with a cloth. The wheel must be completely stopped when this is done. (Lane Community College)

Figure 27. The rough blocks in place with the large vee side up ready to be ground. (Lane Community College)

safety guard, revolves clockwise so that the correct placement of the diamond is a little past the bottom point of the wheel, at about 6:30. This type of dresser has the correct "drag angle" of 10 to 15 degrees built into it.

8. *Dress the wheel (Figure 24).* This operation is to dress the wheel for stock removal so that the diamond can be cross fed rapidly back and forth across the wheel. Coolant is used, since the grinding is done wet, but at greater volume than is shown in the illustration. The coolant was reduced for better visibility, but, in actual dressing, the dresser should be either completely drenched with coolant or no coolant should be used at all.

9. *Wipe off the top of the chuck (Figure 25).* This is to remove any chips or bits of abrasive that

may be on the chuck face. It is done with the wheel at a full stop and with the magnetic chuck turned off. A squeegee is often used before finishing the job with a shop cloth.

10. *Check the chuck for nicks and burrs (Figure 26).* The quickest and best way to do this is to run your hand over the face of the chuck. If the previous step has been done adequately, there should be no slivers or bits of metal to stick in your hand.

11. *Place the two blocks for the first grind (Figure 27).* Note that the sides with the large vees are

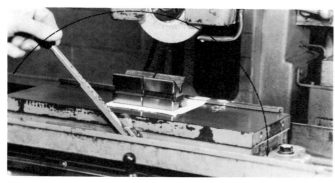

Figure 28. Activating the magnetic chuck. (Lane Community College)

Figure 29. Downfeeding the wheel. (Lane Community College)

Figure 30. Positioning the blocks to set the stroke length. (Lane Community College)

up and that the blocks are placed near the center of the chuck. The paper protects the chuck and the workpieces from each other. For a single job like this, most operators tend to place the workpieces at the center of the chuck. However, if the grinder is in regular use, the paper would probably not be used and each group of parts would be spotted at a different place on the chuck, to equalize wear.

12. *Turn on the chuck (Figure 28)*. Magnetic flux is applied on this chuck by moving a handle from left to right. On other chucks, such as the electromagnetic type, magnetism is activated by an electrical switch.

13. *Downfeed the wheel close to the top of the vee block (Figure 29)*. This is simply a matter of turning the downfeed (sometimes called the infeed) handwheel at the top of the column to lower the wheel to a point an inch or so above the workpieces.

14. *Position the workpieces for grinding (Figure 30)*.

The large wheel, held by the operator's left hand, controls the left and right (traverse) movement of the table and chuck. The smaller wheel, held in the operator's right hand, cross feeds the table and chuck toward or away from the operator. During grinding the table is usually traversed hydraulically with the large handwheel disengaged. The large handwheel is mainly used for positioning work.

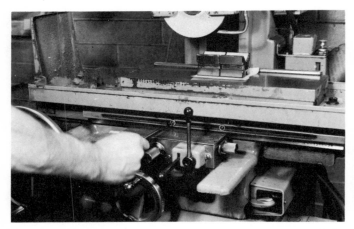

Figure 31. Setting the table stops. (Lane Community College)

Figure 33. Starting the table traverse. The wheel is carefully adjusted to just touch the high point of the work. (Lane Community College)

Figure 32. The wheel is turned on and, after running a minute for safety reasons, the rotating wheel is brought close to the workpiece. (Lane Community College)

Figure 34. The coolant is turned on and an additional downfeed of .002 in. is made. (Lane Community College)

15. *Set the table stops (Figure 31).* Table stops are sometimes called trip dogs. Their purpose is to set the limits within which the table can travel, and these are usually about an inch off either end of the workpiece. All that is needed is to make sure that the wheel clears the end of the workpiece and allows time for cross feeding between traverse passes.

16. *Turn on the grinding wheel spindle.* Also turn on the hydraulic pump motion if your grinder is so equipped. As a safety precaution, let the wheel run for a minute, taking care that you do not place yourself, or allow anyone else, to stand in line with the plane of rotation. The grinding wheel is then brought down close to the work (Figure 32). The high point on the work is sought by manu-

ally cross feeding during the table traverse to find the high spot. When it is found, set the downfeed dial to zero as a reference. Now turn on the traverse (Figure 33), if your machine is equipped with one.

17. *Turn on the coolant (Figure 34).* With the coolant flowing, downfeed about .002 in. With about .015 in. stock to be removed from each dimension and leaving about .003 to .005 in. for the finish cuts, remove about .010 to .012 in. from each of the sides and from the two ends. How much of this total comes off each side or end is usually left up to the operator's judgment, unless it is covered in a job sheet. In this case, perhaps removing a few thousandths of an inch for cleanup would be best until the squaring procedure is finished.

Figure 35. Turning on the table crossfeed; both surfaces are ground on the large vee side. (Lane Community College)

Figure 36. The blocks are turned over and the opposite sides (small vee) are ground. (Lane Community College)

Figure 37. A ground side of the vee block is clamped to a precision angle plate that is turned on its side, and the top side of the vee block is being leveled with a dial indicator that is mounted on a height gage. (Lane Community College)

Figure 38. The precision angle plate and vee block setup is turned with the vee block end up on the magnetic chuck. The end of the vee block is ground square to a ground side. (Lane Community College)

18. *Turn on the table crossfeed (Figure 35).* With power for both traverse and crossfeed, watch the wheel and listen for unusual sounds indicating overloading as it grinds. Downfeed .001 to .002 in. at the end of each complete pass across the two flat surfaces on either side of the large vees on both blocks.

19. *Clean up the opposite sides (Figure 36).* This step repeats Steps 17 and 18 with the large vees turned down. Grinding should continue until the block is just cleaned up. (Steps 20 to 22 are done on each block singly.)

20. *Clamp either finished side of one block to a precision angle plate laid on its side on a surface plate*

(*Figure 37*). Using a dial indicator and either a height or surface gage, clamp the vee block parallel to the surface plate. One end of the block should project slightly beyond the angle plate so it may be ground. This is probably the most critical step in the procedure.

21. *Set up for grinding the end (Figure 38).* Return the angle plate and block to the grinder and place them on the magnetic chuck so that one end of the block is in position to be ground. Turn on the magnetic chuck and grind the end to clean up. The ground end should now be relatively square to the two ground sides.

22. *Grind one of the remaining sides square (Figure*

Figure 39. The other two sides being ground square to the vee block end that was previously ground. (Lane Community College)

Figure 41. The sides that have not been ground are also being ground in one setup to make them parallel to the other sides. (Lane Community College)

Figure 40. Grinding the opposite ends of both blocks in one setup to make them parallel. (Lane Community College)

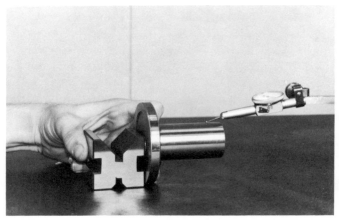

Figure 42. A vee block now being checked for square on all sides using a precision cylindrical square, a dial indicator, and a height gage on a surface plate. (Lane Community College)

39). Use the same procedure as that explained in Step 21, except that the end you have just ground on each block is clamped to the angle plate. The side to be ground must be made parallel to the surface plate. The side to be ground must project above the end of the angle plate so there will be no interference.

23. *Grind the remaining ends at the same time (Figure 40).* Both blocks can be set on the magnetic chuck without additional support and ground separately. Grind to within .003 to .005 in. of finished dimension.

24. *Grind the remaining sides and the two cleaned up sides to .003 to .005 in. oversize (Figure 41).* This is essentially the same setup as for Step 23. These steps can be done without the angle plate after two ajoining sides and one end are square

with each other, because surfaces parallel to square surfaces are also square to each other.

25. *Redress the wheel for finish grinding.* This is essentially the same as Step 8, except that the dresser is cross fed more slowly. Downfeed for the finish grinding should be about .0002 in., about one-tenth of the downfeed used for rough grinding.

26. *Check all sides and ends for square (Figure 42).* This requires a precision cylindrical square and a .0001 in. reading dial indicator supported by a height gage on a surface plate. Correct any errors of squareness by touch-up grinding, using tissue paper shims under one side to make the faulty

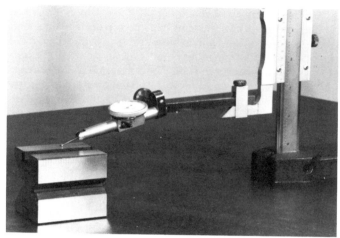

Figure 43. Dimensions being checked using a .0001 in. reading dial indicator on a height gage. Precision gage blocks may be used for comparison measurement for this operation. (Lane Community College)

Figure 44. Setting up the magnetic vee block to grind the angular surfaces on the large vees. (Lane Community College)

surface parallel with the magnetic chuck. Remember to put the tissue paper under the thick side.

27. *Check dimensions.* This can be done by using a ten-thousandths reading micrometer (Figure 43) or by using gage blocks.

Figure 45. The vee blocks are repositioned and the other side of the large vees are being ground. (Lane Community College)

28. *Grind sides and ends to finish dimensions.* If your work to this point has been done carefully, both blocks can be done at the same time, as in Steps 23 and 24. Just be sure that you are grinding corresponding sides on each setup. No tissue paper is used in finish grinding.

29. *Grind one side of the large vee (Figure 44).* Set up both blocks in a magnetic vee block, making sure that the magnetic vee block is aligned parallel with the magnetic chuck. Turn on the magnetic chuck. Turn on the coolant and grind the side for cleanup. Use light downfeed.

30. *Reverse the vee blocks and repeat Step 29 for the other side of the large vee with the same setting (Figure 45).* This will center the vee. Check the large vee for square. The angle between the two sides of the large vee should be exactly 90 degrees. Repeat this procedure on the small vees on the opposite side.

31. *Clean up your machine.* Run the wheel for five minutes to remove coolant, which helps to maintain balance when the wheel is next used. Return the tooling to its proper place. Measure and record the finished dimensions of the vee block. Submit the vee block to your instructor for evaluation.

SELF-TEST

1. The part of the horizontal spindle surface grinder that holds the wheel is called the _____.
2. Workholding on a surface grinder is almost always done on a _____ _____.
3. How can nonparallel (angular) surfaces be ground?

4. What do the finest divisions (marks) on the downfeed handwheel usually represent in inches?
5. What do the marks on the crossfeed handwheel usually represent?
6. On a small surface grinder, the maximum downfeeds

for roughing and finishing are how much?

7. When dressing or trueing the wheel with a diamond, what should the downfeed be per pass?

8. The faster the traverse with diamond dressings, the sharper the wheel will be and a slower dress dulls the abrasive. What affect will this have on the surface of the piece to be ground?

9. When using flood coolant application, why is it not sufficient to just have coolant lying on top of the workpiece?

10. Precision grinding machines, such as cylindrical grinders, centerless grinders, and tool and cutter grinders, are built for special work. The horizontal spindle surface grinder can be considered as the most versatile of grinding machines. Why is this so?

UNIT 4 PROBLEMS AND SOLUTIONS IN SURFACE GRINDING

Producing quality work on a surface grinder bears some resemblance to driving a car. It is not too difficult when everything is going right. It is knowing what to do when something is not going right that separates the skilled from the unskilled. This unit is a discussion of some of the common problems of grinding, how to recognize them, and what to do about them. It should also give you some insight into whether you should try to do something about the problem yourself or whether you should report it to your supervisor.

OBJECTIVES

After completing this unit, you should be able to:

1. Recognize common surface defects resulting from surface grinding, and suggest ways of correcting them.
2. Suggest ways of correcting situations that show up in postinspection such as work that is not flat, parallel, or to form.

Conditions causing problems in surface grinding can conveniently be divided into two groups.

1. *Mechanical.* Those problems having to do with the condition of the machine; for example, worn bearings. Or those problems having to do with the electrical or hydraulic systems of the machine.
2. *Operational.* Those problems having to do with the operation of the machine; for example, the selection and condition of the grinding wheel, selection of coolant, wheel dressing, and other similar responsibilities.

Obviously, there is not a clear line that can be drawn between these two. The division of responsibilities varies between shops; in general, the mechanical condition of the machine, aside from routine daily lubrication, is a shop maintenance responsibility, while the operation of the machine is your responsibility as its operator. Still, there are some close decisions to be made; they are usually decided on the basis of shop policy or practice. In a school shop, it is generally the responsibility of the person using the machine to clean and lubricate it before and after use. If the machine ways are not lubricated, a problem called stick-slip develops as the table traverses, in which the table moves at a varying speed and has a jerky movement. If you are not sure of your responsibilities while using these machines, ask your instructor.

Two other general observations are in order. One is that any surface grinder is limited in the degree of precision and the quality of surface finish that it will produce. A lightly built, inexpensive grinder simply will not produce the quality of finish nor the precision of a more heavily built and more expensive machine.

The second is that any machined surface, even the finest, most mirrorlike surface, is a series of scratches. It is true that on the finer finishes the scratches are finer, closer together, and follow a definite pattern, but there are scratches nonetheless, and they will show up if the surface is sufficiently magnified.

Although you have not studied surface finish in any detail in this section, it is easy to understand that on a reciprocating table surface grinder, the scratches will be parallel and running in the direction of table travel. Anything that differs significantly from this pattern can be considered a surface defect and a problem.

Dirt, heat, faulty wheel dressing, and vibration can cause problems, as indicated earlier. The condition of the wheel's grinding surface can also cause problems. A wheel whose surface is either loaded or glazed will not cut well or produce a good flat surface. **Loading** means that bits of the work material have become embedded in the wheel's face. This usually means that the wheel's structure should be more open. **Glazing** means that the wheel face has worn too smooth to cut. It may result from using a wheel with a grit that is too fine, a structure that is too dense, or a grade that is too hard.

OPERATOR'S RESPONSIBILITIES

In general, it can be said that you, as the operator, are responsible for the daily checking and running of the machine. This includes things such as selecting and dressing the wheel, selecting the coolant, checking to see that the coolant tank is full and the filters are working as they should, and that the coolant is flowing in sufficient volume. You should observe the lubrication of the machine to see whether there is too little or too much. You should be alert for signs that the wheel is not secure or if the bearings are beginning to wear too much. You should not only check the work as instructed, but you should also observe it for surface irregularities. Of course, you will not be working entirely on your own; much of this will be done with the advice and agreement of your

instructor or supervisor, especially when you first begin to use the machine.

GOOD MACHINE CONDITION

It's an old axiom of grinding that you cannot produce quality work on a machine in poor condition. It is true that you can compensate for worn bearings to some degree by substituting a softer wheel, but this is a trap to avoid if at all possible. Not that machine condition is the only factor to be considered, but it is probably the most important single factor in preventing problems.

One frequently hears of unusual causes of trouble, such as the story about a machine that had been performing satisfactorily for years on a given wheel specification. Then suddenly, with the same specification, the operator has nothing but problems. When the manufacturer checked, he found that the customer had moved the machine from the ground floor to the sixth floor, and the added vibration required a new specification. With a wheel that suited the new conditions, there were no further problems. Many problems are not so straightforward. Indeed, the ones that are most difficult are those where there is more than one condition causing the problem. However, in the discussion that follows, the best that can be done is to indicate that there is more than one condition that may cause a particular problem.

Before discussing specific problems, many of which are shown in Table 1, it is worth noting that good dressing practice is probably the best single way an operator can avoid problems. For example, if you cross feed the dresser too fast, you run the risk of causing distinctive spiral scratches on the surface of the workpiece. If you cross feed the diamond too slowly, the wheel is dressed too fine and the effect is similar to that resulting from grit size that is too fine or a wheel that is too hard. A wheel that is too hard also causes burning or burnishing of the work, or hollow spots because an area has become too hot, expanded, been ground off, and then contracted below the surface as it cooled.

Surface finish problems can occur if you have not dressed a small radius on the corner of the wheel, which can be done by touching the rotating wheel with a dressing stick. The first corrective measure you should take, if you have surface finish problems, is to redress the wheel.

SURFACE FINISH DEFECTS

Surface defects usually show up as unwanted scratches in the finish. Figure 1 shows four slightly oversize

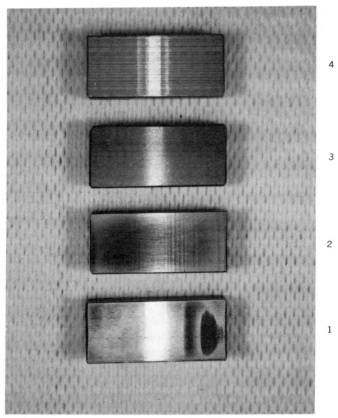

Figure 1. Four specimens of oil hardening steel, approximately 60 RC, with specific surface defects. (Mark Drzewiecki, Surface Finishes Inc.)

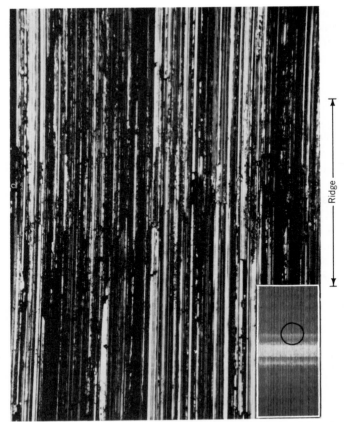

Figure 2. Chatter marks enlarged 80X. Marked inset (1.25X) shows enlarged area. (Mark Drzewiecki, Surface Finishes Inc.)

(1.25X) pieces that have been selected to illustrate common problems of this type. Not all of those discussed in the following paragraphs lend themselves to illustration.

Chatter Marks

These are sometimes referred to as "vibration marks" (Figure 2) because vibration is usually the cause. The reason may be from some outside source such as a punch press operating nearby and transmitting vibration through the machine to the wheel, so that the wheel "slips" instead of cutting for a moment. It may also come from within the machine, a wheel that is not balanced, or a wheel with one side soaked with coolant. It may be from worn wheel bearings, or even from a wheel that loads and/or glazes, and alternately slips, drops its load, and resumes cutting. This slipping and cutting alternation usually produces chatter marks that are close together; the more irregular and wider spaced marks are likely to come from some other source. The remedy may be just redressing, but it could also require a change of grinding wheel, a check on wheel bearings, or even, if none of these

work, a relocation of the machine.

Irregular Scratches

These are random scratches (Figure 3), often called, for obvious reasons, "fishtails." Usually the problem is the recirculation of dirt, bits of abrasive or "tramp" metal in the coolant, or dirt falling from under the wheel guard. If there is not enough coolant, for instance, it may be recirculating too fast for the settling of swarf (grinding particles) to take place. A similar result can occur if you slide a workpiece off a dirty chuck instead of lifting it off, which may be more difficult.

The first thing to do to remedy these problems is to clean out the inside of the wheel guard; then replace or clean the filters in the coolant tank and make sure there is enough coolant in the tank. It should be worthwhile to take some extra time to ensure that the chuck is clean before you load it again.

Discoloration

Discoloration, which is also known as burning or checking, results when a workpiece becomes over-

Figure 3. Grinding marks or "fishtails" also enlarged 80X. Inset (enlarged 1.25X) shows the damaged area. (Mark Drzewiecki, Surface Finishes Inc.)

Figure 4. Discoloration or burning also enlarged 80X, with damaged area marked on inset. (Mark Drzewiecki, Surface Finishes Inc.)

heated. It can be caused by insufficient coolant, improperly applied coolant, a wheel that is too hard or too fine, or by removing too much metal too rapidly from a small area (Figure 4). If carried on too long, it can result in expansion of the metal, probably in the middle of the surface being ground. Then, if the hump is ground off, there will be a low spot when the workpiece cools to normal temperature. However, the burning may be considered in the first place as a surface defect that may or may not be a problem, depending on the final use of the workpiece.

Probably the first remedy is to try to get the wheel to *act* softer by speeding up table traverse, redressing the wheel rougher, or taking lighter cuts and, of course, by checking the supply and the application of the coolant. Whenever there is not enough coolant, it is recirculated before it has had a chance to cool off as it should, and so it simply becomes warmer and warmer.

Burnishing

A burnished surface (Figure 5) is one that is smoothed

by abrasive rubbing instead of by cutting. It often looks good, and it may indeed not be a problem if the surface is not subject to wear. Technically, what happens when a surface is burnished is that the hills of the surface are heated enough so that they can be pushed into the valleys of the scratches. But with any wear at all, the displaced metal breaks loose, and the surface suddenly becomes much rougher. It usually results from using a wheel that is too hard. For that reason, the usual remedy is to get a softer wheel or to change grinding conditions to make the wheel act softer by increasing work speed, redressing the wheel, or taking lighter cuts.

Miscellaneous Surface Defects

As mentioned earlier in this unit, any ground surface is a planned series of scratches, preferably of uniform depth and direction. On the grinder you have been studying, the scratches are parallel and at right angles to the direction of work traverse. Any other pattern, scratches without a pattern, or any discoloration of the work surface can be considered a defect. Some

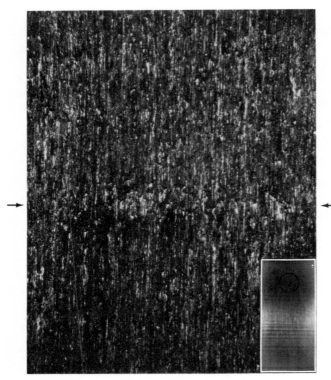

Figure 5. Burnished area enlarged 80X. Inset shows damaged area. (Mark Drzewiecki, Surface Finishes Inc.)

of the causes of these defects are that the wheel may be loose, the bearings may be worn, there may be vibration from some unsuspected source, the wheel may have been dressed too fast or too slow, or the wheel may be too rough or too fine. Sometimes the causes are so remote or obscure as to puzzle even experienced troubleshooters.

Work Not Parallel

It has been said repeatedly that the grinding line of the wheel is parallel to the top of the chuck, and this is true if the grinder is in good condition. However, if the chuck is out of line, dirty, or burred, this parallelism may no longer exist, and any workpiece may be out of parallel. Of the three conditions mentioned above, dirt and burring of the chuck are definitely your responsibility, but the chuck alignment

may or may not be. The remedies are usually obvious, as are those of other possible causes, as shown in Table 1. As you progress, you will develop a sort of routine to be followed for a given machine with this condition.

Work Out of Flat

Lack of flatness in a thick workpiece is likely to be the result of some local overheating, which causes an area of the surface to bulge. Then when the bulge is ground off and the work cools, there is a low spot.

Most flatness problems arise with thin work, and for very obvious reasons. Thin work does not have the bulk to absorb grinding heat without distortion. If it has been rolled, then stresses caused by the passage of the metal through the rolls may have been created, and the grinding may release these stresses on one side, causing the metal to warp or bow out of flat.

Correcting warpage in a thin workpiece is a matter of patience, a right start, and the minimum chuck power required to hold the workpiece in place. The procedure is as follows.

1. Using the least practical amount of chuck power, place the work on the chuck with the bowed side up so that the work rests on the ends.
2. Take a light cut with minimum downfeed. This should grind only the high spots on the work. Cutting should begin near the center.
3. Turn over the workpiece, shim the ends with paper, and take another light cut. This time cutting begins near the ends.
4. Repeat these steps, reducing the shims gradually, until the part is flat within specifications.

Finally, it should only have to be mentioned, as a reminder, that it is impossible to grind work flat if the chuck surface is not flat. When the chuck surface is between .0001 and .0002 in. out of flat, it is time to regrind it.

Perhaps the last word on solving the problems that arise in surface grinding is that care in avoiding the causes of trouble, such as careful and thorough cleaning of the chuck, frequent checking of the coolant level, the condition of the filters, and care in placing and removing workpieces from the chuck, can prevent many of the problems before they happen. It might be called preventive operation of the grinder.

Table 1

Summary of Surface Grinding Defects and Possible Causes

Causes	Burning or Checking	Burnishing of Work	Chatter Marks	Scratches on Work	Wheel Glazing	Wheel Loading	Work Not Flat	Work Out of Parallel	Work Sliding on Chuck
Machine Operation									
Dirty coolant				x		x			
Insufficient coolant	x						x	x	
Wrong coolant					x	x			
Dirty or burred chuck				x			x	x	
Inadequate blocking									x
Poor chuck loading							x	x	x
Sliding work off chuck				x					
Dull diamond					x				
Too fine dress	x				x	x	x		
Too long a grinding stroke								x	
Loose dirt under guard				x					
Grinding Wheel									
Too fine grain size	x				x	x			
Too dense structure					x	x			
Too hard grade	x	x	x		x	x	x		
Too soft grade			x	x					
Machine Adjustment									
Chuck out of line								x	
Loose or cracked diamond				x			x	x	
No magnetism									x
Vibration			x						
Condition of Work									
Heat treat stresses							x		
Thin							x		

Source: *Precision Surface Grinding.* Data courtesy of DoAll Company, 1964.

SELF-TEST

1. List at least three actions of an operator that will help prevent problems in surface grinding.
2. Name at least three general causes of surface grinding problems.
3. What is a surface defect?
4. What is the principal cause of chatter marks? What are some possible remedies?
5. "Fishtails" are another common problem. What are they, what causes them, and what should you do to get rid of them?
6. Name two of the problems that can result from over-heating work. What two or three things might you do if you suspect that a workpiece is getting too hot?
7. What is a burnished surface? Why is it objectionable? What are some remedies?
8. List at least two conditions that could produce out-of-parallel work.
9. List at least two conditions that could cause out-of-flat work.
10. Why is it difficult to grind thin work flat? How do you correct the condition?

APPENDIX I
TABLES

Table 1

Hardness and Tensile Strength Comparison Table

Hardness Conversion Table

Brinell		Rockwell			Brinell		Rockwell		
Indentation diameter (mm)	No.[a]	B	C	Tensile Strength (1000 psi approximately)	Indentation diameter (mm)	No.[a]	B	C	Tensile Strength (1000 psi approximately)
2.25	745		65.3		3.75	262	(103.0)	26.6	127
2.30	712		—		3.80	255	(102.0)	25.4	123
2.35	682		61.7		3.85	248	(101.0)	24.2	120
2.40	653		60.0		3.90	241	100.0	22.8	116
2.45	627		58.7		3.95	235	99.0	21.7	114
2.50	601		57.3		4.00	229	98.2	20.5	111
2.55	578		56.0		4.05	223	97.3	(18.8)	—
2.60	555		54.7	298	4.10	217	96.4	(17.5)	105
2.65	534		53.5	288	4.15	212	95.5	(16.0)	102
2.70	514		52.1	274	4.20	207	94.6	(15.2)	100
2.75	495		51.6	269	4.25	201	93.8	(13.8)	98
2.80	477		50.3	258	4.30	197	92.8	(12.7)	95
2.85	461		48.8	244	4.35	192	91.9	(11.5)	93
2.90	444		47.2	231	4.40	187	90.7	(10.0)	90
2.95	429		45.7	219	4.45	183	90.0	(9.0)	89
3.00	415		44.5	212	4.50	179	89.0	(8.0)	87
3.05	401		43.1	202	4.55	174	87.8	(6.4)	85
3.10	388		41.8	193	4.60	170	86.8	(5.4)	83
3.15	375		40.4	184	4.65	167	86.0	(4.4)	81
3.20	363		39.1	177	4.70	163	85.0	(3.3)	79
3.25	352	(110.0)	37.9	171	4.80	156	82.9	(0.9)	76
3.30	341	(109.0)	36.6	164	4.90	149	80.8		73
3.35	331	(108.5)	35.5	159	5.00	143	78.7		71
3.40	321	(108.0)	34.3	154	5.10	137	76.4		67
3.45	311	(107.5)	33.1	149	5.20	131	74.0		65
3.50	302	(107.0)	32.1	146	5.30	126	72.0		63
3.55	293	(106.0)	30.9	141	5.40	121	69.8		60
3.60	285	(105.5)	29.9	138	5.50	116	67.6		58
3.65	277	(104.5)	28.8	134	5.60	111	65.7		56
3.70	269	(104.0)	27.6	130					

[a] Values above 500 are for tungsten carbide ball; below 500 for standard ball.

Note 1. This is a condensation of Table 2, Report J417b, SAE 1971 Handbook. Values in ( ) are beyond normal range, and are presented for information only.

Note 2. The following is a formula to approximate tensile strength when the Brinell hardness is known:

$$\text{Tensile strength} = BHN \times 500$$

Source: Bethlehem Steel Corporation, *Modern Steels and Their Properties*, Handbook 3310, 1980.

Table 2

Flat Bar Steel—Weight per Linear Foot, Pounds

Thickness (in.)	Width in.											
	$\frac{3}{8}$	$\frac{1}{2}$	$\frac{5}{8}$	$\frac{3}{4}$	$\frac{7}{8}$	1	$1\frac{1}{4}$	$1\frac{1}{2}$	$1\frac{3}{4}$	2	$2\frac{1}{4}$	$2\frac{1}{2}$
$\frac{1}{8}$	.1594	.2125	.2656	.3188	.3719	.4250	.5313	.638	.744	.850	.956	1.063
$\frac{1}{4}$	.3188	.4250	.5313	.6375	.7438	.8500	1.0625	1.275	1.488	1.700	1.913	2.125
$\frac{3}{8}$	.4781	.6375	.7969	.9563	1.1156	1.2750	1.5938	1.913	2.231	2.550	2.869	3.188
$\frac{1}{2}$	.6375	.8500	1.0625	1.2750	1.4875	1.7000	2.1250	2.550	2.975	3.400	3.825	4.250
$\frac{5}{8}$	.7969	1.0625	1.3281	1.5938	1.8594	2.1250	2.6563	3.188	3.719	4.250	4.781	5.313
$\frac{3}{4}$	.9563	1.2750	1.5938	1.9125	2.2313	2.5500	3.1875	3.825	4.463	5.100	5.738	6.375
$\frac{7}{8}$	1.1156	1.4875	1.8594	2.2313	2.6031	2.9750	3.7188	4.463	5.206	5.950	6.694	7.438
1	1.2750	1.7000	2.1250	2.5500	2.9750	3.4000	4.2500	5.100	5.950	6.800	7.650	8.500
$1\frac{1}{8}$	1.4344	1.9125	2.3906	2.8688	3.3469	3.8250	4.7813	5.738	6.694	7.650	8.606	9.563
$1\frac{1}{4}$	1.5938	2.1250	2.6563	3.1875	3.7188	4.2500	5.3125	6.375	7.438	8.500	9.563	10.625
$1\frac{3}{8}$	1.7531	2.3375	2.9219	3.5063	4.0906	4.6750	5.8438	7.013	8.181	9.350	10.519	11.688
$1\frac{1}{2}$	1.9125	2.5500	3.1875	3.8250	4.4625	5.1000	6.3750	7.650	8.925	10.200	11.475	12.113
$1\frac{5}{8}$	2.0719	2.7625	3.4531	4.1438	4.8344	5.5250	6.9063	8.288	9.669	11.050	12.431	13.122
$1\frac{3}{4}$	2.2313	2.9750	3.7188	4.4625	5.2063	5.9500	7.4375	8.925	10.413	11.900	13.388	14.131
$1\frac{7}{8}$	2.3906	3.1875	3.9844	4.7813	5.5781	6.3750	7.9688	9.563	11.156	12.750	14.344	15.141
2	2.5500	3.4000	4.2500	5.1000	5.9500	6.8000	8.5000	10.200	11.900	13.600	15.300	16.150

					Width in.								
$2\frac{3}{4}$	3	$3\frac{1}{4}$	$3\frac{1}{2}$	$3\frac{3}{4}$	4	$4\frac{1}{4}$	$4\frac{1}{2}$	$4\frac{3}{4}$	5	$5\frac{1}{4}$	$5\frac{1}{2}$	$5\frac{3}{4}$	6
1.169	1.275	1.381	1.488	1.594	1.700	1.806	1.913	2.019	2.125	2.231	2.338	2.444	2.550
2.338	2.550	2.763	2.975	3.188	3.400	3.613	3.825	4.038	4.250	4.463	4.675	4.888	5.100
3.506	3.825	4.144	4.463	4.781	5.100	5.419	5.738	6.056	6.375	6.694	7.013	7.331	7.650
4.675	5.100	5.525	5.950	6.375	6.800	7.225	7.650	8.075	8.500	8.925	9.350	9.775	10.200
5.844	6.375	6.906	7.438	7.969	8.500	9.031	9.563	10.094	10.625	11.156	11.688	12.219	12.750
7.013	7.650	8.288	8.925	9.563	10.200	10.838	11.475	12.113	12.750	13.388	14.025	14.663	15.300
8.181	8.925	9.669	10.413	11.156	11.900	12.644	13.388	14.131	14.875	15.619	16.363	17.106	17.850
9.350	10.200	11.050	11.900	12.750	13.600	14.450	15.300	16.150	17.000	17.850	18.700	19.550	20.400
10.519	11.475	12.431	13.388	14.344	15.300	16.256	17.213	18.169	19.125	20.081	21.038	21.994	22.950
11.688	12.750	13.813	14.875	15.938	17.000	18.063	19.125	20.188	21.250	22.313	23.375	24.438	25.500
12.856	14.025	15.194	16.363	17.531	18.700	19.869	21.038	22.206	23.375	24.544	25.713	26.881	28.050
14.025	15.300	16.575	17.850	19.125	20.400	21.675	22.950	24.225	25.500	26.775	28.050	29.325	30.600
15.194	16.575	17.956	19.338	20.719	22.100	23.481	24.863	26.244	27.625	29.006	30.388	31.769	33.150
16.363	17.850	19.338	20.825	22.313	23.800	25.288	26.775	28.263	29.750	31.238	32.725	34.213	35.700
17.531	19.125	20.719	22.313	23.906	25.500	27.094	28.688	30.281	31.875	33.469	35.063	36.656	38.250
18.700	20.400	22.100	23.800	25.500	27.200	28.900	30.600	32.300	34.000	35.700	37.400	39.100	40.800

Table 3
Weights of Round Bars per Linear Foot

Size (diameter in in.)	Weight (pounds per foot)	Size (diameter in in.)	Weight (pounds per foot)	Size (diameter in in.)	Weight (pounds per foot)
$\frac{1}{8}$	.042	$1\frac{3}{4}$	8.178	$3\frac{7}{16}$	31.554
$\frac{3}{16}$	.094	$1\frac{13}{16}$	8.773	$3\frac{1}{2}$	32.712
$\frac{1}{4}$	.167	$1\frac{7}{8}$	9.388	$3\frac{9}{16}$	33.891
$\frac{5}{16}$	.261	$1\frac{15}{16}$	10.024	$3\frac{5}{8}$	35.090
$\frac{3}{8}$	.376	2	10.681	$3\frac{11}{16}$	36.311
$\frac{7}{16}$	.511	$2\frac{1}{16}$	11.359	$3\frac{3}{4}$	37.552
$\frac{1}{2}$	.668	$2\frac{1}{8}$	12.058	$3\frac{13}{16}$	38.814
$\frac{9}{16}$	.845	$2\frac{3}{16}$	12.778	$3\frac{7}{8}$	40.097
$\frac{5}{8}$	1.043	$2\frac{1}{4}$	13.519	$3\frac{15}{16}$	41.401
$\frac{11}{16}$	1.262	$2\frac{5}{16}$	14.280	4	42.726
$\frac{3}{4}$	1.502	$2\frac{3}{8}$	15.062	$4\frac{1}{16}$	44.071
$\frac{13}{16}$	1.763	$2\frac{7}{16}$	15.866	$4\frac{1}{8}$	45.438
$\frac{7}{8}$	2.044	$2\frac{1}{2}$	16.690	$4\frac{3}{16}$	46.825
$\frac{15}{16}$	2.347	$2\frac{9}{16}$	17.535	$4\frac{1}{4}$	48.233
1	2.670	$2\frac{5}{8}$	18.400	$4\frac{5}{16}$	49.662
$1\frac{1}{16}$	3.766	$2\frac{11}{16}$	19.287	$4\frac{3}{8}$	51.112
$1\frac{1}{8}$	3.379	$2\frac{3}{4}$	20.195	$4\frac{7}{16}$	52.583
$1\frac{3}{16}$	3.766	$2\frac{13}{16}$	21.123	$4\frac{1}{2}$	54.075
$1\frac{1}{4}$	4.173	$2\frac{7}{8}$	22.072	$4\frac{9}{16}$	55.587
$1\frac{5}{16}$	4.600	$2\frac{15}{16}$	23.042	$4\frac{5}{8}$	57.121
$1\frac{3}{8}$	5.049	3	24.033	$4\frac{11}{16}$	58.675
$1\frac{7}{16}$	5.518	$3\frac{1}{16}$	25.045	$4\frac{3}{4}$	60.250
$1\frac{1}{2}$	6.008	$3\frac{1}{8}$	26.078	$4\frac{13}{16}$	61.846
$1\frac{9}{16}$	6.520	$3\frac{3}{16}$	27.131	$4\frac{7}{8}$	63.463
$1\frac{5}{8}$	7.051	$3\frac{1}{4}$	28.206	$4\frac{15}{16}$	65.100
$1\frac{11}{16}$	7.604	$3\frac{5}{16}$	29.301	5	66.759
		$3\frac{3}{8}$	30.417		

Table 4
Table of Wire Gages

No. of Wire Gage	Stubs' Steel Wire	Birming-ham or Stubs' Iron Wire	American or B. & S.	New Am. S. & W. Co's Music Wire Gage	Imperial Wire Gage	Washburn & Moen, Worcester, Mass.	W. & M. Steel Music Wire
00000000	...	...	...	...	...	...	0.0083
0000000	...	...	...	...	...	0.490	0.0087
000000	...	...	...	0.004	0.464	0.4615	0.0095
00000	...	...	...	0.005	0.432	0.4305	0.010
0000	...	0.454	0.460	0.006	0.400	0.3938	0.011
000	...	0.425	0.40964	0.007	0.372	0.3625	0.012
00	...	0.380	0.3648	0.008	0.348	0.3310	0.0133
0	...	0.340	0.32486	0.009	0.324	0.3065	0.0144
1	0.227	0.300	0.2893	0.010	0.300	0.2830	0.0156
2	0.219	0.284	0.25763	0.011	0.276	0.2625	0.0166
3	0.212	0.259	0.22942	0.012	0.252	0.2437	0.0178
4	0.207	0.238	0.20431	0.013	0.232	0.2253	0.0188
5	0.204	0.220	0.18194	0.014	0.212	0.2070	0.0202
6	0.201	.0203	0.16202	0.016	0.192	0.1920	0.0215
7	0.199	0.180	0.14428	0.018	0.176	0.1770	0.023
8	0.197	0.165	0.12849	0.020	0.160	0.1620	0.0243
9	0.194	0.148	0.11443	0.022	0.144	0.1483	0.0256
10	0.191	0.134	0.10189	0.024	0.128	0.1350	0.027
11	0.188	0.120	0.090742	0.026	0.116	0.1205	0.0284
12	0.185	0.109	0.080808	0.029	0.104	0.1055	0.0296
13	0.182	0.095	0.071961	0.031	0.092	0.0915	0.0314
14	0.180	0.083	0.064804	0.033	0.080	0.0800	0.0326
15	0.178	0.072	0.057068	0.035	0.072	0.0720	0.0345
16	0.175	0.065	0.05082	0.037	0.064	0.0625	0.036
17	0.172	0.058	0.045257	0.039	0.056	0.0540	0.0377
18	0.168	0.049	0.040303	0.041	0.048	0.0475	0.0395
19	0.164	0.042	0.03589	0.043	0.040	0.0410	0.0414
20	0.161	0.035	0.031961	0.045	0.036	0.0348	0.0434
21	0.157	0.032	0.028462	0.047	0.032	0.03175	0.046
22	0.155	0.028	0.025347	0.049	0.028	0.0286	0.0483
23	0.153	0.025	0.022571	0.051	0.024	0.0258	0.051
24	0.151	0.022	0.0201	0.055	0.022	0.0230	0.055
25	0.148	0.020	0.0179	0.059	0.020	0.0204	0.0586
26	0.146	0.018	0.01594	0.063	0.018	0.0181	0.0626
27	0.143	0.016	0.014195	0.067	0.0164	0.0173	0.0658
28	0.139	0.014	0.012641	0.071	0.0149	0.0162	0.072
29	0.134	0.013	0.011257	0.075	0.0136	0.0150	0.076
30	0.127	0.012	0.010025	0.080	0.0124	0.0140	0.080
31	0.120	0.010	0.008928	0.085	0.0116	0.0132	...
32	0.115	0.009	0.00795	0.090	0.0108	0.0128	...
33	0.112	0.008	0.00708	0.095	0.0100	0.0118	...
34	0.110	0.007	0.006304	...	0.0092	0.0104	...
35	0.108	0.005	0.005614	...	0.0084	0.0095	...
36	0.106	0.004	0.005	...	0.0076	0.0090	...
37	0.103	...	0.004453	...	0.0068	...	...
38	0.101	...	0.003965	...	0.0060	...	...
39	0.099	...	0.003531	...	0.0052	...	...
40	0.097	...	0.003144	...	0.0048	...	...

Note: Dimensions are in inches.

Table 5

Useful Formulas for Finding Areas and Dimensions of Geometric Figures

A = Area; S = Side; D = Diagonal of circumscribed circle; d = Height of circular segment (distance cut into round stock to make 3 flats); h = Height of triangle.

Triangle (equilateral)

$A = Sh/2$
$S = D \times .866$
$d = D \times .25$
$h = D - d$

A = Area; S = Side; D = Diagonal; d = Height of circular segment (distance cut into round stock to make 4 flats).

Square

$A = S^2 = \frac{1}{2}d^2$
$S = D \times .7071$
$D = S \times 1.4142$
$d = D \times .14645$

A = Area; S = Side; D = Diagonal; R = Radius of circumscribed circle; r = Radius of inscribed circle; d = Height of circular segment (distance cut into round stock to make 6 flats).

Hexagon

$A = 2.598S^2 = 2.598R^2 = 3.464r^2$
$S = D \times .5 = R = 1.155r$
$d = D \times .067$
$r = .866S$

A = Area; S = Side; D = Diagonal; R = Radius of circumscribed circle; r = Radius of inscribed circle; d = Height of circular segment (distance cut into round stock to make 8 flats).

Octagon

$A = 4.828S^2$
$S = D \times .3827$
$R = 1.3075 = 1.082r$
$r = 1.207S$
$d = D \times .038$

A = Area; D = Diameter; R = Radius; C = Circumference

Circle

$A = \pi R^2 = 3.1416R^2 = .7854D^2$
$C = 2\pi R^2 = 6.2832R = 3.1416D$
$R = C \div 6.2832$

A = Area; l = Length of arc; C = Chord length; R = Radius; λ = angle (in degrees); d = Height of circular segment (distance cut into round stock to produce a flat of a given width).

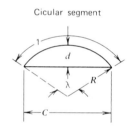

Cicular segment

$A = \frac{1}{2}[Rl - C(R - d)]$
$C = 2\sqrt{d(2R - d)}$
$d = R - \frac{1}{2}\sqrt{4R^2 - C^2}$
$l = .01745R\lambda$
$\lambda = \dfrac{57.296l}{R}$

Note: To find the weight of a metal bar of any cross-sectional shape, find the area of cross section, multiply by the length of the bar in inches and by the weight in pounds per cubic inch of the material.

The weights of some metals in pounds per cubic inch are as follows: steel, .284; aluminum, .0975; bronze, .317; copper, .321; lead, .409; and silver, .376.

Table 6

Inch-Metric Measures

Linear Measures

English	Metric
1 mile = 1760 yards = 5280 feet	10 millimeters (mm) = 1 centimeter (cm)
1 yard = 3 feet = 36 inches	10 centimeters = 1 decimeter (dm)
1 foot = 12 inches	10 decimeters = 1 meter (m)
1 inch = 1000 mils	1000 meters = 1 kilometer (km)

Conversion Factors

1 inch = 2.54 cm = 25.4 mm	1 millimeter = .03937 inch
1 foot = .3048 meter	10 millimeters (cm) = .3937 inch
1 yard = .9144 meter	1 meter = 39.37 inches
	3.2808 feet
	1.0936 yards
1 mile = 1.6093 km	1 kilometer = 1093.6 yards or .62137 mile

Table 7

Decimal and Metric Equivalents of Fractions of an Inch

Fractional Inch	Decimal Inch	Milli-meters	Fractional Inch	Decimal Inch	Milli-meters
$\frac{1}{64}$	0.015625	0.3969	$\frac{33}{64}$	0.515625	13.0969
$\frac{1}{32}$	0.03125	0.7937	$\frac{17}{32}$	0.53125	13.4937
$\frac{3}{64}$	0.046875	1.1906	$\frac{35}{64}$	0.546875	13.8906
$\frac{1}{16}$	0.0625	1.5875	$\frac{9}{16}$	0.5625	14.2875
$\frac{5}{64}$	0.078125	1.9844	$\frac{37}{64}$	0.578125	14.6844
$\frac{3}{32}$	0.09375	2.3812	$\frac{19}{32}$	0.59375	15.0812
$\frac{7}{64}$	0.109375	2.7781	$\frac{39}{64}$	0.609375	15.4781
$\frac{1}{8}$	0.125	3.1750	$\frac{5}{8}$	0.625	15.8750
$\frac{9}{64}$	0.140625	3.5719	$\frac{41}{64}$	0.640625	16.2719
$\frac{5}{32}$	0.15625	3.9687	$\frac{21}{32}$	0.65625	16.6687
$\frac{11}{64}$	0.171875	4.3656	$\frac{43}{64}$	0.671875	17.0656
$\frac{3}{16}$	0.1875	4.7625	$\frac{11}{16}$	0.6875	17.4625
$\frac{13}{64}$	0.203125	5.1594	$\frac{45}{64}$	0.703125	17.8594
$\frac{7}{32}$	0.21875	5.5562	$\frac{23}{32}$	0.71875	18.2562
$\frac{15}{64}$	0.234375	5.9531	$\frac{47}{64}$	0.734375	18.6531
$\frac{1}{4}$	0.25	6.3500	$\frac{3}{4}$	0.75	19.0500
$\frac{17}{64}$	0.265625	6.7469	$\frac{49}{64}$	0.765625	19.4469
$\frac{9}{32}$	0.28125	7.1437	$\frac{25}{32}$	0.78125	19.8437
$\frac{19}{64}$	0.296875	7.5406	$\frac{51}{64}$	0.796875	20.2406
$\frac{5}{16}$	0.3125	7.9375	$\frac{13}{16}$	0.8125	20.6375
$\frac{21}{64}$	0.328125	8.3344	$\frac{53}{64}$	0.828125	21.0344
$\frac{11}{32}$	0.34375	8.7312	$\frac{27}{32}$	0.84375	21.4312
$\frac{23}{64}$	0.359375	9.1281	$\frac{55}{64}$	0.859375	21.8281
$\frac{3}{8}$	0.375	9.5250	$\frac{7}{8}$	0.875	22.2250
$\frac{25}{64}$	0.390625	9.9219	$\frac{57}{64}$	0.890625	22.6219
$\frac{13}{32}$	0.40625	10.3187	$\frac{29}{32}$	0.90625	23.0187
$\frac{27}{64}$	0.421875	10.7156	$\frac{59}{64}$	0.921875	23.4156
$\frac{7}{16}$	0.4375	11.1125	$\frac{15}{16}$	0.9375	23.8125
$\frac{29}{64}$	0.453125	11.5094	$\frac{61}{64}$	0.953125	24.2094
$\frac{15}{32}$	0.46875	11.9062	$\frac{31}{32}$	0.96875	24.6062
$\frac{31}{64}$	0.484375	12.3031	$\frac{63}{64}$	0.984375	25.0031
$\frac{1}{2}$	0.50	12.7000	1	1.000000	25.4000

Table 8
Metric-Inch Conversion Table

Milli-meters	Decimal Inches	Milli-meters	Decimal Inches	Milli-meters	Decimal Inches	Milli-meters	Decimal Inches	Milli-meters	Decimal Inches	Milli-meters	Decimal Inches
.1	.00394	11	.4331	29	1.1417	47	1.8504	65	2.5590	83	3.2677
.2	.00787	12	.4724	30	1.1811	48	1.8898	66	2.5984	84	3.3071
.3	.01181	13	.5118	31	1.2205	49	1.9291	67	2.6378	85	3.3464
.4	.01575	14	.5512	32	1.2598	50	1.9685	68	2.6772	86	3.3858
.5	.01968	15	.5905	33	1.2992	51	2.0079	69	2.7165	87	3.4252
.6	.02362	16	.6299	34	1.3386	52	2.0472	70	2.7559	88	3.4646
.7	.0275	17	.6693	35	1.3779	53	2.0866	71	2.7953	89	3.5039
.8	.0315	18	.7087	36	1.4173	54	2.1260	72	2.8346	90	3.5433
.9	.03541	19	.7480	37	1.4567	55	2.1653	73	2.8740	91	3.5827
1	.0394	20	.7874	38	1.4961	56	2.2047	74	2.9134	92	3.6220
2	.0787	21	.8268	39	1.5354	57	2.2441	75	2.9527	93	3.6614
3	.1181	22	.8661	40	1.5748	58	2.2835	76	2.9921	94	3.7008
4	.1575	23	.9055	41	1.6142	59	2.3228	77	3.0315	95	3.7401
5	.1968	24	.9449	42	1.6535	60	2.3622	78	3.0709	96	3.7795
6	.2362	25	.9842	43	1.6929	61	2.4016	79	3.1102	97	3.8189
7	.2756	26	1.0236	44	1.7323	62	2.4409	80	3.1496	98	3.8583
8	.3150	27	1.0630	45	1.7716	63	2.4803	81	3.1890	99	3.8976
9	.3543	28	1.1024	46	1.8110	64	2.5197	82	3.2283	100	3.9370
10	.3937										

Table 9
Allowances for Fits of Bores in Inches

Diameter (in.)	Running and Sliding Fits (free to rotate and free to slide)	Standard Fits (readily assembled)	Driving Fits (permanent assembly, light drive)	Forced Fits (permanent assembly with hydraulic press)
Up to ½	+.0005 to +.001	+.00025 to +.0005	−.0005	−.00075
½ to 1	+.001 to +.0015	+.0003 to +.001	−.0075	−.0015
1 to 2	+.0015 to +.0025	+.0004 to +.0015	−.001	−.0025
2 to 3½	+.002 to +.003	+.0005 to +.002	−.0015	−.0035
3½ to 6	+.003 to +.004	+.00075 to +.003	−.002	−.0045

Note: In this table, shaft sizes are considered as nominal and bore sizes are varied for fits, thus a negative fit is an interference (press) fit and a positive fit is a free or loose fit.

Table 10
Metric Tap Drill Sizes

| Metric Tap Size | Recommended Metric Drill | | | | Closest Recommended Inch Drill | | | |
	Drill Size (mm)	Inch Equivalent	Probable Hole Size (in.)	Probable Percent of Thread	Drill Size	Inch Equivalent	Probable Hole Size (in.)	Probable Percent of Thread
M1.6 × .35	1.25	.0492	.0507	69	—	—	—	—
M1.8 × .35	1.45	.0571	.0586	69	—	—	—	—
M2 × .4	1.60	.0630	.0647	69	#52	.0635	.0652	66
M2.2 × .45	1.75	.0689	.0706	70	—	—	—	—
M2.5 × .45	2.05	.0807	.0826	69	#46	.0810	.0829	67
*M3 × .5	2.50	.0984	.1007	68	#40	.0980	.1003	70
M3.5 × .6	2.90	.1142	.1168	68	#33	.1130	.1156	72
*M4 × .7	3.30	.1299	.1328	69	#30	.1285	.1314	73
M4.5 × .75	3.70	.1457	.1489	74	#26	.1470	.1502	70
*M5 × .8	4.20	.1654	.1686	69	#19	.1660	.1692	68
*M6 × 1	5.00	.1968	.2006	70	#9	.1960	.1998	71
M7 × 1	6.00	.2362	.2400	70	$\frac{15}{64}$	.2344	.2382	73
*M8 × 1.25	6.70	.2638	.2679	74	$\frac{17}{64}$	.2656	.2697	71
M8 × 1	7.00	.2756	.2797	69	J	.2770	.2811	66
*M10 × 1.5	8.50	.3346	.3390	71	Q	.3320	.3364	75
M10 × 1.25	8.70	.3425	.3471	73	$\frac{11}{32}$	.3438	.3483	71
*M12 × 1.75	10.20	.4016	.4063	74	Y	.4040	.4087	71
M12 × 1.25	10.80	.4252	.4299	67	$\frac{27}{64}$	.4219	.4266	72
M14 × 2	12.00	.4724	.4772	72	$\frac{15}{32}$	.4688	.4736	76
M14 × 1.5	12.50	.4921	.4969	71	—	—	—	—
*M16 × 2	14.00	.5512	.5561	72	$\frac{35}{64}$	.5469	.5518	76
M16 × 1.5	14.50	.5709	.5758	71	—	—	—	—
M18 × 2.5	15.50	.6102	.6152	73	$\frac{39}{64}$	.6094	.6144	74
M18 × 1.5	16.50	.6496	.6546	70	—	—	—	—
*M20 × 2.5	17.50	.6890	.6942	73	$\frac{11}{16}$	.6875	.6925	74
M20 × 1.5	18.50	.7283	.7335	70	—	—	—	—
M22 × 2.5	19.50	.7677	.7729	73	$\frac{49}{64}$	.7656	.7708	75
M22 × 1.5	20.50	.8071	.8123	70	—	—	—	—
*M24 × 3	21.00	.8268	.8327	73	$\frac{53}{64}$	.8281	.8340	72
M24 × 2	22.00	.8661	.8720	71	—	—	—	—
M27 × 3	24.00	.9449	.9511	73	$\frac{15}{16}$	.9375	.9435	78
M27 × 2	25.00	.9843	.9913	70	$\frac{63}{64}$	.9844	.9914	70
*M30 × 3.5	26.50	1.0433						
M30 × 2	28.00	1.1024						
M33 × 3.5	29.50	1.1614						
M33 × 2	31.00	1.2205	Reaming Recommended to the Drill Size Shown					
M36 × 4	32.00	1.2598						
M36 × 3	33.00	1.2992						
M39 × 4	35.00	1.3780						
M39 × 3	36.00	1.4173						

Formula for metric tap drill size: Basic major diameter $-\dfrac{\% \text{ Thread} \times \text{Pitch (mm)}}{76.980}$ = Drilled hole size (mm)
(mm)

Formula for percent of thread: $\dfrac{76.980}{\text{Pitch (mm)}} \times \left[\text{Basic major diameter} - \text{Drilled hole size} \right]$ = Percent of thread
(mm) (mm)

Source: Material courtesy of TRW Inc., *New Greenfield Geometric ISO Metric Screw Thread Manual,* 1973.

BASIC DESIGNATIONS

ISO Metric Threads are designated by the letter "M" followed by the *nominal size* in millimeters, and the *pitch* in millimeters, separated by the sign "×."

Example: M16 × 1.5

Those numbers in the table marked with an asterisk are the commercially available sizes in the United States.

TOLERANCE SYMBOLS

$$\begin{array}{cccc} 3 & 4 & 5 & \boxed{6} \\ & 7 & 8 & 9 \end{array}$$

Numbers are used to define the amount of product tolerance permitted on either internal or external threads. Smaller grade numbers carry smaller tolerances, that is, grade 4 tolerances are smaller than grade 6 tolerances, and grade 8 tolerances are larger than grade 6 tolerances.

$$\text{e} \quad \boxed{\text{H}} \quad \text{G} \quad \boxed{\text{g}}$$

Letters are used to designate the "position" of the product thread tolerances relative to basic diameters. Lower case letters are used for external threads, and capital letters for internal threads.

In some cases the "position" of the tolerance establishes an allowance (a definite clearance) between external and internal threads.

By combining the tolerance amount number and the tolerance position letter, the *tolerance symbol* is established that identifies the actual maximum and minimum product limits for external or internal threads. Generally the first number and letter refer to the pitch diameter symbol. The second number and letter refer to the crest diameter symbol (minor diameter of internal threads or major diameter of external threads).

Example:

Pitch Diameter
Tolerance Symbol ⌐ 5g 6g ⌐ Crest Diameter
Tolerance Symbol

Where the pitch diameter and crest diameter tolerance symbols are the same, the symbol need only be given once.

Example:

6g
⊤
Pitch Diameter and Crest
Diameter Tolerance Symbol

It is recommended that the *coarse series* be selected whenever possible, and that *general purpose grade* 6 be used for both internal and external threads.

Tolerance positions "g" for external threads and "H" for internal threads are preferred.

Other product information may also be conveyed by the ISO metric thread designations. Complete specifications and product limits may be found in the ISO Recommendations or in the B1 report "ISO Metric Screw Threads."

Some examples of ISO Metric Thread designations are as follows:

M10
M18 × 1.5
M6 − 6H
M4 − 6g
M12 × 1.25 − 6H
M20 × 2 − 6H/6g
M6 × 0.75 − 7g 6g

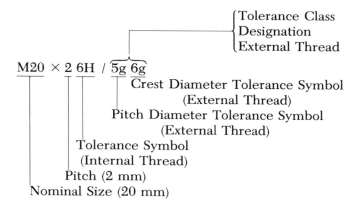

(**Source:** Material courtesy of TRW Inc., *New Greenfield Geometric Screw ISO Metric Thread Manual*, 1973.)

Table 11

Tap Drill Sizes for Unified and American Standard
Series Screw Threads

Thread Size	Threads Per Inch	Series	Tap Drill Diameter		Percent of Full Thread	Thread Size	Threads Per Inch	Series	Tap Drill Diameter		Percent of Full Thread
			Size	Inches					Size	Inches	
0	80	NF	1.25 mm	.0492	66	6	40	NF	31	.1200	55
			1.2 mm	.0472	79				32	.1160	68
			$\frac{3}{64}$"	.0469	81				33	.1130	77
1	64	NC	$\frac{1}{16}$"	.0625	51				34	.1110	83
			53	.0595	66	8	32	NC	28	.1405	58
			54	.0550	88				29	.1360	69
1	72	NF	$\frac{1}{16}$"	.0625	58				3.4 mm	.1338	74
			1.55 mm	.0610	66				3.3 mm	.1299	84
			53	.0595	75	8	36	NF	27	.1440	55
2	56	NC	49	.0730	56				28	.1405	65
			50	.0700	69				29	.1360	77
			1.75 mm	.0689	74				3.4 mm	.1338	83
			51	.0670	82	10	24	NC	22	.1570	61
2	64	NF	1.9 mm	.0748	55				24	.1520	70
			49	.0730	64				25	.1495	75
			1.8 mm	.0709	74				26	.1470	79
			50	.0700	79	10	32	NF	19	.1660	59
3	48	NC	45	.0820	63				20	.1610	71
			46	.0810	66				21	.1590	76
			47	.0785	76				22	.1570	81
			48	.0760	85	12	28	NF	14	.1820	63
3	56	NF	44	.0860	56				15	.1800	66
			2.15 mm	.0846	62				16	.1770	72
			45	.0820	73				17	.1730	79
			46	.0810	77	12	28	NF	12	.1890	58
4	40	NC	42	.0935	57				14	.1820	73
			2.3 mm	.0905	66				15	.1800	77
			43	.0890	71				16	.1770	84
			44	.0860	80	12	32	NEF	$\frac{3}{16}$"	.1875	70
4	48	NF	41	.0960	59				13	.1850	76
			42	.0935	68				14	.1820	84
			2.3 mm	.0905	79	$\frac{1}{4}$"	20	UNC	5	.2055	67
			43	.0890	85			NC	6	.2040	71
5	40	NC	37	.1040	65				7	.2010	75
			38	.1015	72				8	.1990	78
			39	.0995	78	$\frac{1}{4}$"	28	UNF	2	.2210	62
			40	.0980	83			NF	$\frac{7}{32}$"	.2187	67
5	44	NF	36	.1065	62				5.5 mm	.2165	72
			37	.1040	71				3	.2130	80
			38	.1015	79	$\frac{1}{4}$"	32	NEF	5.7 mm	.2244	63
			39	.0995	86				2	.2210	71
6	32	NC	34	.1110	66				$\frac{7}{32}$"	.2187	77
			35	.1100	69				5.5 mm	.2165	82
			36	.1065	77	$\frac{5}{16}$"	18	UNC	$\frac{17}{64}$"	.2656	65
			37	.1040	83			NC	G	.2610	71

Table 11 *(continued)*

Thread Size	Threads Per Inch	Series	Tap Drill Diameter Size	Inches	Percent of Full Thread
			F	.2570	77
			6.4 mm	.2520	84
$\frac{5}{16}''$	24	UNF NF	J	.2770	66
			I	.2720	75
			H	.2660	85
$\frac{5}{16}''$	32	NEF	7.3 mm	.2874	62
			7.2 mm	.2835	71
			$\frac{9}{32}''$	.2812	77
			J	.2770	87
$\frac{3}{8}''$	16	UNC NC	P	.3230	64
			O	.3160	72
			$\frac{5}{16}''$	.3120	77
			7.8 mm	.3071	83
$\frac{3}{8}''$	24	UNF NF	R	.3390	67
			8.5 mm	.3346	74
			Q	.3320	79
			$\frac{21}{64}''$	.3281	86
$\frac{7}{16}''$	14	UNC NC	$\frac{3}{8}''$	.3750	67
			U	.3680	75
			$\frac{23}{64}''$	.3594	84
$\frac{7}{16}''$	20	UNF NF	X	.3970	62
			$\frac{25}{64}''$	.3906	72
			W	.3860	79
$\frac{1}{2}''$	13	UNC NC	$\frac{7}{16}''$	.4375	62
			$\frac{27}{64}''$	.4219	78
			Z	.4130	87
$\frac{1}{2}''$	20	UNF NF	11.75 mm	.4626	57
			$\frac{29}{64}''$	.4531	72
$\frac{9}{16}''$	12	UNC NC	$\frac{1}{2}''$	.5000	58
			$\frac{31}{64}''$	.4844	72
			$\frac{15}{32}''$	.4687	86
$\frac{9}{16}''$	18	UNF NF	$\frac{33}{64}''$	.5156	65
			13 mm	.5118	70
			$\frac{1}{2}''$	.5000	86
$\frac{5}{8}''$	11	UNC NC	14 mm	.5512	63
			$\frac{35}{64}''$	.5469	66
			$\frac{17}{32}''$	.5312	79
$\frac{5}{8}''$	18	UNF NF	$\frac{37}{64}''$	.5781	65
			14.5 mm	.5709	75
			$\frac{9}{16}''$	.5625	87
$\frac{3}{4}''$	10	UNC NC	17 mm	.6693	62
			$\frac{21}{32}''$	.6562	72
			$\frac{41}{64}''$	.6406	84

Thread Size	Threads Per Inch	Series	Tap Drill Diameter Size	Inches	Percent of Full Thread
$\frac{3}{4}''$	16	UNF NF	$\frac{45}{64}''$	.7031	58
			$\frac{11}{16}''$	.6875	77
			$\frac{43}{64}''$	.6719	96
$\frac{7}{8}''$	9	UNC NC	$\frac{25}{32}''$	.7812	65
			$\frac{49}{64}''$	.7656	76
			$\frac{3}{4}''$	.7500	87
$\frac{7}{8}''$	14	UNF NF	$\frac{13}{16}''$	.8125	67
			20.5 mm	.8071	73
			$\frac{51}{64}''$	.7969	84
$1''$	8	UNC NC	$\frac{57}{64}''$	.8906	67
			$\frac{7}{8}''$	.8750	77
			$\frac{55}{64}''$	.8593	87
$1''$	12	UNF N	$\frac{15}{16}''$	.9375	58
			$\frac{59}{64}''$	.9218	72
			$\frac{29}{32}''$	.9062	87

Some Symbols Used for American Threads Are:

Symbol	Reference
NC	American National Coarse Thread Series
NF	American National Fine Thread Series
NEF	American National Extra Fine Thread Series
NS	Special Threads of American National Form
NH	Am. Natl. Hose Coupling and Fire Hose Coupling Thread
NPT	American Standard Taper Pipe Thread
NPTF	American Standard Taper Pipe Thread (Dryseal)
NPS	American Standard Straight Pipe Thread
ACME	Acme Threads—(Acme-C) Centralizing—(Acme-G) General Purpose
STUB ACME	Stub Acme Threads
V	A 60° "V" thread with truncated crests and roots. The theoretical "V" form is usually flattened several thousandths of an inch
SB	Manufacturers Stovebolt Standard Thread

Symbols Used for Unified Threads Are:

Symbol	Reference
UNC	Unified Coarse Thread Series
UNF	Unified Fine Thread Series
UNEF	Unified Extra Fine Thread Series

APPENDIX II
ANSWERS TO SELF-TESTS

SECTION A / UNIT 1 / SHOP SAFETY

SELF-TEST ANSWERS

1. Eye protection equipment.
2. Wear a safety goggle or full face shield. Prescription glasses may be made as safety glasses.
3. Shoes, short sleeves, short or properly secured hair, no rings and wristwatches, shop apron or shop coat with short sleeves.
4. Use of cutting fluids and vacuum dust collectors.
5. They may cause skin rashes or infections.
6. Bend knees and squat, lift with your legs, keeping your back straight.
7. Compressed air can propel chips through the air, implant dirt into skin, and possibly injure ear drums.
8. Good housekeeping includes cleaning oil spills, keeping material off the floor, and keeping aisles clear of obstructions.
9. In the horizontal position with a person on each end.
10. Do I know how to operate this machine?
    What are the potential hazards involved?
    Are all guards in place?
    Are my procedures safe?
    Am I doing something I probably should not do?
    Have I made all proper adjustments and tightened all locking bolts and clamps?
    Is the workpiece secured properly?
    Do I have proper safety equipment?
    Do I know where the stop switch is?
    Do I think about safety in everything I do?
11. No. You should use a hoist or crane for more than 20 or 30 lbs because a lathe chuck is in an awkward location.
12. (b) They are CO_2, dry chemical, and dry chemical multi-use (all purpose).

SECTION A / UNIT 2 / INTRODUCTION TO MECHANICAL HARDWARE

SELF-TEST ANSWERS

1. A bolt goes through parts being assembled and is tightened with a nut. A screw is used where a part is internally threaded and no nut is needed.
2. The minimum recommended thread engagement for a screw in an assembly is as much as the screw diameter; a better assembly will result when $1\frac{1}{2}$ times the screw diameter is used.
3. Class 2 threads are found on most screws, nuts, and bolts used in the manufacturing industry. Cars and machine tools would be good examples.
4. Machine bolts are not machined to the precise dimensions of capscrews. Machine bolts have coarse threads where capscrews may have coarse or fine threads. Machine bolts have many uses in the construction industry, and capscrews are usually used in precision assemblies.
5. The formula is D = number of the machine screw times .013 in. plus .060 in. $D = 8 \times .013$ in. plus $.060 = 0.164$.
6. Set screws are used to secure gears, pulleys, and collars on shafts.
7. Stud bolts can be used instead of long bolts. Stud bolts are used to aid in the assembly of heavy parts by acting as guide pins.

8. Thread forming screws form threads by displacing material. Thread cutting screws produce threads by actually cutting grooves and making chips.
9. Castle nuts can be secured on a bolt with a cotter pin to prevent their accidental loosening.
10. Cap nuts are used because of their neat appearance. They also protect projecting threads from damage.
11. Flat washers protect the surface of parts from being marred by the tightening of screws or nuts. Flat washers also provide a larger contact area than nuts and screw heads to distribute the clamping pressure over a larger area.
12. A helical spring lock washer prevents the unplanned loosening of nut and bolt or screw assemblies. Spring lock washers will also provide for a limited amount of take-up when expansion or contraction takes place.
13. Internal-external tooth lock washers are used on oversized holes or to provide a large bearing surface.
14. Dowel pins are used to achieve accurate alignment between two or more parts.
15. Taper pins give accurate alignment to parts that have to be disassembled frequently.
16. Roll pins are used to align parts. Holes to receive roll pins do not have to be reamed, which is necessary for dowel pins and taper pins.
17. Retaining rings are used to hold bearings or seals in bearing housings or on shafts. Retaining rings have a spring action and are usually seated in grooves.
18. Keys transmit the driving force between a shaft and pulley.
19. Woodruff keys are used where only light loads are transmitted.
20. Gib head keys are used to transmit heavy loads. These keys are installed and removed from the same side of a hub and shaft assembly.

SECTION A / UNIT 3 / SELECTION AND IDENTIFICATION OF METALS

SELF-TEST ANSWERS

1. Carbon and alloy steels are designated by the numerical SAE or AISI system.
2. The three basic types of stainless steels are: martensitic (hardenable) and ferritic (nonhardenable)—both magnetic and of the 400 series—and austenitic (nonmagnetic and nonhardenable, except by work hardening) of the 300 series.
3. The identification for each piece would be as follows:
   (a) AISI C1020 CF is a soft, low carbon steel with a dull metallic luster surface finish. Use the observation test, spark test, and file test for hardness.
   (b) AISI B1140 (G and P) is a medium carbon, resulfurized, free machining steel with a shiny finish. Use the observation test, spark test, and machinability test.
   (c) AISI C4140 (G and P) is a chromium-molybdenum alloy, medium carbon content with a shiny finish. Since an alloy steel would be harder than a similar carbon or low carbon content steel, a hardness test should be used such as the file or scratch test to compare with known samples. The machinability test would be useful as a comparison test.
   (d) AISI 8620 HR is a tough low carbon steel used for carburizing purposes. A hardness test and a machinability test will immediately show the difference from low carbon hot rolled steel.
   (e) AISI B1140 (ebony) is the same as the resulfurized steel in (b), only the finish is different. The test would be the same as for (b).
   (f) AISI C1040 is a medium carbon steel. The spark test would be useful here as well as the hardness and machinability tests.
4. A magnetic test can quickly determine whether it is a ferrous metal or perhaps nickel since they will both be attracted by a magnet. If the metal is white in color, a spark test will be needed to determine whether it is a nickel casting or one of white cast iron, since they are similar in appearance. If a small piece can be broken off, the fracture will show whether it is white or gray cast iron. Gray cast iron will leave a black smudge on the finger. If it is cast steel, it will be more ductile than cast iron and a spark test should reveal a smaller carbon content.
5. O1 refers to an alloy type oil hardening (oil quench) tool steel. W1 refers to a water hardening (water quench) tool steel.
6. (a) No. (b) Hardened tool steel or case hardened steel.
7. Austenitic (having a face centered cubic unit cell in its lattice structure). Examples are aluminum, nickel, 300 series stainless steel, and high manganese alloy steel.
8. Nickel is a nonferrous metal that has magnetic properties. Some alloy combinations of nonferrous metals make strong permanent magnets; for example, the well-known Alnico magnet—an alloy of aluminum, nickel, and cobalt.

9. Some properties of steel to be kept in mind when ordering or planning for a job would be: strength, machinability, hardenability, weldability (if welding is involved), fatigue resistance, and corrosion resistance (especially if the piece is to be exposed to a corrosive atmosphere).
10. *Advantages:* Since aluminum is about one-third lighter than steel, it is used extensively in aircraft. It also forms an oxide on the surface that resists further corrosion. *Disadvantages:* The initial cost is much greater. Higher strength aluminum alloys cannot be welded.
11. The letter "H" following the four digit number always designates strain or work hardening. The letter "T" refers to heat treatment.
12. Magnesium weighs approximately one-third less than aluminum and is approximately one-quarter the weight of steel. Magnesium will burn in air when finely divided.
13. Copper is most extensively used in the electrical industry because of its low resistance to the passage of current when it is unalloyed with other metals. Copper can be hardened or work hardened and certain alloys may be hardened by a solution heat treatment and aging process.
14. Bronze is basically copper and tin. Brass is basically copper and zinc.
15. Nickel is used to electroplate surfaces of metals for corrosion resistance, and as an alloying element with steels and nonferrous metals.
16. Both resist deterioration from corrosion.
17. Alloy.
18. Tin, lead, and cadmium.
19. Die cast metals, sometimes called "pot metal."
20. Wrought aluminum is stronger.
21. Large rake angles (12 to 20 degrees back rake). Use of a lubricant. Proper cutting speeds.
22. No. If a fire should start in the chips, the water based coolant will intensify the burning.
23. The rake should be zero on all cutting tools for brasses or bronzes.
24. Tungsten is combined with carbon to form a tungsten carbide powder that is compressed into briquettes and sintered in a furnace. The resultant tungsten carbide cutting tool is extremely hard and will resist the high temperatures of machining.

SECTION A / UNIT 4 / USE OF PEDESTAL AND PORTABLE GRINDERS

SELF-TEST ANSWERS

1. The shaping and sharpening of tools.
2. Eye protection, no long hair or loose clothing; keep fingers away from the rotating wheel.
3. The soft metals fill the voids in the aluminum oxide wheels and the additional pressure can cause them to shatter. Silicon carbide wheels break down more readily, keeping the wheel from becoming loaded.
4. By ring testing it. A clear ringing sound indicates it is solid, but a dull thud indicates it is cracked.
5. To keep small parts from catching and jamming between the rest and the wheel. This prevents accidents to your fingers where the jammed tool can suddenly force a finger against the wheel.
6. By holding the piece with a locking plier.
7. Overheating will cause carbon tool steel to become softened by tempering.
8. On the periphery (circumference) of the wheel.
9. The possibility of electric shock due to inadequate grounding or a frayed wire.
10. (a) Body and eye protection.
    (b) Avoid grinding around flammable liquid or containers that once held such a liquid. A spark going through a hole in a steel drum containing these fumes can have the explosive force of several sticks of dynamite.
    (c) Let the wheel come to a full stop before setting it down.
    (d) Don't bounce or strike a running wheel.
    (e) Keep the wheel away from your body at all times.

SECTION B / UNIT 1 / WORKHOLDING FOR HAND OPERATIONS

SELF-TEST ANSWERS

1. The vise should be positioned so that a long piece may be held vertically in the jaws without interference from the workbench.

2. The solid base and the swivel base types.
3. By the width of the jaws.
4. Pin vise or toolmaker vise.
5. Insert jaws are hardened and have diamond pattern or criss-cross serrations.
6. Copper, soft metal, or wood may be used to protect finishes from insert jaw serrations.
7. Vises used for sheet metal work have smooth, deep jaws.
8. Vises are used for holding work for assembly and disassembly; for filing, hacksawing, and bending light metal.
9. "Cheater" bars should never be used on the handle. The movable jaw slide bar should never be hammered upon, and excessive heat should never be applied to the jaws.
10. The vise should be taken apart, cleaned, and the screw and nut cleaned in solvent. A heavy grease should be packed on the screw and thrust collars before reassembly.

SECTION B / UNIT 2 / ARBOR AND SHOP PRESSES

SELF-TEST ANSWERS

1. To use the arbor press without instruction is unsafe for the operator; also very expensive equipment and materials may be damaged or ruined.
2. Arbor presses are hand powered and can be mechanical or hydraulic. Large power driven presses do not provide the "feel" needed when pressing delicate parts.
3. The arbor press is used for installing and removing mandrels, bushings, and ball bearings. The hydraulic shop press is also used for straightening and bending.
4. The shaft has seized or welded in the bore because it had not been lubricated with pressure lubricant.
5. A loose and rounded ram could cause a bushing to tilt or twist sideways while pressing and thus be ruined. In any case, the operator should always check to see if a bushing is going in straight. A pressing plug with a pilot would be helpful here.
6. No. Thirty tons would deform the extended end of the bushing. Just enough pressure should be applied to press in the bushing and when it contacts the press plate at the bottom, more pressure will be sensed. At that point it is time to stop.
7. Pressing on the inner race with the outer race supported will damage or break the bearing.
8. Ordinary shafts with press fits are not tapered but have the same dimension along the pressing length. Mandrels taper .006 in./ft, which causes them to tighten in the bore somewhere along their length.
9. The two most important steps, assuming that the dimensions are all correct, are to:
   (a) Make sure the bore has a good chamfer and the bushing should also have a chamfer or "start."
   (b) Apply high pressure lubricant to the bore and the bushing.
10. Five ways to avoid tool breakage and other problems when broaching keyways in the arbor press:
   (a) Make sure the press ram is not loose and check to see that the proper hole in the press plate is under the work so that the broach has clearance to go through the work.
   (b) Clean and lubricate the broach, especially the back edge between each cut.
   (c) Do not use a broach on hard materials (over Rc 35).
   (d) Use the right size bushing for the bore and broach.
   (e) Make sure at least two teeth are continuously engaged in the work.

SECTION B / UNIT 3 / NONCUTTING HAND TOOLS

SELF-TEST ANSWERS

1. Precision layout and grinding work.
2. By the longest opening.
3. Light duty C-clamps are used for holding workpieces as in drilling machines. Heavy duty clamps are used to hold heavy parts such as plates for welding or machining.
4. False. This practice can quickly ruin good machinery so that the proper tool can never be used.
5. The principal advantage of the lever jawed wrench is its great holding power. Most types have hard serrated jaws and so should not be used on nuts and bolt heads.
6. No. Hammers used for layout work ranges from 4 to 10 ounces in weight; a smaller one should be used with a prick punch and for delicate work.

7. Soft hammers and mallets are made for this purpose. When setting down work in a drill press or milling machine vise, for instance, a lead hammer is best because it has no rebound.
8. The box type, either socket or end wrench, would be best as it provides contact on the six points of the capscrew, thus avoiding the damage and premature wear that would be caused by using an adjustable wrench.
9. The hard serrated jaws will damage machine parts. Pipe wrenches should be used on pipe and pipe fittings only.
10. Standard screwdrivers should have the right width blade to fit the screw head. They should be shaped correctly and, if worn, reground and shaped properly.
11. Undertorquing can cause parts to come loose and also cause fatigue failures. Overtorquing can cause warping of parts and weakening of the fastener that leads to failure from breakage or fatigue.
12. By calculation of fastener strength or by using a table that gives inch pounds or foot pounds for standard size bolts with standard grade markings.

SECTION B / UNIT 4 / CUTTING HAND TOOLS: HACKSAWS

SELF-TEST ANSWERS

1. The kerf is the groove produced in the work by a saw blade.
2. The set on a saw blade is the width of the teeth that are bent out from the blade back.
3. The pitch of a hacksaw blade refers to the number of teeth per inch on a saw blade.
4. The first consideration in the selection of a saw blade is the kind of material being cut. For soft materials use a coarse tooth blade and for harder materials use a fine pitch blade. The second point to watch is that at least three teeth should be cutting at the same time.
5. The two basic kinds of saw blades are the all-hard blade and the flexible blade.
6. Generally a speed between 40 and 60 strokes per minute is suggested. It is best to use long and slow strokes utilizing the full length of the blade.
7. Excessive dulling of saw blades is caused by pressure on the saw blade on the return stroke, sawing too fast, letting the saw slide over the workpiece without any cutting pressure, or applying too much pressure.
8. Saw blades break if too much pressure is used or if the blade is not sufficiently tightened in the saw frame.
9. When the saw blade is used, the set wears and makes the kerf cut narrower than the kerf cut with a new blade. If a cut started with a used blade cannot be finished with that blade, but has to be completed with a new blade, the workpiece should be turned over and a new cut started from the opposite end from the original cut. A new blade, when used in a kerf started with a used blade, would lose its set immediately and start binding in the groove.
10. When the blade breaks, it shatters causing blade particles to fly quite a distance with the possibility of injuring someone. Should the blade break while sawing, it may catch the operator off balance and cause him to push his hand into the workpiece while following through with his sawing stroke. Serious cuts or abrasions can be the result of this action.

SECTION B / UNIT 5 / CUTTING HAND TOOLS: FILES

SELF-TEST ANSWERS

1. By its length, shape, cut, and coarseness.
2. Single cut, double cut, curved cut, and rasp.
3. Four out of these: rough, coarse, bastard, second cut, smooth, and dead smooth.
4. The double cut file.
5. To make it possible to file a flat surface by offsetting the tendency to rock a file. To compensate for the slight downward deflection when pressure is applied while filing. To concentrate pressure on fewer teeth for deeper penetration.
6. A blunt file has the same cross-sectional area from heel to point, where a tapered file is larger at the heel than at the point.
7. A mill file is thinner than a comparable size flat file.
8. A warding file is rectangular in shape and tapers to a small point, making filing possible in narrow slots and grooves.

9. Swiss pattern files are more precise in construction; they are more slender and have teeth to the extreme edges.
10. Coarseness is identified by numbers from 00 fine to 6 coarse.
11. Where files touch each other, teeth break off or become dull.
12. Too much pressure will break teeth off a file. It also will cause pinning and scratching of the work surface.
13. Files in contact with each other, files rubbing over the work without any pressure being applied, filing too fast, or filing on hardened materials causes files to dull.
14. As a safety precaution; an unprotected tang can cause serious injury.
15. Measuring the workpiece for flatness and size assures the craftsman that the filing is done in the right place and that the filing is stopped before a piece is undersize. Measuring often is not a waste of time.
16. Touching a workpiece is just like lubricating the workpiece or the file so that it slips over the work without cutting. This also causes a file to dull quickly.
17. A soft workpiece requires a file with coarser teeth because there is less resistance to tooth penetration. A fine toothed file would clog up on soft materials. For harder materials use a fine toothed file in order to have more teeth making smaller chips.
18. The drawfiling stroke should be short enough that the file never slips over the ends of the workpiece. Care should be taken that no hollow surface is created through too short of a stroke.
19. Pressure is only applied on the forward stroke, which is the cutting stroke.
20. Rotating a round file clockwise while filing makes the file cut better and improves the surface finish.

SECTION B / UNIT 6 / HAND REAMERS

SELF-TEST ANSWERS

1. Hand reamers have a square on the shank and a long starting taper on the fluted end.
2. A reamer does its cutting on the tapered portion. A long taper will help in keeping a reamer aligned with the hole.
3. Spiral fluted reamers cut with a shearing action. They will also bridge over keyways and grooves without chattering.
4. The shank diameter is usually a few thousandths of an inch smaller than the nominal size of the reamer. This allows the reamer to pass through the hole without marring it.
5. Expansion reamers are useful to increase hole sizes by a very small amount.
6. Expansion reamers can only be adjusted a small amount by moving a tapered internal plug. Adjustable reamers have a larger range of adjustments, from $\frac{1}{32}$ in. on small diameters to $\frac{5}{16}$ in. on large reamers. Adjustable reamers have removable blades. Size changes are made by moving these blades with nuts in external tapered slots.
7. Taper pin reamers finish holes to receive taper pins that are used to fix the position of parts in relationship to each other.
8. Coolants are used to dissipate the heat generated by the reaming process, but in reaming, coolants are more important in obtaining a high quality surface finish of the hole.
9. Reamers dull rapidly if they should be rotated backwards.
10. The hand reaming allowance is rather small, only between .001 and .005 in.

SECTION B / UNIT 7 / IDENTIFICATION AND USES OF TAPS

SELF-TEST ANSWERS

1. The length of the chamfer.
2. Spiral pointed taps are used on through holes or blind holes with sufficient chip space at the bottom of a hole.
3. Spiral fluted taps draw the chips out of the hole and are useful when tapping a hole that has a keyseat in it bridged by the helical flutes.
4. Thread forming or fluteless taps do not produce chips because they do not cut threads. Their action can be compared to thread rolling in that material is being displaced in grooves to form ridges shaped in the precise form of a thread.
5. Taper pipe taps are identified by the taper of the body of the tap, which is $\frac{3}{4}$ in. per foot of length; also by the size marked on the shank.

6. When an Acme thread is cut, the tap required cuts too much material in one pass. To obtain a quality thread, a roughing pass and then a finishing pass are needed.
7. Taps are driven with tap wrenches or T-handle tap wrenches.
8. A hand tapper is a fixture used to hold a tap in precise alignment while hand tapping holes.
9. A tapping attachment is used when tapping holes in a machine or for production tapping.
10. The strength of a tapped hole is determined by the kind of material being tapped, the percentage of thread used, and the length or depth of thread engagement.
11. Holes should be tapped deep enough to provide 1 to $1\frac{1}{2}$ times the tap diameter of usable thread.
12. Taps break because holes are drilled too shallow, chips are packed tight in the flutes, hard materials or hard spots are encountered, inadequate or the wrong kind of lubricant is used, or the cutting speed used is too great.
13. Tapped holes that are rough and torn are often caused by dull taps, chips clogging the flutes, insufficient lubrication, wrong kind of lubrication, or already rough holes being tapped.
14. Oversize tapped holes can be caused by a loose machine spindle or a worn tap holder, misaligned spindle, oversized tap, a dull tap, chips packed in the flutes, and build up on cutting edges.
15. When using large diameter taps and when fine pitches are used.

SECTION B / UNIT 8 / THREAD CUTTING DIES AND THEIR USES
SELF-TEST ANSWERS

1. A die is used to cut external threads.
2. A die stock is used to hold the die when hand threading. A special die holder is used when machine threading.
3. The size of thread cut can only be changed a very small amount on round adjustable dies. Too much expansion or contraction may break the die.
4. The purpose of the guide is to align the die square to the workpiece to be threaded.
5. When assembling a two-piece die collet, be sure both die halves are marked with the same serial number and that the starting chamfer on the dies are toward the guide.
6. Hexagon rethreading dies are used to clean and recut slightly damaged or rusty threads.
7. The chamfer on the cutting end of a die distributes the cutting force over a number of threads and aids in starting the thread cutting operation.
8. Cutting fluids are very important in threading to achieve thread with a good surface finish, close tolerance, and to give long tool life.
9. Before a rod is threaded, it should be measured to assure its size is no larger than its nominal size. Preferably, it is .002 to .005 in. undersize.
10. The chamfer on a rod before threading makes it easy to start a die. It also protects the starting thread on a finished bolt.

SECTION C / UNIT 1 / SYSTEMS OF MEASUREMENT
SELF-TEST ANSWERS

1. To find in. knowing mm, multiply mm by .03937; $35 \times .03937 = 1.378$ in.
2. To find mm knowing in., multiply in. by 25.4; $.125 \times 25.4 = 3.17$ mm.
3. To find cm knowing in., multiply in. by 2.54; $6.273 \times 2.54 = 15.933$ cm.
4. Length.
5. Degrees, minutes, and seconds.
6. It is checked with an acceptable standard and adjusted if necessary.
7. To find mm knowing in., multiply in. by 25.4; $.050 \times 25.4 = 1.27$ mm.
8. 10 mm = 1 cm; therefore, to find cm knowing mm, divide by 10.
9. To find in. knowing mm, multiply mm by .03937; $.02 \times .03937 = .0008$ in. The tolerance would be ±.0008 in.
10. Yes, by the use of appropriate conversion dials.

SECTION C / UNIT 2 / USING STEEL RULES
SELF-TEST ANSWERS

Fractional Inch Rules
Figure 26 $A = 1\frac{1}{4}$ in.
Figure 26 $B = 2\frac{1}{8}$ in.
Figure 26 $C = \frac{15}{16}$ in.
Figure 26 $D = 2\frac{5}{16}$ in.
Figure 26 $E = \frac{15}{32}$ in.
Figure 26 $F = 2\frac{25}{32}$ in.
Figure 26 $G = \frac{63}{64}$ in.
Figure 26 $H = 1\frac{59}{64}$ in.

Decimal Inch Rules
Figure 27 $A = $ .300 in.
Figure 27 $B = $ .510 in.
Figure 27 $C = 1.020$ in.
Figure 27 $D = 1.190$ in.
Figure 27 $E = 1.260$ in.

Metric Rules
Figure 28 $A = 11$ mm or 1.1 cm
Figure 28 $B = 27$ mm or 2.7 cm
Figure 28 $C = 52$ mm or 5.2 cm
Figure 28 $D = 7.5$ mm or .75 cm
Figure 28 $E = 20.5$ mm or 2.05 cm
Figure 28 $F = 45.5$ mm or 4.55 cm

SECTION C / UNIT 3 / USING VERNIER CALIPERS AND VERNIER DEPTH GAGES
SELF-TEST ANSWERS

Reading Inch Vernier Calipers
Figure 13a. 1.304 in.
Figure 13b. .492 in.
Figure 13c. .532 in.
Figure 13d. .724 in.

Reading Metric Vernier Calipers
Figure 14a. 20.26 mm
Figure 14b. 14.50 mm
Figure 14c. 29.84 mm
Figure 14d. 35.62 mm

Reading Inch Vernier Depth Gages
Figure 15a. .943 in.
Figure 15b. 1.326 in.
Figure 15c. 2.436 in.
Figure 15d. 3.768 in.

SECTION C / UNIT 4 / USING MICROMETER INSTRUMENTS
SELF-TEST ANSWERS

1. Anyone taking pride in his tools usually takes pride in his workmanship. The quality of a product produced

depends to a large extent on the accuracy of the measuring tools used. A skilled craftsman protects his tools because he guarantees his product.

2. Moisture between the contact faces can cause corrosion.

3. Even small dust particles will change a dimension. Oil or grease attract small chips and dirt. All of these can cause incorrect readings.

4. A measuring tool is no more discriminatory than the smallest division marked on it. This means that a standard micrometer can discriminate to the nearest thousandth. A vernier scale on a micrometer will make it possible to discriminate a reading to one ten-thousandths of an inch under controlled conditions.

5. The reliability of a micrometer depends on the inherent qualities built into it by its maker. Reliability also depends upon the skill of the user and the care the tool receives.

6. The sleeve is stationary in relation to the frame and is engraved with the main scale, which is divided into 40 equal spaces each equal to .025 in. The thimble is attached to the spindle and rotates with it. The thimble circumference is graduated with 25 equal divisions, each representing a value of .001 in.

7. There is less chance of accidentally moving the thimble when reading a micrometer while it is still in contact with the workpiece.

8. Measurement should be made at least twice. On critical measurements, checking the dimensions additional times will assure that the size measurement is correct.

9. As the temperature of a part is increased, the size of the part will increase. When a part is heated by the machining process, it should be permitted to cool down to room temperature before being measured. Holding a micrometer by the frame for an extended period of time will transfer body heat through the hand and affect the accuracy of the measurement taken.

10. The purpose of the ratchet stop or friction thimble is to enable equal pressure to be repeatedly applied between the measuring faces and the object being measured. Use of the ratchet stop or friction thimble will minimize individual differences in measuring pressure applied by different persons using the same micrometer.

Exercise Answers

Outside micrometer readings (Figures 28a to 28e):
Figure 28a.   .669 in.
Figure 28b.   .787 in.
Figure 28c.   .237 in.
Figure 28d.   .994 in.
Figure 28e.   .072 in.

Vernier micrometer readings (Figures 29a to 29e):
Figure 29a.   .3749 in.
Figure 29b.   .5377 in.
Figure 29c.   .3123 in.
Figure 29d.   .2498 in.
Figure 29e.   .1255 in.

Metric micrometer readings (Figures 30a to 30e):
Figure 30a.   21.21 mm
Figure 30b.   13.27 mm
Figure 30c.   9.94 mm
Figure 30d.   5.59 mm
Figure 30e.   4.08 mm

Inside micrometer readings (Figures 31a to 31e):
Figure 31a.   1.617
Figure 31b.   2.000
Figure 31c.   2.254
Figure 31d.   2.562
Figure 31e.   2.784

Depth micrometer readings (Figures 32a to 32e):
Figure 32a.   .535 in.
Figure 32b.   .815 in.
Figure 32c.   .732 in.
Figure 32d.   .535 in.
Figure 32e.   .647 in.

SECTION C / UNIT 5 / USING COMPARISON AND ANGULAR MEASURING INSTRUMENTS

SELF-TEST ANSWERS

1. Comparison measurement is measurement where an unknown dimension is compared to a known dimension. This often involves a transfer device that represents the unknown and is then transferred to the known where the reading can be determined.
2. Most comparison instruments do not have the capability to show measurement directly. Direct reading measuring instruments are more reliable than transfer type tools since the measuring "feel" with its potential for error must be made twice with transfer measuring devices.
3. If the axis of measurement is not the same as the axis of the measuring instrument, an error will result. Only dial indicators should be used to make direct measurements; dial test indicators will always have an error in measurement due to the difference between the arc and chord length in the tip travel.
4. Error can be reduced by making sure that the axis of the measuring instrument is exactly in line with the axis of measurement.
5. (c) Adjustable parallel.
   (i) and (j) Dial test indicator and vernier height gage.
   (d) Radius gage.
   (g) Combination square.
   (b) Telescope gage.
   (e) Thickness (feeler) gage.
6. Plate protractor and machinist's combination set bevel protractor.
7. Five minutes of arc.
8. 50° (Figure 52).
   83°45′ (Figure 53).
   34°30′ (Figure 54).
   61°45′ (Figure 55).
   56°25′ (Figure 56).

SECTION D / UNIT 1 / CUTOFF MACHINES

SELF-TEST ANSWERS

1. Reciprocating, horizontal band, universal tilt frame, abrasive, and cold saws.
2. Abrasive.
3. Horizontal band.
4. Reciprocating.
5. Cold saw.
6. Wear eye protection, check blade tensions, and make sure that all guards are in place.
7. Wear face shield or safety glasses with side shields, inspect wheel for chips and cracks, and check speed rating on wheel.
8. Heavy bars of material can pinch fingers or hands.
9. Unplug the machine from the electrical outlet or switch off the circuit breaker and tag with a warning.
10. The abrasive blade may fly apart.

SECTION D / UNIT 2 / USING RECIPROCATING AND HORIZONTAL BAND CUTOFF MACHINES

SELF-TEST ANSWERS

1. Raker, wave, and straight.
2. Workpiece material, cross section shape, and thickness.
3. On the back stroke.

4. The tooth offset on either side of the blade. Set provides clearance for the back of the blade.
5. Standard, skip, and hook.
6. Cutoff material can bind the blade and destroy the set.
7. The horizontal band saw.
8. Cooling, lubrication, and chip removal.
9. Scoring and possible blade breakage.
10. The workpiece must be turned over and a new cut started.

SECTION D / UNIT 3 / VERTICAL BAND MACHINES

SELF-TEST ANSWERS

1. General purpose, power-fed worktable types, high tool velocity machines, and large capacity band machines.
2. High velocity band machines.
3. Filing and band polishing.
4. It is making curving cuts as opposed to straight line cuts.
5. The final shape is produced without converting all unwanted material to chips (material savings). Its contouring ability is the second advantage.
6. Fingers and hands in close proximity to the blade.
7. Changing blades, friction, and high speed sawing.
8. A pusher is a piece of material used to push the workpiece into the blade.
9. Round stock may turn if handheld. This can cause an injury and damage the saw blade.
10. Eye protection is a primary piece of safety equipment.

SECTION D / UNIT 4 / PREPARING TO USE THE VERTICAL BAND MACHINE

SELF-TEST ANSWERS

1. Both ends of the blade should be ground at the same time with the teeth opposed. This will insure that the ends of the blade are square.
2. The blade ends are placed in the welder with the teeth pointed in or just outside of the jaws. The ends must contact squarely in the gap between the jaws. The welder must be adjusted for the band width to be welded. You should wear eye protection and stand to one side during the welding operation. The weld will occur when the weld lever is depressed.
3. The weld is ground on the grinding wheel attached to the welder. Grind the weld on both sides of the band until the band fits the thickness gage. Be careful not to grind the saw teeth.
4. The guides support the band. This is essential to straight cutting.
5. Band guides must fully support the band except for the teeth. A wide guide used on a narrow band will destroy the saw set as soon as the machine is started.
6. The guide setting gage is used to adjust the band guides.
7. Annealing is the process of softening the band weld in order to improve strength qualities.
8. The band should be clamped in the annealing jaws with the teeth pointed out. A small amount of compression should be placed on the movable welder jaw prior to clamping the band. The correct annealing color is dull red. As soon as this color is reached, the anneal switch should be released and then operated briefly several times to slow the cooling rate of the weld.
9. Band tracking is the position of the band as it runs on the idler wheels.
10. Band tracking is adjusted by tilting the idler wheels until the band just touches the backup bearing.

SECTION D / UNIT 5 / USING THE VERTICAL BAND MACHINE

SELF-TEST ANSWERS

1. The three sets are straight, wave, and raker. Straight set may be used for thin material, wave for material with a variable cross section, and raker for general purpose sawing.

2. Scalloped and wavy edged bands might be used on nonmetallic material where blade teeth would tear the material being cut.

3. The job selector provides information about recommended saw velocity, saw pitch, power feed, saw set, and temper. Band filing information is also indicated.

4. Band velocity is measured in feet per minute.

5. The variable speed pulley is designed so that the pulley flanges may be moved toward and away from each other. This permits the belt position to be varied, resulting in speed changes.

6. Speed range is selected by shifting the transmission.

7. Speed range shift must be done with the band speed set at the lowest setting.

8. The upper guidepost must be adjusted so that it is as close to the workpiece as possible.

9. Band pitch must be correct for the thickness of material to be cut. Generally, a fine pitch will be used on thin material. Cutting a thick workpiece with a fine pitch band will clog saw teeth and reduce cutting efficiency.

10. Band set must be adequate for the thickness of the blade used in a contour cut. If set is insufficient, the blade may not be able to cut the desired radius.

SECTION E / UNIT 1 / READING SHOP DRAWINGS

SELF-TEST ANSWERS

1. See Figure 23.

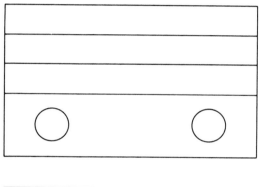

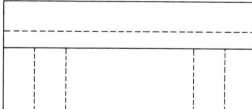 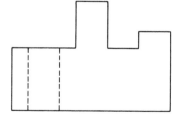

Figure 23.

2. The tolerance of the hole is specified as ±.005. Therefore, the minimum size of the hole would be .750 − .005 or .745.

3. $2\frac{1}{4}$ in.

4. The slot conforms to standard tolerances, width is $\pm\frac{1}{64}$ of an inch.

5. $\frac{1}{8}$ in.

6. 6 in.

7. 1 in.

8. Drilling and reaming.

9. $\frac{1}{8}$ in. by 45 degrees.

10. Note 1 indicates that all sharp corners are to be broken, which means that all burrs and sharp edges left from machining are to be smoothed off.

SECTION E / UNIT 2 / LAYOUT TOOLS

SELF-TEST ANSWERS

1. It is placing reference marks on workpieces for stock cutoff, filing, offhand grinding, drilling, milling, and lathe work.
2. ±.001 in.
3. It is used to make layout marks more visible.
4. Trammel points attached to a bar or rule.
5. The prick punch, also called layout punch, is used only for layout work since it has a sharper point that makes it easier to locate scribe mark locations and to make a slight indentation. The center punch is used to deepen the punch marks prior to a machining operation.
6. An optical center punch.
7. To lay out centerlines on round workpieces.
8. With a beam scale and vernier slide. The discrimination of this tool is to .001 in.; this makes it a precision layout tool.
9. To find and accurately lay out centerlines on round stock.
10. To hold the work perpendicular to the surface plate and give a reference surface. This accessory is often used in conjunction with the vernier height gage.

SECTION E / UNIT 3 / LAYOUT PRACTICE

SELF-TEST ANSWERS

1. The workpiece should have all sharp edges removed by grinding or filing. A thin, even coat of layout dye should be applied.
2. The towel will prevent spilling layout dye on the layout plate or layout table.
3. The punch should be tilted so that it is easier to see when the point is located on the scribe mark. It should then be moved to the upright position before it is tapped with the layout hammer.
4. The combination square can be positioned on the rule for measurements. The square head acts as a positive reference point for measurements. The rule may be removed and used as a straight edge for scribing.
5. The divider should be adjusted until you feel the tip drop into rule engraving.
6. (Figure 37) Inch—5.030 in. Metric—127.76 mm
7. (Figure 38) Inch—8.694 in. Metric—220.82 mm
8. (Figure 39) Inch—5.917 in. Metric—150.30 mm
9. (Figure 40) Inch—4.086 in.
10. (Figure 41) Inch—1.456 in.
11. The position of the vernier scale may be adjusted on height gages to zero the reference lines.
12. By turning the workpiece 90 degrees.

SECTION F / UNIT 1 / HARDENING, TEMPERING, AND CASE HARDENING PLAIN CARBON STEEL

SELF-TEST ANSWERS

1. No hardening would result as 1200°F (649°C) is less than the lower critical point and no dissolving of carbon has taken place.
2. There would be almost no change. For all practical purposes in the shop these low carbon steels are not considered hardenable.
3. They are shallow hardening, and liable to distortion and quench cracking because of the severity of the water quench.
4. Air and oil hardening steels are not so subject to distortion and cracking as W1 steels and are deep hardening.
5. 1450°F (788°C). 50°F (10°C) above the upper critical limit.
6. Tempering is done to remove the internal stresses in hard martensite, which is very brittle. The temperature used gives the best compromise between hardness and toughness or ductility.

7. Tempering temperature should be specified according to the hardness, strength, and ductility desired. Mechanical properties charts give this data.
8. 525°F (274°C). Purple.
9. 600°F (315°C). It would be too soft for any cutting tool.
10. Immediately. If you let is set for any length of time, it may crack from internal stresses.
11. By furnace.
12. (a) Controlled atmosphere furnace.
    (b) Wrapping the piece in stainless steel foil.
    (c) Covering with cast iron chips.
13. An air hardening tool steel should be used when distortion must be kept to a minimum.
14. Since the low carbon steel core does not harden when quenched from 1650°F (899°C), it remains soft and tough, but the case becomes very hard. No tempering is therefore required as the piece is not brittle all the way through as a fully hardened carbon steel piece would be.
15. A deep case can be made by pack carburizing or by a liquid bath carburizing. A relatively deep case is often applied by nitriding or by similar procedures.
16. No. The base material must contain sufficient carbon to harden by itself without adding more for surface hardening.
17. Three methods of introducing carbon into heated steel are roll, pack, and liquid carburizing.
18. Nitriding.
19. The surface decarburizes or loses surface carbon to the atmosphere as it combines with oxygen to form carbon dioxide. An oxide scale forms on the surface.
20. Circulation or agitation breaks down the vapor barrier. This action allows the quench to proceed at a more rapid rate and it avoids spotty hardening.

SECTION F / UNIT 2 / ANNEALING, NORMALIZING, AND STRESS RELIEVING

SELF-TEST ANSWERS

1. Medium carbon steels that are not uniform, have hardened areas from welding or prior heating, need to be normalized before they can be machined. Forgings, castings, and tool steel in the as-rolled condition are normalized before any further heat treatments or machining is done.
2. 1550°F (843°C). 50°F (10°C) above the upper critical limit.
3. The spheroidization temperature is quite close to the lower critical temperature line, about 1300°F (704°C).
4. The full anneal brings carbon steel to its softest condition as all the grains are reformed (recrystallized), and any hardened pieces of steel become soft pearlite as they slowly cool. Stress relieving will only recrystallize distorted ferrite grains and not the hard carbide structures or pearlite grains.
5. Stress relieving should be used on severely cold worked steels or for weldments.
6. High carbon steels (0.6 to 1.7 percent carbon).
7. Process annealing is used by the sheet and wire industry and is essentially the same as stress relieving.
8. In still air.
9. Very slowly. Packed in insulating material or cooled in a furnace.
10. Since low carbon steels tend to become gummy when spheroidized, the machinability is poorer than in the as-rolled condition. Spheroidization sometimes is desirable when stress relieving weldments on low carbon steels.

SECTION F / UNIT 3 / GAS WELDING REPAIRS AND THEIR EFFECTS

SELF-TEST ANSWERS

1. The melting temperature of steel is about 2500°F (1371°C) and it would have to be heated above 3000°F (1649°C) to produce a good weld.
2. Because acetylene becomes unstable at a pressure near 30 PSI and can explode.
3. If the tank should fall over and break off the valve, it becomes a missile that is extremely violent.

4. Fusion welding (melting the parts together) and diffusion welding (braze welding) in which the base metal does not become molten.
5. In braze welding, the parts are joined by using a rod of filler metal, usually brass or bronze. Silver soldering or furnace brazing uses a fixed amount of filler material with flux that goes into place by capillary action when the temperature of the parts reaches the melting point of the brazing material.
6. Temperature. Soldering is done at a temperature under 800°F (426.6°C).
7. Ferrous metals are oxidized (burned) by the stream of oxygen. The edge of the metal must first be brought to the burning (kindling) temperature.
8. Oxygen and acetylene pressures.
9. Too fast.
10. The edge can become very hard because it becomes carburized and is rapidly cooled.

SECTION F / UNIT 4 / ARC WELDING AND ITS EFFECT ON METALS
SELF-TEST ANSWERS

1. The electric arc.
2. AC and DC.
3. Minimum tensile strength in thousand PSI.
4. By "striking" the electode on the base metal like striking a match and then holding the end of the rod about $\frac{1}{8}$ in. away from the base metal.
5. This is called "hot cracking" and is caused by trying to fill in a wide gap on plates that cannot move or by a too high cooling rate. Also porosity or wrong filler metal.
6. By preheating.
7. Underbead or hard cracking.
8. Because it can cause hydrogen embrittlement and underbead cracking.
9. By preheating the base metal.
10. By preheating the base metal.

SECTION G / UNIT 1 / THE DRILL PRESS
SELF-TEST ANSWERS

1. Loose clothing of any kind that could get caught in a machine would be hazardous. Gloves are a danger around rotating machinery. Long loose hair is extremely unsafe around rotating machinery. Loose hair must be tied or restrained when working in the shop. Watches or rings should not be worn.
2. Workpieces that are not clamped can become a danger to the operator. Even work clamped in a drill press vise that is not itself bolted or clamped to the table, can begin to spin if the drill jams. Often this results in a flying workpiece of great mass. The only corrective measure needed is to securely clamp everything that is being drilled.
3. Most machine tools are capable of throwing metal chips at high velocities. Those who have had painful eye injuries can attest to the fact that you cannot get out of the way quickly enough. *Always* wear safety glasses in the machine shop at *all* times.
4. Never clean the spindle taper with the machine running, nor should you change belts, lubricate, or make any other adjustments. The result could be sprained or broken fingers or worse.
5. No. It is not acceptable to leave the chuck key in for even a moment, as this is just the situation in which you are most likely to forget it and turn on the machine. Make it a practice never to let the key leave your hand and you won't have the problem of remembering to remove it from the chuck.
6. No! Slowing or stopping a spindle with the hand can lead to injury. Chips can accumulate around the chuck. The open jaws in an empty chuck could break a finger. Let the machine slow down and stop by itself.
7. This problem can be eliminated by placing a block of wood under the drill on which it can drop. This also protects the cutting edge of the drill.
8. A brush or stick will safely remove chips. A machine table may be wiped down with a cloth only after all of the chips have been removed.
9. The oil can cause anyone walking past or carrying a heavy object to fall. The operator could slip and fall into the machine or against a switch or lever, thus causing an additional accident.

10. Burrs and sharp edges are extremely effective cutting tools for fingers and hands. The sooner the burrs are removed, the less chance there is for you to be cut. Ends of threads, both internal and external, are particularly bad as they are an unexpected hazard.

11. (a) The sensitive, upright, and radial-arm drill presses are the three basic types. The sensitive drill press is made for light duty work and it provides the operator with a sense or "feel" of the feed on the drill. The upright is a similar, but heavy duty, drill press equipped with power feed. The radial-arm drill press allows the operator to position the drill over the work where he needs it, rather than to position the work under the drill as with other drill presses.

(b) The sensitive, upright, and radial-arm drill presses all perform much the same functions of drilling, reaming, counterboring, countersinking, spot facing, and tapping, but the upright and radial machines do heavier and larger jobs. The radial-arm drill can support large, heavy castings and work can be done on them without the workpiece being moved.

12. Sensitive drill press

    __a__ Spindle
    __g__ Quill lock handle
    __e__ Column
    __l__ Switch
    __b__ Depth stop
    __n__ Head
    __o__ Table
    __h__ Table lock
    __f__ Base
    __c__ Power feed
    __j__ Motor
    __d__ Variable speed control
    __k__ Table lift crank
    __i__ Quill return spring
    __m__ Guard

13. Radial drill press

    __b__ Column
    __c__ Radial arm
    __a__ Spindle
    __d__ Base
    __e__ Drill head

SECTION G / UNIT 2 / DRILLING TOOLS

SELF-TEST ANSWERS

1. __t__ Web
    __u__ Margin
    __d__ Drill point angle
    __p__ Cutting lip
    __k__ Flute
    __o__ Body
    __e__ Lip relief angle
    __g__ Land
    __b__ Chisel edge angle
    __j__ Body clearance
    __i__ Helix angle
    __y__ Axis of drill

<u> n </u> Shank length
<u> c </u> Tang
<u> x </u> Taper shank
<u> w </u> Straight shank

2.

	Decimal Diameter	Fractional Size	Number Size	Letter Size	Metric Size
a	.0781	$\frac{5}{64}$			
b	.1495		25		
c	.272			I	
d	.159		21		
e	.1969				5
f	.323			P	
g	.3125	$\frac{5}{16}$			
h	.4375	$\frac{7}{16}$			
i	.201		7		
j	.1875	$\frac{3}{16}$			

SECTION G / UNIT 4 / WORK LOCATING AND HOLDING DEVICES ON DRILLING MACHINES

SELF-TEST ANSWERS

1. The purpose for using workholding devices is to keep the workpiece rigid, from turning with the drill, and for operator safety.
2. Included in a list of workholding devices would be strap clamps, T-bolts, and step blocks. Also used are C-clamps, V-blocks, vises, jigs and fixtures, and angle plates.
3. Parallels are mostly used to raise workpieces off the drill press table or to lift the workpiece higher in a vise, thus providing a space for the drill breakthrough. They are made of hardened steel so care should be exercised in their use.
4. Thin limber materials tend to be sprung downward from the force of drilling until drill breakthrough begins. The drill then "grabs" as the material springs upward and a broken drill is often the result. This can be avoided by placing the support or parallels as near the drill as possible.
5. Angle drilling is done by tilting a drill press table (not all types tilt), or by using an angular vise. If no means of setting the exact angle is provided, a protractor head with level may be used.
6. Vee blocks are suited to hold round stock for drilling. The most frequent use of vee blocks is for cross drilling holes in shafts, although many other setups are used.
7. The wiggler is used for locating a center punch mark under the center axis of a drill spindle.
8. Some odd shaped workpieces, such as gears with extending hubs that need holes drilled for set screws, might be difficult to set up without an angle plate.
9. One of the difficulties with hand tapping is the tendency for taps to start crooked or misaligned with the tap drilled hole. Starting a tap by hand in a drill press with the same setup as used for the tap drilling assures a perfect alignment.
10. Since jigs and fixtures are mostly used for production manufacturing, small machine shops rarely have a use for them.

SECTION G / UNIT 5 / OPERATING DRILLING MACHINES

SELF-TEST ANSWERS

1. The three considerations would be speeds, feeds, and coolants.
2. The RPMs of the drills would be

(a) $\frac{1}{4}$ in. diameter is 1440 RPM.

(b) 2 in. diameter is 180 RPM.

(c) $\frac{3}{4}$ in. diameter is 480 RPM.

(d) $\frac{3}{8}$ in. diameter is 960 RPM.

(e) $1\frac{1}{2}$ in. diameter is 240 RPM.

3. Worn margins and outer corners broken down. The drill can be ground back to its full size and resharpened.

4. The operator will increase the feed in order to produce a chip. This increased feed is often greater than the drill can stand without breaking.

5. The feed is about right when the chip rolls into a close helix. Long, stringy chips can indicate too much feed.

6. Feeds are designated by a small measured advance movement of the drill for each revolution. A .001 in. feed for a $\frac{1}{8}$ in. diameter drill, for example, would move the drill .001 in. into the work for every turn of the drill.

7. The water soluble oil types and the cutting oils, both animal and mineral.

8. Besides having the correct cutting speed, a sulfurized oil based cutting fluid helps to reduce friction and cool the cutting edge.

9. Drill "jamming" can be avoided by a "pecking" procedure. The operator drills a small amount and pulls out the drill to remove the chips. This is repeated until the hole is finished.

10. The depth stop is used to limit the travel of the drill so it will not go on into the table or vise. The depth of blind holes is preset and drilled. Countersink and counterbore depths are set so that several can be easily made the same.

SECTION G / UNIT 6 / COUNTERSINKING AND COUNTERBORING

SELF-TEST ANSWERS

1. Countersinks are used to chamfer holes and to provide tapered holes for flat head fasteners such as screws and rivets.

2. Countersink angles vary to match the angles of different flat head fasteners or different taper hole requirements.

3. A center drill is used to make a 60 degree countersink hole in workpieces for lathes and grinders.

4. A counterbore makes a cylindrical recess concentric with a smaller hole so that a hex head bolt or socket head capscrew can be flush mounted with the surface of a workpiece.

5. The pilot diameter should always be a few thousandths of an inch smaller than the hole, but not more than .005 in.

6. Lubrication of the pilot prevents metal to metal contact between it and the hole. It will also prevent the scoring of the hole surface.

7. A general rule is to use approximately one-third of the cutting speed when counterboring as when using a twist drill with the same diameter.

8. Feeds and speeds when counterboring are controlled to a large extent by the condition of the equipment, the available power, and the material being counterbored.

9. Spotfacing is performed with a counterbore. It makes a flat bearing surface, square with a hole to seat a nut, washer, or bolt head.

10. Counterboring requires a rigid setup with the workpiece being securely fastened and provisions made to allow the pilot to protrude below the bottom surface of the workpiece.

SECTION G / UNIT 7 / REAMING IN THE DRILL PRESS

SELF-TEST ANSWERS

1. Machine reamers are identified by the design of the shank, either a straight or tapered shank and usually a 45 degree chamfer on the cutting end.

2. A chucking reamer is a finishing reamer, the fluted part is cylindrical, and the lands are relieved. A rose reamer is a roughing reamer. It can remove a considerable amount of material. The body has a slight back taper and no relief on the lands. All cutting takes place on the chamfered end.

3. A jobbers reamer is a finishing reamer like a chucking reamer but it has a longer fluted body.

4. Shell reamers are more economical to produce than solid reamers, especially in larger sizes.
5. An accurate hole size cannot be obtained without a high quality surface finish.
6. As a general rule the cutting speed used to ream a hole is about one-third to one-half of the speed used to drill a hole of the same size in the same material.
7. The feed rate, when reaming as compared to drilling the same material, is approximately two to three times greater. As an example, for a 1 in. drill the feed rate is about .010 to .015 in. per revolution. A 1 in. reamer would have a feed rate of between .020 and .030 in.
8. The reaming allowance of a $\frac{1}{2}$ in. diameter hole would be $\frac{1}{64}$ in.
9. Coolants cool the tool and workpiece and act as lubricants.
10. Chatter may be eliminated by reducing the speed, increasing the feed, or using a piloted reamer.
11. Oversized holes may be caused by a bent reamer shank or buildup on the cutting edges. Check also if there is a sufficient amount and if the correct kind of coolant is being used.
12. Bell-mouthed holes are usually caused by a misaligned reamer and workpiece setup. Piloted reamers, bushings, or a floating holder may correct this problem.
13. Surface finish can be improved by decreasing the feed and checking the reaming allowance. Too much or not enough material will cause poor finish. Use a large volume of coolant.
14. Carbide tipped reamers are recommended for long production runs where highly abrasive materials are reamed.
15. Cemented carbides are very hard, but also very brittle. The slightest amount of chatter or vibration may chip the cutting edges.

SECTION H / UNIT 2 / TURNING MACHINE SAFETY

SELF-TEST ANSWERS

1. Both the employer and employee bear the responsibility for worker safety. Also responsible are the manufacturer, reconstructor, and modifier of machinery.
2. A pinch point is where associated parts of a lathe move close together, thus making a hazard to the operator. An example would be a finger caught in gears.
3. When a spindle that has a threaded nose is suddenly reversed, the chuck that was screwed on can unscrew and come out of the lathe. A chuck wrench left in a lathe chuck may fly out when the power is turned on.
4. A lathe operator can get burned from hot chips that fly and strike his body where skin is exposed or by handling hot chips with bare hands. He can also get burned from a part just turned.
5. Sharp edges, projections, stringy chips, and burred edges on workpieces all produce cuts. When measuring a bored hole, cover the tool and back off the carriage.
6. On special setups that are hazardous, signs and verbal instructions are used. The operator should always rotate the setup slowly by hand before turning on the power. He should stand aside when he starts the spindle. A geared, scroll chuck should not be rotated with power when empty. Workpieces should be securely gripped in chuck jaws.
7. Do not turn on the switch yourself. Get the mechanic to do it.
8. The rod will begin to whip immediately and will bend so that its end will describe about a 4 ft diameter circle. At that RPM there would be no time to get out of the way. A trough or tube can be used to guard the extended rod.
9. The bolt that is overstressed may be almost ready to break and when a sudden workload is placed on it, it does break—suddenly. Understressed bolts usually just allow slight slippage of parts out of position.
10. It is probably the result of a wrong attitude. One who acts without thinking, who is careless about work habits, and who does not do the safe thing at all times is a sure prospect for an accident.

SECTION H / UNIT 3 / TOOLHOLDERS AND TOOLHOLDING FOR THE LATHE

SELF-TEST ANSWERS

1. A toolholder is needed to rigidly support and hold a cutting tool during the actual cutting operation. The

cutting tool is often only a small piece of high speed steel or other cutting material that has to be clamped in a much larger toolholder in order to be usable.

2. On a left-hand toolholder, when viewed from above, the cutting tool end is bent to the right.
3. For high speed tools the square tool bit hole is angled upward in relation to the base of the toolholder, where it is parallel to the base for carbide tools.
4. Turning close to the chuck is usually best accomplished with a left-hand toolholder.
5. Tool height adjustments on a standard-type toolholder are made by swiveling the rocker in the tool post ring.
6. Quick-change toolholders are adjusted for height with a micrometer collar.
7. Tool height on turret toolholders is adjusted by placing shims under the tool.
8. Toolholder overhang affects the rigidity of a setup; too much overhang may cause chatter.
9. The difference between a standard type toolholder and a quick-change type toolholder is in the speed with which tools can be interchanged. The tools are usually fastened more securely in a quick-change toolholder and height adjustments on a quick-change toolholder do not change the effective back rake angle of the cutting tool.
10. Drilling machine tools are used in a lathe tailstock.
11. The lathe tailstock is bored with a Morse taper hole to hold Morse taper shank tools.
12. When a series of repeat tailstock operations are to be performed on several different workpieces, a tailstock turret should be used.

SECTION H / UNIT 4 / CUTTING TOOLS FOR THE LATHE

SELF-TEST ANSWERS

1. High speed steel is easily shaped into the desired shape of cutting tool. It produces better finishes on low speed machines and on soft metals.
2. Its geometrical form: the side and back rake, front and side relief angles, and chip breakers.
3. Unlike single point tools, form tools produce their shape by plunging directly into the work.
4. When a "chip trap" is formed by improper grinding on a tool, the chip is not able to clear the tool; this prevents a smooth flow across the face of the tool. The result is tearing of the surface on the workpiece and a possible broken tool.
5. Some toolholders provide a built-in back rake of about 16 degrees; to this is added any back rake on the tool to make a total back rake that is excessive.
6. A zero rake should be used for threading tools. A zero to slightly negative back rake should be used for plastics and brass, since they tend to "dig in."
7. The side relief allows the tool to feed into the work material. The end relief angle keeps the tool end from rubbing on the work.
8. The side rake directs the chip flow away from the cut and it also provides for a keen cutting edge. The back rake promotes smooth chip flow and good finishes.
9. The angles can be checked with a tool grinding gage, a protractor, or an optical comparator.
10. Long, stringy chips or those that become snarled on workpieces, tool post, chuck or lathe dog are hazardous to the operator. Chip breakers and correct feeds can produce an ideal chip that does not fly off but will simply drop to the chip pan and is easily handled.
11. Chips can be broken up by using coarse feeds and maximum depth of cuts for roughing cuts and by using tools with chipbreakers on them.
12. Overheating a tool causes small cracks to form on the edge. When a stress is applied, as in a roughing cut, the tool end may break off.

SECTION H / UNIT 5 / LATHE SPINDLE TOOLING

SELF-TEST ANSWERS

1. The lathe spindle is a hollow shaft that can have one of three mounting devices machined on the spindle nose. It has an internal Morse taper that will accommodate centers or collets.
2. The spindle nose types are the threaded, long taper key drive, and the camlock.
3. The independent chuck is a four-jaw chuck in which each jaw can move separately and independently of the others. It is used to hold odd-shaped workpieces.

4. The universal chuck is most often a three-jaw chuck although they are made with more or less jaws. Each jaw moves in or out by the same amount when the chuck wrench is turned. They are used to hold and quickly center round stock.
5. Combination chucks and Adjust-Tru three-jaw chucks.
6. A drive plate.
7. The live center is made of soft steel so it can be turned to true it up if necessary. It is made with a Morse taper to fit the spindle taper or special sleeve if needed.
8. A hardened drive center is serrated so it will turn the work between centers, but only light cuts can be taken. Face drives use a number of driving pins that dig into the work and are hydraulically compensated for irregularities.
9. Workpieces and fixtures are mounted on face plates. These are identified by their heavy construction and the T-slots. Drive plates have only slots.
10. Collet chucks are very accurate workholding devices. Spring collets are limited to smaller material and to specific sizes.

SECTION H / UNIT 6 / OPERATING THE MACHINE CONTROLS
SELF-TEST ANSWERS

1. Very low speeds are made possible by disengaging the spindle by pulling out the lockpin and engaging the back gear.
2. The varidrive is changed while the motor is running, but the back gear lever is only shifted with the motor off.
3. Levers that are located on the headstock can be shifted in various arrangements to select speeds.
4. The feed reverse lever.
5. These levers are used for selecting feeds or threads per inch for threading.
6. The quick approach and return of the tool; this is used for delicate work and when approaching a shoulder or chuck jaw.
7. Since the cross feed is geared differently (about one-third of the longitudinal feed), the outside diameter would have a coarser finish than the face.
8. The half-nut lever is used only for threading.
9. They are graduated in English units. Some metric conversion collars are being made and used that read in both English and metric units at the same time.
10. You can test with a rule and a given slide movement such as .125 or .250 in. If the slide moves one-half that distance, the lathe is calibrated for double depth and reads the same amount as that taken off the diameter.

SECTION H / UNIT 7 / FACING AND CENTER DRILLING
SELF-TEST ANSWERS

1. A lathe board is placed on the ways under the chuck and the chuck is removed, since it is the wrong chuck to hold rectangular work. The mating parts of an independent chuck and the lathe spindle are cleaned and the chuck is mounted. The part is roughly centered in the jaws and adjusted to center by using the back of a toolholder or a dial indicator.
2. The tool should be on the center of the lathe axis.
3. No. The resultant facing feed would be approximately 0.3 to 0.5 × .012 in., which would be a finishing feed.
4. The compound must be swung to either 30 or 90 degrees so that the tool can be fed into the face of the work by a measured amount. A depth micrometer or a micrometer caliper can be used to check the trial finish cut.
5. A right-hand facing tool is used for shaft ends. It is different from a turning tool in that its point is only a 58 degree included angle to fit in the narrow space between the shaft face and the center.
6. RPM $= \dfrac{300 \times 4}{4} = 300$
7. Center drilling is done to prepare work for turning between centers and for spotting workpieces for drilling in the lathe.

8. Round stock is laid out with a center head and square or rectangular with diagonal lines. Stock can be held vertically in a vise or angle plates and vee blocks for center drilling in a drill press.
9. Center drills are broken as a result of feeding the drill too fast and having the lathe speed too slow. Other causes result from having the tailstock off center, the work off center in a steady rest, or lack of cutting oil.
10. The sharp edge provides a poor bearing surface and soon wears out of round, causing machining problems such as chatter.

SECTION H / UNIT 8 / TURNING BETWEEN CENTERS
SELF-TEST ANSWERS

1. A shaft between centers can be turned end for end without loss of concentricity and it can be removed from the lathe and returned without loss of synchronization between thread and tool. Cutting off between centers is not done as it would break the parting tool and ruin the work. Steady rest work is not done with work mounted in a center in the headstock spindle.
2. The other method is where the workpiece is held in a chuck on one end and in the tailstock center on the other end.
3. Coarser feeds, deeper cuts, and smaller rake angles all tend to increase chip curl. Chip breakers also make the chip curl.
4. Dead centers are hardened 60 degree centers that do not rotate with the work but require high pressure lubricant. Ball bearing centers turn with work and do not require lubricant. Pipe centers turn with the work and are used to support tubular material.
5. With no end play in the workpiece and the bent tail of the lathe dog free to click against the sides of the slot.
6. Because of expansion of the workpiece from the heat of machining, it tightens on the center, thus causing more friction and more heat. This could ruin the center.
7. Excess overhang promotes lack of rigidity. This causes chatter and tool breakage.
8. $RPM = \dfrac{90 \times 4}{1\frac{1}{2}} = 360 \div \dfrac{3}{2} = 360 \times \dfrac{2}{3} = 240$ or $\dfrac{360}{1.5} = 240$
9. The spacing would be .010 in. as the tool moves that amount for each revolution of the spindle.
10. The feed rate should be limited to what the tool, workpiece, or machine can stand without undue stress.
11. For most purposes where liberal tolerances are allowed, .015 to .030 in. can be left for finishing. When closer tolerances are required, two finish cuts are taken with .005 to .010 in. left for the last finish cut.
12. After roughing is completed, .015 to .030 in. is left for finishing. The diameter of the workpiece is checked with a micrometer and the remaining amount is dialed on the cross feed micrometer collar. A short trial cut is taken and the lathe is stopped. This diameter is again checked. If the diameter is within tolerance, the finish cut is taken.

SECTION H / UNIT 9 / ALIGNMENT OF THE LATHE CENTERS
SELF-TEST ANSWERS

1. The workpiece becomes tapered.
2. The workpiece is tapered with the small end at the tailstock.
3. By the witness mark on the tailstock, by using a test bar, and by taking a light cut on a workpiece and measuring.
4. The dial indicator.
5. With a micrometer. The tailstock is set over with a dial indicator.

SECTION H / UNIT 10 / LATHE OPERATIONS
SELF-TEST ANSWERS

1. Drilled holes are not sufficiently accurate for bores in machine parts as they would be loose on the shaft and would not run true.

2. The workpiece is center spotted with a center drill at the correct RPM and, if the hole is to be more than a $\frac{3}{8}$ in. diameter, a pilot drill is put through. Cutting oil is used. The final size drill is put through at a slower speed.

3. The chief advantage of boring in the lathe is the bore runs true with the centerline of the lathe and the outside of the workpiece, if the workpiece has been set up to run true (with no runout). This is not always possible when reaming bores that have been drilled, since the reamer follows the eccentricity or runout of the bore.

4. Ways to eliminate chatter in a boring bar are:
   (a) Shorten the bar overhang, if possible.
   (b) Reduce the spindle speed.
   (c) Make sure the tool is on center.
   (d) Use as large a diameter bar as possible without binding in the bore.
   (e) Reduce the nose radius on the tool.
   (f) Apply cutting oil to the bore.
   (g) Use tuned or solid carbide boring bars.

5. Through boring is making a bore the same diameter all the way through the part. Counterboring is making two or more diameters in the same bore, usually with 90 degree or square internal shoulders. Blind holes are bores that do not go all the way through.

6. Grooves and thread relief are made in bores by means of specially shaped or ground tools in a boring bar.

7. A floating reamer holder will help to eliminate the bell mouth, but it does not remove the runout.

8. Hand reamers produce a better finish than machine reaming.

9. Cutting speeds for reaming are *one-half* that used for drilling; feeds used for reaming are *twice* that used for drilling.

10. Large internal threads are produced with a boring tool. Heavy forces are needed to turn large hand taps, so it is not advisable to use large taps in a lathe.

11. A tap drill can be used as a reamer by first drilling with a drill that is $\frac{1}{32}$ to $\frac{1}{16}$ in. undersized. This procedure assures a more accurate hole size by drilling.

12. A spiral point tap works best for power tapping.

13. The variations in pitch of the hand cut threads would cause the micrometer collar to give erroneous readings. A screw used for this purpose and for most machine parts must be threaded with a tool guided by the lead screw on the lathe.

14. Thread relief and external grooves are produced by specially ground tools that are similar to internal grooving tools except that they have less end relief. Parting tools are often used for making external grooves.

15. Parting tools tend to seize in the work, especially with deep cuts or heavy feeds. Without cutting oil seizing is almost sure to follow with the possibility of a broken parting tool and misaligned or damaged work.

16. You can avoid chatter when cutting off with a parting tool by maintaining a rigid setup and keeping enough feed to produce a continuous chip, if possible.

17. Knurling is used to improve the appearance of a part, to provide a good gripping surface, and to increase the diameter of a part for press fits.

18. Ordinary knurls make a straight or diamond pattern impression by displacing the metal with high pressures.

19. When knurls produce a double impression, they can be readjusted up or down and moved to a new position. Angling the toolholder 5 degrees may help.

20. You can avoid producing a flaking knurled surface by stopping the knurling process when the diamond points are almost sharp. Also use a lubricant while knurling.

SECTION H / UNIT 11 / SIXTY DEGREE THREAD INFORMATION AND CALCULATIONS

SELF-TEST ANSWERS

1. The sharp V thread can be easily damaged while handling if it is dropped or allowed to strike against a hard surface.

2. The pitch is the distance between a point on a screw thread to a corresponding point on the next thread measured parallel to the axis. "Threads per inch" is the number of threads in one inch.

3. American National series and Unified series threads both have the 60 degree included angle and are both

based on the inch measure with similar pitch series. The depth of the thread and the classes of thread fits are different in the two systems.

4. To allow for tolerancing of external and internal threads to promote standardization and interchangeability of parts.

5. This describes a diameter of $\frac{1}{2}$ in., 20 threads per inch, and Unified coarse series external thread with a class 2 thread tolerance.

6. The flat on the end of the tool for 20 threads per inch should be $P = .050 \times .125 = .006$ in.

7. The compound at 30 degrees will move in $.708/20 = .0354$ in. for Unified threads. The compound will move in $.75/20 = .0375$ in. for American National threads.

8. The fit of the thread refers to classes of fits and tolerances, while percent of thread refers to the actual minor diameter of an internal thread, a 100 percent thread being full depth internal threads.

9. The Systéme International (SI) thread and the British Standard ISO Metric Screw Threads are two metric thread systems in use.

10. The tap drill size for a metric thread can be found by subtracting the pitch from the major diameter.

SECTION H / UNIT 12 / CUTTING UNIFIED EXTERNAL THREADS
SELF-TEST ANSWERS

1. A series of cuts are made in the same groove with a single point tool by keeping the same ratio and relative position of the tool on each pass. The quick-change gearbox allows choices of various pitches or leads.

2. The chips are less likely to bind and tear off when feeding in with the compound set at 29 degrees and the tool is less likely to break.

3. The 60 degree angle on the tool is checked with a center gage or optical comparator.

4. The number of threads per inch can be checked with a screw pitch gage or by using a rule and counting the threads in one inch.

5. A center gage is used to align the tool to the work.

6. No. The carriage is moved by the thread on the lead screw when the half-nuts are engaging it.

7. Even numbered threads may be engaged on the half-nuts at any line and odd numbered threads at any numbered line. It would be best to use the same line every time for fractional numbered threads.

8. The spindle should be turning slow enough for the operator to maintain control of the threading operation, about one-fourth turning speeds usually.

9. The lead screw rotation is reversed, which causes the cut to be made from the left to the right. The compound is set at 29 degrees to the left. The threading tool and lathe settings are set up in the same way as for cutting right-hand threads.

10. Picking up the thread or resetting the tool is a procedure that is used to position a tool to existing threads.

SECTION H / UNIT 13 / CUTTING UNIFIED INTERNAL THREADS
SELF-TEST ANSWERS

1. The minor diameter of the thread.

2. By varying the bore size, usually larger than the minor diameter. This is done to make tapping easier.

3. 75 percent.

4. A drill just under the tap drill size should first be used; thus the tap drill acts as a reamer.

5. $500 - \left(\frac{.541}{13} \times 2 \times .6\right) = .450$. Since the nearest fractional drill size is $\frac{29}{64}$ in. $= .453$, the tap drill for 60% threads is $\frac{29}{64}$ in.

6. Large internal threads of various forms can be made, and the threads are concentric to the axis of the work.

7. To the left of the operator.

8. A screw pitch gage should be used.

9. Boring bar and tool deflection cause the threads to be undersize from the calculations and settings on the micrometer collars.

10. The minor diameter equals $D - (P \times .541 \times 2)$.

$\frac{1}{8}$ in. $= .125$ in.

$d = 1$ in. $- (.125 \times .541 \times 2) = .8648$ or $.865$ in.

SECTION H / UNIT 14 / TAPER TURNING, TAPER BORING, AND FORMING

SELF-TEST ANSWERS

1. Steep tapers are quick-release tapers and slight tapers are self-holding tapers.
2. Tapers are expressed in taper per foot, taper per inch, and by angles.
3. Tapers are turned by hand feeding the compound slide, by offsetting the tailstock and turning between centers, or by using a taper attachment. A fourth method is to use a tool that is set to the desired angle and to form cut the taper.
4. No. The angle on the workpiece would be the included angle, which is twice that on the compound setting. The angle on the compound swivel base is the angle with the work centerline.
5. The reading at the lathe centerline index would be 55 degrees, which is the complementary angle.
6. Offset $= \dfrac{10 \times (1.125 - .75)}{2 \times 3} = \dfrac{3.75}{6} = .625$ in.
7. Four methods of measuring the offset on the tailstock are centers and a rule, witness marks and a rule, the dial indicator, and toolholder-micrometer dial.
8. The two types of taper attachments are the plain and the telescopic. Internal and external and slight to fairly steep tapers can be made. Centers remain in line, and power feed is used for good finishes.
9. The taper plug gage and the taper ring gage are the simplest and most practical means to check a taper. Four methods of measuring tapers are the plug and ring gages; using a micrometer on layout lines; using a micrometer with precision parallels and drill rod on a surface plate; and using a sine bar, gage block, and a dial indicator.
10. Chamfers, V-grooves, and very short tapers may be made by the form tool method.

SECTION H / UNIT 15 / LATHE ACCESSORIES

SELF-TEST ANSWERS

1. When workpieces extend from the chuck more than four or five workpiece diameters and are unsupported by a dead center; when workpieces are long and slender.
2. Since they are useful for supporting long workpieces, heavier cuts can be taken or operations such as turning, threading, and grooving may be performed without chattering. Internal operations such as boring may be done on long workpieces.
3. The steady rest is placed near the tailstock end of the shaft, which is supported in a dead center. The steady rest is clamped to the lathe bed and the lower jaws are adjusted to the shaft finger tight. The upper half of the frame is closed and the top jaw is adjusted with some clearance. The jaws are locked and lubricant is applied.
4. The jaws should be readjusted when the shaft heats up from friction in order to avoid scoring. Also soft materials are sometimes used on the jaws to protect finishes.
5. A center punch mark is placed in the center of the end of the shaft. The lower two jaws on the steady rest are adjusted until the center punch mark aligns with the point of the dead center.
6. No.
7. No. When the surface is rough, a bearing spot must be turned for the steady rest jaws.
8. By using a cat head.
9. A follower rest.
10. The shaft is purposely made one or two inches longer and an undercut is machined on the end to clear the follower rest jaws.
11. The tool post grinder may be used for external grinding of shaft diameters that are too hard to machine with a tool. Internal grinding may be done such as trueing up hardened chuck jaws.
12. The ways should be covered to protect them from grinding dust. The operator should stay out of line with the grinding wheel.

13. Only when the needed milling machine is not available.
14. Very light milling jobs such as slotting, making straight and woodruff keyseats and milling flats.
15. By offhand turning and file finishing with a gage or template, by turning with a radius rod, by using a radius tool, and by using a radius attachment.
16. By using a radii-cutter attachment. A motorized attachment is best for extra large diameter balls.
17. An arbor with a cross hole drilled in it is set up and the lathe is set for threading. Friction blocks are arranged so when the wire passes through them, they can be tightened against the wire. Low speeds are used.
18. It affects the diameter to some extent as the spring tends to expand when it is released from the arbor.
19. To convert rotary motion into reciprocating motion.
20. Eccentrics may be made by using offset centers or by turning a short piece that is offset in a four-jaw chuck.

SECTION I / UNIT 1 / MECHANICAL SYSTEMS

SELF-TEST ANSWERS

1. V-belts can deliver much more horsepower for their size than flat belts can.
2. V-belt drives are best suited for the higher speed ranges.
3. High reduction ratios and high efficiencies.
4. Gears provide a positive, constant ratio drive. They can deliver a high torque load for their size and they are highly efficient.
5. Herringbone gears are used to offset the high side or end thrust caused by single helical gears.
6. Bevel and miter gears; helical and worm gears.
7. Steel, cast iron, bronze, die cast, and nonmetallic such as formica and nylon.
8. Hydrostatic drives.
9. A coupling is used to permanently connect two rotating shafts, while a clutch is made to connect and disconnect two rotating shafts.
10. A keyseat is a single groove in a shaft and hub in which a key is fitted. A spline has multiple grooves in the shaft on which similar grooves in the hub are fitted. A spline can transmit a much higher torque load than a key.
11. Drift and pin punches.
12. To connect two rotating shafts that are not in alignment.
13. (a) Clockwise. (b) 4 to 1. (c) 450 RPM.
14. Sleeve bearings have sliding friction and are called friction bearings, while ball or roller bearings have rolling friction and are called antifriction bearings.
15. Pillow blocks and flange mounts are types of bearing housings that are designed to be bolted to machinery frames. They can contain either friction or antifriction type bearings.

SECTION I / UNIT 2 / CAUSES OF MECHANICAL FAILURES

SELF-TEST ANSWERS

1. Wear, overload, metal fatigue, corrosion, and brittle failure.
2. An overload can ruin a precision part if it is stressed beyond its yield point and is permanently deformed because its machined surfaces will no longer be in alignment.
3. Insufficient lubrication, continuous overload, and improper design or selection of material for the wear surface.
4. Dissimilar metals such as steel to bronze. Cast iron to cast iron is an exception.
5. Avoid a sharp change in cross section (undercuts, fillets, and such); utilize proper grain flow in parts as in a forging; increase compressive stress on the surface by surface hardening or shot peening.
6. Welding creates a stress concentration and in carbon rich metals such as springs and bumpers on cars an extremely hard adjacent zone is produced that greatly increases the stress and reduces the fatigue life considerably.
7. Galvanic corrosion which produces rust (iron oxide) on iron.
8. No corrosion can take place without an electrical connector between the anode and cathode. Metal ions

will not detach themselves from the anode unless a respective number of electrons have moved to the cathode.

9. Since zinc is nearer the anode end of the galvanic series, it would form the anode and the nickel would be the cathode.

10. Stainless steel can rust like ordinary iron if it loses its passivity by becoming oxygen starved in a concentration cell. A pile of sand or washers or even lapped plates can cause an oxygen starved area in an electrolyte, which will begin to rust.

11. Anodic coatings, cathodic coatings, organic and inorganic coatings, inhibitive coatings, inhibitors, metal cladding, galvanizing, metal spraying, and metal cementation.

SECTION I / UNIT 3 / MECHANICAL COMPONENT REPAIRS

SELF-TEST ANSWERS

1. Diffusion (carburizing, carbonitriding, cyaniding, nitriding, chromizing, and siliconizing), heat treating (hardening and tempering), surface hardening, hard facing, metallizing, anodizing, and electroplating.

2. Leakage or seepage of lubricant, cooling water or fluid.

3. Welding, spray weld, boring and sleeving, industrial adhesive, or by remachining the part to a new dimension and using a mating part that fits the new dimension.

4. Since welding will always cause some misalignment, the weld for replacement surface must be made oversize to allow for remachining to the original dimension that is in alignment with other machine surfaces.

5. Metal fatigue caused by stress concentration induced by weld passes, especially at the end of the passes. There is little or no stress concentration from metal spray welds.

6. It requires an undercut that slightly weakens the part, and the deposit of metal has a limited bond and is not a very strong material.

7. The millwright's major consideration is to keep the assembly line operating. With 30 workers, a shutdown of even an hour would be very costly. Past experience has taught him not to depend on any kind of shipment to be on time and even if it were on time, it would take time to install it. The locally produced, more expensive shaft would be a better choice in this instance.

8. In a gearbox, one small spalled or broken piece can destroy everything in the box and break the housing.

9. Gears should be match marked in case they are timed or synchronized. Flanges should be marked because the bolt holes are not always symmetrical and will not fit well in a different position.

10. Bearings should be replaced if they have any obvious imperfections; they should always look new. Seals should always be replaced.

SECTION I / UNIT 4 / RETHREADING AND RESTORING DAMAGED
THREADS

SELF-TEST ANSWERS

1. These are carelessness (overtorquing, overloading, and hammering damage), wear, and corrosion.

2. Stripped, galled or welded, corroded, distorted (expanded), and cross threaded.

3. In fine thread series and usually in soft metals.

4. Because the hardened bolt will dull the die or ruin it completely.

5. The universal external thread restorer can be used or the part rethreaded in an engine lathe.

6. By using an engine lathe to recut the damaged threads by thread chasing with a single point tool, coinciding with the original undamaged threads.

7. By using a thread chaser or a tap.

8. (a) Drilling and tapping to a larger size thread.
   (b) Replacing the thread with Heli-Coil[R].
   (c) Drilling and tapping to a larger size, inserting a threaded plug, and drilling and tapping the plug to the original thread size.

9. When the casting or part has a wall thickness that is too thin or the hole is too near an edge or shoulder, an enlarged thread would not work out.

10. Manufactured parts made of soft metals such as aluminum often have these inserts installed at the factory

to increase strength and prevent stripping. They are also often installed where a part is frequently assembled and dismantled to limit wear on the threads.

SECTION I / UNIT 5 / REMOVING BROKEN STUDS, TAPS, AND SCREW EXTRACTORS

SELF-TEST ANSWERS

1. The screw should be soaked with penetrating oil and let stand overnight if possible. If not, a safe amount of heat may be applied while rocking the screw back and forth to break it loose.
2. No; however, drilling out the screw tends to relieve pressure on the threads and less torque than it took to break the screw will often be sufficient to screw it out.
3. When a nut or a metal strap can be welded on the broken end of the stud, it provides a means to apply a torque and the heat of welding often helps to loosen frozen threads.
4. In this case, the stud can be drilled out (in the exact center) and a tap run through the threads to clean them up. If the threads are too severely damaged, they can be restored with Heli-Coil^R.
5. Since the screw may actually be loose in its threads, the first attempt should be to try to screw it out with a small punch and hammer. It may easily screw out without the trouble of drilling and using a screw extractor or other methods.
6. Because of the left-hand twist and the "bite" of the drill, a sufficient torque is sometimes generated on the stud to screw it out.
7. If the screw extractor fails to remove the broken stud, the next step may be to drill out the stud with a tap drill. If the hole is off center, the internal threads will be destroyed on one side of the hole.
8. (a) By using a tap extractor.
   (b) By annealing the tap and drilling it out.
   (c) By embrittling and breaking it up with a punch.
   (d) By using EDM (Electro-Discharge Machining). The entire tap can sometimes be punched out.
9. High speed steel (that cannot be readily softened) and high carbon tool steel (that can be softened by annealing).
10. By spark testing on a grinder. High speed steel has long, straight orange-red carrier lines with globules at the end and no sparklers when ground. High carbon tool steel has many bright yellow to white carrier lines with many bursts of sparklers.

SECTION I / UNIT 6 / NONDESTRUCTIVE TESTING

SELF-TEST ANSWERS

1. Nondestructive testing is used to inspect for cracks or other defects in various metal and nonmetal parts without damaging the part for its intended use.
2. Magnetic particle inspection can only be used for ferromagnetic metals such as iron and steel.
3. The specimen must be magnetized lengthwise in order to find a crosswise crack. A circular magnetic field is needed to find a lengthwise crack.
4. Fluorescent penetrant systems make use of capillary action to draw the penetrating solution into the flaws. The surplus penetrant is removed and a developer is applied to enhance the outline of the defect. The inspection is carried out under black light, which causes the penetrant to fluoresce. This method may be used on almost any dense material.
5. Dye penetrants may be used in remote areas where no power is available for a black light source. Portability, low cost, and ease of application are also advantages.
6. The pulse-echo system and the through transmission system. The pulse-echo system uses one transducer that both sends and receives the pulse. The pulse is reflected and echoed back from the other side or from a flaw. The through transmission system uses a transducer on each side of the test piece. The signal passes through the material and is modified by any flaws present.
7. Internal defects, cracks, porosity, laminations, thicknesses, and weld bonds can be detected with ultrasonic testing systems.
8. X-rays have the ability to pass through solids. A solid steel casting, for example, will have an X-ray source

on one side of the area to be inspected and a photographic plate on the other. A void or crack will appear darker on the negative.

9. Gamma radiation is omnidirectional—it goes in all directions. Therefore a hollow object may be inspected on its entire surface with one exposure. Radium or another radioactive substance is used to produce gamma rays, and extreme care must be taken with its use.

10. Eddy current techniques are used for sorting materials of various alloy compositions and with differing heat treatments. It can be used for detecting seams or variations in thicknesses, mass, and shape. It is limited to testing only materials that conduct electricity.

SECTION I / UNIT 7 / MANUFACTURE OF MECHANICAL PARTS
SELF-TEST ANSWERS

1. Parts are manufactured from bar stock and sheet or plate steel.
2. Forming, forging, machining, casting, weldments, metal spinning, and hand operations.
3. Formed metals are usually of a uniform thickness with cut or sheared edges, sometimes having die marks. Castings and forgings are often variable in thickness and have die or mold marks where the die came together.
4. Heat tends to soften metal that has been work-hardened by a forming operation.
5. Forging causes the fibrous flow lines of the part to follow its contour making it stronger and tougher than an equivalent machined part.
6. Parts made by mass production methods cost less than those made one at a time by a machinist simply because the labor cost per part is much less.
7. Sand casting for engine blocks and die casting for carburetors.
8. This is because an expensive pattern would first have to be made.
9. It is a device or metal part fabricated from standard metal stock by welding and often subsequent machining. Weldments are often made to replace castings.
10. (a) Cleanliness; avoiding dirt and grit.
    (b) Imperfections; remove any burrs, nicks, or dents.
    (c) Alignment; use the correct tools for the job to avoid any damage to the part.

SECTION J / UNIT 1 / HORIZONTAL MILLING MACHINES
SELF-TEST ANSWERS

1. Profile sharpened cutters and form relieved cutters.
2. Light duty plain milling cutters have many teeth. They are used for finishing operations. Heavy duty plain mills have few but coarse teeth, designed for heavy cuts.
3. Plain milling cutters do not have side cutting teeth. This would cause extreme rubbing if used to mill steps or grooves. Plain milling cutters should be wider than the flat surface they are machining.
4. Side milling cutters having side cutting teeth are used when grooves are machined.
5. Straight tooth side mills are used only to mill shallow grooves because of their limited chip space between the teeth and their tendency to chatter. Stagger tooth mills have a smoother cutting action because of the alternate helical teeth; more chip clearance allows deeper cuts.
6. Half side milling cutters are efficiently used when straddle milling.
7. Metal slitting saws are used in slotting or cutoff operations.
8. Gear tooth cutters and corner rounding cutters.
9. To mill V-notches, dovetails, or chamfers.
10. Face mills over 6 in. in diameter.
11. The two classes of taper are self-holding, with a small included angle, and self-releasing, with a steep taper.
12. $3\frac{1}{2}$ IPF.
13. Any small nick or chip between shank and socket or between spacers will cause the cutter to run out and will mar these contact surfaces.
14. Where small diameter cutters are used on light cuts, and where little clearance is available.
15. As close to the cutter as the workpiece and workholding device permit.
16. A style *C* arbor is a shell end mill arbor.

17. Tightening or loosening an arbor nut without the arbor support in place will bend or spring the arbor.

SECTION J / UNIT 2 / USING THE HORIZONTAL MILLING MACHINE

SELF-TEST ANSWERS

1. Eye injuries from flying chips, loose fitting clothing or jewelry can be caught in a cutter, and long or loose hair can be caught.
2. No.
3. $\text{RPM} = \dfrac{\text{CS} \times 4}{D} = \dfrac{90 \times 4}{4} = 90 \text{ RPM}.$
4. Feedrate $= .003 \times 16 \times 90 = 4.32$ IPM.
5. The machine vise.
6. The solid jaw.
7. In a milling vise or clamped to a T-slot.
8. Conventional mode.
9. When climb milling is used on a machine that has backlash.
10. When insufficient coolant is available to flood the cutter and work. Intermittant heating and cooling caused by insufficient coolant causes thermal cracking of carbide insert tools.

SECTION J / UNIT 3 / VERTICAL MILLING MACHINES

SELF-TEST ANSWERS

1. Chips should never be handled with bare hands because they are sharp, possibly hot, and often contaminated from cutting fluids.
2. Eye protection is always worn in a machining facility.
3. Machine guards are there for safety. Guards not in place leave operators or bystanders exposed to dangerous moving parts of machines.
4. A clean work area prevents one from stumbling over parts on the floor and slipping on spilled oil or cutting fluids.
5. The cutting edges of tools are sharp and can cause cuts. Use a rag when handling tools.
6. The motor on top of the workhead is heavy and will flip down if the clamping bolts are loosened completely. The clamping bolts should be kept snug to provide enough drag so that the head only swivels when pushed against.
7. Measurements should only be made when the machine spindle has stopped.
8. When viewed from the cutting end, a right-hand cut end mill will rotate counterclockwise.
9. An end mill has to have center cutting teeth to be used for plunge cutting.
10. End mills for aluminum usually have a fast helix angle and also highly polished flutes and cutting edges.
11. Carbide end mills are very effective when milling abrasive or hard materials.
12. The six major components of a vertical milling machine are the column, knee, saddle, table, ram, and toolhead.

SECTION J / UNIT 4 / USING THE VERTICAL MILLING MACHINE

SELF-TEST ANSWERS

1. Tool material (HS or carbide) workpiece material, condition of the machine, rigidity of the setup, and use of coolant.
2. Thick chips are formed with high feedrates and they often cause a built-up edge on the cutter tooth, leading to rough finishes and high stress on the tool edge. The result is often tool breakage.
3. About one-half the diameter of the end mill.
4. High speed steel tools.
5. No. The previous operator may have been careless in his alignment procedure or perhaps deliberately set an angle for a specific job.

6. An edge finder or dial indicator.
7. By using a paper strip for a feeler gage.
8. All machines have backlash unless they have special antibacklash devices on the feedscrew. A worn feedscrew nut may allow anywhere from .010 to .050 in. error when reversing the machine. Backlash may be eliminated by resetting the micrometer dial.
9. By tilting the toolhead or by setting the work at an angle in a vise.
10. Dividing heads, rotary tables, shaping attachments, and right angle milling attachments.

SECTION K / UNIT 1 / SELECTION AND IDENTIFICATION OF GRINDING WHEELS

SELF-TEST ANSWERS

1. Use silicon carbide for the bronze, aluminum oxide for the steel, either manufactured or natural diamond for the tungsten carbide inserts, and cubic boron nitride or aluminum oxide for the high speed steel. Cubic boron nitride could be eliminated.
2. For a straight wheel the third dimension is the hole size. On a cylinder wheel the third dimension is the wall thickness.
3. Side grinding: cylindrical (Type 2), straight cup (Type 6), flaring cup (Type 11), and dish (Type 12). Peripheral grinding: straight (Type 1) and dish (Type 12). The saucer or dish wheel is the only shape on which both peripheral and side grinding are rated safe.
4. These cutoff wheels would be rubber-bonded, because they are the only ones that can be made so thin.
5. Ball grinding requires the hardest wheels, then cylindrical grinding (cylinder ground with a peripheral wheel), then flat surface grinding with a peripheral wheel, then internal grinding and, finally, the softest, flat surfacing with a side grinding wheel.
6. Wheel 1 is a resinoid-bonded (B) aluminum oxide (A) wheel, very coarse grit (14), very hard (Z) and dense (3) structure. It is a typical specification for grinding castings. Wheel 2 is a vitrified-bonded (V) silicon carbide (C) wheel, same grit size but with medium (J) grade and structure (6), for grinding some soft metal like copper.
7. Wheel 1 is the peripheral grinding wheel, because it is a harder (K) and denser (8) wheel. Both wheels have the same abrasive, silicon carbide, and the same bond, vitrified. The abrasive in wheel 2 is coarser, 24 grit as against 36, and may be used for side grinding.
8. This is a vitrified aluminum oxide wheel, H grade, 8 structure, in a 46 grit size. The 32 indicates a particular kind of aluminum oxide, and the "BE" a particular vitrified bond.
9. In a straight wheel for flat grinding, you would use a softer (J or I) and more open structure (7 or 8) wheel. With a cup or segmental wheel, you would use a still softer grade, like H. For flat grinding, probably a size coarser grit, 46, would be used. The important thing is to go softer.
10. The material to be ground usually determines the abrasive to be used, although it is not practical to stock cubic boron nitride or diamond wheels unless you have a good deal of work to be done with them.

    Hard materials require fine grit sizes and soft grades. With soft, ductile materials, coarser grit sizes and harder, longer-wearing wheels are needed.

    Except for very hard materials, the rule is that grit sizes are coarser as the amount of stock to be removed becomes greater.

    Generally, the finer the grit, the better the finish.

SECTION K / UNIT 2 / GRINDING WHEEL SAFETY

SELF-TEST ANSWERS

1. A 12 in. wheel can be safe on a 2000 RPM grinder. The calculation is either: $12 \times 3.1416 \div 12 \times 2000 = 6283$ SFPM or

$$\frac{6500}{12 \times 3.1416 \div 12} = 2069 \text{ RPM}$$

2. (a) Tap about 45 degrees either side of the vertical centerline.

(b) Use a wooden tapper or mallet.

(c) Suspend the wheel, or place it on a clean, hard floor. Make sure there is nothing to muffle the sound.

3. (a) On a receipt from supplier, to insure getting sound wheels.

(b) Before first mounting, to catch cracks developed in toolroom storage.

(c) Before each remounting, to catch cracks developed in storage at machine.

4. (a) Do not bump or drop wheels.

(b) Do not roll wheel like a hoop.

(c) If you can not carry a wheel, use a hand truck, but support and guard the wheel to prevent cracking.

(d) Never stack anything, particularly metal tools, on top of wheels.

(e) In storage, separate wheels from each other with cardboard or other protection.

5. Flanges must be equal in diameter, clean and flat on the mounting rim, but hollow inside the rim so there is no pressure at the wheel hole. Flanges must be checked to see that there are no burrs and no nicks. They must be at least one-third the wheel diameter.

6. (a) Ring test the wheel.

(b) Check machine RPM against RPM of wheel. Wheel must be equal or higher.

(c) See that wheel fits snugly on spindle or mounting flange. It should not be so tight that is must be forced on, nor should it be loose and sloppy.

(d) Check flanges as above. Blotters should be larger than flanges.

(e) Stand to one side while starting the wheel.

(f) Run the wheel at least a minute without load before starting to grind.

7. (a) Always wear approved safety glasses or other face protection.

(b) Grind on the periphery, except for Type 2 cylinder wheels, cup wheels, and other wheels designed for side grinding.

(c) Never jam work into the wheel.

(d) Always make sure safety guard is in place and properly adjusted.

(e) Never use a wheel that has been dropped.

(f) If using coolant, turn coolant off a minute or so before stopping the wheel. This keeps the wheel from becoming unbalanced.

8. If wheel diameter remains the same, then wheel speed (SFPM) increases as RPM increases. Or to put it another and more practical way, as wheel diameter decreases through wear and dressing, the RPM must be increased to maintain the same speed in surface feet.

9. The fact that a wheel will run at 150 percent of safe speed in a factory test room, under extremely safe conditions with extra protection, and *without any load,* is no guarantee that it would be equally safe *with load under operating* conditions. The 50 percent overspeed has been determined by research to be a reasonable assurance of safe operation at the listed safe speed. It is common manufacturing practice to test products under conditions much more severe than the suppliers ever expect the product to meet in actual use.

10. Long sleeves, neckties, or anything that could be caught in a machine is not acceptable in a grinding shop or in any other shop. This includes finger rings and wristwatches. In fact, you should not wear any watch at all around a surface grinder, where the magnetic chuck could magnetize it and ruin it.

SECTION K / UNIT 3 / USING THE SURFACE GRINDER

SELF-TEST ANSWERS

1. Wheelhead.
2. Magnetic chuck.
3. By mounting it either in a prepared fixture or on a magnetic sine chuck.
4. One ten-thousandths of an inch (.0001 in.,) called a "tenth."
5. One-thousandth of an inch (.001 in.).
6. .002 in. for roughing and .0005 in. for finishing.
7. About .002 in.
8. The sharp wheel will remove stock more rapidly without burning but will give a coarse finish. The duller wheel produces a good finish, but tends to burn the work if too much stock is removed per pass.
9. The wheel simply blows it away and the contact area is not cooled. The coolant ideally should be applied to the wheel so it will be carried to the contact area.
10. Many attachments are available for surface grinders making it possible to perform grinding operations other than flat grinding.

SECTION K / UNIT 4 / PROBLEMS AND SOLUTIONS IN SURFACE GRINDING

SELF-TEST ANSWERS

1. Careful dressing of the wheel, keeping coolant tank full, checking coolant filters, thorough cleaning of chuck before each loading, checking tightness of wheel.
2. Vibration, heat, and dirt, plus poor quality wheel dressing.
3. A surface defect is anything that is out of pattern; any kind of unwanted scratches or discoloration.
4. Vibration is the principal cause. Off-grade wheels and wheels principally too hard or soaked on one side with coolant.
5. Fishtails are random scratches caused most often by dirt and grit in the coolant. Sliding the work off the chuck might also be considered a cause; and less often, a wheel that is very much too soft. Usually, the starting point in getting rid of them is to check the coolant level and filters.
6. Burning, checking, or discoloration is one. Low spots (out of flatness) is the other. Probably you would check first to see that there was enough coolant and that it was actually getting to the cutting area. Speeding up the table speed, if this is possible, could also help.
7. A burnished surface is one in which the hills of the scratches have been rubbed over into the valleys. It is bad because the surface will not resist wear. Probably the wheel is too hard and you should change it.
8. Poor chuck alignment is probably the first cause. You might also look for dirt or swarf on the chuck or insufficient coolant.
9. Overheating, particularly of thin work, is one. Dressing a wheel too fine could also be a cause. A wheel that is too hard is another cause. In fact, although work can be parallel and not flat, or flat and not parallel, the two conditions usually go together.
10. Mostly because thin work cannot absorb the heat, but also because the grinding may release stresses rolled into the workpieces.

GLOSSARY

Abrasive. A substance, such as finely divided aluminum oxide or silicon carbide, used for grinding (abrading), smoothing, or polishing.

Acetylene (C_2H_2). A colorless gas that is very soluble in acetone and has a characteristic odor. It is used with oxygen to produce a very high temperature flame.

Acicular. Needlelike, resembling needles or straws dropped at random.

Acute angle. An angle less than 90 degrees.

Adhesive. A bonding agent or sticky substance used to join similar or dissimilar materials together.

Alignment. The proper positioning or state of adjustment of parts in relation to each other, especially in line as in axial alignment.

Allotropic. Materials that can exist in several different crystalline forms are said to be allotropic.

Alloy. A combination of two or more substances, specifically metals, such as alloy steels or aluminum alloys.

Aluminum oxide. Also alumina (Al_2O_3). Occurs in nature as corundum and is used extensively as an abrasive. Today most aluminum oxide abrasives are manufactured.

Ammonia. A pungent, colorless, gaseous alkaline compound of nitrogen and hydrogen (NH_3). It is very soluble in water.

Amorphous. Having no definite form or outline. Materials such as glass that have no definite crystalline structure.

Angular. Having one or more angles; measured by an angle; forming an angle.

Angularity. The quality or characteristic of being angular.

Angular measure. The means by which an arc of a circle is divided and measured. This can be in degrees (360 degrees in a full circle), minutes (60 min. in one degree), and seconds (60 sec. in one minute), or in radians. (See Radian.)

Anhydrous. Free from water.

Anistropy. Fibrous quality of steel being more pronounced in the direction of rolling.

Anneal. A heat treatment in which metals are heated and then cooled very slowly for the purpose of decreasing hardness. Annealing is used to improve machinability, remove stresses from weldments, forgings, and castings, to remove stresses resulting from cold work, and to refine and make uniform the microscopic internal structures of metals.

Anode. The pole of an electrolytic or galvanic cell that deteriorates or corrodes when current flows to the cathode, which is the other pole.

Anodizing. The formation of a relatively thick corrosion product on some metals such as aluminum to impede further corrosion and to produce a hard wearing surface. The anodized surface is formed in a similar fashion to that of electroplating but with the polarity reversed so that the part to be anodized is the anode.

Arbor. A rotating shaft upon which a cutting tool is fastened. Often used as a term for mandrel.

As rolled. When metal bars are hot rolled and allowed to cool in air, they are said to be in the "as rolled" or natural condition.

Austenite. A solid solution of iron and carbon or iron carbide in which gamma iron, characterized by a face centered cubic crystal, is the solvent.

Axial. Having the characteristics of an axis (that is, centerline or center of rotation); situated around and in relation to an axis as in axial alignment.

Axial rake. An angular cutting surface that is rotated about the axial centerline or a cutting tool such as a drill or reamer.

Axis. Centerline or center of rotation of an object or part; the rotational axis of a machine spindle, which extends beyond the spindle and through the workpiece. Machining of the part imparts the machine axis to that area of metal cutting.

Backlash. Looseness or slack in mechanical parts caused by normal clearance between working parts or wear. Feed and lead screws are common examples of backlash. Some machines have backlash eliminators built into them.

Bearings. Bearings are used on rotating members to give them freedom to rotate with a minimum of friction. Ball and roller type bearings are called antifriction bearings because they were designed to operate with very little friction. In contrast, sleeve bearings of bronze or babbitt metal are called friction bearings since they develop more friction when in service.

Bell mouth. A taper extending inward a short way at the entrance to a bore.

Bezel. A rim that holds a transparent covering, as on a watch or dial indicator.

Blanking. Cutting flat stock in a punch press to a given shape, often circular. Blanking usually precedes forming or drawing.

Blind hole. A hole that does not go completely through an object.

Bolster plate. A structural part of a press designed to support or reinforce the platen (base surface) on which the workpiece is placed for press work.

Boring. The process of removing metal from a hole by using a single point tool. The workpiece can rotate with a stationary bar, or the bar can rotate on a stationary workpiece to bore a hole.

Braze welding. Bonding metals together by use of a filler metal whose melting point is lower than the base metal, which is not melted in the process. The bond is made by diffusion and adhesion. Sometimes capillary action is used in brazing two close fitting parts.

Brinell hardness. The hardness of a metal or alloy measured by hydraulically pressing a hard ball (usually a 10 mm dia. ball) with a standard load into the specimen. A number is derived by measuring the indentation with a special microscope.

Brittleness. That property of a material that causes it to suddenly break at a given stress without bending or distortion of the edges of the broken surface. Glass, ceramics, and cast iron are somewhat brittle materials.

Broaching. The process of removing unwanted metal by pulling or pushing a tool on which cutting teeth project through or along the surface of a workpiece. The cutting teeth are each progressively longer by a few thousandths of an inch to give each tooth a chip load. One of the most frequent uses of broaching is for producing internal shapes such as keyseats and splines.

Buffing wheel. A disc made up of layers of cloth sewn together. Fine abrasive is applied to the periphery of the cloth wheel to provide a polishing surface as the wheel is rotated at a high speed.

Burnish. To make shiny by rubbing. No surface material is removed by this finishing process. External and internal surfaces are often smoothed with high pressure rolling. Hardened plugs are sometimes forced through bores to finish and size them by burnishing.

Burr. (1) A small rotary file. (2) A thin edge of metal that is usually very sharp left from a machining operation. (See Deburr.)

Bushing. A hollow cylinder that is used as a spacer, a reducer for a bore size, or for a bearing. Bushings can be made of metals or nonmetals such as plastics or formica.

Button die. A thread cutting die that is round and usually slightly adjustable. It is held in a diestock or holder by means of a cone point setscrew that fits into a detent on the periphery of the die.

Calibration. The adjustment of a measuring instrument such as a micrometer or dial indicator so it will measure accurately.

Capillary attraction. The force of adhesion between a solid and a liquid when in capillarity, that is, when in a small tube or between close fitting surfaces such as a shaft in a bore having one- or two-thousandths of an inch clearance. Silver soldering or brazing makes use of this phenomenon.

Carburizing compound. A carbonaceous material that introduces carbon into a heated solid ferrous alloy by the process of diffusion.

Caustic soda. Sodium hydroxide, a chemical base material often used to clean metal surfaces. It is harmful to the skin or eyes and must be immediately washed off with cold water.

Cellulose. The cell walls of plants, occurring naturally in such plants as cotton and kapok. Products such as paper and rayon are produced from cellulose. Some welding electrodes have cellulose type coverings.

Celsius. A temperature scale used in the S.I. metric systems of measurement where the freezing point of water is 0° and the boiling point is 100°. Also called centigrade.

Cementite. Iron carbide, a compound of iron and carbon (Fe_3C) found in steel and cast iron.

Centerline. A reference line on a drawing or part layout from which all dimensions are located.

Chamfer. A bevel cut on a sharp edge of a part to improve resistance to damage and as a safety measure to prevent cuts.

Chasing a thread. In machine terminology, chasing a thread is making successive cuts in the same groove with a single point tool. Also when cleaning or repairing a damaged thread.

Chatter. Vibration of workpiece, machine, tool, or a combination of all three due to looseness or weakness in one or more of these areas. Chatter may be found in either grinding or machine operations and is usually noted as a vibratory sound and seen on the workpiece as wave marks.

Checked. A term used mostly in grinding operations indicating a surface having many small cracks (checks). The term "heat checked" or "crazed" is used in reference to friction clutch surfaces.

Chips. The particles that are removed when materials are cut. Also called filings.

Chip trap. A deformed end of a lathe cutting tool that prevents the chip from flowing across and away from the tool.

Circularity. The extent to which an object has the form of a circle. The measured accuracy or roundness of a circular or cylindrical object such as a shaft. A lack of circularity is referred to in shops as out-of-round, egg-shaped, or having a flat spot.

Circumference. The periphery or outer edge of a circle. Its length calculated by multiplying π (3.1416) times the diameter of the circle.

Coarseness. A definition of grit size in grinding or spacing of teeth on files and other cutting tools.

Cold finish. Refers to the surface finish obtained on metal by any of several means of cold working, such as rolling or drawing.

Cold working. Any process such as rolling, forging, or forming a cold metal in which the metal is stressed beyond its yield point. Grains are deformed and elongated in the process, causing the metal to have a higher hardness and lower ductility.

Complementary angles. Two angles whose sum is 90 degrees. Often referred to in machine shop work since most angular machining is done within one quadrant, or 90 degrees.

Concave. An internal arc or curve; a dent.

Concentricity. The extent to which an object has a common center or axis. Specifically, in machine work, the extent to which two or more surfaces on a shaft rotate in relation to each other; the amount of runout on a rotating member.

Convex. An external arc or curve; a bulge.

Coolant. A cutting fluid used to cool the tool and workpiece, especially in grinding operations; usually water based.

Crest of thread. Outer edge (point or flat) of a thread form.

Critical temperatures. The upper and lower transformation points of iron between which is the transformation range in which ferrite changes to austenite as the temperature rises.

Crystallize. Metals, when solidifying from the molten state, form crystals or grains that are characteristic of the type of metal, the alloying elements and cooling rates, or subsequent heat treatment. Metals do not change their crystalline structure at room temperature unless they are cold worked. Metal fatigue does not cause recrystallization or change in the crystal structure.

Cutting fluid. A term referring to any of several materials used in cutting metals: cutting oils, soluble or emulsified oils (water based), and sulfurized oils.

Cyanogen (CN)₂. A colorless, flammable, poisonous gas with a characteristic odor. It forms cyanic and hydrocyanic acids when in contact with water. Cyanogen compounds are often used for case hardening.

Cyclic. To recur in cycles as on a rotating member. A rotating shaft that is slightly bent due to side stress is subject to cyclic stress reversals.

Deburr. The removal of a sharp edge or corner caused by a machining process.

Decarburization. The loss of surface carbon from ferrous metals when heated to high temperatures in an atmosphere containing oxygen.

Decibel. A unit for expressing the relative intensity of sounds on a scale from zero (least perceptible sound) to about 130 (the average pain level).

Deformation. Distortion; alteration of the form or shape as a result of the plastic behavior of a material under stress.

Degrees. The circle is divided into 360 degrees, four 90 degree quadrants. Each degree is divided into 60 minutes and each minute into 60 seconds. Degrees are measured with protractors, optical comparators, and sine bars, to name a few methods. It is also divisions of temperature scales.

Dendrite. A formation that resembles a pine tree in the microstructure of solidifying metals. Each dendrite usually forms a single grain or crystal.

Diagonal. A straight line from corner to corner on a square, rectangle, or any parallelogram.

Diameter. Twice the radius; the length of any straight line going through the center of a figure or body; specifically, a circle in drafting and layout.

Diametral pitch. The ratio of the number of teeth on gears to the number of inches of pitch diameter.

Die. (1) Cutting tool for producing external threads. (2) A device that is mounted in a press for cutting and forming sheet metal.

Die cast metal. Metal alloys, often called pot metals, that are forced into a die in a molten state by hydraulic pressure. Thousands of identical parts can be produced from a single die or mold by this process of die casting.

Diffusion. An absorption of an element such as carbon into the surface of a metal at high temperatures. Carburization

is a form of diffusion in which carbon slowly penetrates the surface grains of steel at high temperature, producing a carbon-rich surface that can be hardened by quenching.

Dimension. A measurement in one direction. One of three coordinates: length, width, and depth. Also thickness, radius, and diameter are given as dimensions on drawings.

Discrimination. The level of measurement to which an instrument is capable within a given measuring system. A .001 in. micrometer can be read to within one-thousandth of an inch. With a vernier, it can discriminate to one ten-thousandth of an inch.

Distortion. The alteration of the shape of an object that would normally affect its usefulness. Bending, twisting, and elongation are common forms of distortion in metals.

Drawing. (1) A method of reducing the diameter of a metal bar or wire by pulling it through successively smaller dies (a form of cold working metals). (2) Drawing the temper of hardened tool steel by reheating (also called tempering or temper drawing).

Ductility. The property of a metal to be deformed permanently without rupture while under tension. A metal that can be drawn into a wire is ductile.

Ebonized. Certain cold drawn or rolled bars that have black stained surfaces are said to be ebonized. This is not the same as the black scaly surface of hot rolled steel products.

Eccentricity. A rotating member whose axis of rotation is different or offset from the primary axis of the part or mechanism. Thus, when one turned section of a shaft centers on a different axis than the shaft, it is said to be eccentric or to have "runout." For example, the throws or cranks on an engine crankshaft are eccentric to the main bearing axis.

Edge finder. A tool fastened in a machine spindle that locates the position of the workpiece edge in relation to the spindle axis.

E.D.M. Electro-discharge machining. With this process, a graphite or metal electrode is slowly fed into the workpiece that is immersed in oil. A pulsed electrical charge causes sparks to jump to the workpieces, each tearing out a small particle. In this way, the electrode gradually erodes its way through the workpiece, which can be a soft or an extremely hard material such as tungsten carbide.

Elasticity. The property of a material to return to its original shape when stretched or compressed.

Elastic limit. The extent to which a material can be deformed and still return to the original shape when the load is released. Deformation occurs beyond the elastic limit.

Electrical conductivity. The relative ability of a material to conduct electricity. Silver is one of the best conductors of electricity, while plastics and ceramics are some of the poorest. Metal alloys are poorer conductors than pure metals.

Electroplating. An electrical deposition of one metal on another in an electrolyte. An electric current is induced through the electrolyte at the cathode (part to be plated) and the anode (the plating material).

Embossing. The raising of a pattern in relief on a metal by means of a high pressure on a die plate.

Embrittle. To cause brittleness. For example, when trapped in steel from welding or plating operations, hydrogen can cause brittleness.

Emulsifying oils. An oil containing an emulsifying agent such as detergent so it will mix with water. Oil emulsions are used extensively for coolants in machining operations.

Eutectic. In double alloys, it is the alloy composition that solidifies at the lowest constant temperature.

Eutectoid. In double alloys such as carbon steel in solid solution, it is a percentage of alloy and the lowest temperature at which the two phases form simultaneously to form a mixture.

Expansion. The enlargement of an object, usually caused by an increase in temperature. Metals expand when heated and contract when cooled in varying amounts depending on the coefficient of expansion of the particular metal.

Extruding. A form of metal working in which a metal bar, either cold or heated, is forced through a die that forms a special cross-sectional shape such as an angle or channel. Extrusions of soft metals such as aluminum and copper are very common.

Fabricated. Manufactured; put together. In metal working the term generally refers to the assembling of standard shapes such as angles, channels, beams, and plates into a useful device by the process of welding; a weldment.

Face. (1) The side of a metal disc or end of a shaft when turning in a lathe. A facing operation is usually at 90 degrees to the spindle axis of the lathe. (2) The periphery or outer cylindrical surface of a straight grinding wheel.

Fahrenheit. A temperature scale that is calibrated with the freezing point of water at 32 degrees and the boiling point at 212 degrees. The Fahrenheit scale is gradually being replaced with the Celsius scale used with the metric system of measurement.

Fatigue. Metal fatigue is a gradual, slow development of a crack that will in time progress through the piece to the point of failure. It is caused by cyclic stress reversals, as on a rotating shaft, and is usually initiated at a point of stress concentration such as a groove, notch, or sharp corner. Fatigue resistance is the property of certain materials to resist fatigue failures under conditions of cyclic stress.

Ferrite. The microstructure of iron or steel that is mostly pure iron and appears light gray or white when etched and viewed with a microscope.

Ferromagnetic. Metals or other substances that have unusually high magnetic permeability, a saturation point with some residual magnetism and high hysteresis. Iron and nickel are both ferromagnetic.

Ferrous. Iron, from the Latin word ferrum, meaning iron. An alloy containing a significant amount of iron.

Fiber. A threadlike material that is strong and tough. Fibers are made of glass, wood, cotton, asbestos, metals, and synthetic materials. When fibers are added to a material to give it added tensile strength, it is called a composite.

Fillet. A concave junction of two surfaces; an inside corner radius of a shoulder on a shaft; an inside corner weld.

Finishing (surface). The control of roughness by turning, grinding, milling, lapping, superfinishing, or a combination of any of these processes. Surface texture is designated in terms of roughness profile in microinches, waviness, and lay (direction of roughness).

Fixture. A device that holds workpieces and aligns them with the tool or machine axis with repeatable accuracy.

Flammable. Any material that will readily burn or explode when brought into contact with a spark or flame.

Flange. A rib for adding strength or for attachment to other objects. An enlargement on the end of a shaft or pipe through which a circle of fasteners (bolts or rivets) hold a mating part.

Flash. (1) Excess material that is extruded between die halves in die castings or forging dies, and also the upset material formed when welding bandsaws. (2) The brilliant light of arc welding that can damage eyes.

Flexible seals. Rubbers, elastomers, and soft plastics are used for sealing material because they are elastic materials and will continue to maintain sealing pressure between two surfaces indefinitely or until the material embrittles from age and begins to crack.

Fluorescence. The property of emitting visible light during or after the absorption of radiation from another source. Fluorescent penetrants or liquids having the ability to emit visible light when absorbing invisible ultraviolet light are used to detect cracks for nondestructive testing of metals.

Flutes. Grooves and lands along a cylindrical tool or shaft such as the flutes of a reamer or drill. The part of a flute that does the cutting is called a tooth.

Flux. A solid or gaseous material that is applied to metal in order to clean and remove the oxides.

Forging. A method of metal working in which the metal is hammered into the desired shape, or is forced into a mold by pressure or hammering, usually after being heated to a more plastic state. Hot forging requires less force to form a given point than does cold forging, which is usually done at room temperature.

Formica. A trademark used to designate several plastic laminated products, especially a laminate used to make gears.

Forming. A method of working sheet metal into useful shapes by pressing or bending.

FPM or SFM. Surface feet per minute on a moving workpiece or tool.

Fracture. A failure of a metal part causing it to separate at a given point. A fracture in metal may be caused by metal fatigue, overload, or stress corrosion.

Friction. Rubbing of one part against another; resistance to relative motion between two parts in contact.

Furnace brazing. A brazing alloy, such as silver solder, that melts at a lower temperature than the base metal is caused to flow into a joint by capillary attraction by heating the parts in a furnace.

Fusion. The joining of two metals by bringing them both to a molten state at the interface so that they become one metal or alloy.

Galling. Cold welding of two metal surfaces in intimate contact under pressure. Also called seizing, it is more severe and more likely to happen between two similar, soft metals, especially when they are clean and dry.

Galvanic corrosion. A slow eroding of a metal in the presence of an electrolyte due to an electrochemical action. In the process, the metal reverts back to an oxide or other chemical combination.

Gas welding. The use of an oxy-acetylene flame to provide the heat to fuse or melt two metals together; also to melt a filler metal to deposit and diffuse it onto a higher melting point metal as with brazing.

Glazing. (1) A work hardened surface on metals resulting from using a dull tool or a too rapid cutting speed. (2) A dull grinding wheel whose surface grains have worn flat causing the workpiece to be overheated and "burned" (discolored).

Globular. Spherical; a ball-like shape.

Graduations. Division marks on a rule, measuring instrument, or machine dial.

Grain. In metals, a single crystal consisting of parallel rows of atoms called a space lattice.

Grain boundary. The outer perimeter of a single grain where it contacts adjacent grains.

Grain growth. Called recrystallization. Metal grains begin to reform to larger and more regular size and shape at certain temperatures, depending to some extent on the amount of prior cold working.

Granular. Made up of grains or crystals.

Graphite. One of the three allotropic forms of carbon. The two others are diamond and amorphous carbon (soot, charcoal).

Grit. (1) Any small, hard particles such as sand or grinding compound. Dust from grinding operations settles on machine surfaces as grit, which can damage sliding surfaces. (2) Diamond dust, aluminum oxide, or silicon carbide particles used for grinding wheels is called grit.

Ground and polished (G and P). A finishing process for some steel alloy shafts during their manufacture. The rolled, drawn, or turned shafting is placed on a centerless grinder and precision ground, after which a polishing operation produces a fine finish.

Gullet. The bottom of the space between teeth on saws and circular milling cutters.

Hard cracking. In welds, usually underbead cracking in the base metal caused by a high carbon content and rapid quenching that produces martensite, a hard constituent of steel.

Hardenability. The property that determines the depth and distribution of hardness in a ferrous alloy induced by heating and quenching.

Hardening. Metals are hardened by cold working and heat treating. Hardening causes metals to have a higher resistance to penetration and abrasion.

Hard faced. Metal that has been surfaced by welding or metal spraying techniques to provide extremely hard wear surfaces.

Harmonic chatter. A harmonic frequency is a multiple of the fundamental frequency of sound. Any machine part, such as a boring bar, has a fundamental frequency and will vibrate at that frequency and also at several harmonic or multiple frequencies. Thus, chatter or vibration of a tool may be noted at several different spindle speeds.

Hazard. A situation that is dangerous to any person in the vicinity. Also a danger to property, as a fire hazard, for example.

Heat treated. Metal whose structure has been altered or modified by the application of heat.

Helix. A line traced around a cylinder at a constant oblique angle (helix angle). Often erroneously referred to as a spiral.

High pressure lube. A petroleum grease or oil containing graphite or molybdenum disulfide that continues to lubricate even after the grease has been wiped off.

Homogenous. A material having a uniform structure throughout its mass is said to be homogenous.

Horizontal. Parallel to the horizon or base line; level.

Hot cracking. In welds, cracking down the center of welds and root passes caused by a poor fit-up or too rapid cooling.

Hot rolled. Metal flattened and shaped by rolls while at a red heat.

Hub. A thickening near the axis of a wheel, gear, pulley, sprocket, and others that provides a bore in its center to receive a shaft. The hub also provides extra strength to transfer power to or from the shaft by means of a key and keyseat.

Hydrogen embrittlement. In welds, entrapped hydrogen from water, oil, or paint on the base metal causes the weld to become brittle.

Increment. A single step of a number of steps. A succession of regular additions. A minute increase.

Induction hardening. A hardening process that uses electrical induction to produce heat above the transformation temperature.

Inert gas. A gas, such as argon or helium, that will not readily combine with other elements.

Infeed. The depth a tool is moved into the workpiece.

Inhibitor. A coating or solution that inhibits corrosion in metals.

Interface. The contact area between two materials or mechanical elements.

Interference fit. Force fit of a shaft and bore, bearings, and housings or shafts. Negative clearance in which the fitted part is very slightly larger than the bore.

Intergranular. Along the grain boundaries of metals.

Internal stress. Also called residual stress. Stress in metals that is built-in by heat treatment or by cold working.

Involute. A profile tooth for gears. A spiral curve that is formed when a point on a thread or string is traced as it is unwound from a circle or cylinder.

Irighizing. A cementation process by which silicon is impregnated into the surface of a metal.

Iron carbide. Also called cementite (Fe_3C), a compound of iron and carbon, which is quite hard.

Jig. A device that guides a cutting tool and aligns it to the workpiece.

Journal. The part of a rotating shaft or axle that turns in a bearing.

Junction zone. In welds, the point at which the weld metal and the base metal join.

Kerf. Width of saw cut.

Key. A removable metal part that, when assembled into keyseats, provides a positive drive for transmitting torque between shaft and hub.

Keyseat. An axially located rectangular groove in a shaft or hub.

Keystock. Square or rectangular cold rolled steel bars used for making and fitting keys in keyseats.

Keyway. Same as keyseat (British terminology).

Knurl. Diamond or straight impressions on a metal surface produced by rolling with pressure. The rolls used are called knurls.

Lamellar. Flat or platelike, such as the alternating plates of pearlite in carbon steel.

Lattice. The space lattice or rows of atoms in a metal crystal.

Lead. The distance a thread or nut advances along a threaded rod in one revolution.

Loading. A grinding wheel whose voids are being filled with metal, causing the cutting action of the wheel to be diminished.

Longitudinal. Lengthwise, as the longitudinal axis of the spindle or machine.

Machinability. The relative ease of machining that is related to the hardness of the material to be cut.

Magnetic. Having the property of magnetic attraction and permeability.

Malleability. The ability of a metal to deform permanently without rupture when loaded in compression.

Mandrel. A cylindrical bar upon which the workpiece is affixed and subsequently machined between centers. Mandrels, often erroneously called arbors, are used in metal turning and cylindrical grinding operations.

Martensite. The hardest constituent of steel formed by quenching carbon steel from the austenitized state to room temperature. The microstructure can be seen as acicular or needlelike.

Mechanical properties (of metals). Some mechanical properties of metals are tensile strength, ductility, malleability, elasticity, and plasticity. Mechanical properties can be measured by mechanical testing.

Metal cementation. Introducing a metal or material into the surface of another by heat treatment. Carburizing is one example of metal cementation.

Metallizing. Applying a coating of metal on a surface by spraying molten metal on it. Also called spray weld and metal spray.

Metal spinning. A process in which a thin disc of metal is rapidly turned in a lathe and forced over a wooden form or mandrel to form various conical or cylindrical shapes.

Metrology. The science of weights and measures or measurement.

Microstructure. Structure that is only visible at high magnification.

Mushroom head. (1) An oversize head on a fastener or tool that allows it to be easily pushed with the hand. (2) A deformed striking end of a chisel or punch that should be removed by grinding.

Neutral. In machine work, neither positive nor negative rake is a neutral or zero rake; a neutral fit is neither a clearance nor interference fit.

Nitriding. A surface hardening treatment for ferrous alloys that is obtained by heating an alloy in the presence of disassociated ammonia gas, which releases nitrogen to the steel. The formation of iron nitride causes the hardened surface.

Nitrogenous gas. Ammonia (NH_3) used in nitriding.

Nominal. Usually refers to a standard size or quantity as named in standard references.

Nonferrous. Metals other than iron or iron alloys; for example, aluminum, copper, and nickel are nonferrous metals.

Normalizing. A heat treatment consisting of heating to a temperature above the critical range of steel followed by cooling in air. Normalizing produces in steel what is called a normal structure consisting of free ferrite and cementite or free pearlite and cementite, depending on the carbon content.

Nose radius. Refers to the rounding of the point of a lathe cutting tool. A large radius produces a better finish and is stronger than a small one.

Obtuse angle. An angle greater than 90 degrees.

Oxide scale. At a red heat, oxygen readily combines with iron to form a black oxide scale (Fe_3O_4), also called mill scale. At lower temperatures 400 to 650°F (204 to 343°C) various oxide scale colors (straw, yellow, gold, violet, blue, and gray) are produced, each color within a narrow temperature range. These colors are used by some heat treaters to determine temperatures for tempering.

Oxidize. Combine with oxygen; to burn or corrode by oxidation.

Oxy-acetylene. Mixture of oxygen and acetylene gases to produce an extremely hot flame used for heating and welding.

Parallax error. An error in measurement caused by reading a measuring device, such as a rule, at an improper angle.

Parting. Also called cutting off; a lathe operation in which a thin blade tool is fed into a turning workpiece to make a groove that is continued to the center to sever the material.

Pearlite. Alternating layers of cementite and ferrite in carbon steel. The microstructure of pearlite sometimes appears under a microscope like mother-of-pearl, hence the name. It is found in steels that have been slowly cooled.

Pecking. A process used in drilling deep holes to remove chips before they can seize and jam the drill. The drill is fed into the hole a short distance to accumulate some chips in the flutes, and then drawn out of the hole, allowing the chips to fly off. This process is repeated until the correct depth of the hole is reached.

Penetrant. A thin liquid that is able to enter small cracks and crevices. Penetrant oils are used to loosen rusted threads, and dye penetrants are used to find hidden cracks.

Periphery. The perimeter or external boundary of a surface or body.

Permanent mold. A reuseable mold for casting molten metals which can usually be used for several thousand castings before discarding.

Perpendicular. At 90 degrees to the horizontal or base line.

Pin. Straight, tapered, or cotter pins are used as fasteners of machine parts or for light drives.

Pinion. The smaller gear of a gear set.

Pitch diameter. For threads, the pitch diameter is an imaginary circle that on a perfect thread occurs at the point where the widths of the thread and groove are equal. On gears, it is the diameter of the pitch circle.

Pitting. A form of corrosion in which anodes form on a metal in many locations, causing a pitted surface.

Plating. The process of applying a thin coating of metal to another metal by electrolysis; electroplating.

Porous. Having pores or voids so that liquids may penetrate or pass through.

Potassium cyanide. A very poisonous crystalline salt (KCN) used in electroplating and for case hardening steel.

Pot metals. Die casting alloys, which can be zinc, lead, or aluminum based (among others).

Precipitation hardening. A process of hardening an alloy by heat treatment in which a constituent or phase precipitates from a solid solution at room temperature or at a slightly elevated temperature.

Precision. A relative but higher level of accuracy within certain tolerance limits. Precision gage blocks are accurate within a few millionths of an inch, yet precision lathe work in some shops may be within a few thousandths of an inch tolerance.

Pressure. Generally expressed in units as pounds per square inch (PSI), and is called unit pressure, while force is the total load.

Pulley. A flat-faced wheel used to transmit power by means of a flat belt. Grooved pulleys are called sheaves.

Quench. A rapid cooling of heated metal for the purpose of imparting certain properties, especially hardness. Quenchants are water, oil, fused salts, air, and molten lead.

Quench cracking. Cracking of heated metal during the quenching operation caused by internal stresses.

Quick-change gearbox. A set of gears and selector levers by which the ratio of spindle rotation to lead screw rotation on the lathe can be quickly set. Many ratios in terms of feeds or threads per inch can be selected without the use of change gears.

Quick-change tool post. A lathe toolholding device in which preset cutting tools are clamped in toolholders that can be placed on the tool post or interchanged with others to an accurately repeatable location.

Radial crack. In welds, cracks that radiate away from the weld bead into the base metal.

Radial rake. On cylindrical or circular cutting tools, such as milling cutters or taps, the rake angle that is off the radius is called the radial rake.

Radian. A unit of angular measurement that is equal to the angle at the center of a circle subtended by an arc equal in length to the radius.

Radioactivity. The property possessed by some elements such as uranium of spontaneously emitting radiation by the disintegration of the nuclei of atoms.

Rake. A tool angle that provides a keenness to the cutting edge.

Rapid traverse. A rapid travel arrangement on a machine tool used to quickly bring the workpiece or cutting tool into close proximity before the cut is started.

Recessing. Grooving.

Reciprocating. A back and forth movement.

Recrystallize. Metal crystals become flattened and distorted as the metal hardens when it is being cold worked. At a particular temperature range (about 950°F or 510°C for low carbon steel) for each metal, called its recrystallization temperature, the distorted grains begin to reform into regular, larger, softer grains or crystals.

Reference point. On a layout or drawing, there must be a point of reference from which all dimensions originate for a part to avoid an accumulating error. This could be a machined edge, datum, or centerline.

Relief angle. An angle that provides cutting edge clearance which allows the cutting action is called the relief angle.

Resulfurized. Sulfur in steel is normally a contaminent that causes steel to be "hot short" (separates while being hot forged) because of iron sulfide inclusions. As much sulfur as possible is therefore removed at the steel mill. Resulfurized steel is free machining because sulfur and manganese (to control the sulfur) are deliberately added to make manganese sulfide (instead of iron sulfide) inclusions that make a sort of lubricant for the chip and do not cause hot shortness.

Right angle. A 90 degree angle.

Ring test. A means of detecting cracks in grinding wheels. The wheel is lightly struck and, if a clear tone is heard, the wheel is not cracked.

Rockwell. A hardness test that uses a penetrator and known weights. Several scales are used to cover the very soft to the very hard materials. The Rockwell "C" scale is mostly used for steel.

Root. The bottom of a thread or gear tooth.

Root crack. In welds, cracking in the root (first) pass.

Root truncation. The flat at the bottom of a thread groove.

Roughing. In machining operations, the rapid removal of unwanted material on a workpiece, leaving a small amount for finishing, is called roughing. Since coarse feeds are used, the surface is often rough.

RPM. Revolutions per minute.

Runout. An eccentricity of rotation, as that of a cylindrical part held in a lathe chuck being off center as it rotated. The amount of runout of a rotating member is often checked with a dial indicator.

Sacrificial anode. A zinc or magnesium mass that is electrically connected to a ship's hull or to a buried pipeline. Since these metals are very low (anodic) on the galvanic series, they will deteriorate instead of the steel hull or pipe.

Scaling. The tendency of metals to form oxides on their surfaces when held at a high temperature in air is called scaling, as the oxides usually form as a loose scale.

Scriber. A sharp pointed tool used for making scratch marks on metal for the purpose of layout.

Semiprecision. Using a method of layout, measurement, or machining in which the tolerances are greater than that capable by the industry for convenience or economy.

Serrated. Small grooves, often in a diamond pattern, used mostly for a gripping surface.

Set. The width of saw tooth. The set of saw teeth is wider than the blade width.

Setup. The arrangement by which the machinist fastens the workpiece to a machine table or workholding device and aligns the cutting tool for metal removal. A poor setup is said to be when the workpiece could move from the pressure of the cutting tool, thus damaging the workpiece or tool, or when chatter results from lack of rigidity.

SFPM. Surface feet per minute.

Shallow hardening. Some steels such as plain carbon steel (depending on their mass), when heated and quenched, harden to a depth of less than $\frac{1}{8}$ in. These are shallow hardening steels.

Shank. The part of a tool that is held in a workholding device or in the hand.

Shearing action. A concentration of forces in which the bending moment is virtually zero and the metal tends to tear or be cut along a transversal axis at the point of applied pressure.

Sheaves. Grooved pulleys such as those used for V-belts or cables.

Sherardized. Zinc inoculated steel; a process by which the surface of steel is given a protective coating of zinc. It is not the same as galvanized or zinc-dipped steel. Zinc powder is packed around the steel while it is heated to a relatively low temperature in the sherardizing process.

Shim. A thin piece of material, usually metal, that is used as a spacer.

Shock loads. Sudden high stresses applied to a machine part far beyond the anticipated design for loading. Shock loads are often the cause of mechanical failures.

S.I. Système Internationale. The metric system of weights and measures.

Silicon carbide. A manufactured abrasive. Silicon carbide wheels are used for grinding nonferrous metals, cast iron, and tungsten carbide, but are not normally used for grinding steel.

Sine bar. A small precision bar with a given length (5 or 10 in.) that remains constant at any angle. It is used with precision gage blocks to set up or to determine angles within a few seconds of a degree.

Sintering. Holding a compressed metal powder briquette at a temperature just below its melting point until it fuses into a solid mass of metal.

Slag. An accumulation through the action of a flux of impurities or nonmetallic constituents of a processed ore usually consisting of oxides.

Slot. Groove or depression as in a keyseat slot.

Snagging. Rough grinding to remove unwanted metal from castings and other products.

Sodium hydroxide (NaOH). A strong caustic used for cleaning metals.

Solder. An alloy of lead and tin that is used for joining parts requiring only low strength bonding.

Solid solution. In certain alloys at certain temperatures below the melting point. Atoms of an immobilized element are able to diffuse into the lattice of the major constituent, thus making a solid solution.

Solidus. The point at which metals solidify on the phase diagram.

Soluble oils. Oils that have been emulsified and will combine with water are called soluble oils.

Solution heat treating. See **Precipitation hardening.**

Solvent. A material, usually liquid, that dissolves another. Dissolved material is the solute.

Spark testing. A means of determining the relative carbon content of plain carbon steels and identifying some other metals by observing the sparks given off while grinding the metal.

Specifications. Requirements and limits for a particular job.

Speeds. Machine speeds are expressed in revolutions per minute; cutting speeds are expressed in surface feet per minute.

Sphericity. The quality of being in the shape of a ball. The extent to which a true sphere can be produced with a given process.

Spherodize anneal. A heat treatment for carbon steels that forms the cementite into spheres, making it softer and usually more machinable than by other forms of annealing.

Spiral. A path of a point in a rotating plane that is continuously receding from the center is called a flat spiral. The term spiral is often used, though incorrectly, to describe a helix.

Sprockets. Toothed wheels used with chain for drive or conveyor systems.

Squareness. The extent of accuracy that can be maintained when making a workpiece with a right angle.

Stepped shaft. A shaft having more than one diameter.

Stick-slip. A tendency of some machine parts that slide on ways to bind slightly when pressure to move them is applied, followed by a sudden release, which often causes the movement to be greater than desired.

Straightedge. A comparison measuring device used to determine flatness. A precision straightedge usually has an accuracy about plus or minus .0002 in. in 24 inches length.

Strain. The unit deformation of a specimen when stress is applied.

Strain hardening. Same as **Work hardening.**

Strength. The ability of a metal to resist external forces. This can be tensile, compressive, or shear strength.

Stress. An external force applied to a specimen.

Stress relief anneal. A heat treatment, usually under the critical range, for the purpose of relieving stresses caused by welding or cold working.

Stroke. A single movement of many, as in a forward stroke with a hacksaw.

Studding. In welding practice, the application of metal plugs for reinforcement on difficult to weld materials.

Sulfides. Sulfur compounds such as metallic ores.

Surface plate. A cast iron or granite surface having a precision flatness that is used for precision layout, measurement, and setup.

Symmetrical. Usually bilateral in machinery where two sides of an object are alike but usually as a mirror image.

Synthetic oils. Artificially produced oils that have been given special properties, such as resistance to high temperatures. Synthetic water soluble oils or emulsions are replacing water soluble petroleum oils for cutting fluids and coolants.

Tang. The part of a file on which a handle is affixed.

Taper. A gradual change in dimension along the length of a part. Taper is measured by the difference in size in one inch length (TPI), one foot length (TPF), and by angular measure (degrees).

Tapered thread. A thread made on a taper, such as a pipe thread.

Tap extractor. A tool that is sometimes effective in removing broken taps.

Tapping. A method of cutting internal threads by means of rotating a tap into a hole that is sufficiently under the nominal tap size to make a full thread.

Telescoping gage. A transfer type tool that assumes the size of the part to be measured by expanding or telescoping. It is then measured with a micrometer.

Temper. (1) The cold worked condition of some nonferrous metals. (2) Also called draw; a method of toughening hardened carbon steel by reheating it.

Temperature. The level of heat energy in a material as measured by a thermometer or thermostat and recorded with any of three temperature scales: Celsius, Fahrenheit, or Kelvin.

Template. A metal, cardboard, or wooden form used to transfer a shape or layout when it must be repeated many times.

Tensile strength. The maximum unit load that can be applied to a material before ultimate failure occurs.

Terminating threads. Methods of ending the thread, such as undercutting, drilled holes, or tool removal.

Test bar. A precision ground bar that is placed between centers on a lathe to test for center alignment using a dial indicator.

Thermal conductivity. The rate of heat transfer through a particular material.

Thermal cracking. Checking or cracking caused by heat.

Thread axis. The centerline of the cylinder on which the thread is made.

Thread chaser. A tool used to restore damaged threads.

Thread crest. The top of the thread.

Thread die. A device used to cut external threads.

Thread engagement. The distance a nut or mating part is turned onto the thread is called the thread engagement.

Thread fit. Systems of thread fits for various thread forms range from interference fits to very loose fits; extensive references on thread fits may be found in machinist's handbooks.

Thread lead. The distance a nut travels in one revolution. The pitch and lead are the same on single lead threads, but not on multiple lead threads.

Thread pitch. The distance from a point on one thread to a corresponding point on the next thread.

Thread relief. Usually an internal groove that provides a terminating point for the threading tool.

Tinning. When soldering or brazing in order to have a bond, it is necessary to first "tin" or wet the base metal with the solder. This is done by capillary attraction on metal that has been cleaned with a flux.

Toe crack. In welding, a crack at one side and at the outer edge of a weld bead.

Tolerance. The allowance of acceptable error within which the mechanism will still fit together and be totally functional.

Tool geometry. The proper shape of a cutting tool that makes it work effectively for a particular application.

Torque. A force that tends to produce rotation or torsion. Torque is measured by multiplying the applied force by the distance at which it is acting to the axis of the rotating part.

Toughness. Toughness in metals is usually measured with an impact test and an Izod-Charpy machine in which a specimen that is notched is struck with a swinging hammer. The amount of energy absorbed by the specimen is the measure of its toughness.

Toxic fumes. Gases resulting from heating certain materials are toxic, sometimes causing illness (as metal fume fever from zinc fumes) or permanent damage (as from lead or mercury fumes).

Transformation temperature. Same as **Critical temperature;** the point at which ferrite begins to transform to austenite.

Transgranular. Through or across the grain; usually refers to cracking in metals.

Transverse. Crosswise to the major or lengthwise axis.

Traverse. To move a machine table or part from one point to another, usually crosswise to the major axis of the machine.

Trueing. In machine work, the use of a dial indicator to set up work accurately. In grinding operations, to dress a wheel with a diamond.

Truncation. To remove the point of a triangle (as of a thread), cone, or pyramid.

Tungsten carbide. An extremely hard compound that is

formed with cobalt and tungsten carbide powders by briquetting and sintering into tool shapes.

Turning. Machine operations in which the work is rotated against a single point tool.

Ultimate strength. Same as tensile strength; or ultimate tensile strength.

Ultrasonic. Sound frequencies above those audible to the human ear.

Vernier. A means of dividing a unit measurement on a graduated scale with a short scale made to slide along the divisions of a graduated instrument.

Vibration. An oscillating movement caused by loose bearings or machine supports, off center weighting on rotating elements, bent shafts, or nonrigid setups.

Viscosity. The property of a fluid or semifluid that enables it to maintain an amount of shear stress and offer resistance to flow, depending on the velocity.

Vise. A workholding device. Some types are bench, drill press, and machine vises.

Vitrified. Fired clay or porcelain.

Wedge angle. Angle of keenness; cutting edge.

Weldability. The extent to which a strong weld can be made on a particular metal without special heat treatments. For example, steel containing .90 percent carbon would have low weldability but could be successfully welded if the correct preheat and postheat techniques were used.

Welding. Fusing or diffusing two metals together.

Weldment. An assembly of parts made by the process of welding.

Wheel dressing. Trueing the grinding surface of an abrasive wheel by means of a dressing tool such as a diamond or Desmond dresser.

Wiggler. A device used to align a machine spindle to a punch mark.

Wire drawing. The process of making wire by pulling a metal rod through a series of consecutively smaller dies by means of capstans.

Witness marks. Index marks used to align two mating parts. Witness marks are often made with a punch, chisel, or number stamp by mechanics when disassembling parts to facilitate reassembly.

Wrought. Hot or cold worked; forged.

Yield point. The amount of stress in PSI at which a certain metal will begin to deform permanently.

Yield strength. The strength in PSI of a particular metal under which point it will behave in an elastic manner and return to its original shape.

Zero back rake. Also neutral rake; neither positive nor negative; level.

Zero index. Also zero point. The point at which micrometer dials on a machine are set to zero and the cutting tool is located to a given reference such as a workpiece edge.

INDEX